Handbook of Metal–Ligand Interactions in Biological Fluids

Bioinorganic Chemistry

Volume 2

Handbook of Metal–Ligand Interactions in Biological Fluids

Bioinorganic Chemistry

Volume 2

edited by

Guy Berthon

Centre National de la Recherche Scientifique (CNRS)
Institut National de la Santé et de la
Recherche Médicale (INSERM)
Toulouse, France

Marcel Dekker, Inc. New York • Basel • Hong Kong

ISBN: 0-8247-9636-5

The publisher offers discounts on this book when ordered in bulk quantities. For more information, write to Special Sales/Professional Marketing at the address below.

This book is printed on acid-free paper.

MARCEL DEKKER, INC.
270 Madison Avenue, New York, New York 10016

Current printing (last digit):
10 9 8 7 6 5 4 3 2

PRINTED IN THE UNITED STATES OF AMERICA

PREFACE TO THE HANDBOOK

In recent years, there has been a growing awareness of the potential contribution of chemistry toward progress in medicine, as is evident from the constant development of related interdisciplinary research. In particular, the area called "medicinal chemistry" has seen enormous growth during the past decade. However, this is largely still the sphere of *organic* chemists, in conjunction with their traditional and, more lately, newer partners in biology. In a recent comprehensive attempt to distinguish it from the larger field known as "bioorganic chemistry," for example, medicinal chemistry has been centered mainly on the understanding of structure–activity relationships and drug design from the mechanistic point of view, with physico-chemical implications hardly being mentioned.

Even though presumably to a lesser extent in quantitative terms, *inorganic* chemistry has also seen a huge development with respect to its contribution to medicine, and the new discipline of "bioinorganic chemistry" has greatly helped toward the understanding of fundamental biological processes in which inorganic elements are involved. Among these elements, metal ions are undoubtedly of the greatest importance. The various roles they play in the above biological processes have thus attracted particular attention regarding the health of living organisms, especially human beings.

In this context, the need for a reference book on the interactions between metal ions and endogenous or exogenous chemical substances occurring in the main body fluids has become evident. This has given rise to the present *Handbook of Metal–Ligand Interactions in Biological Fluids*. In practice, this project took shape from a few common-sense observations.

Keeping human beings healthy is, in the first place, a matter for medicine considered as the nonsurgical treatment of disease. Indeed, except for the strictly "mechanical" repair of physiological injuries, recourse to surgery often amounts to an acknowledgment of medicine's inadequacy. Somewhat paradoxically, however, compared with the gigantic advances in surgery due to recent breakthroughs registered in physics, medicine has progressed much more slowly. This can be attributed at least partly to a lack of adaptation of chemistry to its needs. As a matter of fact, there is no equivalent intermediate field between chemistry and medicine in the way biomedical engineering exists between physics and surgery.

First of all, it seems that full advantage has not been taken of all the potential of chemical research. For example, progress in medicine is mainly expected to ensue from the synthesis of new pharmacologically active substances by organic chemists. The tendency to generalize computer-aided applications of principles such as structure–activity relationships may be effective in the design of new analogs of existing drugs; however, research on how these molecules interact with endogenous substances and how they are translocated in vivo is still in its infancy and is mostly based on animal screening. In this respect, pharmacokinetic profiles are undoubtedly of great value in assessing the extent to which a substance is distributed in vivo, but such macroscopic observations do not provide any quantitative information about the absorption, retention and excretion processes at the molecular level. Also, too often, as it logically ensues from the empirical approach used, the exact site and process of action of new drugs remain obscure whereas their elucidation from a molecular point of view would be essential to allow further progress.

On the other hand, the fundamental activities of inorganic chemists tend to be regarded as being too distant from the biological sciences to be applied to the large variety of medical problems they could in fact help solve. For example, the enormous efforts made in recent years to discover the mechanism of action of the (fortuitously discovered) platinum antitumor compounds has undoubtedly proved useful in that they will eventually help in the design of more specific and less toxic drugs of successive generations. However, the fact that the activity of many centers of excellence has been focussed on this almost exclusive target may have been at the expense of the development of other conceptual approaches. In particular, the possible significance of the systematic variation of parameters such as the level of zinc, copper, histamine and polyamines in tumor cells compared to normal ones has meanwhile been largely overlooked.

Given the extreme complexity of the roles of metal ions in vivo, many objectives are still to be met before these elements can be commonly used in the control of health and disease. Combined efforts from both empirical and theoretical sides are required. In short, these should be directed to (i) provide a more comprehensive knowledge of the physiological functions of essential metals as well as a better understanding of their metabolic processes; (ii) help delineate the origin of the toxicity of harmful elements competing with essential ones and find more efficient detoxication treatments; (iii) develop the use of metal ions for therapeutic and diagnostic purposes; (iv) address the still largely unexplored field of the interactions of metal ions with exogenous organic substances, especially drugs.

Undoubtedly inorganic chemists now have at their disposal extremely powerful techniques to solve many problems with medical applications. The amount of specialized information available to them is also constantly increasing. However, chemists are rarely in a position to appreciate the problems encountered in medicine from their own point of view. For example, while "biological relevance" is commonly put forward to advertise articles of fundamental research relative to substances active in life processes, the corresponding results are too often devoid of real relationships with actual medical or biological problems.

On the other hand, medical researchers are becoming more and more conscious of the importance of the interactions between chemical substances whose effects they investigate in vivo or in vitro, but, having no training in fundamental solution chemistry, they are in turn rarely in a position to tackle the problems they are facing in the proper physico-chemical terms.

In many cases, the information necessary to solve medical problems using chemical techniques may even exist, but it remains so poorly accessible to specialists on both sides that it cannot be successfully exploited. Most of the time, it is so scattered, or described in such specific language, that it is a difficult task even for those working in interdisciplinary sciences to make good progress.

Fortunately, there has recently been a progressive trend to publish at the interdisciplinary border, thus bridging the gap between the chemical and medical communities; however, this has too often resulted in collections of separate, highly specialized contributions. The time has now probably come for a systematic effort to "popularize" the appropriate knowledge from and for the benefit of both sides. In this context, *the objective of the present Handbook is to facilitate contact between medicine and chemistry in the field of the metal–ligand interactions occurring in biological fluids.* It would probably be somewhat overenthusiastic to expect a sudden spectacular advance from bringing together such basically different disciplines; therefore, the first aim of this work is to develop a realistic approach based on a progressive and systematic exchange of information.

For this reason, the Handbook is divided into two two-volume sets that condense this information:

The first, which deals with the physico-chemical aspects of metal–ligand interactions in biological fluids, should provide the fundamental knowledge necessary to understand at the molecular level most of the roles played by metal ions in vivo. Mainly intended to raise chemists' interest in medical applications, it should also introduce medical researchers to the techniques of inorganic chemistry so that they can realize how these techniques can help them solve the problems they face.

The second set, which largely reviews the field of the empirical observations relative to metal ions in medicine, should draw special attention to the relationships between particular metal–ligand interactions and their correlative effects, so as to outline directions for inorganic chemists and to facilitate their transposition of medical problems into physico-chemical terms. This should prompt inorganic chemists to play a more active part in medical research, in particular by taking the lead in predicting physiological or pharmacological effects before they have been fortuitously observed.

Some entries will appear in parallel in both parts of the Handbook, *Bioinorganic Chemistry* and *Bioinorganic Medicine*, which may a priori seem redundant. This, on the contrary, proceeds from the declared will to favor the opening of new avenues in interdisciplinary research, which most of the time results from considering the same notions or facts from distinct perspectives.

The above-mentioned intermediate field between chemistry and medicine remains to be created. In regard to the inorganic side, it is hoped that the Handbook will in the future be considered a determining factor of its emergence.

ACKNOWLEDGMENTS

The editor is grateful to the Institut National de la Santé et de la Recherche Médicale (INSERM, Paris, France), the Centre National de la Recherche Scientifique (CNRS, Paris, France) and the Laboratoires Guerbet (GCA, Aulnay-sous-Bois, France) for financial support to this project.

The editor also thanks the temporary part-time assistants who successively collaborated on the copyediting of the book (Mina Aït' Mbark, Zohra Benzina, Mylène Bielsa, Isabelle Garcia, Sandrine Leulier, Peta Rigaill-Visée, and Ghislaine Touches).

Guy Berthon

PREFACE TO *BIOINORGANIC CHEMISTRY*

Volumes 1 and 2

As an introduction to the second set of volumes in this Handbook—*Bioinorganic Medicine*—which is devoted to the macroscopic effects of metal–ligand interactions in human biology and medicine, this pair of volumes—*Bioinorganic Chemistry*—first deals with the fundamental aspects of these interactions at the molecular level, viewed through the current theories and techniques of inorganic chemistry. Examples of applications of the basic principles of this discipline are then given in relation to the different biomedical implications of metal–ligand interactions, with special attention to the potential uses of chemical speciation in medicine.

In biological fluids, no metal ion can really exist in the free state since the so-called "free" dissociated forms commonly evoked by chemists are in fact solvated. Nevertheless, there is a large consensus to formally consider the notion of metal complexation as corresponding to the exchange of the solvent molecules present in the metal coordination sphere with other ligands. Such exchange processes are of primary importance for the behavior of metal ions in vivo, since the nature of the complexes these locally form determines to a large extent their diverse physiological functions and metabolic pathways.

Of course, the nature of the roles metal ions are led to play in vivo derives in the first place from the chemical properties that differentiate them from one another, but specificity in these different roles always stems from the immediate environment in which each metal ion is placed. Thus, the way a given metal ion is handled or used by a living organism depends mainly on the nature of the ligands it can be associated with, and on the kind of interactions in which it is involved.

Part One addresses these fundamental aspects of metal–ligand interactions in biological fluids. After a first chapter introducing the multiple facets of bioinorganic chemistry, the second one outlines the thermodynamic background against which metal ions and ligands can interact in solution. Then, diverse aspects of the nature and character of metal–ligand interactions are developed: semi-empirical rules of affinity between acceptors and donors, fundamental features of the metal–ligand bond according to current theories, and factors of selectivity and specificity in coordination. Finally, the influence of metal–ligand interactions on the distribution of these reactants in biological fluids is described, and the corresponding physiological implications outlined.

Quantitatively, metal ions are predominantly associated with macromolecules in vivo. In addition, most of their physiological functions originate in their specific interactions with such ligands. Because of this recognized importance, Part Two is devoted to the interactions of metal ions with proteins and nucleic acids.

Starting with a general introduction to all types of metal ion–protein interactions and to techniques of their investigation, the first chapter successively considers the different types of proteins known to interact with each of the essential and beneficial metal ions (respectively defined as being necessary to life or to normal health): work proteins, metallo-enzymes with metal-containing active sites, enzymes with metal ions as structural coordinators, and storage and transport proteins.

Following this review of the individual interactions of metal ions with proteins, the next sections focus on the common roles shared by some of the corresponding metal–protein complexes. For example, almost all metal ions that naturally occur in vivo behave as catalysts in activating different sorts of enzymes. A separate section has thus been devoted to metalloenzyme reactivity and its different aspects. In contrast with inert metalloproteins in which metal ions are firmly incorporated, labile metal–protein complexes carry metal ions in exchangeable forms and play important roles in their storage and transport. Therefore they too are the subject of a particular section. Detrimental metal ions (i.e., those that are exclusively toxic at any level) exert most of their deleterious actions by competing with essential elements for the active sites of proteins. Given the important consequences of these interactions for the toxicity of these elements, they have also been treated separately.

Part Two concludes by reviewing the fundamental aspects of the interactions of metal ions with nucleic acids and their constituents, with a special section devoted to zinc proteins involved in genetic information transfer.

Among the different fractions into which metal ions are distributed in vivo, protein-associated pools represent overwhelming proportions. In the case of transition metal ions, for example, the ultrafiltrable part generally accounts for less than 1% and may thus seem negligible. It is, in fact, no less essential than its quantitatively more important counterpart.

If macromolecular fractions are indeed quite indispensable to fulfill the extremely complex and specific tasks developed in Part Two, small species identifiable as low-molar-mass [l.m.m.] complexes (usually $M < 1$ kg mol^{-1})* are no less necessary since they constitute the "dynamic units" without which the whole system could not work. Among other things, these complexes are the only exchangeable forms in which metal ions can move across membranes from one biological compartment to another and from which metalloproteins are synthesized.

Part Three deals with the interactions through which these small species are formed. After a general overview of the basic thermodynamic notions relative to metal complex equilibria, the first chapter treats the kinetic, as well as the structural, aspects of complex formation in solution.

Examples of interactions of metal ions with different kinds of l.m.m. ligands naturally occurring in vivo are then developed, with amino acids, peptides, neurotransmitters, carbohydrates, polyamines, lipids, aminosugars, and aminophosphonates being successively considered.

Following the description of fundamental principles and experimental features of metal–ligand interactions themselves, Part Four deals with the applications of inorganic chemistry techniques to several biomedical problems in which these interactions are involved.

First, metal–ligand interactions play multiple roles in metal ion metabolism. The

---

* In accordance with IUPAC specifications, molar masses are expressed in SI units, i.e., kg mol^{-1} (equivalent to kDa for the molecular masses used mainly by biochemists).

absorption of a metal from the gastrointestinal tract, its accumulation in specific parts of the body, its urinary as well as its biliary excretions, depend largely on the nature of the complexes it may form in corresponding biofluids. The first chapter deals with the physico-chemical interpretation of this dependence, successively addressing general problems of metal ion translocation and particular aspects of metal absorption, accumulation and excretion.

Metal toxicity and its clinical manifestations are extensively treated in the second set of volumes, *Bioinorganic Medicine*. Some of their physico-chemical bases are developed here, special attention being paid to the competitive processes between essential and beneficial metal ions on the one hand and detrimental metal ions on the other hand.

Recently, there has been a growing awareness about the role of oxygen-free radicals in disease and aging. Given their capacity to exist in different oxidation states, transition metal ions are logically involved in the generation and reactivity of these entities in vivo. Physico-chemical considerations relative to this topic are introduced in the following chapter, which also reviews the effects of complexation on the reactivity of iron and copper ions as well as on the protective role of zinc in this context.

During disease treatment, naturally occurring metal ions interact (intentionally or not) with most of organic drugs being administered. Diverse physico-chemical aspects of these interactions with different drug families have thus been successively considered.

Metal complexes themselves may also be administered as pharmaceuticals. Physico-chemical data underlying the pharmacological effects of such drugs are presented in the next chapter. A large section is naturally devoted to platinum anticancer compounds, but copper and gold complexes are also considered.

In addition to their pharmacological applications, metal ions and their complexes may also be used for diagnostic purposes. For example, radiopharmaceuticals have long been in common use for the detection of tumors and other abnormalities. More recently, paramagnetic metal ions and their complexes have been introduced as contrast agents in magnetic resonance imaging. A special chapter reviews the main physico-chemical aspects of these problems.

Whenever one tries to understand the effects of the interactions between a given metal ion and any substance acting as a ligand in terms of metabolism or mechanism of action of either of these reactants, the first objective should always be to know under which chemical form these reactants are present in the main biofluids and at their presumed physiological target. By definition, this objective falls within the field of speciation. Part Five is devoted to this important domain of application of inorganic chemistry to medicine.

In the present context, the experimental approach to speciation begins with the determination of the overall concentrations of metal ions and ligands in biological media. Specific techniques for these determinations are developed in the first chapter.

Since metal ions in the main biofluids are bound to ligands of different size, the first objective of speciation is to discriminate the different fractions into which each of these is distributed, and whenever possible its different complexes within each of these fractions. The next chapter describes the experimental techniques available, and underlines their limits concerning the identification of small labile species.

Since these small labile species are the above-mentioned "dynamic units" through which most metal ions can be translocated in vivo, it is of particular importance to know which of these predominate under specific circumstances. In the absence of experimental access to their distribution, computer simulation must be used. This approach is thus introduced here: diverse applications of computer-aided speciation to bioinorganic medicine are successively given, and future trends are outlined.

CONTENTS OF *BIOINORGANIC CHEMISTRY*

Volume 2

PART FOUR - BIOMEDICAL IMPLICATIONS OF METAL–LIGAND INTERACTIONS

PHYSICO-CHEMICAL PRINCIPLES UNDERLYING METAL-ION TRANSLOCATION IN VIVO. GENERAL CONSIDERATIONS

by

Constantin Luca

Polytechnic Institute of Bucharest,
Department of Analytical Chemistry and Instrumental Analysis,
Str. Polizu 1-4
77207 - Bucharest 1
Romania

I. INTRODUCTION

One of the main constituents of biological membranes are the lipids. Due to their amphiphilic structure, membrane lipids present distinct polar and non-polar parts. They organize themselves in the biological medium, most of the time as bilayers, with the non-polar part inside and the polar part in contact with the aqueous medium on both sides of the bilayer. This structure is self-organizing, forming an enclosed vesicle which divides the aqueous medium, thus generating a biological membrane to which water has little access. The enclosed structure of the membrane prevents macromolecular species crossing from one compartment to another, and the inner non-polar side of the bilayer effectively prevents the diffusion of metal ions and attenuates that of polar organic species, with the exception of particularly hydrophobic ones.

From the compartmentalization achieved by membrane lipids results the biological cell. However, cells are not simple autonomous systems. Their survival and function involve permanent changes in energy and matter between the two sides of the membrane through which transmission (drive) of information occurs.

An important part of the functions of biological membranes is due to the presence of proteins in the membrane. The functional mechanisms of membrane proteins are related to their structure and to their disposition in the cells, disposition which may be of an integral manner (intrinsic disposition) or of a peripheral manner (extrinsic disposition).

Integral proteins (Fig. 1a) are associated with membranes through hydrophobic interactions with the non-polar centers of membrane lipids. In general, integral proteins adopt a trans-membrane position, the edges of the polypeptidic chain being accessible to both membrane planes. The polypeptidic mass can be located predominantly within the membrane, or be on one or both sides of the membrane. This allows single protein molecules to participate in reaction processes on both sides of the biological membrane, and hence to facilitate the transport of matter and the coupling of energy at the cellular level.

Peripheral proteins (Fig. 1b) are associated with the membrane through electrostatic interactions with the polar centers of membrane lipids or with the polar side of integral

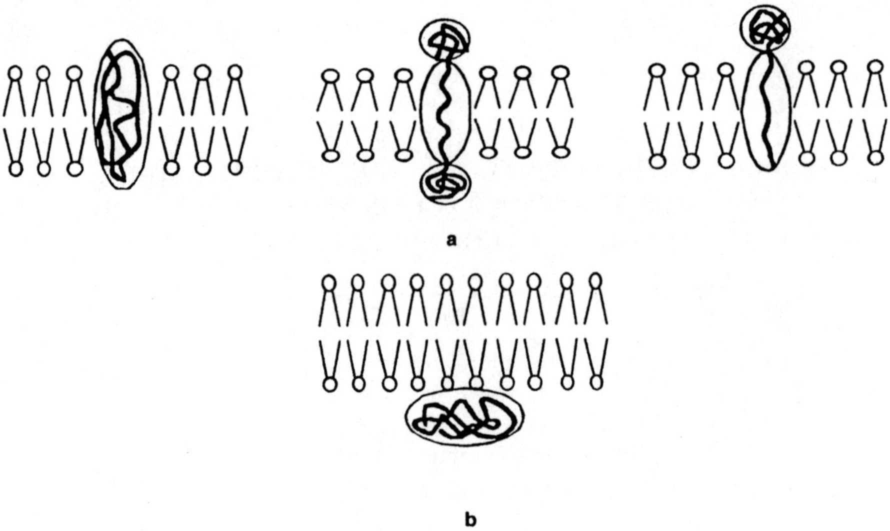

Figure 1.
Disposition of membrane proteins: a. Integral proteins, b. Peripheral proteins.

proteins which occur inside the membrane. Therefore, they establish a connection between the external aqueous medium and the membrane itself. Their peripheral location allows them to take an active part at the interface of the membrane with the surrounding aqueous medium.

The fact that the lipidic and proteic constituents of the membrane are connected only by hydrophobic interactions inside the membrane and by polar interactions at the interfaces with the aqueous phases is very important. This aspect places the membrane structures in the domain of *supramolecular structures*, and the membrane functions in the domain of *supramolecular chemistry* (Lehn, 1988).

The biological membrane has a dynamic structure. The connections between lipidic and proteic components form and open continuously, the whole system being actuated by different movements which provide the membrane with a certain flexibility (Deisenhofer and Michel, 1989; Shechter, 1990).

II. PHYSICO-CHEMICAL PROCESSES IN MEMBRANE TRANSPORT

The transport of metal ions and metabolites through membranes involves diverse mechanisms. Some of these mechanisms are of a spontaneous nature, such as *passive diffusion* and *facilitated diffusion*, whereas others are non-spontaneous, such as *active transport*. Active transport is correlated with numerous energetic couplings, chemi-osmotic, osmo-chemical and osmo-osmotic (Shechter, 1990). The trans-membrane proteins acting as *transporters* (*carriers*) play an essential part in the active transport across membranes and therefore in the membrane functions. Thus, information may be transferred from the extra-cellular medium towards the intra-cellular medium.

Consequently, the carrier allows the cell to respond to the variations of the external medium by modifying its metabolism. The receivers involved in this transfer of information are diverse: duct receivers, receivers coupled with G-proteins, receivers with tyrosine kinase activity, etc.

Soluble proteins capable of crossing membranes play an important part - though poorly elucidated; protein translocation counteracts the main function of the membrane, that is cell compartmentation.

The concentration of any protein in a given biological cell is determined at any moment by transcriptional as well as post-transcriptional mechanisms, which include biochemical modifications, proteolytic process and intracellular transport. Reviews on membrane protein turnover have been published by Thilo (1985), Rechsteiner (1987), Grisolia et al. (1988), and Birnbaumer et al. (1990).

III. MECHANISMS OF PASSIVE TRANSPORT

A. Passive diffusion

The diffusion of a chemical species X through a membrane system is spontaneous in the direction of its decreasing chemical potential, i.e. in that of decreasing concentration. The variation of free energy which accompanies the spontaneous diffusion of one mole of the chemical species X from one compartment in which its concentration is $[X]_1$ into another compartment in which its concentration is $[X]_2$ is given by:

$$\Delta G = \Delta\mu_X = \mu_{X(2)} - \mu_{X(1)} = RT\ln\frac{[X]_2}{[X]_1} \qquad (1)$$

The equilibrium state is reached when $\mu_{X(2)} = \mu_{X(1)}$, therefore when $[X]_2 = [X]_1$. The transport is possible under steady state conditions. In such a state, diffusion takes place with the chemical potentials and consequently the concentrations being constant as a result of a continuous generation of X species in the first compartment (the source phase) and of their involvement in a reaction process in the second compartment (the acceptor phase).

The transport of a species X across a membrane is defined using its flux, J_X, which represents the quantity of X crossing the membrane in a given direction through unit area in unit time, and is expressed in mol $m^{-2}\,s^{-1}$. In some laboratory investigations, instead of the area, the dry membrane mass is used, in which case the flux is expressed in mol $g^{-1}\,s^{-1}$ (g representing the mass of membrane protein or membrane lipid). From a practical point of view, it is interesting to consider the *net flux* ($J_{X_{net}}$) of the transported species from the source phase into the acceptor phase, represented by the difference of the *unidirectional fluxes* in the two opposite senses ($\vec{J}_X, \overleftarrow{J}_X$).

The net flux of the X species is expressed by Fick's first law:

$$J_{X_{net}} = -D\frac{d[X]}{dx} \qquad (2)$$

where the negative sign shows that the concentration of the X species is decreased with the increase of the distance x. The greater the concentration gradient, the greater the net flux, and the net flux is zero when the gradient is zero and the system is at equilibrium. The solubility of the species X may be different in the aqueous medium and in the membrane, and thus a partition coefficient (R) can be defined as the ratio between the solubilities in these two media.

In this case, the net flux can be expressed as:

$$J_{X_{net}} = \frac{RD}{h}([X]_1 - [X]_2) \qquad (3)$$

where h is the membrane thickness (Fig. 2). The term (RD/h) represents the permeability, P, of the membrane for the species X and has the dimension of a rate (m s^{-1}).

Obviously, the net flux depends on time through the concentration of the species X in the two system compartments. The net flux as a function of time can be expressed as:

$$J_{X_t} = P([X]_{1,t} - [X]_{2,t}) \qquad (4)$$

As a function of the initial conditions and the volume of the membrane system phases, the X species can be accumulated in the internal phase of the cell, or can leave the cell.

The cell membrane is in general slightly permeable to inorganic ions and to metabolites, their diffusion being hindered by the central hydrophobic center of the lipidic bilayer. The permeability increases with the strength of the hydrophobic character of the X species through the partition coefficient, R, and decreases with the increase of its molar mass through the diffusion coefficient, D.

Permeability is much decreased by the presence of a positive or negative charge on the X species. Thus, an uncharged species, such as an acid HA, has a greater permeability than

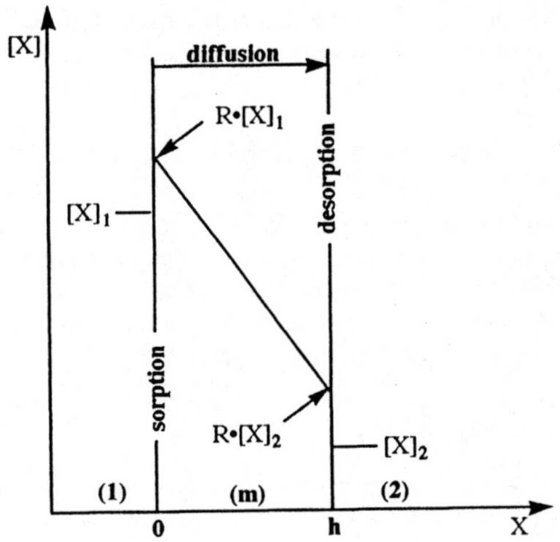

Figure 2.
Passive diffusion: (1) = source phase, (2) = acceptor phase, (m) = membrane.

its conjugated base A^-. Metal ions have very small permeabilities. Thus, the permeability of the Na^+ and K^+ ions through lipids is of the order of 10^{-14} m s^{-1}. The H^+ ions are an exception, their permeability through lipids being of the order of 10^{-6} m s^{-1}. The time to reach the equilibrium state in a lipidic membrane system is very different for the H^+ ions (order of seconds), for glucose (order of minutes) and for metal ions (1 year for K^+). These conclusions derived from in vitro experiments are not always confirmed in vivo. The transport of many organic chemical species and of metallic cations is inconsistent with passive diffusion and consequently with the simple mechanism described above. This is also the case with electrically-charged chemical species.

If the species X is electrically charged and an electric field E is acting on the system, a force

$$F = ZqE \tag{5}$$

will act on this species; here, q is the absolute value of the electric charge and Z the charge of the X species. The displacement rate of the species X at steady state, v_s, will be expressed as:

$$v_s = Zq\gamma^{-1}E = M_X E \tag{6}$$

where γ is the friction coefficient and M_X represents the mobility, i.e. the steady state of an electrically-charged species X in a 1 Vm^{-1} field; the minus sign applies for anions and the plus sign for cations. Under the influence of the electric field E, the X species is transported with a flux

$$J_X = M_X[X]E = -M_X[X]\frac{d\Psi}{dx} \tag{7}$$

where $d\Psi/dx$ represents the electric potential gradient over the distance x in the membrane system. The flux J_X is positive for a cation if $d\Psi/dx$ is negative, i.e. the electric potential decreases with increasing x; the opposite holds for an anion.

If the X species is submitted simultaneously to the action of a chemical gradient and of an electric potential gradient, the total *flux of diffusion-migration* is obtained by summing the diffusion flux (eq. 2) with the migration flux (eq. 7) as follows:

$$J_X = -M_e\frac{RT}{zF}\frac{d[X]}{dx} - M_e[X]\frac{d\Psi}{dx} \tag{8}$$

Equation (8) is known as the Nernst-Planck equation and expresses the *electrodiffusion flux*. Under thermodynamic equilibrium conditions, when the electrochemical potentials of the X species become equal in the phases (1) and (2) of the membrane, the result of (8) is that

$$RT\int_{(1)}^{(2)}\frac{d[X]}{[X]} = -zF\int_{(1)}^{(2)} d\Psi \quad (9)$$

from which, after integration, one obtains

$$[X]_2 = [X]_1\, e^{-\frac{zF}{RT}\Delta\Psi} \quad (10)$$

Consequently, the diffusion-migration couple appears when in addition to the concentration gradient, an electric potential gradient also exists in the membrane system; i.e. there is an electrochemical potential gradient. The origin of the transmembrane electrochemical potential can be either due to *ionic diffusion* or the presence in the phase system of some *non-permeative ions* across the membrane.

B. Ionic diffusion

If in a membrane system ion pairs M^+A^- exist in the source phase and the M^+ ion has a greater mobility than A^-, then the diffusion fluxes of these two ions will be different ($J_{M^+} > J_{M^-}$). As a result, a difference in electrical potential will appear at the interface of the membrane with the source phase (negative potential) and at the membrane interface with the acceptor phase (positive potential). This trans-membrane potential gradient will induce a migration process, modifying both ion fluxes:

$$J_{M^+} = -M_{M^+}\frac{RT}{zF}\frac{d[M^+]}{dx} - M_{M^+}[M^+]\frac{d\Psi}{dx} \quad (11a)$$

$$J_{A^-} = M_{A^-}\frac{RT}{zF}\frac{d[A^-]}{dx} - M_{A^-}[A^-]\frac{d\Psi}{dx} \quad (11b)$$

where M_{M^+} and M_{A^-} represent the mobility of M^+ and A^-, respectively.

In this example, the anion flux increases and the cation flux decreases. As a consequence, a compensation for the difference in mobility of the two ions takes place, and the system soon reaches equilibrium when the net charge flux across the membrane becomes zero ($J_{M^+} = J_{A^-}$).

Obviously, the charge flux at steady state depends on the difference between the mobilities of the two ions as well as on their concentrations in the two compartments (1) and (2). The resulting difference of electrical trans-membrane potential thus established is called the *diffusion potential*.

Now, taking into consideration that at a given point along the membrane the concentrations of the M^+ and A^- ions are equal and consequently their concentration gradients will also be equal,

$$\frac{d[M^+]}{dx} = \frac{d[A^-]}{dx} = \frac{d[X]}{dx} \quad (12)$$

where [X] denotes the concentration $[M^+]=[A^-]$ at the point considered. From equations 11a, 11b and 12,

$$\frac{d\Psi}{dx} = -\frac{RT}{F}\frac{M_{M^+}+M_{A^-}}{M_{M^+}-M_{A^-}}[X]^{-1}\frac{d[X]}{dx} \quad (13)$$

Multiplying the terms of equation (13) by dx and integrating between x=0 and x=h (h being the membrane thickness), one obtains an expression for the diffusion potential:

$$\Delta\Psi = -\frac{RT}{F}\frac{M_{M^+}+M_{A^-}}{M_{M^+}-M_{A^-}}\ln\frac{[X]_2}{[X]_1} \quad (14)$$

where $[X]_1$ and $[X]_2$ are the concentrations of the ion pair, M^+A^-, in the source phase and in the acceptor phase of the cell, respectively. In conclusion, the diffusion potential is determined by the difference between the mobilities of the permeative ions and by the ratio of the concentrations of these ions in the two aqueous phases of the membrane system. As long as diffusion takes place, the concentrations of the ions in both phases vary, increasing in the acceptor phase and diminishing in the source phase. Therefore, the diffusion potential decreases, accompanying the diffusion. Consequently, $\Delta\Psi$ is not an equilibrium potential.

C. Non-permeative ions

The presence of non-permeative ions in the membrane system can also produce a gradient of electric trans-membrane potential. Let us consider for example the simultaneous presence of an ion pair, M^+A^-, in the source phase and of another ion pair, M^+B^-, in the acceptor phase (M^+ being a common ion), the concentrations of these two ion pairs being equal and the B^- anion being unable to cross the membrane. The following apply:

(i) at time t=0, the electric potential has the same value in phases (1) and (2) of the membrane system;

(ii) the net flux of the common cation M^+ is zero ($J_{M^+_{net}} = 0$);

(iii) the non-permeative anion B^- does not diffuse, its concentration being constant in the acceptor phase;

(iv) the permeative anion A^- diffuses from (1) to (2) which induces the appearance of a trans-membrane potential gradient;

(v) the electric potential, negative in the acceptor phase and positive in the source phase, inhibits the diffusion of the A^- anion from the source phase to the acceptor phase and at the same time the migration of the M^+ cation in the same direction;

(vi) the fluxes of the M^+ and A^- ions cancel out as the electrochemical potentials of these permeative ions reach equal values in the two phases.

Taking account of equation (10), after reaching the equilibrium state, one obtains:

$$\frac{[M^+]_2}{[M^+]_1} = \frac{[A^-]_1}{[A^-]_2} = e^{-\frac{zF}{RT}\Delta\Psi} \tag{15}$$

where $\Delta\Psi = \Psi_2 - \Psi_1$. If more than one permeative ion (anions and cations) and the non-permeative anion B^- are present in the membrane system, and imposing the conditions of electroneutrality of phases and of concentration balance from (15), the so-called Donnan potential is obtained:

$$\Delta\Psi = \frac{RT}{F}\left[\left[1 + \frac{[B^-]_2}{[2[Y]_1]^2}\right]^{1/2} + \frac{[B^-]_2}{2[Y]_1}\right] \tag{16}$$

where $[Y]_1$ is the sum of the cation concentrations in phase (1), and is equal to the sum of the anion concentrations in the same phase. Relation (16) is written for ionic species with unit charge. The trans-membrane, Donnan-type potential is an equilibrium potential and depends on the concentration of the permeative species in the source phase and on the concentration of the non-permeative species in the acceptor phase.

The total value of the trans-membrane potential gradient is obtained by summarizing the Donnan potential with the diffusion potential. Both the diffusion transport and the migration transport are based on *mechanisms of passive transport*.

IV. FACILITATED DIFFUSION

There exist membrane systems in which membrane phase contains a chemical species which is able to complex the transported ion(s), this making the transport mechanism different from passive diffusion.

Such a chemical species, involved in the transport process, is called a *carrier*. The carrier interacts specifically or selectively with the transported ion. In such systems, the transported ion is translocated from the source phase into the acceptor phase with a rate that is characteristically greater than would be predicted from passive diffusion. Moreover, its rate of translocation reaches saturation at high concentrations of the transported species in the acceptor phase, and the presence in the source phase of other species which are analogous to that transported inhibits the translocation process.

Transport that involves the participation of such carriers is called *facilitated transport*. As in the case of passive diffusion, the diffusion process in facilitated transport occurs in the direction of decreasing electrochemical potentials. The transport which takes place is spontaneous and the net flux of the transported ion becomes zero when its electrochemical potential becomes equal in the acceptor phase and the source phase.

The carrier must present a center of specific binding of the transported ion. The facilitated transport mechanism (Fig. 3) involves several steps:

(i) The *binding step* of the transported species, X, at the interface between the membrane where the carrier T is located and the source phase in which the concentration of the X species is $[X]_1$. At this interface the equilibrium reaction

$$T_{(1)} + X_{(1)} \underset{k'_1}{\overset{k_1}{\rightleftharpoons}} TX_{(1)} \qquad (17)$$

is established, characterized by the equilibrium constant, K_1

$$K_{(1)} = \frac{k'_{(1)}}{k_{(1)}} = \frac{[T]_1[X]_1}{[TX]_1} \qquad (18)$$

in which k'_1 and k_1 represent the rate constants of the reaction process (17). If the carrier T binds several ions from the source phase, the equilibrium constants of the TX-type complexes allow one to estimate the selectivity of the carrier towards a group of similar X ions.

(ii) The *translocation step*, which involves the actual transport of the species X across the membrane in the form of the complex TX from the interface with the phase source to the interface with the acceptor phase. This transport is achieved by the carrier T, which after binding the species X, delivers it at the interface with the acceptor phase.

Other translocation mechanisms are possible in biological membranes, where the protein carriers of large molecular size can effect the transport by rotational movements of the binding center from one interface to the other or by specific conformational changes which lead to the formation of hydrophilic channels, which may sometimes be particular for metallic cations (see Chapter 1-A2).

(iii) The *dissociation step* of the X species at the membrane interface with the acceptor phase. This step is controlled by an equilibrium reaction process which is the opposite of

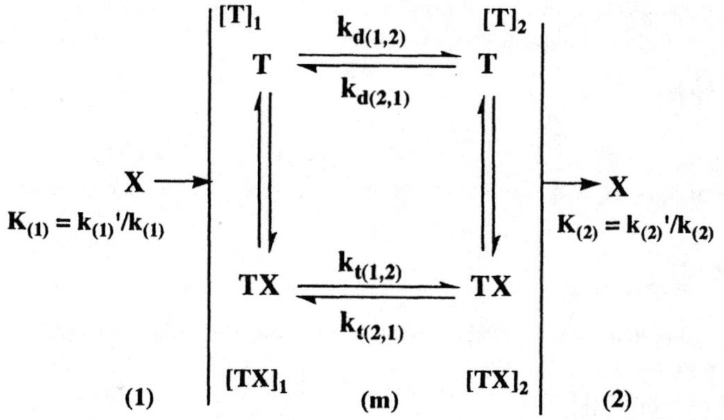

Figure 3.
Facilitated diffusion.

process (17):

$$TX_2 \underset{k_2}{\overset{k'_2}{\rightleftharpoons}} T_2 + X_2 \tag{19}$$

and is characterized by the equilibrium constant K_2 (written with the same meaning as in (17), i.e. as a dissociation constant):

$$K_{(2)} = \frac{k'_{(2)}}{k_{(2)}} = \frac{[T]_2[X]_2}{[TX]_2} \tag{20}$$

The diffusion of the free carrier T from the interface (1) to interface (2) is under the control of the rate constant $k_{d(1,2)}$, and the translocation of the species X between the same two interfaces by diffusion of the complex TX is controlled by the rate constant $k_{t(1,2)}$. The rate constants $k_{d(2,1)}$ and $k_{t(2,1)}$ apply to the same processes working in the opposite direction. Expressing the rate transport constant by fluxes and taking into account that the net flux is given by the difference of the unidirectional fluxes, the equation of the net flux of the X species can be written as:

$$J_{X_{net}} = k_{t(1,2)}\,[TX]_1 - k_{t(2,1)}\,[TX]_2 \tag{21}$$

Similar equations can be written for the diffusion of the free carrier, with corresponding rate constants $k_{d(1,2)}$ and $k_{d(2,1)}$.

One can deduce that at time t=0, an accumulation of carrier molecules takes place at the interface with the source phase, but, after a time necessary to reach the steady state, the carrier concentration becomes equal at both interfaces. Then, the equation corresponding to the steady state can be written as:

$$k_{d(1,2)}[T]_1 + k_{t(1,2)}[TX]_1 = k_{d(2,1)}[T]_2 + k_{t(2,1)}[TX]_2 \tag{22}$$

which in combination with equation (21) and with the distribution of the carrier in the system,

$$[T]_{total} = [T]_1 + [T]_2 + [TX]_1 + [TX]_2 \tag{23}$$

allows to obtain the rate equation of the transport in the steady state. Taking into consideration the simplified case of a *symmetrical carrier*, for which $K_1 = K_2 = K$ and that the rate constants of species T and TX are also equal in both directions,

$$\vec{J}_X = \frac{k[T]_{total}}{2}\left[\frac{[X]_1}{K+[X]_1}\right] \tag{24}$$

The unidirectional flux of the X species from (1) to (2) is therefore a hyperbolic function of the concentration of this species in the source phase, and tends to a maximum value, $\vec{J}_{X_{max}}$, for a given value of $[X]_1$. From equation (24) it follows that for $[X]_1 = K$

$$\vec{J}_X = \frac{k[T]_{total}}{4} = \frac{\vec{J}_{X_{max}}}{2} \tag{25}$$

Thus, an apparent transport constant, $K_{1/2}$, is defined which corresponds to the concentration of species X in the source phase when the flux $\vec{J}_X$ equals half its maximum value:

$$K_{1/2} = [[X]_1]_{\vec{J}_X = \vec{J}_{X_{max}}/2} \tag{26}$$

Similarly, under the same conditions one can deduce the unidirectional flux expression in the opposite direction:

$$\overleftarrow{J}_X = \frac{k[T]_{total}}{2}\left[\frac{[X]_2}{K+[X]_2}\right] \tag{27}$$

From equations (24) and (27) it follows that $(\vec{J}_X)_{[X]_2=0} = (\vec{J}_X)_{[X]_2=0}$ and therefore the flux $(\vec{J}_X)_{[X]_2=0}$ can be expressed by:

$$(\vec{J}_X)_{[X]_2=0} = \frac{[X]_1(K+\rho[X]_2)}{(K+[X]_1)(K+\rho[X]_2)+(K+\rho[X]_1)(K+[X]_2)} \rho k_{d(1,2)}[T]_{total} \quad (28)$$

where $\rho = k_{t(1,2)}/k_{d(1,2)}$, and the net flux $J_{X_{net}}$

$$J_{X_{net}} = \frac{K([X]_1-[X]_2)}{(K+[X]_1)(K+\rho[X]_2)+(K+\rho[X]_1)(K+[X]_2)} \rho k_{d(1,2)}[T]_{total} \quad (29)$$

The flux of the species X from (1) to (2) evidently depends on the concentration of this species in both phases and also on the total carrier concentration, $[T]_{total}$. If the rate constants of the free carrier, T, and of the TX complex are different, it results that $\vec{J}_{X_{max}}$ increases as a function of $[X]_1$ along with the ratio value, ρ.

On the other hand, the value of the apparent constant $K_{1/2}$ decreases when ρ is increased. The dependence of the flux on $[X]_1$ at a given value of the concentration $[X]_2$ is always hyperbolic.

There are two cases which are related to the action of the mobile carrier T on the translocation process of the X species from the source phase into the acceptor phase as a function of the ratio value, ρ:

- *trans-stimulation*, when $\rho > 1$;
- *trans-inhibition*, when $\rho < 1$.

If the carrier T also recognizes and transports a second species, X', present together with X in the source phase, a competitive inhibition may appear. Thus, when both the species X and X' are present in the system, their fluxes are often different from their values determined separately. When the TX' complex is more stable than TX, the carrier is largely required to form TX' and consequently the transport process of X is inhibited.

When the dissociation constants of the TX complex have different values at the membrane interfaces with phases (1) and (2), the carrier has an asymmetric action and is called an *asymmetric carrier*. The expressions of the unidirectional and net fluxes can be deduced by similar reasoning to that given for symmetric carriers.

A. The coupled transport: cation-anion

If the transported species is a cation, X^+, and the carrier is neutral, the TX^+ complex which is formed is also positively charged, a fact which reduces the hydrophobic character of the carrier. In this case, the association of an anion A^- with which the cationic complex forms an ion pair, TX^+A^-, is possible. If the A^- anion exists in the source phase, it may be transported together with X^+ in the acceptor phase. Consequently, a simultaneous coupled transport cation-anion takes place. This transport mechanism is governed by the same principles as presented above for the X species. The transport takes place in the direction of decreasing concentration of species X^+ and A^- ("down-hill" transport) and depends in large measure on the nature of the coupling anion A^-, especially its hydrophobicity.

V. ACTIVE TRANSPORT

By coupling the transport processes of two or more chemical species, the movement of one can be mediated by the chemical or electrical gradient due to other species in the system. The transport of one is said to "pump" the transport of the other against its concentration gradient. Such a process always involves the coupling of a counter-flux of ions or electrons. The following example illustrates this mechanism (Fig. 4).

An anionic carrier, T^-, having a basic character (proton acceptor) and present in the membrane, fixes the cationic species, X^+, at the interface with the source phase (aqueous solution at a pH which stabilizes the carrier in its T^- form), forming the ion pair T^-X^+ which passes through the membrane by diffusion:

$$(T^-)_m + (X^+)_1 \rightarrow (T^-X^+)_m \quad (30)$$

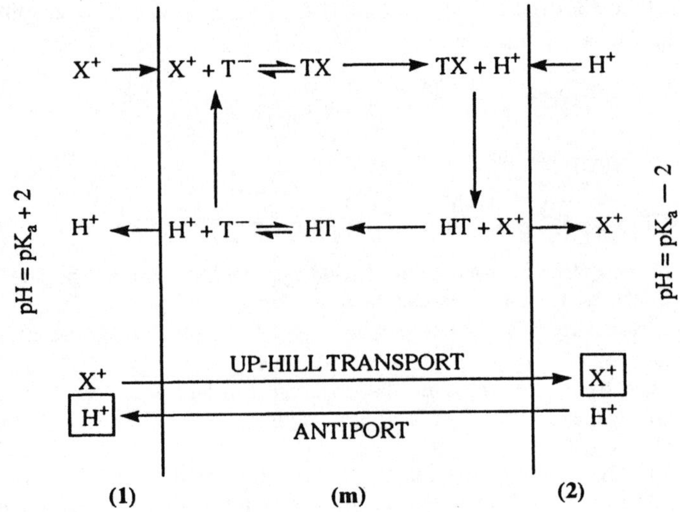

Figure 4.
Active transport. T^- = anionic carrier (proton acceptor).

At the membrane interface with the acceptor phase (aqueous solution with a pH value at which the carrier T^- is transformed into TH), the carrier T^- from the ion pair T^-X^+ fixes the hydrogen ion, a process which releases the free cationic species X^+ in the acceptor phase:

$$(T^-X^+)_m + (H_3O^+)_2 \rightarrow (TH)_m + (X^+)_2 + (H_2O)_2 \tag{31}$$

The acidic conjugate TH of the carrier thus formed at the interface with the acceptor phase then diffuses toward the interface with the source phase where it is deprotonated:

$$(TH)_m + (OH^-)_1 \rightarrow (T^-)_m + (H_2O)_1 \tag{32}$$

The displacement of equilibrium (32) increases the concentration of the active form, T^-, of the carrier, which in turn produces a simultaneous displacement of equilibrium (30) to the right and consequently permits movement of a new X^+ ion in the transport process toward the acceptor phase.

Hydrogen ions are transported across the membrane in the reverse direction relative to the transport of the X^+ species. The hydrogen ion transport is determined by the pH gradient that exists between the two aqueous phases of the system. The backward transport - *antiport* - of H^+ ions produces chemical changes of the carrier, which takes the form T^- (able to fix the X^+ species) at the interface with the source phase and the TH form at the interface with the acceptor phase. By this mechanism, the transport of the X^+ species becomes an *active transport*; the X^+ species could be transported against its concentration gradient ("up-hill" transport). Such an active transport is also called "ionic pump" (Fyles and Malix-Diemer, 1982).

The example considered is a "pump" of a X^+ species, mediated by a pH gradient between the aqueous phases of the membrane system. The pH values in the source and acceptor phases needed for the pump to function are determined by the value of the acidic constant for the couple TH/T^-:

$$TH + HOH \rightarrow T^- + H_3O^+ \tag{33}$$

$$K_a = \frac{[T^-][H_3O^+]}{[TH]} \tag{34}$$

Thus, transforming the expression of the constant K_a to a Henderson-type relationship,

$$pH_{(1)} = pK_a + \log\frac{[T^-]}{[TH]} \tag{35}$$

one can see that in the case of a predominant transport (99%) in the basic form T^-, the pH of the source phase at the interface must have the value $pH_{(1)} = pK_a + 2$. By a similar procedure, for a predominant transport in the TH form, the pH must be $pH_{(2)} = pK_a - 2$ at the interface with the acceptor phase.

Consequently, the optimum pH gradient between phases (1) and (2) must have values of

$$pH \geq (pK_a \pm 2) \tag{36}$$

As the pump functions, the pH of the source phase decreases and the pH of the acceptor phase increases so the pH gradient decreases, tending to zero, which would stop the pump. Therefore, a pH gradient must be maintained in the membrane system to keep the pump functioning; this is achieved by means of a buffered medium or a competitive reaction process.

Besides H^+, other ions may play a mediatory role in mechanisms similar to active transport, facilitated by carriers with adequate reaction properties (Shinkai and Kinda, 1982; Schechter, 1984).

Generally the carrier is involved in reaction processes at the membrane interface with both the source phase and the acceptor phase. Thus, differences of electrochemical potential appear between the phases, which induces the transport of one or more chemical species either in the same sense - *symport* - or in opposite directions - *antiport*.

REFERENCES

Birnbaumer, L., Abramowitz, I., and Brown, A.M. (1990). *Biochim. Biophys. Acta, 1031*: 163.

Deisenhofer, I., and Michel, H. (1989). *Science, 245*: 1463.

Fyles, T.M., and Malix-Diemer, V.A. (1982). *Can. J. Chem., 60*: 2259.

Grisolia, S., Hernandez-Yago, I., and Knecht, E. (1988). *Curr. Top. Cell Regul., 27*: 387.

Hare, I.F. (1990). *Rev. Biomembranes, 1131*: 71.

Lehn, I.M. (1988). *L'Actualité Chimique*, Nov.-Déc., 265.

Rechsteiner, M. (1987). *Annu. Rev. Cell. Biol., 3*: 1.

Shechter, E. (1984). *Membranes Biologiques*, Masson, Paris.

Shechter, E. (1990). *Biochimie et Biophysique des Membranes*, Masson, Paris.

Shinkai, S., and Kinda, K.J. (1982). *J. Chem. Soc. Chem. Commun.*, 125.

Thilo, L. (1985). *Biochim. Biophys. Acta, 822*: 243.

MOLECULAR DYNAMICS SIMULATIONS OF METAL ION INTERACTIONS WITH CHANNELS

by

Osvaldo Alvarez,

Departamento de Biologia,
Facultad de Ciencias,
Universidad de Chile,
Casilla 653
Santiago
Chile

Frederick S. Lee and George Eisenman

Department of Physiology,
UCLA School of Medicine,
Los Angeles, CA 90024
U.S.A.

I. SUMMARY

We describe some solved and many unsolved problems in applying Free Energy Perturbation / Molecular Dynamics (FEP/MD) simulations to explore the permeation properties of three structures postulated for the Shaker K channel by Guy and Durrell: a ***short beta barrel***, an ***intermediate beta barrel***, and a ***random coil***. We accept and enumerate some of the many problems in starting from a conjectured, rather than measured, 3-dimensional structure. We also introduce such limitations as starting from strictly 4-fold symmetrical structures whose ionizable residues are left un-ionized for simplicity. We do this because we believe that such simulations can provide complementary information to that obtained from genetic manipulation, can narrow down the range of acceptable structures, and will provide experience of use in deducing the permeation properties of reliable 3-dimensional structures when these ultimately become known from crystallographic and other structural techniques. We also suggest that useful insights into the fundamentals of ionic selectivity can be gleaned from such simulations.

Within these limitations we find that the short beta barrel is the only structure of the three that is compatible with a K-permeable channel. The energy profile for intermediate beta barrel is incompatible with ion conduction, and the random coil structure forms an Na-permeable channel. In all cases the energy profile is dominated by ion hydration: unfavorable energies occur where water molecules are lost, favorable energies where the ion is surrounded by many water molecules.

We have used spherical boundary conditions and King and Warshel's surface constrained all-atom solvent procedure for hydrating the channel. This procedure was developed for globular proteins and has not yet been critically tested in a narrow channel within a protein. Some of our findings suggest that artifactually favorable energies may result from these boundary conditions. We point this out by way of a warning, noting that this particular problem will not arise when using periodic boundary conditions.

Though, for simplicity, we have restricted our simulations to occupancy by no more than one cation in a structure whose acidic and basic residues are un-ionized, we discuss the necessity of removing this restriction in the future (for example, allowing acidic residues to ionize in order to balance the charge when several cations at a time are present simultaneously). If this is done to allow for as many as 4 cations to find acceptable "single-filing" positions, it seems highly likely that the carboxylate residues would distribute themselves along the channel axis; and the strict 4-fold symmetry of the present Guy-Durrell models would be lost.

II. INTRODUCTION

Ion channels are integral membrane proteins that create pathways for ions across biological membranes. For a protein to work as a channel it must be folded to create polar tunnels connecting both faces of the membrane. These tunnels must have a size wide enough to allow ions to move freely and must be lined with chemical groups able to interact with ions to maintain a low potential energy. Multi-ion channels must also provide several charged sites to allow simultaneous occupancy by several ions. Ion channels not only provide a transmembrane pathway but also can be exquisitely selective among different ions. They can be gated by a wide variety of stimuli such as chemical agonists and transmembrane electric fields. Understanding of how a channel works requires a knowledge of its detailed three-dimensional structure at the atomic level. Since this information is still missing, we can only report here our progress in developing strategies to evaluate the functionality of different structures deduced from conformation prediction procedures. Checking the implied conduction properties of alternative structures should help focus the efforts in searching for correct 3-D structures.

We have chosen as subject the Shaker K-channel, which is a K-selective channel that is defective in the "shaker" *Drosophila Melanogaster* mutants, and was the first K channel whose DNA was cloned (Papazian et al., 1987; Iverson et al., 1988; Pongs et al., 1988). Shaker is a representative example of a large family of voltage-gated channels (Caterall, 1988); its physiological properties have been extensively studied, its amino-acid sequence is known, its gene has been cloned and subject to point mutations which have been correlated with functional consequences (MacKinnon and Yellen, 1990; Yellen et al., 1991; Yool and Schwarz, 1991; Heginbotham and MacKinnon, 1992). As a result of these studies, a widely accepted general model of the protein folding has emerged (Miller, 1991), and plausible structures detailed to the atom level have been proposed (Durell and Guy, 1992; Bogusz et al., 1992). In these structures several potential ion-binding ligands occur in the lining of the channel, such as hydroxyl groups of tyrosine, tryptophan and threonine, carboxylate groups from aspartate, carbonyls from the protein backbone, and aromatic rings. We will present estimations of the conduction and selectivity properties of various of these models and describe a strategy for estimating free energy profiles of ions in channels.

III. METHODS

A. General

Our methods are designed to evaluate the potential energy of transfer of an ion from water to the channel. A Born-Haber thermodynamic cycle enables this energy to be calculated from four separate processes, namely: (i) discharging the ion in water, (ii) removing the "uncharged ion" from water, (iii) inserting the "uncharged ion" into the channel, (iv) charging the ion in the channel. The same steps in reverse are thermodynamically equivalent. Among these steps, the charging and discharging contributions are the largest. The energy to discharge and then remove the ion from water can be obtained from thermodynamic data on hydration energies. The work of inserting the ion into the channel and charging it there must be calculated, which we do using the MD/FEP method (see e.g. Warshel, 1991). Calculations involve the simulations of the system at room temperature and consider the effects of dynamic fluctuations of the structure. The charging process is performed as a series of steps in which the charge of the ion is incremented by a small amount while the channel is allowed to rearrange and fluctuate around the ion. The time-average difference in energy of the charged and uncharged states of the ion placed into that fluctuating channel configuration is collected for each step. The total energy of the charging process is obtained by the integration of all the steps.

Ideally, all the atoms of the protein that make up the channel, the atoms of the membrane that contains the channel, and the water molecules and components of the surrounding solutions should be considered in the simulation. This is materially impossible and the system has to be reduced to tractable dimensions. King and Warshel (1989) successfully estimated the free energy of solvation of ions by cutting the system to a sphere centered at the ion for the case of an ion in pure water. We have followed this idea and used a sphere of radius 11.5 Å centered at the ion at various positions in the channel. This sphere contains a small fraction of the protein atoms (about 2000), no membrane atoms, and a finite number (20 to 200) of water molecules. The rest of the protein atoms are immobilized. Our calculations compare the difference of free energy calculated using MD/FEP for such a reduced system with that of an ion in an 11.5-Å droplet of SPC water molecules. The present knowledge of the structures does not justify the use of more complete systems. Some shortcomings of this strategy are discussed.

B. Choosing a force field

Of the various Force Fields available for proteins, we chose GROMOS (van Gunsteren and Berendsen, 1987) because it has been widely tested; but we also carried out some comparisons with MOLARIS (Warshel and Creighton, 1989) which is better able to handle properly the distant interactions using Local Reaction Field approach (Lee and Warshel, 1992) and with which we can include ion-induced dipole interactions (see Part Two, Chapter 1-B6 of this volume) when this becomes appropriate.

C. Ion model

Ions are described by their charge and a set of 6-12 Lennard-Jones parameters for van der Waals forces. We used a self-consistent set of parameters that describe simultaneously the correct hydration energy and radial distribution functions of water (Aqvist, 1990; 1991; 1992).

D. The water model and boundary conditions

We use spherical boundary conditions and the Surface Constrained All Atom Solvent (SCAAS) procedure of King and Warshel (1990) with a flexible SPC water, which has been calibrated to simulate the physical properties of water and ion solvation. Radial constraints are used to keep the water molecules on the surface from evaporating, and angular constraints are used to prevent the surface waters from over-polarizing. An alternative procedure would have been to use Periodic Boundary Conditions, which might have circumvented certain problems in hydrating the channel that are discussed below but might have introduced other complications. The best procedure to be used for a channel is still an open question.

E. Adding water to the channel

To calculate energies of ions through the channel we create an ion at different places in the structure, and the empty space not occupied by protein atoms within a 11.5-A sphere from the ion is filled with water molecules. At confined and tortuous locations in the channel it is not easy to fill the space with the correct water density, because water molecules are discrete. This is not trivial because errors in the density can introduce pressure or tension that distorts ion-water and protein-water interactions. Possible artifacts that can arise from this source are discussed in Section VI.A.1. Our procedure creates a 2-Å cubic grid of possible water locations. Each location is examined for contacts with other water, protein atoms, or ions and discarded if these exist. We tested 50 random grid centers near the ion to select the one giving the best packing. Thermal shaking under surface constraints is then used to relax the cubic grid to the final water radial distribution function.

F. Holding the ion in place
For mapping the free energy path it is necessary to take readings at various z locations along the channel axis, and this is accomplished by constraining the ion to stay in place during the MD/FEP procedure. We have found that in the charging steps where the ion is uncharged or weekly charged, it tends to drift out of the 11.5-Å water drop. To avoid this, which would invalidate the FEP results, we used an harmonic constraint of 50 kcal/Å^2 on x, y and z directions to hold the ion in place. In addition, we so far have restricted the exploration to the path along the channel axis.

G. Reducing the number of atoms
We treat CH, CH_2 and CH_3 groups as single super-atoms. In an average protein where all the 20 amino acids are present in the same proportion, the atom count is reduced by 47%.

H. Holding the protein atoms in place
Since the system has been truncated, atoms may be able to move more freely than in the complete system because of losing interactions with the missing parts. To make the reduced system retain its original structure, atoms contained in a 11.5-Å sphere centered at the ion coordinates are subject to a restraining force of 0.001 kcal/Å^2, and the rest of the structure is held in place by a 30 kcal/Å^2 restraining force.

I. Distance cutoffs for ion-protein interactions
Since the atoms outside the 11.5-Å sphere are not free to move, a problem arises in calculating the contribution of these atoms to the free energy of the ion. The MD/FEP method is based on calculating the work of charging the ion in small increments while the surrounding protein and water atoms move around, driven by thermal motion and the emerging electric field generated by the increasing charge of the atom. If the immobilized atoms happen to be in a configuration that has favorable interactions with the ion, the free energy will be overestimated. To avoid this, we have chosen not to include the energy from inmobilized atoms by using a radial cutoff of 11.5-Å in calculating ion-protein interactions.

J. Simulating the membrane environment
Another problem is created by using an incomplete system. The MD/FEP calculations give a detailed treatment of the interaction of the ion with protein and solvent atoms located within a 11.5-Å sphere, and the rest of the system is treated as a continuum, using Born's energy. Still another problem arises here because the dielectric constant of the surroundings varies as the ion moves to various locations within the channel. Born's energy will be less favorable as waters are shed within the channel. Since this energy is unknown, we have chosen to compare the energy of charging the ion in a 11.5-Å sphere carved into the channel to charging the ion in a 11.5-Å sphere of water. Born energies are left out, and this will give an extra energy within the channel equal to the difference in Born's energy; in the worst case this can be as much as -14.5 kcal for a 11.5-Å sphere. If the whole protein were used, imbedded in a phospholid bilayer, and with a very long cutoff radius, the problem would be solved.

K. Ionization state of acidic and basic residues
For simplicity, none of the acidic or basic residues were considered as ionized in the initial exploration reported here. Allowing the groups to ionize is a formidable problem which will only be worth tackling when the possible structures have been narrowed down considerably. This will be discussed in Section VI. B.

IV. THE SHAKER CHANNEL

The Shaker channel is a tetramer of four identical subunits that form a single pore at the axis of assembly of the subunits. The Guy-Durrell models assume that each subunit makes exactly one fourth of the channel wall and that 4-fold symmetry exists. Each monomer has six putative transmembrane alpha-helix segments named S1 to S6. There is compelling evidence that the channel is formed by the segment linking S5 and S6, specifically by the so called P-segment that starts on PRO430 and ends on PRO450. The effect of point mutations on TEA binding demonstrate that THR441 is accessible from the internal (intracellular) side of the channel and THR449 is accessible from the external side (MacKinnon and Yellen, 1990; Yellen et al., 1991). The positions of these two THR residues impose a severe constraint that must be met by all models of the pore. In addition, the 9 aminoacid residues ranging from THR441 and THR449 are more conserved than the rest of the P-segment, as revealed by comparison of the structures of different K channels. This 9-residue stretch is thus a good candidate to be the lining of the pore.

A. Choosing starting structures

There are several models of the detailed structure of the pore, constructed by different folding, that vary in the number of residues that face the inside of the pore. All model structures should have PHE443 and THR449 at the external side and THR441 at the internal side of the channel. Therefore the P-segment must be configured in a U-shape with the turn at the inside and the ends at the outside. We have examined three Durrell-Guy structures (Durrell and Guy, 1992; and personal communication) for the P segment of the Shaker channel: a short beta barrel, an intermediate beta barrel and a random coil model. All of these were assumed to be 4-fold symmetrical and then optimized in vacuo and in the absence of ions by Guy and Durrell using CHARMM. There is no experimental evidence that 4-fold symmetry actually exists; and some relaxation of this restriction, which causes all ionizable residues to lie in one plane, may be called for ultimately in trying to model a file of sites in the channel when fully ionized and occupied by more than one ion at a time. More will be said about this in Section VI.2.

The ***short beta barrel*** starts at the external side on VAL437, turns in the inside at VAL437 and ends on the external side on THR449. Among the foldings explored here, this is the one that presents the least obvious relation with the mutation affecting ion permeation. In the pore lining of this structure, PHE433 and TRP434 are not included and VAL 438 points away from the pore; however, TYR445 and ASP447 do face in. We have chosen it because it is the structure that allows the best overall folding of the protein consistent with Guy's and Durrell's models of gating. The ***intermediate beta barrel*** starts at the external side on PHE433, turns at THR441 and ends on residue 449 that has been mutated to TYR449. It has ASP431, TRP434, VAL438, TYR445, and ASP 447 projecting into the pore. The ***random coil*** starts at ASP439, turns at THR442 and ends also on TRY449. It has TRP434, VAL438, TYR445, and ASP447 facing in.

B. Preparation for MD/FEP

Our starting atom coordinates are those of the energy-minimized four-fold symmetrical tetramers supplied by Durrell and Guy. To keep the number of atoms manageable, we trimmed the structure leaving only the P segment and whatever other protein residues had any atoms within 15 Å the channel axis. In effect, we kept only the residues from VAL345 to PHE404 and from SER424 to HIS486. We move the ion along the z-axis and we calculate energies only for the atoms located in a sphere within 11.5 Å of the ion. Thus, for all the z locations, we in effect use all atoms located inside a cylinder of 11.5 Å radius surrounding the axis of the channel. The first operation on the structures was a short 0.2-ps (100 steps) annealing run of dynamics at 1K to relieve any bad atom contacts that may became apparent when passing from CHARMM to GROMOS force fields.

C. Equilibration

We then carried out 10 ps (5000 steps) of equilibration at 298K, with either a non-charged or fully charged ion at a given z-location. This equilibration gives time for water molecules, initially arranged in a cubic grid, to randomize and approach to the correct radial distribution function. The protein configuration is also relaxed to optimize water-protein and ion-protein interaction. We compared the results obtained when equilibrating with the charged or the uncharged ion. We found very different results, that will be presented in section V.B.2. When the structure is equilibrated with the charged ion in place, we get larger solvation energies than equilibrating with the uncharged ion. This is probably because during the long equilibration with the charged ion the system can jump over high energy barriers to reach a very stable state, from which it will not be able to exit during the discharge process. We trust more the charging procedure because the system has less chance to get trapped in energy minima, and a more representative sample of the energy states can be acquired during the (necessarily) limited simulation time. This is why we choose to use equilibration with the uncharged ion for the present exploration.

D. Ion charging

The charging process consists of eleven 2-ps long steps at 298K, each with the ion increasingly charged. In the first step the ion is totally discharged, and in the succeeding steps, 10% of the charge is added to have the ion fully charged on the 11th step. At each charge the system is allowed to fluctuate and reach a quasi-static state. We allow 1 ps (500 steps of dynamics) for equilibration at each partial charge and we take energy readings during another 1 ps. At each partial charge, the Boltzman's average of the potential energy of the fully charged and fully discharged ion forms are computed. The potential energy is composed of the Coulomb and van der Waals interactions of the ion with the protein atoms and the water atoms present within the 11.5-Å sphere. The difference of these two averages is the slope of the free energy vs charge curve at full charge. The total free energy of charging is obtained by integration of the 11 steps. As discussed in section E, we also tested a discharging protocol, obtaining different results. The discrepancy between charging and discharging results should be resolved if longer simulation times were used, but we do not know at present how long is long enough.

V. RESULTS

A. Short beta barrel

1. *Structure*

The original Durell-Guy short beta barrel structure of the part of the P-segment that makes the narrow part of the pore is drawn in Figure 1 (Molscript drawings (Kraulis, 1991)). Atoms are represented as balls with a radius 1/4 of the atom radius, and bonds are represented as sticks. Only two of the four monomers are drawn. Peptide bond C and N atoms of the inward segment are highlighted on the upper monomer, and those of the outward segment are highlighted on the lower monomer. In this beta structure, the side chain of every other residue points into the pore. These are VAL437, THR439, THR441, on one limb, TYR445 and ASP447 on the other.

2. *Energy profile and ligands*

The Free Energy of transfer of an ion from water into the channel is plotted in Figure 2. It can be described as a three-barrier four-site profile; there are no entry barriers. As the ion approaches from either side of the channel, it gets attracted by the channel as shown by the negative values of energy. For example, approaching from the left (intracellular side) the ion reaches a minimum at z = -4. To analyze the ligands we have collected all the oxygen atoms located within a 5-Å sphere with center at the ion. These data are summarized in Table 1 where the ligands located at 3 Å or closer are also listed.

3. *Selectivity*

Comparison of the energy profiles for the various ions gives information about ion selectivity. The highest peak occurs at z = +2 and, since permeabilities are inversely related to peak height, this structure is a K-selective channel with a permeability sequence of K> Rb> Cs> Na. On the other hand, the lower the energy level is on these plots, the stronger is the binding. Examining the energies at the various loci, it is seen that only two binding selectivity sequences occur. The sequence Na> K> Rb> Cs is found at z = -4, +4, +6 and +8; while the sequence K> Rb> Cs> Na is found at z = -2, 0, +2 and +10. It is noteworthy that both of these are regular Eisenman Selectivity Sequences (Eisenman, 1961; 1962).

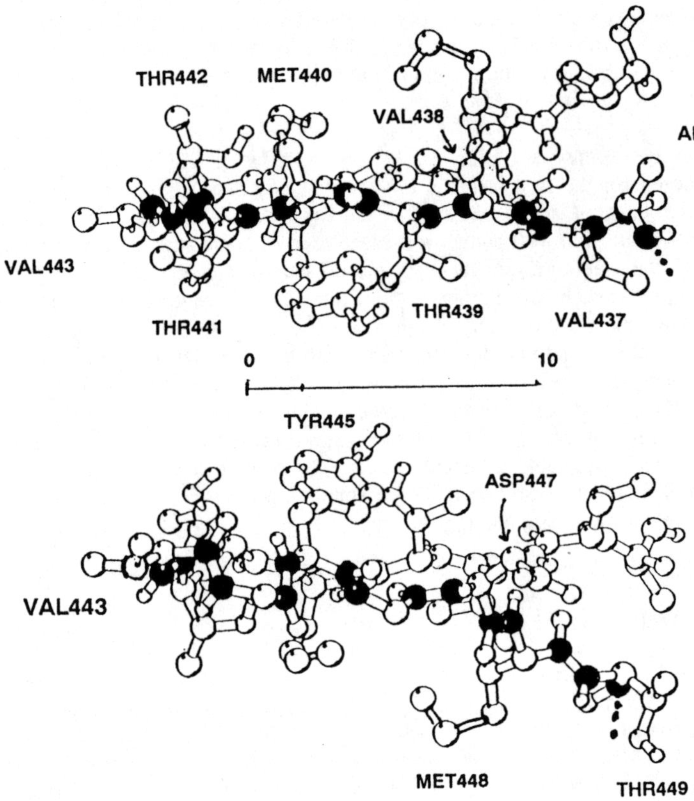

Figure 1.
Short Beta Barrel Structure. Durell and Guy's, Short Beta Barrel structure of the section of the Shaker channel aminoacid sequence forming the pore. Only two of the four monomers are shown for clarity of the drawing. The intracellular side of the channel is at the left. Balls radii are 1/4 of the relative size of the corresponding atom. CH, CH_2 and CH_3 are displayed as a single sphere. On the upper-monomer, peptide bond C and N atoms are highlighted from ALA436 to VAL443, which is the section of the P-segment that runs from the outside to the inside side of the channel, and that is closer to the viewer. On the bottom monomer, peptide bond C and N's are highlighted from VAL 443 to THR449, which is the section that runs from inside to outside, and is closer to the viewer.

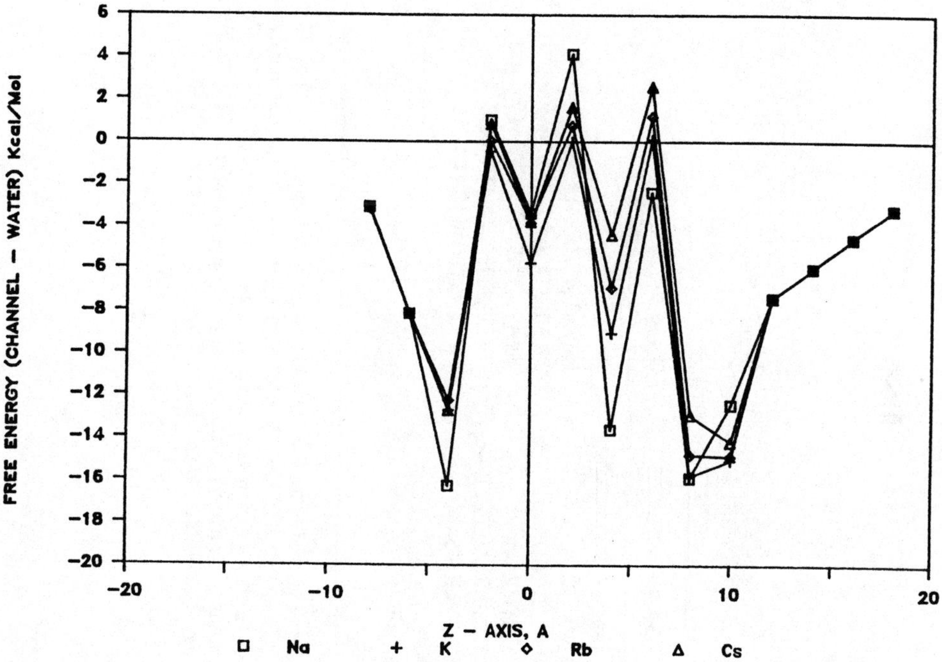

Figure 2.
Energy Profile for the Short Beta Barrel. The energy difference of charging the ion in the channel and charging it in water is plotted for various z locations for Na, K, Rb and Cs.

Table 1

Short Beta Barrel

z	Na energy	O's < 3Å	O's < 5Å	Selectivity
-8	-3.1	6WAT	15WAT	Na>K>Rb>Cs
-4	-16.3	6WAT	16WAT,4THR441	Na>K>Rb>Cs
-2	1.0	3WAT,1THR441	9WAT,3THR441,1TYR445	K>Rb>Cs>Na
0	-3.4	2WAT,3THR441	4WAT,4THR441,4TYR445	K>Rb>Cs>Na
2	4.1	1WAT,2THR441,2TYR445	3WAT,3THR441,4TYR445	K>Rb>Cs>Na
4	-13.6	3WAT,2TYR445	6WAT,4TYR445,2THR439	Na>K>Rb>Cs
6	-2.4	5WAT	8WAT,2TYR445,3THR439,1ASP447	Na>K>Rb>Cs
8	-15.9	6WAT,1TYR445	10WAT,1TYR445,2THR439,1ASP447	Na>K>Rb>Cs
10	-12.5	6WAT	16WAT,1ASP447	K>Rb>Cs>Na

Summary of free energy of transfer (kcal mol^{-1}) for Na ion from water to the axis of the Short Beta Barrel channel structure, at the z coordinate listed under the column "z". Column "O's < 3 Å" identifies the residues having oxygen atoms in close contact with the ion with are those contained into a 3-Å sphere centered at the ion. Column O's < 5 Å lists the residues with oxugen atom within 5 Å of the ion. The column "selectivity" lists, for each z-location, the selectivity sequence obtained from comparison of the free energies of transfer from water to the channel for each ion. The most preferred ion is listed first and the least prefered ion the last.

B. Intermediate beta barrel

1. *Structure*

The structure of Durell-Guy intermediate beta barrel equilibrated with water and a Na ion at z = +2 is drawn in Figure 3. Since the hairpin is longer than in the short beta barrel model and it wraps the pore as it runs along z, we show only one of the monomers whose inward and outward segments are shown separately in the upper and lower parts of the figure, respectively. On the inward segment TRP434, VAL438 and THR441 point into the channel. On the outgoing segment, TYR445, ASP447 and TYR449 are the residues projecting into the pore.

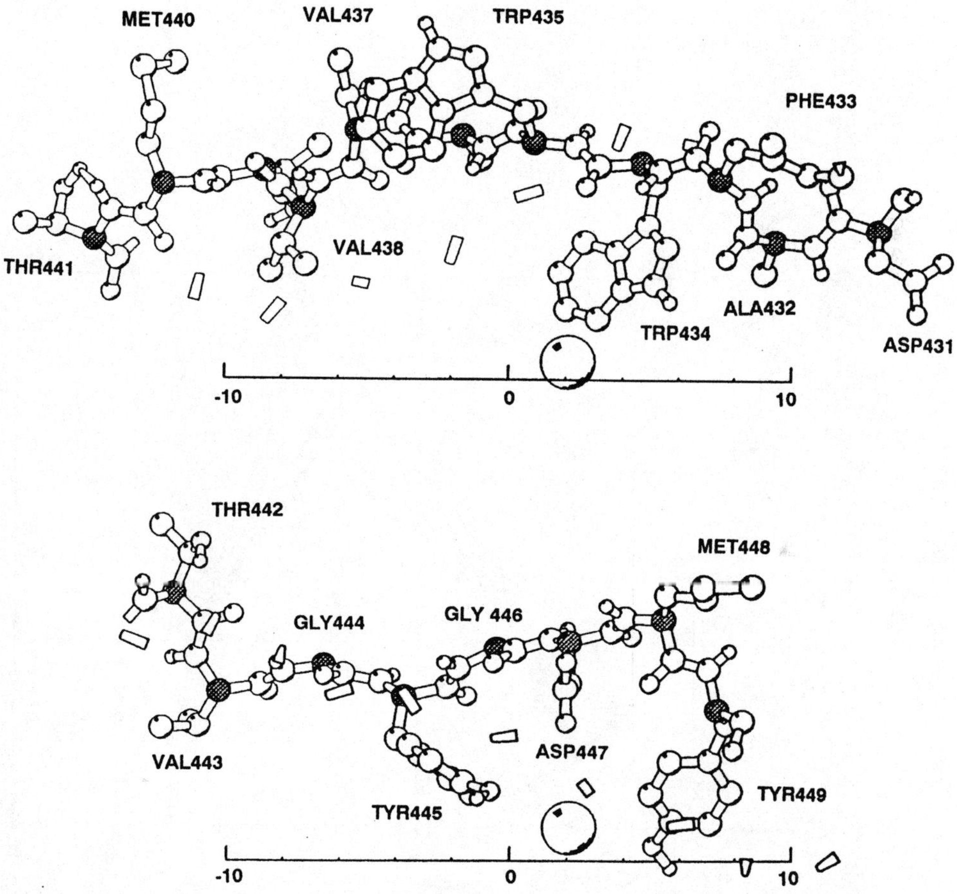

Figure 3.
Intermediate Beta Barrel Structure. Stick and ball representation of the Intermediate lenght Beta Barrel of Durell and Guy, equilibrated with a charged Na ion at z = +2 and with the space within 11.5 Å of the ion filled with water molecules, which are not shown. The upper part of the figure shows residues ASP431 through THR441 of one monomer, and the lower part shows from THR442 to TYR449 of the same monomer, rotated 45 degrees around the z axis. For each part of the figure, the scale is drawn along the center of the pore, i.e. x = 0 and y = 0. Alpha carbons are highlighted, and the scattered squares seen correspond to the peptide bonds connecting the residues not drawn on each panel.

2. *Energy profile and ligands*

The energy profile plot in Figure 4 shows two minima at the channel entrances and a very high peak which suggests that this structure is incompatible with ion conduction. The result of the analysis of the ligands is in Table 2. The low energy sites occur where the ion is surrounded by many water molecules and the high energy regions are those were the ion looses hydration. At the peak, only two TYR449 oxygens interact closely with the ion. The only protein ligands that are able to get close to the ion are TYR445 and TYR449, neither TRP434 nor ASP447 interact closely.

3. *Selectivity*

Potassium is the ion with the lowest energy at the peak, followed by Rb, Cs and Na. Although the energy at the peak is so high that no ion conductance is expected, K would exhibit the highest permeability. (Of course, if the carboxylate residues lining the channel were allowed to ionize, the energy level could drop to become quite favorable, as discussed in Section VI.2.). Four different binding selectivity sequences occur in this structure. K> Rb> Cs> Na at z = 0 -10, -2 and 0; K> Rb> Na> Cs at z = -8; Na> K> Rb> Cs at z = -6, -4 , 6, 8 and 10; and Cs> Rb> K> Na at z = 2.

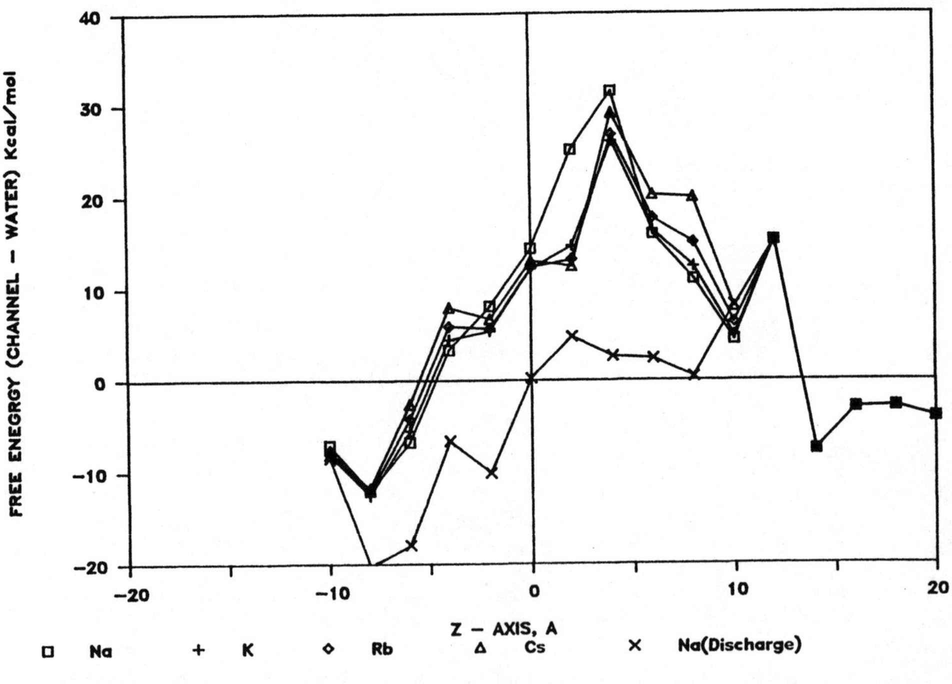

Figure 4.
Energy Profile for the Intermediate Beta Barrel. The energy difference of charging the ion in the channel and charging it in water is plotted for various z locations for Na, K, Rb and Cs. The trace labeled Na (discharge) shows the data for the Na profile calculated equilibrating the structure with the charged ion and simulating a discharge process.

Table 2

Intermediate Beta Barrel

z	Na energy	O's < 3 Å	O's < 5 Å	Selectivity
-10	-7.1	5WAT	17WAT	K>Rb>Cs>Na
-8	-12.0	6WAT	12WAT	K>Rb>Na>Cs
-6	-6.7	5WAT	9WAT	Na>K>Rb>Cs
-4	3.4	3WAT,1TYR445	6WAT,1TYR445	Na>K>Rb>Cs
-2	8.1	3WAT	5WAT,1TYR445	K>Rb>Cs>Na
0	14.4	1WAT,2TYR445	2WAT,3TYR445,2TYR449	K>Rb>Cs>Na
2	25.1	1TYR445,2TYR449	3TYR445,3TYR449	Cs>Rb>K>Na
4	31.5	2TYR449	1WAT,3TYR449	K>Rb>Cs>Na
6	15.9	4TYR449	1WAT,4TYR449	Na>K>Rb>Cs
8	11.0	2WAT,3TYR449	5WAT,4TYR449	Na>K>Rb>Cs
10	4.4	5WAT,1TYR449	10WAT,2TYR449	Na>K>Rb>Cs

Summary of free energy of transfer (kcal mol^{-1}) for Na ion from water to the axis of the Intermediate Beta Barrel channel structure, at the z coordinate listed under the column "z". See also legend of Figure 1.

C. Random coil

1. *Structure*

The random coil structure equilibrated with water and a Na ion is drawn in Figure 5. On the inward segment, TRP434, VAL438 and THR441 point into the channel. On the outgoing segment, TYR445, ASP447 and TYR449 are the residues projecting into the pore.

2. *Energy profile*

The energy profile, plotted in Figure 6, shows a sharp energy barrier at z = -8 to -6 and a wide energy well located from z = 4 to 8. The analysis of the ligands is summarized in Table 3. Throughout this structure, 10 or more water molecules are found within 5 Å of the ion. This explains why the energy is relatively favorable at all points. Among all the polar groups in the channel, only TYR445 oxygen gets closer than 3 Å from the ion. The peak observed at -8 to -6 corresponds to the ring of TYR445 and is the narrowest part of the channel.

3. *Selectivity*

In this structure, Na is the most permeable ion since its energy at the peak is the lowest, K and Rb energies are 4 kcal higher and Cs is one more kcal higher than Rb. This sieve is made by the ring of four TYR445 located at the left side entrance of the narrow region of the pore. On other locations, although all the closest ligands are water molecules, many selectivity sequences occur.

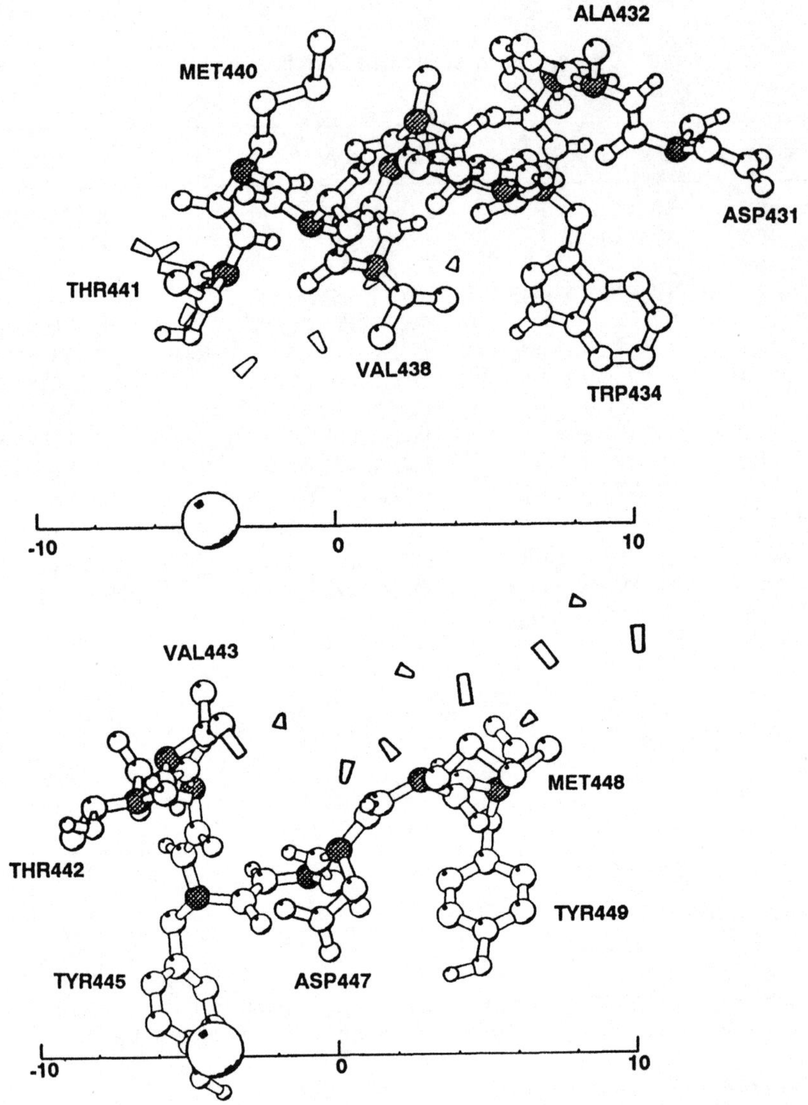

Figure 5.
Random Coil Structure. Durell and Guy's random coil structure equilibrated with a charged Na ion at z = -4 and with the space within 11.5 Å of the ion filled with water molecules, which are not shown. The upper part of the figure shows residues ASP431 through THR441 and the lower part shows residues THR442 to TYR449 of the same monomer, rotated 45 degrees around the z axis. The scale is drawn along the center of the pore, i.e. x = 0 and y = 0. Alpha carbons are highlighted.

Table 3

Random Coil

z	Na energy	O's < 3Å	O's < 5Å	Selectivity
-10	1.0	5WAT	15WAT,1TYR445	K>Na>Rb>Cs
-8	6.4	4WAT,1YR445	12WAT,3TYR445	Na>>K>Rb>Cs
-6	5.4	2WAT,2TYR445	10WAT,4TYR445	Cs>Rb>K>Na
-4	-8.3	5WAT	10WAT,2TYR445,1ASP447	K>Na>Rb>Cs
-2	-7.3	5WAT	12WAT,2TYR445,3ASP447	Na>>K>Rb>Cs
0	-6.4	4WAT	10WAT,1TYR445,1GLY446,4ASP447,1TYR449	Na>K>Rb>Cs
2	-5.9	5WAT	13WAT,2ASP447,3TYR449	Na>Rb>K>Cs
4	-3.7	4WAT	12WAT,4TYR449,1ASP447	Na>K>Rb>Cs
6	-5.7	4WAT	14WAT,3TYR449	Na>K>Cs>Rb
8	-5.7	6WAT	17WAT	K>Rb>Na>Cs
10	-1.9	6WAT	18WAT	K>Rb>Cs>Na

Summary of free energy of transfer (kcal mol^{-1}) for Na ion from water to the axis of the Random Coil channel structure, at the z coordinate listed under the column "z". See also legend of Figure 1.

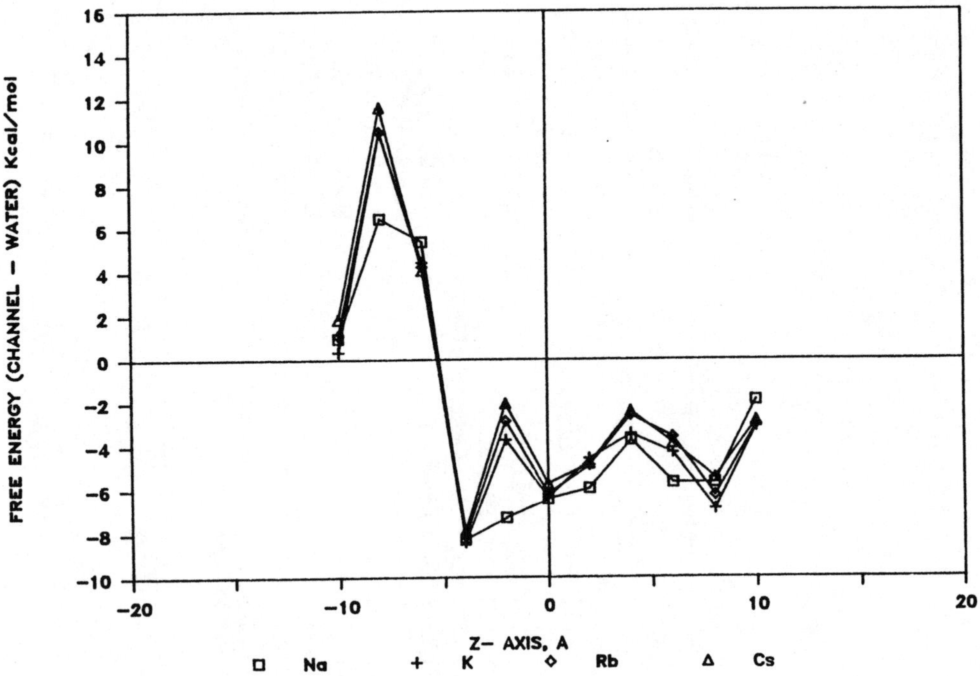

Figure 6.
Energy Profile for the Random Coil Structure. The energy difference of charging the ion in the channel and charging it in water is plotted for various z locations for Na, K, Rb and Cs.

VI. DISCUSSION

We have compared the energy profiles for various ions to cross three alternative structures proposed for the Shaker K channel by Durell and Guy. We have done this initially for the simplest situation, namely only one ion in the channel and with no acidic or basic groups ionized. This does not correspond to either the Physiological or expected Biochemical state where multi-ion occupancy is known and ionization of neighboring acidic and basic residues is likely. Nevertheless, we believe the differences in permeation properties revealed by the simulations may be helpful in refining the folding further. Thus, the Intermediate Beta Barrel has such a high peak (30 kcal mol^{-1}) that it makes this structure incompatible with ionic conduction unless appropriate acidic residues were ionized. The Random configuration has a lower peak (6 kcal mol^{-1}) but would be a Na channel. By contrast, the energy profile of the Short Beta Barrel has no barriers incompatible with ion conduction and would be selectively permeable to K because this ion has the lowest peak to surmount.

It is of interest that the energy profile of the short beta barrel model shows two very deep (15 kcal mol^{-1}) binding sites located at the entrances of the narrowest part of the pore. An ion binding to one of these sites would get "trapped" because of the high activation energy needed to leave the energy well. Thus this structure will be "ion-avid" and it would be occupied most of the time by at least one ion, and probably two ions, even at very low concentrations. Ion conduction in this structure would thus be consistent with the known high ion affinity and multiple ion conduction mechanism of K-conducting channels. It is quite surprising to find such a situation in a postulated structure in the absence of having to ionize any of its acidic residues. This will be discussed further in the next two sections.

A. Origin of the energy profile features

Comparison of the energy profiles computed for the Intermediate Beta Barrel and the Random Coil foldings suggests that hydration dominates those situations where relatively favorable free energy of the ion within the channel is computed. Thus, in the shorter and more open Random Coil structure, the ion is always surrounded by water and energy readings are favorable over most of the channel length (the barrier at z = -8 Å being due to replacing several water molecules by tyrosine hydroxyl ligands). For comparison, the Intermediate Beta Barrel, which is much less hydrated than the Random Coil, exhibits an unfavorable energy in all regions where significant dehydration occurs.

A common feature for all three structures is an unusually favorable energy for an almost fully hydrated ion within the channel. Such a finding is counterintuitive and is particularly surprising in the very deep wells of the Short Beta Barrel. This might be due to dipolar fields from non-neighboring atoms within the 11.5-Å cutoff for computing ion-protein interactions. Certainly part of this extra energy comes from the assumption that the dielectric constant within the channel and in bulk water are the same, which adds an unknown amount of energy to that calculated within the channel. But we suspect that most of this energy comes from a subtle difficulty in using the Surface Constrained All-Atom Solvent Procedure (King and Warshel, 1990) when modelling water within the channel, as discussed further in the next section.

1. *A possible problem with applying the SCAAS procedure to water in narrow channels*

The Surface Constrained All-Atom Solvent Procedure (King and Warshel, 1990) has not been critically tested in a confined system such as the present narrow channel carved into a protein. It is important to point out that serious artifacts can arise if the following problems are not addressed correctly. One major problem is to fill the empty space with water molecules properly. The problem arises because constraints are only applied to the outer 1.5-Å shell of water molecules. If the empty space is not well-filled with water molecules (e.g. too few water molecules) their tendency to be attracted by the charged ion will remove water molecules from the outer 2.5-Å shell. Therefore no water molecules will be constrained and surface water molecules will become overpolarized (King and Warshel, 1990). This will cause an excess of solvation energy to be calculated. To improve the procedure for filling the empty space with waters we reduced the spacing of the water grid from 3 to 2 Å. We then tested 50 random centers for the grid and chose the tightest packing obtained. These refinements have, indeed, reduced the extra energy but we are not yet sure that it has been reduced to zero. In part, this has been difficult to do because, unlike the situation for valinomycin that we describe elsewhere in this volume (Part Two, Chapter 1-B6), we have not had a model system on which to compare computations with actual data. We suspect that a large part of the surprisingly favorable energy for ions within channel water is due to an artifactual water polarization, but we cannot prove this yet.

B. Limitations and future enhancements of the calculations

It is important to realize that the largest uncertainty in everything done so far is in the choice of the starting structures. We have chosen to try to test which of various alternative structures suggested by others is most compatible with the observed permeation properties of the channel. While we have used Molecular Dynamics techniques to refine the initial structures slightly, and we have allowed the structures to hydrate and to rearrange in order to calculate Free Energy realistically, we do not believe that these methods are the appropriate way to try to improve the folding until far better guesses as to initial structure are available.

Our present MD/FEP level of simulations goes beyond our previous frozen-coordinate studies in channels (Eisenman et al., 1988; Eisenman and Alvarez, 1991) even though the number of atoms allowed to move is quite limited by the decision to essentially freeze all atoms further than 11.5 Å from the ion. This decision was taken because we found that placing an ion into the channel and letting the atoms move freely under its force field led to such severe changes of the structures that the original structure was essentially lost. At present, the initial structural guesses are not close enough to the "real" structure for MD procedures to have any chance of ever finding it within the astronomical conformational space that exists. We were, therefore, forced to limit the goal of our study to the restricted question of what energy profile is associated with each of the proposed alternative structures.

We have restricted our exploration to the energy profile for a single ion in the channel. Since K channels are known to be multi-ion channels, it would be interesting to simulate systems with more than one ion. To compensate the electric charge of these ions, acidic groups may have to be ionized. Recent models for ion conduction call for concerted ion movements in a single file (Schumaker and MacKinnon, 1990) which need the presence of several sites along the channel rather than a single site that can accept several ions. In the present level of structural model building, the ***assumption*** that the channel has a four-fold symmetry results in making a ring of four ASP447 at one position along the channel axis. Such an arrangement of carboxylate groups is not as well suited to producing a multi-K ion channel as, for example, structures where the monomers might be slightly displaced along the channel axis to produce a chain of ion binding sites (e.g. in a spiral).

VII. CONCLUSIONS

Since no high resolution 3-dimensional structure of the K channel exists, we could not address the relatively straightforward question of what permeation properties such a structure would have. Instead, we were forced to use three conjectured foldings of the Shaker K-channel to gain experience in the kinds of problems that will be encountered when one tries to ask this question for a known structure. We hope that our results contribute a useful first step toward the overall goal of inferring channel function from structure.

ACKNOWLEDGEMENT

The research was supported by USPHS Grant GM 24749 and FONDECYT Grant 1134-1992.

REFERENCES

Aqvist, J. (1990). *J. Phys. Chem.*, *94*: 8021.

Aqvist, J. (1991). *J. Phys. Chem.*, *95*: 4587.

Aqvist, J. (1992). *J. Mol. Struct. (THEOCHEM)*, *256*: 135.

Aqvist, J., and Warshel, A. (1989). *Biophys. J.*, *56*: 171.

Catterall, W.A. (1988) *Science*, *242*: 50.

Bogusz, S., Boxer, A., and Busath, D. D. (1992). *Protein Engineering*, *5*: 285.

Durell, S.R., and Guy, H. (1992). *Biophys. J.*, *62*: 238.

Eisenman, G. (1961). *Symposium on Membrane Transport and Metabolism* (A. Kleinzeller and A. Kotyk, eds.), Academic Press, New York, p. 163.

Eisenman, G. (1962). *Biophys. J.*, 2: 259.

Eisenman, G., and Alvarez, O. (1991). *J. Membrane Biol.*, *119*: 109.

Eisenman, G., Oberhauser, A., and Bezanilla, F. (1988). *Transport Through Membranes: Carriers, Channels and Pumps* (A. Pullman, J. Jortner and B. Pullman, eds.), Kluwer Academic, Dordrecht-Boston-London, p. 329.

Heginbotham, L., and MacKinnon, R. (1992). *Neuron*, *8*: 483.

Iverson, L.E., Tanouye, M.A., Lester, H.A., Davidson, N., and Rudy, B. (1988). *Proc. Nat. Acad. Sci. USA*, *85*: 5732.

King, G., and Warshel, A. (1990). *J. Chem. Phys.*, *93*: 8682.

Kraulis, P.J. (1991). *J. Appl. Cryst.*, *24*: 946.

Lee, F.S., and Warshel, A. (1992). *J. Chem. Phys.*, *97*: 3100.

MacKinnon, R. (1991). *Nature*, *350*: 232.

MacKinnon, R., and Yellen, G. (1990). *Science*, *250*: 276.

Miller, C. (1991). *Science*, *252*: 1092.

Papazian, D.M., Schwarz, T.L., Tempel, B.L., Jan, Y.N., and Jan, L.Y. (1987). *Science*, *237*: 749.

Pongs, O., Kecskemethy, N., Muller, R., and Krah-Jentgens, I. (1988). *EMBO J.*, *7*: 1087.

Schumaker, M.F., and MacKinnon, R. (1990). *Biophys. J.*, *58*: 975.

van Gunsteren, W.F., and Berendsen, H.J.C. (1987). *Groningen Molecular Simulation (GROMOS) Library Manual*, Biomos B. V., Nijenborgh 16, Groningen, The Netherlands.

Warshel, A. (1991). Computer Modeling of Chemical Reactions, in *Enzymes and Solutions*, John Wiley and Sons, Inc. New York-Chichester-Brisbane-Toronto-Singapore.

Warshel, A., and Creighton, S. (1989). *Computer Simulation of Biomolecular System* (W.F. van Gunsteren and P.K. Weiner, eds.), ESCOM, Leiden.

Yellen, G., Jurman, M.E., Abramson, T., and MacKinnon, R. (1991). *Science*, *251*: 939.

Yool, A.J., and Schwarz, T.L. (1991). *Nature*, *349:* 700.

INFLUENCE OF METAL INTERACTIONS ON MECHANISMS OF METAL TRANSPORT ACROSS MEMBRANES

by

Constantin Luca

Polytechnic Institute of Bucharest,
Department of Analytical Chemistry and Instrumental Analysis,
Str. Polizu 1-4
77207 - Bucharest 1
Romania

I. INTRODUCTION

Cell membranes are only weakly permeable to metal ions, mainly by passive diffusion mechanisms (see Part Four, Chapter 1-A1). The transport of metallic cations across these membranes occurs by means of "ionic channels" (see Chapter 1-A2), or by way of an active transport mechanism with the aid of "ionic pumps" which involve the presence of a mobile carrier in the membrane (see Chapter 1-A1).

In physiological conditions, the carrier is submitted to permanent changes in which its binding center is alternatively placed at the interfaces with the source phase and the acceptor phase of the membrane system, respectively. Often, these changes constitute the limitation step of the transport fluxes and explain the inferior transport rates for the transport facilitated by mobile carriers compared to the transport by channels.

Research on the transport mechanisms in cell membranes often utilizes the artificial permeability of membranes towards synthetic ionophores which simulate physiological species (Painter and Pressman, 1982). For example, valinomycin makes cellular membranes permeable to K^+ ions, playing the part of a carrier which catalyzes the facilitated diffusion of this ion by complexation. Valinomycin, with a macrocyclic structure, is a cyclodepsipeptide having a hydrophobic exterior which makes it soluble in membranes (Hilgenfeld and Saenger, 1982). The membrane specifically fixes the K^+ ions inside its intramolecular cavity where six oxygen atoms form an octahedral configuration around the cation. In this way, the K^+ ion can be transported as a K^+-valinomycin complex from the phase in which its electrochemical potential is smaller. The presence of the valinomycin carrier in the membrane produces a decrease of the trans-membrane electric potential gradient due to the facilitated transport of K^+ ions.

Nigericin, a monocarboxylic polyalcohol-polyether, can diffuse across cellular membranes, either in acidic form (H^+), or in salt form (K^+), allowing an ionic exchange between phases of H^+ and K^+ ions. Such a type of carrier catalyzes an electroneutral exchange, which is no longer accompanied by a net charge transfer. Like nigericin, some other ionophores produce electroneutral ionic exchanges, of Na^+-H^+ (monensin) or Ca^{2+}-$2H^+$ (A23187) - type.

Consequently, cell membranes are the center of transport processes and energy couplings (Cevc, 1990; Moule, 1990; Shechter, 1990). An energy coupling associates a spontaneous process and a non-spontaneous process. The mechanical work liberated by the first can stimulate the second. Evidently, such a coupling requires a common species, which usually is of proteic nature and plays the part of a carrier. The proteic transporter must have a fixation center, more or less specific for the transported ions.

II. DIFFERENT TYPES OF MEMBRANE COUPLINGS

The membrane couplings may be variously classified depending on the spontaneous or non-spontaneous nature of the two coupled transformations:

(i) a spontaneous chemical reaction is coupled to an active, non-spontaneous transport of an ion or, more generally, of a solute;

(ii) a facilitated spontaneous transport of an ion is coupled to an non-spontaneous chemical reaction;

(iii) a facilitated spontaneous transport of a solute is coupled to an active, non-spontaneous transport of a second solute.

These couplings could explain the role and the influence of the metal-ligand interactions on the transport mechanisms across cell membranes.

Type (i) coupling, called ***chemio-osmotic coupling***, involves spontaneous chemical reaction, such as a hydrolysis or an oxidation-reduction change of a chemical species present in the membrane. These reactions are coupled with the non-spontaneous transport of some cations, such as H^+, Na^+, K^+, Ca^{2+}, forming an ionic pump which performs the charge transport across the biological membrane. Examples of such pumps are respectively the action of some proton-ATPases or Ca^{2+}-ATPases which carry H^+ or Ca^{2+} ions, or the Na^+/K^+ ATPase which transports Na^+ and K^+ ions with different stoichiometries in both directions. If two different ions, but of the same electrical charge, are carried in opposite directions, the ionic pump is electroneutral. This is the case of the H^+/K^+ ATPase.

The hydrolysis of ATP in the cell cytoplasm, catalyzed by various ATPases, can couple the active transport of some ions and organic solutes to move these from one phase to the other of the membrane system (Flemming, 1991). Thus, for example, the Na^+/K^+ ATPase in the cytoplasm membrane of animal cells produces the ATP hydrolysis, a process which couples the electrogenic transport of Na^+ and K^+ ions; namely, to three Na^+ ions transported from the internal phase to the external phase of the cell correspond two K^+ ions transported in the opposite direction. This couple imparts to the cell membrane a series of functions, such as:

- the control of the intra-cellular volume;
- formation of a trans-membrane electrochemical potential gradient of the Na^+ ions, which can induce the active transport of certain solutes by secondary active transport mechanisms;
- maintenance of a low concentration of Na^+ ions in the intracellular liquid;
- formation of a trans-membrane electrical gradient, providing the driving force for the excitable cells.

The presence in the cytoplasmic membranes of gastric cells of an ATPase which allows the electroneutral coupling of the ATP hydrolysis with the H^+ and K^+ ions transport, leads to the transport of the H^+ ions from the interior towards the exterior of the cell, and to the transport of the K^+ ions in the opposite direction. This chemio-osmotic coupling is responsible for stomach acidic secretion and the maintenance of a gradient of about six pH units between the interior and exterior of the gastric cell.

The Ca^{2+} ATPase in the cytoplasmic membranes of animal cells and also in the endoplasmic and sacroplasmic reticula can couple the ATP hydrolysis with the transport of the Ca^{2+} ions towards the exterior of the cell. The low intracellular concentration of Ca^{2+} ions is maintained by this coupling.

Type (ii) coupling, called ***osmo-chemical***, involves the presence of a facilitated spontaneous transport system in the membrane which couples with a non-spontaneous chemical reaction. In some membrane systems with chemio-osmotic coupling, an electron transfer is coupled to the active carrier of Na^+ ions against H^+ ions, and the presence in such a membrane of an ATPase can couple the facilitated diffusion of Na^+ ions with the synthesis of ATP. In *Propionigenium modestum*, an anaerobic bacterium, decarboxylation of methymalonyl-CoA is catalyzed by a membrane decarboxylase which couples the decarboxylation with the active transport of Na^+ ions from the cell's interior towards the exterior. As a result, a trans-membrane electrochemical potential gradient appears in respect of Na^+ ions . The presence of an ATPase in this bacterium also leads to the ATP synthesis, which in turn decreases the electrochemical potential gradient for Na^+ ions. This example, therefore, represents a transport of Na^+ ions coupled to ATP synthesis.

Type (iii) coupling, called ***osmo-osmotic*** or "secondary active transport", involves a facilitated spontaneous transport, coupled with an active non-spontaneous transport. As has already been shown, an electrochemical potential gradient of the transported ions is produced in the chemio-osmotic coupling. The presence of this gradient can bring about:

- spontaneous diffusion of the corresponding ion in the direction of the decreasing electrochemical potential, which can be either passive or facilitated;
- spontaneous diffusion of the corresponding ion coupled with the active non-spontaneous transport of other chemical species present in the membrane system (ions or molecules). In this case, the coupling is osmo-osmotic and requires a carrier which can recognize and transport two chemical species simultaneously, one representing the driving ion which diffuses spontaneously, the second being the species actively transported.

The carrier that ensures the coupling between the facilitated diffusion and the active transport is called "port", i.e. entrance gate. The action of a port carrier can be:

- ***cotransport*** (or ***symport***), when the diffusion and the active transport take place in the same direction;
- ***countertransport*** (or ***antiport***), when the diffusion and the active transport take place in opposite directions.

Both symport and antiport can be electrogenic or electroneutral. If in a membrane system only an "uniport" process takes place, it is evidently electrogenic.

In the cytoplasm membrane of *Escherichia coli*, there exists hydrocarbonate chains of membrane lipids which permit an electron transfer and the appearance of an electrochemical potential gradient of the H^+ ions. In such a membrane also exist proteic carriers which ensure an active secondary transport of lactase, an osmo-osmotic transport coupled with the facilitated diffusion of H^+ or Na^+ ions. Thus, the lactose permease carrier (a polytopical protein) of *Escherichia coli* produces a coupling of the active carrier of β-galactosides with the facilitated diffusion of H^+ ions. The gradient of the electrochemical potential of H^+ ions plays an important part on the active cotransport mechanism (Ghazi and Shechter, 1981) which can follow different paths.

The cotransport of lactose and H^+ ions in the above example may involve the following phenomena:

- the fixation of the carrier on the trans-membrane protein chain at the interface with the external phase and of the H^+ ion followed by lactose fixation;
- the fixation at the same interface of lactose, followed by H^+ ion fixation;
- the increase of the carrier affinity for lactose, as a consequence of the H^+ ion fixation, or the increase of the H^+ carrier affinity as a consequence of lactose fixation;
- the appearance of some conformational changes of the carrier as a consequence of the fixing centers crossing the cell membrane from one interface to the other.

As this phenomenon may have rate constants which depend on the conformational state of the protein carrier, modification of one of the constituents of the electrochemical potential of the driving ion may change the transport scheme. The precise mechanism by which lactose is carried through by lactose permease is not known so far, although intensive mutagenic research (Kaback, 1987) has attempted to elucidate this mechanism.

In the eucaryote cytoplasm membranes where a Na^+/K^+ ATPase exists, the following phenomena can appear: an ATP hydrolysis is coupled to the active transport of Na^+ ions from the membrane towards its exterior, and of K^+ ions from the exterior into the membrane; a coupling of the facilitated diffusion of Na^+ ions from the exterior towards the interior, with an active transport of an organic species if an adequate carrier is present, for example of the glycoprotein type. In this way, a cotransport is achieved; examples are D-glucose with Na^+, and aminoacids with Na^+.

As an example of countertransport one may consider the case of the cytoplasm membranes (of procaryotes or eucaryotes) where a H^+/Na^+ or Na^+/H^+ antiport couples the facilitated diffusion either of the H^+ or of the Na^+ ion, respectively, to an active secondary transport of various solutes.

III. CONCLUSION

This summary of the transport mechanisms and of the various couplings between spontaneous and non-spontaneous phenomena emphasizes the important role played by the interactions between protein carriers as the ligands and metal ions in the membrane functions of biological cells. However, it is well to remember that a cell is not a completely autonomous entity. Its function involves permanent changes of matter, energy and information, changes based on these metal-ligand interactions. Therefore, its compartmentation constitutes an essential condition. Despite the fact that the biological membrane, which is an essentially lipid bilayer, constitutes an impermeable barrier for most solutions from the internal and external cell compartments, protein carriers, if they are present, operate the solution transport across the membrane and thus ensure the cellular functions. Moreover, the cells can respond to various external factors by changing their metabolism to match modifications in their external phase.

A more detailed description of various particular transport mechanisms of metal ions across cell membranes as a function of the nature of the metal ion and of the type of interactions between metal ion and the carrier ligand, natural or artificial, are presented in *Bioinorganic Medicine*, Part One, Chapter 4.

REFERENCES

Cevc, G. (1990). *Biochim. Biophys. Acta, 1031*: 31.

Flemming, C. (1991). *Biochim. Biophys. Acta, 1071*: 19.

Ghazi, A., and Shechter, E. (1981). *Biochim. Biophys. Acta, 645*: 305.

Hilgenfeld, R., and Saenger, W. (1982). *Host Guest Complex Chemistry II* (F.L. Boschke, ed.), Akademie Verlag, Berlin, p. 1.

Kaback, H.R. (1987). *Biochemistry, 26*: 2071.

Moule, S.K., and McGivan, J.D. (1990). *Biochim. Biophys. Acta, 1031*: 383.

Painter, G.R., and Pressman, B.C. (1982). *Host Guest Complex Chemistry II* (F.L. Boschke, ed.), Akademie Verlag, Berlin, p. 83.

Shechter, E. (1990). *Biochimie et Biophysique des Membranes*, Masson, Paris.

IONOPHORE ANTIBIOTIC COMPLEXES AND TRANSPORT OF METAL IONS ACROSS MEMBRANES

by

K.R.K. Easwaran

Molecular Biophysics Unit,
Indian Institute of Science,
Bangalore-560012
India

I. INTRODUCTION

The problem of transport of ions and other hydrophilic substances across biological membranes, crucial to all living cells and organisms, has been of great concern to cellular and electrophysiologists for the last several decades (Grell et al., 1975; Yagi and Pullman, 1985; Jain, 1988). The transmembrane cation transport is generally facilitated by large membrane proteins spanning the bilayer (Scarpa and Azzone, 1970; Giebisch et al., 1987). Several channel proteins such as acetylcholine receptor, sodium, potassium and calcium channel proteins, Gap Junction, etc, have been identifed as macromolecular pores, and extensive studies on the structure, aggregation and interaction of these proteins to understand their mechanisms of channel action are going on in different laboratories around the world (Hille, 1984; Keeling and Benham, 1989). Ionic channels are fundamental elements in gating the flow of ions across cells and in maintaining the membrane potential. The translocation of ions through these channels is modulated by discrete movements of the so-called gates which are controlled either by ligand binding, i.e. chemical transmitter (e.g. acetyl choline receptor), or by membrane electric field (e.g. sodium channel), or by both (e.g. calcium channel). However, our present knowledge on the mechanism of ion transport by membrane channel proteins is limited because of the problems related to the isolation, characterization and structure elucidation of these relatively large-sized membrane proteins. However, the discovery of a large number of specialized transporter molecules (ionophores) that function as carriers or pores in artificial and biological membranes, and extensive structural studies on these have to some extent helped us in the understanding of the ion-transport mechanism in molecular terms (Meuller and Rudin, 1967; Szabo et al., 1969; Ovchinnikov et al., 1974; Ivanov, 1975; Pressman, 1976; Bakkar, 1979; Dobler, 1981).

Ionophores are compounds produced by microorganisms (a majority of these by various species of *streptomyces*) which have the ability to promote transfer of ions relatively from an aqueous medium into a hydrophobic one, i.e. across artificial or biological membrane (Meuller and Rudin, 1967; Szabo et al., 1969; Ovchinnikov et al., 1974; Ivanov, 1975; Pressman, 1976; Bakkar, 1979; Dobler, 1981). They are classified as natural neutral carrier ionophores, natural carboxylic ionophores, synthetic carrier ionophores and model channel formers (Table 1). The neutral carriers (among which is Valinomycin, a 12-membered depsipeptide highly selective for K^+ ions) move as charged cation-ionophore

Table 1

Ion Selectivity of Ionophores

Class of ionophores	Compound	Monovalent cations	Divalent cations
Neutral carriers (natural)	Valinomycin	Rb>K>NH_4> Na>Li	Weak complexation
	Enniatin B	NH_4>K>Rb>Na	Weak complexation
	Beauvericin	Cs~Rb>K>Na>Li	Weak complexation
	Nactins-Nonactin	K>Cs> Rb>Na>Li	Complexation with Ca, Ba
	Boromycin	K>Na>Li	
Carboxylic carriers	Nigericin	H:K>Rb>Na>Li	—
	Monensin	H:Na>K>Rb>Li	—
	Dianemycin	H:Na>K>Rb>Li	—
	Lasalocid (X537 A)	H:K, Rb>Na>Li	Ca>Mn>Sr>Mg
	Calcimycin (A23187)	H:Li>>Na>K	Mn>>Ca>Mg> Sr>Br strong divalent specific
Synthetic carriers	Crown ethers (e.g.18-crown-6-dibenzo)	K>Na>Cs>NH_4>Na	—
	Cryptates(e.g. (222)-Cryptate	K>Rb>Na>Cs	Ba>Sr>Ca>Mg
Channel formers (voltage independent)	Gramicidin A	Rb~K>Cs>Na>Li	—
	Polyene antibiotics (Amphotericin B)	—	—
(voltage dependent)	Alamethicin	K>Rb>Na>Li	Ca
	Monoazomycin	Na~K	Ca

complex across the membrane, while the carboxylic carriers are believed to move across as neutral cation-anion pairs. Interestingly, ionophores are only found in prokaryotes, but are effective on membranes of both prokaryotic and eukaryotic cells, which raises very interesting questions about their role in microorganisms. Most of the ionophores listed in Table 1 have no special role in biological transport, but are excellent model compounds for delineating information on the mechanisms of transmembrane cation-transport at the molecular level. Several studies have been reported in the literature on the selectivity of ionophores to particular cations (Table 1). The elucidation of the structure and conformation of various ionophores and their complexes with different cations when incorporated into model membranes using biophysical techniques such as NMR, CD, fluorescence, X-ray diffraction, etc has provided information on the selectivity of the ionophores to a particular cation, on the nature of ligands and, in general, a molecular basis for the ion capture and release of these transporter molecules (Grell et al., 1973; 1975; Ovchinnikov et al., 1974; 1975; Ivanov, 1975; Easwaran, 1985; 1987). Among the carrier ionophores, valinomycin, a K^+ ion carrier, is the one studied extensively (Easwaran, 1985). Typical chemical structures of some carrier ionophores and model channel formers are given in Figure 1.

a

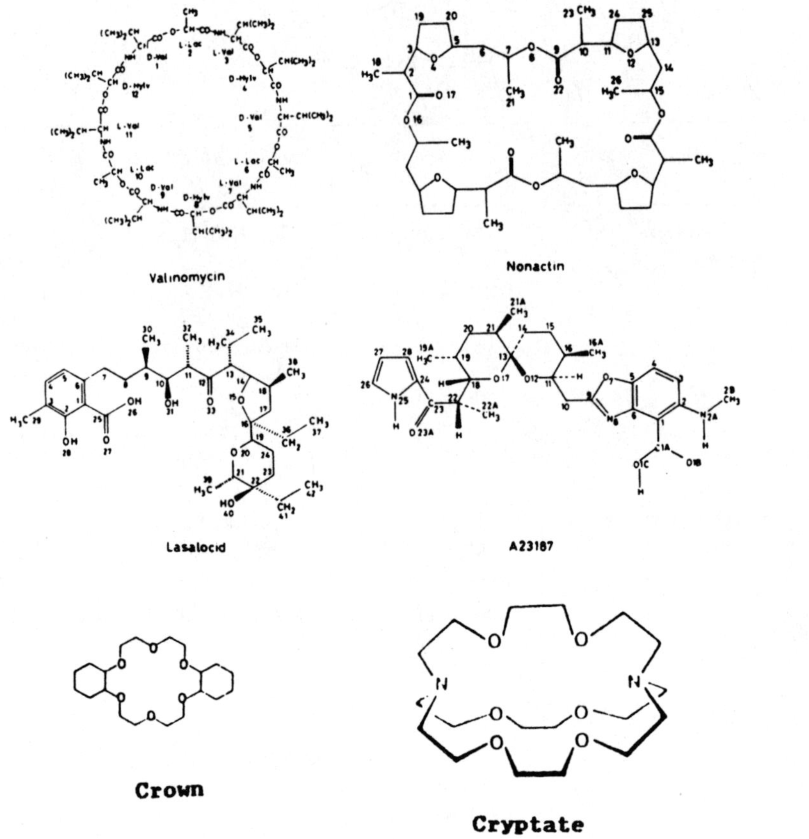

b

gramicidin

HCO-NH-L-Val-Gly-L-Ala-D-Leu-L-Ala-D-Val-L-Val-D-Val-L-Trp-
D-Leu-L-Trp-D-Leu-L-Trp-D-Leu-L-Trp-CO-NHCH$_2$CH$_2$OH

Alamethicin (AL30, main component)
Ac-Aib-Pro-Aib-Ala-Aib-Ala-Gln-Aib-Val-Aib-Gly-Leu-Aib-
1 2 6 7 11
-Pro-Val-Aib-Aib-Glu-Gln-Pheol]
14 18 19 20

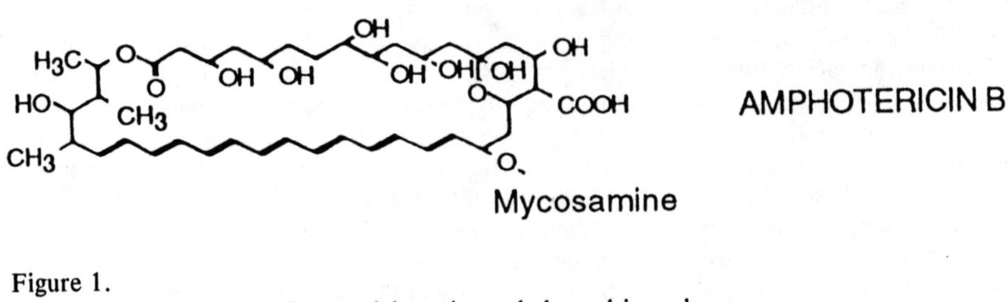

Figure 1.
Chemical structures of a few model carrier and channel ionophores.

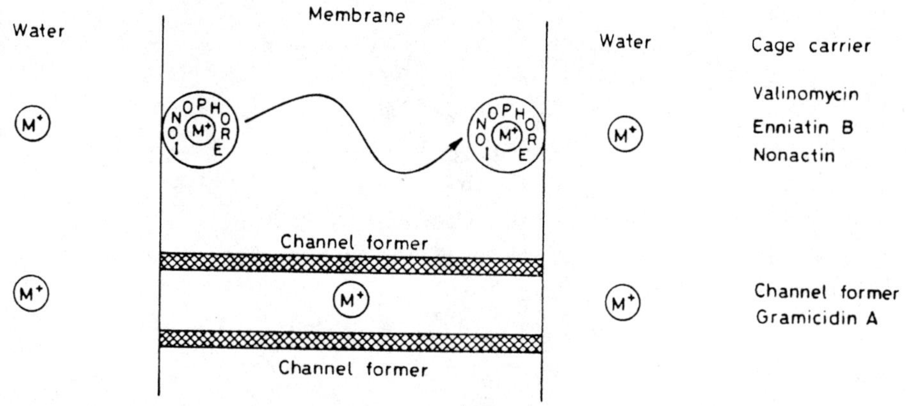

Figure 2.
Mechanisms of ion transport.

In understanding the possible ion-transport pathway, two mechanisms have been postulated: carrier, and channel mechanisms (Fig. 2). A carrier forms a complex with the ion and transports it across membrane as a carrier-ion complex with decomplexation taking place at the interface, whereas channel forming ionophores span the lipid membrane bilayer and form hydrophilic pathway for the ions. Understanding the structure and conformation of various ionophores and their cation complexes as well as their interaction with model and biological membranes is a prerequisite to work out the detailed mechanism of ion transport across membranes mediated by these transporter system. In this review, a brief summary on the structures of a few carrier ionophores, their interaction with model membranes and the kinetics of cation transport across bilayer membranes is given: major part will focus on the data on the two important highly cation-selective ionophores, namely valinomycin (K^+ selective) and the carboxylic ionophore, A23187 (divalent Ca^{2+}/Mg^{2+}-selective). A brief account of the channel-forming ionophores is also presented.

II. STRUCTURES OF IONOPHORES AND THEIR CATION COMPLEXES

A very large number of papers have been published in the literature on the structure, conformations and interaction of various ionophores and their possible cation complexes (Grell et al., 1973; Ovchinnikov et al., 1974; 1975; Ivanov, 1975; Easwaran, 1985; 1987). These results have shed light on the origin of cation selectivity in terms of the structural flexibilities of the molecule, nature of ligands, etc. A few representative results are given below.

A. Valinomycin

Among the carrier ionophores, valinomycin, a cyclic dodecadepsipeptide, cyclo (L-Valine-D-Hydroxyisovaleric acid-D-Valine-L-Lactic acid)$_3$ which has been found to be highly selective for K^+ ions and used specifically in several potassium ion permeability-dependent processes in biological systems, is the one most well studied (Ovchinnikov et al., 1974; Devarajan et al., 1980; Devarajan and Easwaran, 1981; Sankaram and Easwaran, 1982; Vishwanath and Easwaran, 1982; Easwaran, 1985). The conformation of this molecule in solution has been found to depend on the polarity of the solvent (Ovchinnikov et al., 1974). For example, in non-polar solvents such as hexane, chloroform, etc the molecule assumes a 'bracelet' structure with six intramolecular hydrogen bonds, the conformation having a C_3 symmetry and six $1 \rightarrow 4$ type IIβ turns (Fig. 3A), whereas in medium polar solvents, such as a mixture of 3:1 CCl_4:DMSO, it assumes a propellor conformation having only 3 intramolecular hydrogen bonds (Fig. 3B). In polar solvents such as DMSO the molecule has no fixed conformation, the average conformation is the result of several interconverting conformations (Fig. 3C).

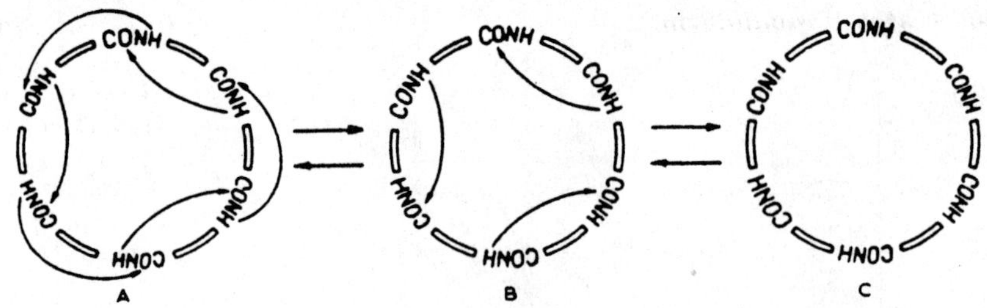

Figure 3.
Conformation of valinomycin in (A) non-polar (B) medium polar (C) polar solvents (from Ovchinnikov et al., 1974).

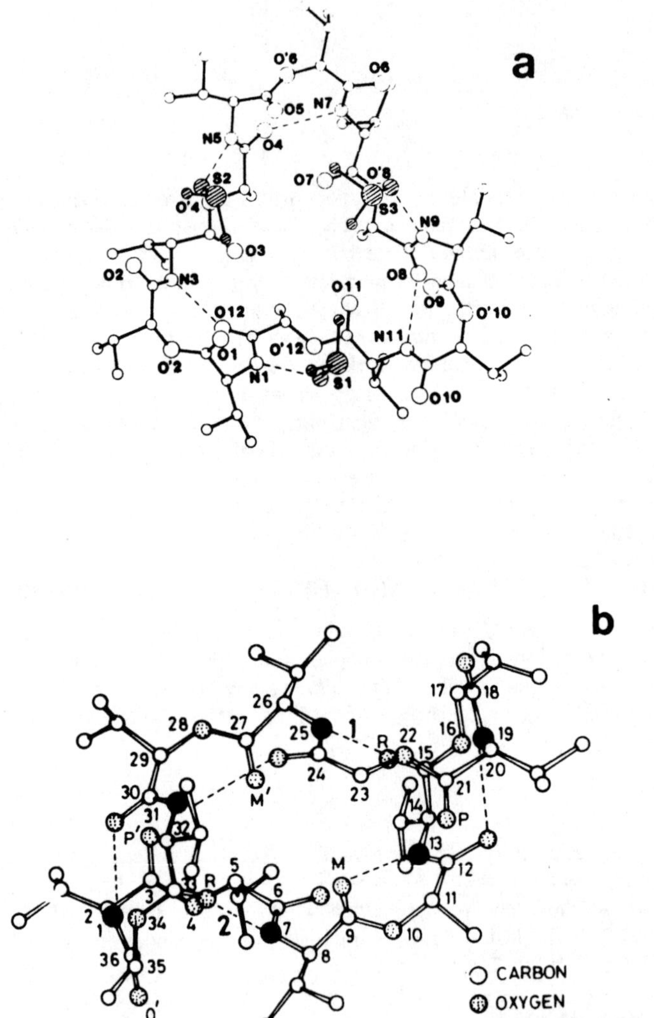

Figure 4.
a) Crystal structure of valinomycin crystallized in a medium polar solvent, $(CH_3)_2SO$ (from Karle and Flippen-Anderson, 1988); b) Crystal structure of free valinomycin crystallized in a non-polar solvent (from Smith et al., 1975).

The single crystal structure of uncomplexed valinomycin (Smith et al., 1975) (Fig. 4b) did not resemble any of the forms proposed in solution. The crystal structure showed no 3-fold symmetry. There are six intramolecular hydrogen bonds, four of the 4 → 1 type (two between L-Val and D-Hylv and two between D-Val and L-Lac) and two 5 → 1 type (between D-Val and L-Val residues). The crystal structure of valinomycin crystallized in a medium polar solvent (Karle and Flippen-Anderson, 1988) showed very similar conformation as proposed in solution (Fig. 4a). Studies on the valinomycin-K^+ complex showed that the conformation of the complex in solution and solid state (Easwaran, 1985) were very similar. The molecule assumed a bracelet type structure with six turns and having all six amide hydrogen atoms involved in intramolecular hydrogen bonds. The six ester carbonyls, three from each side of the bracelet, point inward and coordinate with K^+ ion in an octahedral geometry. The complexes formed with the other monovalent cations Rb^+ and Cs^+ are very similar. The Na^+ complex differs from the VM-K^+ complex in terms of the anion and solvent dependence and nature of complexation. A 1:1 peripheral complex with the cation complexed on the D-Val-L-Lac side of the bracelet structure has been observed for the valinomycin-Na^+ picrate complex in solid state (Steinrauf et al., 1982). Detailed NMR studies on valinomycin-Li^+ complex gave a similar conformation to that of the Na^+ complex (Sankaram and Easwaran, 1982). Very interesting structures for the valinomycin-divalent complexes have also been reported. For example, valinomycin has been shown to form complexes with barium and strontium ions with stoichiometries 2:1, 1:1, and 1:2, ionophore:cation in solution with the most stable being 1:2 (Devarajan and Easwaran, 1981). The single crystal X-ray structure of valinomycin-barium showed very novel conformation for the complex (Devarajan and Easwaran, 1981; Devarajan et al., 1980; 1984). The molecule assumed a flat open conformation devoid of any intramolecular hydrogen bonds and having two barium ions, each barium coordinated to 3 amide carbonyls, the remaining coordinating ligands being anion, solvent and water molecule oxygens. The two barium ions in the structure are separated by 4.57Å. Figure 5 shows a

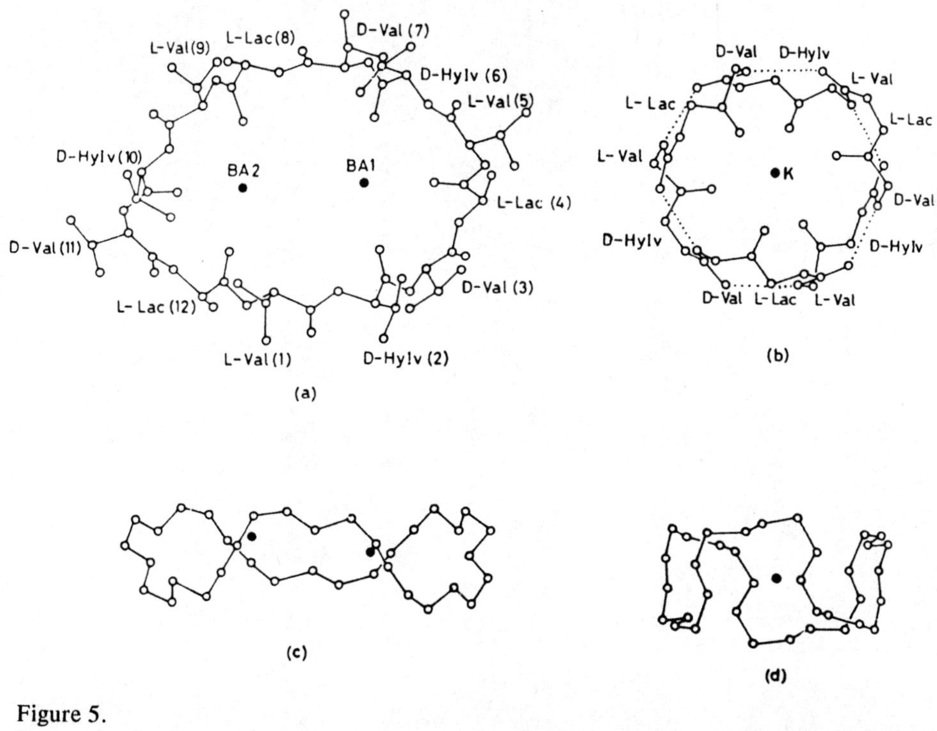

Figure 5.
Structure of valinomycin molecule: a) barium perchlorate complex, b) potassium iodide complex; c) and d) represent the corresponding side views (from Devarajan et al., 1984).

comparison of the structures of valinomycin-K^+ and Ba^{2+} ion complexes. The differences in the conformation of the two complexes in spite of the similar ionic radii of K^+ and Ba^{2+} ions have been explained as due to their difference in hydrated radii (Easwaran, 1987). The solution conformation of valinomycin-calcium complex derived from NMR showed that the molecule form a 2:1 ionophore:calcium complex at lower salt concentration and a 2:1 and 1:1 peripheral complex at high salt concentration (Vishwanath and Easwaran, 1982).

B. Other neutral carriers

The solution conformation and solid state structures of other carrier ionophores such as Enniatin, Beauvericin and nactin series and their cation complexes have been reported (Dobler, 1981; Vishwanath and Easwaran, 1981). These molecules, however, do not exhibit selectivity to a particular cation as shown by valinomycin. Crystal structures of the complexes of non-actin with Na^+, K^+, Cs^+, $NH_4{}^+$ and Ca^{2+} ion have been determined (Dobler, 1981; Vishwanath et al., 1983). Figure 6 shows a typical structure of nonactin-calcium complex. The structures of all complexes are similar. The cation sits at the center of a more or less perfect cube formed by four ether oxygen atoms from tetrahydrofuran rings and four carbonyl oxygen atoms from ester groups. The conformation resembles the seam of a tennis ball. A detailed discussion on these structures and other neutral carrier ionophores is given in Dobler (1981), Vishwanath and Easwaran (1981) and Vishwanath et al. (1983).

C. Carboxylic ionophores

Structural data on several carboxylic ionophores Nigercin, Monensin, Grisorixin, A-204A, Septamycin, Ionomycin, X-206, Alborixin, Lasalocid (X-537A) and Calcimycin (A23187)

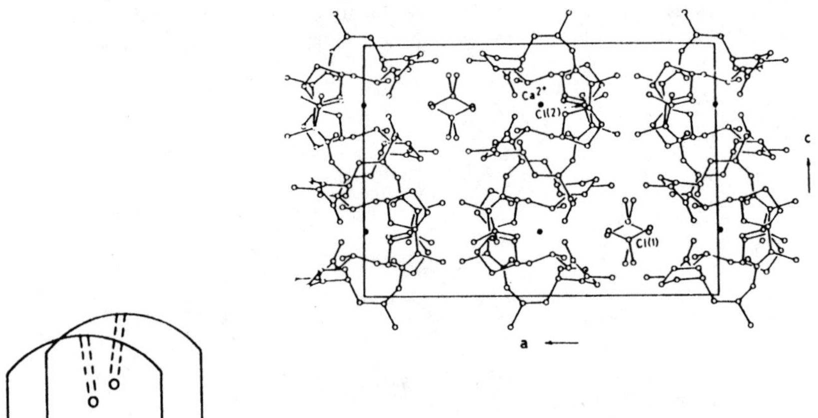

Figure 6.
Crystal structure of nonactin-Ca^{2+} perchlorate complex as viewed down the b axis (upper right); perspectives of the nonactin-Ca^{2+} complex (lower right). Schemative diagram of nonactin-Ca^{2+} perchlorate complex (from Easwaran, 1987).

and their cation complexes have been documented in the literature (Dobler, 1981; Westley, 1982). Among them, the two most well studied are X-537A and A23187. A23187 is by far the only divalent calcium ion-selective ionophore with potential applications (Caswell and Pressman, 1972; Pfeiffer et al., 1974; Deber and Pfeiffer, 1976; Patel and Shen, 1976; Puskin et al., 1981; Westley, 1982; Balasubramanian and Easwaran, 1989; Antenuis, 1977; Easwaran and Balasubramanian, 1992).

D. Structure of X-537A (Lasalocid) and its cation complexes

The chemical structure of X-537A, a member of the nigercin type of carboxylic ionophores, consists of a terminal salicylic acid group, a phenolic hydroxyl, a keto carbonyl group as well as a tetrahydrofuran and a tetrahydropyran ring (see Fig. 1). The antibiotic was found to form crystalline complexes with both mono and divalent cations. The single crystal structures of free form and complexes with Ba^{2+}, Ag^{+} and Na^{+} ions crystallized in different solvents have been reported (Dobler, 1981). In the lasalocid-Ba^{2+} ion complex, two lasalocid molecules coordinate to the cation in an unsymmetrical way. The Ba^{2+} ion is coordinated to six ligands of one lasalocid molecule and two ligands of the second molecule. The two monomer units are placed 'head-to-tail' (salicylate and tetrahydropyran moieties are referred to as the 'head' and the 'tail' of the molecule, respectively) with no hydrogen bonds connecting the two lasalocid molecules (Dobler, 1981). However, each of the lasalocid molecule is folded into a pseudo-cyclic conformation stabilized by an intramolecular hydrogen bond between the carboxylate and the terminal hydroxy group. The monovalent Ag^{+} ion has also been found to form a dimeric 'head-to-tail' complex, but two Ag^{+} ions are complexed by the ligand molecules as a 2:2 Ag^{+}:lasalocid complex. While the Na^{+}-lasalocid complex crystallized from non-polar solvent (carbon tetrachloride) gave a dimeric structure similar to that of the Ag^{+}-lasalocid complex, those crystallized from a solvent of medium polarity such as acetone showed a head-to-head association. Overall, the solid state structures of free lasalocid and lasalocid-cation complexes exhibited dimeric structures with variation in the way the two lasalocid molecules arranged themselves in the dimer, but the backbone conformation remains invariant.

Extensive studies on lasalocid and its complexes with mono and divalent cations in solution using a variety of spectroscopic techniques have been reported (Alpha and Brady, 1973; Degani and Friedman, 1974; Pfeiffer et al., 1978; Vishwanath and Easwaran, 1983;

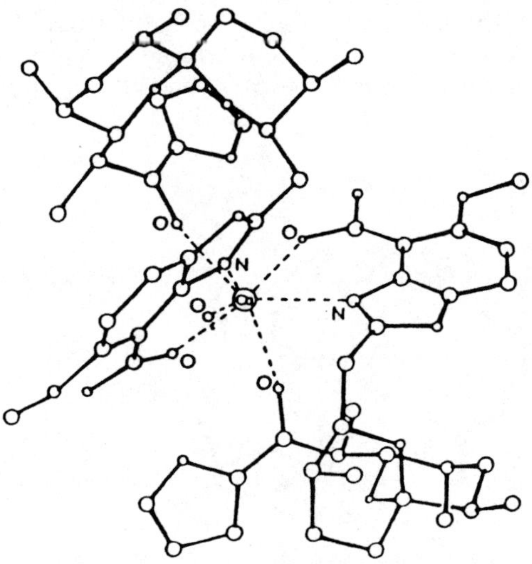

Figure 7.
Proposed model for the 2:1 complex of X-537A with Ca^{2+} in solution (from Vishwanath and Easwaran, 1985).

1985; Shastri and Easwaran, 1984). Detailed NMR studies indicated that lasalocid forms charged complexes with calcium with 2:1 and 1:1 stoichiometries (Shastri and Easwaran, 1984; Vishwanath and Easwaran, 1985). The conformational model for the complex based on NMR data showed that the calcium ion is preferentially bound to one end of the molecule, and ligands to three oxygen atoms, the other end (salicylic acid part) being relatively free. In the 2:1 lasalocid:calcium dimeric complex, the calcium ion is sandwiched between the two X-537A molecules with three oxygen atoms binding to it from each molecule (Fig. 7).

E. A23187 (Calcimycin) and its divalent cation complex

A23187 has been shown to be highly selective to divalent cations such as Ca^{2+} and Mg^{2+} over monovalent cations and has been used extensively in studies of the role of the calcium ion in several cellular phenomena (Pfeiffer et al., 1978; Pressman and Fahim, 1982; Westley, 1982). It is isolated from *Streptomyces chartreusensis* and has a structure consisting of a ketopyrrole group, a spiroketal portion and a benzoxazole ring (Fig. 1). Extensive spectroscopic (UV, CD, fluorescence and NMR) studies on free A23187 and its Ca^{2+} and Mg^{2+} complexes have been carried out by various groups (Deber and Pfeiffer, 1976; Puskin et al., 1981; Balasubramanian and Easwaran, 1989; Antenuis, 1977). The solution conformation of free A23187 gave a monomeric structure for the molecule similar to that reported for its structure in the solid state (Chaney et al., 1974). In the conformation of the Ca^{2+}- and Mg^{2+}-A23187 complexes, the two monomeric units aggregate around the divalent cation to form a 2:1 (ionophore:cation) complex with a cavity of approximately 3.5Å in diameter and exhibiting C-2 symmetry.

In this conformation, the intermolecular hydrogen bonding was observed between the ketopyrrole NH of one molecule and the COOH of benzoxazole of the second molecule and vice versa. The carboxyl and carbonyl oxygen, and benzoxazole nitrogen, of each molecule form ligands to the cation, providing an octahedral environment around the divalent cation. The structure derived from X-ray crystal structure analysis of A23187-Ca^{2+} complex shown in Figure 8 is very similar to that obtained in solution (Caswell and Pressman, 1972; Deber and Pfeiffer, 1976; Smith and Duax, 1976).

However, recent studies on the structure and conformation of free A23187 in solution using IR, UV, CD, fluorescence and NMR methods indicated that the most preferred stable conformation in non-polar solvents is a dimer and that it aggregates further to form a dimeric pore structure in phospholipid vesicles (Balasubramanian and Easwaran, 1989). The CD and fluorescence spectra of free A23187 showed remarkable time and concentration dependence indicating aggregation of the molecule. The involvement of the carboxyl C=O group of one molecule in forming an intermolecular hydrogen bond with the NH of the ketopyrrole group is seen from the IR data as a function of ionophore concentration

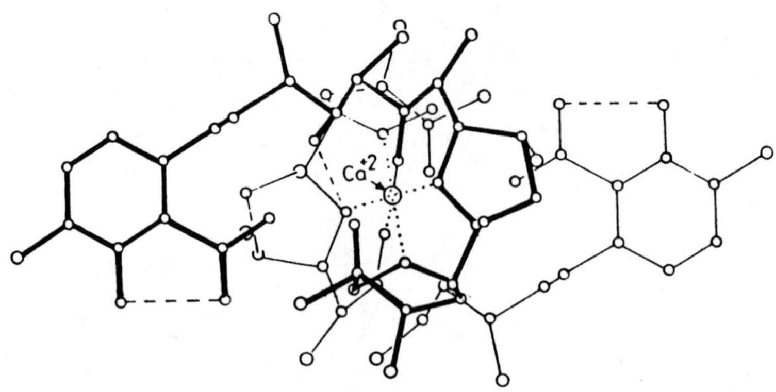

Figure 8.
Structure of A23187-Ca^{2+} complex (2:1) (from Smith and Duax, 1976).

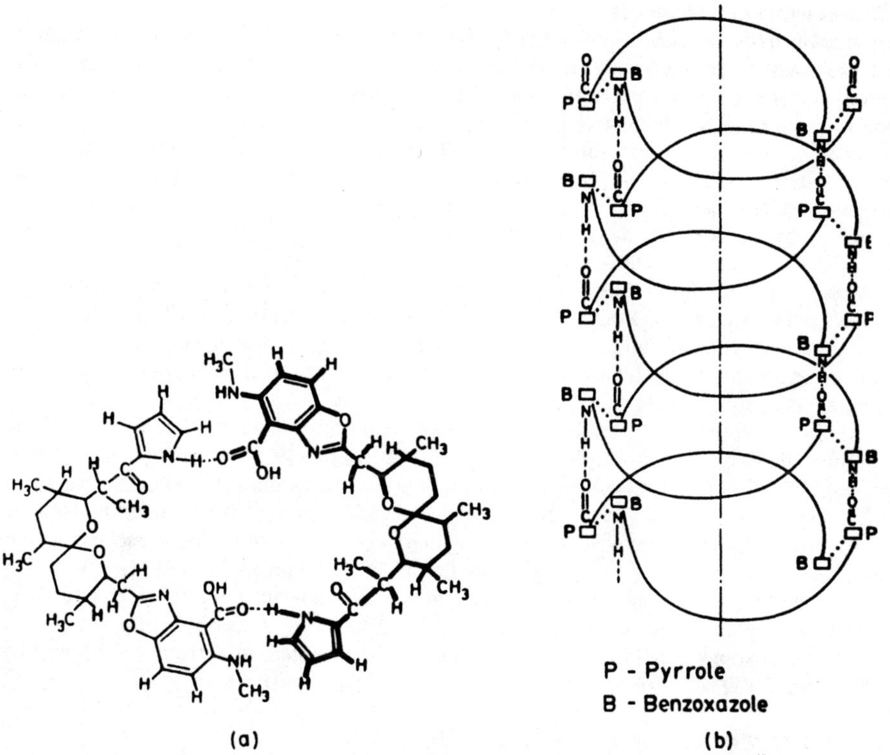

Figure 9.
a) Dimer structure of A23187 free acid in non-polar solvents. b) Model of stacked dimeric pore of A23187 in phospholipid vesicle (from Balasubramanian and Easwaran, 1989).

(Easwaran and Balasubramanian, 1992). Also, the one-dimensional NOE (Nuclear Overhauser Effect) and two-dimensional NOESY (Nuclear Overhauser Effect Spectroscopy) data showed usual NOE connectivities between certain protons (for e.g. 26 NHMe to 11, 12 protons and 15 proton to 8 proton) which could not be explained as arising from a single molecule (Patel and Shen, 1976). All the results clearly pointed to a dimeric structure (Fig. 9a) for free A23187 in solution. Spectroscopic (UV, CD and fluorescence) studies on A23187 incorporated into phospholipid vesicles showed further aggregation of the dimeric structure. Analysis of the results established the formation of inter dimeric hydrogen bond between the NH group of the benzoxazole portion and the C=O of the ketopyrrole group. A model of the stacked dimeric structure proposed for A23187 in phospholipid is shown in Figure 9b. For the first time it was suggested that A23187, believed to be a carrier, could possibly aggregate in phospholipids and transport calcium ion by a pore or channel mechanism.

F. Synthetic ionophores

Among the synthetic ionophores designed to mimic the behavior of natural ionophores and to optimize the reactivity to a particular cation, crown ethers (with varying number of ether oxygen atoms in the ring, and derivatives where some of the oxygen atoms are replaced by sulphur or nitrogen) and cryptates are the most prominent ones (Dobler, 1981) (Fig. 1). The complexation abilities of these with different cations, selectivity to a particular cation and the structure of the free and cation complexes are discussed in detail by Dobler (1981). In addition, several synthetic cyclic peptides and bicyclic peptides with selective cation-binding properties and ionophoric properties and their structures have been reported in the literature.

G. Model channel ionophores

As indicated in Table 1, this class of compounds do not fall in the category of ionophores in the sense that they do not act as ion carriers. They form a channel or pore across membrane for transporting cations selectively. Gramicidin, a hydrophobic linear pentadecapeptide model channel (Fig. 1), is the one physiologically well characterized, and extensive studies have been reported in the literature on the conformation and functionally related structural features (Anderson, 1984; Hladky and Haydon, 1984; Hinton and Koeppe, 1985; Yagi and Pullman, 1985). A large amount of studies have also been carried out to understand the channel properties of this molecule using the BLM technique to delineate the channel conductance under a variety of conditions, and different structural models have been proposed to account for its electrophysiological properties (Anderson, 1984; Hladky and Haydon, 1984; Hinton and Koeppe, 1985; Yagi and Pullman, 1985). Alamethicin, a second example of a polypeptide model channel (Fig. 1), is a voltage-dependent channel, and its structure, aggregation and channel characteristics in reconstituted membranes have been studied extensively.

Among the other pore-forming compounds, some of the naturally occurring polyene antibiotics display channel-forming activity in cell membranes (Medoff et al., 1965; Lampen, 1966; Bolard, 1986). Amphotericin B (Fig. 1), an antifungal polyene, is the most studied and clinically used for treating systemic fungal infections. While extensive studies are reported in the literature, the action of polyene antibiotics at the cellular and membrane level (Medoff et al., 1965; De Kruijff and Demel, 1976; Kobayashi and Medoff, 1977; Bolard, 1986), the mechanism of their action at the molecular level, their structure, aggregation properties and role of different sterols are not fully understood. Recent studies have addressed some of these questions in the case of Amphotericin B (Rajini Balakrishnan and Easwaran, 1993a; 1993b).

III. ION TRANSPORT OF CHANNELS AND CARRIERS

The kinetic properties of well studied examples of carriers and channels, namely translatory ion carriers of the valinomycin type and model ion channels such as gramicidin A, are well documented in the literature (Stark, 1978; Lauger, 1980). The bilayer membrane conductance (BLM) method is the most extensively used to study the kinetics of ion transport (Hladky and Haydon, 1984).

A. Carrier conductance

Experiments on the stationary conductivity measurements across bilayer membranes incorporated with ionophores such as valinomycin established the general scheme of the transport model, namely (i) formation of a cation carrier ionophore complex at the interface, (ii) translocation of this complex through the membrane interior, (iii) decomplexation at the interface liberating the ion, and (iv) back diffusion of the ionophore, with each step characterized by a rate constant (Stark, 1978) (Fig. 10). For a typical simple carrier mechanism (valinomycin) for a 1:1 ionophore:cation complex, a strict proportionality was observed between conductance of a bilayer membrane and ionophore concentration (Fig. 11).

B. Channel conductance

Most direct evidence for the existence of a pore or channel in bilayer membranes comes from experiments in which the electrical current through the bilayer membrane is monitored as a function of time, while voltage across the membrane is kept constant. Under these conditions, discrete conductance changes can be seen in the current-time traces. These quantized conductance changes correspond to rapid flow of ions through the pore. Large ion fluxes (10^7 to 10^9 ions per second) can be observed in a typical channel, while the ion fluxes induced by a carrier such as valinomycin are diffusion limited. The discrete current jumps in a BLM experiment incorporated with gramicidin is shown in Figure 12.

Each channel-forming molecule is characterized by three important parameters, namely, unit conductance, channel opening, frequency, and the average time the channel spends in the open state. Gramicidin is one of the model channel transporter system on

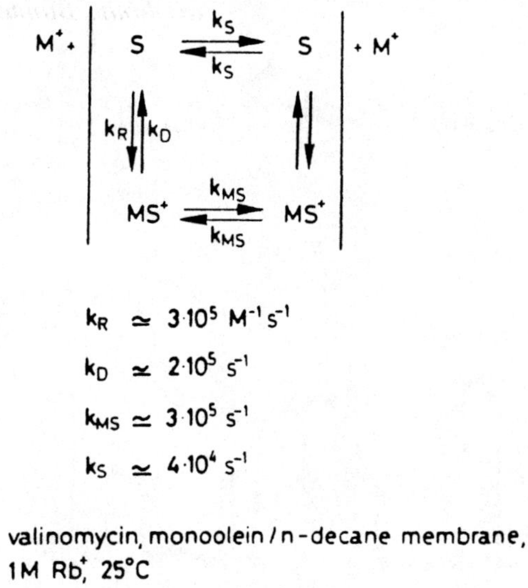

Figure 10.
Kinetics of the cation (M) mediated by a neutral carrier (from Stark, 1978; and Lauger, 1980).

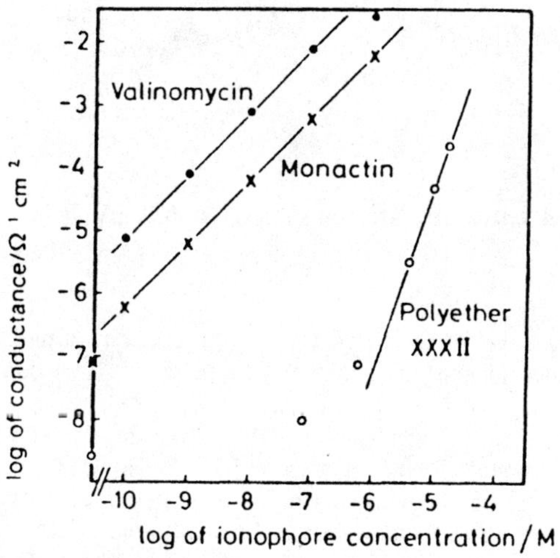

Figure 11.
Membrane conductance as a function of concentration of neutral macrocyclic ionophores (from Stark, 1978).

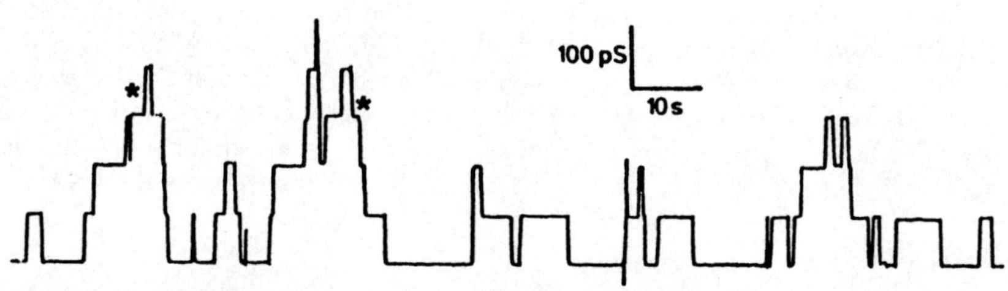

Figure 12.
Current steps due to fluctuations of two gramicidin A channels. Indicates incorporation of more than one channel and opening of them simultaneously.

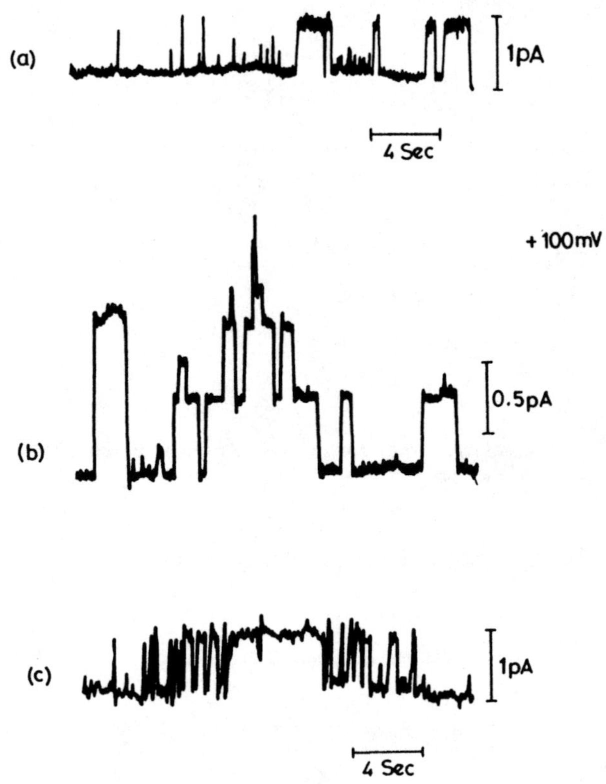

Figure 13.
A23187-induced channel activity in different bilayers at 100 mV: a) DPhPC bilayer, b) egg lecithin, c) DPPC.

which extensive studies have been carried out on the understanding of the structure and interaction of the peptides on the one hand, and the BLM channel conductance profile on the other.

Most of the ionophores (listed in Table 1) fall in the category of a typical carrier (e.g. valinomycin) or typical model channel (e.g. gramicidin). However, as mentioned earlier, recent studies on some of the ionophores which are believed to be typical carriers have indeed shown these to exhibit channel activity. An example of this type is calcimycin, A23187, a divalent (Ca^{2+} and Mg^{2+}) specific carrier ionophore. Detailed studies using several spectroscopic methods such as NMR, UV, CD, fluorescence and DSC (Differential Scanning Calorimetry) have established a stacked pore structure of A23187 consisting of several dimers of the molecule aggregated in lipid bilayers by formation of intermolecular hydrogen bonds (Antenuis, 1977) (Fig. 9). The conductance measurements across BLM's incorporated with A23187 gave a step-like current profile typical of a pore and similar to the gramicidin channel (Balasubramanian et al., 1992) (Fig. 13), and detailed studies on the current profile, effect of different lipids, cholesterol etc on the channel conductance have been reported. For the first time it has been shown that A23187, believed to be a divalent cation-selective carrier, does indeed form an aggregated pore-like structure and transports Ca^{2+} and Mg^{2+} ions by a pore or channel mechanism (Balasubramanian et al., 1992).

IV. SUMMARY

Maintenance of ionic balance and related transmembrane ion transport is crucial for all living cells and organisms. Understanding the structure, interaction and dynamics of the ion transporters (carriers, model channels, channel proteins, etc) is important to get an insight into the mechanisms of ion transport across membranes at the molecular level. This review briefly describes the structural aspects of a few model transporter systems, namely carrier ionophores, and discusses their importance in understanding the selectivity and ion transport across membranes.

REFERENCES

Alpha, S.R., and Brady, A.H. (1973). *J. Amer. Chem. Soc.*, *95*: 7043.

Anderson, O.S. (1984). *Annu. Rev. Physiol.*, *46*: 531.

Antenuis, M.J.O. (1977). *Bioorg. Chem.*, *6*: 1.

Bakkar, E.P (1979). *Antibiotics*, Vol. 5 (F.E. Hahn, ed.), Springer-Verlag, Berlin, p. 67.

Balasubramanian, S.V., and Easwaran, K.R.K. (1989). *Biochem. Biophys. Res. Commun.*, *158*: 891.

Balasubramanian, S.V., Sikdar, S.K., and Easwaran, K.R.K. (1992). *Biochem. Biophys. Res. Commun.*, *189*: 1038.

Bolard, J. (1986). *Biochim. Biophys. Acta*, *864*: 257.

Caswell, A.H., and Pressman, B.C. (1972). *Biochem. Biophys. Res. Commun.*, *49*: 292.

Chaney, M.O., Demaro, P.V., Jones, N.D., and Occolouctz, J.C. (1974). *J. Amer. Chem. Soc.*, *96*: 1932.

Deber, M., and Pfeiffer, D.R. (1976). *Biochemistry*, *15*: 132.

Degani, H., and Friedman, H. (1974). *Biochemistry*, *13*: 5022.

De Kruijff, B., and Demel, R.A. (1976). *Biochim. Biophys. Acta*, *457*: 109.

Devarajan, S., and Easwaran, K.R.K. (1981). *Biopolymers*, *20*: 891.

Devarajan, S., Nair, C.M., Vijayan, M., and Easwaran, K.R.K. (1980). *Nature*, *286*: 640.

Devarajan, S., Vijayan, M., and Easwaran, K.R.K. (1984). *Int. J. Peptide Prot. Res.*, *23*: 324.

Dobler, M. (1981). *Ionophores and their Structures*, John Wiley & Sons, New York.

Easwaran, K.R.K. (1985). *Metal Ions in Biological Systems*, Vol. 19 (H. Sigel, ed.), Marcel Dekker, Inc., New York, p. 109.

Easwaran, K.R.K. (1987). *Ion Transport through Membranes* (K. Yagi and B. Pullman, eds.), Academic Press, Tokyo, p. 17.

Easwaran, K.R.K., and Balasubramanian, D. (1992). *Membrane Structure and Function - The State of the Art* (B.P. Gaber and K.R.K. Easwaran, eds.), Adenine Press, New York.

Giebisch, G., Tosteson, D.C., and Ussing, H.H., eds. (1987). *Membrane Transport in Biology*, Vol. 1, Springer-Verlag, Berlin.

Grell, E., Funck, Th., and Santer, H.P. (1973). *Eur. J. Biochem.*, *34*: 415.

Grell, E., Funck, T., and Eggers, F. (1975). *Membranes - A Series of Advances*, Vol. 3 (G. Eisenman, ed.), Marcel Dekker, Inc., New York.

Hladky, S.B., and Haydon, D.A. (1984). *Curr. Top. Membrane Transport*, *21*: 327.

Hille, B. (1984). *Ionic Channels of Excitable Membranes*, Sinauer Associates, Sunderland.

Hinton, J.F., and Koeppe, R.E., II (1985). *Metal Ions in Biological Systems*, Vol. 19 (H. Sigel, ed.), Marcel Dekker, New York, p. 173.

Ivanov, V.T. (1975). *Ann. N.Y. Acad. Sci.*, *264*: 221.

Jain, M.K (1988). *Introduction to Biological Membranes*, John Wiley & Sons, New York.

Karle, I., and Flippen-Anderson, J.L. (1988). *J. Amer. Chem. Soc.*, *110*: 3253.

Keeling, D., and Benham, C., eds. (1989). *Ion-Transport*, Academic Press, London.

Kobayashi, G.S., and Medoff, G. (1977). *Annu. Rev. Microbiol.*, *31*: 291.

Lampen, J.O. (1966). *Biochemical Studies of Antimicrobial Drugs* (B.A. Newton and P.E. Reynolds, eds.), Cambridge University Press, p. 111.

Lauger, P.J. (1980). *Membrane Biol.*, *57*: 163.

Medoff, G., Bratyburg, J., Falkowski, L., and Borowski, E. (1965). *Ann. Soc. Chim. Polon.*, *39*: 497.

Meuller, P., and Rudin, D.O. (1967). *Biochem. Biophys. Res. Commun.*, *26*: 398.

Ovchinnikov, Y.A., and Ivanov, V.T. (1975). *Tetrahedron*, *31*: 2177.

Ovchinnikov, Y.A., Ivanov, V.T., and Shkrob, A.M. (1974). *Membrane Active Complexors*, Elsevier, Amsterdam.

Patel, D.J., and Shen, S. (1976). *Proc. Nat. Acad. Sci. USA*, *73*: 1796.

Pfeiffer, D.R., Reed, R.W., and Lardy, H.A. (1974). *Biochemistry*, *13*: 4007.

Pfeiffer, D.R., Taylor, R.W., and Lardy, H.A. (1978). *Ann. N.Y. Acad. Sci.*, *307*: 402.

Pressman, B.C. (1976). *Annu. Rev. Biochem.*, *45*: 501.

Pressman, B.C., and Fahim, M. (1982). *Annu. Rev. Pharmacol. Toxicol.*, 22: 465.

Puskin, J.S., Visities, M.F., and Wene, M.T. (1981). *Arch. Biochem. Biophys.*, *206*: 164.

Rajini Balakrishnan, A., and Easwaran, K.R.K. (1993a). *Biochemistry*, *32*: 4139.

Rajini Balakrishnan, A., and Easwaran, K.R.K. (1993b). *Biochim. Biophys. Acta*, *1148*: 296.

Sankaram, M.B., and Easwaran, K.R.K. (1982). *Biopolymers*, *21*: 1557.

Scarpa, A., and Azzone, G.F. (1970). *Eur. J. Biochem.*, *12*: 328.

Shastri, B.P., and Easwaran, K.R.K. (1984). *Int. J. Biol. Macromol.*, *6*: 219.

Smith, G.D., and Duax, W.L. (1976). *J. Amer. Chem. Soc.*, *98*: 1578.

Smith, G.D., Duax, W.L., Long, D.A., deTitta, G.T., Edmonds, J.W., Rohrer, D.C., and Weeks, C.W. (1975). *J. Amer. Chem. Soc.*, *97*: 7242.

Stark, G. (1978). *Membrane Transport in Biology* (G. Giebisch, D.C. Tosteson and H.H. Ussing, eds.), Springer Verlag, Berlin, p. 447.

Steinrauf, L.K., Hamilton, J.A., and Sabesau, H.N. (1982). *J. Amer. Chem. Soc.*, *104*: 4085.

Szabo, G., Eisenman, G., and Ciani, S. (1969). *J. Membrane Biol.*, *1*: 346.

Vishwanath, C.K., and Easwaran, K.R.K. (1981). *Biochemistry*, *20*: 2018.

Vishwanath, C.K., and Easwaran, K.R.K. (1982). *Biochemistry*, *21*: 2612.

Vishwanath, C.K., and Easwaran, K.R.K. (1983). *FEBS Lett.*, *153*: 320.

Vishwanath, C.K., and Easwaran, K.R.K. (1985). *J. Chem. Soc., Perkin Trans.*, *II*: 65.

Vishwanath, C.K., Shamala, N., Easwaran, K.R.K., and Vijayan, M. (1983). *Acta Cryst.*, *C 39*: 1640.

Yagi, K., and Pullman, B., eds. (1985). *Ion Transport through Membranes*, Academic Press, Tokyo.

Westley, J.W., ed. (1982). *Polyether Antibiotics*, Vols. I and II, Marcel Dekker, Inc., New York.

INTESTINAL ABSORPTION AND METAL-LIGAND INTERACTIONS

by

Nicholas J. Birch

Biomedical Research Laboratory,
University of Wolverhampton,
School of Health Sciences,
62-68 Lichfield Street,
Wolverhampton, WV1 1DJ
U.K.

I. INTRODUCTION

The contents of the gastrointestinal tract are external to the body and though they may be modified in composition to a limited extent by the activities of the body, they are subject to the vagaries of dietary intake and the environment in which an animal exists. Digestive tracts throughout the animal kingdom are all based upon the principle of maximizing the intestinal surface in contact with the nutrient medium. In this chapter, we shall deal primarily with mammalian intestine. The jejunum and ileum are the primary sites of absorption of many nutrients, metals and drugs in man.

A. Intestinal structure

The gastrointestinal tract of mammals, from the oesophagus to the anus, shares a common structure, of which certain features may be enlarged or modified in the various regions. In addition to the absorptive and motile activities, the intestine contains a large variety of exocrine glands which secrete into the lumen of the intestine or its associated organs, such as the pancreas, enzymes which may be specific for particular chemical groupings; carbohydrate, fats, protein. Glands also secrete acids and bases which adjust the intestinal conditions to provide an optimum sequence of activity of the secreted enzymes so that digestion may be continuous, controlled and efficient. Other glands secrete substances which act as surfactants to enhance the dispersion of the fatty constituents of the diet.

B. Factors affecting intestinal absorption

1. *The intestinal barrier*

The intestinal mucosa is comprised of epithelial cells in the form of a thin sheet which are attached to a basement membrane. Underlying this is the submucosa in which are embedded blood capillaries and lacteals, the fine vessels of the lymphatic system. The epithelial structures are folded firstly at the level of microvilli and then to form the villus. Encasing the whole structure are bands of longitudinal and circular muscles. The normal passage of absorptive products is thus across the epithelial membrane and into the extracellular supportive tissue to diffuse into the closely interwoven blood vessels and lacteals. The epithelial cells in the mucosa are embedded in the underlying connective tissue and are attached to each other at the lateral surfaces near to their apices by *zonulae occludentes*. The permeability of the resulting junctional complexes determines whether an epithelium is described as "tight", as in colon or urinary bladder, or "leaky", as in small intestine.

Two major transport routes across intestinal epithelia are possible: transcellular and extracellular. Transcellular routes include lipid and aqueous routes for non-ionic and water-soluble solutes respectively, and a carrier route for hydrophilic substances too large for the aqueous route. The principal extracellular route is the paracellular route via the junctional complexes between the cells.

Paracellular transport depends on water flow; a passive phenomenon that is dependent on hydrostatic or osmotic pressure gradients across membranes. The flow of fluid through a membrane will influence the diffusion of solutes using the same pathway through the membrane. When water flows, solutes are transported along the channels. Passage of solutes in the same direction as the water flow will be enhanced by solvent drag (Behar, 1974). The energy for such bulk flow of water comes from the osmotic potential generated by the blood flow in the capillary and this may originate either from the activities of the sodium pump locally, or by the maintenance of a substantial blood flow by the heart. Pappenheimer and his colleagues proposed that active, sodium-coupled transport of organic solutes from lumen to intercellular space provides the principal osmotic force for fluid absorption and triggers the widening of the intercellular junctions, thus promoting absorption of nutrients by solvent drag (Pappenheimer and Reiss, 1987).

Transcellular absorption of a substance via the epithelial cells requires the passage of the substance across both the mucosal and the serosal membrane of the cell, together with translocation within the cell interior. It is known that a number of metal ions may be absorbed in this way by the activity of the various ATPase pumps and that these pumps are interlinked with mechanisms for absorption of vital cellular fuels such as glucose and amino acids. Many transcellular transport processes involve a reaction with a membrane protein which is saturable and linked to the transport of other ions travelling in the same (symport) or opposite directions (antiport) (see Chapter 1-A). The membrane is permeable, however, to lipophilic substances by a process of diffusion into the cell.

Some epithelial cells contain proteins which may be shown to bind particular metals. One particular example of this is the calcium-binding protein in the intestine (Wasserman and Fullmer, 1989; Wasserman et al., 1992). However, the fact that these binding substances exist in intestinal cells does not mean that they are the major route of absorption of these metals and it is possible that the various binding proteins may exist purely for the defence of the homeostasis of the intestinal cells themselves against high and low concentrations of the metal in their mucosal environment (Birch et al., 1991). In principle, it would seem more likely that the less energetically expensive paracellular route might be preferred for metal ion absorption in the intestine.

2. *Ligands in intestinal transport and absorption*

Ligands which may be involved in transport of substances across the membrane may feature in a number of different mechanisms:

(i) they may form complexes in the intestinal lumen which facilitate the uptake of the whole complex by the intestinal cell;

(ii) they may reside in the mucosal membrane of the cell to enhance the uptake of substances into the cell interior;

(iii) they may reside in the cell cytoplasm or organelles in order to translocate the absorptive product to the basement membrane;

(iv) they may reside in the basement membrane on the serosal side of the cell to enhance the expulsion of the substance from a cell towards the associated mesenteric capillary;

(v) if the transport is paracellular, ligands which occur in the lumen may form complexes which will be carried via bulk flow of fluid through the pericellular spaces to be passively absorbed by the capillaries.

The intestinal concentration of metals, as with other products of digestion, may vary from very high to very low and it would seem biologically simpler for the body to avoid transcellular flux of metals which are either potentially metabolically disruptive or toxic or which may cause dysregulation of other metabolic processes within the cell. For instance, this would be the case with magnesium whose concentration is fairly closely regulated by all cells and whose concentration in turn regulates the activity of many enzymes (Padgham et al., 1993). Were unrestricted flow of magnesium to occur, the metabolic consequences would be wide ranging and unfortunate for the cell. "Transport proteins" and other ligands which have been identified in blood and in cells may have more to do with the defence of individual cells against either high or low concentrations of the metal rather than be part of the intestinal uptake process (Birch et al., 1991).

II. EXPERIMENTAL STUDIES OF METAL ION ABSORPTION

We have studied lithium and magnesium absorption in rodents and man, and zinc absorption in rodents. Lithium is an example of a metal which is not usually present in quantity in the diet and therefore might not be expected to have specialized transport mechanisms (Davie, 1991). Magnesium (Padgham et al., 1993) and zinc (Davie et al., 1991a) are both carefully controlled by cells because they have important regulatory effects themselves on cell metabolism. It is thus essential for the cell to prevent either excess or deficit within the cell itself.

A. Lithium transport in the intestine

1. *Isolated cell preparations*

The absorption of lithium by the gut has not been widely studied and we have studied both cellular transport in isolated intestinal cells disaggregated from their epithelial structure and in intact, isolated epithelial sheets obtained ex vivo from animal and human tissues. Lithium uptake into, and efflux from, guinea pig isolated epithelial cells is passive (Birch et al., 1985). Isolated intestinal epithelial cells (enterocytes), suspended in Krebs-Henseleit bicarbonate buffer, were obtained from guinea pigs and prepared by the method of Towler (Towler et al., 1978; Birch et al., 1985). Cell viability was confirmed by trypan blue dye exclusion and oxygen consumption using a Clarke oxygen electrode. Ouabain, phloretin and DNP/NaF inhibitors had no effect on lithium efflux. There was no significant difference between efflux into sodium chloride or potassium chloride media, indicating that lithium-sodium countertransport (LSC) does not contribute to lithium efflux (Birch and Phillips, 1992). However, LSC might operate at the basolateral membrane where phloretin is known to have its effect on glucose transport (Kimmich, 1983). Lithium transport in enterocytes in situ might be regulated at the basolateral membrane. This can be investigated more reliably in a model in which the mucosal epithelium remains intact.

2. *The isolated stripped mucosal preparation in rodent and man*

The guinea pig intestinal mucosal preparation was introduced by Lauterbach (Lauterbach, 1977) and modified by Davie (Davie, 1988; Partridge et al., 1987). Further modifications were necessary when the technique was adapted for human tissue (Davie, 1991). Sections of intestine were isolated by sequential mesenteric clamping which allowed the blood supply to be maintained intact until immediately prior to excision of tissue. Once excised, the mucosal epithelium was stripped and placed mucosal side down onto a Nylon mesh and mounted between two prepunched PVC membranes in a flux chamber so that it occluded a circular porthole of 5 mm diameter. The tissue was incubated in bicarbonate buffer solutions based on that of Krebs and Henseleit (1932) and was aerated throughout the 45 minute experiment. Polyethylene Glycol (PEG 900) was used as an extracellular marker to measure water transport and to indicate the viability of the preparation. Tissue preparation, i.e. removal from animal to beginning of incubation, took less than two minutes. At the end of the experiment, buffer solutions were withdrawn simultaneously from the two half chambers by suction and the area of tissue exposed to the bathing medium was removed for analysis.

Transepithelial lithium permeation was found to be linear up to 50 mM Li. No evidence of transport saturation was seen and apparent tissue lithium uptake was also linear with no evidence of saturation. These results in guinea-pig are typical of a passive diffusion process. The overall rate of lithium transfer was proportional to the initial lithium concentration.

In human distal ileum, there was a significant correlation between luminal buffer lithium concentration and amount of lithium transported (Davie, 1991). This suggests that lithium transport is primarily paracellular via the tight junctions and confirms earlier work using everted sacs (Birch et al., 1983; 1984). Passive lithium uptake into both kidney slices and isolated guinea-pig intestinal epithelial cells has also been reported (Birch et al., 1985).

We have shown that, when lithium associated with the extracellular space was taken into account, acute cellular uptake of lithium was negligible (Phillips et al., 1988; 1990). This is consistent with studies that we have carried out using ^{7}Li NMR to investigate the uptake of lithium in a number of different cell types. Lithium uptake was shown to be less than about 10 percent of the extracellular lithium concentration in Swiss Mouse 3T3 fibroblasts (Hughes, 1989) and rat hepatocytes (Thomas et al., 1988) and less than 15 percent in naive rat and human erythrocytes (Hughes et al., 1988). Long-term treatment in humans leads to inhibition of lithium countertransport and to a rise in the proportion of lithium which is taken up by erythrocyte to a maximum of about 30 percent of the external concentration (Ehrlich et al., 1981; 1983).

B. Magnesium transport in the intestine

1. *Transepithelial transport of magnesium*

Transepithelial transport of magnesium can occur through one or both transport routes (Ebel, 1990; Aldor and Moore, 1970). In our laboratory, the isolated epithelial technique has been used to study magnesium kinetics in guinea-pigs (Birch et al., 1991; Padgham et al., 1993). The efflux of magnesium from the serosal side was significantly greater than that from the mucosal side. Metabolic inhibitors (dinitrophenol and sodium fluoride) did not affect serosal efflux but equalized the mucosal efflux rates. Mucosal magnesium efflux thus may be masked by an active magnesium uptake mechanism at the brush border on the mucosal side.

Mucosal magnesium concentration was demonstrated to affect differentially the tissue uptake of magnesium. At low mucosal magnesium concentrations, there was a small apparent net efflux of magnesium from the epithelial cells. At intermediate mucosal magnesium concentrations (above 8 mmol l^{-1}) there was an apparent net accumulation of magnesium in the tissue. From these experiments it appears that total intracellular magnesium in mucosal cells is maintained at a relatively constant level, despite wide variations in extracellular mucosal magnesium concentration. There could be a homeostatic control mechanism regulating intracellular magnesium within narrow limits in gut tissue and this may occur through a cation selective transporter (Birch et al., 1991). In the absorptive direction, no evidence of a saturable system was found and the transepithelial movement of magnesium was not affected by the addition of inhibitors. We conclude that transepithelial magnesium transport in guinea-pig isolated mucosa was passive (Birch et al., 1991).

Magnesium transport studies in the human ileum agree with the findings in the guinea-pig (Phillips et al., 1991; Birch et al., 1991). There was no evidence of an active transport process for magnesium in the isolated human ileal mucosa, nor of a ouabain-sensitive Na-Mg exchange process. Tissue magnesium was maintained within fairly narrow limits despite wide fluctuations in external mucosal magnesium concentration. In the ileum, other workers have confirmed that the major magnesium transport process is paracellular in nature (Karbach and Feldmeier, 1991; Karbach and Rummel, 1990; Hardwicke et al., 1991; Karbach, 1989; see also *Medicine*, Part Three, Chapter 2-F2). At usual magnesium intakes, most studies suggest that magnesium is absorbed primarily by intercellular diffusional and solvent drag mechanisms. In the rat colon, however, there is evidence of saturable uptake of magnesium (Karbach and Ewe, 1987; Meneely et al., 1982; Karbach, 1989).

Lutz et al. (1991) examined the effects of short chain fatty acids and pH on the absorption of magnesium by the rat colon and caecum using a luminal perfusion technique. Short chain fatty acids, particularly butyrate, stimulated the absorption of magnesium and potassium in the distal colon and of sodium and water in the proximal colon and caecum. A decrease in pH from 6.4 to 5.0 in the presence of short chain fatty acids produced similar results. This is consistent with the presence of a magnesium/proton and a potassium/proton exchanger in the membrane of the distal colon and a sodium/proton exchanger in the proximal colon and caecum. Short chain fatty acids act as intracellular proton donors and enhance the cation absorption by proton exchange mechanisms.

C. Zinc transport in the intestine

Davie has investigated in detail the mechanisms of absorption of zinc at the isolated intestinal mucosa (Logue et al., 1990; Davie, 1988; Davie et al., 1991a; 1991b). This provides an example of yet another type of intestinal mechanism, this time involving possible ligand-metal interactions. At a physiological pH the free zinc ion is present only at insignificant concentrations. However, the overall process of zinc absorption from the small intestine is primarily a passive process: uptake kinetics were directly related to initial buffer concentration, which suggests that both zinc uptake and transfer are passive. However, the absorption kinetics were significantly influenced by the nature of the zinc-binding ligands which were present simultaneously. Certain ligands, particularly phytate, are capable of making the luminal zinc unavailable for absorption (see Chapter 1-B2).

Using the isolated, stripped, intestinal mucosal preparation (described above) transport was studied using zinc and histidine in a variety of ratios. The relative amount of zinc-ligand complex to free ligand was also shown to be of importance. High free ligand concentrations were shown to reduce total zinc absorption, presumably by a competitive inhibition interaction. At the concentrations used, the most favourable histidine-zinc ratio was shown to be 3:1.

Much zinc absorption may be occurring by a process of solvent drag of the zinc-ligand complex. In this respect, the presence of ligands in solution may have a significant effect on zinc absorption. Glucose or amino acids promote zinc absorption by stimulating the sodium exchange mechanisms required to produce the standing osmotic potential in the lateral intercellular spaces. However, only a minority of zinc, less than 2%, which binds to the mucosal membrane is subsequently delivered to the serosal surface. This avid binding of zinc to the mucosal tissue has been reported by other workers (Sahagian et al., 1966; 1967; Pearson and Schwink, 1966). However, it is unclear whether this is true tissue uptake into the enterocytes or simply non-specific binding due to the presence of metallothionein which might block absorption at the mucosal surface. Only a small proportion of the tissue-bound zinc might then be allowed to cross the membrane. Fine control of zinc homeostasis may be achieved by resecretion of zinc into the gut lumen.

Oberleas has suggested that zinc tissue binding is more than simple adsorption, as zinc accumulation by intestinal tissue was not easily removed by washing (Oberleas and Murer, 1966). It is possible that the enterocyte zinc binding may be the rate-limiting step in zinc intestinal absorption. A relationship between increased metallothionein-zinc binding in response to raised zinc levels in the diet has been shown (Richards and Cousins, 1975; 1977; Smith et al., 1978).

Although Davie has shown that zinc absorption is passive, whole body zinc homeostasis has been demonstrated by many workers (Cotzias and Papavasiliou, 1964; Weigand and Kirchgessner, 1978; Steel and Cousins, 1985) and this must indicate an active component in zinc transport at some point in the process. The location of this homeostatic mechanism has not been determined. Zinc secretion into the intestine has been demonstrated by a number of workers (Antonson et al., 1979; Davies, 1980) and it is suggested that total body zinc is regulated by intestinal secretion of excess zinc rather than by blocking total absorption. Both processes may be occurring simultaneously. The time relations of the zinc uptake are interesting: there appears to be a fast mucosal zinc uptake phase followed by a slower process which may involve the transport of zinc across the basolateral membrane. It is possible that this second phase of zinc absorption represents the active component.

III. CONCLUSION

The recognition that intestinal uptake and transport of lithium, magnesium and zinc may not involve transcellular transport of the metal is in accord with proposed transport mechanisms for other metals and should be borne in mind when considering mucosal transport of other metals. Metal-ion carrier proteins are ***not*** essential for rapid absorption to occur: their function is to assist when existing equilibrium conditions are unfavorable.

REFERENCES

Aldor, T.A.M., and Moore, E.W. (1970). *Gastroenterology, 59*: 745.

Antonson, D.L., Barak, A.J., and Vanderhoof, J.A. (1979). *J. Nutr., 109*: 142.

Behar, J. (1974). *Amer. J. Physiol., 227*: 334.

Birch, N.J., and Phillips, J.D. (1992). *Lithium Kinetics (Pharmacotherapy Monographs, Vol. 1)* (M. Thellier and J.-C. Wissocq, eds.), Marius Press, Carnforth, Lancashire, U.K., p. 69.

Birch, N.J., Coleman, I.P.L., and Karim, A.R. (1983). *Brit. J. Pharmacol., 80*: 443P.

Birch, N.J., Coleman, I.P.L., Hilburn, M.E., and Karim, A.R. (1984). *Brit. J. Pharmacol., 81*: 158P.

Birch, N.J., Coleman, I.P.L., Karim, A.R., and Mann, C.J. (1985). *Biochem. Soc. Trans., 13*: 250.

Birch, N.J., Davie, R.J., and Phillips, J.D. (1991). *Magnesium - A Relevant Ion?* (B. Lasserre and J. Durlach, eds.), John Libbey, London, p. 125.

Cotzias, G.C., and Papavasiliou, P.S. (1964). *Amer. J. Physiol., 206*: 787.

Davie, R.J. (1988). *The Uptake and Transport of Zinc in Isolated Intestinal Mucosa*, PhD Thesis (CNAA), The Polytechnic, Wolverhampton.

Davie, R.J. (1991). *Lithium and the Cell* (N.J. Birch, ed.), Academic Press, London, p. 243.

Davie, R.J., Phillips, J.D., and Birch, N.J. (1991a). *J. Inorg. Biochem., 43*: 686.

Davie, R.J., Phillips, J.D., and Birch, N.J. (1991b). *Trace 89*, Proceedings of the Third International Congress on Trace Metals in Health and Disease, Adana, Turkey, 1989 (G.T. Yuregar, O. Donma and L. Kayrin, eds.), Cukerova University Publishing Company, Cukerova, Turkey, p. 325.

Davies, N.T. (1980). *Brit. J. Nutr., 43*: 189.

Ebel, H. (1990). *Metal Ions in Biological Systems*, Vol. 26 (H. Sigel and A. Sigel, eds.), Marcel Dekker, New York and Basel, p. 227.

Ehrlich, B.E., Diamond, J.M., and Gosenfield, L. (1981). *Biochem. Pharmacol., 30*: 2539.

Ehrlich, B.E., Diamond, J.M., Fry, V., and Meier, K. (1983). *J. Membrane Biol., 75*: 233.

Hardwicke, L.L., Jones, M.R., Brautbar, N., and Lee, D.B. (1991). *J. Nutr., 121*: 13.

Hughes, M.S. (1989). *Lithium Interactions with Erythrocytes Studied by NMR and AAS*, PhD Thesis, The Polytechnic, Wolverhampton.

Hughes, M.S., Flavell, K., and Birch, N.J. (1988). *Biochem. Soc. Trans., 16*: 827.

Karbach, U. (1989). *Gastroenterology, 96*: 1282.

Karbach, U., and Ewe, K. (1987). *Digestion, 37*: 35.

Karbach, U., and Feldmeier, H. (1991). *Magnesium Res., 4*: 9.

Karbach, U., and Rummel, W. (1990). *Gastroenterology, 98*: 985.

Kimmich, G.A. (1983). *Intestinal Transport* (M. Gillies-Ballien and R. Gillies, eds.), Springer Verlag, New York, p. 87.

Krebs, H.A., and Henseleit, K. (1932). *Hoppe-Seyler Z., 210*: 33.

Lauterbach, F. (1977). *Arch. Pharm., 297*: 201.

Logue, A.M., Davie, R.J., and Birch, N.J. (1990). *Biochem. Soc. Trans., 18*: 1006.

Lutz, T., Würmli, R., and Scharrer, E. (1991). *Magnesium - A Relevant Ion?* (B. Lasserre and J. Durlach, eds.), John Libbey, London, p. 131.

Meneely, R., Leeper, L., and Ghishan, F.K. (1982). *Pediat. Res., 16*: 295.

Oberleas, D., and Murer, M. (1966). *J. Nutr., 90*: 56.

Padgham, C., Birch, N.J., Davie, R.J., and Phillips, J.D. (1993). *Magnesium and the Cell* (N.J. Birch, ed.), Academic Press, London, p. 263.

Pappenheimer, J.R., and Reiss, K.Z. (1987). *J. Membrane Biol.*, *100*: 123.

Partridge, S., Davie, R.J., and Birch, N.J. (1987). *Magnesium-Bull.*, *9*: 151.

Pearson, W.N., and Schwink, T.R. (1966). *J. Nutr.*, *90*: 239.

Phillips, J.D., Davie, R.J., and Birch, N.J. (1988). *Brit. J. Pharmacol.*, *95*: 836P.

Phillips, J.D., Davie, R.J., and Birch, N.J. (1990). *Biochem. Soc. Trans.*, *18*: 653.

Phillips, J.D., Davie, R.J., Keighley, M.R., and Birch, N.J. (1991). *J. Amer. Coll. Nutr.*, *10*: 200.

Richards, M.P., and Cousins, R.J. (1975). *Bioinorg. Chem.*, *4*: 215.

Richards, M.P., and Cousins, R.J. (1977). *Biochem. Biophys. Res. Commun.*, *75*: 286.

Sahagian, B.M., Harding-Barlow, I., and Perry, H.M. (1966). *J. Nutr.*, *90*: 259.

Sahagian, B.M., Harding-Barlow, I., and Perry, H.M. (1967). *J. Nutr.*, *93*: 291.

Smith, K.T., Cousins, R.J., Siblon, B.L., and Failla, M.L. (1978). *J. Nutr.*, *108*: 1849.

Steel, L., and Cousins, R.J. (1985). *Amer. J. Physiol.*, *248*: G46.

Thomas, G.M.H., Hughes, M.S., Partridge, S., Olufunwa, R.I., Marr, G., and Birch, N.J. (1988). *Biochem. Soc. Trans.*, *16*: 208.

Towler, C.M., Pugh-Humphreys, G.P., and Porteous, J.W. (1978). *J. Cell Sci.*, *29*: 53.

Wasserman, R.H., and Fullmer, C.S. (1989). *Adv. Exp. Med. Biol.*, *249*: 45.

Wasserman, R.H., Chandler, J.S., Meyer, S.A., Smith, C.A., Brindak, M.E., Fullmer, C.S., Penniston, J.T., and Kumar, R. (1992). *J. Nutr.*, *122*: 662.

Weigand, E., and Kirchgessner, M. (1978). *Nutr. Metab.*, *22*: 101.

METAL-PHYTATE INTERACTIONS: INTERFERENCE WITH METAL ION ABSORPTION

by

F.M. Clydesdale

Professor and Department Head,
Food Science Department,
Chenoweth Laboratory,
University of Massachusetts,
Amherst, MA 01003
U.S.A.

I. CHEMISTRY OF PHYTIC ACID

Chemically, phytic acid is designated as myo-inositol 1,2,3,4,5,6-hexakis (dihydrogen phosphate) and its titration curve displays two inflection points at pH 4.9 and 8.1 (Reddy et al., 1982). It has 12 replaceable protons, six strongly dissociated with a pK of approximately 1.5, one weakly dissociated with a pK of about 5.7, two weakly dissociated with a pK of about 6.8-7.6, and three very weakly dissociated with a pK greater than 10.

II. METAL ION COMPLEXES

This provides phytic acid with a strong negative charge in the pH range encountered in foods, making likely complex formation with positively charged species in foods, such as metal ions (Camire, 1982). Such complexation could have a positive, negative, or no effect on metal ion absorption depending on the solubility of the complex formed and the stability constant which exists within the complex. Solubility is critical for absorption since a metal must be in a soluble form to be absorbed, but not all soluble metals are necessarily absorbed. If a metal is in a soluble complex with a ligand and the complex has a stability constant such that the metal may be transferred to a serosal or mucosal receptor, the ligand is known as an enhancer of absorption whereas ligands which do not allow transfer are known as inhibitors.

Phytates have often been relegated to the role of inhibitor but the data is controversial and in many instances they have been shown to act as enhancers. The reason for this is that phytates can form complexes with metals that include a range of molar ratios from 1:1 to 6:1. They can also complex with more than one metal and bind to another compound, like protein, simultaneously (Clydesdale, 1988). Further, phytate and its various complexes are pH dependent, changing from an axial to an equatorial form outside a pH range of 5-12 (Martins and Evans, 1986). As a result, a number of structures have been postulated for phytates and metal ions (Weingartner and Erdman, 1978; Kratzer and Vohra, 1986; Nolan et al., 1987).

III. SOLUBILITY OF PHYTATE COMPLEXES

The major determinant of metal ion absorption from phytate complexes in food is probably solubility rather than the strength of the metal-phytate bond (Champagne, 1988). Although there is little thermodynamic evidence upon which to base this conclusion, there is evidence that lignin, a common component of food systems, will remove 35-70% of the iron from an iron-phytate complex rendering it insoluble (Platt and Clydesdale, 1984). This indicates that the iron-phytate bond is not excessively strong. Supporting this conclusion are other studies which show that iron from some forms of phytate is absorbed depending on the molar ratio of iron to phytate (Morse and Ellis, 1976). Apparently the various molar ratios of iron to phytate produce varying bond strengths and those soluble forms with weak bonds can release iron to a mucosal acceptor.

Phytate does, however, bind metals to different degrees. Using titrimetric methods, Vohra et al. (1965) showed that the capacity of sodium phytate to form complexes with metal is pH dependent and that at pH 7.4 it bound metal in the following decreasing order: Cu, Zn, Ni, Co, Mn, Fe, Ca. However, in food, where absorption is most relevant, conflicting results have been found. Water washing did not remove iron from wheat bran (Ellis and Morris, 1979) nor from soy isolates at pH 4.3 (Davies et al., 1962) but did remove calcium, magnesium and zinc. Obviously the phytate in food was bound differently than that observed in the model systems, which would of course influence absorption. Such differences were also shown in another study (Platt and Clydesdale, 1987) where phytic acid, phosphorous, protein, endogenous calcium, and soluble iron decreased significantly when both zinc and iron were added to a wheat bran fraction. However, when iron and zinc were added to sodium phytate alone, all remained soluble indicating the need for another component, such as protein or calcium which were present in the wheat bran, to cause precipitation and thus inhibit absorption.

With or without multi-complexes, consideration of pH is essential prior to drawing conclusions concerning phytate-metal solubility and subsequent absorption.

Jackson and Black (1951) found that the magnesium and calcium salts of phytate were soluble at low and insoluble at high pH. Al salts were soluble at extremes of pH with a minimum solubility at pH 3.5, and ferric salts were insoluble at pH 1.0-3.5, but became progressively more soluble as the pH was raised. Champagne (1985) has shown that the solubility of calcium in rice bran decreases above pH 5 and Gifford and Clydesdale (1990) found that an increase in pH significantly reduced zinc solubility in the presence of soy concentrate, casein or torula yeast. In order to account for physiological pH effects, Platt and Clydesdale incorporated a sequential pH treatment simulating gastrointestinal changes. Using this technique with phytate-rich wheat bran containing added iron, it was found that both calcium and zinc caused a decrease in the solubility of phytate and iron, with zinc also causing a decrease in protein. This indicated that calcium does not require protein in its phytate-metal complex to cause precipitation. Copper had no effect on the solubility of iron, phytate, or protein, indicating that a soluble complex is formed. This is in accord with previous work (Vohra et al., 1965) on single metal-phytate complexes and could explain why phytate inhibits zinc but not copper absorption (Turnlund et al., 1985).

Mills (1985) has suggested that the arguments made for the inhibitory effect of phytate on zinc utilization do not consider the restricted physiological relevance of the simplified aqueous systems within which such evidence was first obtained. The major variables interfering with extrapolations are the stability of phytic acid to hydrolysis both in the food and in vivo, the presence of other intraluminal ligands which bind zinc and prevent it from reacting with phytic acid (this phenomenon was also found to occur in food systems with iron, Platt and Clydesdale, 1984), and the high concentration of calcium in rat bioassay diets as compared to human diets. In fact, Mills (1985) suggested that the rate of weight gain of rats ΔW (gm/day) can be related to the (molar) concentration of dietary calcium, zinc, and phytate by the expression:

$$\Delta W = 6.213 - 0.331\ X$$

where X = [Ca][phytate]/[Zn] moles/kg diet; with the critical threshold ratio of [Ca][phytate]/[Zn] above which available zinc supply was inadequate to maintain growth being approximately 3.5.

Although the predictive value of the [Ca][phytate]/[Zn] ratio may prove to be superior to the more commonly used phytate/zinc ratio, it must be realized that either one will have limitations in diets which differ in respect of amino acids, proteins, organic acids, pH, and redox potential. All of these factors will affect the formation and stability of Ca-phytate-Zn complexes, since some compounds will bind zinc prior to its uptake by calcium phytate, while others may remove it from the complex (Clydesdale, 1988). For instance there is evidence that histidine, cysteine, and methionine readily desorb zinc from an insoluble Zn-Cu-phytate complex (Wise and Gilburt, 1982). It was further shown that the yield of copper or zinc from phytic acid complexes was proportional to the amino-N content of the intestinal soluble phase, and the effectiveness of release of a range of metals from their complexes with phytate was typically Cu>Cd>Mn>Zn. This fact, along with the differences in solubility of Zn- versus Cu-phytate complexes mentioned previously, might be used to explain the bioavailability of copper, as compared to the decreased bioavailability of zinc, in the presence of phytate. Magnesium deficiency can also be induced by a high fiber diet, and the magnesium phytate complex formed may be as effective in inhibiting zinc absorption as calcium phytate (Mills, 1985).

IV. EFFECT OF PHYTATE ON CALCIUM ABSORPTION

Although the inhibitory effect of phytate on calcium absorption is well documented in the literature, the numerous chemical reactions involved with its interaction with phytate (Wise, 1983) must be considered. Graf and Eaton (1985) have shown that phytate does not impair the absorption of calcium when fed to mice by gavage, indicating that the reactions which occur both in vitro and in vivo should be considered in this phenomenon.

V. CONCLUSIONS

It is clear that phytate has an effect on metal ion absorption. Nearly half a century ago, Mellanby, McCance and Widdowson, and others advanced the thesis that the fiber and phytic acid in unrefined cereals and other plant foods could interfere with the bioavailability of calcium and other mineral salts. The highly practical issue at that time, and now, is whether negative balances persist or satisfactory adaptation occurs (Walker et al., 1992). Opposite conclusions have been reached in several studies. This is probably due to the complexity of reactions which phytate undergoes and which have been briefly alluded to previously. Therefore it would be unwise to assume that phytate will contribute negatively to metal absorption in all diets and certainly dietary recommendations should not be based on concerns for reasonable levels of phytate in a well fed population.

REFERENCES

Camire, A.L., Jr. (1982). *Interaction of Metal Ions with Food Constituents with Particular Emphasis on Phytic Acid*, Ph.D. Thesis, Univ. of Mass., Amherst.

Champagne, E.T. (1988). *J. Amer. Coll. Nutr.*, *7*: 499.

Champagne, E.T., Rao, R.M., Luzzo, J.A., Robinson, J.W., Gale, R.J., and Miller, F. (1985). *Cereal Chem.*, *62*: 218.

Clydesdale, F.M. (1988). *Trace Minerals in Food* (K.T. Smith, ed.), Marcel Dekker Inc., New York, p. 57.

Davies, P.N., Norris, L.C., and Kratzer, F.H. (1962). *J. Nutr.*, *77*: 217.

Ellis, R., and Morris, E.R. (1979). *Nutr. Pept. Inter.*, *20*: 739.

Gifford, S.R., and Clydesdale, F.M. (1990). *J. Food Sci.*, *55*: 1720.

Graf, E., and Eaton, J.W. (1985). *Nutritional Bioavailability of Calcium* (C. Kies, ed.), Symp. Series No. 275, American Chemical Society, Washington, D.C.

Jackson, R.H., and Black, C.A. (1951). *Soil Sci.*, *72*: 179.

Kratzer, F.H., and Vohra, P. (1986). *Chelates in Nutrition*, CRC Press, Boca Raton, p. 69.

Martin, C.J., and Evans, W.J. (1986). *J. Inorg. Biochem.*, *27*: 17.

Mills, C.F. (1985). *Annual Review of Nutrition* (R.E. Olson, ed.), Annual Rev. Inc., Palo Alto, CA, p. 173.

Morse, E.R, and Ellis, R. (1976). *J. Nutr.*, *106*: 53.

Nolan, K.B., Duffin, P.A., and McWieny, D.J. (1987). *J. Sci. Food Agric.*, *40*: 79.

Platt, S.R., and Clydesdale, F.M. (1984). *J. Food Sci.*, *49*: 531.

Platt, S.R., and Clydesdale, F.M. (1987). *Cereal Chem.*, *64*: 102.

Reddy, N.R., Sathe, S.K., and Salunkhe, D.K. (1982). *Advances in Food Research*, Vol. 28 (C.O. Chickester, E.M. Mrak and G.F. Stewart, eds.), Academic Press, New York, p. 1.

Turnlund, J.R., King, J.C., Gong, B., Keyes, W.L., and Mitchell, M.C. (1985). *Amer. J. Clin. Nutr.*, *42*: 18.

Vohra, P., Gray, G.A., and Kratzer, F.H. (1965). *Proc. Soc. Exp. Med.*, *120*: 447.

Walker, A.R.P., Walker, B.F., Vorster, H.H., and Glatthaar, I.I. (1992). *Amer. J. Clin. Nutr.*, *56*: 608.

Weingartner, K.E., and Erdman, J.W. (1978). *Illinois Res.*, *20*: 4.

Wise, A. (1983). *Nutr. Abs. Rev. Rev. Clin. Nutr.*, *53*: 791.

Wise, A., and Gilburt, D. (1982). *Toxicol. Lett.*, *11*: 49.

MOLECULAR INTERACTIONS IN TRANSPORT, ACCUMULATION AND EXCRETION OF METALS

by

A. Mas* and B. Sarkar

Department of Biochemistry Research,
The Hospital for Sick Children and
the Department of Biochemistry,
University of Toronto,
Toronto, Ontario
Canada

I. INTRODUCTION

Most of the transition metals are bound to proteins in biological systems. Although other components (nucleic acids, sugars, amino acids) bind these metals and relevant functions are assigned to them, the binding to proteins is of paramount importance from pathological, toxicological and nutritional points of view. In fact, most of the transport and accumulation of transition metals is in the metal-protein form, whereas the excretion is in the form of metals bound to peptides and other low-molar-mass (l.m.m.) components. Consequently, there is a great surge of interest in the studies of the metal-protein interactions in transport, accumulation and excretion of metals (Sarkar, 1987). However, the molecular mechanisms of these processes still remain unclear.

For essential metals, provision is made for their materno-fetal transfer during gestation and their utilization during the postnatal period. It is known that placenta is highly selective. Essential metals such as zinc (Feaster et al., 1955; Evans and Reis, 1976) and copper (Terao and Owen, 1977) are easily transferred through the placenta, whereas "toxic" metals such as cadmium (Webb and Samawickrama, 1981) and nickel (Mas et al., 1985) find a "placental barrier" that protects the fetus from their toxic effects. The mechanisms underlying such selective action of the placenta have not been clearly understood. Kidney is known to be the target organ for a number of metals including nickel (Sarkar, 1980), and urinary excretion is the major route of its elimination. The metal-binding components in kidney and urine are not completely elucidated. In this article, we will focus on what is known about the molecular mechanisms of trace element transport in blood, their materno-fetal transfer across the placenta, and their interaction in kidney and excretion in urine, with special reference to the differences between essential (copper and zinc) and "toxic" (nickel and cadmium) elements in the process of transport, accumulation and excretion.

Present address:
* Bioquímica. Univ. Rovira i Virgili. 43005-Tarragona, Spain

II. TRANSPORT

Most of the metals are transported in blood bound to specific ligands in plasma. Although there is a considerable amount of metals in red blood cells, this compartment is not in equilibrium with plasma and, thus, it should not be considered as a transport form of metals.

A. Serum

The status of metals in serum is better known for copper and zinc. Copper in serum is mostly associated with ceruloplasmin (60-90% of total copper in serum) and the rest of copper is associated with high-molar-mass (h.m.m.) proteins, albumin and amino acids (Sass-Kortsak et al., 1967). Copper associated with ceruloplasmin is tightly bound and its release is associated with the changes of either the redox status of the metal or the protein conformation. This fact, coupled with its other physiological properties (mostly derived from its oxidase activity) raises the question of ceruloplasmin being a copper transport protein. However, copper is known to be released effectively from ceruloplasmin and incorporated and metabolized in the peripheral tissues (Marceau and Aspin, 1971). Because of the strong association of copper with ceruloplasmin, the rest of the copper in serum is considered to be in an exchangeable form. Upon equilibration of copper with adult or cord sera, most of the metal is found to be associated with albumin, although 5-7% is associated with h.m.m. proteins as well (Lau and Sarkar, 1984). Identification of these h.m.m. components has not yet been accomplished. Although a copper-binding protein given the name of transcuprein has been reported (Weiss and Linder, 1985), it still remains to be characterized. This protein was found in portal circulation of some systems after intestinal perfusion with high specific activity copper-67 (Weiss and Linder, 1985). However, it was not always found to be the case (Gordon et al., 1987). A small fraction of copper is associated with amino acids in which copper-histidine was found to be the major component (Sarkar and Kruck, 1966). Copper has a specific transport site at the NH_2-terminal of albumin, involving the α-NH_2, two intervening peptide nitrogens, imidazole nitrogen of histidine residue in the third position and carboxyl side chain of the aspartyl residue (Laussac and Sarkar, 1984). It is also known that copper bound to albumin coexists with ternary complexes of albumin-copper-amino acids (mostly with histidine) (Lau and Sarkar, 1971). These ternary complexes are in equilibria with copper bound to amino acids, which has been proposed to play an important role in the exchange and transfer of copper between albumin and amino acids (Lau and Sarkar, 1971; Tabata and Sarkar, 1992).

Zinc in serum is mostly associated with albumin and h.m.m. components, mostly α_2-macroglobulin. Both cord and adult sera were fractionated by means of column chromatography of Sephadex G-100 after equilibration with zinc-65 in serum. Most of the zinc (75%) was bound to albumin and 15% to α_2-macroglobulin (Lau and Sarkar, 1984). These results are consistent with those obtained when serum was analyzed by Atomic Absorption Spectrophotometry (Cousins, 1985). Recent studies show that the zinc transport site of albumin is located within the polypeptide chain of the protein and not at the NH_2-terminal as in the case of copper (Goumakos et al., 1992). Beside these proteins, the rest is associated with l.m.m. components, mostly amino acids (Giroux and Henkin, 1972). Zinc in these forms is freely exchangeable, and is thus considered to play an important role in the transport of this essential metal.

Cadmium and nickel are not normally found in plasma in appreciable amounts. These two metals are found in plasma only after exposure or administration. Their half-life in plasma is short and they are either rapidly accumulated in the tissues or excreted in urine. When equilibrated with serum, great differences are observed between adult and cord sera. Both metals show higher amounts of albumin-bound forms in adult sera than those in cord sera. In these equilibration studies (Lau and Sarkar, 1984), nickel was shown primarily bound to albumin and l.m.m. components in a ratio of 2:1 in adult sera. Reverse is the case of cord sera. Characterization of these l.m.m. binding components is still lacking, although they could have similarities with those found in kidneys and urine. For cadmium, a similar distribution is observed, although it is predominantly bound to albumin and h.m.m. components. The variation in nickel distribution in cord serum could be due to the presence of specific nickel-binding components in cord serum.

B. Placenta

Mammalian placentae provide a unique system to study both metal transport and its accumulation. Placenta is a typical transport epithelium, acting as an active organ in transporting nutrients from the mother to the fetus. Furthermore, placenta acts as a very complex organ with a broad array of functions such as hormone production, fetal excretion and accumulation of non-transferable metabolites. Regarding accumulation and transfer, it is worth mentioning the differences between two essential elements (e.g. zinc and copper) and two "toxic" elements (e.g. cadmium and nickel). Zinc and copper are easily transferred to the fetus. The fetal liver has a higher content of these metals than the maternal liver 24 h after the administration of zinc-65 or copper-67. On the other hand, cadmium and nickel accumulate in the placenta and very little is found in the fetus (Mas and Sarkar, 1988). All these placental functions lie on a single layer of specialized cells, the trophoblast which, at term, is formed almost exclusively by the syncytiotrophoblast.

By using the isolated trophoblast from human placentae at term, it has been possible to elucidate some of the characteristics of the processes involved in the placental transport. Uptake studies of copper by these cells have clearly shown that ceruloplasmin releases its copper, which is subsequently taken up by the cells in a similar manner as that by copper-histidine complexes or ionic copper. The release of copper from ceruloplasmin involves the loss of oxidase activity of ceruloplasmin (Mas and Sarkar, 1992). These results support the role of ceruloplasmin as a copper-transport protein, in agreement with results obtained by other laboratories (Orena et al., 1986; Dameron and Harris, 1987). This function of ceruloplasmin may be carried out by a receptor-mediated process. The interaction between the receptor and ceruloplasmin probably results in the release of copper near the cell surface from where the uptake may take place. Although an endocytotic pathway is also present for ceruloplasmin in these cells (Kennedy et al., 1989), most of the copper transported across the placenta appears to originate in the ceruloplasmin-receptor interaction at the cell surface. Once inside the cell, copper is mostly in the cytosol where it binds to metallothionein and other h.m.m. proteins (Mas and Sarkar, 1988).

In the absence of serum, zinc is rapidly removed from the media by binding to the syncytiotrophoblast; yet, in the presence of serum, the binding is greatly reduced. However, most of this binding in the presence of serum results in internalization, whereas in the absence of serum it is not. Zinc internalization is dependent upon the availability of free SH groups whereas the binding process is not, as demonstrated by the reduction of zinc uptake in presence of N-ethylmaleimide (Mas and Sarkar, 1991). Nevertheless, the binding to SH-containing proteins should be of low affinity to allow the release of zinc to the fetal circulation.

III. ACCUMULATION

Potentially, all the tissues have the ability to accumulate metals, although only few tissues use this ability for large storages. Among them, liver, kidney and bone could play an important role in removing the excess of metals in the organism before their excretion (liver and kidney) or in long term immobilization (bone). Placenta plays a very different role compared to those mentioned above, in the way that the organ itself is eliminated at the time of delivery.

A. Placenta

It is known that the effect of cadmium administration in pregnant animals results in zinc deficiency. This is due to the increase of metallothionein synthesis in placenta by cadmium which in turn causes the retention of zinc in placenta (Webb and Samawickrama, 1981). The exposure of the syncytiotrophoblast to cadmium shows a sharp increase of metallothionein and this increase is clearly correlated to a decreased efflux of zinc from the cells (Torreblanca et al., 1992), a step necessary to free this metal to the fetal side of the placenta. However, without induction of metallothionein, zinc would not normally be bound to this placental protein (Mas and Sarkar, 1988). Metallothionein is found only in low concentration in human placenta (Clough et al., 1986). Furthermore, the transfer of zinc across the placenta is dependent on the presence of ligands in the fetal side, basically albumin and amino acids (Paterson et al., 1991).

As for nickel and cadmium, they are known to be selectively retained by the placenta and only minute amounts are transported to the fetus (Webb and Samawickrama, 1981; Mas et al., 1985; Mas and Sarkar, 1988). As previously stated, cadmium induces and binds to metallothionein, as well as other h.m.m. proteins in the placenta (Mas and Sarkar, 1988).

The retention of nickel by the placenta does not necessarily imply internalization. Thus, nickel binding to isolated human syncytiotrophoblast was analyzed in the presence of albumin and histidine, which are the most likely ligands in plasma. The kinetics of nickel binding were similar to those observed in vivo (Mas et al., 1986) when nickel was administrated as nickel-histidine complexes, whereas nickel-albumin complexes showed a different pattern (Mas et al., 1990). Analysis of nickel-binding components in placenta showed that most of the nickel is bound to l.m.m. components in the trophoblast and in rat placenta (Mas and Sarkar, 1988).

B. Kidney

Similarly, in kidney, metallothionein plays a key role in accumulation akin to what is observed with placenta. Furthermore, in kidney cells, it has been shown that metallothionein is induced by zinc and cadmium in an additive fashion (Harford and Sarkar, 1991). This indicates that induction of metallothionein by the two metals occurs by different mechanisms. However, the accumulation of metals in the kidney could be an intermediary step toward the final excretion from the body, which would be through the urine.

IV. EXCRETION

Mammals dispose off the unwanted metals through different routes. The main excretory routes are considered to be the urine, bile and sweat. For essential metals, however, excretion and losses are difficult to distinguish. For instance, the presence of zinc in the pancreatic secretion should be considered more as a loss than an excretion. Bile has been considered the main route of excretion for copper, yet the status of this element in bile is not known. Although urinary excretion of copper and zinc are not considered as the main route of excretion, a consistent amount of both metals is found in urine, and changes of these urinary levels reflect different pathological states.

Nickel is rapidly cleared in urine and most of the characterization studies of the urinary excretion have been possible with this element. Distribution of nickel following intraperitoneal administration of its soluble salts in rats clearly showed the highest concentration of nickel in kidney and urine (Sarkar, 1980). Subcellular distribution of nickel in kidneys showed most of the metal to be present in the soluble postmicrosomal fraction. About 70% of the nickel in this fraction is associated with a carbohydrate-rich fraction, and 30% is bound to an acidic polypeptide (Templeton and Sarkar, 1985). Characterization of these two components showed that the polypeptide is both intra- and extracellular with little variation from various species, whereas the oligosaccharide is intracellular, derived from the degradation of internalized heparan sulfate with two to four overall negative charges (Predki et al., 1992). Both compounds are able to bind different metals such as copper, manganese, cadmium and zinc. In physiological conditions, the higher affinity of these elements for other cellular components reduces their binding to the peptide and oligosaccharides.

Similar compounds are also found in urine, although they are not fully characterized. Most of the nickel found in urine is bound to a polypeptide, similar to the one reported in kidney. Small amounts of substances similar to the oligosaccharides are also found. The polypeptides in urine, although heterogeneous, have compositions similar to that of the renal polypeptide. Zinc and copper appear to be present in urine also bound to this heterogeneous pool of acidic polypeptides (Sarkar et al., 1989).

V. CONCLUSION

Metal-binding proteins and polypeptides play an important role in transport, accumulation and excretion of metals. Transport is carried out mostly by large proteins. Low-molar-mass components, mainly amino acids, play an important role in the delivery of metals to the cells. Accumulation and excretion, instead, involve smaller proteins (e.g. metallothionein) or polypeptides. However, the characterization of all these processes and the molecular interactions between metals and the different molecules involved are still far from being completely understood. Further studies of these processes are clearly warranted.

REFERENCES

Clough, S.R., Mitra, R.S., and Kulkarni, A.P. (1986). *Biol. Neonate, 49*: 241.

Cousins, B. (1985). *Physiol. Rev., 65*: 238.

Dameron, C.T., and Harris, E.D. (1987). *Biochem. J., 248*: 663.

Evans, G.W., and Reis, B.L. (1976). *Amer. J. Clin. Nutr., 29*: 814.

Feaster, J.P., Hansard, S.L., McCall, J.T., and Davis, G.K. (1955). *Amer. J. Physiol., 181*: 287.

Gordon, D.T., Leinart, A.S., and Cousins, R.J. (1987). *Amer. J. Physiol., 252*: E327.

Goumakos, W., Laussac, J.P., and Sarkar, B. (1992). *Biochem. Cell Biol., 69*: 809.

Giroux, E.L., and Henkin, R.I. (1972). *Biochim. Biophys. Acta, 273*: 64.

Harford, C., and Sarkar, B. (1991). *Biochem. Biophys. Res. Commun., 177*: 224.

Kennedy, M., Douglas, G., King, B., and Lönnerdal, B. (1989). *FASEB J., 3*: A560.

Lau, S.-J., and Sarkar, B. (1971). *J. Biol. Chem., 246*: 5938.

Lau, S.-J., and Sarkar, B. (1984). *Can. J. Biochem. Cell. Biol., 62*: 449.

Laussac, J.P., and Sarkar, B. (1984). *Biochemistry, 23*: 2832.

Marceau, N., and Aspin, N. (1971). *Amer. J. Physiol., 222*: 106.

Mas, A., and Sarkar, B. (1988). *Biol. Tr. Elem. Res., 8*: 105.

Mas, A., and Sarkar, B. (1991). *Biochim. Biophys. Acta, 1092*: 35.

Mas, A., and Sarkar, B. (1992). *Biochim. Biophys. Acta, 1135*: 123.

Mas, A., Holt, D., and Webb, M. (1985). *Toxicology, 35*: 47.

Mas, A., Peligero, M.J., Arola, L., and Alemany, M. (1986). *Clin. Exp. Pharmacol. Physiol., 13*: 91.

Mas, A., Yeger, H., and Sarkar, B. (1990). *Res. Commun. Chem. Pathol. Pharmacol., 67*: 271.

Orena, S.J., Goode, C.A., and Linder, M.C. (1986). *Biochem. Biophys. Res. Commun., 139*: 822.

Paterson, P.G., Mas, A., Sarkar, B., and Zlotkin, S.H. (1991). *J. Nutr., 121*: 338.

Predki, P.F., Whitfield, D.N., and Sarkar, B. (1992). *Biochem. J., 281*: 835

Sarkar, B. (1980). *Nickel Toxicology* (F.W. Sunderman and S.S. Brown, eds.), Academic Press, New York, p. 81.

Sarkar, B. (1987). *Prog. Food Nutr. Sci., 11*: 363.

Sarkar, B., and Kruck, T.P.A. (1966). *Biochemistry of Copper* (J. Peisach, P. Aisen and W. Blumberg, eds.), Academic Press, New York, p. 183.

Sarkar, B., Abdulwajid, A.W., Mukherjee, S., Khandelwal, S., and Mas, A. (1989). *Elements in Health and Disease* (M. Said, ed.), Hamdard Foundation, Karachi, p. 274.

Sass-Kortsak, A., Clarke, R., Harris, D.I.M., Neuman, P.Z., and Sarkar, B. (1967). *Progr. Neurogen., 175*: 625.

Tabata, M., and Sarkar, B. (1992). *J. Inorg. Biochem., 45*: 93.

Templeton, D.M., and Sarkar, B. (1985). *Biochem. J., 230*: 35.

Terao, T., and Owen, C.A., Jr. (1977). *Amer. J. Physiol., 232*: E172.

Torreblanca, A., Del Ramo, J., and Sarkar, B. (1992). *Toxicology, 72*: 167.

Webb, M., and Samawickrama, G.P. (1981). *J. Appl. Toxicol., 1*: 270.

Weiss, K.C., and Linder, M.C. (1985). *Amer. J. Physiol., 249*: E77.

METAL-LIGAND INTERACTIONS IN BIOMINERALIZATION

by

Brigid R. Heywood

Division of Chemical Sciences,
Science Research Institute,
University of Salford,
Salford M5 4WT
U.K.

I. INTRODUCTION

A. Biomineralization

Of the many roles ascribed to metals in biological systems, perhaps none is more challenging to the imagination than the formation of solid phase inorganic deposits. Biological minerals formed from calcium phosphates, carbonates, oxalates and sulfates are widespread amongst plants and animals, as are deposits of iron oxides and sulfides, barium and strontium sulfates and silica (Lowenstam and Weiner, 1989). Even precipitates of non-essential elements (silver, gold and uranium) have been identified in the cell walls of some bacteria (McClean and Beveridge, 1989). In addition to the more obvious examples of biomineral in structural/defensive applications (bones, teeth, shells, etc), devices for inertial detection, magnetic sensing, optical transmittance and homeostatic regulation are included amongst the many varied uses for which they are utilized in vivo. Thus, biological minerals arrest our interest not only because of their novel and elegant forms but also because they are employed for specialist functions which require materials of specific size, structure, shape, orientation and organization to optimize performance.

Clearly, 'nature' has evolved unique crystallization strategies which circumvent the accepted thermodynamic directives normally controlling crystal nucleation and growth to facilitate the fabrication of inorganic materials tailored for precise functional use. The elucidation of these strategies so that they can be mimicked, adapted and controlled to produce novel materials with enhanced properties (and the added benefit of environmental compatibility) is now the common goal of materials scientists, gene technologists, microbiologists, protein chemists, structural engineers and crystal chemists (Heuer et al., 1992; Mann, 1993).

B. Solid state bio-engineering

To the crystal chemist, a review of biomineralization shows that while a complex array of strategies have evolved to regulate the formation of bio-crystalline phases, one feature is common to the biological paradigm: interactions between organized biopolymeric assemblies (e.g. collagen-based matrices, β-pleated polypeptide sheets, proteolipid membranes), and the nascent inorganic solids play a fundamental role in the mineralization process (Weiner, 1986). The events of nucleation and growth, the crystal morphology and the aggregation (assembly) of the mineral units into functional hierarchical microarchitectures, are all regulated by these bio-organic matrices.

On the microscopic level, this control may be through the chemical manipulation of the reaction microenvironment and/or the spatial constraint of the evolving mineral structure. At the molecular level, the unique chemical and structural identity of these bio-organic templates regulates the crystallochemical events at the inorganic-organic interface which in turn governs crystallographic selectivity in the precipitating mineral phases (Mann and Heywood, 1989; Addadi et al., 1991). Thus, specific molecular processes dependent upon the recognition and translation of chemical information at the interface between the bio-organic macromolecules and the inorganic crystal faces can be credited with mediating the energy cascade responsible for stabilizing nuclei and directing crystal growth.

The nature and identity of these molecular recognition processes is the subject of the present work. Circumstantial evidence suggests that the selective interaction of the precipitating cations and anions with a unique catalogue of bio-organic molecules which cycle through during mineralization (Slavkin and Price, 1992) simulates the lattice geometry (coordination chemistry, hydration profile, etc) of the ions in the mineral phase and potentiates the mineralization event. The observation that bio-organic molecules (e.g. proteins, phospholipids, etc) with functionalities which mimic the oxyanions of the contiguous biomineral are frequently isolated from mineralized tissues gives further support to this argument. For example, carboxylate-rich acidic proteins are commonly associated with calcium carbonate biominerals, whereas the species isolated from calcium phosphate biominerals are heavily phosphorylated and the organic matrices associated with silica or ice formation are distinguished by the abundance of polar -OH groups. On the basis of these observations and evidence obtained from comparative structural analyses of several biominerals, a purely epitactic relationship was proposed since the structural studies revealed a direct geometric correspondence between the lattice coordinates defining the principal face of the biomineral and specific conformational features of the organic matrix. However, the failure of in vitro crystallization assays with soluble polypeptides isolated from mineralizing matrices to verify this model, coupled with recent genetic and biochemical studies relating to the primary structure of matrix constituents (principal functional residues, degree and type of post-translational modifications, viz phosphorylation, glycosylation, sulphation, etc) prompted the evolution of an interfacial (organic-inorganic) model encompassing a more complex interplay of molecular parameters to include geometric complementarity, charge localization, the chemistry of the hydration layers and the stereochemical motifs defining both the organic and inorganic surfaces.

The next question was how to verify and expand this model ? One approach involves the direct probing of mineralizing systems in vivo using the sophisticated repertoire of techniques currently available to molecular biologists (cDNAs, expression cloning, in situ hybridization and developmental switching of gene expression) (Wade et al., 1991; Slavkin and Price, 1992). For solid state chemists, engineers and material scientists, the challenge rests with combining in depth studies of the mineral phase with the development of effective and reproducible biomimetic crystallization strategies (Mann et al., 1992; Birchall, 1993). The purely synthetic challenge of designing polymers containing surface functionalities which are active with respect to crystal nucleation and growth has been successfully taken up by some groups (Calvert, 1992; Brisdon et al., 1993). The further development of this work from the phenomenological to a predictive science has proven more challenging since the molecular identity of the active sites is difficult to characterize.

As an alternative, the feasibility of using supramolecular assemblies of organic molecules as model templates to control nucleation and growth is under consideration (Mann and Heywood, 1989; Heuer et al., 1992; Heywood, 1993). This biomimetic approach to crystallization has the advantage of providing a route to control the interfacial chemistry by molecular recognition through the synthesis of tailored organic templates (head group identity, polarity, spatial ordering, etc), in order to initiate and regulate crystal growth. A certain measure of success has already been achieved with respect to size control (often nanoscale) through the use of synthetic surfactant aggregates (vesicles, micelles, etc)

(Mann et al., 1986; Bhandarkar and Bose, 1990; Heywood and Eanes, 1992; Meldrum et al., 1993) native phytochelatins and their synthetic analogues (Dameron et al., 1989) and protein cages (Meldrum et al., 1991; 1992), all of which function as spatially constrained micro-environments for controlled crystal growth. The advent of programmed crystal habit modification through the use of surface-specific additives (one dimensional templates) which interact only with designated crystal faces has been another key development (Davey et al., 1991). The facility for orienting the nucleation and controlling the growth of crystals has evolved from the design and application of two-dimensional films of ordered surfactants, Langmuir monolayers (Gavisch et al., 1990; Heywood and Mann, 1992; 1993). By adopting ideas and concepts derived from biomineralization, this novel interdisciplinary approach to crystal engineering has brought about the successful integration of organic supramolecular chemistry and molecular self-assembly with the classical elements of inorganic synthesis and materials science to open new horizons in biomimetics and materials research.

II. THE ORGANIC-INORGANIC INTERFACE

A. Matrix-mediated biomineralization - Issues of substrate identity, surface recognition and oriented nucleation

As discussed earlier, analyses of well mineralized tissues (e.g. bones, oyster shells, magnetotactic bacteria) (Fig. 1) demonstrate clearly that the mineral crystallites are not randomly distributed throughout the tissue/organism, but are aligned in an ordered array, complementary to the structure of the underlying host matrix.

In bone for example, small platelike single domain mineral particles (35 x 20 x 5 nm) of the calcium phosphate mineral, hydroxyapatite ($Ca_5(PO_4)_3(OH)$) are embedded within the ordered polymeric matrix of collagen (Fig. 1) (Heywood et al., 1990). The small size of the crystallites coupled with (i) the knowledge that their crystallographic *c* axis is consistently co-aligned with the host protein lattice, and (ii) an awareness of the quasi-three dimensional packing model proposed for collagen fibrils (Katz and Li, 1973), leads to the conclusion that the process of nucleation may be directed by interactions between the organic matrix and nascent crystal lattice with the imposition of spatial constraints rather than specific matrix-crystal interactions regulating particle maturation. Whilst pure native collagen evinces no nucleation potential in controlled crystallization assays, several specific bone

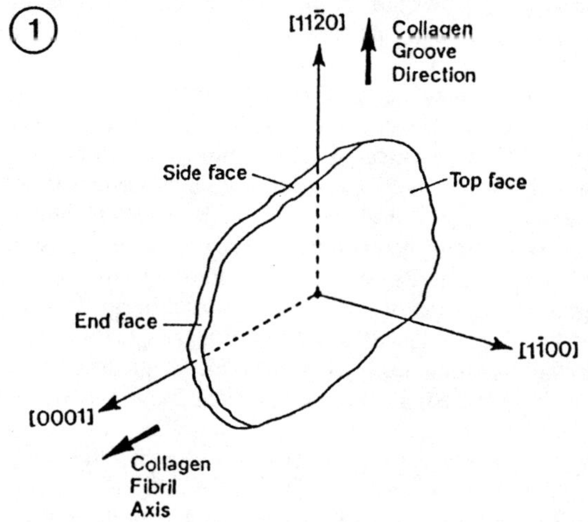

Figure 1.
Schematic showing the proposed relationship between the crystallographic structure of the plate-shaped apatite crystals extracted from calcified turkey tendon, their morphology and the collagen fibrils of the surrounding matrix (Heywood et al., 1991d).

proteins have now been isolated from bone and may prove to be significant to the mineralization process (Slavkin and Price, 1992).

The chemistry and biology of the organic matrix in mollusc shells is equally of interest since the elaborate nacreous (aragonite) and foliated (calcite) structures of $CaCO_3$ which characterize the mineralized conch (shell) of the adult and juvenile forms are noted for their unusual toughness and fracture resistance which arises directly from the unique crystallographic microstructure of the bulk material (Lowenstam and Weiner, 1989). Each crystal unit is preferentially elongated along the [001] axis and oriented with respect to their nearest neighbors such that their [100] and [010] crystallographic axes were coincident. A preferred alignment of the aragonite *c* axis with the underlying protein substrate is also evident.

As with bone, the formation of this unified mineral layer argues strongly in favor of biological control over the crystallochemical sequelae of nucleation and growth. It has been postulated that the role of the organic matrix is to lower the activation energy for nucleation. Simple catalysis involving complementarity of electrostatic binding at the inorganic-organic interface would be sufficient, but a more complex model dependent upon the stabilization of the inorganic nuclei in a specific transition state by structural and chemical interfacial recognition processes may also apply.

B. Biomineralization in restricted reaction volumes - Issues of surface charge and spatial constraint

Biological control of crystal nucleation and growth is perhaps best exemplified by the formation of intracellular magnetic deposits in magnetotactic bacteria (Fig. 2). These aquatic bacteria are susceptible to the earth's geomagnetic field and employ magnetotaxis to direct their motion towards suitable habitats. A feature common to them all is the presence of discrete intracellular magnetic inclusions, magnetosomes, aligned in chains along the motility axis of the organism. The size and orientation of these individual magnets imparts a permanent magnetic dipole moment which is responsible for the magnetotactic response. In each species, the magnetic particles are well ordered single domain membrane-bound crystals which exhibit a narrow size distribution (50-90 nm) and unique crystallographic morphologies (e.g. cubo-octahedral, rectangular prismatic) (Frankel and Mann, 1989). Each crystal is preferentially elongated along the easy axis of magnetization ($<111>$ for magnetite, [100] for greigite). Until recently, only crystals of the mixed valence iron oxide magnetite had been identified in these prokaryotes; now, bacteria which contain crystals of the ferrimagnetic iron sulphide, greigite (Fe_3S_4), non-magnetic pyrite (FeS_2) (Heywood et al., 1991b) and some which synthesize both magnetite and greigite (Bazylinski et al., 1993) have been reported.

The expression of such unique crystal morphologies in the bacterial magnetosomes gives evidence of a sophisticated hierarchy of crystallochemical mechanisms controlling the bioprecipitation of this mineral. Critical to these biologically-mediated processes is the trilaminate membrane which provides (i) an enclosed microenvironment for the precipitation reactions, (ii) the potential for chemical specificity via selective ion transport, and (iii) imposes geometric constraint upon the growing crystals through the asymmetric imposition of physiological stresses (e.g. pH, degree of supersaturation, solution composition, redox gradients). The preferred association of a designated crystallographic axis with the direction of elongation (e.g. $<111>$ of magnetite, $<100>$ of greigite) also attests to the role of biomolecular engineering directing nucleation with the magnetosome membrane strongly implicated as the active surface controlling the chemical, interfacial and spatial aspects of the process.

It is interesting to note that an ion-translocating proteolipid capable of initiating membrane-mediated mineralization has been isolated from the cell membrane of a calcifying bacteria, *Bacterionema matruchotii* (Swain and Boyan, 1989). It is also noteworthy that a series of unique proteins has been isolated from the magnetosome membrane of a magnetite-forming bacteria (Gorby et al., 1988). Now that pure cultures of some magnetotactic bacteria have been established, it is widely anticipated that progress in this field will be rapid (Bazylinski et al., 1993).

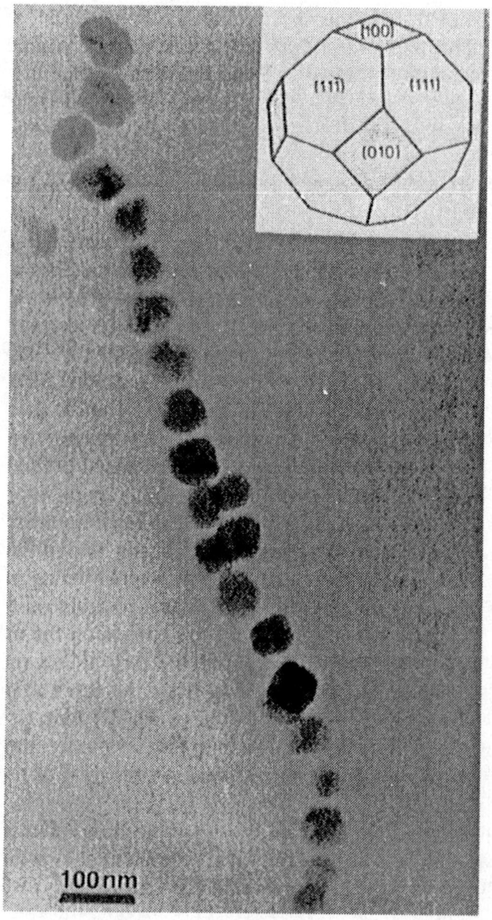

Figure 2.
Transmission electron micrograph of two types of magnetotactic bacteria collected from sulphide-rich aquatic environments.
(a) Unicellular rod-shaped bacteria containing electron dense cubo-octahedral inclusions of greigite.
(b) Large unicellular rod-shaped organism containing discrete prismatic rectangular greigite crystals (Heywood et al., 1990).

III. CRYSTAL ENGINEERING

A. Biomimetic crystal engineering

The identity of many of the organic polymeric species (e.g. glycoproteins, acidic phospholipids, polyanionic proteins) associated with biomineralization have now been identified and it is known that many of them will enhance the rate of nucleation and alter crystal morphologies in controlled crystallization assays. However, the complexity of these organic molecules coupled with the loss of their native configuration once isolated and purified compromises the extrapolation of data from such studies to the in vivo process.

Parallel studies have sought to provide adequate explanations for the processes that result in the controlled crystallization of inorganic solids in biological systems by using a ***biomimetic*** approach based upon the integration of supramolecular organic chemistry with inorganic synthesis (Mann and Heywood, 1989; Mann et al., 1992). The possibility of using host-guest systems to confine the synthesis of inorganic materials within small reaction volumes (e.g. protein cages, synthetic phospholipid vesicles) has been fruitfully explored. Also, the potential of ordered one and two-dimensional organic surfaces to orient the nucleation of inorganic materials and control their subsequent growth has been investigated (Heywood and Mann, 1993).

B. Solid state reactions in self-assembling supramolecular assemblies

1. *Protein cages (Ferritin)*

The integration of supramolecular organic chemistry with inorganic materials fabrication is perhaps best exemplified by the iron storage protein, ferritin. This unique protein which is found in plants, bacteria and animals comprises a spherical polypeptide shell (apoferritin) of 24 ordered subunits encapsulating an inorganic core of the hydrated iron oxide ferrihydrite ($5Fe_2O_3.9H_2O$). The mechanism for Fe uptake and biomineralization inside the core is well documented (Harrison et al., 1989) and is illustrative of the controlled synthesis of an inorganic phase within a constrained reaction space to generate particles of nanometer (6-9nm) dimensions; the potential value of these nano-solids lies in the fact that particles of 1 - 100 nm size can exhibit unique atomic, or surface-influenced properties, which are distinct from those found in the bulk material.

Efforts to interpret the molecular details of the ferritin mineralization process have been complicated by the heteropolymeric nature of the two different subunit types, H (heavy) and L (light). The H-subunit bears a specific metal-binding oxidation site which lies in close proximity to a unique motif of three glutamate residues on the inner surface of the protein cage (Lawson et al., 1991). Mineralization depends on the uptake of Fe(II) through the intermolecular channels and metal binding to the ferroxidase center followed by rapid oxidation. The charged molecular niche generated by the three acidic residues provides a localized gradient which favors the accumulation of Fe(III) ions. It is possible, therefore, that the electrostatic center functions to reduce the encounter times between molecular collisions and thus potentiate nucleation by kinetic stabilization of the critical nuclei (Wade et al., 1991).

These findings suggested the possibility of synthesizing other inorganic phases within this supra molecular organic cage by exploiting the remarkable stability of the polypeptide assembly to a wide range of reaction conditions (20 - 85°C; pH 7 - 9.0) (Fig. 3) (Meldrum et al., 1991; 1992). It is now possible to form a range of inorganic materials inside apoferritin using either (i) controlled thermoregulated reconstitution (magnetite), (ii) in situ reaction of the native core (iron sulphide), (iii) redox driven reactions (manganese oxide), or (iv) ion-binding coupled with hydrolytic polymerization (uranyl oxyhydroxide) (Fig. 3). The successful utilization of the bio-protein cage for the controlled formation of nanodimensional ferrimagnetic particles is of particular interest given its potential uses in biomedical imaging, cell labelling and separation procedures.

2. *Polypeptide clusters (Phytochelatins)*

Another exciting biomimetic route to the fabrication of inorganic nanophases has been the exploitation of the detoxification chemistry that many plant species engage in response to toxic heavy metal stress (e.g. Cd^{2+}, Cu^{2+}, Hg^{2+}, Pb^{2+}, Pb^{4+}, Zn^{2+}). Exposure to these materials stimulates the immediate production of metal-chelating peptides (phytochelatins) which sequester the ions and induce the precipitation of nanoscale crystallites (McClean and Beveridge, 1989). Phytochelatins are sulphur-rich, heavy-metal binding peptides composed of only three amino acids cysteine, glutamic acid and glycine. The glutamic acid is linked to the cysteine residues by a γ linkage and the general structure of the complex is $(\gamma\text{-Glu-Cys})_n\text{Gly}$ (n = 2 - 7) (Dameron et al., 1989). As noted already, ultrasmall (< 100 nm) semiconductor particles are of interest because size and/or dimensionality reduction results in substantially altered chemical, mechanical, electrical, electro-optical and magnetic properties.

Crystallization assays performed in vitro with isolated peptides have confirmed that they are capable of sulfide-mediated CdS crystallite formation. The resulting peptide-stabilized CdS crystallites are remarkable for their size (2 - 3 nm longest physical

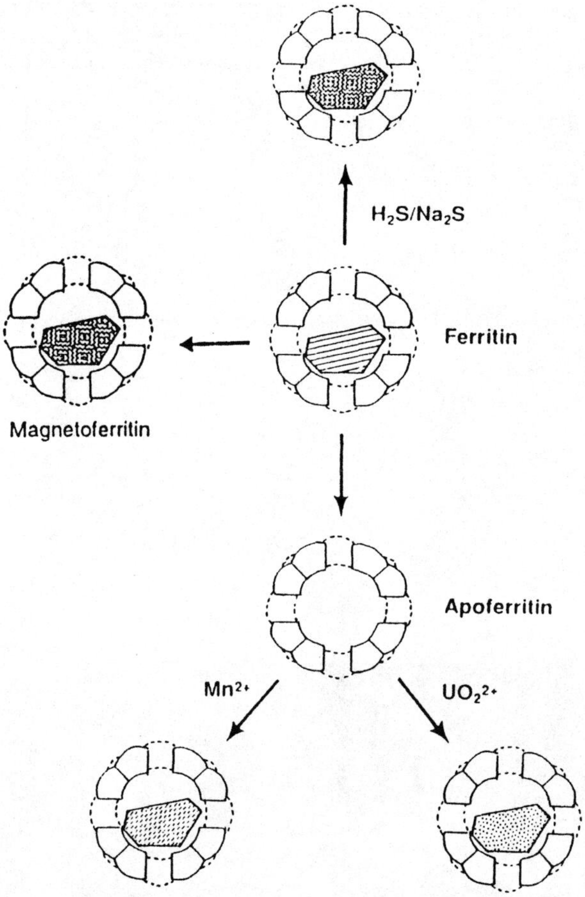

Figure 3.
Schematic showing the various synthetic strategies that have evolved for the formation of novel inorganic phases inside the protein cavity of ferritin (Meldrum et al., 1992).

dimension) and the novel blue-shifted near UV transition which they display. Changes in the molecular structure (chain length) and linkage motif (γ to α) of the peptide complex produce some minor changes in CdS particle size whereas the habit and morphology remains essentially unchanged (Heywood, 1993).

It is evident from these crystallization studies with biological molecules, (phytochelatins and ferritin) that the chemical organization of a polymeric microenvironment operates at the molecular level to control certain aspects of the nucleation, growth and stabilization of inorganic particles. The studies also uphold the belief that the use of biological and synthetic templates of this type constitutes a promising approach to the crystal engineering of nanodimensional solids of defined habit.

3. *Membrane-mimetic chemistry*
This promise is also highlighted in recent work on the controlled crystallization of inorganic solids on model organic surfaces. Supramolecular organic assemblies prepared from surfactants, phospholipids, etc offer a greater degree of versatility with respect to substrate identity and geometry and, hence, the option to engineer the interfacial chemistry directing crystallization on a molecular scale. A potential reaction volume is defined by the closed bilayer structures which form when amphiphilic macromolecules undergo spontaneous self-assembly in aqueous media.

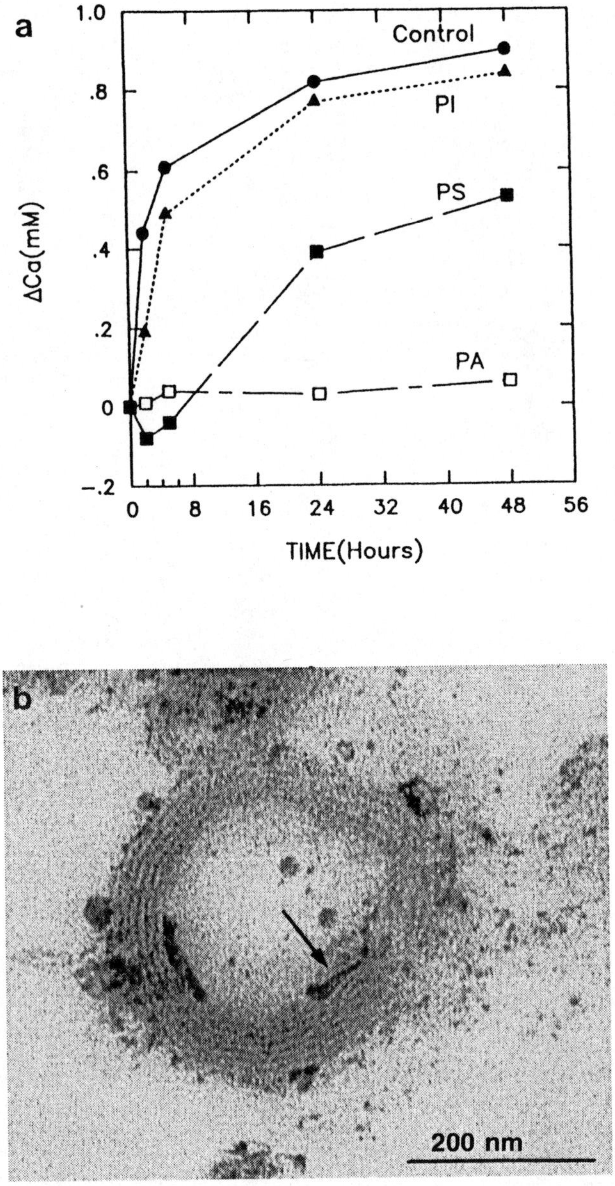

Figure 4.
(a) Schematic showing crystallization strategy for membrane-mediated calcium phosphate formation in phospholipid liposomes.
(b) Kinetic plots of intraliposomal precipitation of calcium phosphate in phosphatidylcholine vesicles (PC) and for liposomes doped with other acidic phospholipids (Heywood and Eanes, 1992).

Biomimetic studies of mineral formation in synthetic vesicles, for example, have shown that a dramatic reduction in particle size can be achieved because of the size range of the reaction environment (1 - 100nm) with the bilayer membrane which encloses each crystal limiting particle-particle interactions (Heywood et al., 1990; Meldrum et al., 1993). It has also been established that the molecular identity and geometry of the amphiphile in the bilayer membrane can have a significant effect upon the precipitation kinetics and matrix-mineral interactions associated with crystal formation (Heywood et al., 1990; Heywood and Eanes, 1992).

For example, phosphatidylcholine (PC) vesicles doped with phosphatidic acid (PA) or phosphatidylserine (PS) exhibit retarded kinetics for calcium phosphate deposition and a reduced particle size when crystallization is initiated within the intraliposomal volume (Fig. 4); an effect which relates to the specific interactions between the lattice ions of the mineral phase and the polar head group of the lipid moiety during crystal nucleation and growth. More specifically, the interaction of Ca^{2+} with either of these two lipid species (PA, PS) induces lateral phase separations within the bilayer and promotes the transient formation of enriched domains as the lipids segregate with the concomitant dehydration of the head group region and the generation of Ca-acidic phospholipid complexes. The formation of these enriched domains facilitates the crystal-lipid interactions observed in calcified PA- and PS-doped vesicles and extends the lipid-mineral contact time, thus inhibiting growth in selected crystallographic directions.

IV. ORIENTED NUCLEATION AND CONTROLLED GROWTH

A. Template-directed crystal engineering

A further refinement of the biomimetic approach has been the adoption of *Langmuir monolayers* as two-dimensional templates to control crystallization events. The tendency for insoluble amphiphilic molecules to accumulate at a solution surface and form highly organized monomolecular films is a physical phenomenon which has been widely exploited (Gaines, 1966; Addadi et al., 1985; Roberts, 1990). Indeed, investigations of Langmuir monolayers have benefited researchers with interests in molecular science ranging from biochemistry to condensed matter physics.

As an approximation to half of the bilayer structure of biomembranes, Langmuir monolayers present a controllable model system (Ringsdorf et al., 1988) and, since the transfer of monolayers to solid supports for the formation of multilayers (Langmuir-Blodgett films) is facile there is considerable interest in their use for electronics applications (Roberts, 1990). For crystal engineers, interest has centered upon the use of monolayers with interfaces which mimic the surfaces of three-dimensional crystal because the adsorption of an amphiphile with the appropriate head group at the gas/liquid interface will induce an oriented nucleation event.

B. Crystallization under compressed Langmuir monolayers

The potential of polyelectrolyte monolayers to promote three-dimensional crystallization events was first realized by researchers at the Weizmann Institute, Israel who used amphiphilic films of resolved amino acids to induce the crystallization of organic crystals from a saturated subphase (Landau et al., 1985; Jacquemain et al., 1992). The monolayer catalyzes the nucleation of the crystalline phase and stabilizes the resulting nuclei in a preferred orientation relative to the interface. It should be noted, however, that the subsequent growth of these oriented nuclei is unaffected by the template with the result that the mature crystals express an equilibrium morphology commensurate with the prevailing solution conditions. The inductive influence of ordered organic surfaces can be related to one of three interfacial processes, namely (i) the accumulation of charges species and stabilization of specific electrostatic interactions, (ii) the translation of precise geometric information, or (iii) the provision of stereochemical complementarity between the monolayer and nascent crystal.

In a systematic study of the template-directed crystallization of inorganic solids, Mann, Heywood and co-workers have established that compressed Langmuir monolayers of the appropriate molecular design will selectively nucleate inorganic crystals in a preferred crystallographic orientation (Heywood and Mann, 1993). An added dimension to their approach has been the syntemporal use of soluble one-dimensional templates in the subphase to elicit further control over the growth process and achieve complete morphological tailoring (Heywood and Champ, 1993).

1. *Molecular recognition and the inorganic-organic interface*

The effect of engineering the head group identity, polarity and packing conformation of the amphiphilic monolayer upon the crystallization of calcium carbonate has been investigated now in some detail (Heywood et al., 1991a; 1991b; 1991c; Heywood and Mann, 1993). For example, compressed Langmuir monolayers of saturated long alkyl chain carboxylates, sulfates, amines and alcohols have all been reviewed as potential molecular templates for the oriented nucleation of calcium carbonate. The choice of these surfactants was based upon the similarity of their head group functionalities to those identified in the organic matrix proteins associated with calcium carbonate biominerals.

With the exception of aliphatic alcohols which inhibit nucleation, the pattern of precipitation under compressed monolayers is reproducible; the crystals nucleate exclusively at the monolayer/solution interface, the particles so-formed always exhibit a unimodal size distribution and the induction time to nucleation is significantly reduced. More importantly, all the crystals are crystallographically oriented relative to the monolayer.

Consider, for example, compressed monolayers of a saturated long alkyl chain carboxylate. Molecular templates of this surfactant will nucleate $CaCO_3$ from a supersaturated solution; calcite crystals oriented with the [1-1.0] axis perpendicular to the monolayer form at subphase conditions of $[Ca^{2+}] > 10$ mM (Fig. 5a) whereas crystals of the metastable polymorph vaterite nucleate on a (00.1) face at $[Ca^{2+}] < 5$mM (Rajam et al., 1991). If the polarity of the monolayer is reversed (e.g. positively charged films of octadecylamine) then only vaterite forms regardless of the subphase conditions (Heywood et al., 1990) (Fig. 5b). In this case, however, two distinct orientations of the vaterite crystals are observed; type I crystals oriented with the *c* axis perpendicular to the monolayer and type II crystals with the *a* axis aligned perpendicular to the monolayer/solution interface (Heywood et al., 1991c).

2. *Molecular requirements for oriented nucleation*

From two-dimensional crystallographic studies (using surface specific X-ray and neutron diffraction techniques) of amphiphilic molecules at the air/water, it is known that aliphatic primary acids pack in a marginally distorted hexagonal unit cell (area per molecule, $A = a^2 \sin 60°$) with the molecules aligned vertically (Landau et al., 1989a; 1989b). Additional evidence from compression isotherms (Gavisch et al., 1990) indicates the formation of a diffuse layer of metal ions beneath the charged surfactant surface although recent data from X-ray and neutron reflectivity measurements of arachidate monolayers cast on a Ca^{2+} subphase suggest that the density of this layer is low with one or perhaps two water molecules screening each calcium ion and forming OH.....O hydrogen bonds with the monolayer carboxylates (Landau et al., 1989b). Under these conditions, it is unlikely that the preferential accumulation of calcium ions in a diffuse two-dimensional network beneath the monolayer would be sufficient to catalyze nucleation; a supposition which is supported by the observation that oriented vaterite nucleation does not require the formation of a Ca^{2+} Stern layer. Since nucleation is a three- rather than two-dimensional process, the translation of other structural information must be significant.

The (1-10) face of calcite contains rows of carbonate anions oriented with their C_{2v} axis perpendicular to the surface, a motif which mimics the orientation of the COO^- groups in the monolayer. Moreover, the carbonate-carbonate spacings (0.496 nm) in this face are commensurate with the interhead-group spacing of the monolayer (0.5 nm), suggesting a role for both geometric correspondence and stereochemical complementarity in the interfacial events mediating nucleation. Similar criteria can be invoked to explain the preferential nucleation of vaterite under carboxylate monolayers. No structural relationship exists between the (00.1) face of vaterite and the monolayer, but the coplanarity of the proposed hydrogen bonded OH......OCO_2^{2-} system would favor the orientation of bound

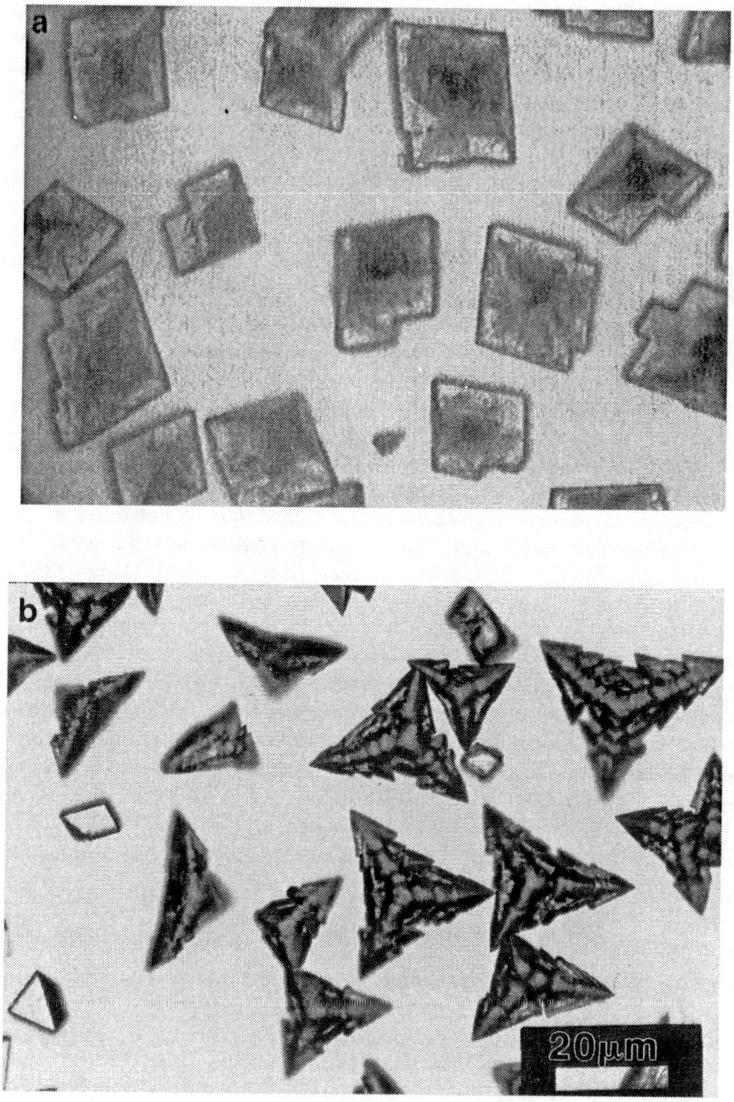

Figure 5.
(a) SEM micrographs of calcite crystals formed under a compressed monolayer of eicosanoic acid. The crystals orient with the [10.0] crystallographic axis perpendicular to the monolayer/solution interface.
(b) SEM micrographs of calcite crystals formed under a compressed monolayer of n-eicosyl sulphate. The crystal are oriented with the [0001] crystallographic axis perpendicular to the monolayer/solution interface (Heywood and Mann, 1993).

carbonate ions from the subphase perpendicular to the monolayer, an orientation which mimics the crystal structure of vaterite. The exclusive selection of both the (00.1) and {11.0} faces of vaterite under amine monolayers is perhaps more difficult to rationalize although both have carbonates aligned perpendicular to the crystal face nucleated under the organic surface. In this instance, solution conditions (pH 6.0) may favor the binding of HCO_3^- ions orthogonal to the $-NH_3^+$ head group which would in turn facilitate some degree of stereochemical recognition between the monolayer and the nascent crystals.

The generic requirement for stereochemical complementarity as the key determinant of oriented nucleation was confirmed in studies where the bidentate motif of the carboxylate head groups was replaced with the trigonal pattern of a sulphate moiety (Heywood and Mann, 1993). Here the tridentate template of three oxygens favors the nucleation of calcite on a (001) face as it mimics the stereochemical disposition of the planar carbonates in this face. It must be assumed that the failure of alcohol monolayers to induce the oriented nucleation of calcium carbonate relates to the neutrality of the head group limiting cation binding and the non-complementary stereochemical motif (Heywood and Mann, 1992b).

3. *Structural requirements for controlled growth*

In all of the systems discussed above, the influence of the organic template is confined solely to the events associated with nucleation. Studies with barium sulphate highlighted another aspect of template-directed crystallization; the controlling influence of the organic surface could be extended to the growth process (Heywood and Mann, 1992a; 1992b; 1992c). Compressed monolayers of an aliphatic long chain sulphate will induce barium sulphate crystals to nucleate with the (100) face parallel to the plane of the monolayer. Once again the nucleation events are critically dependent upon the template which mimics both the lattice geometry of the cations and stereochemistry of the oxyanions in the (100) face and translates this information to the forming crystals. In addition to their preferred orientation relative to the monolayer a unique feature of the $BaSO_4$ crystals grown under the sulphate monolayers is their unusual morphology. Each crystal forms as a thin anisometric plate distinguished by nano-dimensional texture comprising an interconnecting mosaic of some 20-40 nm diamond-shaped subunits bounded by four well-defined {011} side-faces. Despite this complex nano-dimensional architecture each plate is a crystallographically continuous single crystal.

V. TEMPLATE-DIRECTED MODELLING OF CRYSTAL MICROSTRUCTURES

The observation that two-dimensional organic template could at one and the same time direct nucleation events through a series of molecular recognition events at the organic-inorganic interface and subsequently influence the growth of the crystals as a result of selected interactions with specific crystal faces raises a number of interesting issues. One in particular is the possibility that the syntemporal use of both insoluble two-dimensional surfaces and soluble one-dimensional additives might be a route to achieving both an oriented nucleation event and control of the crystal habit during growth. As noted previously (see I.B), soluble tailor-made additives are used routinely to selectively stabilize specific crystal faces and effect the programmed habit modification of both inorganic and organic crystals. The essential process is again one of molecular recognition mediated by electrostatic interactions and charge neutralization events, and the close lattice matching and stereochemical complementarity of the guest additive molecule with specific surfaces in the host crystal.

The presence of soluble organic additives (e.g. α,ω-dicarboxylates) is known to have a significant effect upon the morphology of calcite crystals (Mann et al., 1990). Malonic acid, for example, transforms calcite from its normal rhombohedral ({10.4}) habit into spindle-shaped crystals elongated along the *c* axis with curved prismatic {10.0} faces. The planar bidentate COO^- groups of the dicarboxylate mimics the surface incorporation of HCO_3^- into the (1-1.0) face; this stereochemical relationship is compounded by a potential geometric match between adjacent layers of carbonates in the {10.0} faces of the crystal lattice and the two carboxylate residues in the additive molecules. The summative effect of these molecular interactions is the selective stabilization of the first order prismatic {1-1.0} faces. The

Figure 6.
(a) Crystals of calcite nucleated under a compressed monolayer of eicosanoic acid from a subphase doped with malonic acid.
(b) Crystals of calcite nucleated under a compressed monolayer of n-eicosyl sulphate from a subphase doped with lithium ions (Heywood and Mann, 1993).

presence of malonic acid in the solution subphase under carboxylate monolayers stabilizes the symmetry-related prismatic face on the ventral surface with the result that the arrow-head morphology of the immature particles is sustained in the mature crystals (Fig. 6a).

Soluble inorganic additives can be equally effective as crystal habit modifiers (Rajam and Mann, 1990). Consider, for example, rhombohedral calcite crystals grown in the presence of lithium ions. Doped crystals exhibit well-defined basal faces suggesting that these relatively high energy faces are neutralized by the incorporation of Li^+ (0.59Å, four-fold coordination) into the surface interstitial sites of the carbonate layer. If Li^+ is introduced into the subphase under a compressed monolayer of a long alkyl chain sulphate, crystal growth along the [001] axis is inhibited and leads to the formation of a thin (0.5 - 1.0 mm) triangular tablet of calcite (Fig. 6b).

VI. SUMMARY

The facility to tailor the crystallographic form of inorganic solids in a predictable and reproducible manner is attractive to material scientists, crystal chemists and chemical engineers seeking to optimize and/or exploit selected physicochemical features of crystalline materials. The use of water-soluble additives to regulate particle size and morphology is a common strategy, whereas the potential of organized assemblies of insoluble organic molecules to control the nucleation and growth has yet to be fully realized. This is largely due to our incomplete understanding of the molecular processes governing the crystallization of inorganic crystals on organized substrates. On the horizon are a range of new physical techniques (atomic force microscopy, surface sensitive analytical techniques such as grazing angle neutron scattering) for probing molecular processes at the atomic level that will clearly make a significant impact on this area.

A wide range of molecular template have now been investigated and it is clear that one theme is common to them all; the oriented nucleation of an inorganic solid is dependent upon a sequence of molecular recognition events between the two-dimensional organic template and the pre-critical aggregates in solution which mediate nucleation. To date these studies, and related work with one dimensional templates, have focused attention upon the importance of structural, stereochemical and dynamical relationships to the host-guest chemistry of these processes. As with the majority of such interactions which operate on a molecular scale, specificity is critically dependent upon the complementarity in size, charge and molecular shape of the active species in both the host organic template and the opposing inorganic (guest) surface.

We are only at the beginning of these explorations but already the many possible benefits of this biomimetic approach to materials fabrication are unfolding. With the progress of knowledge through the continued development of interdisciplinary research into the biological process of crystal growth and strategies for modelling and exploiting the crystallochemical paradigms of biomineralization, there is every hope that the potential of this work will be realized in the production of novel nanophase materials, ceramics, etc with enhanced functional performance.

ACKNOWLEDGEMENTS

In reporting these results I note with thanks the collaborative support of Professor R. Frankel, Dr D. Bazylinski, Mr S. Champ, Dr E.D. Eanes, Professor D. Winge, Professor N. Watabe, Professor S. Mann and Dr F.C. Meldrum.

The financial support of the SERC and BP Research International Ltd is gratefully acknowledged.

REFERENCES

Addadi, L., Berkovitch-Yellin, Z., Weissbuch, I., van Mil, J., Shimon, L. J., Lahav, M., and Leiserowitz, L. (1985). *Angew. Chem., Int. Ed. Engl.*, *24*: 466.

Addadi, L., Moradian-Oldak, J., and Weiner, S. (1991). *Surface Reactive Peptides and Polymers* (C.S. Sikes and A.P. Wheeler, eds.), ACS Washington, DC, p. 13.

Bhandarkar, S., and Bose, A. (1990). *J. Colloid Interface Sci.*, *135*: 531.

Barbas, J., Santhanagopolan, V., Blaszczynski, M., Elllis, W.R., and Winge, D.R. (1992). *J. Inorg. Biochem.*, *48*: 95.

Bazylinski, D.A., Heywood, B.R., Mann, S., and Frankel, R.B. (1993). *Nature*, *366*: 218.

Birchall, J.D. (1993). *Materials Res. Soc. - Symposium Proceedings*, *286*: 297.

Brisdon, B., Heywood, B.R., Hodson, A., Mann, S., and Wong, K. (1993). *Advanced Materials*, *5*: 49.

Calvert, P.D. (1992). *Materials Res. Soc. Bull.*, *17*: 37.

Dameron, C.T., Reese, R.N., Mehre, R.K., Kortan, A.R., Carroll, P.J., Steigerwald, M.L., Brus, L.E., and Winge, D.R. (1989). *Nature*, *338*: 596.

Davey, R.J., Black, S.N., Bromley, L.A., Cottier, D., Dobbs, B., and Rout, J.E. (1991). *Nature*, *353*: 549.

Frankel, R., and Mann, S. (1989). *Biomineralization: Chemical and Biochemical Perspectives* (S. Mann, J. Webb and R.J.P. Williams, eds.), VCH, Weinheim, Germany, p. 296.

Gaines, G.L. (1966). *Insoluble Monolayers*, Wiley-Interscience, New York.

Gavisch, M., Popovitz-Biro, R., Lahav, M., and Leiserowitz, L. (1990). *Science*, *250*: 973.

Gorby, Y.A., Beveridge, T.J., and Blakemore, R.P. (1988). *J. Bacteriol.*, *170*: 834.

Harrison, P.M., Artymuik, G.C., Ford, D.M., Lawson, J.M.A., Smith, A., Treffry, A., and White, J.L. (1989). *Biomineralization: Chemical and Biochemical Perspectives* (S. Mann, J. Webb and R.J.P. Williams, eds.), VCH, Weinheim, Germany, p. 257.

Heuer, A.H., Fink, D.J., Laraia, V.J., Arias, J.L., Calvert, P.D., Kendall, K., Messing, G.L., Blackwell, J., Rieke, P.C., Thompson, D.H., Wheeler, A.P., Veis, A., and Caplan, A.I. (1992). *Science*, *255*: 1098.

Heywood, B.R. (1993). *Microsc. Res. Technique*, *27*: 376.

Heywood, B.R., and Champ, S. (1993). Unpublished data.

Heywood, B.R., and Eanes, E.D. (1992). *Calcif. Tissue Int.*, *50*: 149.

Heywood, B.R., and Mann, S. (1992a). *Advanced Materials*, *4*: 278.

Heywood, B.R., and Mann, S. (1992b). *J. Amer. Chem. Soc.*, *114*: 4681.

Heywood, B.R., and Mann, S. (1992c). *Langmuir*, *8*: 1492.

Heywood, B.R., and Mann, S. (1993). *Advanced Materials*, *6*: 9.

Heywood, B.R., Bazylinski, D., Garratt-Reed, A., Frankel, R., and Mann, S. (1990). *Naturwissenschaften*, *77*: 536.

Heywood, B.R., Frankel, R., and Mann, S. (1991a). *Materials Res. Soc. - Symposium Proceedings*, *218*: 93.

Heywood, B.R., Fendler, J., and Mann, S. (1991b). *J. Colloid Interface Sci.*, *138*: 295.

Heywood, B.R., Rajam, S., and Mann, S. (1991c). *J. Chem. Soc., Faraday Trans.*, *5*: 7735.

Heywood, B.R., Sparks, N.H.C., Shellis, R.P., Weiner, S., and Mann, S. (1991d). *Connect. Tissue Res.*, *25*: 1.

Jacquemain, D., Grayer-Wolf, S., Leveiller, F., Deutsch, M., Kjaer, K., Als-Nielson, J., Lahav, M., and Leiserowitz, L. (1992). *Angew. Chem., Int. Ed. Engl.*, *31*: 130.

Katz, E., and Li, S.T. (1973). *J. Mol. Biol.*, *80*: 1.

Landau, E.M., Levanon, M., Leiserowitz, L., Lahav, M., and Sagiv, J. (1985). *Nature*, *318*: 353.

Landau, E.M., Grayer-Wolf, S., Levanon, M., Leiserowitz, L., Lahav, M., and Sagiv, L., (1989a). *J. Amer. Chem. Soc.*, *111*: 1436.

Landau, E.M., Grayer-Wolf, S., Deustch, M., Kjaer, K., Als-Nielson J., Leiserowitz, L., and Lahav, M. (1989b). *Pure Appl. Chem.*, *61*: 673.

Lawson, D.M., Artymuik, P.J., Yewdall, S.J., Smith, J.M.A., Livinstone, J.C., Treffry, A., Luzzageo, S., Levi, S., Arosio, P., Cesareni, G., Thomas, C.D., Shaw, W.V., and Harrison, P.M. (1991). *Nature*, *349*: 541.

Lowenstam, H., and Weiner, S. (1989). *On Biomineralization*, Oxford University Press, London.

Mann, S. (1993). *Nature, 365*: 499.

Mann, S., and Heywood, B.R. (1989). *Chem. Brit., 38*: 23.

Mann, S., Hannington, J.P., and Williams, R.J.P. (1986). *Nature, 324*: 565.

Mann, S.J., Sparks, N.H.C., Frankel, R.B., Bazylinski, D.A., and Jannasch, H.W. (1990). *Nature, 334*: 258.

Mann, S., Archibald, D., Didymus, J., Heywood, B.R., Meldrum, F.C., and Wade, V.J. (1992). *Materials Res. Soc. Bull., 17*: 32.

McClean, R.J.C., and Beveridge, T.J. (1989). *Microbial Mineral Recovery* (C. Brierly, J. Brierly and H. Ehrlich, eds.), McGraw, New York, p. 185.

Meldrum, F.C., Nimmo, D., Wade, V.W., Heywood, B.R., and Mann, S. (1991). *Nature, 349*: 684.

Meldrum, F.C., Heywood, B.R., and Mann, S. (1992). *Science, 257*: 533.

Meldrum, F.C., Heywood, B.R., and Mann, S. (1993). *J. Colloid Interface Sci., 161*: 66.

Rajam, S., Heywood, B.R., Walker, J.B.A., Mann, S., Davey, R.J., and Birchall, J.D. (1991). *J. Chem. Soc., Faraday Trans., 87*: 727.

Rajam, S., and Mann, S. (1990). *J. Chem. Soc. Chem. Commun., 24*: 1789.

Ringsdorf, H., Schlarb, B., and Venzmer, J. (1988). *Angew. Chem., Int. Ed. Engl., 27*: 113.

Roberts, G.G. (1990). *Langmuir-Blodgett Films* (G.G. Roberts, ed.) Plenum Press, New York, p. 317.

Slavkin, H., and Price, P. (1992). *The Chemistry and Biology of Mineralised Tissues*, Excerpta Medica, Elsevier, Amsterdam.

Swain, L.D., and Boyan, B.D. (1989). *J. Dent. Res., 67*: 526.

Wade, V.J., Levi, S., Arosio, P., Treffry, A., Harrison, P.M., and Mann, S. (1991). *J. Mol. Biol., 221*: 1443.

Weiner, S. (1986). *CRC Crit. Rev. Biochem., 20*: 365.

LIGAND DESIGN RULES FOR CHELATION THERAPY

by

Mark M. Jones

Department of Chemistry and Center in Molecular Toxicology,
Box 1583, Station B,
Vanderbilt University,
Nashville, TN 37235
U.S.A.

I. INTRODUCTION

The purpose of chelation therapy is the removal of a toxic metal from a person or animal suffering from metal intoxication via the administration of a chelating agent which will enhance the urinary or fecal excretion of the toxic metal. In the design of new ligands for potential use in chelation therapy, there is what may be called a hierarchy of design rules, whose number and exact sequence is as yet imperfectly understood. Nevertheless, these rules can be conveniently divided into two categories. The first category is one which can be designated as *thermodynamic* and relates to the requirement that the overall process in which the toxic metal ion is removed from the body must be thermodynamically spontaneous. The second, and more complex category deals with the manipulation of various rates and equilibria which determine the *pharmacokinetics* of the overall process.

II. THERMODYNAMIC REQUIREMENTS

The thermodynamic requirements which must be incorporated into successful ligand design can be simply formulated. The reaction in which the chelating agent removes the toxic metal ion from its in vivo binding site must be spontaneous. Thus K for the reaction:

$$MB + C = MC + B$$

(where M represents the toxic metal ion, B represents the in vivo binding site of the toxic metal ion and C the chelating agent) must have an equilibrium constant greater than 1 under physiological conditions. In fact, it is generally assumed that as this equilibrium constant increases to values increasingly greater than 1, the effectiveness of the metal removal process will also increase, though this is not always true if such increases in equilibrium constant are achieved via structural changes which drastically and unfavorably change basic processes involved in the pharmacology of the metal removal process. This requires C to bind effectively to M in the presence of a variety of competing essential metal ions and biological ligands such as proteins and DNA.

The selection of structures which will satisfy this requirement must be successfully carried out prior to any other structural manipulations. This is most readily approached using a knowledge of the donor atoms preferred by the toxic metal ion and its relative affinity for a variety of chelating groups. At this point, the nature of the competition with endogenous essential metal ions must be given some consideration. The donor atom preferences can be approached most directly via the Hard and Soft acid/base approach of Pearson (Pearson, 1968a; 1968b), which allows the selection of the donor atom types forming the most stable linkages with the toxic metal. In this search, it is very advantageous to consider the biological data available on those chelating agents commonly used in the clinic as antagonists for toxic metals. The structures of several of these are shown in Figure 1,

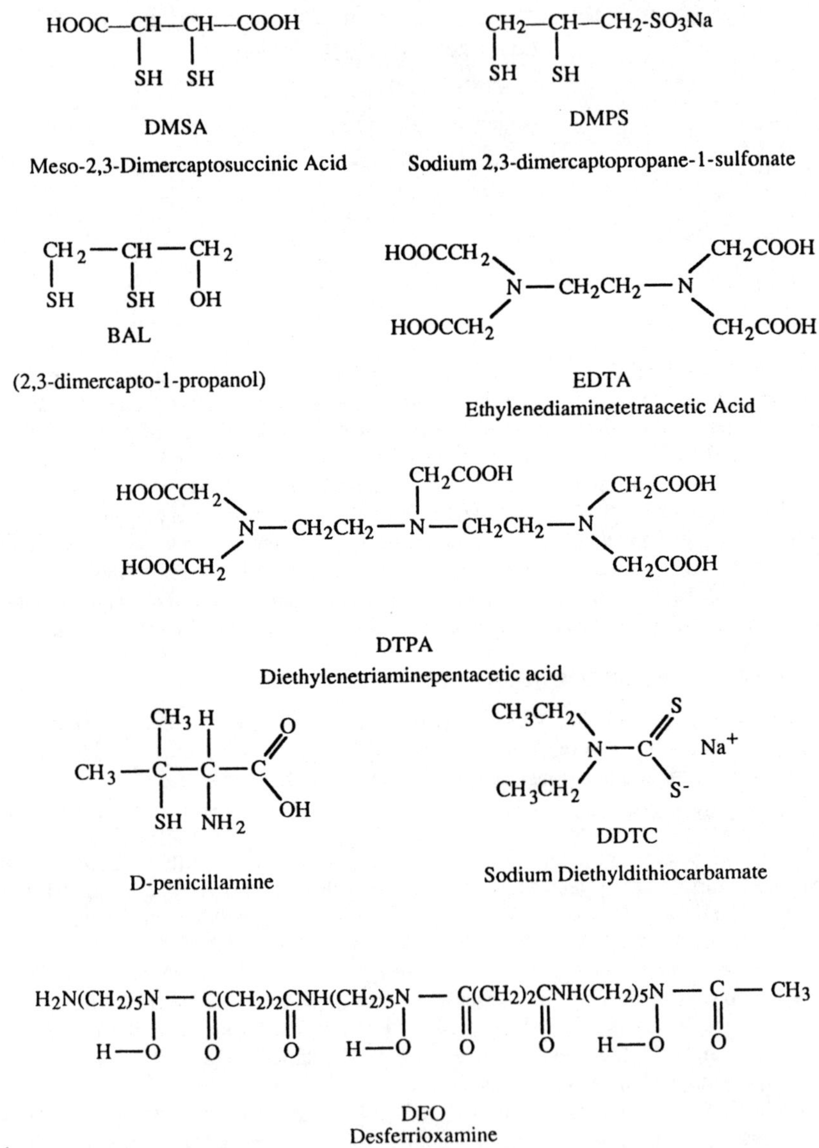

Figure 1.
Structures of clinically utilized chelating agents.

Table 1

Clinical Use of Chelating Agents[a]

Chelating agent	Metal ions antagonized
BAL	As(III)*, Hg(II), Pb(II), Cu(II), Au(I)
DMSA	Pb(II)*, As(III)*, Hg(II)*, Cu(II), Au(I)
DMPS	Hg(II)*, Cu(II), Pb(II), As(III), Au(I)
$Na_2CaEDTA$	Pb(II)*, Pu(IV)
$Na_3ZnDTPA$	Pu(IV)*
D-Penicillamine	Cu(II)*, Pb(II)
Triethylenetetramine·2 HCl	Cu(II)*
Desferrioxamine	Fe(III)*, Al(III)*, Ga(III)
Sodium diethyldithiocarbamate	$Ni(CO)_4$

[a] Structures of these compounds are shown in Figure 1; data from Jones (1991b).
* Indicates a superior antagonist for this toxic metal.

and some of the metal ions for which they have been used as antagonists are summarized in Table 1. It must be remembered that the administration of any chelating agent which firmly binds calcium can lead to a rapid drop in serum calcium levels and possible death; this is the reason for the administration of EDTA as its calcium complex, and DTPA as its calcium or zinc complex.

The subsequent design of the chelating agent to achieve an optimized binding constant can be achieved by following the patterns of structural development which have been used in previous studies of the factors which can affect the stability constant of a specific donor atom collection with a given metal ion. These have been recently reviewed (Hancock and Martell, 1988; 1989; Hancock, 1990). The usual procedures involve the addition of donor groups on an organic framework in such a fashion as to put the donor atoms at the sites preferred by the metal ion in its most stable coordination geometry (e.g. Martell et al., 1981; Raymond and Smith, 1981; Garrett et al., 1989). The advantages to be gained from an increased stability constant by the addition of chelate rings can be seen by a comparison of the stability constants (the equilibrium constant for the formation of the complex from its constituent parts in the absence of competing metal ions or ligands) of the equal numbers of uncharged donors ammonia (four donors and no chelate rings), ethylenediamine (en) (four donor atoms and two chelate rings), and triethylenetetramine (trien) (four donor atoms and three chelate rings) and in the series with equal numbers of donors on charged ligand groups: two glycines (four donors and two chelate rings) and ethylenediaminediacetic acid (EDDA) (four donors and three chelate rings). The structures of these are shown in Figure 2 and data on the stability constants shown in Table 2.

There are, of course, also problems which arise due to electrostatic repulsion when a large number of negatively charged donor groups are to be arranged around a small cation of low positive charge, as such an arrangement may result in a significant reduction in the number of such charged groups bonded to the small cation.

The use of molecular mechanics programs to determine the conformations of optimum stability for both chelating agents and metal complexes can be expected to be an area of growing importance in the design of chelating agents as such calculations should provide very useful information on the effects of relatively small structural changes on the relative stability of the metal complexes which are to be formed. A very promising start in this area has already been made (Hancock, 1990; Bourari et al., 1990; Bernhardt and Comba, 1992).

It must be noted that simple chelating agents may possess considerable advantages over more complex structures with many chelate rings because of the superiority of

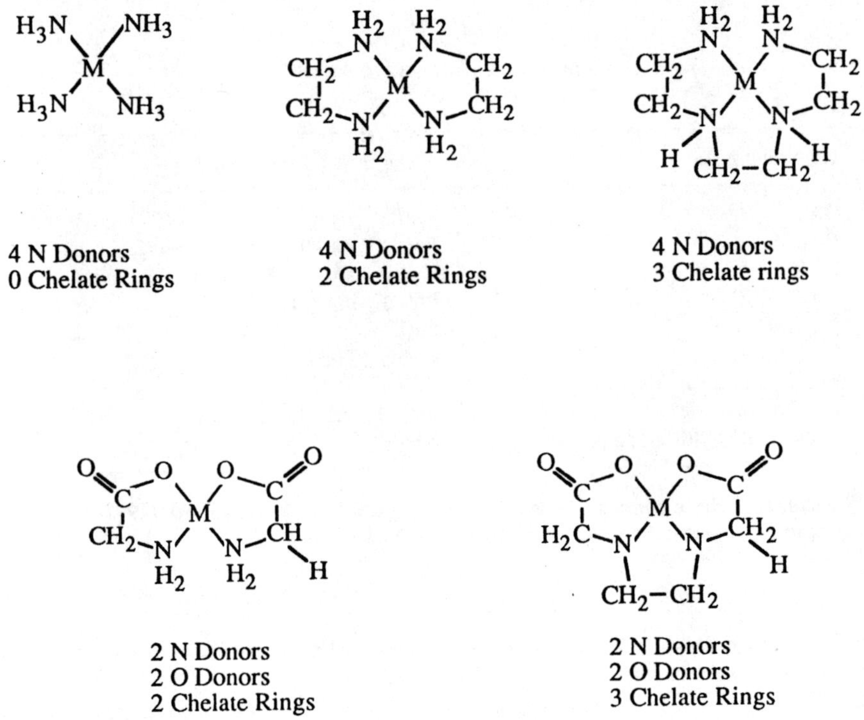

Figure 2.
Donor atoms and chelate rings.

Table 2

Dependence of Stability Constants on Number of Chelate Rings[a]

Ligand	Donor atoms	Chelate rings	Log $K_{stab}(Zn^{2+})$	Log $K_{stab}(Cu^{2+})$
Uncharged donors				
Ammonia	4	0	9.65(β_4)	13.00(β_4)
en	4	2	10.8(β_2)	19.9(β_2)
trien	4	3	12.0	20.1
Charged donor atoms				
glycine	4	2	9.19(β_2)	15.03(β_2)
EDDA	4	3	11.22	16.2

[a] Values of stability constants from Martell and Smith (1974).

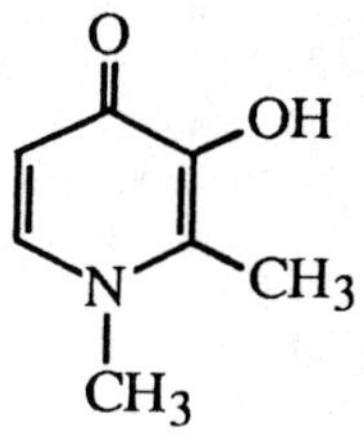

Figure 3.
1,2-Dimethyl-3-hydroxy-pyridin-4-one.

pharmacological properties or ease of preparation or administration. A clear example of this is seen in the new iron chelating agent, 1,2-methyl-3-hydroxy-pyridin-4-one (Fig. 3) which is easily prepared and can be given orally, and is superior in many respects to the much more elaborate chelating agent desferrioxamine (Fig. 1) (Porter et al., 1988; Hider and Hall, 1991; Gale et al., 1991).

III. EXTRACELLULAR VS INTRACELLULAR MOBILIZATION OF TOXIC METALS

After the initial judgement has been made about the donor atoms to be used and the chelate systems which are most promising, it is necessary to decide whether the final chelating agent is to be used to remove extracellular or intracellular deposits of the toxic metal. Even if a chelating agent has a very high effective stability constant for a toxic metal (i.e. the equilibrium constant for the formation of the complex in the presence of competing metal ions and ligands of the sort found in a physiological milieu at pH 7.4), its administration will not result in the mobilization of an appreciable amount of the toxic metal if the chelating agent cannot gain access to the intracellular sites in the organs where the toxic metal ion is concentrated. For chelating agents which are confined to the serum, the sites from which the toxic metal can be removed will be limited to the serum and those surface sites on tissues which are exposed to the serum (Fig. 4).

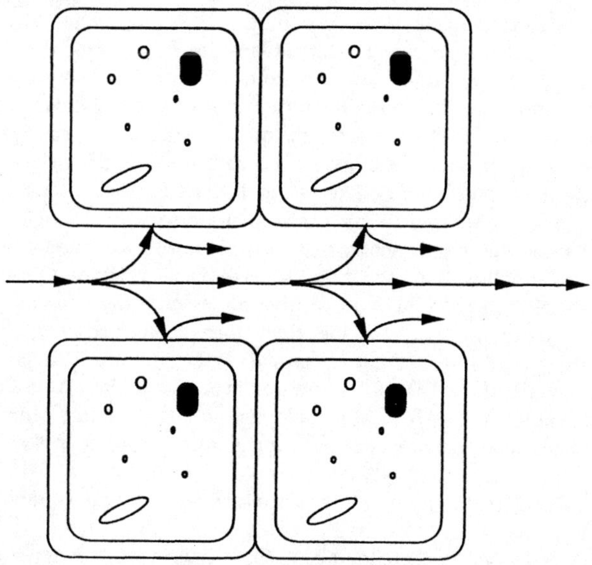

Figure 4.
In vivo pathway of chelating agents restricted to extracellular spaces.

These chelating agents can be expected to be useful primarily for the removal of toxic metals which are present in sites available to the serum. For some toxic metals which rapidly re-equilibrate among biological binding sites, repeated treatments with such chelating agents may be able to remove an appreciably larger fraction of the toxic metal than would otherwise be removed by a single treatment which reduces the toxic metal present in the serum via normal processes of excretion. After treatment with such a chelating agent, one generally finds a "rebound effect" in which the toxic metal present in sites inaccessible to the chelating agent re-equilibrates with the serum (Haust et al., 1980). Such chelating agents, however, will generally be much less effective than those chelating agents which can gain access to the major intracellular sites at which a toxic metal is deposited. There are three ways in which this can be achieved. *First,* by the use of a lipid soluble chelating agent (Shaikh and Lucis, 1972; Cherian et al., 1982), *second* by the encapsulation of the chelating agent in a liposome (Rahman et al., 1979), and *third,* by designing the chelating agent so that it will be actively transported to the interior of the cells in the organ in which the toxic metal is concentrated (Stewart and Diamond, 1987).

IV. MOLECULAR PARAMETERS WHICH GOVERN PHARMACOLOGICAL BEHAVIOR

If it is desired to design a chelating agent which will penetrate the cells of a particular organ and mobilize the intracellular toxic metal ions for excretion, it is necessary to insure that the overall molecular structure selected will be transported into the cells of that organ which contains the toxic metal. This can often be done via the manipulation of the variables upon which the pharmacological behavior of the chelating agent is dependent. If one summarizes the variables upon which the effectiveness of a chelating agent in the mobilization of a toxic metal is dependent, one can present this as a functional dependence of the sort (Jones, 1991a):

$$E = E(K_{eff}, \Sigma\pi, M, Z, \chi, R/S, \ldots\ldots\ldots\ldots) \quad (1)$$

where K_{eff} is the effective stability constant for the toxic metal in a biological medium at pH 7.4, $\Sigma\pi$ is the sum of the Hansch constants (Hansch and Leo, 1979) for the molecular constituents (which is closely related to the partition coefficient), M is the molar mass, Z is the ionic charge, χ is the degree of chain branching as represented by the appropriate molecular connectivity, R/S is the chirality and the remaining variables are as yet incompletely specified. The variables listed above, other than K_{eff}, are ones which determine the pharmacological properties of the chelating agent. The manner in which variations in one of these parameters, $\Sigma\pi$, affects the behavior of a chelating agent can be seen in the data of Porter, Hider and their colleagues on iron mobilizing agents (Porter et al., 1988), where the most effective of a series of similar compounds are those with values of $\Sigma\pi$ (and the partition coefficient) with intermediate values, a common pattern. The role of M in determining the in vivo behavior of the compound is the result of the fact that the route of excretion of molecules is determined in part by their molar mass. While water-soluble low-molar-mass molecules are excreted preferentially through the kidney into the urine, those of somewhat higher molar mass and/or more hydrophobic in nature are usually excreted through the liver into the bile (Klaassen and Watkins, 1984). Thus, chelating agents which are either lipid soluble or of high molar mass are generally more effective for the removal of toxic metals from the liver than molecules with analogous chelating structures which are of lower molar mass and water soluble. This is clearly seen in the differences between BAL and DMSA in the removal of cadmium from the liver (Shinobu et al., 1983). Here BAL, which is lipid soluble and able to penetrate cellular membranes, can remove appreciable amounts of cadmium from its aged deposits in the liver while DMSA, which has an identical vicinal dithiol group to coordinate the cadmium, is ineffective because it is largely restricted to the extracellular spaces and is rapidly excreted in the urine.

The total charge on the chelating agent, Z, is important as it determines which, if any, of the membrane transport systems will be available to the chelating agent. In general, values of Z of 0 or -1 give the most useful chelating agents; Z values as great as -4, such as is found for EDTA, limit the sphere of action of the chelating agent almost exclusively to the extracellular space, the surface of cells and supporting tissue and to excretion via simple glomerular filtration in the kidney.

The degree of chain branching or the connectivity in an organic chain has a significant effect on the biological properties of molecules (Kier and Hall, 1976), including chelating agents. Thus, chelating agents which have isomeric alkyl groups but are otherwise identical may be found to differ significantly in effectiveness (Jones et al., 1991).

The chirality of the chelating agent, R/S, can be very important in governing the degree of interaction with the living organism. This is most clearly seen in the case of penicillamine, which occurs in D and L forms. The L-form is much more toxic because of its action as a vitamin B_6 antagonist, and the form used in the clinic is the much less toxic D-penicillamine (Walshe, 1981; Planas-Bohne, 1981; Perrett, 1981).

V. BIOLOGICAL PARAMETERS OF SIGNIFICANCE

In addition to the molecular variables indicated above, there are several biological parameters of which the most important deal with the acute and chronic toxicity of the chelating agent. The toxicity is determined by the nature of the groups present and the nature of their metabolic products. For many chelating agents, a part of their toxicity is due to their ability to enhance the excretion of essential metal ions. With such chelating agents, the continuous administration of elevated doses on a daily basis to an animal will result in chronic toxicity and ultimately death because of the loss of essential metal ions such as zinc, copper, etc.

Other biological parameters of significance in governing the effectiveness of chelating agents include whether the compound can be absorbed to an appreciable extent from the gastrointestinal tract, the relative rates and pathways of metabolism and the manner in which structural changes affect these, the rates of excretion, and the ability to enter into the cells of various organs. While these are dependent on the chemical structures of the chelating agent, this dependence is not always easily quantified.

VI. CONCLUSION

The ligand design rules for chelation therapy may be summarized as follows:
(i) First make a chelating agent which is thermodynamically capable of removing the metal ion from the biological sites (i.e. at thermodynamic equilibrium) using the steps:
- Select those donor atoms which are known to bond preferentially to the toxic metal ion of interest.
- From the donor atoms selected, develop a chelating agent with a sufficiently large stability constant that it is capable of removing the metal from biological binding sites under in vitro conditions.

(ii) Adjust the parts of the molecule other than the chelating moiety to achieve the desired biological properties:
- Adjust the values of $\Sigma\pi$, M and Z to allow access to intracellular sites in the main organs in which the toxic metal is present. If this organ is the kidney, this is most easily achieved with a chelating agent with Z = -1 with subsequent fine-tuning of $\Sigma\pi$ and M values to optimize. For mobilization from hepatic sites again, Z = -1 is preferred, but a larger molar mass favors excretion through the cells of the liver into the bile, with $\Sigma\pi$ values allowed over a somewhat wider range.
- Reduce the toxicity to the desired level, which can usually be achieved via increases in the polarity, if necessary via the introduction of non-ionic polar groups such as the hydroxy group or by the selection of the least toxic of enantiomeric forms.
- Fine tune the structure for any biological properties which are still to be adjusted via changes in molecular structures which preserve overall values of a variable such as $\Sigma\pi$, M, χ, R/S, etc in an acceptable range.

REFERENCES

Bernhardt, P.V., and Comba, P. (1992). *Inorg. Chem.*, *31*: 2638.

Bourari, A., Fathallah, M., Blaive, B., and Gallo, R. (1990). *J. Chem. Soc., Perkin Trans.*, *2*: 1211.

Cherian, M.G., Onosaka, S., Carson, G.K., and Dean, P.A.W. (1982). *J. Toxicol. Environ. Health*, *9*: 389.

Gale, G.R., Lichtenberg, W.H., Smith, A.B., Singh, P.K., Campbell, R.A., and Jones, M.M. (1991). *Res. Commun. Chem. Pathol. Pharmacol.*, *73*: 299.

Garrett, T.M., Miller, P.M., and Raymond, K.N. (1989). *Inorg. Chem.*, *28*: 128.

Hancock, R.D. (1990). *Accounts Chem. Res.*, *23*: 253.

Hancock, R.D., and Martell, A.E. (1988). *Comments Inorg. Chem.*, *6*: 237.

Hancock, R.D., and Martell, A.E. (1989). *Chem. Rev.*, *89*: 1875.

Hansch, C., and Leo, A. (1979). *Substituent Constants for Correlation Analysis in Chemistry and Biology*, John Wiley & Sons, New York, p. 49.

Haust, H.L., Ali, H., Haines, D.S.M., and Forret, C. (1980). *Int. J. Biochem.*, *12*: 897.

Hider, R.C., and Hall, A.D. (1991). *Perspect. Bioinorg. Chem.*, *1*: 209.

Jones, M.M. (1991a). *J. Coord. Chem.*, *23*: 187.

Jones, M.M. (1991b). *Crit. Rev. Toxicol.*, *21*: 209.

Jones, M.M., Singh, P.K., Jones, S.G., Mukundan, C.R., Banton, J.A., Gale, G.R., Atkins, L.M., and Smith, A.B. (1991). *Chem. Res. Toxicol.*, *4*: 27.

Kier, L.B., and Hall, L.H. (1976). *Molecular Connectivity in Chemistry and Drug Research*, Academic Press, New York, 1976.

Klaassen, C.D., and Watkins, J.B., III (1984). *Pharmacol. Rev.*, *36*: 1.

Martell, A.E., Anderson, W.F., and Badman, D.G. (1981). *Development of Iron Chelators for Clinical Use*, Elsevier North-Holland, New York.

Martell, A.E., and Smith, R.M. (1974-1989). *Critical Stability Constants*, Plenum Press, New York.

Pearson, R.G. (1968a). *J. Chem. Educ.*, *45*: 581.

Pearson, R.G. (1968b). *J. Chem. Educ.*, *45*: 643.

Perrett, D. (1981). *J. Rheumatol.*, *8 (Suppl.7)*: 41.

Planas-Bohne, F. (1981). *J. Rheumatol.*, *8 (Suppl. 7)*: 35.

Porter, J.B., Gyparaki, M., Burke, L.C., Huehns, E.R., Sarpong, P., Saez, V., and Hider, R.C. (1988). *Blood*, *72*: 1497.

Rahman, Y.E., Rosenthal, M.W., and Cerny, E.A. (1979). *Science*, *180*: 300.

Raymond, K.N., and Smith, W.L. (1981). *Struct. Bonding*, *43*: 159.

Shaikh, Z.A., and Lucis, O.J. (1972). *Arch. Environ. Health*, *24*: 410.

Shinobu, L.A., Jones, M.M., Basinger, M.A., Mitchell, W.M., Wendel, D., and Razzuk, A. (1983). *J. Toxicol. Environ. Health*, *12*: 757.

Stewart, J.R., and Diamond, G.L. (1987). *Amer. J. Physiol.*, *252*: F800.

Walshe, J.M. (1981). *J. Rheumatol.*, *8 (Suppl. 7)*: 3.

PHYSICOCHEMICAL ASPECTS OF THE INTERFERENCE OF DETRIMENTAL METAL IONS WITH NORMAL METAL METABOLISM

by

Detmar Beyersmann

Fachbereich Biologie/Chemie,
Universität Bremen,
D-28334 Bremen
Germany

I. INTRODUCTION

The range of metal ions used by biological systems is very large: it reaches from the alkaline metal ions to the transition metal ions. "Normal" metal metabolism is defined as the uptake, distribution, binding and biochemical effects of metals that are essential for the maintenance of life. "Detrimental" metal ions are defined as substances which impair the regular course of life functions at all concentrations. In addition, metals that are essential or beneficial nutrients at low levels may also be toxic at more elevated doses (see *Medicine*, Part One, Chapter 1-C).

The chemical forms of a metal, i.e. its oxidation state and its complexed forms (its "speciation"), are determining for its behavior with respect to life. For example, chromium is essential for mammals in μg/day quantity; higher doses of trivalent chromium are tolerable, but hexavalent chromium, which is anionic, causes cancer. Another example is copper. This element is highly toxic to mammals as the free ion, whereas copper chelated to appropriate carrier molecules is essential. In addition, the route of uptake and the intracellular localization are crucial for the beneficial or toxic character of a metal ion. Nickel compounds have caused cancer in industrial environments when inhaled, but nickel can be tolerated by the gastrointestinal tract in mg quantities without observable health effects. Similarly for iron the intracellular distribution is crucial: iron is taken up by humans in mg/day quantity and reaches millimolar total concentrations in the cytoplasm of blood and liver cells, but micromolar concentrations of free iron ions in the cell nucleus may catalyze the formation of reactive oxygen species that damage the genetic material. Physicochemical criteria greatly help us to understand observed toxic interactions and to formulate predictions about possible detrimental effects. These predictions cannot, however, be quantitative calculations of biological effects since living organisms are extremely complex and we cannot calculate much more than model reactions.

This chapter will mainly deal with the effects of detrimental metals on mammalian systems relevant to human health, since metals are often used in a different way in lower organisms. The following material is organized according to essential elements and their normal metabolism following the classification of the periodic table from alkaline metals to transition metals via earth alkaline elements. Interactions of detrimental ions will be discussed within the sections on essential metals.

Texts and reviews relating to the subject of this chapter are the textbook on "The Biological Chemistry of the Elements" by Fraústo da Silva and Williams (1991), reviews on metal interactions in general toxicity (Nordberg et al., 1986; Beyersmann, 1991) and in carcinogenesis (Nordberg and Pershagen, 1984; Furst, 1987). Relevant up-to-date reviews can also be found in this Volume (see next Sections in the present Chapter 2), and in *Bioinorganic Medicine*, Volume 1, Part Three, and Volume 2, Part Four).

II. INTERFERENCE WITH SODIUM AND POTASSIUM FUNCTIONS

Fundamental life processes depend on sodium and potassium ions, and they are highly vulnerable to toxic interactions. Na^+ and K^+ are used to control the osmotic pressure and thus the integrity and the shape of living cells. The transport of these ions through biomembranes is interwoven with the transport of other ions like H^+, Ca^{2+} and anions. Na^+ and K^+ neutralize negative charges of intracellular anions and serve as charge carriers in neurophysiological processes. The most important biochemical devices of control of alkaline metal ion concentrations in life are active, energy-consuming pumping and passive transport across biomembranes, both catalyzed by integral membrane proteins. The prominent enzyme activity involved in the osmotic balance of cells as well as in the generation of nerve axon potentials is Na^+,K^+-ATPase, which requires Na^+, K^+ and Mg^{2+} for its activity. This enzyme couples the hydrolysis of one ATP with the export of three Na^+ and the import of two K^+ through cell membranes. In consequence, living cells accumulate potassium relative to their exterior medium and maintain a lower intracellular sodium concentration. To evaluate toxic interactions it is important to note that cation channels are highly selective. For example, the K^+ channels permit Na^+ to pass only with 1% of the rate of K^+ transport, and they totally exclude anions and multivalent cations (Cantley, 1981).

Na^+ and K^+ transport is sensitive to the toxic effects of other metal ions, mainly Tl^+, Ba^{2+} and VO_3^-. In spite of its much larger atomic mass, Tl^+ is only slightly larger (1.5 Å) than K^+ (1.4 Å) and penetrates the potassium channels even faster (2.3-fold) than K^+ itself. Furthermore, Tl^+ interferes with the K^+ binding site of Na^+, K^+-ATPase (Gehring and Hammond, 1967), where it can form stronger bonds than K^+ due to its higher polarizability (4-fold that of K^+). These molecular properties of Tl^+ may well explain its toxic effects in humans and experimental animals: thallium is rapidly absorbed in the gastrointestinal tract and is widely distributed in the body. However, in accordance with its higher polarizability and protein affinity as compared to K^+, Tl^+ is less readily excreted. If taken up orally in accidents or suicidal efforts, it causes acute gastrointestinal problems and subsequently polyneuropathic symptoms. It has still to be elucidated whether the neurotoxic properties of Tl^+ are directly related to interference with K^+ transport and Na^+, K^+-ATPase activity. An experimental result supporting this interpretation is the observation that infusion of K^+ in experimental animals caused increased mobilization of Tl^+ from tissues and increased renal clearance with relief of the toxic symptoms (Kazantzis, 1986).

Ba^{2+} is much heavier than K^+ but has almost the same size (1.35 Å). This ion binds to Na^+, K^+ channels and K^+ channels, blocking Na^+-K^+ exchange and K^+ transport, respectively (Nielsen, 1981). Like Tl^+, Ba^{2+} causes gastrointestinal pain. In addition, it has strong effects on muscle functions: first, it stimulates cardiac and skeletal muscle contraction and later the muscles are paralyzed. Furthermore, disorders of heart rhythm appear (Reeves, 1986). Barium intoxication of humans has been reported from accidental exposure, but there may also be health problems with welders using barium-rich electrodes and showing increased blood plasma and urine levels of barium (Zschiesche et al., 1992).

An anionic metal compound, vanadate, also interferes with the functions of Na^+ and K^+. VO_3^- competes with phosphate in binding to Na^+, K^+-ATPase, inhibits this enzyme and hence disturbs the osmotic equilibria and electrophysiological functions of the alkaline metal ions (Nechay, 1984). VO_3^- also binds to and inhibits other ATPases, ATP synthases and other phosphate-metabolizing enzymes. Among other consequences, the intracellular ATP concentration as well as the biosynthesis of phospholipids and coenzyme A are diminished, and oxidative phosphorylation is uncoupled. In accordance with this wide spectrum of metabolic effects of VO_3^-, the medical symptoms in vanadate intoxication are not uniform. If inhaled in the processing of vanadium-containing ores or alloys, local effects are noted in the respiratory tract (irritation, bronchitis and pneumonia). The wide distribution in the body after inhalation results in impairments of the functions of liver, kidney, nervous and cardiovascular systems, the blood forming organs and others (Lagerkvist et al., 1986; see also *Medicine*, Part Four, Chapter 1-B).

Other metal ions may also interfere with sodium and potassium metabolism. Many transition metal ions are bound to proteins at other sites than alkali metal ligands and modify or block enzymic activities. As an example, Na^+, K^+-ATPase is inhibited by CrO_4^{2-}, Co^{2+}, Ni^{2+}, Cu^{2+}, Zn^{2+} and Hg^{2+} (Bader et al., 1970). However, these ions interfere with sodium and potassium metabolism at relatively high concentrations compared to other targets, and it would be an oversimplification to conclude that these effects also are the cause of human health impairment by these metal ions.

III. INTERFERENCE WITH MAGNESIUM FUNCTIONS

Magnesium ions are required for nearly every reaction in the production and consumption of biochemical energy. The general function of magnesium is to serve as a cofactor in the biochemistry of phosphates. There is a prominent exception to this role: magnesium is located in the heart of chlorophyll, bound to five nitrogen atoms, stabilizing this light harvesting molecule without participating in electron transfers.

In the majority of magnesium-dependent biochemical reactions, Mg^{2+} is bound to oxygen atoms of phosphates and bridges these to the side groups of enzymes. Table 1 lists a selection of Mg^{2+}-dependent enzymes and reactions. These reactions comprise nearly every aspect of energy-rich phosphate-bond formation and degradation as well as macromolecular synthesis and degradation. The common denominator of magnesium functions in enzymic reactions is its role as Lewis acid catalyst with phosphate turnover. Li^+, Be^{2+}, Al^{3+}, Mn^{2+} and many other transition metal ions are capable of interfering with the biological functions of magnesium.

Table 1

A Selection of Mg^{2+}-dependent Enzymes and Reactions

Reactions catalyzed	Enzymes involved
ATP synthesis	Mitochondrial ATP Synthase
ATP hydrolysis	Myosin ATPase
Pyrophosphate hydrolysis	Pyrophosphatase
Transmembrane pumping	Na^+, K^+-ATPase
Metabolite phosphorylation	Creatine kinase Phosphoglycerate kinase
Phosphate ester hydrolysis	Alkaline phosphatase
Second messenger formation	Adenylate cyclase
Protein phosphorylation	Protein kinase
Nucleic acid biosynthesis	DNA polymerase RNA polymerase Reverse transcriptase
Nucleic acid degradation	DNase Ribozymes (catalytic RNAs)
Protein biosynthesis	Ribosomal proteins
CO_2 assimilation	Ribulose bisphosphate carboxylase

Lithium is chemically the closest relative to magnesium according to the rule of diagonal relations in the periodic table. The radii of lithium and magnesium ions in hexacoordination are 0.76 and 0.72 Å, respectively. Lithium carbonate is used as a remedy against certain forms of depression, and its toxic dose is only slightly above the therapeutic dose (corresponding to 1 mM in blood plasma). Both kinds of actions might be attributed to competition of Li^+ for Mg^{2+} in the Na^+,K^+-ATPase of neurons and/or interactions of Li^+ with the metabolism of inositol phosphates which act as second messengers in the transduction of biochemical signals (Berridge et al., 1989). However, Li^+ forms weaker complexes than Mg^{2+}, and it will bind to sites not occupied by Mg^{2+}. This fact is difficult to reconcile with hypotheses assuming a competitive interaction of Li^+ for Mg^{2+} sites.

The ***beryllium*** ion is much smaller (0.45 Å in hexacoordination) than Mg^{2+}, and it has a correspondingly much higher charge density. Be^{2+} polarizes its ligands to such an extent that its ligand bonds tend to covalency. Be^{2+} may interfere with Mg^{2+}-dependent processes, since it carries the same charge, but forms bonds with enzymes and substrates that are more stable kinetically. Beryllium is highly toxic when ingested and it is carcinogenic when inhaled. The mechanism of beryllium carcinogenicity is still obscure, but there is strong evidence that points to a disturbance of immunochemical reactions (Reeves, 1983). At the very high concentration of 10 mM $BeCl_2$, also the error rate (infidelity) of DNA synthesis in vitro is increased and DNA replication is inhibited (Zakour et al., 1981).

Aluminum follows magnesium in the periodic table. Al^{3+} is smaller (0.54 Å in hexacoordination) than Mg^{2+} and is not bioavailable from neutral solutions due to the formation of hydroxides. Al^{3+} may substitute for Mg^{2+} where charge neutralization is required but the size control is not stringent. Once bound, Al^{3+} exchanges much slower than Mg^{2+}, a factor that may cause inactivation of Mg^{2+}-dependent reactions. The concentration of aluminum is highest in the brain, and it has been proposed as a cofactor in the development of Alzheimer's dementia (Shore and Wyatt, 1983; see *Medicine*, Part Four, Chapter 5-B).

Manganese is an essential trace element and is part of the active sites of many metalloenzymes, e.g. superoxide dismutases, glycosyl transferases or acid phosphatase. When taken up in excess, manganese is toxic and by inhalation may even cause cancer. Mn^{2+} is larger (0.83 Å in hexacoordination) than Mg^{2+}, but still has been shown to substitute for Mg^{2+} in enzymes and in many instances even maintains the enzyme activity (Schramm and Wedler, 1986). Hence Mn^{2+} has been used as a paramagnetic and NMR probe for Mg^{2+} in enzyme studies.

Other transition metal ions also may substitute for Mg^{2+} with respect to its charge interactions but cause functional alterations. A most relevant example is the substitution of Mg^{2+} by other transition metal ions in DNA polymerase assays. If Ag^+, Be^{2+}, Cd^{2+}, Co^{2+}, Cr^{3+}, Cu^{2+}, Ni^{2+} or Pb^{2+} ions are used instead of Mg^{2+} in enzymic DNA synthesis, increases in DNA mispairing are observed. Interestingly, compounds of these metals with the exception of silver and copper are also involved in carcinogenesis (Zakour et al., 1981). In some instances, the ion concentrations required to cause mispairing in DNA synthesis are relatively high, i.e. above 1 mM (of Be^{2+}, Cr^{3+}, Co^{2+}, Mn^{2+}, Ni^{2+} and Pb^{2+}), and other toxic symptoms probably are dominant. Eichhorn (1979) discusses increasing errors introduced into the genetic information as a cause of ageing. Free transition metal ions at elevated concentrations disturb not only DNA synthesis and repair (Hartwig and Beyersmann, 1989) but also interfere with the roles of Mg^{2+} in RNA and protein synthesis. In certain cases the rate of RNA synthesis may even be enhanced by toxic metal ions as in the case of trivalent chromium which increases the rate of nucleolar RNA synthesis (Okada et al., 1989). This effect is interpreted as being a result of complexation of Cr^{3+} to DNA causing enhanced template activity of DNA for in vitro RNA synthesis. Furthermore, the biosynthesis of nucleotides is disturbed by many transition metal ions at toxic doses, resulting in detrimental effects both on macromolecular syntheses and bioenergetics.

The chemical interpretation of the interference of toxic metal ions with the functions of Mg^{2+} in nucleic acid synthesis is not as simple as it may seem, since toxic ions may (i) substitute for Mg^{2+} in the biosynthesis of the precursors, i.e. the nucleotides; (ii) substitute for Mg^{2+} in the enzymes involved and modify their activity and/or specifity; (iii) bind to the phosphates of the nucleic acid backbone and/or to the bases. The binding of the genotoxic ions Cd^{2+} (Jacobson and Turner, 1980) and Cr^{3+} (Kortenkamp et al., 1992) to DNA induces a more condensed conformation of the macromolecule which in turn may change the interaction of DNA with DNA polymerases, nucleases, histones and/or transcription factors.

The similar structural requirements of Mg^{2+} and Ni^{2+} may explain that Mg^{2+} protects from nickel toxicity in rats (Kasprzak et al., 1986) and carcinogenicity (Kasprzak et al., 1985), and that magnesium deficiency increases the incidence of malignant lymphoma in rats (Averdunk et al., 1982).

IV. INTERFERENCE WITH CALCIUM FUNCTIONS

Calcium is the element that is used as a signal and regulator of nearly every major biological process in multicellular organisms and its biological functions are highly vulnerable to toxicants. The calcium ion often acts as second messenger in intracellular signal transduction when cells are receiving messages in the form of hormones, growth factors, and neurotransmitters. Table 2 lists a selection out of nearly 100 processes known to be influenced by intracellular calcium ions. These include control of intermediary metabolism, cell proliferation and differentiation, motion and membrane transport. A further specialized function of Ca^{2+} is the mineralization in bone and teeth. It is important to note that it is the free intracellular Ca^{2+} concentration and not the total calcium that controls cellular events. The basic level of free intracellular Ca^{2+} is very low, i.e. about 10^{-7} M, and is elevated to about 10^{-5} M when the calcium system is triggered. The free Ca^{2+} concentration is controlled by a balance between calcium pumping out of the cytoplasm by Ca^{2+}-ATPase and Ca^{2+} influx through controlled channels. When calcium is mobilized into the cytoplasm upon response to extracellular signals, it may activate enzymes directly by binding to specific regulatory sites, but in many instances Ca^{2+} is first bound to special calcium binding proteins which subsequently modulate enzymic functions. Examples of signalling calcium proteins are calmodulin, troponin C, and osteocalcin.

The physicochemical factors which allow the specific biochemical recognition of Ca^{2+} are the relatively large size (about 1.0 Å depending on the number of bonding atoms), the hard character of the ion and the very high rate of ligand exchange (rate constants 10^8 to 10^9 s^{-1}). In contrast to Mg^{2+}, Ca^{2+} has the capacity to form complexes with up to eight ligands with irregular geometry. Hence Ca^{2+} is able to fit into flexible binding sites, an important feature for its regulatory functions. Also, the fast ligand exchange rate makes Ca^{2+} an ideal controlling element for immediate biological response. Ca^{2+} preferentially binds to oxygen atoms, with highest affinity for uncharged oxygen like in alcohols or carbonyl compounds.

Toxic metal ions interfering with the metabolism of calcium typically are Pb^{2+}, Cd^{2+} and Hg^{2+}. Since these metal ions are "softer" than Ca^{2+} ions, they preferentially bind to sulfur rather than to oxygen and tend to form covalent bonds. The size of their divalent ions in hexacoordinate compounds is, however, close to that of Ca^{2+} and the exchange rates for inner sphere water are similarly high as those for calcium complexes. The lanthanum(III) ion also has a similar size and preferentially binds to oxygen, but has a somewhat lower exchange rate for bound water than Ca^{2+}. The interference of heavy metals with calcium functions has a major impact on human health, since these elements are widely polluting the environment. Detrimental metal ions may either mimic calcium ions and bind to calcium sites in regulatory proteins or may bind to other sites (preferentially cysteinyl-sulfur in proteins) and inactivate enzymes and regulatory proteins. In the former case, detrimental metal ions may not only bind to calcium sites but also functionally substitute for Ca^{2+} as in the case of calmodulin activation (Table 3; Cheung, 1984; Chao et al., 1984).

Lead has been most thoroughly studied as toxic to calcium-mediated reactions. Pb^{2+} may be taken up by Ca^{2+} channels, and in human erythrocytes Pb^{2+} may even be taken up actively (Simons, 1988). Pb^{2+} interferes with all levels of calcium-regulated physiological processes. Table 4 summarizes examples of direct effects of Pb^{2+} on Ca^{2+} functions and their biological consequences. Perhaps the most sensitive and early physiological effect of

Table 2

Some Major Calcium-mediated Biological Processes and Toxic Interferences

General process	Ca^{2+} mediated reaction	Possible consequences of interference by toxic metals
Intermediary metabolism	Glycolysis oxidative phosphorylation	Altered response to hormonal signals
Cell proliferation	Phosphorylation of transcription factors DNA replication and cell division	Altered transcription Impaired cell proliferation
Differentiation	Control of transcription by phosphorylation of transcription factors	Altered transcription
Cell death (Apoptosis)	Activation of DNase and protease	Interference with programmed cell death
Signal transduction	Second message after hormone reception Intracellular response to neurotransmitters	Calmodulin activity diminished Voltage-gated Ca^{2+} channels inhibited
Blood clotting	Phosphorylation of clotting factors	Impaired blood clotting
Fiber concentration	Actomyosin ATPase mediated muscle contraction	Impaired muscle function
Membrane transport	Ca^{2+} transport through biomembranes	Disturbed Ca^{2+} homeostasis
Biomineralization	Control of bone formation via osteocalcin	Impaired bone formation

Table 3

Physicochemical Properties of Some Metal Ions Interfering with Calcium Metabolism and Effects of Ca^{2+} Substitution in Calmodulin

Ion	Ionic radius (Å)[a] hexacoordin.	Ionic radius (Å)[a] octacoordin.	Preferred donor	Water exchange rate constant[b]	Relative amounts bound to calmodulin[c]	Relative calmodulin activity (cyclic phosphodiesterase)[c]
Ca^{2+}	1.00	1.12	O	$10^8 s^{-1}$	1.00	1.00
Cd^{2+}	0.95	1.10	S	$10^8 s^{-1}$	0.81	0.90
Pb^{2+}	1.19	1.29	S	$10^8 s^{-1}$	0.47	0.92
Hg^{2+}	1.10		S	$10^8 s^{-1}$	0.75	0.73
La^{3+}	1.06	1.18	O	$10^7 s^{-1}$	0.70	0.94

[a] Shannon (1976); [b] Diebler et al. (1969); [c] Cheung (1984)

lead at low body burden is the impairment of nerve development and function. However, it is not known up to now which is the primary molecular target of lead toxicity. A plausible hypothesis assumes that it is the calcium messenger system (Pounds, 1984). Besides its early and specific effects on the calcium signalling system, lead inhibits numerous biochemical reactions due to its high affinity for protein sulfhydryl groups. Health aspects of lead have been reviewed by Ewers and Schlipköter (1991).

Cadmium like lead interferes at several levels with calcium-mediated processes. Its close chemical relationship to zinc explains its toxic effects also on the function of zinc proteins (see below the section on zinc). Cd^{2+} competes for Ca^{2+} in the uptake through cell membranes and intracellular binding sites (Planas-Bohne and Klug, 1988; Verbost et al., 1989). Vice versa, extracellular calcium protects cells against cadmium, since it interferes with the uptake of the toxic ion (Sorensen, 1988). Cadmium also perturbs the intracellular homeostasis of calcium which is extremely carefully controlled. Cd^{2+} inhibits the transport of Ca^{2+} into intracellular vesicles (Zhang et al., 1990; Hechtenberg and Beyersmann, 1991). The result is an increase in free cytosolic Ca^{2+} concentration, which in turn may switch on a series of calcium-dependent processes (Tsien, 1987; Suszkiw et al., 1984). The biochemical competition of Cd^{2+} for Ca^{2+}-binding sites in cells offers a mechanistic interpretation for the so-called Itai-itai-disease (pain disease) that occurred in Japan in an area highly polluted by industrial emissions of cadmium. The symptoms were severe renal tubular damage and osteomalacia causing great pain in back and legs. These effects were observed predominantly in women who had a low daily intake of calcium (Nordberg et al., 1986). Cadmium and calcium seem to compete for calcium channels in the kidney which reabsorb calcium and phosphate. The resulting mineral deficiency causes a depletion of calcium from the bone. Cadmium like lead interferes also with the calcium-calmodulin messenger system (Cheung, 1984) and causes neurotoxic symptoms. As with lead, many detrimental effects of cadmium may also be due to its high affinity for protein thiol groups and inhibition of enzymic activities. Medicinal aspects of cadmium toxicity have been reviewed by Friberg et al. (1986a).

The ***Lanthanum(III)*** ion has an ionic radius close to that of Ca^{2+} and (like Ca^{2+} but in contrast to Pb^{2+} and Cd^{2+}) prefers oxygen as a ligand. Therefore, La^{3+} has been used as a spectroscopic probe for Ca^{2+}-binding proteins. La^{3+} has a lower exchange rate for bound water than Ca^{2+}. This may be the cause for the observation that La^{3+} blocks Ca^{2+} channels in experimental settings. However, environmental lanthanum intoxications have not occurred due to the low environmental concentration of this element.

V. INTERFERENCE WITH ZINC FUNCTIONS

Zinc is essential for all forms of life. It occurs in several grams in the human body. It is a constituent of nearly 300 proteins (at least 200 work enzymes, dozens of regulatory proteins and some peptide toxins). Its small ionic radius (0.74 Å) and its high charge density make Zn^{2+} an ideal Lewis acid catalyst. In many hydrolases (e.g. proteins catalyzing protein degradation) and dehydrogenases (e.g. alcohol dehydrogenase) Zn^{2+} serves to polarize oxygen. An outstanding feature of Zn^{2+} is the variety of atoms it binds and its flexibility in coordination geometry: it binds nitrogen, sulfur and oxygen and it forms tetrahedral, pyramidal and octahedral coordination with four to six ligands. Therefore, zinc is able to act as a structural element in the folding of a large variety of proteins like enzymes and transcription factors (proteins controlling the expression of specific genes). In the latter case, Zn^{2+} fixes loops in the protein structure by coordinating four amino acid side chains (cysteine sulfur and histidine nitrogen atoms) in the so-called zinc fingers (see Part Two, Chapters 1-B3 and 2-C). Many enzymes contain zinc atoms like alcohol dehydrogenase which has structural zinc (coordinating four sulfur atoms) and catalytic zinc (coordinating a sulfur, two nitrogens and an oxygen) (Bertini et al., 1986). A key regulatory enzyme, protein kinase C, contains structural zinc in zinc-finger like motifs (Hubbard et al., 1991) and regulatory zinc which modulates the enzyme activity by modifying its affinity for critical sites within the cell (Block, 1992).

The supply of zinc to various organs and the homeostatic control of free Zn^{2+} are believed to be regulated by metallothionein, a small (7 to 8 kg mol^{-1}) sulfur-rich protein

Table 4

Interference of Pb^{2+} with Some Calcium-dependent Processes

Biochemical process	Mode of interference by Pb^{2+}	Toxic effect	Refs.
Ca^{2+} homeostasis	Inhibition of Ca^{2+} ATPase	Sustained elevation of free Ca^{2+}	1-3
	Binding to ligand-operated Ca^{2+} channels	Interruption of signalling by Ca^{2+}	4
Calmodulin activation	Functional substitution of Ca^{2+} by Pb^{2+}	Ca^{2+}-independent activation of cyclic phosphodiesterase	5-7
Neurotransmission	Binding to voltage-gated Ca^{2+} channels	Inhibition of neurotransmission	8
Gene expression	Binding to proteins controlling transcription	Altered gene expression	9
Blood clotting	Competition for Ca^{2+} in clotting factors	Inhibition of blood clotting	10
Programmed cell death	Inhibition of Ca^{2+} activated endonuclease	Retarded DNA fragmentation	11
Bone mineralization	Substitution of Ca^{2+} by Pb^{2+} in osteocalcin	Diminished binding of osteocalcin to bone mineral	12-13

1. Zhang et al. (1990)
2. Mas-Oliva (1989)
3. Hechtenberg and Beyersmann (1991)
4. Pounds (1988)
5. Cheung (1984)
6. Fullmer et al. (1985)
7. Chao et al. (1984)
8. Pounds (1984)
9. Fowler (1989)
10. Aseni et al. (1979)
11. Lohmann and Beyersmann (1993)
12. Rosen and Pounds (1989)
13. Schanne et al. (1990)

(Schäffer and Kägi, 1991), which binds Zn^{2+} with high affinity. A striking feature of zinc biochemistry is the high zinc content of reproductive cells like sperm and pollen, the high zinc concentration of cell nuclei and the requirement for zinc enzymes and regulatory proteins for cell proliferation (DNA polymerase, RNA polymerase, transcription factors, protein kinases). It is tempting to speculate that zinc may function as a "master hormone" for nuclear activities, differentiation and proliferation (Fraústo da Silva and Williams, 1991). The broad variation in biochemical functions of zinc correlates with the considerable variation in ligand exchange rates of Zn^{2+} in proteins. Depending on the protein environment, the half-time of ligand exchange at zinc centers is between an hour in metallothionein (a zinc finger protein with all-sulfur coordination) and a few days in zinc enzymes with nitrogen and oxygen coordination (Bertini et al., 1986).

Regarding toxic ions interfering with zinc functions, Co^{2+}, Cd^{2+} and Pb^{2+} are of major concern. In animal nutrition, zinc protects from adverse effects of these elements, and zinc defiency increases their toxicity (Mills, 1987), Furthermore, zinc prevents the induction of tumors by cadmium (Gunn et al., 1963; Waalkes and Rehm, 1987) and the teratogenicity of cadmium (Ferm and Carpenter, 1967). The ability of detrimental metal ions to substitute for Zn^{2+} depends on their preferred ligand environment, their ligand exchange rates and, of course, on their concentration.

Cobalt is essential for mammals at doses of a few micrograms per kg and day, but may be toxic at elevated concentrations, and may even cause cancer when inhaled (IARC, 1991). Co^{2+} substitutes for Zn^{2+} in many zinc proteins, and it has been widely used as a spectroscopic probe for the non-absorbing and non-magnetic zinc ion (see Part Two, Chapter 1-B3). Co^{2+} has the same size (0.74 Å) as Zn^{2+}, it prefers the octahedral coordination to six ligands in suitable zinc proteins and it binds to the same broad variety of atoms as Zn^{2+}, i.e. oxygen, nitrogen and sulfur. However, Co^{2+} exchanges ligands much slower than Zn^{2+} with the consequence of lower or absent biological activity of Co^{2+}-substituted zinc enzymes (Bertini et al., 1986). Interference of Co^{2+} with zinc enzymes could also be the mechanism of its carcinogenic and genotoxic effects: Co^{2+} interferes with enzymes that repair damaged DNA (Beyersmann and Hartwig, 1992).

Cadmium ions have an analogous electron configuration ($4d^{10}$) with Zn^{2+} ($3d^{10}$), and they have a closely related chemical reactivity. However, Cd^{2+} is considerably larger (0.95 Å in hexacoordination) than Zn^{2+}. Due to the flexibility of zinc ligation in many proteins, cadmium often can substitute for zinc in spite of its larger size. In some instances, cadmium-substituted enzymes even have full activity, e.g. in the case of porphobilinogen synthase (Schlösser and Beyersmann, 1987). In other cases, cadmium substitutes for zinc but confers an altered enzyme specifity. This is the case with protein kinase C, an enzyme regulating cellular differentiation and proliferation by protein phosphorylation (Block et al., 1992). One must be warned against oversimplification: the metabolic toxicity of Cd^{2+} is not always caused by substitution for Zn^{2+} but often is due to preferential binding of this ion to functional sulfhydryl groups of enzymes rendering them inactive. Interference with zinc occurs also at the level of genetic control: Cd^{2+} is an effective inducer of biosynthesis of the protein metallothionein which subsequently binds Cd^{2+} with a nearly thousand-fold higher affinity than Zn^{2+} (Schäffer and Kägi, 1991). This reaction is regarded as a major mechanism of cadmium capturing and detoxification. Cadmium salts are also carcinogenic, but only weakly mutagenic. The interference of Cd^{2+} with the Mg^{2+}-dependent DNA replication resulting in increasing errors (mentioned above in section IV) cannot explain the carcinogenicity of cadmium satisfactorily. A stimulating hypothesis has been proposed by Sunderman (1990): Cd^{2+} could substitute for Zn^{2+} in the zinc finger structures of transcription factors, thus changing the pattern of RNA synthesis towards an expression of oncogenes. In consequence, the proliferation of tissues would be deregulated and tumor growth promoted. However, cadmium also interferes with several other levels of cellular control (Beyersmann et al., 1992), and it is premature to assign the carcinogenic effect of this metal to a single interaction.

Lead like cadmium is a general metabolic poison. Like cadmium it also binds to functional sulfhydryl groups of proteins and inactivates enzymes. One of the biochemical pathways most sensitive to lead is the biosynthesis of heme, where Pb^{2+} inhibits the formation of pyrrole by the zinc enzyme porphobilinogen synthase (Doss, 1978). This enzyme contains eight zinc atoms per octameric protein; four of these contain zinc bound to a coordination sphere with three or four sulfur atoms and exchange Zn^{2+} for Pb^{2+} (Dent et al., 1990). These physicochemical data correspond to the observed protection by zinc against hematopoietic symptoms of lead intoxication in animals (Willoughby et al., 1972). At variance, the neurotoxicity and disturbation of bone formation of lead are interpreted by interference with the metabolism of calcium (see above section on calcium).

VI. INTERFERENCE WITH BIOCHEMICAL FUNCTIONS OF IRON AND COPPER

Iron and copper are elements capable of electron transfer under physiological conditions, and their ions are involved in biochemical catalysts. In prokaryotic organisms iron-sulfur proteins act as electron transmitters in the metabolism of energy-supplying substances like sulfur, hydrogen sulfide, nitrate etc. In higher organisms, iron-sulfur proteins are carrying out redox reactions in mitochondria and chloroplasts, i.e. in organelles believed to be derived from prokaryotic origin in evolution (Beinert and Kennedy, 1989). Mitochondrial membranes also contain iron in the form of cytochromes, which are iron-porphyrins carrying electrons in the electron transport chain of respiration. In the cytoplasm cytochromes occur as the catalytic groups of monooxygenases which detoxify hydrocarbons by hydroxylation. Red blood cells, of course, use iron in hemoglobin as a reversible oxygen carrier. Iron cannot be transported in living cells as the free Fe^{3+} ion because it forms unsoluble hydroxides. Hence special iron-transporting proteins, e.g. transferrin, are used for transmembrane transport of iron (Harrison et al., 1987).

Little is known about interference of toxic metal ions with the transport and biological functions of iron. Trivalent chromium binds to transferrin and may be taken up by cells via the iron way, but there is no report on an impairment of iron transport by chromium(III). Lead(II) interferes with the insertion of iron(II) into protoporphyrin by the enzyme ferrochelatase (Granick et al., 1978). This effect, in addition to the inhibitory action of lead(II) on porphobilinogen synthase, is the cause for the lead-induced anaemia at elevated body-burdens of lead (Doss, 1978). At variance, the inhibition of hepatic hydrocarbon hydroxylation by various transition and heavy metal ions (Mn^{2+}, Cu^{2+}, Zn^{2+}, Cd^{2+}, Hg^{2+} and Pb^{2+}) is not due to competition for cytochrome iron, but to inactivation of the enzyme protein component (Tephly et al., 1979).

Similar to iron, copper is the active part of some electron transferring enzymes in pro- and eukaryotic organisms (Lontie, 1984). Copper is part of cytochrome *c* which transports electrons to molecular oxygen in respiration. Copper(II) also is at the active center of the enzyme superoxide dismutase which detoxifies superoxide when it is produced as a side product of oxygen metabolism. In chloroplasts, copper is part of the photosystem II which produces molecular oxygen. Since free copper ions are highly toxic, copper is transported in the blood stream bound to a protein, ceruloplasmin, and excess copper(II) in tissues is reduced to Cu(I) and bound extremely tightly to metallothionein (Hamer, 1986). There are no reports on toxic competition of detrimental metal ions with the electron-carrying functions of copper so far. But there may be interactions with zinc and cadmium ions in metallothionein: if zinc or cadmium ions induce the synthesis of metallothionein at elevated concentrations, Cu(I) may by captured by metallothionein since its affinity for this protein is 100 times that for Cd(II), which in turn is bound 1000 times tighter than Zn(II). Due to its high affinity to metallothionein, copper then is no longer bioavailable (Hamer, 1986).

VII. SUMMARY AND CONCLUSIONS

Detrimental metal ions interfere with normal metal functions either by interacting competitively with the specific binding sites of the essential elements or by modifying metal binding proteins at other sites which in turn modulate the functions of essential metal ions. Insights into physicochemical mechanisms of toxic metal interactions have been gained mainly from studies on chemical model systems which are of reduced complexity compared to the conditions in living systems. This reduction has the consequence that model studies at best allow qualitative predictions about possible detrimental interactions in life but that quantitative conclusions cannot be drawn. A major obstacle to quantitative calculations is the difficulty to estimate the concentrations of free metal ions in a multi-ligand system. In some instances, free metal ions can be measured by suitable fluorescence indicators as in the case of measurement of the intracellular free calcium concentration with fluorescent chelators. It is highly desirable to develop suitable in situ analytic methods for free metal ions, e.g. NMR techniques (Schanne et al., 1990).

With regard to the individual interactions, much knowledge has been accumulated on the uptake, distribution, binding and metabolic and health consequences of the "classical" heavy metal ions Pb^{2+}, Cd^{2+} and Hg^{2+} (Friberg et al., 1986a). Also much research activities have been devoted to the elucidation of mechanisms of carcinogenesis by genotoxic metal compounds like those of chromium and nickel (IARC, 1990). There is, however, much to be learnt in the future about metal interactions resulting in more subtle health effects like subclinical neurotoxicity, impairment of immune response and reproductive toxicity. A lot of work will be necessary to interpret and understand these types of interactions in physicochemical terms. What we need from inorganic biochemistry are the kinetic and equilibrium data on the interactions of toxic metals with those biomolecules which are relevant to biological and medicinal effects.

ACKNOWLEDGEMENT

I wish to thank Professor Julio Pedrosa de Jesus of the University of Aveiro, Portugal, for critical and helpful comments. I am also grateful to Mrs. Brigitte Lübkemann for expert assistance in the preparation of the manuscript.

REFERENCES

Aseni, P., Bossa, R., Galatulas, I., and Perrone, G. (1979). *IRCS Medial Science*, *7*: 235.

Averdunk, R., Bippus, P.H., Günther, T., and Merker, H.J. (1982). *J. Cancer Res. Clin. Oncol.*, *104*: 63.

Bader, H., Wilkes, A.B., and Jean, D.H. (1970). *Biochim. Biophys. Acta*, *198*: 583.

Berridge, M.J., Downes, C.P., and Hanley, M.R. (1989). *Cell*, *59*: 411.

Beinert, H., and Kennedy, M.C. (1989). *Eur. J. Biochem.*, *186*: 5.

Bertini, I., Luchinat, C., Maret, W., and Zeppezauer, M., eds. (1986). *Zinc Enzymes*, Birkhäuser, Basel.

Beyersmann, D. (1991). *Metals and Their Compounds in the Environment* (E. Merian, ed.), VCH, Weinheim, p. 491.

Beyersmann, D., and Hartwig, A. (1992). *Toxicol. Appl. Pharmacol.*, *115*: 137.

Beyersmann, D., Hechtenberg, S., Block, C., and Kirchherr, H. (1992). *Metal Compounds in Environment and Life*, *4*: 187.

Block, C. (1992). *Funktionelle Differenzierung der Zink-Bindungsstellen von Proteinkinase C*, Dissertation, Universität Bremen.

Block, C. Freyermuth, S., Beyersmann, D., and Malviya, A.N. (1992). *J. Biol. Chem.*, *267*: 19824.

Cantley, L.C. (1981). *Curr. Top. Bioenerget.*, *11*: 201.

Chao, S.H., Suzuki, Y., Zysk, J.R., and Cheung, W.Y. (1984). *Mol. Pharmacol.*, *26*: 75.

Cheung, W.Y. (1984). *Fed. Proc.*, *43*: 2995.

Dent, A., Beyersmann, D., Block, C., and Hasnain, S. (1990). *Biochemistry*, *29*: 7822.

Diebler, H., Eigen, M., Ilgenfritz, G., Maass, G., and Winkler, R. (1969). *Pure Appl. Chem.*, *20*: 93.

Doss, M., ed. (1978). *Diagnosis and Therapy of Porphyrias and Lead Intoxication*, Springer, Berlin.

Eichhorn, G.L. (1979). *Mech. Ageing Dev.*, *9*: 291.

Ewers, U., and Schlipköter, H.-W. (1991). *Metals and Their Compounds in the Environment*, VCH, Weinheim, p. 971.

Fowler, B.A. (1989). *Comments Toxicol.*, *3*: 27.

Ferm, V.H., and Carpenter, S.J. (1967). *Nature, 216*: 1123.

Fraústo da Silva, J.J.R., and Williams, R.J.P. (1991). *The Biological Chemistry of the Elements*, Clarendon Press, Oxford.

Friberg, L., Nordberg, G.F., and Vouk, V., eds. (1986a). *Handbook on the Toxicity of Metals*, Elsevier, Amsterdam.

Friberg, L., Elinder, C.-G., Kjellstroem, T., and Nordberg, G.F., eds. (1986b). *Cadmium and Health*, Vol. II, CRC Press, Boca Raton, Florida.

Fullmer, C.S., Edelstein, S., and Wassermann, R.H. (1985). *J. Biol. Chem., 260*: 6816.

Furst, A. (1987). *Genotoxic and Carcinogenic Metals* (L. Fishbein, A. Furst and M.A. Mehlman, eds.), Scientific Publishing Co., Princeton, New Jersey, p. 279.

Gehring, P.J., and Hammond, P.B. (1967). *J. Pharmacol. Exp. Ther., 155*: 187.

Granick, J.L., Sassa, S., and Kappas, A. (1978). *Adv. Clin. Chem., 20*: 287.

Gunn, S.A., Gould, T.C., and Anderson, W.A.D. (1963). *J. Nat. Cancer Inst., 31*: 745.

Hamer, D. (1986). *Annu. Rev. Biochem., 55*: 913.

Harrison, P.M., Andrews, S.C., Ford, G.C., Smith, J.M.A., Teffry, A., and White, J.L. (1987). *Iron Transport in Microbes, Plants and Animals* (G. Winkelmann, D. Van der Helm and J.B. Neilands, eds.), VCH, Weinheim, p. 445.

Hartwig, A., and Beyersmann, D. (1989). *Biol. Tr. Elem. Res., 21*: 359.

Hechtenberg, S., and Beyersmann, D. (1991). *Enzyme, 45*: 109.

Hubbard, S.R., Bishop, W.R., Hamer, D. (1986). *Annu. Rev. Biochem., 55*: 913.

Kirschmeier, P., George, S.J., Cramer, P., and Hendrickson, W.A. (1991). *Science, 254*: 1776.

IARC (1990). *Chromium, Nickel, and Welding*, IARC Monographs, Vol. 49.

IARC (1991). *Chlorinated Drinking-Water; Chlorination By-Products; Some other Halogenated Compounds; Cobalt and Cobalt Compounds*. IARC Monographs, Vol. 52.

Jacobson, K.B., and Turner, J.E. (1980). *Toxicology, 16*: 1.

Kasprzak, K.S., Quander, R.V., and Poirier, L.A. (1985). *Carcinogenesis, 6*: 1161.

Kasprzak, K.S., Waalkes, M.P., and Poirier, L.A. (1986). *Toxicology, 42*: 57.

Kazantzis, G. (1986). *Handbook on the Toxicity of Metals* (L. Friberg, G.F. Nordberg and V. Vouk, eds.), Elsevier, Amsterdam, p. 549.

Kortenkamp, A., O'Brien, P., and Beyersmann, D. (1992). *Toxicol. Environ. Chem., 35*: 149.

Lagerkvist, B., Nordberg, G.F., and Vouk, V. (1986). *Handbook on the Toxicity of Metals* (L. Friberg, G.F. Nordberg and V. Vouk, eds.), Elsevier, Amsterdam, p. 638.

Lohmann, R., and Beyersmann, D. (1993). *Biochem. Biophys. Res. Commun., 190*: 1097.

Lontie, R., ed. (1984). *Copper Proteins and Copper Enzymes*, Vol. 1, CRC Press, Boca Raton, Florida.

Mas-Oliva, J. (1989). *Mol. Cell. Biochem., 89*: 87.

Mills, C.F. (1987). *Trace Element Analytical Chemistry in Medicine and Biology*, Vol. 4 (P. Brätter and P. Schramel, eds.), W. de Gruyter, Berlin, p. 23.

Nechay, B.R. (1984). *Annu. Rev. Pharmacol. Toxicol., 24*: 501.

Nielsen, R. (1981). *Adv. Physiol. Sci., 3*: 135.

Nordberg, G.F., and Pershagen, G. (1984). *Toxicol. Environ. Chem., 9*: 63.

Nordberg, G.F., Parizek, J., Pershagen, G., and Gerhardsson, L. (1986). *Handbook on the Toxicity of Metals* (L. Friberg, G.F. Nordberg and V. Vouk, eds.), Elsevier, Amsterdam, p. 175.

Okada, S., Tsukada, H., and Tezuka, M. (1989). *Biol. Tr. Elem. Res., 21*: 35.

Planas-Bohne, F., and Klug, S. (1988). *Toxicol. Environ. Chem., 18*: 229.

Pounds, J.G. (1984). *Neurotoxicology, 5*: 295.

Pounds, J.G. (1988). *Toxicol. Appl. Pharmacol., 94*: 331.

Reeves, A.L. (1983). *The Immunotoxicity of Beryllium* (G.G. Gibson, R. Hubbard and D.V. Parke, eds.), Academic Press, New York, p. 261.

Reeves, A.L. (1986). *Handbook on the Toxicity of Metals* (L. Friberg, G.F. Nordberg and V. Vouk, eds.), Elsevier, Amsterdam, p. 175.

Rosen, J.F., and Pounds, J.G. (1989). *Toxicol. Appl. Pharmacol.*, *98*: 530.

Schäffer, A., and Kägi, J.H.R. (1991). *Metals and Their Compounds in the Environment* (E. Merian, ed.), VCH, Weinheim, p. 523.

Schanne, F.A.X., Dowd, T.L., Gupta, R.K., and Rosen, J.F. (1990). *Biochim. Biophys. Acta*, *1054*: 250.

Shannon, R.D. (1976). *Acta Crystallogr.*, *A 32*: 751.

Schlösser, M. and Beyersmann, D. (1987). *Biol. Chem. Hoppe-Seyler*, *368*: 1469.

Shore, D., and Wyatt, R.J. (1983). *Ment. Dis.*, *171*: 553.

Schramm, V.L., and Wedler, F.C. eds. (1986). *Manganese in Metabolism and Enzyme Function*, Academic Press, Orlando, Florida.

Simons, T.J.B. (1988). *J. Physiol.*, *405*: 105.

Sorensen, E.M.B. (1988). *Toxicol. Lett.*, *41*: 39.

Sunderman, F.W., Jr. (1990). *Toxicol. Environ. Chem.*, *27*: 131.

Suszkiw, J., Toth, G., Muransky, M., and Cooper, G.P. (1984). *Brain Res.*, *323*: 31.

Tephly, T.R., Wagner, G., Sedman, R., and Piper, W. (1979). *Fed. Proc.*, *38*: 346.

Tsien, R.W. (1987). *Annu. Rev. Biophys. Bioengin.*, *16*: 265.

Verbost, P.M., Flik, G., Pang, P.K.T., Lock, R.A.C., and Bonga, S.E.W. (1989). *J. Biol. Chem.*, *264*: 5613.

Waalkes, M.P., and Rehm, S. (1987). *The Toxicologist*, *7*: 71.

Willoughby, R.A., MacDonald, E., McSherry, B.J., and Brown, G. (1972). *Can. J. Comp. Med.*, *36*: 348.

Zakour, R.A., Kundel, T.A., and Loeb, L.A. (1981). *Environ. Health Perspect.*, *40*: 197.

Zhang, G.h., Yamaguchi, M., Kimura, S., Higham, S., and Kraus-Friedmann, N. (1990). *J. Biol. Chem.*, *265*: 2184.

Zschiesche, W., Schaller, K.-H., and Weltle, D. (1992). *Int. Arch. Occup. Envir. Health*, *64*: 13.

METAL COMPLEXES IN METAL TOXICITY

by

R. Bruce Martin

Chemistry Department,
University of Virginia,
Charlottesville, Virginia 22903
U.S.A.

I. INTRODUCTION

For the purposes of this chapter these two volumes classify metal ions into four groups:
(i) *Essential* - those in whose absence an organism cannot live.
(ii) *Beneficial* - those in whose absence an organism can survive but without optimized health. Considered as beneficial metal ions are those of nickel, tin, chromium, and vanadium discussed in *Bioinorganic Medicine*, Part Two, Chapter 3-B.
(iii) *Neutral* - those usually present in an organism at benign levels for which neither a beneficial nor toxic role has been found.
(iv) *Detrimental* - those which are exclusively toxic. Although even essential elements may become toxic at high levels, by this definition they do not belong to the exclusive detrimental group.

Only 3% by weight of the human body is composed of metal ions and of that amount over 80% is calcium in bones. These gross figures belie the fact that there are ten essential metal ions for humans and most mammals, most of them present only in trace amounts: two alkali metal ions, sodium and potassium; two alkaline earth metal ions, magnesium and calcium; five first transition row metal ions, manganese, iron, cobalt, copper, and zinc (all the metal ions from manganese through zinc except nickel); and one second transition-row metal ion, molybdenum. Other classes of organisms need additional elements. For more discussion of essential metal ions, see the *Bioinorganic Medicine* volumes.

Virtually all heavy metal ions are toxic if available in sufficient amounts. Since the essential metal ions seldom occur at toxic levels, this chapter focuses on complexes of the detrimental heavy metals cadmium, mercury, and lead, and the light metal aluminum, which acts very differently than the heavy metal ions. For greater details on toxicities of detrimental metals, see *Medicine*, Part Four, Chapter 1-C, and for their metabolism see *Medicine*, Part Three, Chapter 8. An extended discussion of metabolism and toxicity of several elements appears in the book by Ochiai (1987). For element-by-element details on toxicities of all inorganic compounds, see the authoritative handbook edited by Seiler and Sigel (1988).

II. INTERDEPENDENCIES

Metal ion toxicity does not relate to essentiality. Toxicity and essentiality do share, however, one feature: frequent interdependencies among metal ions, between metal ions and

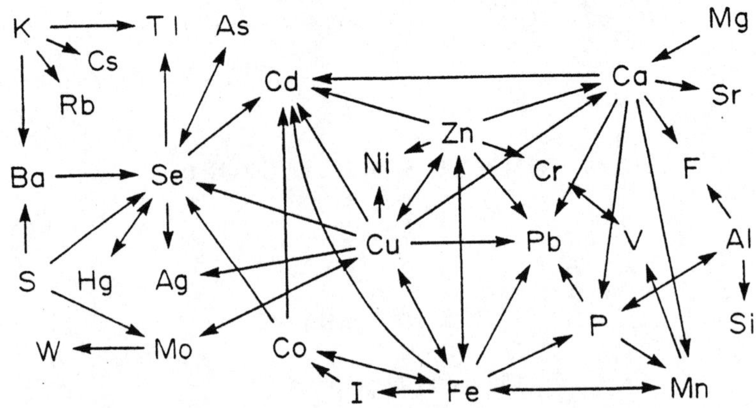

Figure 1.
Interdependencies among thirty elements in mammals. An arrow from element A to B indicates that administration of element A may reduce toxicity of element B, low levels of element A may heighten toxicity of element B, or high levels of element B may inhibit salutary effects of element A. From Martin (1988b).

non-metals, and between both and organic molecules such as vitamins (Micronutrient Interactions, 1980; Underwood, 1979). For example, excess Zn^{2+} may reduce absorption of essential Cu^{2+}, while a deficiency of Zn^{2+} worsens Cd^{2+} toxicity. Such antagonisms and mutual dependencies produce variable toxicities for comparable concentrations of a designated element, and complicate any attempt to formulate simple guidelines for essentiality or toxicity.

Interdependencies among thirty elements in mammals are diagrammed in Figure 1. The figure is not exhaustive of all possibilities but serves to indicate both the commonality and complexity of the interactions among elements. Figure 1 points up that heavily supplementing a diet with one mineral may upset the balances with others.

III. HEAVY METAL ION COMPLEXES

Figure 1 shows that both Cd^{2+} and Pb^{2+} share interdependencies with several other elements accounting for some sources of their toxicities. Cd^{2+} interferes with activities of essential Ca^{2+} and Zn^{2+}, and a low Ca^{2+} diet enhances Cd^{2+} absorption. Pb^{2+} interacts with systems that use Ca^{2+}, iron, and Zn^{2+}. Ca^{2+}, Cd^{2+}, and Pb^{2+} possess similar ionic radii (Table 1 in Part One, Chapter 3-A), so the pair of detrimental metal ions may substitute for Ca^{2+}.

Fortunately, many metal ions are less harmful than they might be because of their intrinsic chemical behavior, such as formation of insoluble compounds in the gut and subsequent elimination of the metal ion. Other fractions of metal ions travel in the blood to other tissues where they are immobilized like lead in bone, or to the liver or kidney, where they are complexed in a less toxic or disposable form.

Sulfhydryl groups of proteins furnish the vehicle for both the toxicity and detoxification of most heavy metal ions. In the stability ruler discussion in Part One, Chapter 3-A, we noted the special avidity of heavy metal ions for interaction with sulfhydryl groups. Both Cd^{2+} and Pb^{2+} exhibit enhanced binding to sulfhydryl groups compared to N or O donor ligands. Still stronger is the binding of both Hg^{2+} and CH_3Hg^+ to sufhydryl groups. For more detailed information on protein interactions with detrimental metal ions see Part Two, Chapter 1-E.

In response to the presence of detrimental Cd^{2+}, Hg^{2+}, Pb^{2+}, and other heavy metal ions, the human liver and kidneys synthesize more *metallothionein*, an unusual small protein in which approximately one-third of the 61 amino acid residues are cysteine. The frequency and juxtaposition of sulfhydryl groups provide strong binding sites for heavy metal ions. For a detailed discussion of metallothionein see Part Two, Chapter 1-D.

Though not as profusely as metallothionein, many proteins contain sulfhydryl groups that may become metalated by toxic heavy metal ions such as Cd^{2+}, Hg^{2+}, and Pb^{2+}, and it is widely believed that this complex formation is the mode of toxicity of these metal ions. The exact proteins at which the most consequential damage occurs remains uncertain. Toxic metal ions enter many tissues, and the greatest damage does not always occur where the metal ion is most plentiful. For example, more than 90% of Pb^{2+} may become immobilized in bone, but the toxic effects result from the remaining 10% distributed among other tissues, where it may react with sufhydryl groups on any number of proteins. The immobilization of Pb^{2+} in bone may act as a detoxifying mechanism even though it is more likely the result of an incidental property of the metal ion rather than a deliberate strategy of the organism.

IV. MERCURY (Martin, 1986a)

Mercury is detrimental to mammals in all of its forms. The global release of mercury by degassing from the earth's crust and oceans exceeds that produced by humans by a factor of at least five, but industrial discharges are more local and concentrated. The average person carries 13 mg of mercury in the body, none of which serves any useful purpose, although various mercuric salts have been used as therapeutic agents: for example, mercuric benzoate to treat syphilis and gonorrhea. Use of mercury reagents as insecticides and fungicides has led to many small-scale and several disastrous large-scale poisonings involving thousands of people. Mercury poisoning is a world-wide problem.

Mercury occurs in three common forms and one uncommon form as mercurous ion that disproportionates to two of the common forms, elemental mercury and mercuric ion:

$$Hg_2{}^{2+} \rightleftharpoons Hg^0 + Hg^{2+} \quad K = [Hg^{2+}]/[Hg_2{}^{2+}] = 1/170$$

The equilibrium constant value for the reaction indicates that the favored direction of the reaction is actually from right to left. It is driven to the right by strong complexation of Hg^{2+} with any of a large number of ligands. The third common form of mercury is as organic mercurials such as methyl mercury, CH_3Hg^+.

Mercury is one of the few elements that remains liquid at room temperature. Though its boiling point is 357°C, it is more volatile and hence more dangerous than generally appreciated. At saturation at 25°C a cubic meter of air holds 20 mg of Hg. Spilled mercury is a health hazard in rooms with poor ventilation. Elemental mercury is almost insoluble in water, dissolving to the extent of 0.29 μM at 25°C (Glew and Hames, 1971; Spencer and Voigt, 1968), 58 μg/liter, or 58 ppb.

Both mercuric ion, Hg^{2+}, and methyl mercury, CH_3Hg^+, prefer linear 2-coordination. They form stronger complexes than most metal ions with a single donor atom, especially N and S (see Part One, Chapter 3-A). Rare among metal ions, Hg^{2+} in basic solutions substitutes for a hydrogen on amines (in contrast to ammonium ions) as in mercuric amidochloride, $ClHgNH_2$. At high pH values, excess CH_3Hg^+ substitutes for hydrogen on the -NH groups at C4 in cytidine and C6 in adenosine (Martin, 1979). Hg^{2+} may add third and fourth donor atoms to give a distorted tetrahedron. The last two ligands are bound much more weakly than the first two and exhibit longer Hg^{2+}-donor atom bond lengths. The ability of linear, 2-coordinated Hg^{2+} and CH_3Hg^+ to associate weakly with additional donor atoms accounts for the rapid metal ion exchange among donor atoms, discussed below.

Because they prefer linear, 2-coordination, there is not much chelate effect with Hg^{2+} and CH_3Hg^+. In contrast to most metal ions, the stability constant logarithm for Hg^{2+} and CH_3Hg^+ binding to 1,2-diaminoethane is less than the sum of the values for binding to two ammonia molecules, that to glycine is less than the sum of ammonia and acetate, and that to histidine is less than the sum of ammonia and imidazole. However, owing to the great strength of binding to a single nitrogen donor atom, the stability constants for the weakly chelated Hg^{2+} and CH_3Hg^+ with diaminoethane, glycine, and histidine still exceed those for most metal ions (see Part One, Chapter 3-A).

In pure water, HgO precipitates at pH > 2 and CH_3HgOH forms at pH > 5. Chloride ion binds with high stability. The first two Cl^- to give linear $HgCl_2$ are bound 10 times more strongly than the last two Cl^- to give tetrahedral $HgCl_4^{2-}$ (Baes and Mesmer, 1976). In seawater, with its high 0.56 M Cl^- background, the predominant species are $HgCl_4^{2-}$ and CH_3HgCl.

In almost any biological system, an excess of sulfhydryl donors is expected to bind all Hg^{2+} and CH_3Hg^+. The name mercaptan ("mercury capture") derives from the strong Hg^{2+} binding to thiols. In red blood cells, Hg^{2+} links glutathione and hemoglobin sulfhydryl groups in a mixed complex (Rabenstein and Isab, 1982). Only a fraction of body Hg^{2+} remains in the blood. Though the molecular basis of Hg^{2+} toxicity by interaction at sulfhydryl groups is generally understood, the specific proteins undergoing consequential damage remain uncertain. Many proteins contain reactive sulfhydryl groups. Hg^{2+} also forms inert bonds at C5 in pyrimidines (Dale et al., 1975). The long retention times of mercury in tissues may result from formation of inert aromatic carbon-mercury bonds involving nucleic bases or amino acid side chains.

In spite of being very strongly bound, ligands show rapid exchange in and out of coordination to Hg^{2+} and CH_3Hg^+. Substitution of inner-sphere water in Hg^{2+} occurs with a half-life of less than 10^{-9} seconds, as fast as any metal ion (see Part Three, Chapter 1-B). For CH_3Hg^+ binding to the sulfhydryl group of glutathione, the stability constant log K = 15.9 (Rabenstein, 1978). For such strong binding, the half-life for exchange by a dissociation mechanism is about 10 days; yet the average lifetime for the CH_3HgSR complex in the presence of excess glutathione is less than 0.01 sec (Rabenstein et al., 1982). This comparison shows that exchange does not occur by a dissociation mechanism. The clue to the exchange mechanism in Hg^{2+} and CH_3Hg^+ compounds is the weak coordination of third and fourth ligands (mentioned near the beginning of this section). Excess ligand participates in a nucleophilic attack at an uncoordinated site on the metal ion, with rearrangement in the coordination geometry and release of a formerly bound ligand. This addition-elimination mechanism accounts for the very rapid exchange in Hg^{2+} and CH_3Hg^+ complexes.

The rapid exchange of Hg^{2+} and CH_3Hg^+ among excess donor ligands such as sulfhydryl groups plays a key role in toxicology. It accounts for the fast partitioning of the mercury among sulfhydryl groups in tissues. In the blood, CH_3Hg^+ partitions in the same proportion as the populations of sulfhydryl groups, about 10% in the plasma and 90% in the erythrocytes, which contain both hemoglobin and glutathione sulfhydryl groups. To be avoided as an antidote for mercury poisoning is simple administration of British anti-Lewisite or BAL (2,3-dimercaptopropanol), since it promotes the distribution of mercury throughout the body.

V. ALUMINUM

Locked in soils and minerals as oxides and complex aluminosilicates and comprising 8% of the earth's crust, aluminum conceals its status as the most abundant metal in the crust. For the first time in the history of the earth, this metal is being released by acid rain into water supplies. Modern industrial society has unwittingly set in motion an epic experiment to test the ability of living creatures -including man- to cope with this release. Aluminum in antacids and in waters used for dialysis is known to cause dementias and result in bone deterioration (*Medicine*, Part Three, Chapter 8-A, and Part Four, Chapter 1-C). Elevated aluminum levels have been found in the brains of Alzheimer's disease victims (Martyn et al., 1989), and its role in the etiology of the disease is debated (McLachlan et al., 1991). Appreciation of the basic chemistry of aluminum leads to understanding and prediction of its potential roles in more complex systems (Martin, 1988a; 1991a; 1992).

The chemistry of aluminum is relatively simple. It reacts 10^6 times faster than Cr^{3+} (see Part Three, Chapter 1-B, II), its hydroxide is much more soluble than that of Fe^{3+}, and it exhibits only one oxidation state in biological systems, Al^{3+}. There is no oxidation-reduction chemistry to Al^{3+} in biology. Al^{3+} interactions are primarily ionic, and in the Table 1 of Part One, Chapter 3-A, we find its ionic radius to be the second smallest ion listed (after Be^{2+}), and smaller but comparable to Fe^{3+} and Mg^{2+}, both of which compete with it in biological systems. Both Al^{3+} and Mg^{2+} favor oxygen donor ligands, especially phosphate groups. Al^{3+} is 10^7 times more effective than Mg^{2+} in promoting polymerization of tubulin to microtubules (Macdonald et al., 1987).

In solutions more acid than pH < 5, aluminum occurs as the octahedral hexahydrate, $Al(H_20)_6{}^{3+}$, usually abbreviated as Al^{3+}. As a solution becomes less acidic, $Al(H_2O)_6{}^{3+}$ undergoes successive deprotonations to yield $Al(OH)^{2+}$, $Al(OH)_2{}^+$, and soluble $Al(OH)_3$, with a decreasing and variable number of water molecules. Neutral solutions give an $Al(OH)_3$ precipitate that redissolves, owing to formation of tetrahedral aluminate, $Al(OH)_4{}^-$, the primary soluble Al^{3+} species at pH > 6.2. The four successive deprotonations squeeze into a narrow range 5.5 < pH < 6.2; the deprotonations from ocatahedral hexahydrate to tetrahedral aluminate occur cooperatively (Martin, 1988a; 1991b). Formerly locked in soils as the hydroxide or aluminosilicates, Al^{3+} is now being released from these minerals by acid rain.

The most likely Al^{3+} binding sites are oxygen atoms, especially if they are negatively charged (Martin, 1988a; 1991a; 1992). Carboxylate, catecholate (Kiss et al., 1989), and phosphate (Kiss et al., 1991) groups are the strongest Al^{3+} binders. In contrast to their reaction with heavy metal ions discussed above, even when part of a potential chelate ring, sulfhydryl groups do not bind Al^{3+} (Toth et al., 1984). Amines do not bind Al^{3+} strongly except as part of multidentate ligand systems such as nitrilotriacetate (NTA) and EDTA. The nitrogenous bases of DNA and RNA do not bind Al^{3+} (Martin, 1986b; 1988a). Fluoride binds Al^{3+} strongly, and at the 1 ppm level where it is added to acidic drinking water, most aluminum appears as $AlF_2{}^+$ and as neutral AlF_3 (Martin, 1988c). In mixed complexes of ADP and F^-, the ternary complex appears with the frequency expected statistically on the basis of binary complex stabilities (Nelson and Martin, 1991).

Citrate facilitates incorporation of Al^{3+} into mammals. Al^{3+} levels were elevated in both the brain and bones of rats fed a diet containing aluminum citrate or even just citrate (Domingo et al., 1991; Slanina et al., 1984; 1985). Evidently citrate alone chelates trace Al^{3+} in the diet. Dosing lambs with aluminum citrate promotes absorption of Al^{3+} and alters the balance of other minerals (Allen et al., 1990; 1991). Increased levels of serum Al^{3+} were found in patients with chronic renal failure taking an Al^{3+} containing phosphate binder with citrate (Hewitt et al., 1988). A rapidly fatal encephalopathy in human patients with chronic renal failure has been attributed to concomitant ingestion of $Al(OH)_3$ and citrate (Bakir et al., 1986; Kirschbaum and Schoolwerth, 1989). Moreover, healthy adults taking $Al(OH)_3$-based antacids along with citric acid, citrate salts, or citrus fruits showed substantial increases in Al^{3+} levels of blood (Slanina et al., 1986; Weberg and Berstad, 1986) and urine (Bakir et al., 1989; Coburn et al., 1991; Walker et al., 1990).

Citrate exists mainly in the form of the tricarboxylate anion at pH > 6, and at 0.1 mM in the blood plasma, it is the leading small molecule Al^{3+} binder (Martin, 1986b; 1986c). In neutral solutions the main species is $HOAlLH_{-1}{}^{2-}$ followed by $MLH_{-1}{}^-$ with $pK_a \sim 6.5$ for the loss of a proton from metal ion bound water (Martin, 1988a; Öhman, 1988). Citrate solubilizes Al^{3+} from both insoluble $Al(OH)_3$ and $AlPO_4$ (Martin, 1986b; 1986c; 1988a). Based on information from equilibrium constants, it was strongly urged in 1986 that people not take Al^{3+} and citrate together (Martin, 1986b; 1986c). At 0.1 mM, citrate is the main small molecule Al^{3+} binder in the plasma.

Transferrin is the main protein carrier of Al^{3+} in the plasma. Displacement of the 10^9 times stronger binding Fe^{3+} is unnecessary because plasma transferrin is about 50 μM in unoccupied sites. Recent experiments confirm the conclusion based on stability constants (Martin, 1986b; Martin et al., 1987) that citrate is the low-molar-mass and transferrin the high-molar-mass carriers of Al^{3+} in rat serum (Van Ginkel et al., 1990). Proton nuclear magnetic resonance spectroscopy furnishes direct evidence for Al^{3+} binding to citrate in human blood plasma (Bell et al., 1993).

The bioavailability of Al^{3+} is discussed further in Part Five, Chapter 4-G.

REFERENCES

Allen, V.G., Fontenot, J.P., and Rahnema, S.H. (1990). *J. Anim. Sci.*, *68*: 2496.

Allen, V.G., Fontenot, J.P., and Rahnema, S.H. (1991). *J. Anim. Sci.*, *69*: 792.

Baes, C.F., and Mesmer, R.E. (1976). *The Hydrolysis of Cations*, Wiley-Interscience, New York.

Bakir, A.A., Hryhorczuk, D.O., Berman, E., and Dunea, G. (1986). *Trans. Amer. Soc. Artif. Intern. Organs*, *32*: 171.

Bakir, A.A., Hryhorczuk, D.O., Ahmed, S., Hessl, S.M., Levy, P.S., Spengler, R., and Dunea, G. (1989). *Clin. Nephrol.*, *31*: 40.

Bell, J.D., Kubal, G., Radulovic, S., Sadler, P.J., and Tucker, A. (1993). *Analyst*, *118*: 241.

Coburn, J.W., Mischel, M.G., Goodman, W.G., and Salusky, I.B. (1991). *Amer. J. Kidney Dis.*, *17*: 708.

Dale, R.M.K., Martin, E., Livingston, D.C., and Ward, D.C. (1975). *Biochemistry*, *14*: 2447.

Domingo, J.L., Gomez, M., Llobet, J.M., and Corbella, J. (1991). *Kidney Int.*, *39*: 598.

Glew, D.N., and Hames, D.A. (1971). *Can. J. Chem.*, *49*: 3114.

Hewitt, C.D., Poole, C.L., Westervelt, F.B., Savory, J., and Wills, M.R. (1988). *Lancet*, *ii*: 849.

Kirschbaum, B.B., and Schoolwerth, A.C. (1989). *Amer. J. Med. Sci.*, *297*: 9.

Kiss, T., Sóvágó, I., and Martin, R.B. (1989). *J. Amer. Chem. Soc.*, *111*: 3611.

Kiss, T., Sóvágó, I., and Martin, R.B. (1990). *Inorg. Chem.*, *30*: 2130.

Macdonald, T.L., Humphreys, W.G., and Martin, R.B. (1987). *Science*, *236*: 183.

Martin, R.B. (1979). *Metal Ions in Biological Systems*, Vol. 8 (H. Sigel, ed.), Marcel Dekker, New York, p. 57.

Martin, R.B. (1986a). *Metal Ions in Biological Systems*, Vol. 20 (H. Sigel, ed.), Marcel Dekker, New York, p. 21.

Martin, R.B. (1986b). *Clin. Chem.*, *32*: 1797.

Martin, R.B. (1986c). *J. Inorg. Biochem.*, *28*: 181.

Martin, R.B. (1988a). *Metal Ions in Biological Systems*, Vol. 8 (H. Sigel, ed.), Marcel Dekker, New York, p. 1.

Martin, R.B. (1988b). *Handbook on Toxicity of Inorganic Compounds* (H.G. Seiler and H. Sigel, eds.), Marcel Dekker, New York, p. 9.

Martin, R.B. (1988c). *Biochem. Biophys. Res. Commun.*, *155*: 1194.

Martin, R.B. (1991a). *Aluminium in Chemistry, Biology, and Medicine* (M. Nicolini, P.F. Zatta and B. Corain, eds.), Cortina International-Verona, Raven Press, New York, p. 3.

Martin, R.B. (1991b). *J. Inorg. Biochem.*, *44*: 141.

Martin, R.B. (1992). *Aluminium in Biology and Medicine*, Ciba Foundation Symposium 169, John Wiley, Chichester, p. 5.

Martin, R.B., Savory, J., Brown, S., Bertholf, R.L., and Wills, M. (1987). *Clin. Chem.*, *33*: 405.

Martyn, C.N., Osmond, C., Edwardson, J.A., Barker, D.J., Harris, E.D., and Lacey, R.F. (1989). *Lancet*, *i*:59.

McLachlan, D.R.C., Kruck, T.P., Lukiw, W.J., and Krishnan, S.S. (1991). *Can. Med. Assn. J.*, *145*: 793.

Micronutrient Interactions: Vitamins, Minerals, and Hazardous Elements (1980). *Ann. N.Y. Acad. Sci.*, *355*.

Nelson, D.J., and Martin, R.B. (1991). *J. Inorg. Biochem.*, *43*: 37.

Ochiai, E. (1987). *General Principles of Biochemistry of the Elements*, Plenum, New York.

Öhman, L.O. (1988). *Inorg. Chem.*, *27*: 2565.

Rabenstein, D.L. (1978). *Accounts Chem. Res.*, *11*: 100.

Rabenstein, D.L., and Isab, A.A. (1982). *Biochim. Biophys. Acta*, *721*: 374.

Rabenstein, D.L., Isab, A.A., and Reid, R.S. (1982). *Biochim. Biophys. Acta*, *696*: 53.

Seiler, H.G., and Sigel, H., eds. (1988). *Handbook on Toxicity of Inorganic Compounds*, Marcel Dekker, New York.

Slanina, P., Falkeborn, Y., Frech, W., and Cedergren, A. (1984). *Food Chem. Toxicol.*, *22*: 391.

Slanina, P., Frech, W., Bernhardson, A., Cedergren, A., and Mattsson, P. (1985). *Acta Pharmacol. Toxicol.*, *56*: 331.

Slanina, P., Frech, W., Ekström, L., Lööf, L., Slorach, S., and Cedergren, A. (1986). *Clin. Chem.*, *32*: 539.

Spencer, J.N., and Voigt, A.F. (1968). *J. Phys. Chem.*, *72*: 464.

Toth, I., Zekany, L., and Brucher, E. (1984). *Polyhedron*, *7*: 871.

Underwood, E.J. (1979). *Toxicity of Heavy Metals in the Environment*, Part 2 (F.W. Oehme, ed.), Marcel Dekker, New York, p. 641.

van Ginkel, M.F., van der Voet, G.B., van Eijk, H.G., and de Wolff, F.A. (1990). *J. Clin. Chem. Clin. Biochem.*, *28*: 459.

Walker, J.A., Sherman, R.A., and Cody, R.P. (1990). *Arch. Intern. Med.*, *150*: 2037.

Weberg, R., and Berstad, A. (1986). *Eur. J. Clin. Invest.*, *16*: 428.

MOLECULAR TARGETS OF METAL TOXICITY

by

E.A. Conner and B.A. Fowler

Program in Toxicology,
University of Maryland at Baltimore,
660 West Redwood Street
Baltimore, MD 21201
U.S.A.

I. INTRODUCTION

Molecular targets of metal toxicity include DNA, RNA and proteins. The detrimental effects of toxic metals through their interactions with proteins such as enzymes, membrane and cytoskeletal proteins and high-affinity metal-binding proteins have already been covered in this volume (see Part Two, Chapter 1-E). This chapter will examine nucleic acids and attempt to delineate the molecular factors that regulate these mechanisms.

The interactions of metals with DNA and RNA are often associated with carcinogenesis. Before discussing the various interactions, certain underlying observations of metal carcinogenesis and fundamental principles will be mentioned to give some insight into the complexities of the subject. First, most data suggest that the metal or its complexes are the ultimate carcinogen. An agent of this type is a chemically and biologically reactive entity by virtue of its specific structure. Such a metal or complex can interact specifically with sensitive target elements of tissues, cells and cellular component macromolecules to yield a neoplastic state. Secondly, for most carcinogenic metals there is generally not a strong correlation between mutagenicity and carcinogenicity. The problem here is that the standard mutagenicity assays using *Salmonella typhimurium* or *Esherichia coli* may not be appropriate when using metals. Negative results are often associated with metal precipitation in cell culture, use of genetically incompatible human and bacterial cell lines, metals being co-mutagenic, and the possibility that there is no correlation between carcinogenicity and mutagenicity for metals. Lastly, the molecular mechanisms describing the roles of metals in carcinogenesis remain unclear and subject to change as more data are accumulated.

The basic chemical properties of the metal also add to the complexity of the problem. Thermodynamic principles describing ligand-dissociation constants, stability constants and free energies need to be considered as well as how a metal traverses a membrane. The question of whether the metal diffuses actively or passively or does it use an already existing transport. Arsenate and chromate are believed to diffuse across membranes by sulfate-phosphate exchange (Jennette, 1981). In the case of lead, it is hypothesized that the proximal tubule uptake of lead in the circulation is not due to endocytosis and/or passive diffusion but rather that the lead in the circulation is bound to diffusible proteins. Studies in rodents have shown that lead binds to the rat protein α_2u-globulin (Fowler and DuVal, 1991). The rat renal lead-binding protein has also been shown to facilitate the cell free nuclear translocation of lead (Mistry et al., 1985; 1986). Recently, lead-binding proteins similar in size and ion-exchange properties but that did not cross-react with the polyclonal antibody to α_2u-globulin have been isolated in the kidney and liver of humans (Kahng et al., 1992; Smith et al., 1993), suggesting an analogous mechanism of transport and cellular handling of lead in these different species. Another diffusible protein, namely metallothionein, is responsible for the transport of cadmium and other transition metals. Please refer to Part Two, Chapter 1-E for further discussion on the roles of these high-affinity metal-binding proteins in mediating the molecular mechanisms of metal toxicity.

II. MODES OF ACTION

Metals can act directly or indirectly at the molecular level. Metals which react directly with nucleic acids are called direct-acting and those metals which affect some other aspect of DNA processing are considered indirect-acting. Metals such as Cu, Pt, Mn, Ni, and Cr are generally considered to be direct-acting carcinogens. Metals of this type are often demonstrated to form cross-links between the DNA strands by binding to the bases of DNA (Marzilli et al., 1980; Eichhorn, 1973) or bring about mispairing of the bases by binding to the phosphate groups in the deoxyribosephosphate backbone. Indirect modes of action by metals on DNA include effects on DNA repair, modification of metabolic and biochemical processes and metal-metal interactions.

III. SPECIFIC INTERACTIONS OF METALS WITH NUCLEIC ACIDS

The remainder of this review will concentrate on the most recent issues concerning metals and genotoxicity. Please refer to the following for further discussion (Kasprzak, 1991; Skroch et al., 1993; Rossman et al., 1992; Costa, 1991; Martell et al., 1981).

Cr(III) creates DNA protein cross-links by binding to the reactive amino acids (i.e. cysteine, tyrosine or histidine) and linking these to the phosphate backbone of DNA (Salnikow et al., 1992). Ni(II), like Cr(III), also produces stable DNA-protein cross-links (Lin et al., 1992). Oxidative mediated Ni(II)/Ni(III) redox cycling, however, is thought to be the cause and not direct metal involvement (Lin et al., 1992). Cr(III), on the other hand, was found to be involved directly in the formation of the DNA-protein complexes (Salnikow et al., 1992). It has been suggested that both DNA-protein as well as amino acid DNA cross-links be used as biomarkers of human exposure (Toniolo et al., 1993; Costa et al., 1993).

Cr(VI), Co(II) and complexes of Ni(II) with certain peptides and proteins are capable of reacting with H_2O_2, organic peroxides and/or O_2 to produce oxygen-derived radicals (Shi et al., 1992; 1993a; 1993b). Hydroxyl free radical generation is also implemented in site-specific DNA damage induced by hydrazine in the presence of Mn and Cu ions (Yamamoto and Kawanishi, 1991). The modified DNA bases identified in some of these studies are suggested to play a major role in metal ion-induced carcinogenicity.

Cadmium can bind to the bases as well as the phosphate groups of DNA (Jacobson and Turner, 1980) but whether or not it acts directly or indirectly remains to be determined. Recent findings have described the ability of Cd-MT complex to cause DNA strand breaks (Muller et al., 1991). Other studies have shown cadmium treatment inducing a "pro-oxidant state" by causing a depletion of cellular glutathione (Ochi et al., 1987). Cadmium can also inhibit DNA repair (Mandel and Ryser, 1984; Synder, 1990), and can therefore act as a comutagen (Hartwig and Beyersmann, 1989). For further discussion on the genotoxicity of cadmium, refer to recent review by Rossman et al. (1992).

IV. ROLES OF METAL-TRANSCRIPTIONAL FACTOR COMPLEXES

The induction of metallothionein by heavy metal ions is mediated by four metal response elements (MREs) which are recognized by a metal response factor (MRF). Using genetic selection in yeast, Skroch et al. (1993) have recently isolated two potential cDNA clones that could encode the human MRF. The primary structure predicted by these cDNAs suggest that the MFR is a metal-binding protein.

Studies by Mistry et al. (1985; 1986) suggested that the high-affinity lead-binding proteins in kidney are also capable of acting in a manner similar to receptor proteins to facilitate the intranuclear translocation of lead and chromatin binding of lead, resulting in altered renal gene expression (see Fowler et al., 1994 for a review). The demonstrated existence of these proteins in specific target cell populations of the renal proximal tubule (Fowler and DuVal, 1991) suggest that these proteins may play a role in mediating the preferential toxicity of lead for these cells. Further studies are needed to delineate the exact DNA regions that are the molecular targets for these proteins.

Studies with arsenite (Van Wijk et al., 1993; van Delft et al., 1993) have shown that this agent is capable of stimulating cell cycle progression and increasing the expression of the HSP 68, HSP 84 and the protooncogene c-fos through a presently unidentified serum response element (SRE). In addition, these studies (van Delft et al., 1993) have shown that casein kinase II may also be involved in the SRE induction of c-fos by arsenite and cadmium. Further studies are needed to identify the SRE and determine the mechanism by which it mediates intracellular induction of these gene products following exposure to arsenite and cadmium. The role(s) of ancillary factors such as casein kinase II in mediating genetic responses to metals in target tissues also needs to be delineated.

V. CONCLUSION

From the above discussion, it is clear that metals have a number of intracellular molecular targets which may play major roles in mediating not only central mechanisms of cell injury/cell death, but also carcinogenicity. Data from some studies suggest direct metal ion interactions with molecules such as DNA as a primary mechanism while others indicate the involvement of intermediary transcriptional factors that alter gene expression within target cell populations. More recent studies have also demonstrated that metals capable of producing oxygen radicals withing cells may damage molecular targets through this mechanism. It is clear that the number of molecular targets for metal toxicity will continue to grow as our understanding of these processes increases. The possibility that more than one mechanism may be operating must also be considered.

REFERENCES

Costa, M. (1991). *Annu. Rev. Pharmacol. Toxicol.*, *31*: 321.

Costa, M., Zhitkovich, A., Taioli, E., and Toniolo, P. (1993). *J. Toxicol. Environ. Health*, *40*: 217.

Eichhorn, G.L. (1973). *Inorganic Chemistry*, Elsevier Scientific Publ. Co., Amsterdam.

Fowler, B.A., and DuVal, G.E. (1991). *Environ. Health Perspect.*, *91*: 77.

Fowler, B.A., Kahng, M.W., and Smith, D.R. (1994). *Environ. Health Perspect.*, in press.

Hartwig, A., and Beyersman, D. (1989). *Biol. Tr. Elem. Res.*, *21*: 359.

Jacobson, K.B., and Turner, J.E. (1980). *Toxicology*, *16*: 1.

Jennette, K.W. (1981). *Environ. Health Perspect.*, *40*: 233.

Kahng, M.W., Conner, E.A., and Fowler, B.A. (1992). *The Toxicologist*, *12*: 788.

Kasprzak, K.S. (1991). *Chem. Res. Toxicol.*, *4*: 604.

Lin, X., Zhuang, Z., and Costa, M. (1992). *Carcinogenesis*, *13*: 1763.

Mandel, R., and Ryser, H.J.P. (1984). *Mutat. Res.*, *18*: 9.

Martell, A.E. (1981). *Environ. Health Perspect.*, *40*: 27.

Marzilli, L., Kistenmacher, T., and Eichhorn, G.L. (1980). *Nucleic Acid Metal-Ion Interactions*, John Wiley and Sons, New York.

Mistry, P., Lucier, G., and Fowler, B.A. (1985). *J. Pharmacol. Exp. Ther.*, *232*: 462.

Mistry, P., Mastri, C., and Fowler, B.A. (1986). *Biochem. Pharmacol.*, *53*: 711.

Muller, T., Schuckelt, R., and Jaenicke, L. (1991). *Arch. Toxicol.*, *65*: 20.

Ochi, T., Takahashi, K., and Oshawa, M. (1987). *Mutat. Res.*, *180*: 257.

Rossman, T.G., Roy, N.K., and Lin, W. (1992). *Cadmium in the Environment : Toxicity and Carcinogenicity*, Publ. No. 118 (G.F. Nordberg, R.F.M. Heber and L. Alesso, eds.), IARC Lyon, p. 367.

Salnikow, K., Zhitkovich, A., and Costa, M. (1992). *Carcinogenesis*, *13*: 2341.

Shi, X., Dalal, N.S., and Kasprzak, K.S. (1992). *Arch. Biochem. Biophys.*, *299*: 154.

Shi, X., Dalal, N.S., and Kasprzak, K.S. (1993a). *J. Inorg. Biochem.*, *50*: 211.

Shi, X., Dalal, N.S., and Kasprzak, K.S. (1993b). *Arch. Biochem. Biophys.*, *302*: 294.

Skroch, P., Buchman, C., and Karin, M. (1993). *Prog. Clin. Biol. Res.*, *380*: 113.

Smith, D., Kahng, M.W., Conner, E.A., and Fowler, B.A. (1993). *The Toxicologist*, *13*: 440.

Synder, R.D. (1990). *Biological Effects of Heavy Metals*, Vol. II, CRC Press, Boston, p. 77.

Toniolo, P., Zhitkovich, A., and Costa, M. (1993). *Int. Arch. Occup. Envir. Health*, *65*: S87.

Van Delft, S., Coffer, P., Kruijer, W., and Van Wijk, R. (1993). *Biochem. Biophys. Res. Commun.*, *197*: 542.

Van Wijk, R., Welters, M., Souren, J.E.M., Ovelgonne, H., and Weigant, A.C. (1993). *J. Cell Physiol.*, *155*: 265.

Yamamoto, K., and Kawanishi, S. (1992). *Chem. Res. Toxicol.*, *5*: 440.

CHEMICAL MECHANISMS OF METAL CARCINOGENESIS

by

Max Costa

New York University Medical Center,
Nelson Institute of Environmental Medicine,
550 First Avenue
New York, NY 10016
U.S.A.

I. INTRODUCTION

The differences in chemical reactions that probably contribute to the carcinogenesis of relevant metals and their compounds are more diverse than has been previously realized. Metals that are carcinogenic to humans include chromate and its compounds as well as compounds containing nickel and cadmium (Bhamra and Costa, 1992; Cohen et al., 1990). There is also evidence that beryllium is carcinogenic to humans although very little work has been done on the mechanisms (Bhamra and Costa, 1992; Cohen et al., 1990). Lead produced renal cancers at very high doses in rats. However, recent speculation has suggested that these cancers developed in this organ as a result of the toxicity produced by lead (Cohen et al., 1990). Two metals could not be more different in terms of their chemistry than nickel and chromate (Cohen et al., 1990). Therefore, it is probably best to confine a discussion of the chemical mechanisms of metal carcinogenesis to individual metals and their compounds.

The study of metal carcinogenesis challenges basic knowledge and understanding of mechanisms involved in DNA damage, mutagenesis, DNA replication, DNA repair, DNA methylation protein-DNA and protein-protein interaction (Bhamra and Costa, 1992; Cohen et al., 1990). Metals interact with many tissue and cellular components, membrane, organelles, proteins, metabolic enzymes, replicative and repair enzymes, nucleic acids, amino acids and nucleotides and it is, therefore, difficult to point to a critical interaction that is more important in the carcinogenesis of each metal compound (Klein et al., 1991a). One can only speculate which effect is more important than the others based upon the selectivity of the injury and its relevance to carcinogenesis. Mounting experimental evidence in recent years has pointed to oxidative processes in metal carcinogenesis and thus, many metal compounds may indirectly produce their genotoxic effects by generating oxygen radicals (Klein et al., 1991a). A number of recent reviews have been written on this topic and certainly it is an area that has received considerable interest and attention in recent years (Standeven and Wetterhahn, 1991; Kasprzak, 1991).

The metal carcinogenesis process can be divided into a number of steps for general consideration. The first involves exposure to carcinogenic metals which, as shown in Table 1, may be occupational as well as environmental. Occupational exposure to the carcinogenic metals is primarily by inhalation whereas environmental involves both inhalation and

Table 1

Exposures Relevant to Carcinogens

Occupational	Environmental	
Ni, Cr, Pb, Cd, Be	Cr, Pb, Cd	
↓	↙	↘
Inhalation	Inhalation,	Ingestion

ingestion. The ability of metals to penetrate cells in the body at the site of exposure (i.e. by inhalation) is of the utmost importance in determining the carcinogenic potential of the metal. A metal that is potentially genotoxic must have entered the cell and reacted with chromatin in order to produce its carcinogenic response. Metal carcinogenesis almost solely depended upon the bioavailability of the metal ion to sites such as chromatin (Cohen et al., 1990).

Table 2 describes some general cellular uptake phenomenon by different types of carcinogenic metal compounds. For example, metal compounds with low cellular uptake are weakly carcinogenic. It is the case of the water-soluble Ni salts which are not easily taken up into cells (Coogan et al., 1989), of trivalent Cr compounds that do not effectively penetrate cells, and of certain particulate water-insoluble Ni compounds that are not well phagocytized and thus do not enter the cell (Heck and Costa, 1986; Costa, 1991a). In contrast, potent metal carcinogens are those that enter the cell well, such as phagocytized crystalline nickel sulfides, water-soluble hexavalent Cr compounds, and many of the other metal compounds of carcinogenic concern that are well taken up by cells (Table 2).

Once a metal enters the cell, it changes oxidation states. Table 3 illustrates examples of metabolism of metal compounds after they have entered the cell. Very often, it is the metabolism of the metal compound that yields the genotoxic species. For example, hexavalent Cr compounds are taken up by the sulphate transport system and are reduced in the cell to trivalent Cr compounds and during this reduction process, reactive oxygen and reactive Cr intermediates are formed (Standeven and Wetterhahn, 1991). There are other examples of redox reactions occurring inside the cell that generate reactive species. This will be discussed further in a separate section dealing with each metal. It should be noted that metal metabolism is not limited to intracellular sites, since rapid extracellular reduction

Table 2

General Cellular Uptake

Low	Adequate or high
1) Highly water soluble Ni salt	1) Phagocytized crystalline nickel sulfides water soluble
2) Trivalent Cr	2) Hexavalent Cr compounds
3) Non-phagocytized nickel containing particles	3) Phagocytized trivalent Cr containing particles
	4) Cd compounds
	5) Pb compounds

Table 3

Examples of Metabolism of Carcinogenic Metals

1)	Hexavalent Cr reduction to trivalent by ascorbic acid, glutathione, etc
2)	$Ni^{2+} \rightarrow Ni^{3+}$ by very strong oxidants (H_2O_2) when Ni^{2+} is bound to certain ligands (i.e. histidine)

of chromate to the trivalent state can attenuate its uptake. Following phagocytosis of Ni_3S_2, the particles are contained in a cytoplasmic vacuole that is highly acidified, which promotes the intracellular dissolution of Ni^{2+} ions from Ni_3S_2 particles. Dissolution is another example of intracellular metabolism.

Table 4 lists the various genetic events that may be involved in metal carcinogenesis, along with the metal ions (in order of potency) that have been shown to have these genetic effects in cells (Snow, 1992). As can be seen from an examination of this table, Cr compounds appear to be active at many of the genetic endpoints described, whereas some metal ions such as beryllium are not often mentioned simply because they have not been well studied with respect to all these endpoints. These genotoxic events lead to altered gene expression or altered genetic information, which is thought to be one of the important events in metal carcinogenesis. Recent evidence has implicated tumor-suppressor genes as extremely important in the carcinogenesis process in general, and recent studies indicated that Ni compounds can, in fact, induce mutations in tumor-suppressor and senescense genes (Klein et al., 1991b). The fact that metals transformed cells directly in vitro indicated that they were directly carcinogenic and even though some metals may have had indirect carcinogenic effects such as suppressing the immune system, their ability to directly transform cells was an important indicator that they were a direct-acting type of carcinogen (Klein et al., 1991a). The fact that so many different genetic processes were affected by metals indicated that they interfered with cellular genetic transfer at many levels (Snow, 1992).

Essential metal ions play a very important role in regulatory enzymes dependent on genetic transfer as well as maintaining chromatin structure (Bhamra and Costa, 1992). These essential metal ions include zinc, magnesium, manganese, and calcium as well as others, and Table 5 lists the carcinogenic metal ions that interfere with the function of the indicated essential metal ions because they have similar chemical properties (Bhamra and Costa, 1992). Consequently, the actions of carcinogenic metals are at a number of levels

Table 4

Genetic Events in Metal Carcinogenesis

1)	DNA binding (Cr >> Ni, Be, Cd)
2)	Effects on DNA replication (Cr, Ni, Be, Cd)
3)	Effects on specific DNA-protein interaction (Ni, Cd, Pb)
4)	DNA strand breaks (Cr, Ni, Cd)
5)	DNA-protein crosslinks (Cr, Ni)
6)	Oxidative DNA damage (Cr, Ni, Cd)
7)	Bacterial mutagenesis (Cr)
8)	Mammalian cell mutagenesis (Cr, Ni, Cd)
9)	Chromosomal aberrations (Cr, Ni, Cd)
10)	Inhibition of DNA repair (Cr, Ni)
11)	Direct transformation of cells in vitro (Cr, Ni, Be, Cd)
12)	Effects on the immune system (Cr, Ni, Be, Cd)

Table 5

Similarity of Carcinogenic Metals with Essential Metals and Potential Antagonism

Carcinogenic metal ion	Essential metal ion	Antagonism of genotoxicity by essential ions
Ni^{2+}	Mg^{2+}, (Mn^{2+})	Yes
Cd^{2+}	Zn^{2+}, Ca^{2+}	Yes?
Pb^{2+}	Ca^{2+}	?
CrO_4	SO_4^{2-}	?

since essential metal ions perform so many diverse functions in the cells, serving as cofactors for enzymes or being involved in the structure of proteins such as the zinc fingers. For example, DNA polymerase activity is highly dependent upon magnesium and manganese. The substitution of nickel for magnesium affects the ability of the polymerase to replicate properly (Snow, 1992).

II. CHROMATE

The most toxic and carcinogenic form of chromium (Cr) is the hexavalent form which at physiological pH exists as an oxyanion (Snow, 1992; Klein et al., 1991b). The other stable form of Cr is the trivalent form (Miller and Costa, 1989a). The difference in uptake into many different types of cells between the hexavalent and the trivalent form is approximately 1000-fold (Miller and Costa, 1989a; Costa, 1991b). The reason for this is that the hexavalent form being an oxyanion uses anion transport systems to be actively transported into cells whereas the trivalent form of Cr by comparison exhibits little cell penetration (Miller and Costa, 1989a). The trivalent form of Cr exhibits some interesting chemical properties. It binds very tightly to a number of ligands such as sulfur and oxygen and once it is bound to these ligands, it is very resistant to ligand exchange reactions. This is in contrast to many other metal compounds which undergo extensive ligand exchange reactions. Thus, subcellular localization of trivalent Cr is more conclusive in view of its rather stable interactions with ligands. This is not, however, true of other forms of chromate. Once the hexavalent form of Cr is actively transported into the cell, it undergoes reduction reactions, ultimately forming the highly stable trivalent Cr through intermediates of Cr(V) and Cr(IV) (Miller and Costa, 1989a).

In the past, a number of enzymatic and non-enzymatic reductants such as microsomal enzymes, gluthathione and ribonucleotides have been proposed as major pathways to hexavalent chromate reduction (Standeven and Wetterhahn, 1991). However, ascorbic acid has recently been suggested as the major intracellular reductant of hexavalent Cr (Standeven and Wetterhahn, 1991; Snow, 1992). Various intermediate oxidation states of hexavalent Cr have been described. Based upon electron spin resonance, pentavalent Cr is known to be formed during the reduction of hexavalent Cr (Standeven and Wetterhahn, 1991). The formation of Cr(IV) has also been proposed (Standeven and Wetterhahn, 1991). The reduction of hexavalent Cr to trivalent Cr also results in the generation of oxygen radicals, and oxidative damage to DNA has been found following treatment with hexavalent Cr (Standeven and Wetterhahn, 1991). However, in addition to this oxidant damage, the ultimate reductant of hexavalent Cr is the trivalent form which binds to DNA primarily at the phosphate backbone and coordinates the crosslinking of proteins and glutathione to the DNA phosphate backbone (Costa, 1991b). Recent studies suggest that trivalent Cr binds to terminal phosphates on DNA, not internal phosphates.

The DNA-protein crosslinks produced by chromate have been found to be relatively stable lesion and not readily repaired (Costa, 1991b). This lesion has been studied by a variety of techniques including alkaline elution, filter binding, Western transfer, and immunological methods (Costa, 1990). An antibody has been made to proteins crosslinked to DNA by Cr and this antibody recognized a 95 kg mol^{-1} protein, which was not one of the original proteins identified as part of the crosslink by silver staining (Costa, 1990; 1991b). The reason for this was that this protein was not stained by silver, but it elicited antibody formation. DNA-protein crosslinks were detected by Western blotting and by slot blotting (Miller and Costa, 1989b). It has been very difficult to study the DNA-protein crosslink lesion induced by Cr compounds because of background problems of proteins associated with the DNA. There are a number of proteins that are covalently bound to the DNA in DNA in normal cells (Costa, 1990; 1991b; Miller and Costa, 1989b). My laboratory developed a new method to analyze DNA-protein crosslinks based upon the selective precipitation of DNA containing crosslinked proteins with potassium SDS (Zhitkovich and Costa, 1992). This represented a simpler assay for DNA-protein crosslinks but, due to background, analysis of proteins crosslinked to DNA continued to be a problem. There were a number of residual proteins that remained tightly associated with DNA isolated by proteinase K and phenol extraction.

Another approach taken in my laboratory examined residual amino acids or peptides associated with DNA in intact cells treated with chromate. These studies revealed that cysteine, tyrosine, histidine, methionine and threonine were complexed more than other amino acids to DNA in intact cells treated with chromate (Lin et al., 1992). Interestingly, when trivalent Cr was reacted with these amino acids, the same ones that showed higher activity in intact cells were also more active in being complexed to DNA in vitro (Salnikow et al., 1992). The exact site on the amino acid that was complexed through trivalent Cr has not been studied. However, one can speculate that the sulfur groups of cysteine and the imidazole nitrogen of histidine, as well as the hydroxyl groups of tyrosine and threonine, were probably important since Cr(III) binds to oxygen, nitrogen and sulfurs (Salnikow et al., 1992). Cr(III) in turn was bound to DNA through terminal phosphates (Salnikow et al., 1992). These amino acids reacted with Cr directly while they were on proteins and also by direct interaction between free amino acids and DNA through trivalent Cr as well. Since glutathione was present in many cells at mM concentrations and this tri-peptide contained one molecule of cysteine, it reacted with chromate forming DNA-Cr(III) glutathione adducts (Borges and Wetterhahn, 1989). These crosslinks were extremely stable due primarily to the relative inertness of Cr(III) to ligand substitution reaction. As with proteins, there were also residual amino acids tightly associated with DNA in untreated cells (Lin et al., 1992).

Crosslinks can also be formed between proteins and DNA through the generation of oxygen radicals. The radicals oxidized DNA bases and proteins that became reactive towards each other, or superoxide or ·OH radicals generated lipid peroxidation which resulted in the formation of malondialdehyde which in turn produced crosslinking of proteins to DNA (Klein et al., 1991a; Standeven and Wetterhahn, 1991; Kasprzak, 1991). In several isolated instances, the stability of DNA-protein crosslinks has been measured and results indicated that chromate-induced DNA-protein crosslinks were extremely stable and not readily repaired (Costa, 1990; 1991b). They actually accumulated with time after exposure to chromate. Thus, they were very slow to form and very slow to disappear in cells, making them a potentially important lesion in the carcinogenesis process. Since the DNA-protein crosslink is an extremely bulky lesion, it is likely to lead to deletions of DNA sequences during DNA replication (Snow, 1992). This lesion was also utilized as a biomarker of Cr exposure (Zhitkovich and Costa, 1992).

The generation of oxygen radicals has also been shown to produce DNA strand breaks. However, strand breaks produced by chromate were alkaline labile and not readily repaired. These lesions may also have contributed to chromate carcinogenic activity (Klein et al., 1991a; Standeven and Wetterhahn, 1991).

Cr(III) does not form adducts with any particular base specificity. Instead, it reacts with the phosphate backbone of the DNA and thus, specific DNA adducts formed by trivalent Cr are not likely to be an important factor in chromate carcinogenesis (Salnikow et al., 1992). In view of all the different reactive intermediates that can be formed and the varying types of genetic lesions produced, hexavalent Cr compounds have been shown to be more positive in more genetic test endpoints than any other human carcinogen (IARC, 1990).

In studying the reactions of trivalent Cr with DNA and protein, it was noted that cysteine was a major component in the DNA-protein crosslinks. This was done by radiolabeling cells with amino acids, treating cultured cells with chromate compounds, and subsequently examining DNA isolated following proteinase K phenol extraction with subsequent study of amino acids complexed to DNA (Lin et al., 1992). Cysteine was one of the major amino acids whose complexing to DNA was increased by chromate exposure, and experiments using cyclohexamide to inhibit protein synthesis still found that cysteine was complexed to DNA, suggesting that most of the cysteine complexed to DNA originated from glutathione and not proteins (Lin et al., 1992). Since cysteine is usually buried in proteins and not accessible for crosslinking reactions, this explanation is quite feasible.

This discussion of the molecular mechanisms of chromate carcinogenesis left out the important understanding of the appropriate target tissues for chromate-induced cancer. In human epidemiological studies, lung cancers have been associated with chromate exposure since most exposures occur by inhalation (IARC, 1990). However, the ability of a tissue to reduce hexavalent chromate obviously play a very important role in mediating its genotoxicity. It is not, however, only the reduction of hexavalent chromate that is important but also its rate of reduction since this rate influenced the type of genetic lesions found (Snow, 1992).

III. NICKEL

The ability of nickel (Ni) ions to gain entry inside the cell seems to have been one of the most important determinants of its carcinogenic potency (Costa, 1991a). Thus, water-insoluble Ni compounds are generally not carcinogenic in mammalian systems because they do not readily enter cells (Costa, 1991a). However, certain water-insoluble Ni compounds following inhalation exposure are very potent carcinogens because they are phagocytized by cells (Costa, 1991a). Thus, it has been proposed that the carcinogenic activity of particulate Ni compounds is proportional to their cellular uptake by phagocytosis (Costa, 1991a). There have been extensive studies examining the important determinants of the cellular uptake of particulate Ni compounds. The crystalline NiS particles that were potently carcinogenic and actively taken up had a negative surface charge while compounds that were not taken up, such as amorphous NiS, had a positive surface charge (Costa, 1991a). By changing the surface charge from positive to negative, the phagocytosis of the amorphous NiS compounds was activated and these negatively charged particles exhibited a carcinogenic potency similar to that of the crystalline NiS compound (Costa, 1991a).

NiO compounds were also taken up by phagocytosis. However, their uptake was not as easy to study as that of the crystalline NiS particles. Since NiO compounds exhibited a dissolution halftime of > 11 years in water and months to several years in serum or renal cytosol, Ni ions were likely to be lower in cells containing phagocytized NiO. Their carcinogenic activity may have depended less upon the dissolution of Ni and its delivery to DNA as did the carcinogenic activity of other more soluble Ni compounds (Ni_3S_2) (Sunderman et al., 1987).

A popular notion with regard to the mechanism of Ni carcinogenesis has been that the carcinogenic activity depended upon the concentration of Ni^{2+} ions inside the cell. In the past, Ni^{2+} oxidation/reduction reactions have not been considered important in the mechanisms of Ni carcinogenesis. This has been a rather perplexing problem since Ni ions do not form strong covalent bonds with DNA and while the binding constant for protein is considerably higher than that for DNA, its interaction with protein is not stable. The affinity constant for Ni^{2+} binding to DNA bases is approximately $K_a = 10^2$, whereas the affinity constant of Ni^{2+} for the imidazole nitrogen of histidine is 10^9 (Sillén and Martell, 1971). The affinity constant for many other amino acids is from 10^5 to 10^9, indicating that there is anywhere from 3 to 7 orders of magnitude more affinity of Ni for amino acids on proteins than for DNA (Sillén and Martell, 1971). Therefore, the interaction of Ni with proteins on DNA was probably a most important chemical reaction in the carcinogenic response rather than the interaction of Ni with DNA directly (Costa, 1991a). This is in contrast to many other types of carcinogens which interact with DNA. In particular, Ni ions have been shown to have a very high affinity for heterochromatin regions (Costa, 1991a). These regions contained a high concentration of highly condensed DNA-protein structure and perhaps the more compact nature of the proteins present in this structure and/or the binding of magnesium to this structure for which Ni has very similar ionic radii, may have caused enhanced Ni interaction with heterochromatin. The binding of Ni^{2+} in place of magnesium for these proteins or the interaction of Ni with these proteins at Non + Mg^{2+} binding sites probably led to heterochromatin decondensation (Costa, 1991a; Klein et al., 1991b).

Another important effect of Ni binding to protein (i.e. imidazole nitrogen of histidine), is that it was oxidized from Ni^{2+} to Ni^{3+} by various oxidizing agents such as hydrogen peroxide (Klein et al., 1991a; Kasprzak, 1991). The oxidation of Ni^{2+} to Ni^{3+} generated hydroxyl and other oxygen radicals which produced genetic damage either directly by oxidizing the DNA bases or indirectly through lipid peroxidation and the formation of a secondary intermediate, i.e. malondialdehyde (Klein et al., 1991a; Kasprzak, 1991). Malondialdehyde is a potent crosslinker of protein to protein, and protein to DNA. Consequently, it has been hypothesized that oxygen reactive intermediates which in turn produce oxidized DNA damage and form DNA-protein crosslinks indirectly subsequent to the formation of malondialdehyde were mediators of nickel's genotoxicity. In fact, many of the DNA-protein crosslinks formed by Ni have been shown to involve covalent linkages and not directly involve the metal ion participating in the crosslink (Klein et al., 1991b). Thus, it was believed that Ni was inducing its genetic damage through interactions with proteins, generating oxygen radicals by its binding to such proteins. The enhanced formation of oxygen radicals produced oxidized bases such as 8-hydroxy-guanine, which has been shown to increase following Ni exposure (Kasprzak, 1991). These reactive oxygen intermediates also elevated malondialdehyde levels subsequent to lipid peroxidation resulting in covalent crosslinks of proteins to DNA (Kasprzak, 1991).

A DNA-protein crosslink represents a bulky lesion that, if DNA replicates in its presence, deletions of DNA sequences will ensue. It has been known in recent years that Ni induced deletions of DNA sequences such as inactivation of a senescence gene (Klein et al., 1991b). In Chinese hamster cells, the longest and most contiguous region of heterochromatin was found on the long arm of the X chromosome. Figure 1 below shows the selective damage produced by Ni in the heterochromatin long arm of the X chromosome, while Figure 2 shows a deletion of this region in Ni-transformed male Chinese hamster cells (Conway and Costa, 1989; Sen and Costa, 1986). The same region that was damaged in 1-2% of the cells treated with Ni compounds led to a deletion of this region in over half of the male Ni-transformed cells. Subsequent studies revealed that the X chromosome region deleted was involved in senescence of cells (Klein et al., 1991b). Thus, Ni-induced immortalization of cells involved the deletion of this senescence gene.

IV. CADMIUM

Cadmium (Cd) ions are extremely soft metal ions with very high reactivity towards anions. In fact, the chemistry of Cd and mercury (Hg) are quite similar. However, Hg has not been sufficiently implicated as a carcinogen (Bhamra and Costa, 1992). Similarly, the data for Cd-inducing human cancers is very weak and inconclusive but is discussed here primarily because of its ability to induce tumors in experimental animals (Bhamra and Costa, 1992). The ability of Cd to complex anions is so strong that complexed anionic species form in solution. In experimental animals and humans, Cd has been associated with cancers of the prostate, lung and testes, as well as at the sites of injection in experimental animals (Bhamra and Costa, 1992). The ability of Cd to induce cancers in the testes and prostate is related in part to the lack of metallothionein in these organs (Bhamra and Costa, 1992). While Cd has been shown to produce some genotoxic lesions in cells, these generally occur at very high concentrations (Rossman et al., 1992). These concentrations are well beyond the levels needed, for example, to induce metallothionein and produce other cytotoxic effects in cells. Therefore, the carcinogenic activity of Cd is puzzling and perplexing and is lacking in explanation. Cd is a very toxic metal ion and its carcinogenic effects are probably less important than its toxicity in vivo which primarily involves the kidney.

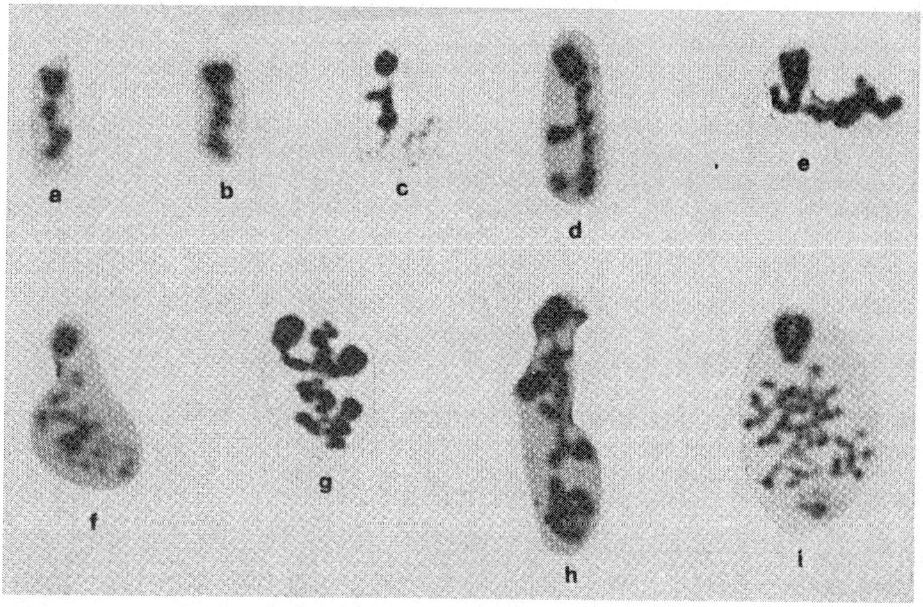

Figure 1.
Photograph of Chinese hamster ovary cell X chromosome after cells have been treated with crystalline NiS particles. Cells were treated with crystalline NiS particles ranging in concentration from 10 to 20 mg/ml for 24 to 48 hrs. The figure shows increasing degree of fragmentation of the long arm of the X chromosome with higher dosages and longer exposure times (a-i). Note the absence of damage to the euchromatic short arm and the selective damage produced in the heterochromatic long arm. Reprinted from Sen and Costa (1986), with permission.

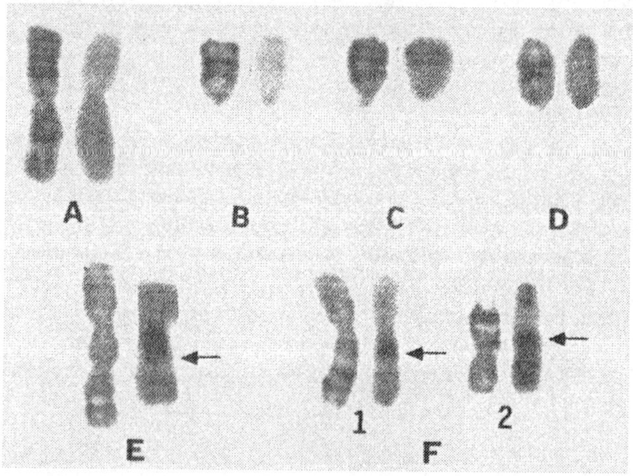

Figure 2.
G- (left) and C-banding (right) profiles of the altered X chromosomes from the nickel-transformed SA clones. (*A*) normal X chromosome; (*B-D*) abnormal X chromosomes from the male clones, Ni-2/9, Ni-3/1, and Ni-4/1, respectively, exhibiting complete deletions of the heterochromatic X long arm; (*E*) abnormal X chromosome from the male clone, Ni-1/10, showing the partial Xq deletion and subsequent X:5 translocation; (*F*, 1 and 2) female clone, Ni-10/1, showing the two derivative chromosomes resulting from the X:5 reciprocal translocation. Translocation breakpoints in (*E*) and (*F*) are designated by *arrows*. The breakpoint on chromosome 5 in (*E*) and (*F*) is at band 5q3. Reprinted from Conway and Costa (1989), with permission.

Cd is similar to Ni in that it has a very low affinity for DNA interactions and a very high affinity for proteins. Thus, it is likely that Cd produced genetic damage through interactions with proteins in chromatin. Alternatively, Cd may also have been able to increase the production of oxygen radicals in cells, probably secondarily to binding to proteins in chromatin or to depletion of glutathione creating a pro-oxidant state (Rossman et al., 1992).

V. BERYLLIUM

Although beryllium has been implicated as a human carcinogen, the delayed immunotoxic response in the lung associated with beryllium disease was a much more prevalent concern than its ability to induce cancers (Kuschner, 1981; Leonard and Lauwerys, 1987; ATSDR, 1987). The mechanism by which beryllium interacts with DNA and other proteins is not well studied. The chemistry of beryllium is extremely complicated. Its ability to elicit a strong immune response may be the important factor in its carcinogenesis in vivo or its effects on the immune system causing hypersensitivity reaction such as occurred during beryllium disease in the lung (Cohen et al., 1990). It can inhibit various enzymes, such as alkaline phosphatase and other phosphate-binding enzymes (Cohen et al., 1990). However, the mechanism of beryllium toxicity and carcinogenicity certainly needs further investigation (Cohen et al., 1990).

VI. LEAD

There have been a few studies that have implicated lead in an increased incidence of cancers in humans. However, most studies tend to show a less than significant or marginally significant carcinogenic response of lead in humans (Cooper and Gaffey, 1975; Cooper, 1976; McMichael and Johnson, 1982). Lead is very similar to calcium and thus, its ability to antagonize calcium effects and to interact with calcium-dependent processes is probably responsible in large part for its diverse toxicological effects in vivo (Cohen et al., 1990). Lead affects nerve transmission, inhibits enzymes such as those that are important in porphyrin biosynthesis (ALA-dehydratase). Lead even caused the precipitation of nuclear proteins in cells forming intranuclear inclusion bodies at very high concentrations where tumors were seen in vivo, such as in kidney cells of rats treated with high doses of metal (Cohen et al., 1990).

The carcinogenic response of lead has been largely attributed to its renal toxicity (Koller et al., 1985; Schroeder et al., 1965; Zollinger, 1953). Thus, lead only produced cancers in the kidney in rats at very high levels that themselves produced overt renal damage. The effects of lead on various endpoints related to carcinogenicity were perplexing (Cohen et al., 1990). Lead increased cell transformation in vitro (Cohen et al., 1990). There was some mutagenic response of lead although genetic lesions were low with many lead compounds (Cohen et al., 1990). Lead did, in fact, increase the incidence of chromosomal aberrations slightly but these effects, again, were not striking and not consistent with its carcinogenic response. I think the major conclusion with lead is that it is far more toxic than it is carcinogenic and thus, toxicity to the nervous system and to the HEME biosynthetic system and to other systems in the body overshadowed its carcinogenic response (Cohen et al., 1990).

REFERENCES

Agency for Toxic Substances and Disease Registry (ATSDR) (1987). *Toxicology Profile for Beryllium*.

Bhamra, R.K., and Costa, M. (1992). *Environment Toxicants, Human Exposures* (M. Lippmann, ed.), Van Nostrand Reinhold, New York, p. 575.

Borges, K.M., and Wetterhahn, K.E. (1989). *Carcinogenesis, 10*: 2165.

Cohen, M.D., Latta, D., Coogan, T., and Costa, M. (1990). *Biological Effects of Heavy Metals, Mechanisms of Metal Carcinogenesis* (E. Foulkes , ed.), CRC Press, Boca Raton, Florida, p. 19.

Conway, K., and Costa, M. (1989). *Cancer Res., 49*: 6032.

Coogan, T.P., Latta, D.M., Snow, E.T., and Costa, M. (1989). *CRC Crit. Rev. Toxicol.*, *19*: 341.

Cooper, W.C., and Gaffey, W.R. (1975). *J. Occup. Med.*, *17*: 100.

Cooper, W.C. (1976). *Ann. NY Acad. Sci.*, *271*: 250.

Costa, M. (1990). *J. Cell. Biol.*, *44*: 127.

Costa, M. (1991a). *Annu. Rev. Pharmacol. Toxicol.*, *31*: 321.

Costa, M. (1991b). *Environ. Health Perspect.*, *92*: 45.

Heck, J.D., and Costa, M. (1986). *Metal Ions in Biological Systems*, Vol. 20 (H. Sigel, ed.), Marcel Dekker, New York, p. 279.

IARC (1990). *Monograph on the Evaluation of Carcinogenic Risks to Humans, Chromium, Nickel and Welding*, Vol. 49, Lyon, France.

Kasprzak, K.S. (1991). *Chem. Res. Toxicol.*, *4*: 604.

Klein, C.B., Frenkel, K., and Costa, M. (1991a). *Chem. Res. Toxicol.*, *4*: 592.

Klein, C.B., Conway, K., Wang, X.W., Bhamra, R.K., Lin, X., Cohen, M.D., Annab, L., Barrett, J.C., and Costa, M. (1991b). *Science*, *251*: 796.

Koller, L.D., Karkvliet, N.L., and Exon, J.H. (1985). *Toxicol. Pathol.*, *13*: 50.

Kuschner, M. (1981). *Carcinogenesis*, *40*: 101.

Leonard, A., and Lauwerys, R. (1987). *Mutat. Res.*, *186*: 35.

Lin, X., Zhuang, Z., and Costa, M. (1992). *Carcinogenesis*, *13*: 1763.

McMichael, A.J., and Johnson, H.M. (1982). *J. Occup. Med.*, *24*: 375.

Miller, C.A., III, and Costa, M. (1989a). *Carcinogenesis*, *10*: 667.

Miller, C.A., III, and Costa, M. (1989b). *Mol. Toxicol.*, *2*: 11.

Rossman, T.G., Roy, N.K., and Lin, W.-C. (1992). *IARC Publication No. 118* (G.F. Nordberg, L. Alessio and R.S.M. Herber, eds.), Lyon, France.

Salnikow, K., Zhitkovich, A., and Costa, M. (1992). *Carcinogenesis*, *13*: 2341.

Schroeder, H.A., Balassa, J.J., and Vinton, W.H., Jr. (1965). *J. Nutr.*, *86*: 51.

Sen, P., and Costa, M. (1986). *Toxicol. Appl. Pharmacol.*, *84*: 278.

Sillén, L.G., and Martell, A.E. (1971). *Stability Constants Supplement No. 1*, Special Publication 25, Chemical Society, London.

Snow, E.T. (1992). *Pharmacol. Ther.*, *53*: 31.

Standeven, A.M., and Wetterhahn, K.E. (1991). *Chem. Res. Toxicol.*, *4*: 616.

Sunderman, D.W., Jr., Hopfer, S.M., Knight, J.A., McCully, K.S., Cecutti, A.G., Thornhill, P.G., Conway, K., Miller, C., Patierno, S., and Costa, M. (1987). *Carcinogenesis*, *8*: 305.

Zhitkovich, A., and Costa, M. (1992). *Carcinogenesis*, *13*: 1485.

Zollinger, H.U. (1953). *Virchows Arch. Pathol. Anat. Physiol.*, *323*: 694.

OXYGEN, TRANSITION METAL IONS AND FREE RADICAL FORMATION

by

John M.C. Gutteridge

Oxygen Chemistry Laboratory,
Department of Anaesthesia & Intensive Care,
Royal Brompton Hospital and National Heart & Lung Institute,
Sydney Street
London, SW3 6NP
U.K.

I. SOME HISTORY AND CHEMISTRY OF OXYGEN

The element oxygen (O) exists in air as a molecule (O_2) known as dioxygen or molecular oxygen. It was first isolated and characterized between 1772 to 1774 by the individual skills of the great European scientists Priestley, Lavoisier and Scheele. Dioxygen, hereafter referred to as oxygen, appeared in significant amounts on the surface of the Earth some 2.5 x 10^9 years ago, and geological evidence suggests that this was due to the photosynthetic activity of micro-organisms (blue-green algae). As they split water to obtain their essential requirement for hydrogen, blue-greens released tons of oxygen into the atmosphere creating the worst case of environmental pollution ever recorded on this planet. The slow and steady rise in atmospheric oxygen concentration was accompanied by the formation of the ozone layer in the stratosphere. Both oxygen and the ozone layer acted as critical filters against the intense solar ultraviolet light reaching the surface of the Earth. Ascending in altitude from the Earth's atmosphere into outer space, we see a change from relatively heavy molecules such as O_2, N_2, H_2O to lighter molecules, atoms, and ions such as $H^{\bullet}$, H^+, $^{\bullet}OH$, and electrons which are dominant above 800 km. The universe exists predominantly of hydrogen (H) and helium (He) with the Earth as a unique center of oxidation in an otherwise reducing Universe.

The percentage of oxygen in dry air is now around 21% making it, after nitrogen (78%), the second most abundant element in the atmosphere. However, this amount of oxygen in the air is negligible when compared with that present as part of the water molecule in oceans, lakes and rivers, and that present as part of mineral reservoirs in the Earth's crust where it is by far the most abundant element. When the Earth's atmosphere changed from a highly reducing state to the oxygen-rich state that we know today, anaerobic life forms ceased to exist or retreated to places where oxygen was excluded. The slow change from anaerobic to aerobic life necessitated the evolution of specialized antioxidants to protect against the toxic properties of oxygen. Aerobic life uses oxygen to oxidize (burn) carbon and hydrogen rich substrates (foods) to obtain the chemical energy and heat essential for life. Unfortunately, when we oxidize molecules with oxygen, the oxygen molecule itself becomes reduced and forms intermediates; two of which are free radicals (Equations 1-4)

$O_2 + e + H^+ \rightarrow HO_2^\bullet$ hydroperoxyl radical (1)

$HO_2^\bullet \rightarrow H^+ + O_2^-$ superoxide radical

$O_2^- + 2H^+ + e \rightarrow H_2O_2$ hydrogen peroxide (2)

$H_2O_2 + e \rightarrow OH^- + {}^\bullet OH$ hydroxyl radical (3)

${}^\bullet OH + e + H^+ \rightarrow H_2O$ (4)

Under normal conditions oxygen is a stable, odourless, tasteless and colourless gas of limited solubility in water; three volumes of gas dissolving in 100 volumes of water. This limited solubility is vital to aquatic life, and essential for normal respiratory functions in man. The air dissolved in water contains a higher percentage of oxygen (34% - because nitrogen has a solubility around 2 volumes per 100 volumes) than does dry air (21%). Oxygen is considerably more soluble in organic solvents than it is in water. For example, the fat solvent chloroform ($CHCl_3$) at 10°C can dissolve up to 219.5 ml of oxygen per litre at one atmosphere, whereas under the same conditions water dissolves only 38.2 ml of oxygen. These differences in solubility are important when considering the availability of oxygen for chemical reactions inside biological membranes which predominantly consist of lipids.

When compared with the other elements, oxygen has the third highest affinity for electrons and should be considerably more reactive than it is observed to be. The reason why its reactivity is masked involves the electronic structure of oxygen and only when this is overcome can the 'true' reactivity of oxygen be expressed.

Chemical molecules of interest to biologists and chemists differ in their tendency to accept or donate electrons, and the voltage required to oppose this tendency in aqueous solution under defined conditions is known as the "redox potential".

When oxygen is reduced by the stepwise addition of electrons (eqs. 1-4) two free radicals ($HO_2^\bullet$, ${}^\bullet OH$) are formed, together with hydrogen peroxide (H_2O_2). At a physiological pH value of 7.4, the hydroperoxyl radical ($HO_2^\bullet$) with a pKa of 4.8 (the pH value at which equal concentrations of both acid ($HO_2^\bullet$) and base (O_2^-) are present) dissociates to give the superoxide anion radical (O_2^-).

$$HO_2^\bullet \rightarrow H^+ + O_2^- \quad (5)$$

It is now considered unnecessary to show the unpaired electron (bold dot $^\bullet$) on the superoxide anion radical ($O_2^{\bullet -}$) since it is less of a radical than molecular oxygen (which has two unpaired electrons but is never shown as $O_2^{\bullet\bullet}$). At this point, before going on to discuss free radicals, we need a simple definition of a free radical.

'A free radical may be defined as any chemical species capable of independent existence that contains one or more unpaired electrons'.

This definition is a broad biological one (Halliwell and Gutteridge, 1989) which does not specify exactly where the unpaired electron is. It is preferred because it allows us to classify most of the transition metal ions as free radicals, and so better understand the close inter-relationship between oxygen and reactive metal ions.

A. Oxygen and is reduction intermediates

1. *The superoxide anion*

Superoxide (O_2^-) is a radical anion formed when one electron enters one of the π*2p orbitals of oxygen. The chemistry of superoxide differs greatly depending on its solution environment. In aqueous solution, O_2^- is a weak oxidizing agent able to oxidize molecules such as ascorbic acid and thiols. However, O_2^- is a much stronger reducing agent, and is able to reduce several iron complexes such as cytochrome *c* and ferric-EDTA. Superoxide rapidly disappears in aqueous solution due to its dismutation reaction in which hydrogen peroxide and oxygen are formed :

$$O_2^- + O_2^- + 2H^+ \rightarrow H_2O_2 + O_2 \quad (6)$$

Superoxide dismutase (SOD) discovered by McCord and Fridovich in 1968 greatly accelerates the above reaction. Several different forms of the enzyme are known to exist in plant, microbial and mammalian cells, differing in structure and transition metal ions at their active centers, but catalyzing exactly the same chemical reaction shown in equation 6 (Fridovich, 1982). The protonated form of O_2^-, the hydroperoxyl radical ($HO_2^\bullet$), is both a more powerful oxidant and reductant than O_2^-, but little $HO_2^\bullet$ will be present at pH 7.4.

2. *Hydrogen peroxide*

Any system producing superoxide will, as a result of the dismutation reaction, also produce hydrogen peroxide (H_2O_2). Many enzymes such as urate oxidase, glucose oxidase and D-amino acid oxidase produce H_2O_2 directly by the transfer of two electrons to oxygen. Hydrogen peroxide is a weak oxidant and weak reducing agent that is relatively stable in the absence of transition metal ions. The molecule has an uncharged covalent structure. It readily mixes with water, and is treated as a water molecule by the body, rapidly diffusing across cell membranes. The redox properties of H_2O_2, and its ability to form highly reactive free radicals in the presence of transition metal ions has necessitated the evolution of body defenses against it. Unwanted hydrogen peroxide is removed from cells by the action of enzymes such as catalase, glutathione peroxidase (selenium containing), and certain other peroxidases.

$$\text{Catalase} + 2H_2O_2 \rightarrow O_2 + 2H_2O$$

$$\text{Peroxidase} + H_2O_2 \xrightarrow[\text{+ cofactor}]{} \text{cofactor oxidation} + H_2O$$

3. *Hydroxyl radical*

The hydroxyl radical ($^\bullet OH$) is a major product arising from the high energy ionization of water (radiolysis)

$$H_2O \rightarrow {}^\bullet OH + H^\bullet + E_{aq} + H_2O_2 \qquad (7)$$

Most of our definitive knowledge about the hydroxyl radical comes from studies by radiation chemists who, by convention, show the unpaired electron on the hydrogen atom, i.e. $OH^\bullet$. This is, incorrect since the unpaired electron is on the oxygen atom. The $^\bullet OH$ radical is an extremely aggressive oxidant that can attack most biological molecules at an almost diffusion controlled rate. Three types of chemical reactions commonly occur with the $^\bullet OH$ radical (see Fig. 1):

(i) Hydrogen abstraction
(ii) Aromatic hydroxylation
(iii) Electron transfer

Use is often made of these reactions in detecting damage caused by $^\bullet OH$ radicals. Some of the ways in which $^\bullet OH$ radicals can be formed during the handling of biological materials are shown in Table 1 below.

4. *Singlet oxygen*

Singlet oxygen ($^1\Delta g\ O_2$) is not a free radical as it does not contain an unpaired electron. Nevertheless, it is a highly reactive form of oxygen in which the spin restriction of oxygen

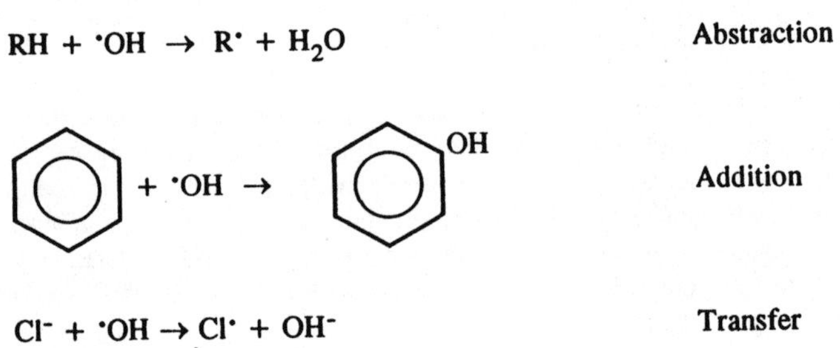

Figure 1 Types of radical reactions.

Table 1

Mechanisms by which Oxygen Radicals Can Be Formed During the Handling of Biological Molecules

(i)	Transition metal ion-dependent Fenton chemistry
(ii)	Radiolysis of water (X-rays, γ-rays)
(iii)	Drying reactions in air
(iv)	Freeze-thawing
(v)	Milling, grinding and cutting
(vi)	Subjecting to ultrasound
(vii)	Freeze-drying

(two unpaired electrons with parallel spins) is removed thereby increasing its oxidizing ability. In addition to $^1\Delta g\ O_2$, another form of singlet oxygen exists ($^1\Sigma g^+$). However, this state is extremely energetic and rapidly decays to the $^1\Delta g$ form in biological systems. Formation of singlet oxygen is extremely important when considering photo-chemical reactions.

5. *Ozone*
The pale blue gas ozone (O_3) provides an important protective shield (global antioxidant) in the stratosphere against solar radiation. At ground level, however, ozone is a toxic and unwanted pollutant. Ozone is produced in polluted urban air and by intensive light sources used in scientific equipment and some photocopying machines. It is extremely damaging to the lung readily oxidizing proteins, DNA and lipids.

6. *Oxides of nitrogen*
Nitric oxide ($NO^\bullet$) and nitrogen dioxide ($NO_2^\bullet$) contain odd numbers of electrons, and are therefore free radicals, whereas nitrous oxide (N_2O) does not. Nitrogen dioxide is a dense brown poisonous gas, and a powerful oxidizing agent. Nitric oxide, on the other hand, is a colourless gas and a weak reducing agent. it was first recognized as a distinct gas by Joseph Priestly in 1772 who prepared an iron complex of it. Recently, biological interest in nitric oxide has centered around the observation that the vascular endothelium and other cells in the body produce small amounts of the gas from the amino acid L-arginine (reviewed in Moncada and Higgs, 1991). At present nitric oxide is indistinguishable from the vasodilator Endothelium Derived Relaxing Factor (EDRF) which can react with another endogenous free radical, superoxide, to produce a reactive intermediate peroxynitrite ($ONOO^-$) (Beckman et al., 1990; Saran et al., 1990). Peroxynitrite is a powerful oxidant, able to damage many biological molecules, and can decompose to release hydroxyl radicals independent of metal catalysis (Radi et al., 1991) (eq. 8). Nitrous oxide is a colorless gas with a sweetish odor and taste that is used as an anaesthetic ("laughing gas").

$$ONOO^- + H^+ \rightarrow {}^\bullet OH + NO_2^\bullet \qquad (8)$$

II. TRANSITION METAL IONS AND FREE RADICAL FORMATION

When the Earth's atmosphere changed to an aerobic state, iron (the fourth most abundant element, present in the reduced ferrous Fe(II) state) oxidized to the insoluble ferric state, Fe(III). Poor solubility placed considerable restrictions upon its availability to early life forms, such as microorganisms, that required iron as a catalyst for aerobic processes. To overcome the limited bioavailability of iron, microorganisms adapted by synthesizing iron-chelators (siderophores) that were able to chelate, solubilize and capture ferric ions, and internalize them by receptor recognition to provide the iron required. Today, microbial siderophores, such as desferrioxamine, are among some of the most effective iron chelators used to treat iron-overload diseases. Hereafter, a close relationship evolved between oxygen and iron, and together with copper these metals form the major catalysts for aerobic metabolism often found at the active center of oxidases, oxygenases, antioxidants, electron transport and oxygen transport proteins.

A. The reaction of iron with oxygen

During the 1890's the Cambridge chemist H.J.H. Fenton described a reaction between iron salts and hydrogen peroxide that caused oxidative damage to organic molecules such as tartaric acid (Fenton, 1894). This reaction is known as the 'Fenton reaction' and widely represented as follows :

$$Fe^{2+} + H_2O_2 \rightarrow Fe^{3+} + OH^- + {}^{\bullet}OH \quad (9)$$

$$Fe^{3+} + H_2O_2 \rightarrow Fe^{2+} + HO_2^{\bullet} + H^+ \quad (10)$$

overall reaction

$$\text{Iron salt} + H_2O_2 \rightarrow 2H_2O + O_2 \quad (11)$$

Reactions 9 and 10 are gross oversimplifications of the chemistry involved, particularly when such reactions are considered in biological systems.

At physiological pH values (7.4) in the presence of oxygen and phosphate ions (PO_4^{3-}), ferrous ions, Fe(II), exist only transiently before auto-oxidizing to the ferric state Fe(III). In the process of changing from the ferrous to the ferric state, an electron is transferred from iron to oxygen to make superoxide as follows:

$$Fe^{2+} + O_2 \rightarrow Fe^{2+}O_2 \leftrightarrow Fe^{3+}O_2^- \rightarrow Fe^{3+} + O_2^- \quad (12)$$

The intermediates $Fe^{2+}O_2$ and $Fe^{3+}O_2^-$ are known as the perferryl ion, in which iron has an oxidation state of V.

Discovery of the superoxide dismutase enzymes (McCord and Fridovich, 1968) changed biological thinking concerning oxygen radicals, previously considered the domain of radiation chemists. It is not surprising, therefore, that many theoretical aspects of radiation chemistry were transferred into biological Fenton chemistry without carefully considering the full implications. One such example is the protection afforded by 'hydroxyl radical scavengers'. Hydroxyl radical scavengers are usually low-molar-mass chemicals with known second order rate constants for reaction with the hydroxyl radical, that are determined by competition kinetics using techniques such as pulse radiolysis (Anbar and Neta, 1967).

The homolysis (radiolysis) of water produces a variety of products; a major species being the hydroxyl radical (${}^{\bullet}OH$).

$$H_2O \rightarrow E_{aq}, H^{\bullet}, H_3O^+, {}^{\bullet}OH, H_2O_2 \quad (7)$$

$$H^{\bullet} + O_2 \rightarrow H^+ + O_2^- \quad (12)$$

$$E_{aq} + O_2 \rightarrow O_2^- \quad (13)$$

The hydroxyl radical produced by X or γ irradiation of water is formed free in solution, and can react with any added chemical at a rate calculated in moles per second as its second order rate constant ($M^{-1}\ s^{-1}$). Most of the 'scavengers' used in radiation chemistry react at high rates with ${}^{\bullet}OH$ ($1 - 6 \times 10^9\ M^{-1}\ s^{-1}$). The ability to protect a selected detector molecule from damage by ${}^{\bullet}OH$ will be a function of concentration present, and of the second order rate constant. Do these considerations still hold true for ${}^{\bullet}OH$ radicals formed in Fenton chemistry?

The answer is 'no' because biological Fenton chemistry is dependent upon iron ions, which are insoluble and unreactive at physiological pH. Biological systems overcome the problem of solubility by binding iron ions to a ligand. If the bound iron still remains redox active, any ${}^{\bullet}OH$ radicals generated by its participation in Fenton chemistry will be site-directed to the ligand holding the iron. Iron therefore determines which biological molecules will receive ${}^{\bullet}OH$ radical damage, and this tends to overrule predictions for scavengers based on rate constants and concentrations present.

High Energy Radiation → $^{\bullet}OH$ (high reactivity, low specificity)

Fenton Chemistry (Fe) → $^{\bullet}OH$ (high reactivity, high specificity)

The feeding of hydroxyl radical scavengers to animals is therefore unlikely to achieve simple competitive scavenging since in most cases insufficient scavenger will enter cells to compete with fast site-specific chemistry. Many hydroxyl radical scavengers will, of course, have a multitude of pharmacological properties in their own right, which may explain why some remarkable responses have been reported. Even in simple test tube 'Fenton' experiments, we do not see hydroxyl radical scavengers behaving as predicted from their concentrations and second order rate constants. To explain these discrepancies, others have suggested that the oxidizing species formed during Fenton chemistry is not the $^{\bullet}OH$ radical but an oxo-iron species such as the ferryl ion (FeO^{2+}) (Rush and Koppenol, 1986; Sutton and Winterbourn, 1989), in which iron has an oxidation number of IV. So far, ferryl ions have not been chemically or spectroscopically identified outside a haem ring and, although it is expected to have considerable oxidizing power, its 'footprints' of attack on scavengers or more complex biological molecules such as DNA do not support the view that it is an absolute alternative to the $^{\bullet}OH$ radical (reviewed in Gutteridge and Halliwell, 1989).

It seems more likely that the ferryl ion is often formed as an intermediate in Fenton chemistry leading to $^{\bullet}OH$:-

$$Fe^{2+} + H_2O_2 \rightarrow FeO^{2+} \rightarrow {}^{\bullet}OH \qquad (14)$$

and that its reaction with detector molecules or added scavengers will depend upon the experimental conditions used.

B. Molecular damage by oxygen

Oxygen damage to living cells was originally attributed to the oxygen molecule itself. Although isolated examples of this do exist in biology, it was soon realized that the rates of enzyme inactivation by oxygen were too slow to account for the extreme toxicity caused by oxygen to plant and animal systems.

Gerschman et al., (1954) first proposed that the damaging effects of oxygen were caused by oxygen radicals. Their theory was latter expanded and extended, following the discovery of the superoxide dismutases by McCord and Fridovich (1968), into the 'Superoxide Theory of Oxygen Toxicity'. This theory proposes that superoxide is a major biological toxin, and that the superoxide dismutase enzymes provide a key biological defense against oxygen toxicity. Twenty five years on, the theory still holds true although there are still some intriguing puzzles to solve. As previously discussed, much of our early biological knowledge of radicals came from radiation studies which led to the proposal that hydrogen peroxide reacted directly with superoxide to yield hydroxyl radicals. It soon became obvious that this reaction could not occur in biological systems, and that a catalyst was required much along the lines (involving iron) proposed by Haber and Weiss in 1932.

Table 2

Requirements for an Oxidant to Be the Same Species as $^{\bullet}OH$

(i) Must attack spin traps to give correct "$^{\bullet}OH$" signal
(ii) React with $^{\bullet}OH$ scavengers at similar second order rate constants to $^{\bullet}OH$
(iii) Require H_2O_2 for its formation (inhibited by catalase)
(iv) Hydroxylate aromatic compounds to give the same "pattern" as $^{\bullet}OH$
(v) Give the same pattern of DNA base damage as $^{\bullet}OH$

The 'Haber-Weiss reaction' shown below (eqs. 15, 6, 9) can also be described as a 'Superoxide-driven Fenton reaction'.

$$O_2^- + Fe^{3+} \leftrightarrow Fe^{2+} + O_2 \quad (15)$$

$$2O_2^- + 2H^+ \rightarrow H_2O_2 + O_2 \quad (6)$$

$$Fe^{2+} + H_2O_2 \rightarrow Fe^{3+} + OH^- + {}^\bullet OH \quad (9)$$

C. How does Fenton chemistry occur in vivo?

Superoxide is produced continuously in the body during aerobic metabolism, and by the dismutation reaction will form H_2O_2 both in the presence or absence of the superoxide dismutase enzymes. When O_2^- and H_2O_2 are produced in quantities that exceed the ability of protective enzymes such as superoxide dismutase, catalase and glutathione peroxidase to remove them, they can facilitate the release of iron from iron-containing proteins. Redox active iron + O_2^- + H_2O_2 make hydroxyl radicals; thus the excessive production of O_2^- can provide all the necessary ingredients for Fenton chemistry to take place (reviewed in Gutteridge et al., 1985).

D. If superoxide dismutases only make H_2O_2, why do we need them?

Superoxide-dismutase deficient bacterial mutants have confirmed the essentiality of SOD for aerobic life. The poor reactivity of O_2^- has led to the identification of only a few examples, in plant and microbial systems, which suggest a direct toxic effect of O_2^-. Yet, we know that the generation of O_2^- in biological systems can lead to extensive molecular damage, and protection can be afforded by SOD. The only plausible explanation we have, at present, is that SOD's protect by preventing the O_2^--dependent reduction of redox active iron complexes (i.e. SOD's are anti-reductants). Our knowledge of the redox active iron-complexes present in biological systems is still, however, sadly lacking (Gutteridge, 1991).

As previously mentioned, superoxide can react with another endogenous free radical, nitric oxide ($NO^\bullet$), to form peroxynitrite. Superoxide dismutase prevents this reaction and the subsequent formation of damaging $ONOO^-$ and $^\bullet OH$ as well as prolonging the biological activity of $NO^\bullet$.

E. Iron-containing proteins as Fenton catalysts

Several claims have been made that iron storage, transport and functional proteins are Fenton catalysts (reviewed in Halliwell and Gutteridge, 1990). Under normal physiological conditions, however, this is extremely unlikely. Iron correctly loaded into transferrin, lactoferrin and into the core of ferritin do not facilitate Fenton chemistry. A small amount of labile iron associated with ferritin, but not at its ferric core, is often available (depending on the preparation of the sample) to participate in Fenton chemistry. Similarly, iron in the plasma of patients with haemochromatosis and in the synovial fluid of patients with rheumatoid arthritis appears to be more labile under conditions of pH, or oxidant stress, than that present in normal control patients (Gutteridge, 1988). This iron can be detected by using the ability of bleomycin to degrade DNA in the presence of the iron that it chelates from the biological sample. Haem proteins have been observed to stimulate Fenton chemistry, and the reason is that iron leaves the haem protein and not the $^\bullet OH$ radical (Gutteridge, 1986). Incubation of haem proteins with H_2O_2 causes release of chelatable iron that is available for Fenton chemistry (Gutteridge, 1986).

F. Iron distribution in the body

An average adult human male contains some 4.5 g of iron, absorbs about 1 mg of iron per day, and when in iron-balance excretes the same amount. Only slight disturbances to the delicate balance between iron intake and iron loss can push the body into conditions of iron overload or iron deficiency. It has been estimated that in the world today some 500 million people are iron deficient and several millions are iron overloaded. No specific mechanisms exist for iron excretion; loss occurring by the turnover of intestinal epithelial cells, in sweat, faeces, urine and by menstrual bleeding in women.

Most of the body's iron (some two thirds) is found in the oxygen-carrying protein haemoglobin, with smaller amounts present in myoglobin, various enzymes, the iron transport proteins transferrin and lactoferrin, and the iron storage proteins ferritin and hemosiderin. Transferrin binds with high affinity ($Ka=10^{28}$) two moles of iron per mole of protein and is normally only one-third loaded with iron, retaining a considerable metal-binding capacity. Metals other than iron, for example aluminum, can bind to transferrin and enter cells via the transferrin receptor. Lactoferrin shares many similarities with transferrin, but is immunologically distinct and can hold on to bound iron down to pH values of 4.0.

The iron storage protein ferritin consists of a protein shell surrounding an iron core of ferric ions, comprising some 4,500 iron ions per mole of protein (see Part Two, Chapter 1-D4). Iron enters ferritin in the reduced ferrous state and is laid down in the core by a process of oxidation involving the 'ferroxidase' activity of the protein, and the growth of ferric oxide crystal. For it to leave the core, iron has to be reductively mobilized back to the ferrous state by either the entry of a reductant to the ferritin core or by long-range electron transfer through the coat of ferritin. Hemosiderin is an insoluble iron-protein complex that probably arises as a lysosomal degradation product of ferritin, although similar conversions can be made by subjecting ferritin to oxidizing free radicals (O'Connell et al., 1986).

Most of our dietary intake of iron is in the form of non-haem ferric ion. This requires reduction and solubilization before absorption from the intestine can occur. Gastric HCl and ascorbate (vitamin C) facilitate absorption in this way. Dietary haem-iron is mostly derived from red-meat products, and can be directly absorbed by the intestine. Only a small, and carefully controlled, fraction of the total iron ingested is absorbed and allowed to enter the circulation bound to transferrin. Cells requiring iron for essential aerobic processes have transferrin receptors which allow uptake of iron into the cell. Once inside the cell, reductive release of iron from transferrin occurs which allows the apotransferrin molecule (devoid of iron) to return to the circulation to continue iron transport functions.

G. Copper as a biological catalyst

Copper is a trace element in the Earth's crust, and some sixty times less abundant than iron in the human body. Copper salts are usually more reactive than iron in oxygen radical reactions, but poor bioavailability and avid protein-binding often mask this increased reactivity. Absorption of dietary copper takes place in the stomach, or upper small intestine, from whence it is transported to the liver as an albumin complex; human albumin having one high-affinity binding site for copper (see Part Two, Chapter 1-D2, and *Medicine*, Vol. 1, Part Three, Chap. 2-F5). The liver incorporates most of the copper into the plasma protein ceruloplasmin; said to show sequence homology with lactoferrin and coagulation factor V. Ceruloplasmin has a molar mass of around 130 kg mol^{-1}, and contains some 6 or 7 copper ions per mole of protein (see Part Two, Chapter 1-B2). Its biological function has been considered as an acute phase reactant. Proteins showing an increase in plasma concentration after trauma or in the acute phase of an inflammatory response are often called acute phase reactants. The acute phase proteins have been defined as those plasma proteins which increase in concentration by 25% or more in the first 7 days following tissue damage. These proteins are synthesized by the liver and induced by cytokines derived from the inflammatory focus. The purpose of the acute phase response appears to be the activation of a protective screen against further tissue injury caused by release of proteolytic enzymes and biological pro-oxidants from damaged cells.

Ceruloplasmin does not donate or specifically bind copper ions whilst in the circulation. However, it is able to enter cells and release copper for the synthesis of copper-containing proteins, undergoing proteolytic degradation in the process. Total plasma copper levels are essentially a measure of ceruloplasmin, although many established texts report that up to 10% or even more of plasma copper is labile, 1% being bound to amino acids. Using a molecular recognition assay similar to that of the bleomycin assay but specific for copper (based on 1,10-phenanthroline), we have been unable to detect labile copper in freshly taken plasma (Gutteridge, 1984). Whenever plasma is stored or mishandled, however, labile copper readily appears due to rapid proteolytic degradation of ceruloplasmin. Copper released from ceruloplasmin in this way causes the peroxidation of plasma lipoproteins. The resulting oxidized lipid-protein-copper complex has a ferroxidase activity (Ferroxidase II) similar to that of ceruloplasmin (Ferroxidase I), and so the potential for artefacts when studying plasma copper biochemistry are enormous (reviewed in Gutteridge and Stocks, 1981).

When copper ions are non-specifically attached to aminoacids and proteins, they often generate oxygen radicals, although the radicals will not be seen by an added detector molecule because the reaction is highly site-specific.

SUMMARY

Oxygen is constantly reduced in the body to reactive intermediates which damage biological molecules. Life continues by a process of reactive oxygen elimination, protection, and by the constant repair of oxidative damage.

ACKNOWLEDGEMENTS

JMCG holds the first BLF/BOC Senior Research Fellowship in Respiratory Critical Care and gratefully thanks the British Lung Foundation, British Oxygen Group, and the British Heart Foundation for their generous support.

REFERENCES

Anbar, M., and Neta, P. (1967). *Int. J. Appl. Radiat. Isotopes*, *18*: 493.

Beckman, J.S., Beckman, T.W., Chen, J., Marshall, P.A., and Freeman, B.A. (1990). *Proc. Nat. Acad. Sci. USA*, *87*: 1620.

Fenton, H.J.H. (1894). *J. Chem. Soc.*, *65*: 899.

Fridovich, I. (1982). *Superoxide Dismutase*, Vol. 1 (L.W. Oberley, ed.), CRC Press, Boca Raton, p. 1.

Gerschman, R., Gilbert, D.L., Nye, S.W., Dwyer, P., and Fenn, W.O. (1954). *Science*, *119*: 623.

Gutteridge, J.M.C. (1984). *Biochem. J.*, *218*: 983.

Gutteridge, J.M.C. (1986). *FEBS Lett.*, *201*: 291.

Gutteridge, J.M.C. (1988). *Free Radical Methodology and Concepts* (C. Rice-Evans and B. Halliwell, eds.), Richelieu Press, London, p. 429.

Gutteridge, J.M.C. (1991). *J. Trace Elem. Electr. Hlth Dis.*, *5*: 279.

Gutteridge, J.M.C., and Halliwell, B. (1989). *Clinical Haematology. Iron Chelating Therapy*, Vol. 2 (C. Hershko, ed.), No 2, Bailliere Tindall, London, p. 195.

Gutteridge, J.M.C., and Stocks, J. (1981). *CRC Crit. Rev. Clin. Lab. Sci.*, *14*: 257.

Gutteridge, J.M.C., Westermarck, T., and Halliwell, B. (1985). *Free Radicals, Aging and Degenerative Diseases* (J E. Johnson, R. Walford, D. Harman and J. Miquel, eds.), Alan R. Liss, New York, p. 99.

Haber, F., and Weiss, J. (1932). *Naturwissenschaften*, *20*: 948.

Halliwell, B., and Gutteridge, J.M.C. (1989). *Free Radicals in Biology and Medicine*, Oxford University Press, Oxford, p. 10.

Halliwell, B., and Gutteridge, J.M.C. (1990). *Methods in Enzymology*, Vol. 186 (L. Packer and A.N. Glazer, eds.), Academic Press, San Diego, p. 1.

McCord, J.M., and Fridovich, I. (1968). *Fed. Proc.*, *28*: 361.

Moncada, S., and Higgs, E.A. (1991). *Eur. J. Clin. Invest.*, *21*: 361.

O'Connell, M.J., Baum, H., and Peters, T.J. (1986). *Biochem. J.*, *240*: 297.

Radi, R., Beckman, J.S., Bush, K.M., and Freeman, B.A. (1991). *J. Biol. Chem.*, *266*: 4244.

Rush, J.D., and Koppenol, W.H. (1986). *J. Biol. Chem.*, *261*: 6730.

Saran, M., Michel, C., and Bors, W. (1990). *Free Rad. Res. Commun.*, *10*: 221.

Sutton, H.E., and Winterbourn, C.C. (1989). *Free Rad. Biol. Med.*, *6*: 53.

IRON-LIGAND INTERACTIONS AND THE FENTON REACTION

by

R.A. Addy and B.C. Gilbert

Department of Chemistry,
University of York,
Heslington, York, YO1 5DD
U.K.

I. INTRODUCTION

The reaction between iron(II) and hydrogen peroxide in aqueous solution is generally referred to as the Fenton reaction, to acknowledge the discovery about a hundred years ago that the oxidation of polycarboxylic acids by hydrogen peroxide is accelerated by the presence of ferrous salts (Fenton, 1984). More specifically the title is often used to describe the occurrence of reaction (1), which was proposed as the crucial first stage and the source of the key oxidant by Haber and Weiss (1934), in a reaction scheme subsequently modified by Uri (1952).

$$Fe(II) + H_2O_2 \rightarrow Fe(III) + HO^{\cdot} + HO^- \qquad (1)$$

Although there has been considerable support for the claim that hydroxyl radicals are indeed involved in this reaction (Barb et al., 1951a; 1951b; Merz and Waters, 1947; 1949; and Walling, 1975), it has also been proposed that the oxidizing intermediate produced in the initial step is actually an Fe(IV) or ferryl species (FeO^{2+}) (see e.g. Bray and Gorin, 1932; Rahhal and Richter, 1988; Shiga, 1965) or a complex $Fe(II)H_2O_2$ (Rush and Koppenol, 1987).

We will first briefly review our present understanding of the nature of the intermediates formed in the reaction of aquo-Fe(II) with H_2O_2 (note that reactions of $HO_2^{\cdot}$ and Fe(III) must also be considered in the reaction scheme) with emphasis on the approaches used to characterize the species involved and to determine the rate constants for their reaction. For a more detailed discussion of relevant research, the reader is referred to several excellent reviews (Walling, 1975; Edwards and Curci, 1993; Brown et al., 1970) and including that with a particular biochemical relevance (Halliwell and Gutteridge, 1990). Secondly, a more detailed review will be presented of the effect of different ligands on the nature and rate of reaction (1), with emphasis on recent results and a variety of techniques. Finally, mention will be made of particular examples where good evidence exists for the mediation of high-valent iron-oxygen species rather than hydroxyl radicals.

II. THE FENTON REACTION BETWEEN AQUO-FE(II) AND HYDROGEN PEROXIDE

When ferrous iron reacts with hydrogen peroxide in aqueous solution in the absence of other substrates, there is considerable kinetic and stoichiometric evidence to support the occurrence of a free-radical decomposition as indicated by reactions (1) - (3) (for iron in excess) and, when H_2O_2 is in excess, by the reactions shown in Scheme 1 (Edwards and Curci, 1993; Walling, 1975; Haber and Weiss, 1934; Barb et al., 1951a; 1951b). The rate constant for reaction (1) at 20°C in perchloric acid-sulphuric acid mixtures is 41.5 dm^3 mol^{-1} s^{-1} (Hardwick, 1957), and other rate constants have been reported as follows: $k_{(2)} \sim 3 \times 10^8$ dm^3 mol^{-1} s^{-1}, $k_{(3)} \sim 3 \times 10^7$ dm^3 mol^{-1} s^{-1} (Buxton et al., 1988), $k_{(4)} \sim 1.2 \times 10^6$ dm^3 mol^{-1} s^{-1}, and $k_{(5)} \sim 1 \times 10^5$ dm^3 mol^{-1} s^{-1} (at pH 2) (Bielski et al., 1985).

$$Fe(II) + H_2O_2 \rightarrow Fe(III) + HO^{\bullet} + OH^- \quad (1)$$
$$Fe(II) + HO^{\bullet} \rightarrow Fe(III) + OH^- \quad (2)$$
$$HO^{\bullet} + H_2O_2 \rightarrow H_2O + HO_2^{\bullet} \quad (3)$$
$$HO_2^{\bullet} + Fe(II) \rightarrow Fe(III) + HO_2^- \quad (4)$$
$$HO_2^{\bullet} + Fe(III) \rightarrow Fe(II) + H^+ + O_2 \quad (5)$$

Scheme 1: The Fe(II) - H_2O_2 reaction in the absence of substrate

Studies of the stoichiometry and products in experiments in which organic substrates were added (see e.g. Merz and Waters, 1947; 1949; Walling, 1975 and references therein) could then be satisfactorily interpreted in terms of the scavenging of the hydroxyl radical by the substrate(s) (by hydrogen abstraction and addition), with subsequent further reactions of organic radicals (e.g. dimerization, or oxidation by Fe(III) as appropriate). For example, the data obtained by Walling and Kato (1971) for the oxidation of propan-2-ol are clearly compatible with the occurrence of preferential $^{\bullet}OH$ attack on the α-position and subsequent oxidation of $^{\bullet}CMe_2OH$ by Fe(III) (see Scheme 2); for this and many other substrates, the results obtained are in excellent agreement with predictions based on the known behavior of the hydroxyl radical generated radiolytically or photolytically from H_2O_2 (see e.g. Walling and Kato, 1971; Walling and El Taliawi, 1973a; 1973b; Walling et al., 1974). Reliable relative reactivities of the hydroxyl radical towards different types of C-H bond have been rationalized and examples given of aromatic hydroxylation and of radical-metal ion reactions [including the rapid oxidation of electron-rich radicals by Fe(III) and the reduction of certain carboxyl-conjugated radicals by Fe(II) (Walling 1975)].

$$Fe(II) + H_2O_2 \rightarrow Fe(III) + HO^{\bullet} + OH^- \quad (1)$$
$$HO^{\bullet} + Me_2CHOH \rightarrow H_2O + {}^{\bullet}CMe_2OH \quad (6)$$
$$HO^{\bullet} + Me_2CHOH \rightarrow H_2O + {}^{\bullet}CH_2CHMeOH \quad (7)$$
$${}^{\bullet}CMe_2OH + Fe(III) \rightarrow Me_2CO + H^+ + Fe(II) \quad (8)$$
$$2\ {}^{\bullet}CH_2CHMeOH \rightarrow MeCH(OH)CH_2CH_2CHMeOH \quad (9)$$

Scheme 2: The Fenton Reaction ($Fe(II)/H_2O_2$) in the presence of Propan-2-ol

The corresponding reaction of ferric iron and hydrogen peroxide (wherein the former catalyzes the decomposition of the latter to give oxygen) is also generally believed to be a closely-related free-radical reaction as outlined in Scheme 3 (see e.g. Haber and Weiss, 1934; Barb et al., 1951b; Walling, 1975; Edwards and Curci, 1993). Following formation of $HO_2^{\bullet}$ in reaction (10), in which a complex $FeOOH^{2+}$ may well be involved, oxygen results from reaction (5). Other reactions which occur, depending on conditions, are (1) - (4): the combination of reactions (5) and (1) is often referred to as the iron-catalyzed Haber-Weiss cycle, whereby $HO_2^{\bullet}$ brings about the decomposition of H_2O_2 to $^{\bullet}OH$. Organic substrates retard the decomposition via reaction with $^{\bullet}OH$ (Walling 1975).

$$H_2O_2 + Fe(III) \rightarrow Fe(II) + H^+ + HO_2^{\bullet} \quad (10)$$
$$HO_2^{\bullet} + Fe(III) \rightarrow Fe(II) + O_2 + H^+ \quad (5)$$
$$Fe(II) + H_2O_2 \rightarrow Fe(III) + HO^{\bullet} + OH^- \quad (1)$$
$$Fe(II) + HO^{\bullet} \rightarrow Fe(III) + OH^- \quad (2)$$
$$HO^{\bullet} + H_2O_2 \rightarrow H_2O + HO_2^{\bullet} \quad (3)$$
$$HO_2^{\bullet} + Fe(II) \rightarrow HO_2^- + Fe(III) \quad (4)$$

Scheme 3: The Fe(III)-H_2O_2 Reaction

III. THE EFFECT OF LIGANDS ON THE FENTON REACTION IN AQUEOUS SOLUTION

There is considerable contemporary interest in the influence of inorganic ligands, especially polyaminocarboxylate complexes (e.g. EDTA) on the reaction of Fe(II), especially, and Fe(III) with H_2O_2; recent discussion has largely focussed on the reaction between Fe(II) and H_2O_2 and in particular the rate of this reaction, and on the nature of the intermediate(s) involved ($^{\bullet}OH$ or ferryl ion). The following discussion will focus first on the results obtained for Fe(II)-EDTA (at pH ca 7) with an introduction to the methodologies employed for study of rates, products, intermediates and mechanisms; towards the end of the section, results for other ligands - and a comparison of different mechanistic interpretations - will be attempted.

A. Kinetic measurements

There is little disagreement amongst a variety of researchers in this field that the chelation of iron(II) to EDTA at pH ca 7 dramatically accelerates the rate constant of the Fenton reaction (to ca 7×10^3 dm^3 mol^{-1} s^{-1}). This rate constant has been measured by a number of methods including polarography (Borggaard et al., 1971), spectrophotometry (Rush and Koppenol, 1987), product analysis of Fe(II)/H_2O_2 reactions (e.g. after generation of Fe(II) from Fe(III) with paraquat$^{+\bullet}$; Sutton and Winterbourn, 1984) and both direct ESR investigations (Croft et al., 1992a; 1992b; Gilbert and Jeff, 1988) and spin-trapping ESR studies (Yamazaki and Piette, 1990; 1991). Since the use of ESR spectroscopy also enables information to be obtained (relatively directly) concerning the intermediates involved, special mention will be made here of the basis of experiment and intepretation of results.

In continuous-flow experiments, three solutions (typically of Fe(II), H_2O_2 and a substrate) are mixed a very short time (usually ca 20-50 ms) before the combined solution passes through a flattened aqueous sample cell held in the cavity of the spectrometer. For the Fe(II)-H_2O_2 and related redox systems (including Ti(III)-H_2O_2; Fe(III)-$S_2O_8^{2-}$; Cu(I)-$S_2O_8^{2-}$ etc; Gilbert et al., 1988; Gilbert and Stell, 1991) signals can be detected from relative low concentrations (ca micromolar) of short-lived organic free radicals: the absolute and relative concentrations of these organic radicals can be measured and the ESR spectra reveal unambiguously the precise nature and structure of the radicals detected (see e.g.

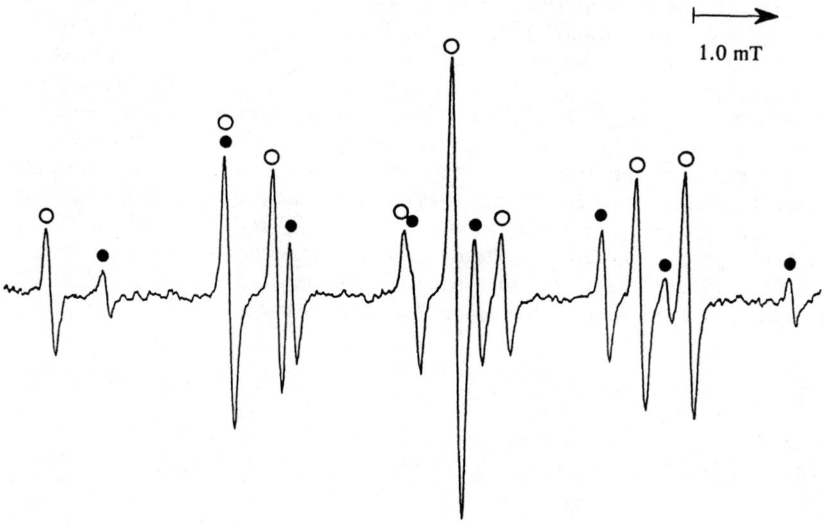

Figure 1.
ESR spectra of $^{\bullet}CHMeOH$(●) and $^{\bullet}CH_2CH_2OH$(○) obtained at pH 7 with the Fe(II)EDTA/H_2O_2 couple and ethanol. $[Fe(II)]_o$ 6.7×10^{-4}, $[EDTA]_o$ 1.3×10^{-3}, $[EtOH]_o$ 0.6, $[H_2O_2]$ 1.0×10^{-2} mol dm^{-3}.

Norman and Gilbert, 1967). An example is provided in Figure 1 which shows the ESR spectra of α- and β- radicals obtained from ethanol with the $Fe(II)/H_2O_2$ redox couple.

Since the life-time of such radicals is of the order of microseconds, it can be shown (Czapski, 1971) that organic radicals (R•) detected in such experiments are those formed in the cavity (where the metal-peroxide reaction still proceeds) and then a pseudo steady-state is established. For a system involving reactions (1), (11) and (12), and in which (RH) is high enough to scavenge •OH, appropriate kinetic analysis gives equation (13) (where t represents the time after mixing).

$$Fe(II) + H_2O_2 \xrightarrow{k_{in}} Fe(III) + HO^{\bullet} + OH^- \quad (1)$$

$$HO^{\bullet} + RH \rightarrow R^{\bullet} \quad (11)$$

$$R^{\bullet} + R^{\bullet} \xrightarrow{2k_t} \text{molecular products (e.g. } R_2) \quad (12)$$

$$k_{in}[Fe(II)]_t[H_2O_2]_t = 2k_t[R^{\bullet}]^2 \quad (13)$$

From analysis of systems in which $[H_2O_2]$ is kept in excess ($[H_2O_2]_t \sim [H_2O_2]_o$) it can be shown (Czapski 1971; Croft et al., 1992a; 1992b) that k_{in} can be obtained (and analysis verified) by experiments in which [R•] is monitored as a function of $[H_2O_2]$ [the maximum value of [R•] is given by equation (14)] or via variation of t and analysis according to equation (15). A variety of applications of this method (supported by computer simulations) have been described (Czapski, 1971; Gilbert and Jeff 1988; Croft et al., 1992a; 1992b). When extra reactions contribute to the fate of [R•] [e.g. with Fe(II) and Fe(III), reaction (16)], computer-fitting of the variation of e.g. [R•] with $[M^{n+}]_o$ and $[H_2O_2]_o$ can be used to determine the rate constants for k_{in} and the extra reactions (e.g. reaction (16)).

$$[H_2O_2]_o \text{ for } [R^{\bullet}]_{max} = 1/k_{in}t \quad (14)$$

$$[R^{\bullet}] = \left[\frac{k_1[Fe(II)]_o[H_2O_2]_o}{2k_t}\right]^{1/2} \exp(-k_1[H_2O_2]_o/2t) \quad (15)$$

$$R^{\bullet} + Fe(III) \rightarrow Fe(II) + R^+ \quad (16)$$

An alternative ESR approach - suitable for small-scale non-flow work - involves the addition of a spin-trap to a reaction in which radicals are believed to be involved (see e.g. Janzen and Haire, (1990); Buettner (1987); Davies (1987)). These molecules, normally nitroso compounds or nitrones, are chosen so that, in their addition reaction with radicals, short-lived organic radicals become trapped as relatively long-lived nitroxides, which can be studied at leisure. In particular, in many cases it is possible to choose a trap so that the spectrum of the resulting nitroxide provides information about the first-formed radical. Whilst nitroso compounds provide an excellent method of distinguishing different types of carbon-centred radical (e.g. primary, secondary, tertiary), nitrones possess the advantage that carbon-, oxygen-, and sulphur-centred radicals can be distinguished (via the magnitude of the appropriate β-proton-splitting). An excellent example of the latter category is DMPO, 5,5-dimethyl pyrroline-N-oxide (1); reaction (17) shows the process where •OH is trapped and recognized via its characteristic adduct (2). Yamazaki and Piette (1990; 1991) have exploited this approach in studies of Fenton chemistry; in particular, they have investigated the stoichiometry and kinetics of the $Fe(II)/H_2O_2$ reaction in the presence of EDTA and similar ligands by monitoring the increase with time in the concentration of (2).

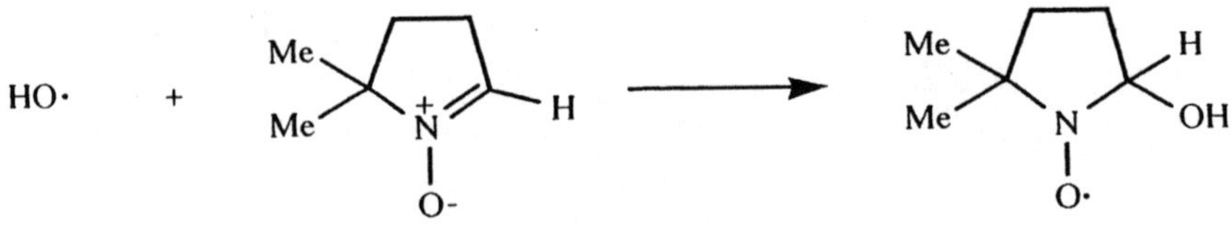

(17)

$a_N = a_H = 14.9$ mT

(1) (2)

B. The nature of the intermediates

The evidence that the Fe(II) EDTA-H_2O_2 reaction does indeed produce $^{\bullet}OH$ (reaction (18)) as the reactive intermediate has been very effectively summarized by Halliwell and Gutteridge (1990). As noted above, this conclusion is based on the result of spin-trapping studies, the detailed analysis of products from aromatic hydroxylation, scavenging experiments with a variety of substrates, and kinetic analysis (and comparison with data for other, independent methods for $^{\bullet}OH$ production).

Recent continuous-flow ESR experiments provide considerable support for this interpretation and also throw light upon the earlier observations of Shiga (1965). The latter compared the ESR spectra of organic radicals from the Ti(III)-H_2O_2 system (which is believed to produce $^{\bullet}OH$) under steady-state conditions with those from Fe(II)EDTA-H_2O_2 under identical conditions. Results from the two systems are not identical; for example in the reaction of EtOH, the predominant radical from Ti(III)-H_2O_2 is from $^{\bullet}CHMeOH$ (the α -radical) as expected from $^{\bullet}OH$ (reaction (19)) whereas $^{\bullet}CH_2CH_2OH$ (the β-radical) predominates for Fe(II)EDTA-H_2O_2, an observation interpreted in terms of the formation and reaction of a ferryl species. However, we have shown (Gilbert and Jeff, 1988; Croft et al., 1992a; 1992b) that these discrepancies can be explained if both systems serve as an effective method to generate $HO^{\bullet}$ but that in iron-containing systems the α-alcohol radicals are destroyed by rapid reaction with Fe(III) (as would be expected: see e.g. Walling, 1975). Detailed kinetic analysis allows the derivation of not only the initiation rate constant but also the rate constant (1 - 1.5 x 10^9 dm^3 mol^{-1} s^{-1}) for the rapid oxidation of $^{\bullet}CHMeOH$ by Fe(III) (reaction (20)), as well as the rate constant of the appropriate oxidation of $^{\bullet}CHMeOH$ by H_2O_2 $k_{(21)}$ = 1.5 x 10^5 dm^3 mol^{-1} s^{-1} (see Croft et al., 1992b)]. Comparable experiments with ^tBuOH (to give detectable signals from $^{\bullet}CH_2CMe_2OH$), for EDTA and a number of related complexes (including those from some aminophosphonic acids) indicate that rapid radical reduction by Fe(II) can also occur [(reaction (22): Croft et al., 1992a].

$$\text{Fe(II)EDTA} + H_2O_2 \rightarrow \text{Fe(III)EDTA} + HO^{\bullet} + OH^- \quad (18)$$

$$HO^{\bullet} + \text{EtOH} \rightarrow H_2O_2 + \underset{\text{major}}{^{\bullet}\text{CHMeOH}} + \underset{\text{minor}}{^{\bullet}CH_2CH_2OH} \quad (19)$$

$$^{\bullet}\text{CHMeOH} + \text{Fe(III)EDTA} \rightarrow \text{Fe(II)EDTA} + \text{MeCHO} + H^+ \quad (20)$$

$$^{\bullet}\text{CHMeOH} + H_2O_2 \rightarrow \text{MeCHO} + HO^{\bullet} + H_2O \quad (21)$$

$$^{\bullet}CH_2Me_2OH + \text{Fe(II)L} \rightarrow \text{Fe(III)L} + CH_2{=}CMe_2 + OH^- \quad (22)$$

It can also be shown in related ESR experiments that for a variety of chelates (including EDTA or DTPA) the selectivity of attack by the Fe(II)L/H_2O_2 redox couple on other molecules with several sites available for C-H abstraction (e.g. propanoic acid) or addition (e.g. crotonic acid) matches that observed for Ti(III)-H_2O_2; the observation of *hydroxyl*-radical adducts to alkenes in the Fenton system is of course particularly significant. Further, as has been pointed out recently (Croft et al., 1992b) the fact that these experiments provide simultaneous direct evidence for hydroxyl-radical formation and kinetic results close to those obtained with other techniques is compelling evidence that the hydroxyl radical is the major (if not sole) oxidant formed under conditions of the ESR experiment (pH 7; [Fe(II)] typically less than [H_2O_2]).

Use of the spin-trapping technique provides considerable support of the notion that reactions of Fe(II)EDTA and Fe(II)DTPA as well as Fe(II)ADP, with H_2O_2 also produce $^{\bullet}OH$: experiments involved trapping of $^{\bullet}OH$ (with DMPO) together with, e.g., an ethanol-derived radical when the substrate was also present (Yamasaki and Piette, 1990). It was concluded, however, that with Fe(II)DTPA only the hydroxyl radical was formed, probably via a peroxo complex, but that with Fe(II)EDTA both $^{\bullet}OH$ and an extra ethanol-oxidizing species (probably FeO^{2+}) result (the latter becoming particularly effective at higher concentrations of Fe(III) (see reactions (23) and (24)). However, this view has been challenged by Burkitt (1993) who has shown that under the reaction conditions employed secondary reactions occur involving ligand-dependent oxidation and reduction of the adduct (2) by Fe(III) and Fe(II); he claims that detailed quantitative conclusions are invalid.

$$Fe(II) + H_2O_2 \nearrow Fe(II)\ldots O(H){-}OH \rightarrow Fe(III) + HO^{\bullet} + OH^- \quad (23)$$

$$Fe(II) + H_2O_2 \searrow Fe(II)\ldots \begin{matrix} OH \\ | \\ OH \end{matrix} \rightarrow Fe(IV){=}O + H_2O \quad (24)$$

Sutton and Winterbourn (1984) find that Fe(II)EDTA and Fe(II)DTPA catalyze the $^{\bullet}OH$-producing reaction between H_2O_2 and $PQ^{+\bullet}$, via the occurrence of the Fenton reaction and the lower-valence iron complex (obtained via reaction of Fe(III) with $PQ^{+\bullet}$: EDTA and DTPA have comparable efficiency). In contrast, DTPA is a much less effective catalyst than EDTA for the Haber-Weiss cycle (i.e. in the absence of $PQ^{+\bullet}$) not because the Fenton reaction is significantly retarded but rather because reaction of Fe(III) with O_2^- is dramatically slowed. This conclusion is also supported by product studies of iron-induced hydroxylation using different ligands; see e.g. Winterbourn and Sutton (1986); Burkitt and Gilbert (1990). For the reaction of O_2^- with Fe(III)EDTA at pH 7 - 8, values of this rate constant in the range 10^5 - 6×10^6 $dm^3\ mol^{-1}\ s^{-1}$ have been recorded (Bielski et al., 1985) whereas for Fe(III)DTPA the rate constant is $< 10^4$ $dm^3\ mol^{-1}\ s^{-1}$ (Buettner et al., 1983).

$$Fe(III)L + O_2^- \rightarrow Fe(II)L + O_2 \quad (25)$$

Other ligands for iron whose presence encourages the Haber-Weiss cycle, with clear implications for biological damage, are NTA and citrate, though not phytate or succinate (Burkitt and Gilbert, 1990). However, ESR-stop flow students establish that in the presence of ascorbate, where the possibility exists for reduction of complexed Fe(III) back to Fe(II) by *ascorbate ion* (to give $Asc^{-\bullet}$) a number of other ligands (including DTPA and succinate) induce effective hydroxylation (via the so-called *ascorbate-driven* Fenton reaction).

Sutton et al. (1987) have also used an approach involving deoxyribose oxidation to contrast the non-chelated Fe(II)-H_2O_2 reaction (which is believed to occur via a peroxo complex or tetravalent iron) and the Fe(II)EDTA-H_2O_2 reaction, for which it is suggested that 90% of the pathway at pH 7.3 gives 1e-transfer to give $^{\bullet}OH$, with 2e-oxidation accounting for the remaining 10%. Other potential chelates (including citrate, ADP, and ATP) have also been explored (Vile et al., 1987). Evidence has also been obtained that iron bound to xanthine oxidase catalyzes production of a strong oxidant (not $^{\bullet}OH$) from H_2O_2 and O_2^- (Winterbourn and Sutton, 1986).

Evidence that a ferryl species, rather than $^{\bullet}OH$, might be involved in Fe(II)L-H_2O_2 reactions has been summarized and discussed by Rush et al. (1990a). These claims are largely based on stoichiometric studies of the reaction in the presence of different (radical) scavengers for the reactive intermediate, and in particular, on the way in which substrates affect the ratio $\Delta[Fe(II)]/\Delta[H_2O_2]_{lim}$ - i.e. the amount of Fe(III) produced per H_2O_2 consumed. The distinction can in principle be made between the behavior expected for potential oxidizable ($^{\bullet}$C-OH), reducible ($^{\bullet}$C-CO), and "inert" ($R^{\bullet}$) radicals (for which different stoichiometries are expected). The method has been applied for a variety of complexes including EDTA, DTPA and HEDTA (Rush and Koppenol, 1986; 1987) and it has been pointed out that, with ^tBuOH for example, the ratio of 1.7-2.0 for $\Delta(Fe(II)/H_2O_2)$ is best interpreted in terms of oxidation via a higher oxidation state of iron [on the assumption that $^{\bullet}OH$ is involved, to give $^{\bullet}CH_2CMe_2OH$, a ratio of ca 1.0, would be expected: reduction of $^{\bullet}CH_2CMe_2OH$ with Fe(II) (see earlier) to give $CH_2{=}CMe_2$, detected in *low* yields, would not be expected to account for the difference]. This relative *lack* of effective scavenging (see also Rahhal and Richter, 1988) is also typical of several other studies (cf reactions in the presence of Br^-, with ABTS as marker (Rush and Koppenol, 1988)) which serve to emphasize the conclusion that a complete knowledge of the reactions is lacking (Rush et al., 1990a).

C. Comparison of rates of reaction

Rate data taken from a variety of sources referred to above are collected together in Table 1; conditions are believed to be broadly comparable (pH 7, room temperature). As can be

Table 1

Rate Constants for the Fe(II)L-H_2O_2 Reaction (pH 7, room temperature)
($k/10^3\ dm^3\ mol^{-1}\ s^{-1}$)

(a) Aminocarboxylate ligands

	Borggaard et al.[a]	Rahhal and Richter[b]	Rush et al.[c,d]	Sutton and Winterburn[e]	Yamazaki and Piette[f]	Gilbert et al.[g,h]
EDTA	9		7	7	14	7
HEDTA	17		42			10
EDDA			78			23
DTPA		1.3	0.8	0.5	0.4	27
NTA	18		30			9.7
Citrate			4.9			4.9
ADP			11		8.2	
ATP			6.6			

(b) Aminophosphonic acid ligands

Ethylenediaminetetrakis (methylenephosphonic acid)	$(H_2O_3PCH_2)_2NCH_2CH_2N(CH_2PO_3H_2)$	200
Hexamethylenediaminetetrakis (methylenephosphonic acid)	$(H_2O_3PCH_2)_2N(CH_2)_6N(CH_2PO_3H_2)_2$	10
Diethylenetriaminepentakis (methylenephosphonic acid)	$(H_2O_3PCH_2)_2NCH_2CH_2N(CH_2PO_3H_2)CH_2CH_2N(CH_2PO_3H_2)_2$	40
Nitrilotris (methylenephosphonic acid)	$N(CH_2PO_3H_2)_3$	200

a Borggaard, O.K., Farver, O., and Andersen, V.S. (1971). *Acta Chem. Scand.*, *25*: 3541.
b Rahhal, S., and Richter, H.W. (1988). *J. Amer. Chem. Soc.*, *110*: 3126.
c Rush, J.D., and Koppenol, W.H. (1987). *J. Inorg. Biochem.*, *29*: 199.
d Rush, J.D., Maskos, Z., and Koppenol, W.H. (1990b). *FEBS Lett.*, *261*: 121.
e Sutton, H.C., and Winterbourn, C.C. (1984). *Arch. Biochem. Biophys.*, *235*: 106.
f Yamazaki, I., and Piette, L.H. (1990). *J. Biol. Chem.*, *265*: 13589.
g Croft, S., Gilbert, B.C., Lindsay Smith, J.R., Stell, J.K., and Sanderson, W.R. (1992). *J. Chem. Soc., Perkin Trans.*, *2*: 153.
h Croft, S., Gilbert, B.C., Lindsay Smith, J.R., and Whitwood, A.C. (1992). *Free Rad. Res. Commun.*, *17*: 21.

seen, there is good agreement over the appropriate value for EDTA (ca $7 \times 10^3\ dm^3\ mol^{-1}\ s^{-1}$). The rate constants for HEDTA and NTA appear to be broadly similar, perhaps a little higher, whereas those of citrate and particularly DTPA are reduced (there is a surprising spread in the values of the latter). The aminophosphonic acids clearly accelerate the reaction.

In comparing the rate constants for different ligands, it might be noted that the fact that the rate constants for Fe(II)EDTA and some related ligands are so much greater than that for $Fe(II)_{aq}$ is as expected if an outer-sphere mechanism is involved, for which the rate constant would be largely governed by thermodynamic factors (compare E° values of 0.12V and 0.71V, respectively; cf E° for $H_2O_2/{}^{\bullet}OH$ 0.46 V: Koppenol and Butler (1985)). On the other hand, the observation of a considerably lower rate constant for Fe(II)DTPA (for which E° is 0.03V) than Fe(II)EDTA and its close analogues suggests that steric hindrance might be involved - and hence that an inner-sphere reaction, involving formation of a peroxo-complex, is likely to be the appropriate mechanism (reaction (23)).

$$Fe(II) + H_2O_2 \rightarrow Fe(II)\cdots\overset{\overset{\large H}{|}}{O}\text{—}OH \rightarrow Fe(III) + HO^{\bullet} + OH^- \qquad (23)$$

D. Reactions of DFO and its protective effects

For an iron complex to catalyze the generation of the hydroxyl radical from hydrogen peroxide, it must permit both the efficient reduction of iron(III) and the subsequent reaction of iron(II) in the Fenton reaction (e.g. EDTA, citrate or NTA). On the basis of oxygen-uptake, spin-trapping and hydroxylation reactions it has been shown (Burkitt and Gilbert, 1990; 1991) that desferrioxamine, DFO [which has been used in the treatment of iron overload (N.F. Olivieri et al., 1986)] prevents both reactions of the Haber-Weiss cycle and also the ascorbate-driven Fenton reaction. Thus the iron-reduction step (reaction (25)) is retarded because the metal-ion's redox potential [E° Fe(III)/Fe(II)-DFO, -0.45V] is reduced below that of the O_2/O_2^- couple; and the Fenton reaction (1) does not occur to a significant extent because the Fe(II) complex is very rapidly oxidized to Fe(III) by oxygen (Burkitt and Gilbert, 1990; 1991). It is also believed that organic radicals fail to reduce Fe(III)-DFO and hence stimulate Fenton chemistry.

IV. HIGH-VALENCE STATE IRON INTERMEDIATES

Whilst the hydroxyl radical has been implicated in the majority of aqueous Fe(II)- hydrogen peroxide reactions, there are examples where it is evident that a different oxidizing intermediate is involved. For example, Groves and Van der Puy (1974; 1976) have shown that when hydrogen peroxide is added slowly to a solution containing cyclohexanol and ferrous iron in 90% MeCN/10% H_2O, cyclohexanone and a family of diols are formed: high regio- and stereo-selectivity is observed, with 70% of the product being the *cis*-1,3 - diol. This selectivity is lost when 50% aqueous acetonitrile is used as the solvent. These results are clearly inconsistent with the generation and reaction of $^{\bullet}OH$ via the Fenton mechanism, and the involvement of an iron-bound oxidizing intermediate (Scheme 4) has been proposed.

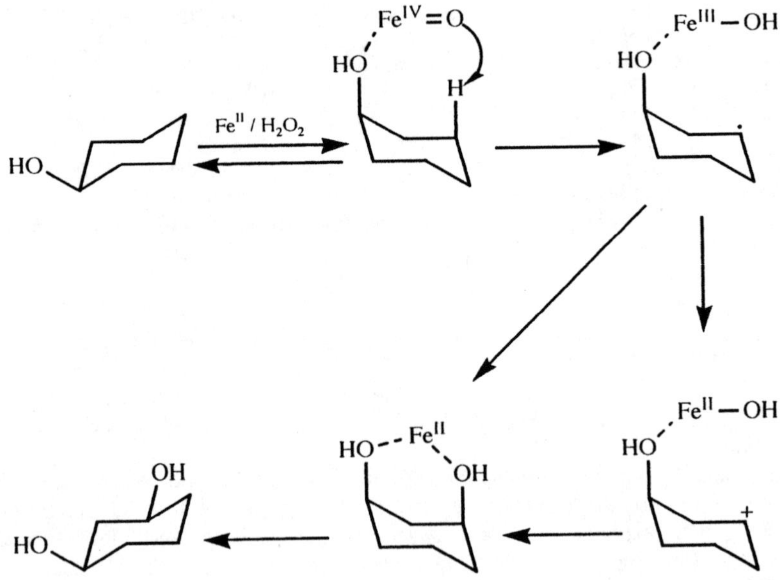

Scheme 4: Oxidation with Fe(II)/H_2O_2 in MeCN/H_2O

Sawyer and Sugimoto (1984; 1985) have also reported that when dilute hydrogen peroxide is added to dry MeCN containing Fe(II) and RH, mono-oxygenation (PPh_3) or dehydrogenation ($PhCH_2OH$) reactions occur, depending on the substrate: $FeCl_3$ in dry MeCN activates hydrogen peroxide for the epoxidation of alkenes. These results are also inconsistent with the production of the hydroxyl radical, and an iron-oxygen intermediate has been suggested.

Stein and Weiss (1951) have shown that the Fenton oxidation of phenol produces a different isomer ratio to that observed when phenol is oxidized by hydroxyl radicals generated from the radiolysis of water. Hamilton and his co-workers have studied this system in detail and propose a mechanism which is consistent with the observed kinetics and products: the crucial intermediate is thought to be an iron oxide which is formed by the elimination of water from a complex involving ferric iron, hydrogen peroxide, and the aromatic diol (Hamilton et al., 1966).

Finally, the reactions of H_2O_2 with iron-containing porphyrins is of great interest, not least because of their relevance to enzyme behavior. It is generally believed that the reaction of ferric porphyrins with hydrogen peroxide produces the high-valent iron oxo species ($^{+\bullet}$porph)Fe(IV)O as an oxidizing intermediate (Zipplies et al., 1986; Panicucci and Bruice, 1990), though the exact mechanism depends on pH and upon the nature of the porphyrin. The reader is referred to these excellent reports and the references therein.

REFERENCES

Barb, W.G., Baxendale, J.H., George, P., and Hargrave, K.R. (1951a). *Trans. Faraday Soc.*, *47*: 462.

Barb, W.G., Baxendale, J.H., George, P., and Hargrave, K.R. (1951b). *Trans. Faraday Soc.*, *47*: 591.

Bielski, B.H.J., Capelli, D.E., Arudi, R.L., and Ross, A.B. (1985). *J. Phys. Chem. Ref. Data*, *14*: 1041.

Borggaard, O.K., Farver, O., and Andersen, V.S. (1971). *Acta Chem. Scand.*, *25*: 3541.

Bray, W.C., and Gorin, M.H. (1932). *J. Amer. Chem. Soc.*, *54*: 2124.

Brown, S.B., Jones, P., and Suggett, A. (1970). *Prog. Inorg. Chem.*, *13*: 159.

Buettner, G.R. (1987). *Free Rad. Biol. Med.*, *3*: 259.

Buettner, G.R., Doherty, T.P., and Patterson, L.K. (1983). *FEBS Lett.*, *158*: 143.

Burkitt, M. (1993). *Free Rad. Res. Commun.*, *18*: 43.

Burkitt, M., and Gilbert, B.C. (1990). *Free Rad. Res. Commun.*, *10*: 265.

Burkitt, M., and Gilbert, B.C. (1991). *Free Rad. Res. Commun.*, *14*: 107.

Buxton, G.V., Greenstock, C.L., Hetman, W.P., and Ross, A.B. (1988). *J. Phys. Chem. Ref. Data*, *17*: 513.

Croft, S., Gilbert, B.C., Lindsay Smith, J. R., Stell, J.K., and Sanderson, W.R. (1992a). *J. Chem. Soc., Perkin. Trans.*, *2*: 153.

Croft, S., Gilbert, B.C., Lindsay Smith, J.R., and Whitwood, A.C. (1992b). *Free Rad. Res. Commun.* *17*: 21.

Czapski, G. (1971). *J. Phys. Chem.*, *75*: 2957.

Davies, M.J. (1987). *Chem. Phys. Lipids*, *44*: 149.

Edwards, J.O., and Curci, R. (1993). *Activation of Hydrogen Peroxide* (G. Struker, ed.), Kluwer Academic Press, Dordrecht, p. 97.

Fenton, H.J.H. (1894). *J. Chem. Soc.*, *65*: 899.

Gilbert, B.C., and Stell, J.K. (1990). *J. Chem. Soc., Perkin Trans.*, *2*: 1281.

Gilbert, B.C., and Stell, J.K. (1991). *J. Chem. Soc., Faraday Trans.*, *86*: 3261.

Gilbert, B.C., and Jeff, M. (1988). *Free Radicals in Chemistry, Biology and Medicine* (C. Rice-Evans and T. Dormandy, eds.), Richelieu Press, p. 25.

Gilbert, B.C., Stell, J.K., and Jeff, M. (1988). *J. Chem. Soc., Perkin Trans.*, *2*: 1867.

Groves, J.T., and Van der Puy (1974). *J. Amer. Chem. Soc.*, *96*: 5274.

Groves, J.T., and Van der Puy, M. (1976). *J. Amer. Chem. Soc.*, *98*: 5290.

Haber, F., and Weiss, J. (1934). *Naturwiss.*, *20*: 948.

Halliwell, B., and Gutteridge, J.M.C (1990). *Meth. Enzymol.*, *186*: 1.

Hamilton, G.A., Hanifin, J.W., and Friedman, J.P. (1966). *J. Amer. Chem. Soc.*, *88*: 5269.

Hardwick, T.J. (1957). *Can. J. Chem.* *35*: 428.

Janzen, E.G., and Haire, L.D. (1990). *Adv. Free Rad. Chem.* *1*: 253.

Koppenol, W.H., and Butler, J. (1985). *Adv. Free Rad. Biol. Med.*, *1*: 91.

Mertz, J.H., and Waters, W.A. (1947). *Disc. Faraday Soc.*, 2: 179.

Mertz, J.H., and Waters, W.A. (1949). *J. Chem. Soc.*, 2427.

Norman, R.O.C., and Gilbert, B.C. (1967). *Adv. Phys. Org. Chem.*, *5*: 53.

Olivieri, N.F., Buncic, J.R., Chew, E., Gallent, T., Harrison, R.V., Keenan, N., Logan, W., Mitchell, D., Ricci, G., Skarf, B., Taylor, M., and Freedman, M.H. (1986). *N. Engl. J. Med.*, *314*: 869.

Panicucci, R., and Bruice, T. (1990). *J. Amer. Chem. Soc.*, *112*: 6063.

Rahhal, S., and Richter, H.W. (1988). *J. Amer. Chem. Soc.*, *110*: 3126.

Rush, J.D., and Koppenol, W.H. (1986). *J. Biol. Chem.*, *261*: 6730.

Rush, J.D., and Koppenol, W.H. (1987). *J. Inorg. Biochem*, *29*: 199.

Rush, J.D., and Koppenol, W.H. (1988). *J. Amer. Chem. Soc.*, *110*: 4957.

Rush, J.D., Maskos, Z., and Koppenol, W.H. (1990a). *Meth. Enzymol.*, *186*: 148.

Rush, J.D., Maskos, Z., and Koppenol, W.H. (1990b). *FEBS Lett.*, *262*: 121.

Sawyer, D.T., and Sugimoto, H. (1984). *J. Amer. Chem. Soc.*, *106*: 4283.

Sawyer, D.T., and Sugimoto, H. (1985). *J. Org. Chem.*, *50*: 1784.

Shiga, T. (1965). *J. Phys. Chem.*, *69*: 3805.

Stein, G., and Weiss, J. (1951). *J. Chem. Soc.*, 3265.

Sutton, H.C., and Winterbourn, C.C. (1984). *Arch. Biochem. Biophys.*, *235*: 106.

Sutton, H.C., Vile, G.F., and Winterbourn, C.C. (1987). *Arch. Biochem. Biophys.*, *256*: 462.

Uri, N. (1952). *Chem. Rev.*, *50*: 375.

Vile, G.F., Winterbourn, C.C., and Sutton, H.C. (1987). *Arch. Biochem. Biophys.*, *259*: 616.

Walling, C. (1975). *Accounts Chem. Res.*, *8*: 125.

Walling, C., and Kato, S. (1971). *J. Amer. Chem. Soc.*, *93*: 4275.

Walling, C., and El Taliawi, E.M. (1973a). *J. Amer. Chem. Soc.*, *95*: 844.

Walling, C., and El Taliawi, E.M. (1973b). *J. Amer. Chem. Soc.*, *95*: 848.

Walling, C., El Taliawi, E.M., and Johnson, R.A. (1974). *J. Amer. Chem. Soc.*, *96*: 133.

Winterbourn, C.C., and Sutton, H.C. (1986). *Arch. Biochem. Biophys.*, *244*: 27.

Yamazaki, I., and Piette, L.H. (1990). *J. Biol. Chem.*, *265*: 13589.

Yamazaki, I., and Piette, L.H. (1991). *J. Amer. Chem. Soc.*, *113*: 7588.

Zipplies, M.F., Lee, W.A., and Bruice, T. (1986). *J. Amer. Chem. Soc.*, *108*: 4433.

COPPER COMPLEXES AND FREE RADICALS

by

U. Weser, C. Steinkühler

Anorganische Biochemie,
Physiologish-Chemisches Institut der Universität Tübingen,
Hoppe Seyler Straße 4, D-72076 Tübingen
Germany

and

G. Rotilio

Department of Biology,
University of Rome "Tor Vergata", Rome
Italy

I. INTRODUCTION

The pronounced tendency of copper to coordinate with biologically relevant chelators and the pronounced positive redox potentials of the resulting compounds are features that render copper an ideal cofactor for enzymes involved in the catalysis of oxygen-based redox reactions. Considering the chelation properties of the metal, it is not surprising that an overwhelming portion of the total body copper is bound to proteinaceous ligands under physiological conditions. In mammalian serum for instance, which contains about 15 μM copper, roughly 93% are found bound to caeruloplasmin while the residual portion is bound to albumin and only 2-4 nM are chelated by low-molar-mass (l.m.m.) serum components (Deuschle and Weser, 1985). The two major complexes are copper-albumin, a related complex called transcuprein (Wirth and Linder, 1985) and a ternary Cu-albumin-histidine complex (Lau and Sarkar, 1971). Likely, over 90% of the erythrocyte copper is bound to the protein Cu,Zn-SOD (Gärtner and Weser, 1984).

On the other hand, in copper-resistant hepatomas over 60% of the cytoplasmic copper was isolated as a l.m.m. copper-glutathione complex (Freedman et al., 1989), and data obtained with cells cultured in the presence of micromolar copper concentrations indicate that a minor, as yet unquantified, amount of intracellular copper may also exist in a glutathione-associated form under physiological conditions (Steinkühler et al., 1991). Thus, although to a small extent, l.m.m. copper compounds are present in organisms both extra- and intracellularly, and their possible interaction with oxygen might be of physiological importance - especially in hereditary diseases affecting copper metabolism such as Wilson's or Menkes syndrome which eventually lead to abnormal copper distribution within the organism (Ettinger, 1984).

The aim of this chapter is to review some general characteristics of the interaction of copper complexes with oxygen and its reduction products, and to discuss their possible physiological consequences.

II. ASPECTS OF COPPER REDOX CHEMISTRY IN PHYSIOLOGICAL MILIEUX

The redox behaviour of ligated copper ions is strongly influenced by the molecular architecture of the ligating species. Thus, Cu $3d^9$ (Cu(II)) is unstable in symmetric tetrahedral or octahedral ligand fields, whereas it is usually found in Jahn-Teller distorted 6-coordinate tetragonal or 5-coordinate square pyramidal coordination geometries. Quite frequently, the metal-ligand distances are conspicuously longer for the axial ligands, which are usually kinetically labile. This particular situation confers high exchange and reaction rates to complexes with free or solvent-coordinated axial positions (Schechinger et al., 1988).

Cu(I) is preferentially found in tetrahedral binding geometries. Moreover, this species, unless sufficiently stabilized by appropriate ligands, will undergo disproportionation to Cu(0) and Cu(II) in aqueous solutions. Among biologically relevant ligands, Cu(II) is found to be coordinated most strongly to nitrogen (either aliphatic or heteroaromatic) or oxygen (mainly carboxylate) as donor atoms whereas Cu(I) is generally coordinated to "soft" thiolate sulfur or unsaturated nitrogen groups.

As a consequence of these differences in coordination geometry, any factor(s) affecting the symmetry of the ligand field is also expected to have effects on the redox behavior of the ligated copper ion. In fact, to be able to bind both Cu(I) and Cu(II), a ligand must be flexible enough to shuttle between distorted tetragonal and tetrahedral coordination structures. Thus, complexes lacking flexibility and free coordination sites, such as Cu(II)-EDTA, will render the metal almost redox-inert (Beyer and Fridovich, 1988). In this context, however, it should be pointed out that most copper complexes with flexible ligands will prove unstable in physiological milieux due to the presence of excess competing unspecific copper-binding sites on proteins, which usually readily sequester copper from l.m.m. complexes (Deuschle and Weser, 1985). The redox chemistry of this unspecifically protein-bound copper is still poorly explored.

The stability of a given complex can vary considerably as a function of the physiological milieu in which it is investigated. This can be related to differences in the redox state of the metal: whereas Cu(II) can apparently exist in blood (Ettinger, 1984; Sarkar, 1981), intracellular copper is likely to be mainly Cu(I).

Bleomycin, for example, has a high affinity for Cu(II) and sequesters the metal in the blood (Kanao et al., 1973). Upon entry into the cell, the Cu(II)-Bleomycin complex is reduced and the resulting copper (I) is transferred to a cellular protein, presumably to metallothionein (Takahashi et al., 1977).

Intracellular reduction of copper complexes is apparently accomplished mainly by thiol compounds (Byrnes et al., 1990; Steinkühler et al., 1990). In the case of thiosemicarbazonato-Cu(II) complexes, ternary adducts with thiols have been characterized in cells by low temperature EPR spectroscopy (Saryan et al., 1981). The main intracellular thiol involved appears to be glutathione, as evidenced by studies using cultured cells or isolated erythrocytes and a number of copper compounds (Byrnes et al., 1990; Caffrey et al., 1986; Fernandes et al., 1988; Steinkühler et al., 1991a). Ascorbate (Shinar et al., 1989), H_2O_2 (Ferreira and Toma, 1988) and presumably also electron transport chains (Werringloer et al., 1979) are also able to reduce a number of copper coordination compounds. NADH or NADPH will not produce any detectable reduction of Cu(II)-diSchiff-base (Steinkühler et al., 1991a), Cu(II)-thiosemicarbazonato (Byrnes et al., 1990) or Cu(II)-phenanthroline (Que et al., 1980) complexes in the absence of H_2O_2. In the presence of H_2O_2, however, pyridine dinucleotides are able to reduce Cu(II)-phenantroline, albeit by an indirect mechanism postulated to proceed via formation of paramagnetic NAD species (Reich et al., 1981).

In the presence of oxygen, the Cu(I)-species formed by one of these reactions are able to cause damage to proteins (Marx and Chevion, 1985), lipids (Stacey and Klaassen, 1981), nucleic acids (Reed and Douglas, 1989), and to oxidize alcohols (Johnson and Nazhat, 1987), deoxyribose (Gutteridge and Halliwell, 1982) and salicylic acid (Rowley and Halliwell, 1983). The exact nature of the oxidizing species is, in many cases, still object of debate (Sutton and Winterbourn, 1989) and may vary according to the system observed. It is usually assumed that the oxidant is $^{\bullet}OH$, produced in a Fenton type reaction (with Red = reduced glutathione or other thiol compounds or ascorbate):

$$Red + Cu(II) \rightarrow Cu(I) + Red^{-} \quad (1)$$
$$2\,Cu(I) + 2\,H^{+} + O_2 \rightarrow 2\,Cu(II) + H_2O_2 \quad (2)$$
$$Cu(I) + H_2O_2 \rightarrow Cu(II) + OH^{-} + {}^{\bullet}OH \quad (3)$$

Cyclic voltammetry studies done with Cu(II)-imidazole complexes lead to the conclusion that reaction (3) is slow as compared to the reoxidation of the reduced complex by oxygen (Ferreira and Toma, 1988), indicating that the reoxidation of Cu(I) will be mainly accomplished by O_2 as long as $[O_2] >> [H_2O_2]$. This means that $^{\bullet}OH$ generation by Cu(I) and H_2O_2 is favored under hypoxic conditions.

The $^{\bullet}OH$ radical was invoked to be the ultimate oxidant in chemical cleavage of DNA by glutathione (Reed and Douglas, 1991) or ascorbate and H_2O_2 (Stoewe and Prütz, 1987) in the presence of Cu(II) ions. Generally, the involvement of $^{\bullet}OH$ is inferred from protection by scavengers. This procedure may not always give unequivocal results. Thus, Sagripanti and Kraemer (1989) reported the involvement of hydroxyl free radical in the cleavage reaction of supercoiled DNA by Cu(II) and H_2O_2. However, spin trapping experiments suggest that the oxidant involved in DNA damage by Cu(II) and H_2O_2 is not $^{\bullet}OH$ itself, but rather a copper-peroxo species with a reactivity similar to singlet oxygen or $^{\bullet}OH$ radical (Yamamoto and Kawanishi, 1989). Its formation was suggested to be triggered by a previous reduction of copper by H_2O_2 according to equation (4):

$$Cu^{2+} + H_2O_2 \rightarrow Cu^{+} + HO_2 + H^{+} \quad (4)$$
$$Cu^{+} + H_2O_2 \rightarrow [Cu\text{-}OOH] + H^{+} \quad (5)$$

The cleavage of DNA by Cu(II)-phenantroline in the presence of thiols and H_2O_2 was suggested not to proceed via a diffusible species, as $^{\bullet}OH$ scavengers used at a concentration of 10 mM failed to inhibit it (Marshall et al., 1981). However, Que et al. (1980) were able to observe protection by addition of 50-100 mM $^{\bullet}OH$ scavengers. The suggestion (Goldstein and Czapski, 1985) that in the case of Cu(II)-phenanthroline, a copper-peroxo complex is first formed which then decomposes to yield approximately equal amounts of $^{\bullet}OH$ and Cu(III), both of which are capable to react with DNA, may reconcile these apparently contradictory results. The ability to observe a protection by scavenging molecules may in these cases be strictly dependent on the ratios between target and scavenger, and on the accessibility of the scavenger to the microenvironment in which the radical is formed. This may be of particular importance in the case of so-called site-specific damage (Czapski and Aronowitch, 1984), where the oxidation reaction occurs at the metal binding site of a macromolecule, either because a metal-linked oxidant (e.g. a Cu(III) species or a copper-peroxo complex) is formed that reacts with neighboring nucleophilic groups on the macromolecule, or because a diffusible species is formed (e.g. the $^{\bullet}OH$) which, due to its high reactivity, will react preferentially with groups of the macromolecule that are adjacent to its formation site.

There are many examples for site-directed cleavage of DNA (Pope and Sigman, 1984; Spassky and Sigman, 1985; Francois et al., 1988; Veal and Rill, 1988; Yoon et al., 1990) or proteins (Marx and Chevion, 1985) by copper ions or complexes in the presence of reductants and oxygen (or partial reduction products thereof). It generally appears that the tendency to produce site-directed damage to DNA is greater for copper complexes than for iron (Celander and Cech, 1990), which has an analogous redox chemistry. Probably this finding reflects the more pronounced steric requirements of copper for binding sites on nucleic acids. In fact, Cu(II) ions preferentially bind to the DNA bases A, G and C at N7 of purines and N1 of the pyrimidine (Martin and Mariam, 1979; Tkeshelashvili et al., 1991). Furthermore, Cu binding to DNA may depend also on specific conformations of the macromolecule. In fact, single strand DNA with the capacity to form secondary hairpin structures that accomodate Cu bound to a G residue is specifically cleaved at this single site

(Kazakov et al., 1988). Moreover, comparison of the interaction of Cu(II)bleomycin, Cu(II)-2-formylpyridine monothiosemicarbazonato and the tripeptide complexes Cu(II)-GHK and Cu(II)-GHG with DNA, using EPR spectroscopy, indicates that ternary complexes are formed with the macromolecule whose spatial orientation strongly depends on the ligating species (Chikira et al., 1991).

III. SCAVENGING OF PARTIALLY REDUCED OXYGEN SPECIES BY COPPER COMPOUNDS

A. Reactivity with hydrogen peroxide

Many copper compounds catalytically decompose H_2O_2 (Sigel, 1969). As copper undergoes cyclic changes in its redox state during catalysis, the reactivity of a given complex depends on its coordination geometry according to what has been discussed above. The catalytic activity is considerably enhanced by ligands such as pyridine (Otto et al., 1981), 2,2'-bipyridine (Sigel et al., 1979), ethylenediamine (Kaden and Sigel, 1968) and 1,10-phenanthroline (Goldstein and Czapski, 1983). Binuclear complexes with appropriate geometries react faster than mononuclear planar compounds (Nishida et al., 1982). In the case of Cu(II)-imidazolate, the rate-limiting step appears to be the formation of a Cu(II)-peroxo species that decomposes yielding the Cu(I)-complex (Ferreira and Toma, 1988). Generally, Cu(I)-species thus formed are able to react with a second molecule of H_2O_2 in a copper(I)-driven Fenton reaction. From a biological standpoint, the copper-catalyzed H_2O_2 decomposition is therefore deleterious and can cause oxidative damage to physiologically relevant target molecules.

B. Reactivity with superoxide

Pulse radiolysis experiments have shown that copper ions and a series of copper coordination compounds catalyze the dismutation of the superoxide radical very efficiently (Table 1). The mechanism of the dismutation reaction is apparently the same as for the enzyme Cu,Zn-superoxide dismutase (Cu,Zn-SOD), i.e. a Ping-Pong mechanism by which copper is alternately reduced and oxidized by the superoxide anion radical, O_2^- (Fielden et al., (1974):

$$Cu^{2+} + O_2^- \rightarrow Cu^+ + O_2 \qquad (6)$$
$$Cu^+ + O_2^- + 2H^+ \rightarrow Cu^{2+} + H_2O_2 \qquad (7)$$

Although many copper complexes catalytically remove superoxide in pulse radiolysis experiments, they fail to do so in vivo because of their already discussed instability in the presence of competing biological ligands (Deuschle and Weser, 1985). On the other hand, ligand flexibility, which is the cause of kinetic lability, is a prerequisite for a superoxide dismutase mimic that has to efficiently accomplish a redox-shuttle. In an attempt to solve this problem, tetradentate Schiff-base complexes bridged by an aliphatic C4-chain were

Table 1

SOD Activity of Copper Complexes Determined by Pulse Radiolysis

Copper compound	k_{cat} ($M^{-1}s^{-1}$)/eq. of Cu
Cu,Zn SOD	1.3×10^9
Cu(II) aq.	2.7×10^9
Cu(II)-Lys	0.56×10^9
Cu(II)-(Gly-His)	0.29×10^9

Data from Fielden et al. (1974) and Brigelius et al. (1974).

synthesized (Fig. 1). The most efficient SOD-mimic thus obtained, CuPuPy, was sufficiently stable in the presence of albumin (Linss and Weser, 1987) and had an SOD activity about 3% of the activity of bovine Cu,Zn-SOD (Linss and Weser, 1986), which is in the order of magnitude of many proteinaceous counterparts. Analogous with native Cu,Zn-SOD, CuPuPy protected cells from the toxic effects of superoxide radicals produced by xanthine oxidase (Fig. 2). Furthermore, the compound inhibited the superoxide production by 50% during the oxidative burst of activated polymorphonuclear leukocytes (Miesel et al., 1990).

Although it thus appears possible to synthesize model compounds that survive in the extracellular milieu and display SOD activity in pulse radiolysis experiments, most complexes studied up to now are of undesirable toxicity (Steinkühler et al., 1988). This toxicity is closely related to a lack of specificity towards superoxide as substrate. In fact, the enzyme Cu,Zn-SOD reacts at high rates and very specifically with its substrate, through pre-collision guidance of the negatively charged O_2^- by an electrostatic gradient at the active site produced by positively charged amino acid residues (Getzhoff et al., 1983).

In the case of l.m.m. analogs of Cu,Zn-SOD, the lack of a protein moeity renders the metal much more accessible to reductants or oxidants other than superoxide. Considering that the intracellular concentrations of reductants like glutathione are in the order of 10^{-3} M (Meister, 1988) against 10^{-11} M O_2^- (Cadenas et al., 1983), it is not surprising that the copper in these complexes, instead of performing reaction (6) of the catalytic cycle, is reduced by reductants other than O_2^- upon entry into a cell. The reduced complex on the other hand is much more likely to be reoxidized by molecular oxygen, whose physiological concentration is estimated to be in the order of 10^{-4} M (Cadenas et al., 1983), than by O_2^-. The result is a redox cycle in which copper catalyzes the oxidation of a given reductant (e.g. glutathione (GSH)) with oxygen as an electron acceptor:

$$2\ GSH + 2\ Cu^{2+} \rightarrow 2\ Cu^{+} + GSSG + 2\ H^{+} \quad (8)$$

$$2\ Cu^{+} + O_2 + 2\ H^{+} \rightarrow 2\ Cu^{2+} + H_2O_2 \quad (9)$$

$$\text{net: } 2\ GSH + O_2 \rightarrow GSSG + H_2O_2 \quad (10)$$

In the case of Cu(I)-pyridine complexes, the mechanism of the reoxidation reaction (9) seems to proceed via a ternary complex between two copper(I) and one O_2, yielding H_2O_2 in a two-electron transfer reaction (Stevanato et al., 1984):

$$2\ Cu^{+} + O_2 \rightarrow Cu^{+}\text{--O-O--}Cu^{+} \xrightarrow{2H^{+}} 2\ Cu^{2+} + H_2O_2 \quad (11)$$

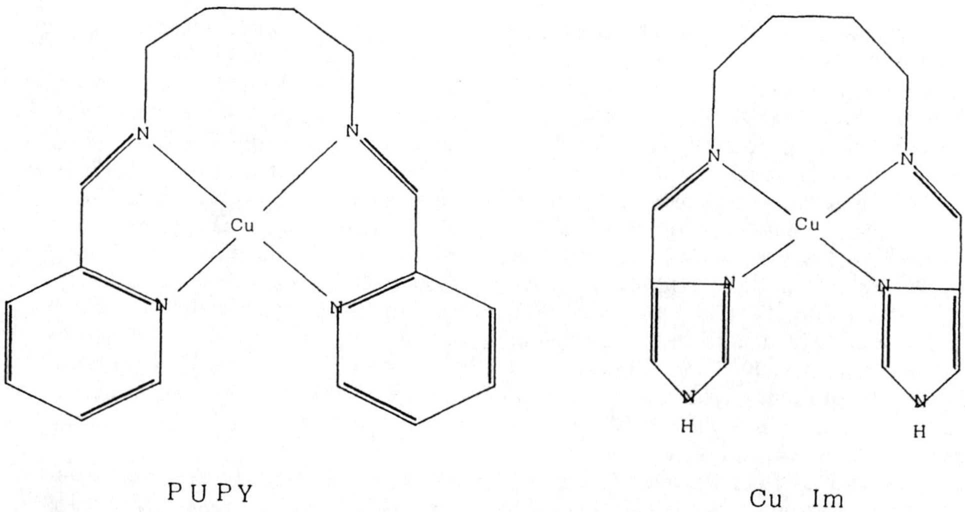

Figure 1.
Flexible di-Schiffbase copper complexes with superoxide dismutase-like activity.

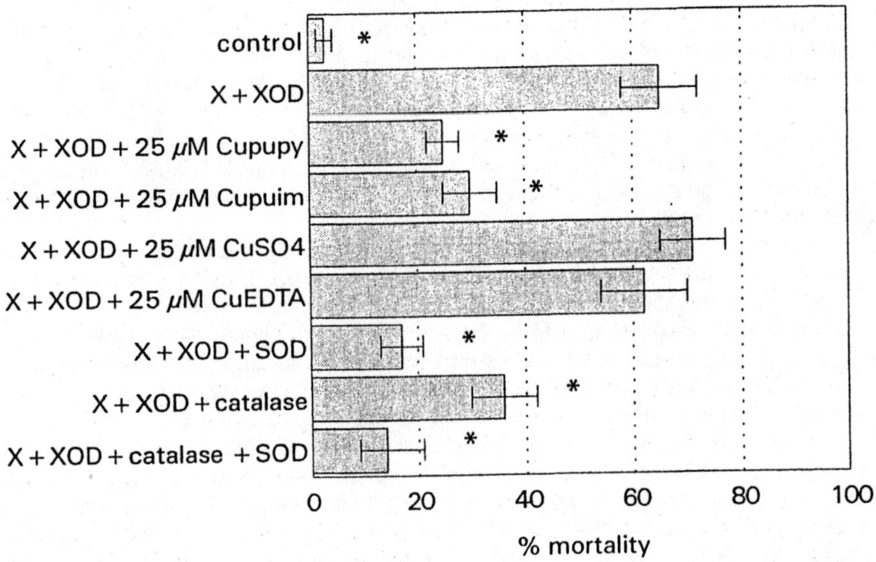

* significantly different from x + xod-treated cells as evaluated by Student's t-test ($P < 0.005$)

Figure 2.
2 x 10^6 K562 erythroleukemia cells were incubated in two ml Eagle's minimal essential medium supplemented with 10% heat-inactivated fetal calf serum at 37°C in humidified atmosphere containing 5% CO_2. 20 mU xanthine oxidase (XOD) and 1 mM xanthine (X) were added to the cell suspension containing copper compounds or 12.5 µM Cu,Zn-SOD. Where indicated, catalase was added at a final concentration of 1000 U/ml. After 21 h aliquots of the cell suspensions were collected and viability was determined by trypan blue exclusion. No decrease in viability was observed in the presence of 25 µM ligands of any copper compound tested. Data are mean values ± standard deviation of three different experiments.

This reaction is not possible with Cu,Zn-SOD as the distance between the two copper ions in the protein is about 30 Å (Tainer et al., 1983), and an intermolecular mechanism appears improbable due to steric hindrance by the protein moeity. Reduced Cu,Zn-SOD is therefore reoxidized by oxygen only via a much slower one-electron tranfer on oxygen yielding O_2^- (Viglino et al., 1986). Table 2 illustrates that for Cu,Zn-SOD $k_{(7)} >> k_{(9)}$, whereas for SOD mimics this difference is generally smaller (Czapski and Goldstein, 1988), which causes the macroscopic catalytic constant to become $k_{cat} << k_{(6)}$ at physiological $[O_2^-]$. These circumstances render l.m.m. SOD analogs effective only at O_2^- concentrations $> 10^{-8}$ M (Czapski and Goldstein, 1991), e.g. those produced by pulse radiolysis, during which in fact both Cu,Zn-SOD and its l.m.m. analogs show comparable activities (Table 1).

As depicted above, the toxicity of SOD mimics results from their ability to perform redox cycles with concomitant production of H_2O_2. As many tumor cells are deficient in enzymes that detoxify H_2O_2 (Mavelli et al., 1987), some SOD mimics show selective toxicity toward tumor cells both in vitro and in vivo (Kensler et al., 1983; Leuthauser et al., 1981; Oberley et al., 1983; Oberley et al., 1984; Steinkühler et al., 1988; Steinkühler et al., 1990; Miesel and Weser, 1990).

Whether this peculiar reactivity will turn out to be pharmacologically exploitable greatly depends on the possibility to design complexes with improved overall stability and tumor cell selectivity. This and the attempt to design an efficient l.m.m. superoxide dismutase mimic appear to be the most promising task.

Table 2

Kinetic Parameters of the Superoxide Dismutation Reaction Catalyzed by Copper Compounds

Compound	$k_{(6)}$ ($M^{-1}s^{-1}$)	$k_{(7)}$ ($M^{-1}s^{-1}$)	$k_{(9)}$ ($M^{-1}s^{-1}$)	k_{cat} ($M^{-1}s^{-1}$)
Cu,Zn SOD	2.6×10^9	2.6×10^9	4.4×10^{-1}	2.6×10^9
Cu(II)-Phenanthroline	1.9×10^9	3.0×10^8	5.0×10^4	1.0×10^5
Cu(II)-Histidine	3.4×10^8	3.4×10^8	4.0×10^4	2.4×10^5

Data from Fielden et al. (1974), Czapski and Goldstein (1988), and Viglino et al. (1986).

ACKNOWLEDGEMENTS

C.S. was recipient of a post-doctoral fellowship awarded by the Deutsche Forschungsgemeinschaft. Previous support by the same foundation (DFG We 401-24/1-3) and by the Consiglio Nazionale delle Ricerche special project "Chimica fine II" for experimental work is gratefully acknowledged.

REFERENCES

Beyer, W.F., and Fridovich, I. (1988). *Anal. Biochem.*, *173*: 160.

Brigelius, R., Spöttl, R., Bors, W., Lengfelder, E., Saran, M., and Weser, U. (1974). *FEBS Lett.*, *47*: 72.

Byrnes, R.W., Mohan, M., Antholine, W.E., Xu, R.X., and Petering, D.H. (1990). *Biochemistry*, *29*: 7046.

Cadenas, E., Brigelius, R., Akerboom, T., and Sies, H. (1983). *34. Kolloquium Moosbach: Biological Oxidations*, Springer Verlag, Heidelberg, Berlin, p. 288.

Caffrey, J.M., Dasmahapatra, A., Smith, H.A., Hede, K., and Frieden, E. (1986). *Biol. Tr. Elem. Res.*, *11*: 19.

Celander, D.W., and Cech, T.R. (1990). *Biochemistry*, *29*: 1355.

Chikira, M., Sato, T., Antholine, W.E., and Petering, D.H. (1991). *J. Biol. Chem.*, *266*: 2859.

Czapski, G., and Aronowitch, J. (1984). *Oxygen Radicals in Chemistry and Biology* (W. Bors, M. Saran and D. Tait, eds.), Walter de Gruyter, Berlin, New York, p. 225.

Czapski, G., and Goldstein, S. (1988). *Free Rad. Res. Commun.*, *4*: 225.

Czapski, G., and Goldstein, S. (1991). *Free Rad. Res. Commun.*, *12-13*: 167.

Deuschle, U., and Weser, U. (1985). *Prog. Clin. Biochem. Med.*, *2*: 99.

Ettinger, M. (1984). *Copper Proteins and Copper Enzymes* (R. Lontie, ed.), CRC Press, Boca Raton, Fl., p. 175.

Fernandes, A., Mira, M.L., Azevedo, M.S., and Manso, C. (1988). *Free Rad. Res. Commun.*, *4*: 291.

Ferreira, A.M.C., and Toma, H.E. (1988). *J. Coord. Chem.*, *18*: 351.

Fielden, E.M., Roberts, P.B., Bray, R.C., Lowe, D.J., Mautner, G.N., Rotilio, G., and Calabrese, L. (1974). *Biochem. J.*, *139*: 49.

Francois, J.C., Saison-Behmoaras, T., Chassignol, M., Thuong, N.T., Sun, J.S., and Helene, C. (1988). *Biochemistry*, *27*: 2272.

Freedman, J.H., Ciriolo, M.R., and Peisach, J. (1989). *J. Biol. Chem.*, *264*: 5598.

Gärtner, W., and Weser, U. (1984). *Biochem. J.*, *221*: 549.

Getzhoff, E.D., Tainer, J.A., Weiner, P.K., Kollman, P.A., Richardson, J.S., and Richardson, D.C. (1983). *Nature*, *306*: 287.

Goldstein, S., and Czapski, G. (1983). *J. Amer. Chem. Soc.*, *103*: 7276.

Goldstein, S., and Czapski, G. (1985). *Free Rad. Biol. Med.*, *1*: 373.

Gutteridge, J.M.C., and Halliwell, B. (1982). *Biochem. Pharmacol.*, *31*: 2801.

Johnson, G.R.A., and Nazhat, N.B. (1987). *J. Amer. Chem. Soc.*, *109*: 1990.

Kaden, T., and Sigel, H. (1968). *Helv. Chim. Acta*, *51*: 947.

Kanao, M., Tomita, S., Ishida, A., Murakami, A., and Okada, H. (1973). *Chemotherapy*, *21*: 1305.

Kazakov, S.A., Astashkina, T.G., Mamaev, S.V., and Vlassov, V.V. (1988). *Nature*, *335*: 186.

Kensler, T.W., Bush, D.M., and Kozumbo, W.J. (1983). *Science*, *221*: 75.

Lau, S., and Sarkar, B. (1971). *J. Biol. Chem.*, *246*: 5938.

Leuthauser, S.W.C., Oberley, L.W., Oberley, T.D., Sorenson, J.R.J., and Ramakrishna, K. (1981). *J. Nat. Cancer Inst.*, *66*: 1077.

Linss, M., and Weser, U. (1986). *Inorg. Chim. Acta*, *125*: 117.

Linss, M., and Weser, U. (1987). *Inorg. Chim. Acta*, *138*: 175.

Marshall, L.E., Graham, D.R., Reich, K.A., and Sigman, D.S. (1981). *Biochemistry*, *20*: 244.

Martin, R.B., and Mariam, Y.H. (1979). *Metal Ions in Biological Systems*, Vol. 8 (H. Sigel, ed.), Marcel Dekker, Inc., New York, p. 57.

Marx, G., and Chevion, M. (1985). *Biochem. J.*, *236*: 397.

Mavelli, I., Rotilio, G., Ciriolo, M.R., Melino, G., and Sapora, O. (1987). *Human Tumor Markers* (S. Cimino, S. Birkmayer, P. Klavious, P. Pimentel and F. Salvatore, eds.), Walter de Gruyter, Berlin, p. 883.

Meister, A. (1988). *J. Biol. Chem.*, *263*: 17205.

Miesel, R., and Weser, U. (1990). *Free Rad. Res. Commun.*, *11*: 39.

Miesel, R., Hartmann, H.J., Li, Y., and Weser, U. (1990). *Inflammation*, *14*: 409.

Nishida, Y., Oishi, N., Kuramoto, H., and Kida, S. (1982). *Inorg. Chim. Acta*, *57*: 253.

Oberley, L.W., Rogers, K.L., Schutt, L., Oberley, T.D., Leuthauser, S.W.C., and Sorenson, J.R.J. (1983). *J. Nat. Cancer Inst.*, *71*: 1089.

Oberley, L.W., Leuthauser, S.W.C., Pasternack, R.F., Oberley, T.D., Schutt, L., and Sorenson, J.R.J. (1984). *Agents Actions*, *15*: 535.

Otto, M., Lerchner, J., Pap, T., Zwanziger, H., Hoyer, E., Inczedy, J., and Werner, G. (1981). *J. Inorg. Nucl. Chem.*, *43*: 1101.

Pope, L.E., and Sigman, D.S. (1984). *Proc. Nat. Acad. Sci. USA*, *81*: 3.

Que, B., Downey, K.M., and So, A.G. (1980). *Biochemistry*, *19*: 5987.

Reed, C.J., and Douglas, K.T. (1989). *Biochem. Biophys. Res. Commun.*, *162*: 1111.

Reed, C.J., and Douglas, K.T. (1991). *Biochem. J.*, *275*: 601.

Reich, K.A., Marshall, L.E., Graham, J., and Sigman, D.S. (1981). *J. Amer. Chem. Soc.*, *103*: 3582.

Rowley, D.A., and Halliwell, B. (1983). *Arch. Biochem. Biophys.*, *225*: 279.

Sagripanti, J.L., and Kraemer, K.H. (1989). *J. Biol. Chem.*, *264*: 1729.

Sarkar, B. (1981). *Metal Ions in Biological Systems*, Vol. 12 (H. Sigel, ed.), Marcel Dekker Inc., New York, p. 233.

Saryan, L.A., Mailer, K., Krishnamurti, C., Antholine, W., and Petering, D.H. (1981). *Biochem. Pharmacol.*, *30*: 1595.

Schechinger, T., Hiller, W., Maichle, C., Strähle, J., and Weser, U. (1988). *Biol. Metals*, *1*: 112.

Shinar, E., Rachmilewitz, E.A., Shifter, A., Rahamim, E., and Saltman, P. (1989). *Biochim. Biophys. Acta*, *1014*: 66.

Sigel, H. (1969). *Angew. Chem. Int. Ed.*, *8*: 167.

Sigel, H., Wyss, K., Fischer, B.E., and Prijs, B. (1979). *Inorg. Chem.*, *18*: 1354.

Spasski, A., and Sigman, D.S. (1985). *Biochemistry, 24*: 8050.

Stacey, N.H., and Klaassen, C.D. (1981). *Toxicol. Appl. Pharmacol., 58*: 211.

Steinkühler, C., Mavelli, I., Melino, G., Rossi, L., Weser, U., and Rotilio, G. (1988). *Ann. N. Y. Acad. Sci., 551*: 133.

Steinkühler, C., Mavelli, I., Rossi, L., Pedersen, J.Z., Melino, G., Weser, U., and Rotilio, G. (1990). *Biochem. Pharmacol., 39*: 1473.

Steinkühler, C., Sapora O., Carri, M.T., Nagel, W., Marcocci, L., Ciriolo, M.R., Weser, U., and Rotilio, G. (1991). *J. Biol. Chem., 266*: 24580.

Steinkühler, C., Pedersen, J.Z., Weser, U., and Rotilio, G. (1991a). *Biochem. Pharmacol., 42*: 1821.

Stevanato, R., Argese, E., Viglino, P., and Rigo, A. (1984). *Inorg. Chim. Acta, 92*: 177.

Stoewe, R., and Prütz, W.A. (1987). *Free Rad. Biol. Med., 3*: 97.

Sutton, H.C., and Winterbourn, C.C. (1989). *Free Rad. Biol. Med., 6*: 53.

Tainer, J.A., Getzhoff, E.D., Richardson, J.S., and Richardson, D.E. (1983). *Nature, 306*: 284.

Takahashi, K., Yoshioka, O., Matsuda, A., and Umezawa, H. (1977). *J. Antibiotics, 30*: 861.

Tkeshelashvili, L.K., McBride, T., Spence, K., and Loeb, L.A. (1991). *J. Biol. Chem., 266*: 6401.

Veal, J.M., and Rill, R.L. (1988). *Biochemistry, 27*: 1822.

Viglino, P., Scarpa, M., Coin, F., Rotilio, G., and Rigo, A. (1986). *Biochem. J., 237*: 305.

Werringloer, J., Kawano, S., Chacos, N., and Estabrook, R.W. (1979). *J. Biol. Chem., 254*: 11839.

Wirth, P.L., and Linder, M.C. (1985). *J. Nat. Cancer Inst., 75*: 277.

Yamamoto, K., and Kawanishi, S. (1989). *J. Biol. Chem., 624*: 15435.

Yoon, C., Kuwabara, M.D., Spasski, A., and Sigman, D.S. (1990). *Biochemistry, 29*: 2116.

ZINC-LIGAND INTERACTIONS AND OXYGEN FREE RADICAL FORMATION

by

Alain Emile Favier

Groupe de Recherche et d'Etude des Pathologies Oxydatives (GREPO),
Faculté de Pharmacie de Grenoble,
Domaine de la Merci
38700 La Tronche
France

I. INTRODUCTION

Zinc is an essential trace element for many living species, from bacteria to man. Its essentiality results from its capacity to bind to a great number of biologically active molecules and to modify their activity by changing their conformation. Almost every stage in the metabolism or activity of cells is zinc dependent, at least for one of its steps such as gene expression, multiplication, protein synthesis, membrane stability, cell moving, hormone secretion and immunity. With so many different effects on cells, it is not really surprising that zinc may be involved in free radical metabolism (Bray and Bettger, 1990; Willson, 1989). Although neither usually considered a strong biological antioxidant when compared with selenium or copper nor apparently necessary for any major antioxidant enzyme, zinc is an important factor in oxidation phenomena. Its ways of acting are numerous and diffuse as it plays several roles inside the cell, acting as both antioxidant and (in some circumstancies) prooxidant. It is often difficult to pinpoint the precise mechanism by which zinc acts in a process: enzymatic, hormonal, antioxidant, as these generally operate in combination. The biphasic effect of zinc in many biochemical or physiological events such as enzyme activity, phagocytosis, and platelet clotting also complicates the data interpretation. In such cases, zinc deficiency or excess induce the same effect. Nevertheless, mild zinc deficiency is very common in western countries, where it seems to play an important part in the occurrence and clinical evolution of oxidative diseases.

II. CHARACTERIZATION OF THE ANTIOXIDANT EFFECT OF ZINC

Since the first observation by Chvapil et al. (1972) of a protective effect of zinc supplementation on CCl_4-induced lipid peroxidation, many authors have reported an antioxidant action of zinc on organelles, cells, and in animal experiments. Two different ways are commonly used to characterize this action: zinc supplementation (often with very high levels) and experimental zinc deficiency. In both cases, great care should be taken to avoid artefacts. Zinc-deficient diets and culture media are difficult to prepare without modifying other nutrients and changing animal behavior. On the other hand, zinc excess can act in non-physiological pathways.

In **isolated organelles**, zinc addition decreases the production of MDA and peroxides due to Fe-xanthine oxidase-xanthine induced radicals on erythrocyte membranes (Girotti et al., 1985). However, in some experiments using rat erythrocyte membranes (Bettger et al., 1978) or frog rod outer segments (Pasantes-Morales and Cruz, 1984), zinc addition decreases lysis but not lipid peroxidation.

In **cells,** zinc produces metabolic changes and induction of proteins like metallothioneins (see Part Two, Chapter 1-D5). Compared to organelles, cells can, in some way, regulate their zinc metabolism to adapt to experimental deficiency or supplementation. Cultured hepatocytes treated with iron or t-butyl hydroperoxide decrease their MDA and hydroxyl radical production when incubated with zinc (Coppen et al., 1988). The induction of metallothionein (MT) protects against UV-A induced photooxidative damage (Dukek et al., 1990). Incubating fibroblasts with an increased amount of zinc in the culture medium produced a 34% decrease in MDA production by UV-A and total cell protection from UV-induced cytolysis (Richard et al., 1992). In contrast, zinc does not protect the same cell line against xanthine oxidase-xanthine or H_2O_2 cytotoxicity.

In **animals,** Chvapil et al. (1972) found that zinc supplementation decreased lipid peroxidation in rats poisoned with CCl_4. Poisoned rats supplemented with zinc also accumulate less collagen in their liver (Anttinen et al., 1984). CCl_4 toxicity is mainly due to its ability to form free radicals inside the body (Poyer et al., 1980), and the protective effect of zinc is closely linked to MT induction in the liver (Clarke and Lui, 1986). Zinc supplementation also protects against the toxicity of other free radical producing agents such as alcohol (Dar et al., 1986; Floresheim, 1987) or irradiation (Floresheim and Floresheim, 1986; Matsubara et al., 1986; Bienvenu and Kergonou, 1990). Simultaneous injection of zinc decreases lipid peroxidation induced by administration of iron and salicylate by an effect not related to MT induction but to an antagonism between zinc and iron (Günther et al., 1991). Zinc protects rats against galactosamine-induced liver injury. For example, zinc supplementation (50 mg/kg) decreases mortality, restores liver pathomorphological changes, maintains zinc content, improves liver function and inhibits lipid peroxidation (Huh et al., 1992).

During experimental zinc deficiency in animals, we observed an increase in both plasma peroxides and lipoprotein susceptibility to in vitro oxidation by copper ascorbate (Faure et al., 1991b). Zinc-deficient rats have higher MDA level in plasma during pregnancy compared to pair-fed pregnant controls (Dreostii and Patrick, 1987). This effect could be due to an overproduction of free radicals by Zn-deficient animals. Microsomes from the lungs of Zn-deficient rats produce more free radicals (Bray et al., 1986; Kubow et al., 1986). Several authors have found a similar enhanced production of free radicals in liver microsomes of Zn-deficient rats. However, in some experiments, zinc deficiency manifests as a general tendency towards lipid peroxidation. In our own experiments, an inconstant MDA increase was obtained. More often, we observed an increased susceptibility to an in vitro oxidative stress without peroxidation in the control Zn-deficient but non-stressed group. Sullivan et al. (1980) found the liver of Zn-deficient rats more sensitive to in vitro peroxidation by iron ascorbate. Zinc-deficient rats are also more sensitive to peroxidation during alcohol intoxication (Coudray et al., 1991). Interestingly, Burke and Fenton (1985) observed no increase of conjugated drenes in liver microsomes or mitochondria and no increase of peroxides in plasma of Zn-deficient mice, but organelles where more sensitive to oxidation by Fe-ADP-NADPH. Erythrocytes of Zn-deficient rats are not more sensitive to in vitro lipid peroxidation than erythrocytes of Zn-sufficient rats, unless vitamin E has been omitted from their diet (Record et al., 1990).

Clinical and biological signs of zinc deficiency are unrepresentative of symptoms resulting from an oxidative process. Zinc deficiency leads mainly to skin and intestinal lesions, impairment of immunity and vision. With selenium or vitamin E deficiency, more cardiovascular and muscles abnormalities are seen. However, this difference in expression may be the consequence of a different localization of the Zn-dependent enzymes or zinc biological targets. Zinc seems to exert most of its effects on the liver where it induces high amounts of MTs. Nevertheless, some of the signs of zinc deficiency can be corrected by enrichment of the diet with another antioxidant. For example, vitamin E can prevent skin lesions and tissue lipid peroxidation of Zn-deficient chicken (Bettger et al., 1980).

III. HOW DOES ZINC ACT AS AN ANTIOXIDANT ?

Compared with other metals, zinc has a very diffuse action in a multiplicity of biochemical oxidative mechanisms.

A. Direct antioxidant mechanism of zinc

1. *Action on copper, zinc-superoxide dismutase (Cu,Zn-SOD)*

Cu,Zn-SOD is an important but somewhat ambiguous enzyme involved in the chain of detoxifying oxygen radicals. Generally its activity is beneficial and the longevity of animal species is proportional to their SOD levels. However, SOD transforms superoxide that is usually not a very reactive species per se into hydrogen peroxide that is more of an oxidant. Consequently, an increase in SOD concentration can be prooxidant if there is a defect in catalase and glutathione peroxidase activity. Transfection of SOD in cells of mice produces an overexpression resulting in an increased sensitivity to oxygen in cell culture (Amstad et al., 1991), and in increased glutathione peroxidase (Ceballos et al., 1988) and TBA-reactants in the brain (Nicole et al., 1991).

Cu,Zn-SOD exists in the cytosol, and consists of a dimeric protein with a molar mass of 16 kg mol^{-1} per unit. Each unit contains one atom of copper and one atom of zinc (Fridovich, 1986; see Part Two, Chapter 1-C7).

In non-hemolyzed plasma, the low SOD activity comes from different enzymes, but mainly from a glycosylated extracellular SOD with adhesive properties towards endothelial cells (Oyanagui and Sato, 1991). EC-SOD binding to the endothelial cell surface might be an efficient protection against extracellular superoxide production. Another type of Cu,Zn-SOD has been found to be present with catalase inside peroxisomes, whose transport is disturbed in the same way as catalase in Zellweger syndrome (Keller et al., 1991).

Copper plays a far more important part in the catalytic process of SOD than zinc. Active site copper is alternately oxidized and reduced during the enzymatic dismutation reaction (Roberts et al., 1991) (see equations below).

$$\text{E-Cu(II)} + O_2^- \rightarrow \text{E-Cu(I)} + O_2$$
$$\text{E-Cu(I)} + O_2^- + 2H^+ \rightarrow \text{E-Cu(II)} + H_2O_2$$
$$\overline{\quad 2O_2^- + 2H^+ \rightarrow O_2 + H_2O_2 \quad}$$

Zinc stabilizes and protects the enzyme, and at basic pH values takes some part in catalysis by stabilizing the imidazole of histidine 63 located between the two metals (Gutteridge and Bannister, 1986).

Zinc alone cannot restore the activity of the SOD apoenzyme in vitro (Weser et al., 1972). Copper can restore activity by occupying not only the copper site but also the zinc-vacant place at the active site (Valentine et al., 1979). However, the copper-loaded enzyme needs zinc to entirely recover its full activity (Koukay, 1989).

In vivo results are more complex since zinc possesses many other actions. For example, zinc can modify the synthesis of the enzyme (synthesis of all proteins is defective in Zn-deficient cells), can modify the resistance of the enzyme to proteases and so increase its half-life. In addition, zinc can induce a defect in copper absorption at the intestinal level when given in large amounts, which can create a defect in the activity of Cu,Zn-SOD. In a cellular model of cultured human fibroblasts, we have observed an increase in cellular zinc and a 26% decrease in Cu,Zn-SOD activity after increasing the zinc content of the culture medium (Leccia et al., 1992). In contrast, Fischer et al. (1991) found that increasing zinc in the diet of rats (150 mg/kg zinc) increased Cu,Zn-SOD in erythrocytes, whereas decreasing copper in the diet decreased enzyme activity.

In animal or human studies, the activity of SOD is often found to be related to the zinc status (Coudray et al., 1992). For example, in hemodialyzed patients with renal failure there is a linear relationship between zinc status measured by plasma zinc and erythrocyte Cu,Zn-SOD activity (Richard et al., 1991) (Fig. 1).

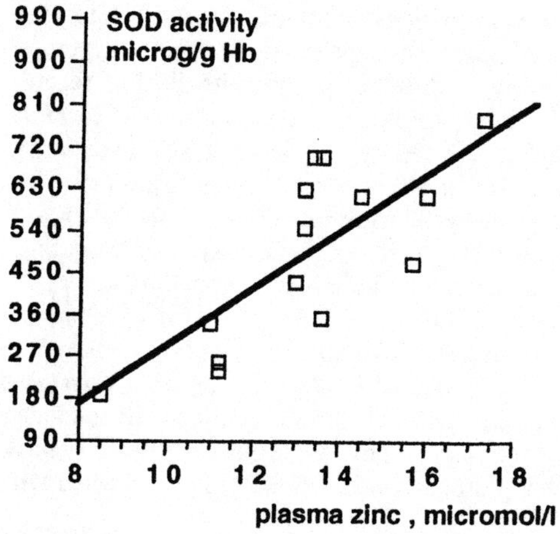

Figure 1.
Relationship between plasma zinc and erythrocyte Cu,Zn-SOD activity in hemodialyzed patients ($r = 0.58$, $p < 0.02$).

2. *Action against iron-induced production of oxygen radicals*
Iron mainly on the form of Fe^{2+} is a prooxidative agent in cells, driving the Fenton reaction and lipid peroxidation. Potentially dangerous iron is often described as "free" iron and is believed to be involved in the appearance of many oxidative diseases (Gutteridge et al., 1979; 1984). However, other forms of iron in the body, even when linked to proteins, are potentially dangerous. Ferritin, whose synthesis is induced in response to oxidative stress through a decreased free iron pool (Klausner, 1988), may loose "free" iron when attacked by an excess of superoxide whose small size permit it to enter the protein shell. Saturated transferrin, but not the unsaturated form of this protein (which inhibits the Fenton reaction), and abnormal forms of hemoglobin or myoglobin (Misrah and Fridovich, 1972) can also generate oxygen free radicals.

Zinc interacts with iron at different levels: with iron absorption, which alters the size of the iron pool in the body, with the proposed very reactive "free" iron pool, with the Fenton reaction catalyzed by "free" iron, and with iron linked to thiol groups of proteins.

a. Effect of zinc on iron metabolism
Experimentally iron and zinc exert a mutual inhibitory effect on their mechanism of absorption (Aggett et al., 1983; Solomons, 1988; see also *Bioinorganic Medicine*, Part Three, Chapter 2-G). This effect is responsive for a deficiency in the status of one of the metals during long-term supplementation with high doses of the other metal. For instance, infants (Friel et al., 1988) or adult females (Yadrick et al., 1989) receiving a zinc supplement show a decrease in serum ferritin. The exact mechanism of this interaction has not yet been elucidated. It is suggested that a common carrier exists in the enterocyte, making iron and zinc have to compete to enter or exit the intestinal cell.

b. Effect of zinc on the Fenton reaction
Iron(II) is able to produce hydroxyl radicals from superoxide in the Fenton reaction, and can change its oxidation state and catalyze the Haber Weiss cycle producing hydroxyl radical from O_2^- and H_2O_2. Even if the exact mechanism and the nature of the iron is not known (Fontecave and Pierre, 1991), the reaction is very important for explaining damage in oxidative diseases, and its importance has been stressed by Halliwell and Gutteridge (1989). In such a reaction, the form of iron is very important. For example, iron is more efficient in some chelated forms, as with EDTA. In contrast, other complexes such as those with desferrioxamine or apotransferrin are unable to drive this reaction.

Zinc decreases the amount of •OH trapped by DMPO in Fenton reactions involving iron, cysteine and oxygen. Zinc is supposed to complex with cysteine in the place of iron and so decrease the ability of iron to catalyze the Fenton reaction (Searle and Tomasi, 1982). Zinc complexes are redox-inactive. Zinc is observed to inhibit lipid peroxidation of small unilamellar liposomes induced by iron-ascorbic acid (Szebeni et al., 1988).

We have observed a similar effect of zinc on the Fenton reaction, using iron citrate and H_2O_2: here zinc decreased the production of ethylene after attack of keto-methylthiobutyrate by •OH (Coudray et al., 1993). This observation is very important because, in vivo, iron citrate is supposed to be the non-protein-bound iron that occurs in plasma and tissues in pathological circumstances. This has been demonstrated by computer simulation (May and Linder, 1976), by isolation of the complex using HPLC (Grooteveld et al., 1989) and NMR. Iron citrate was found able to produce MDA from deoxyribose and to hydroxylate benzoate as well as to undergo autoreduction to produce ferrous ions (Gutteridge, 1991).

Zinc can also displace iron when complexed with toxic agents. Günther et al. (1991) have demonstrated such a protective effect of zinc in rats injected with iron and salicylate.

c. Effect of zinc on the process of lipid peroxidation

The degradation of polyunsaturated fatty acids by free radicals is a complex multistep process that needs iron to catalyze some of the reaction (Minotti and Aust, 1987). Even though it has not yet been proven experimentally, zinc may act as a competitor with iron in the process at protecting fatty acids.

d. Protective effect of zinc on thiol groups

The iron-catalyzed oxidation of thiols is a rapid reaction, known for a long time, as is the inhibitory effect of zinc on this reaction (Mathews and Walker, 1989). Iron chelated to thiol groups can react with oxygen because it can change its oxidation state. When zinc takes the place of iron (because of its stronger affinity for thiol groups) the new zinc chelate is unable to react with oxygen. This effect has been observed by Willson on thiol groups such as in metronidazole, a radio-protective agent. Willson has proposed a general scheme for this mechanism (Fig. 2) (Willson, 1976).

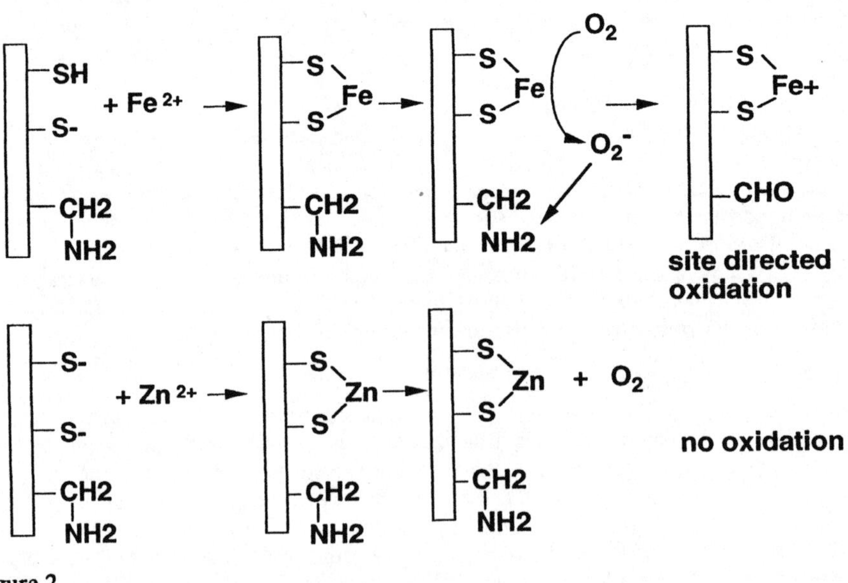

Figure 2.
Zinc protection mechanism of thiol groups from iron-induced oxidation, and consequences on protein damage (adapted from Willson, 1989).

In vivo this mechanism is probably very important for protecting the thiol groups of bioactive proteins. Binding of iron to thiols is likely to produce free radicals resulting in oxidation and damage to the protein (Davies et al., 1991). In the same way, enzymes with thiol goups at their active sites are sensitive to oxidation. Zinc protects such enzymes from oxidation. Delta amino levulinate dehydratase possesses four cysteines on each sub-unit. Zinc protects enzyme activity in the presence of oxygen by preventing these thiol groups forming internal disulphide-bridges (Gibbs et al., 1985). Zinc similarly protects thiol groups of other active thiol proteins such as dihydroorotase (Washabaugh and Collins, 1986), and tubulin (Hesketh, 1983).

3. *Antagonistic effect of zinc on other prooxidant metals*

a. Aluminum

Aluminum has only one oxidation state (+3) and so cannot drive the Fenton reaction. However, it accelerates iron-induced lipid peroxidation (Gutteridge et al., 1985), and peroxidation of erythrocyte membranes by H_2O_2 (Quinlan et al., 1988). Aluminum also inhibits the activity of Cu,Zn-SOD in the brains of rats (Ohtawa et al., 1983) or in vitro (Shainkin-Kestenbaum et al., 1989), without binding to SOD protein (Serra et al., 1991). The effect of aluminum is possibly due to in vivo mechanisms involving displacement of iron from neutralizing proteins such as transferrin or changing structural integrity, thereby promoting iron damage. Zinc is reported to decrease the Al-induced production of free radicals by microglial cells (Evans et al., 1991).

b. Copper

Copper, like iron in the presence of oxygen, can undergo redox cycling with reducing agents to form reactive oxygen species (Halliwell and Gutteridge, 1989). Cu^{2+} is unable to modify the production of •OH by radiolysis, but can modify, in a biphasic way, the production of •OH in the Fenton reaction, measured by ESR or by the production of hydroxysalicylate or MDA (Maestre et al., 1992). In vitro, copper is more reactive than iron in damaging DNA in the presence of H_2O_2 and ascorbic acid (Aruoma et al., 1991). However, in vivo copper ions have not the same potential as iron to generate free radicals. They bind to proteins such as albumin which prevents formation of •OH radicals in free solution from H_2O_2. However, •OH radicals are still generated at a site-specific location on the protein molecule (Gutteridge and Wilkins, 1983). On the other hand, copper is very useful for antioxidant systems such as SOD and ceruloplasmin. Copper deficiency therefore exacerbates peroxidative damage produced in rats by carbon tetrachloride (Lawrence and Jenkinson, 1987).

The same interaction that exists between zinc and iron also exists between zinc and copper. The effect of zinc on copper absorption occurs at an equimolar level, and is exacerbated by increasing the ratio of histidine in the diet (Wapnir and Balkman, 1991). This antagonistic effect of zinc is beneficial in the therapy of Wilson's disease (Brewer et al., 1987; Hoogenraad et al., 1987; see also *Medicine*, Part Five, Chapter 1-I). Zinc is also an antagonist of copper in several mechanisms of peroxidation. For instance, thiol-rich proteins of the lens are able to produce large amounts of H_2O_2 in the presence of copper and this effect is suppressed by zinc (Hunt et al., 1989). Survival and cell growth of HepG2 cells are reduced by addition of copper chloride to the media, and incubation of cells with zinc acetate before exposure to copper raises the LD_{50}, improves protein synthesis and increases MT synthesis. Zinc protects *Escherichia coli* against free-radical mediated copper-paraquat toxicity. The protective mechanism of zinc involves either a direct displacement of copper by zinc or the formation of a ternary complex, paraquat-copper-zinc, unable to undergo redox cycling (Korbashi et al., 1989).

B. Indirect antioxidant action of zinc

1. *Induction of metallothioneins*

The MT molecule contains 50% cysteine and can bind different metals (Cu, Zn, Cd, Mn..). MTs are antioxidants by direct scavenging (through their thiol groups) of •OH but not of superoxide (Thornally and Vašak, 1985), by binding Cu^+ and preventing it from driving

Fenton reactions inside the cell. MTs are induced by prooxidants, inflammation, injection of LPS, Interleukin 1 (Karin, 1985), heat shock (Brady, 1982) and UV radiation. The antioxidant effect of MTs explains the increase in X-ray resistance of cells which contain high levels of these proteins (Bakka and Webb, 1981). Addition of MT to the perfusate of ischaemic rat hearts decreased ESR signals corresponding to radicals generated during reperfusion, thereby protecting myocytes from oxidative stress (Cheng and Tang, 1990).

Different metals can induce MT synthesis, but zinc is by far the most effective. Conversely, Zn-deficient rats show decreased concentrations of liver MT (Taylor et al., 1988). When Zn-MT is attacked by $^{\bullet}OH$, the reaction produces a disulphide bridge and releases zinc (Thornally and Vašak, 1985).

2. *Influence of zinc status on other antioxidant systems*

a. Glutathione transferase

This class of enzymes contains a great number of isoenzymes whose total in rat liver represent 20 % of all proteins. These enzymes detoxify numerous electrophilic compounds (drugs, carcinogens, xenobiotics...) through covalent linking to reduced glutathione (Jakoby, 1985). Glutathione transferases, mainly GT1-2 have also a peroxidase-like activity towards organic hydro-peroxides but not H_2O_2. They are also named non-seleno glutathione peroxidases. Experimental zinc deficiency decreases the activity of glutathione transferase in rat by an unknown mechanism (Lee and Fong, 1986).

b. Tocopherol

Tocopherols, the most important of which is α-tocopherol, are present inside the lipid bilayer of cells and mitochondrial membranes as well as in lipoproteins (Chazan and Szalc, 1987). Tocopherol reacts with O_2, $O_2^{\bullet-}$, $^{\bullet}OH$, but its main biological role is to react with peroxyl radicals to form a more stable tocopheryl radical (Jore et al., 1991). Zinc deficiency modifies the metabolism of tocopherol. Zinc-deficient animals present with a lower tocopherol level in serum, and a slower replenishment after supplementation with tocopherol-rich diet (Bunk et al., 1989).

3. *Zinc-induced decrease in cell free radical production*

a. Phagocytes

Zinc ions inhibit the NADPH oxidase of leukocytes (Chvapil et al., 1977), decrease cell stimulation and protect cell membranes (see *Bioinorganic Medicine* 2, Part Two, Chapter 2-D). Zinc can protect PMNs against cytolysis induced by dehydroabietic acid (Sunzel et al., 1991). Addition of zinc, however, gives different results in vitro and in vivo. In vitro, zinc decreases the production of superoxide whatever the stimulant but its in vivo effect varies with the agent, from no change using phorbol myristate to augmentation of superoxide production using zymosan (Srinivas et al., 1988) or concanavallin A (Nakamura et al., 1987).

b. Microsomal production

Zinc decreases lipid peroxidation in liver microsomes by inhibition of NADPH oxidase activity. Observed in vitro (Chvapil et al., 1975) and in vivo (Sullivan et al., 1980), this effect results from the formation of a 2:1 zinc:NADPH complex (Chvapil et al., 1975), and from the activation by zinc of a nucleotide pyrophosphatase that hydrolyzes the NADPH-depleting cofactor essential for NADPH oxidase activity (Jefferey, 1983).

c. TNF

The cytokine aTNF is cytolytic by a mechanism that makes the target cell become apoptotic. Free radicals are involved in the killing process that is important in the inflammatory process or in the destruction of cancerous cells. Overproduction of Mn-SOD by transfection as well as treatment with various antioxidant agents induces a resistance to TNF. Addition of zinc protects sensitive WEH 1164-S cells from aTNF-induced lysis, but since it prevents DNA fragmentation it may act by inhibition of endonuclease (an important factor in the programmed cell death in response to TNF), and not by acting as an antioxidant (Flieger et al., 1989).

4. *Effect of zinc on the substrates of peroxidation*
a. Membrane stabilization
Zinc plays an important role in the maintenance of membrane structure and function (Bettger and O'Dell, 1981). Zinc deficiency, like zinc excess, decreases membrane fluidity and so influences free radical production by phagocytes and prostaglandin secretion by cells.

b. Polyunsaturated fatty acids (PUFA)
Zinc deficiency changes fatty acid levels and thereby the quality of platelet membrane (Faure et al., 1991a) as well as tissues (Cunnane, 1988a) or plasma lipids (Faure, 1992). Table 1 shows the changes in fatty acids observed in rats after 4 weeks of experimental zinc deficiency: as described by Cunnane (1988b), zinc deficiency increases the ratio 18:2/ 20:4 in different tissues. Sullivan et al. (1980) observed an increase in liver phospholipids in Zn-deficient animals. As PUFAs are the main substrates for lipid peroxidation, these changes can explain the increased sensitivity to in vitro peroxidation of Zn-deficient animals. This effect of zinc to alter fatty acid distribution also changes the metabolism of prostaglandins, decreasing prooxidant metabolites produced by cyclooxygenase.

IV. PROOXIDANT EFFECT OF ZINC

In some conditions, an excess of zinc can increase damage due to free radicals. A decrease in cell viability and adhesion after stress with H_2O_2 has been observed in cultured human fibroblasts (Fig. 3). In such a model, the effect of zinc addition varies from protection, when the oxidative stress is induced by UV-B or UV-A, to exacerbation of damage when induced by xanthine-oxidase/xanthine and H_2O_2. The same exacerbation in cytotoxicity was observed when high doses of zinc were added to the medium of U 929 leukocytes before stress induction by H_2O_2. This effect of zinc may not be an actual prooxidative effect. Zinc may inhibit the activity of an enzyme necessary for cell repair after such a stress. We have often observed a dose-dependent biphasic activity of zinc metalloenzymes, with a progressive increase in activity when small amounts of zinc are added to the apoenzyme, and a progressive decrease in activity when large amounts are added.

In vitro, very large, far from physiological concentrations of zinc increased the peroxidation of erythrocyte membranes by xanthine oxidase-xanthine and iron-citrate systems. In this experiment, zinc may displace iron from its complexes and facilitate its action for decomposition of lipid peroxides (Koukay, 1989).

Table 1

Fatty Acid Distribution (%) in Platelets of Rats Fed during 4 Weeks with a Zn-deficient Diet (ZD), or Pair-fed with a Normal Diet (PF).
From Faure (1992)

Fatty acids	After 4 weeks		After 8 weeks	
	ZD	PF	ZD	PF
C16 : 0	21.51 ± 1.03	23.11 ± 1.86	30.03 ± 1.19	32.44 ± 2.21
C16 1: (n -7)	2.55 ± 0.56*	2.03 ± 0.41	1.02 ± 0.07	1.08 ± 0.02
C18 2 : (n - 6)	20.81 ± 3.90*	14.91 ± 1.96	8.71 ± 0.47	7.81 ± 0.43
C18 3: (n -3)	0.37 ± 0.02	0.52 ± 0.24	0.31 ± 0.29	0.51 ± 0.21
C 20 4: (n - 6)	21.64 ± 0.61*	21.55 ± 1.66	28.68 ± 0.26	30.51 ± 0.32
C 22 5: (n - 3)	0.28 ± 0.09***	0.44 ± 0.10	0.45 ± 0.16**	0.23 ± 0.04
C 22 6 : (n - 3)	0.97 ± 0.24**	1.19 ± 0.24	0.66 ± 0.12	0.62 ± 0.11

* $p < 0.5$
** $p < 0.01$
*** $p < 0.001$

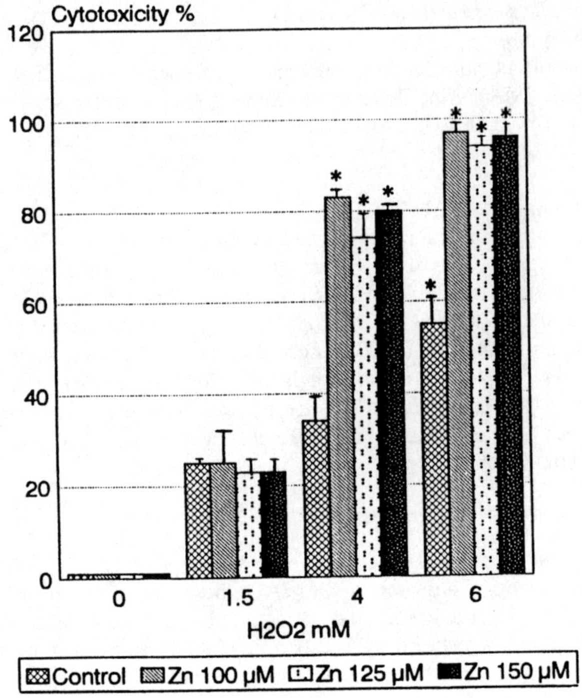

Figure 3.
Exacerbation by zinc of damage induced by addition of H_2O_2 to fibroblasts culture medium (* $p < 0.05$ vs control cells).

V. CONCLUSIONS

Zinc has different and sometimes conflicting ways of acting in oxidative stress. Its main antioxidant effect seems to be its capacity to displace iron, not only free iron in the Fenton reaction but also, more importantly, iron bound to the surface of proteins. Zinc has also a protective effect on thiols and other reactive groups of active proteins against local damage when redox potential inside cells changes. The role of zinc in such phenomena, not only in cellular damage but also in normal functions such as cellular cycles, is a rewarding direction for future investigations.

REFERENCES

Aggett, P.J., Crofton, R.W., Khin, C., Gvozdanovic, S., and Gvozdanovic, D. (1983). *Zinc Deficiency in Human Subjects*, Alan R. Liss, New York, p. 117.

Amstad, P., Peskin, A., Shah, G., Mirault, M., Moret, R., Zbinden, I., and Cerutti, P. (1991). *Biochemistry*, *30*: 9305.

Anttinen, H., Ryhanen, L., Puistola, U., Arranto, A., and Oikarinen, A. (1984). *Gastroenterology*, *86*: 532.

Aruoma, I., Halliwell, B., Gajewski, E., and Dizdaroglu, M. (1991). *Biochem. J.*, *273*: 601.

Bakka, A., and Webb, M. (1981). *Biochem. Pharmacol.*, *30*: 721.

Bettger, W.J., and O'Dell, B.L. (1981). *Life Sci.*, *28*: 1425.

Bettger, W.J., Fish, T.J., and O'Dell, B.L. (1978). *Proc. Soc. Exp. Biol. Med.*, *158*: 279.

Bettger, W.J., Reeves, P.G., Savage, J.E., and O'Dell, B.L. (1980). *Proc. Soc. Exp. Biol. Med.*, *163*: 432.

Bienvenu, P., and Kergonou, J.F. (1990). *Metal Ions in Biology and Medicine* (Ph. Collery, L.A. Poirier, M. Manfait and J.C. Etienne, eds.), John Libbey Eurotext, Paris, p. 545.

Brady, F.O. (1982). *Trends Biochem.*, *7*: 143.

Bray , T.M., and Bettger, W.J. (1990). *Free Rad. Biol. Med.*, *8*: 281.

Bray, T.M., Kubow, S., and Bettger, W.J. (1986). *J. Nutr.*, *116*: 1054.

Brewer, G., Hill, G., Dick, R., Norstrant, T., Sams, J., Wells, J., and Prasad, A. (1987). *J. Lab. Clin. Med.*, *109*: 526.

Bunk, M., Dnistrian, A., Schwartz, M., and Rivlin, R. (1989). *Proc. Soc. Exp. Biol. Med.*, *190*: 379.

Burke, J.P., and Fenton, M.R. (1985). *Proc. Soc. Exp. Biol. Med.*, *179*: 187.

Ceballos, I., Delabar, J.M., Nicole, A., Lynch, R.E., Hallewell, R.A., Kamoun, P., and Sinet, P.M. (1988). *Biochim. Biophys. Acta*, *58*: 59.

Cheng, S., and Tang, C.S. (1990). *Metal Ions in Biology and Medicine* (Ph. Collery, L.A. Poirier, M. Manfait and J.C. Etienne, eds.), John Libbey Eurotext, Paris, p. 24.

Chazan, J.B., and Szalc, M. (1987). *Cah. Nutr. Diet.*, *23*: 291.

Chvapil, M., Ryan, J.N., and Zukoski, C.F. (1972). *Proc. Soc. Exp. Biol. Med.*, *141*: 150.

Chvapil, M., Sipes, I.G., Ludwing, J.C., and Halladay, S.C.I. (1975). *Biochem. Pharmacol.*, *24*: 917.

Chvapil, M., Stankova, L., Zukoski, C., IV, and Zukoski, C., III (1977). *J. Lab. Clin. Med.*, *89*: 1;135.

Clarke, I.S., and Lui, E.M.K. (1986). *Can. J. Physiol. Pharmacol.*, *64*: 1104.

Coppen, D.E., Richardson, D.E., and Cousins, R.J. (1988). *Proc. Soc. Exp. Biol. Med.*, *189*: 100.

Coudray, C., Boucher, P., Richard, M.J., Arnaud, J., De Leiris, J., and Favier, A. (1991). *Biol. Tr. Elem. Res.*, *30*: 103.

Coudray, C., Richard, M.J., Laporte, F., Faure, P., Roussel, A.M., and Favier, A. (1992). *J. Nutr. Med.*, *3*: 13.

Coudray, C., Rachid, S., and Favier, A. (1993). *Biol. Tr. Elem. Res.*, *38*: 273.

Cunnane, S.C. (1988a). *Brit. J. Nutr.*, *53*: 273.

Cunnane, S.C. (1988b). *Prog. Food Nutr. Sci.*, *12*: 151.

Dar, M.S., Towsend, S.M., and Wooles, W.R. (1986). *J. Toxicol. Environ. Health*, *18*: 41.

Davies, M., Gilbert, B., and Haywood, R. (1991). *Free Rad. Res. Commun.*, *15*: 111.

Dreostii, E., and Patrick, E.J. (1987). *Biol. Tr. Elem. Res.*, *14*: 179.

Dukek, E., Peak, J., Roth, R., and Peak, M. (1990). *J. Free Rad. Biol. Med.*, *9*: 76.

Evans, P., Klinowski, J., and Yano, E. (1991). *Med. Hypotheses*, *34*: 209.

Faure, P. (1992). *Zinc et Diabète Sucré: Implications dans la Peroxydation Lipidique et la Sensibilité à l'Insuline*, PhD. Thesis, Grenoble, France.

Faure, P., Roussel, A.M., and Favier, A. (1991a). *Trace Elements in Man and Animals*, Vol. 7 (B. Momcilovic, ed.), IMI, Zagreb, p. 18.

Faure, P., Roussel, A.M., Foulon, T., Groslambert, P., Hadjian, A., and Favier, A. (1991b). *Biol. Tr. Elem. Res.*, *28*: 135.

Fischer, P., Campbell, J., and Giroux, A. (1991). *Biol. Tr. Elem. Res.*, *30*: 65.

Flieger, D., Riethmuller, G., and Ziegler-Heitbrock, H.W. (1989). *Int. J. Cancer*, *44*: 315.

Floresheim, G.L. (1987). *Agents Actions*, *21*: 217.

Floresheim, G.L., and Floresheim, P. (1986). *Brit. J. Radiol.*, *59*: 597.

Fontecave, M., and Pierre, J.L. (1991). *Biometals*, *4*: 133.

Fridovich, I. (1986). *Adv. Enzymol.*, *58*: 61.

Friel, J., Cox, M., Cornel, A., Matthew, D., and Andrews, W. (1988). *Trace Elements in Man and Animals*, Vol. 6 (L.S. Hurley, C.L. Keen, B. Lönnerdal and R.B. Rucker, eds.), Plenum Press, New York, p. 571.

Gibbs, P.N.B., Gore, M.G., and Jordan, P.M. (1985). *Biochem. J.*, *225*: 573.

Girotti, A.W., Thomas, J.P., and Jordan, J.E. (1985). *J. Free Rad. Biol. Med.*, *1*: 395.

Grooteveld, M., Bells, J., Halliwell, B., Aruoma, O., Bomford, A., and Sadlers, S.P. (1989). *J. Biol. Chem.*, *264*: 4417.

Günther, T., Grosrau, R., Hollriegl, V., and Vormann, J. (1991). *J. Trace Elem. Electr. Hlth Dis.*, *5*: 95.

Gutteridge, J.M.C. (1991). *Free Rad. Biol. Med.*, *11*: 401.

Gutteridge, J.M.C., and Bannister, J.V. (1986). *Biochem. J.*, *234*: 225.

Gutteridge, J.M.C., and Wilkins, S. (1983). *Biochim. Biophys. Acta*, *759*: 38.

Gutteridge, J.M.C., Richmond, R., and Halliwell, B. (1979). *Biochem. J.*, *184*: 469.

Gutteridge, J.M.C., Halliwell, B., and Rowley, D.A. (1984). *Life Chem. Rep.*, *Suppl. 2*: 15.

Gutteridge, J.M.C., Quinlan, G.J., Clark, I., and Halliwell, B. (1985). *Biochim. Biophys. Acta*, *835*: 441.

Halliwell, B., and Gutteridge, J.M.C. (1989). *Free Radicals in Biology and Medicine*, Oxford University Press, Oxford, U.K., p. 1.

Hesketh, J.E. (1983). *Int. J. Biochem.*, *15*: 743.

Hoogenraad, T., Van Hattum, J., and Van Den Hammer, J. (1987). *J. Neurol. Sci.*, *77*: 137.

Huh, L., Chen, R.D., and Ma, L. (1992). *Biol. Tr. Elem. Res.*, *34*: 27.

Hunt, J., Jiang, Z., and Wolff, S. (1989). *Free Rad. Biol. Med.*, *13*: 319.

Jakoby, W. (1985). *Methods Enzymol.*, *113*: 495.

Jefferey, E. (1983). *Mol. Pharmacol.*, *23*: 467.

Jore, D., Ferradini, C., Madden, K., and Patterson, L.K. (1991). *Free Rad. Biol. Med.*, *11*: 349.

Karin, M. (1985). *Cell*, *41*: 9.

Keller, G., Warner, T., Steimer, K., and Hallewell, R.A. (1991). *Proc. Nat. Acad. Sci. USA*, *88*: 7381.

Klausner, R. (1988). *Science*, *246*: 870.

Korbashi, P., Katzhendlers, J., Saltman, P., and Chevion, M. (1989). *J. Biol. Chem.*, *264*: 8479.

Koukay, N. (1989). *Rôle du Zinc et du Sélénium dans la Génération des Radicaux Libres lors des Accidents Ischémiques du Myocarde*, PhD Thesis, Grenoble, France.

Kubow, S., Bray, T.M., and Bettger, W.J. (1986). *Can. J. Physiol. Pharmacol.*, *64*: 1281.

Leccia, M.T., Richard, M., Beani, J.C., Faure, H., Monjo, A., Cadet, J., Amblard, P., and Favier, A. (1993). *Photochem. Photobiol.*, *58*: 548.

Lee, I.S.K., and Fong, L.Y.Y. (1986). *Carcinogenesis*, *7*: 1111.

Maestre, P., Lambs, L., Thouvenot, J.P., and Berthon, G. (1992). *Free Rad. Res. Commun.*, *15*: 305.

Mathews, A.P., and Walker, S. (1989). *J. Biol. Chem.*, *6*: 299.

Matsubara, J., Shida, T., Ishioka, K., Egawa, S., Inada, T., and Machida, K. (1986). *Environ. Res.*, *41*: 558.

May, P.M., and Linder, P.W. (1976). *J. Chem. Soc., Dalton Trans.*, 588.

Misrah, P., and Fridovich, I. (1972). *J. Biol. Chem.*, *247*: 6960.

Minotti, G., and Aust, S.D. (1987). *Chem. Phys. Lipids*, *44*: 191.

Nakamura, T., Shiraishi, N., and Aono, K. (1987). *Phys. Chem. Phys. Med. NMR*, *19*: 147.

Nicole, A., Porier, M., Agid, F., Lafon, M., Thevenin, M., Sinet, P.M., and Ceballos-Picot, I. (1991). Communication to "Free Radical and Aging", Paris.

Ohtawa, M., Seko, M., and Takayama, F. (1983). *Chem. Pharm. Bull.*, *31*: 1415.

Oyanagui, Y., and Sato, S. (1991). *Free Rad. Res. Commun.*, *12/13*: 229.

Pasantes-Morales, H., and Cruz, C. (1984). *J. Neurol. Sci. Res.*, *11*: 303.

Poyer, J.L., Mac Cay, P.B., Lai, E.K., Janzen, E.G., and Davis, E.R. (1980). *Biochem. Biophys. Res. Commun.*, *94*: 1154.

Quinlan, G.J., Halliwell, B., Moorhouse, C.P., and Gutteridge, J.M.C. (1988). *Biochim. Biophys. Acta*, *962*: 196.

Record, I., Mac Queen, S., and Dreostii, E. (1990). *Biol. Tr. Elem. Res.*, *23*: 89.

Richard, M.J., Arnaud, J., Jurkovitz, C., Hachache, T., Meftahi, H., Laporte, F., Foret, M., Favier, A., and Cordonnier, D. (1991). *Nephron*, *57*: 10.

Richard, M.J., Guiraud, P., Arnaud, J., Monjo, A.M., Beani, J.C., and Favier, A. (1993). *Biol. Tr. Elem. Res.*, *37*: 187.

Roberts, V., Fisher, C., Redford, S., Mc Ree, D., Parge, H., Getzoff, D., and Tainer, J. (1991). *Free Rad. Res. Commun.*, *12-13*: 269.

Searle, A.I.F., and Tomasi, A. (1982). *J. Inorg. Biochem.*, *17*: 161.

Serra, M.A., Barassi, V., Canavese, C., and Sabbioni, E. (1991). *Biol. Tr. Elem. Res.*, *31*: 79.

Shainkin-Kestenbaum, R., Adler, A.J., Berlyne, G.M., and Caruso, C. (1989). *Clin. Sci.*, *77*: 463.

Solomons, N.W. (1988). *Current Topics in Nutrition and Disease*, Vol. 18 (A.S. Prasad, ed.), Alan R. Liss, New York, p. 509.

Srinivas, U., Braconier, H., Jeppsson, B., Abdulla, M., Akesson, B., and Ockerman, P.A. (1988). *Scand. J. Clin. Lab. Invest.*, *48*: 495.

Sullivan, J.F., Jetton, M.N., Hahn, H.K.J., and Burch, R.E. (1980). *Amer. J. Clin. Nutr.*, *33*, 51.

Sunzel, B., Soderberg, T.A., Reuterving, C.O., Hallman, G., Holm, S.E., and Hanstrom, L. (1991). *Biol. Tr. Elem. Res.*, *31*: 33.

Szebeni, J., Eskel Son, C., and Chvapil, M. (1988). *Phys. Chem. Phys. Med. NMR*, *20*: 205.

Taylor, C., Bettger, W.J., and Bray, T. (1988). *J. Nutr.*, *118*: 613.

Valentine, J.S., Pantoliano, M.W., Mc Donnell, P.J., Burger, A.R., and Lippard, S.J. (1979). *Proc. Nat. Acad. Sci. USA*, *76*: 4245.

Wapnir, R., and Balkman, C. (1991). *Biol. Tr. Elem. Res.*, *29*: 193.

Washabaugh, M.W., and Collins, K.D. (1986). *J. Biol. Chem.*, *261*: 5920.

Weser, U., Barth, G., Djerassi, C., Hartmann, H.J., Krauss, P., Voelcker, G., Voelter, W., and Voetsch, W. (1972). *Biochim. Biophys. Acta*, *278*: 24.

Willson, R.L. (1976). *Ciba Found. Symp.*, *51*: 331.

Willson, R.L. (1989). *Zinc in Human Biology* (C.F. Mills, ed.), Springer-Verlag, London, p. 147.

Yadrick, M.K., Kenney, M., and Winterfeldt, E.I. (1989). *Amer. J. Clin. Nutr.*, *49*: 145.

STRUCTURAL ASPECTS OF METAL ION - DRUG INTERACTIONS

by

Frederick T. Greenaway

Gustaf H. Carlson School of Chemistry,
Clark University,
Worcester, MA 01610
U.S.A.

I. INTRODUCTION

In recent years it has become increasingly apparent that interactions with metal ions are important factors in the modes of action of many drugs. In most cases the details of why the metal-drug complex is more effective than the drug alone remain a matter of conjecture, but the importance of understanding the nature of the interactions of drugs with metals which might be present in body fluids is now beyond dispute.

The pharmacology of metal-drug interactions can depend on the method of administration; a metal-drug complex can be administered as the therapeutic agent, or the drug can be administered and interact with metal ions in the body. From a *thermodynamic* point of view, there is no reason to anticipate significant differences in the structures of the final products found in body fluids for the two methods of administration. From a *kinetic* point of view, however, there may be substantial differences for the two forms of administration. Although many ligand exchange reactions are fast on a pharmacologically relevant time scale, which is typically longer than seconds, some complexes are quite *inert*, i.e. they exchange their ligands slowly. Thus some complexes may decompose sufficiently slowly that the therapeutic action can be due to the administered complex. One example is of insoluble complexes which often owe their insolubility to polymerization. Although it is highly improbable that polymers will form in dilute solutions because the low concentrations of both metal ions and drug make aggregation unlikely, polymeric solids commonly undergo very slow ligand exchange reactions and, when administered, may persist longer in vivo than monomeric species.

The examples in this review will be restricted to studies of interactions between organic molecules which are administered as drugs and for which interaction with metals has been implicated in the therapeutic action. In many of these cases, it is often not yet known whether the metal is essential to the local drug action, or whether the metal serves instead to modify drug transport mechanisms. Frequently a metal-drug complex provides an uncharged moiety which should have enhanced lipophilicity, facilitating adsorption and passage through membranes, but only if the complex is sufficiently long-lived in the presence of other potential ligands found in body fluids to remain as a neutral species until adsorption or membrane transport has occurred.

Although there have been structural studies of complexes of many drugs with a wide variety of metal ions, including those with little or no physiological relevance, and although these often lead to a good understanding of the chemistry of the drugs which can be extrapolated to other metal complexes, the therapeutic action of metal-drug complexes is presumably significant only for the metals normally present in body fluids, which are (Hughes, 1987) the alkali metals (Na^+, K^+) which form very weak complexes, the alkaline earth metals (Mg^{2+}, Ca^{2+}) which form complexes with rather low stability constants, and some of the 3d transition metal ions (Cu^{2+}, Fe^{2+}, Fe^{3+}, Zn^{2+}). Vanadium, molybdenum, Mn^{2+}, Ni^{2+}, and Co^{2+} are also present in vivo but are unlikely to be available to interact with drugs. Other metal ions, of uncertain physiological importance, are also present from dietary intake (Al^{3+}, Cd^{2+}, Hg^{2+}, etc). Several of these are detrimental.

II. SPECIATION AND KINETIC FACTORS

Under normal circumstances, almost all metal ions capable of forming stable complexes with drugs, especially the transition metal ions, are already present as chelate complexes in the body (Howard-Lock and Lock, 1987), and significant interactions between metal ions and drugs therefore depend on ligand substitution reactions. In most cases administration of a metal-drug complex results in rapid ligand exchange with the formation of ternary complexes (metal-drug-other ligands) and metal removal (metal-other ligand complexes + free drug). One should remember that both the free drug and the metal-other ligand complexes are potential therapeutic agents, thus the metal-drug complex may be a source of metal ions, or a source of drug, as well as a source of a metal-drug complex. If equilibrium is attained, which is by no means assured in vivo, the most stable complexes will persist. In these cases speciation is probably the most important consideration. Traditionally speciation studies have involved determination of formation constants of all possible metal-ligand complexes by potentiometric titration techniques (Wilson and Wilson, 1964) followed by development of a computer-generated model based on thermodynamic competition between the available ligands for the metal (Jackson et al., 1978; Martell and Motekaitis, 1989a). Under these conditions, metal-drug complexes are usually minor components of the mixture (May and Williams, 1981) since many biologically-available ligands are present in higher concentrations than the drugs (Laurie, 1987). In addition, many proteins contain multidentate binding sites which, due to the chelate effect, have higher stabilities than simple metal-drug complexes which frequently only have bidentate ligands. Speciation studies are discussed in detail in Parts Three and Five.

Most speciation studies published to date have generally involved in vitro studies where solutions of the competing ligands were studied over long periods when equilibrium was likely to have been attained. In pharmacological situations, however, kinetic factors may be at least as important as thermodynamic (stability) factors although they have been much less studied. Thus although equilibrium speciation studies may indicate that an administered metal-drug complex will decompose in the body, this decomposition may be sufficiently slow that the therapeutic action may still be due to the complex.

The variation of complex structure with pH (and sometimes with ionic strength) is too often neglected. Because all ligands are Lewis bases, it is essential to study metal ion - proton competition for systems subjected to low pH, as for example in the gut. It is important to carry out solution studies over the full physiologically relevant pH range and to evaluate stability constants as Pettit et al. (1992) did in the study of Cu(II) binding to a 1-hydroxy-4-(Gly-His-Lys)-anthraquinone as a model for anthraquinone antitumor agents. In this study, the metal-ligand interactions were investigated by potentiometric titration and by CD and EPR as a function of pH, stability constants were evaluated, and a pH-dependent speciation model was developed.

It seems appropriate to caution about extrapolating the results of metal-drug studies carried out in high concentrations and in the absence of competing biologically-relevant ligands. Nevertheless, structural studies of metal-drug complexes prepared in even such a manner can indicate the detailed nature of the metal-drug interaction: the charge on the species, which is important for solubility and transport considerations; the metal-ligand stoichiometry; the stability of the complex; and the geometry of the metal coordination, which will frequently lead to an understanding of kinetics of ligand substitution by utilization of basic principles of inorganic chemistry. For redox-active metals, structural studies can lead to an understanding of reduction potentials which often change significantly with geometry and ligand environment.

Structural studies of metal-drug interactions are commonly performed by preparing and isolating pure metal-drug complexes and utilizing standard spectroscopic techniques to elucidate structure. In many cases solution studies are carried out in addition to or in place of studies of the isolated solids and these are undoubtedly more relevant than the solid-state studies, although not infrequently the results are similar. Since most spectroscopic techniques are sensitive only above 100 micromolar concentrations, extrapolations to the lower concentrations of pharmacological relevance are important. It is probable that the importance of metal-drug interactions has been overstated, since the stability constants of many complexes are not high enough for there to be appreciable concentrations under equilibrium conditions in biological fluids. There is a particularly great need for more work on the kinetics of ligand exchange under biologically appropriate conditions.

III. METAL-ASSISTED DEGRADATION AND COMBINATION REACTIONS

One significant action of metal ions is to facilitate degradation of bound ligands. While the kinetics of degradation are usually studied in detail, structural studies can also lead to an understanding of metal-assisted hydrolysis mechanisms (Black, 1987; Hay and Morris, 1976). In some cases the role of the metal may be to facilitate degradation of the administered drug to an active or more active metabolite. It is often unknown whether it is the drugs or their metabolites which are of primary physiological significance, and the importance of metal-assisted degradation reactions of drugs remains little investigated. Structural studies are often required to identify the new product(s), which may remain coordinated to the metal. One example where such hydrolysis has been shown, although its physiological significance is unknown, is the Cu(II)-promoted hydrolysis of the nitrile of coordinated cimetidine (Greenaway et al., 1980b). Metal ion catalyzed decomposition may be an important factor affecting the stability of drugs in vivo. For example, Iley et al. (1991) have shown that metals catalyze decomposition of monoalkyltriazenes. Addition reactions can also be facilitated by metal ions, either because of the entropic gain from coordination of the reacting species, or from electronic effects (Black, 1987). The review by Leussing (1976) gives many examples of Schiff base degradation and formation reactions of relevance to drug action which are facilitated by metal ions.

IV. SPECTROSCOPIC TECHNIQUES FOR STRUCTURE ELUCIDATION

Many metal ion - drug complexes have been prepared in recent years, following the realization that such complexation may occur in vivo. The usual method of study is to prepare metal complexes in solution and attempt to isolate solids, preferably crystalline, by precipitation or by removal of solvent. In general, structural studies are carried out on both solids and solutions using a wide variety of complementary structural techniques. Typical studies, for example those of metal complexation to phenobarbital (Sevilla et al., 1992; Pezeshk, 1986; Pezeshk et al., 1983; Hänsel, 1982; Caira et al., 1973; Wang and Craven, 1971), utilize a wide range of complementary techniques, most commonly IR, UV-visible absorption, EPR, and ^{1}H and ^{13}C NMR spectroscopies, X-ray crystallography, thermogravimetric analyses, microanalyses, and magnetic studies. Many excellent and readable general summaries of these techniques are available (Drago, 1991; Parrish, 1990; Ebsworth et al., 1987; Darnall and Wilkins, 1980; Brown, 1980).

The metal-ligand stoichiometry is the most basic form of structural data, and can be readily determined by titrating metal ions into drug solutions (or the reverse) at constant pH and monitoring the system by almost any convenient spectroscopic or analytical technique which can detect changes upon metal-drug coordination. An alternative, and more common, technique is potentiometric titration, where pH is monitored as base (or acid) is added to solutions containing both metal ion and drug (Martell and Motekaitis, 1990). Increasingly such titrations are also being monitored by spectroscopic techniques which may allow more direct structural information (Jackson and Kelly, 1989; Lincoln and Stephens, 1991). Although such titrations have been commonly applied for decades in systems where metal-ligand interactions are strong, the availability of computer programs has increased the information available from these studies and allowed their extension to systems where interactions are weaker (Martell and Motekaitis, 1989a).

A. Electronic absorption spectroscopy

Most studies utilize electronic (UV-visible) absorption spectroscopy (Lever, 1984) to monitor metal-drug interactions. Many drugs, especially those with aromatic groups, absorb strongly in the ultraviolet and these bands are often perturbed upon coordination of the ligand to a metal ion. While the high extinction coefficients of these intraligand bands allow work at low concentrations, unfortunately such studies seldom enable the exact nature of the coordinating atoms to be determined since the bands are commonly due to transitions of delocalized electrons and the origin of the shift in wavelength is not obvious. The metal-centered d-d bands of transition metal ions are frequently utilized in a similar manner and, although these are weaker since the transitions are Laporte forbidden, they also give more information about coordination number and the types of coordinating atoms (Figgis, 1987). Correlations between λ_{max} and crystal field strength are quite well understood (Billo, 1974) and enable the ligating atoms to be determined. In addition to the d-d bands, metal → ligand or, less frequently, ligand → metal charge transfer (CT), bands are observed in many cases, especially when the metal has unoccupied d-orbitals to accept nonbonding or π-electrons from the ligand. Metal→ligand CT bands are most commonly observed with sulfur ligand atoms (Jørgensen, 1968), since the sulfurs have empty d-orbitals which can accept metal d-electrons, or with ligands such as CN^- which have empty π^* orbitals. Although the presence of CT bands confirms formation of a metal-drug complex, the bands are difficult to assign and the band positions generally are not especially helpful in giving structural details.

B. Fluorescence spectroscopy

Fluorescence spectroscopy (Harris and Bashford, 1987) has been utilized for drugs which have rigid structures, especially delocalized multi-ring systems. Fluorescence maxima are affected by binding of diamagnetic metals, and fluorescence is commonly quenched by paramagnetic ions. In either case, stability constants and metal-drug stoichiometries can be determined. In addition, binding sites can be deduced. For example, Stoel et al. (1976) studied the interaction of several metal ions with anhydrotetracycline by UV-visible spectroscopy and also by fluorescence spectroscopy, and were able to identify both the primary metal binding site and a secondary site which is occupied when the first site is blocked.

Fluorescence of lanthanide metal ions has been utilized in model studies of calcium and magnesium binding to drugs. Lenkinski et al. (1980) showed that the visible luminescence of Tb^{3+} is enhanced on binding to bleomycin and were able to evaluate binding constants which were in excellent agreement with constants obtained from potentiometric titrations. Lanthanide fluorescence has also been utilized extensively to study metal binding to ionophores (Horrocks, 1982).

C. ORD, CD and MCD spectroscopy

In addition to absorption spectroscopy, ORD, CD, and MCD can be utilized to provide additional information. Optical rotatory dispersion, ORD, is the variation of optical rotation with the wavelength of plane polarized light. Circular dichroism, CD, is the difference between the extinction coefficients for left and right handed circularly polarized light ($\varepsilon_L - \varepsilon_R$).

CD can be useful when the drug (ligand) is itself optically active. Ligand-based CD was utilized by Newman and Frank (1976) to show 2:1 metal ion : ligand complex formation between tetracycline and calcium and 1:1 formation with magnesium. By contrast, relatively few complexes display metal-based optical activity and CD studies of metal-drug complexes are usually based on the optical activity of the ligand which, since the d-d transitions have partial ligand character, may be observed in the visible spectrum where the metal-based d-d bands occur (Martin et al., 1983).

For molecules which are optically inactive, such as $[Co(H_2O)_6]^{2+}$, magnetic circular dichroism, MCD, can be induced by a magnetic field and thus this technique is a complement to CD (Dooley and Dawson, 1984) but with the advantage that it is generally more sensitive than UV-visible studies, and is not confused by ligand-centered transitions. MCD studies are most commonly utilized with paramagnetic metal ions to confirm assignments of bands as d-d transitions.

D. Vibrational spectroscopy

When a complex can be isolated as a solid, vibrational spectroscopy (including IR, FTIR, and Raman) frequently enables the coordinating groups to be identified. Most vibrational studies use an empirical analysis based on comparison of structures of known metal-ligand complexes, although in principle more detailed geometric information is possible from normal coordinate analysis of the data (Nakamoto, 1986). Whenever solid complexes are prepared, it is standard practice to acquire an IR spectrum, since it is usually possible to determine the chelating groups by studying shifts of the ligand-based absorption peaks from their positions in the ligand itself. For example, Preti and Tosi (1979) were able to show metal-nitrogen bonding in several metal diazepam complexes. There is an exhaustive body of IR data in the literature and qualitative interpretations are usually possible. Difficulties exist for larger ligands when there are several similar binding sites and it is usually not possible to distinguish which of several similar groups is ligated to the metal. However, coordination details can commonly be deduced. For example, differences between unidentate, bidentate, and bidentate-bridging carboxylates are quite obvious (Nakamoto, 1986; Melnik, 1982).

In addition to ligand-based vibrations, metal-ligand vibrations can be of assistance. Such vibrations usually occur at frequencies below 600 cm^{-1}, and not all IR spectrophotometers extend to this range. Many of the common mulling or diluting agents also absorb in this region, obscuring the rather weak metal-ligand bands. For these reasons, most published work in this field has not utilized metal-ligand bands in structural elucidation. A useful introduction to this area is the book by Ferraro (1971).

Vibrational spectroscopy has been much less utilized in solution studies. Water is the solvent of choice for studying drug chelation and, although modern FTIR techniques are sufficiently sensitive for the technique to be useful, the strong H_2O bands provide only a small window. Use of D_2O as well as H_2O, and solvent subtraction techniques enable solution FTIR studies, but experimental problems remain and the technique has been not utilized as extensively as it might.

Resonance Raman spectroscopy is a technique where the molecule is excited electronically and the vibrational fine structure is observed. The technique provides greater sensitivity since the greater intensity of the Raman transition is utilized, and simplifies the vibrational assignments since only vibrations involving the absorbing moieties are observed (Clark and Dines, 1986). Solution Resonance Raman spectroscopy is useful for interpreting vibrational bands of ligand groups which coordinate to metals having d-d bands, but has not been widely used for this purpose with small molecules to date.

E. EPR spectroscopy

Although restricted to paramagnetic systems, EPR has been much utilized for work with some metal ions, particularly Cu(II). The g-values and copper hyperfine coupling constants can commonly be correlated with the ligand atoms that coordinate to the copper (Peisach and Blumberg, 1974) in a manner similar to that utilized for interpretation of d-d transition λ_{max} values (Billo, 1974) and the analyses are, in fact, based on the same theoretical principles. In addition, nitrogen superhyperfine structure enables direct verification of the number of coordinating nitrogen atoms although spectral simulation is desirable for interpreting frozen solution or powder spectra (Hyde et al., 1986). EPR has been utilized for many other paramagnetic metal ions, but often the information content is lower than for copper, and structural details less readily deduced. The technique is also particularly useful (although under-utilized) in evaluating stabilities, since overlapping spectra from several species can frequently be separated and the relative concentrations of the various species quantitated. Added advantages are that EPR is commonly more sensitive than electronic absorption spectroscopy, and spectra are uncluttered by ligand-based transitions. The high sensitivity has also enabled some experiments to be carried out in biologically relevant systems, for example studies of copper nirvanol and copper diisopropylsalicylate in tumor cells (Pezeshk et al., 1987). In addition, some studies of reactions between metal-drug complexes and other biologically active species have been carried out; for example the reaction of superoxide with Cu(II) salicylate as a function of pH showed that the superoxide reduced the Cu(II) of $Cu(HSal)_2$ to Cu(I) with the evolution of O_2 (O'Young and Lippard, 1980).

Nuclear Quadrupole Resonance Spectroscopy, NQR, although less common, can also provide useful information about complexes of paramagnetic metal ions (Buslaev and Kravcenko, 1987; Parrish, 1990).

F. NMR spectroscopy

NMR techniques (Harris, 1986; Parrish, 1990) have been applied in many cases to studies of metal-drug interactions in solution. The technique is not useful for low-molar-mass complexes with metal ions which are paramagnetic unless metal-ligand exchange rates are high, but many studies have been carried out with diamagnetic ions where analysis of chemical shifts and spin-spin interactions can lead to detailed structural information. Concentrations need typically to be of the order of 0.1 mM for ^1H NMR, and higher for NMR of other nuclei. An early utilization of NMR techniques was the determination of the metal binding site of tetracyclines (Williamson and Everett, 1975; D'Arcy and McElnay, 1982) which helped explain the observation that absorption of tetracyclines is inhibited by interaction with Ca^{2+}, Fe^{2+}, Mg^{2+}, and Al^{3+}. A number of NMR studies have been made of interactions between paramagnetic metal ions and drugs which are large enough that they have atoms an appreciable distance (>5Å) from the metal ion. By utilizing the effect of the unpaired electrons in enhancing spin-lattice relaxation rates, it is possible to determine distances between the metal ion and the NMR-active nucleus (e.g. Camparini et al., 1983, with Mn(II) and morphine), although most workers (e.g. Lee and Everett, 1981, with Mn(II) and tetracycline) have utilized a more empirical approach since distance calculations require careful temperature-dependent studies of relaxation times and incorporate a number of theoretical approximations which are sometimes difficult to justify (Bertini and Luchinat, 1986). In addition to providing information about the metal-binding site, NMR techniques also provide some information about kinetics of ligand exchange, since the technique can to some extent distinguish exchange rates. Since oxygen is a common drug ligand atom, studies utilizing drugs where the ^{16}O has been replaced by ^{17}O should yield additional information (Boykin, 1990), but as yet this has not been widely utilized. It is clear that NMR techniques, with their improving sensitivity and sophistication will be utilized increasingly in this field, particularly for higher molar mass drugs.

An example of a simple 1-D NMR study of metal binding is provided by a series of studies with the bleomycins, glycopeptides of about 1.5 kg mol^{-1}. NMR has been extensively utilized to monitor the binding sites of several metals. Complete ^1H and ^{13}C NMR spectral assignments of all resonances has enabled the observed shifts on metal ion binding to be analyzed in some detail. The earliest ^1H and ^{13}C NMR studies of the diamagnetic Zn(II) complex confirmed the nature of the metal binding site (Dabrowiak et al., 1978; Greenaway et al., 1980a). The shortening of the relaxation time of the NMR-observable nuclei on binding of the paramagnetic Cu(II) ion was observed to be so great that the effect was observed primarily as loss of resonances due to atoms close to the ligating groups (Dabrowiak et al., 1978). In the Fe(II) and Fe(III) complexes, where the effect on the relaxation time was somewhat smaller, some bleomycin resonances were lost, but others were observed to both shift and broaden on metal ion binding (Dabrowiak et al., 1980). Several Co(III) complexes of bleomycin have also been analyzed in some detail (Tan et al., 1992).

NMR detection of metal ions themselves is becoming increasingly common, if they are diamagnetic (Johansson and Drakenberg, 1990). ^{23}Na NMR has been used extensively to study Na^+ complexes of ionophores and also to show the effect of anesthetics on Na^+ binding at specific sites of the channel-forming peptide gramicidin A (Hunt and Veiro, 1986). Since most metals have at least one isotope with a non-zero nuclear spin, metal NMR is feasible for most diamagnetic species, but to date has been most utilized for ^{23}Na, ^{25}Mg, and ^{27}Al (Laszlo, 1983).

Modern 2-D and 3-D NMR spectroscopic techniques which enable significantly more detail to be extracted by measuring cross-relaxation have yet to be extensively utilized for metal-drug systems, but should be of considerable value for larger drugs since they are capable, in conjunction with computer modelling techniques, of providing full three-dimensional solution structures, including distances and geometries (Oppenheimer and James, 1989).

G. Mössbauer spectroscopy

Mössbauer spectroscopy (Parrish, 1990) is limited to the study of γ-emitting nuclei and is not readily available to the average chemist, and thus has been little utilized as a structural elucidation technique for many metal ion-drug complexes, except for a few studies utilizing ^{57}Fe, ^{61}Ni, ^{67}Zn, and ^{197}Au. It is, however, particularly useful in the study of iron complexes, and clear differences in both quadrupole splittings and isomer shifts enable characterization of the valence and spin state of both Fe(II) and Fe(III). Since EPR spectroscopy is generally more available to most researchers, does not require the use of ^{57}Fe, and yields similar information for Fe(III) more readily than Mössbauer spectroscopy, it is the preferred technique for studying Fe(III) complexes. However, Mössbauer techniques are especially helpful in the study of Fe(II) complexes. Examples of the utility of Mössbauer spectroscopy are the studies of the binding of Fe(II) to bleomycin (Dabrowiak et al., 1982) and of Au(I) and Au(III) to 2,3-dimercaptopropanol (Brown et al., 1979).

H. X-ray absorption spectroscopy

X-ray absorption spectroscopy is a technique which utilizes synchrotron radiation to ionize core electrons of the metal. The absorption edge position depends on the oxidation state of the metal, and X-ray absorption fine structure, EXAFS, on the absorption band provides information about the ligand atoms attached to the metal (Sham, 1986). The technique is particularly useful because metal-ligand atom distances, information analogous to X-ray crystallography for the first (and to some extent the second) coordination sphere of the metal ion, can be obtained for complexes *in solution*. Limited access to instrumentation, and the importance of good model compounds in the interpretation of the data, however, have limited application of the technique in studies of metal-drug complexes.

V. NON-SPECTROSCOPIC TECHNIQUES FOR STRUCTURE ELUCIDATION

A. Potentiometric titration

Although not technically a "structural" technique, potentiometric titration remains the most utilized technique for determining stabilities of complexes and elucidation of speciation diagrams, and its importance has been discussed earlier. A large body of data exists on stability constants of metal complexes (Martell and Smith, 1989b). In addition, much work has been carried out on metal complexes of amino acids and peptides, and the results have been used to predict metal ion speciation in plasma (e.g. Jackson and Kelly, 1988). Details can be found in Parts Three and Five of this volume. Recent interest has been shown in testing these predictions using a variety of spectroscopic techniques.

B. Conductivity

Solution conductivity is a readily measured quantity and since the conductivity depends on the number and charges of the ions in solution, such measurements have long been utilized to measure ionic stoichiometries.

C. Thermogravimetric analysis

Thermogravimetric analysis, TGA, studies the energetics of the decomposition of solid complexes as the temperature is raised. It is utilized to help characterize solids, especially regarding solvation, but the results are less readily interpreted than most spectroscopic alternative methods.

D. Magnetic susceptibility

This is frequently utilized to determine valence, spin states, and especially the presence of metal-metal interactions. Simple apparatus is available for measuring magnetic moments of solids at room temperature, and results are routinely reported for solid paramagnetic complexes. Variable temperature measurements of magnetic susceptibility are normally carried out only when metal-metal interactions are indicated by EPR or by abnormal room temperature magnetic moments. Binuclear structures, especially with carboxylate ligands (Melnik, 1982), and antiferromagnetic interactions in Cu(II)-drug complexes are quite common in the solid state, one example being the study of interactions in the Cu(II) complex of bromazepam (Real et al., 1987).

E. Crystallography

X-ray (or neutron) diffraction (crystallography) has become increasingly common with the advances of modern instrumentation and computer-assisted structure solving techniques, and allows complete characterization of the structures of crystallizable metal-drug complexes (Stout and Jensen, 1989). However, in many cases, the solid-state structures are not the same as the solution structures. Since water is present in high concentration (55 M) and is a significantly strong Lewis base, it can displace some or all of the coordinating ligand atoms in aqueous solution, resulting in major changes in geometry and structure. Many structural studies of metal-drug complexes thus have little relevance to pharmacology, since the structures found to exist in the solid are not retained in aqueous solution, even in the absence of other chelating agents. For example, Tajmir-Riahi and Boghai (1982) showed by crystallography that vitamin C forms salts with alkali metal ions in the solid state, but ^{13}C NMR and FTIR experiments in aqueous solution, showed that no significant metal-vitamin C interactions occur. Dimeric copper carboxylate structures are common in the solid state (Melnik, 1982; Weser et al., 1980), but few persist in aqueous solution, as has been shown with copper diisopropylsalicylate complexes (Greenaway et al., 1988). Solid-state studies of metal-ligand complexes which have been prepared in non-aqueous solutions, or in solutions at very high pH are similarly of doubtful relevance to pharmacological studies. Nevertheless, crystallographic methods provide far more definitive structures than can be deduced from most spectroscopic studies, and in many cases have rendered spectroscopic studies of solids largely redundant. In many cases the solid-state structures determined by crystallography also enable a more definitive interpretation of solution state structures determined by spectroscopic means, particularly when the solution and solid state structures are similar. For example, a study by Cini et al. (1990) of the binding of a variety of metal ions to the antiinflammatory drug piroxicam used the crystallographic results to interpret the absorption, ^{13}C NMR, and magnetic results. Similarly Tan et al. (1992) showed that the 1-D and 2-D solution NMR results for a Co(III) bleomycin model compound were consistent with the complex structure determined for the solid by crystallographic methods.

F. Cyclic voltammetry

Cyclic voltammetry is a technique where a varied potential is applied to a solution and the current flow is measured (Bard and Faulkner, 1980). The technique is a sensitive procedure for determining oxidation-reduction potentials. In the future spectroelectrochemical techniques will become more common for characterizing intermediates (McCreery, 1987).

REFERENCES

Bard, A.J., and Faulkner, L.R. (1980). *Electrochemical Methods*, Wiley-Interscience, New York.

Bertini, I., and Luchinat, C. (1986). *NMR of Paramagnetic Molecules in Biological Systems*, Benjamin Cummings, Reading, Mass.

Billo, E.J. (1974). *Inorg. Nucl. Chem. Lett.*, *10*: 613.

Black, D. St C. (1987). *Comprehensive Coordination Chemistry*, Vol. 1 (G. Wilkinson, R.D. Gillard and J.A. McCleverty, eds.), Pergamon Press, Oxford, p. 415.

Boykin, D.W., ed. (1990). *^{17}O NMR Spectroscopy in Organic Chemistry*, CRC Press, Florida.

Brown, D.H., McKinlay, G.C., and Smith, W.E. (1979). *J. Inorg. Biochem.*, *10*: 275.

Brown, S.B. (1980). *An Introduction to Spectroscopy for Biochemists*, Academic Press, Inc. London.

Buslaev, Yu. B., and Kravcenko, E.A. (1987). *Coord. Chem. Rev.*, *82*: 1.

Caira, M.R., Fazakerley, G.V., Linder, P.W., and Nassimbeni, L.R. (1973). *Inorg. Nucl. Chem. Lett.*, *9*: 1101.

Camparini, I.B., Gaggelli, E., and Valensin, G. (1983). *J. Inorg. Biochem.*, *18*: 349.

Cini, R., Giorgi, G., Cinquantini, A., Rossi, C., and Sabat, M. (1990). *Inorg. Chem.*, *29*: 5197.

Clark, R.J.H., and Dines, T.J. (1986). *Angew. Chem.*, *98*: 131.

Dabrowiak, J.C., Greenaway, F.T., and Grulich, R. (1978). *Biochemistry*, *17*: 4090.

Dabrowiak, J.C., Greenaway, F.T., Santillo, F.S., Crooke, S.T., and Essery, J.M. (1980). *Inorganic Chemistry in Biology and Medicine*, ACS Symposium Series, Vol. 140 (A.E. Martell, ed.), p. 237.

Dabrowiak, J.C., Greenaway, F.T., and Reiff, W.M. (1982). *J. Inorg. Biochem.*, *16*: 161.

D'Arcy, P.F., and McElnay, J.C. (1982). *Metal Ions in Biological Systems*, Vol. 14 (H. Sigel, ed.), Marcel Dekker, Inc., New York, p. 1.

Darnall, D.W., and Wilkins, R.G. (1980). *Methods for Determining Metal Ion Environments in Proteins: Structure and Function of Metalloproteins*, Elsevier/North Holland, Inc., New York.

Dooley, D.M., and Dawson, J.H. (1984). *Coord. Chem. Rev.*, *60*: 1.

Drago, R.S. (1991). *Physical Methods for Chemists*, 2nd ed., W.B. Saunders Co., Philadelphia.

Ebsworth, E.A.V., Rankin, D.W.H., and Cradock, S. (1987). *Structural Methods in Inorganic Chemistry*, Blackwell Scientific Publications, Oxford.

Ferraro, J.R. (1971). *Low Frequency Vibrations of Inorganic and Coordination Compounds*, Plenum Press, New York.

Figgis, B.N. (1987). *Comprehensive Coordination Chemistry*, Vol. 1 (G. Wilkinson, R.D. Gillard and J.A. McCleverty, eds.), Pergamon Press, Oxford, p. 213.

Greenaway, F.T., Dabrowiak, J.C., Grulich, R., and Crooke, S.T. (1980a). *Org. Magn. Resonance*, *13*: 270.

Greenaway, F.T., Brown, L.M., Dabrowiak, J.C., Thompson, M.R., and Day, V.M. (1980b). *J. Amer. Chem. Soc.*, *102*: 7782.

Greenaway, F.T., Norris, L.J., and Sorenson, J.R.J. (1988). *Inorg. Chim. Acta*, *145*: 279.

Hänsel, W. (1982). *Metal Ions in Biological Systems*, Vol. 14 (H. Sigel, ed.), Marcel Dekker, Inc., New York, p. 243.

Harris, D.A., and Bashford, C.L. (1987). *Spectrophotometry and Spectrofluorimetry - A Practical Approach*, IRL Press, Oxford.

Harris, R.K. (1986). *Nuclear Magnetic Resonance Spectroscopy*, Longman Scientific and Technical, Essex.

Hay, R.W., and Morris, P.J. (1976). *Metal Ions in Biological Systems*, Vol. 5 (H. Sigel, ed.), Marcel Dekker, Inc., New York, p. 173.

Horrocks, W. DeW. (1982). *Advances in Inorganic Biochemistry*, Vol. 4, Elsevier Science Publishing Co., Inc., New York, p. 201.

Howard-Lock, H.E., and Lock, C.J.L. (1987). *Comprehensive Coordination Chemistry*, Vol. 6 (G. Wilkinson, R.D. Gillard and J.A. McCleverty, eds.), Pergamon Press, Oxford, p. 755.

Hughes, M.N. (1987). *Comprehensive Coordination Chemistry*, Vol. 6 (G. Wilkinson, R.D. Gillard and J.A. McCleverty, eds.), Pergamon Press, Oxford, p. 541.

Hunt, G.R.A., and Veiro, J.A. (1986). *Biochem. Soc. Trans.*, *14*: 603.

Hyde, J.S., Antholine, W.E., Froncisz, W., and Basosi, R. (1986). *Progr. Inorg. Biochem. Biophys.*, *2*: 363.

Iley, J., Moreira, R., and Rosa, E. (1991). *J. Chem. Soc., Perkin Trans. II*, 81.

Jackson, G.E., and Kelly, M.J. (1988). *Inorg. Chim. Acta*, *152*: 215.

Jackson, G.E., and Kelly, M.J. (1989). *J. Chem. Soc., Dalton Trans.*, 2429.

Jackson, G.E., May, P.M., and Williams, D.R. (1978). *J. Inorg. Nucl. Chem.*, *40*: 1227.

Johansson, C., and Drakenberg, T. (1990). *Annual Reports in NMR Spectroscopy*, Vol. 22 (G.A. Webb, ed.), Academic Press, London, p. 1.

Jørgensen, C.K. (1968). *Inorg. Chim. Acta Rev.*, *2*: 65.

Laszlo, P., ed. (1983). *NMR of Newly Accessible Nuclei*, Academic Press, New York.

Laurie, S.H. (1987). *Comprehensive Coordination Chemistry*, Vol. 2 (G. Wilkinson, R.D. Gillard and J.A. McCleverty, eds.), Pergamon Press, Oxford, p. 959.

Lee, J.Y., and Everett, G.W. (1981). *J. Amer. Chem. Soc.*, *103*: 5221.

Lenkinski, R.E., Peerce, B.E., Pillai, R.P., and Glickson, J.D. (1980). *J. Amer. Chem. Soc.*, *102*: 7808.

Leussing, D.L. (1976). *Metal Ions in Biological Systems*, Vol. 5 (H. Sigel, ed.), Marcel Dekker, Inc., New York, p. 1.

Lever, A.P.B., ed. (1984). *Inorganic Electronic Spectroscopy*, 2nd ed., Elsevier Science, Amsterdam.

Lincoln, S.F., and Stephens, A.K.W. (1991). *Inorg. Chem.*, *30*: 3529.

Martell, A.E., and Motekaitis, R.J. (1989a). *Determination and Use of Stability Constants*, VCH Publishers, New York.

Martell, A.E., and Motekaitis, R.J. (1990). *Coord. Chem. Rev.*, *100*: 363.

Martell, A.E., and Smith, R.M. (1989b). *Critical Stability Constants*, Plenum, New York.

Martin, M., Krogh-Jespersen, M.B., Hsu, M., Tewksbury, J., Laurent, M., Viswanath, K., and Patterson, H. (1983). *Inorg. Chem.*, *22*: 647.

May, P.M., and Williams, D.R. (1981). *Metal Ions in Biological Systems*, Vol. 12 (H. Sigel, ed.), Marcel Dekker, Inc., New York, p. 283.

McCreery, R.L. (1987). *Physical Methods in Chemistry*, Vol. 2 (B. Rossiter, ed.), Wiley, New York.

Melnik, M. (1982). *Coord. Chem. Rev.*, *42*: 259.

Nakamoto, K. (1986). *Infrared and Raman Spectra of Inorganic and Coordination Compounds*, 4th ed., Wiley-Interscience, New York.

Newman, E.C., and Frank, C.W. (1976). *J. Pharm. Sci.*, *65*: 1728.

Oppenheimer, N.J., and James, T.L. (1989). *Methods in Enzymology*, Vol. 176, Academic Press, Inc., San Diego.

O'Young, C.-L., and Lippard, S.J. (1980). *J. Amer. Chem. Soc.*, *102*: 4920.

Parrish, R.V. (1990). *NMR, NQR, EPR, and Mössbauer Spectroscopy in Inorganic Chemistry*, Ellis Horwood, New York.

Peisach, J., and Blumberg, W.E. (1974). *Arch. Biochem. Biophys.*, *165*: 691.

Pettit, L.D., Ueda, J., Morier-Teissier, E., Helbecque, N., Bernier, J.-L., Henichart, J.-P., and Kozlowski, H. (1992). *J. Inorg. Biochem.*, *45*: 203.

Pezeshk, A. (1986). *J. Amer. Chem. Soc.*, *108*: 8285.

Pezeshk, A., Greenaway, F.T., and Sorenson, J.R.J. (1983). *Inorg. Chim. Acta*, *80*: 191.

Pezeshk, A., Antholine, W.E., and Petering, D.H. (1987). *J. Inorg. Biochem.*, *31*: 283.

Preti, C., and Tosi, G. (1979). *J. Inorg. Nucl. Chem.*, *41*: 1079.

Real, J.A., Borrás, J., Munoz, M.C., Mosset, A., and Galy, J. (1987). *J. Inorg. Biochem.*, *31*: 221.

Sevilla, I., Perelló, L., and Borrás, J. (1992). *J. Inorg. Biochem.*, *45*: 221.

Sham, T.K. (1986). *Accounts Chem. Res.*, *19*: 99.

Stoehl, L.J., Newman, E.C., Mariano, S.L., and Frank, C.W. (1976). *J. Pharm. Sci.*, *65*: 1794.

Stout, G.H., and Jensen, L.H. (1989). *X-Ray Structure Determination. A Practical Guide*, 2nd ed., Wiley-Interscience, New York.

Tajmir-Riahi, H.A., and Boghai, D.M. (1982). *J. Inorg. Biochem.*, *45*: 73.

Tan, J.D., Hudson, S.E., Brown, S.J., Olmstead, M.M., and Mascharak, P.K. (1992). *J. Amer. Chem. Soc.*, *114*: 3841, and references therein.

Wang, B.C., and Craven, B.M. (1971). *Chem. Commun.*, 290.

Weser, U., Sellinger, K.-H., Lengfelder, E., Werner, W., and Strähle, J. (1980). *Biochim. Biophys. Acta, 631*: 232.

Williamson, D.E., and Everett, G.W. (1975). *J. Amer. Chem. Soc.*, *97*: 2397.

Wilson, C.L., and Wilson, D.W., eds. (1964). *Comprehensive Analytical Chemistry*, Vol. IIA, Elsevier, Amsterdam.

ELECTROCHEMICAL STUDIES ON METAL ION - DRUG INTERACTIONS

by

J.-C. Viré and J.-M. Kauffmann

Département d'Analyse Instrumentale et Bioélectrochimie,
Université Libre de Bruxelles (ULB),
Institut de Pharmacie,
Campus Plaine, C.P. 205/6
Boulevard du Triomphe
B-1050 Brussels
Belgium

I. INTRODUCTION

Numerous drugs are capable of forming chelate complexes with metals ions, mainly transition and heavy metal ions. This may in some instances bear a direct relation to the major action of the drug, or be responsible for significant side effects.

Drug - metal ion interactions have often been evoked to explain certain mechanisms of drug action but have also been exploited for a variety of purposes, including: (i) transient combination of a ligand with metalloenzymes without removal of the metal ion from the enzyme, (ii) stabilization of either a chelating molecule or a valence state of a metal ion, (iii) complex formation to improve the delivery of a ligand or metal ion to a target site, (iv) chelation of toxic metal ions allowing their excretion (antidotes), and (v) formation of "unsaturated" chelates to induce or increase the toxic effects of a metal ion (antibacterial activity).

A knowledge of the reactivity of drugs towards metal ions appears to be of prime importance (Foye, 1961; Wilson et al., 1977; Chappuis, 1991) not only because it may affect the therapeutic activity of drugs, but also because it may modify the concentration and the distribution of essential elements in living organisms.

Most of the physico-chemical properties of the ligand or of the metal are modified by complexation. These modifications can be used to detect complex formation, to identify the species formed and to determine the thermodynamic parameters of the involved equilibria.

Numerous techniques have been proposed to investigate complexation reactions. Among them, electrochemical methods appear to be well suited due to their ability to detect the modifications of the redox properties resulting from the chelation process. A mutual interest exists between complexation studies and electrochemical techniques because the latter allow to determine the structure and the stability constants of the chelates formed while the former have given rise to several electroanalytical procedures.

II. INSTRUMENTAL METHODS

A. Complexation studies

1. *Complexometric titrations*

Complexometric titrations of metal ions or chelating agents (De Robertis et al., 1987) as well as potentiometric pH titrations of solutions containing different metal-to-ligand ratios (Pavlovich et al., 1988; Singh et al., 1991; Sigel et al., 1991) are well-known. In order to enhance selectivity, ion-selective electrodes have been proposed to detect the end point, but with some restrictions (Cosofret, 1980; Vander Linden, 1981; Midorikawa et al., 1990). A detailed overview of these methods is proposed by Linder in Volume 1 (Part Three, Chapter 1-A2).

2. *Polarographic methods*

In addition to the above titration methods, polarography constitutes a versatile technique for studying complexes in solution (Amatore, 1990). Different procedures have been developed to make polarography applicable to both reversible (fast uptake *and* release of electron) and irreversible (reduction *or* oxidation with slow electron transfer) processes, broadening its use considerably. Linder has described the basic principles of the method in Part Three, Chapter 1-A2, and useful informations can be found in the detailed monograph of Crow (1969).

B. Analytical procedures

Metal ion - drug interaction studies may also lead to analytical procedures allowing the quantitative determination of pharmaceutical compounds.

1. *Volumetric methods*

Complexometric precipitation and redox titrations are often used when selectivity is not required (raw material, pharmaceutical forms) and numerous applications have been described, involving the assay of drugs available in their hydrochloride or hydrobromide form with a silver nitrate solution (Davis, 1964; Zapotocky, 1966; Jenkins et al., 1967). These methods are included in several pharmacopoeia.

2. *Coulometric methods*

The possibilities of volumetric methods have been considerably extended by the use of coulometric titration. Constant current coulometry, which is based on Faraday's laws, allows to perform determinations identical to volumetric techniques, but exhibits a better sensitivity and does not require the standardization of the reactants. Moreover, in addition to precipitating cations or complexing agents, unstable powerful oxidizing (Ag(II), Mn(III)) or reducing (Cr(II), Ti(III)) reactants can be electrogenerated in situ. The principles, possibilities and limitations of the technique have been extensively described (Milner and Phillips, 1967; Bishop, 1975).

3. *Ion-selective electrodes*

Ion-selective electrodes have found some applications for the detection of the end point of volumetric titrations (Ross and Frant, 1969; Baumann and Wallace, 1969; Suarez et al., 1986) but few metal ion - drug interactions have been studied using ISE's and only valinomycine has been proposed to construct potassium-selective electrodes (Kuan and Guilbault, 1987; Eyal and Rechnitz, 1971).

4. *Voltammetric methods*

In addition to complexation studies, polarography and more generally voltammetric techniques have been applied to metal ion - drug interaction studies. These methods are often concerned with the direct reduction or oxidation of the molecule at the electrode surface. This includes mechanistic and quantitative analysis of organometallic drugs. Numerous compounds exhibiting complexing properties or forming insoluble salts with mercury ions give rise to an anodic polarographic wave corresponding to the easier oxidation of the mercury electrode.

5. *Stripping voltammetric methods*

The behavior evoked above has permitted to develop a very sensitive stripping technique (10^{-10} M) which involves a preconcentration step at a potential selected in the diffusion current of the anodic wave. The so-called cathodic stripping voltammetry allows to accumulate a drug as its insoluble mercury salt. The recorded current results from a negative going potential scan where reduction of mercury ions and dissolution of the deposited film occur (Vydra et al., 1976; Wang, 1985).

Such stripping techniques can also be applied to flow injection analysis (Luque de Castro and Izquierdo, 1991).

6. *Chemically modified electrodes*

In an approach which is similar to the development of ion-selective electrodes, different modifications have been brought to the surface of voltammetric electrodes in order to improve sensitivity or selectivity, but also, in the present context, to make easier the redox process occurring at the electrode - solution interface (faster electron transfer, lower redox potentials). This catalytic effect can be achieved by immobilization of an organometallic compound (cobalt, ruthenium, osmium, iridium, iron derivatives) which exhibits reversible redox properties, by (i) chemical binding to the electrode surface, (ii) incorporation in a polymeric film which is deposited on the surface, or (iii) mixing it with the electrode matrix, i.e. carbon paste and carbon composite electrodes. Theoretical and practical considerations have been detailed (Murray, 1984; Kalcher, 1990; Forster and Vos, 1992).

III. APPLICATIONS

A. Complexation studies

Most electrochemical investigations of complexation processes deal with titration procedures and polarographic measurements. They mainly concern electrochemically reversible systems involving copper, lead and cadmium ions.

Table 1 shows that antibiotics are mainly concerned with complexation studies, and few other drugs have been investigated by means of electrochemical techniques. The numbers appearing in the remarks column refer to some additional informations given below:

1) Prewave of kinetic - catalytic nature corresponding to the reduction of the metal of the adsorbed Ni-Amoxicillin complex.

2) The electroinactive antibiotic cephalexin forms 1:1 complexes with Ni and Co (Kapetanovic et al., 1990a) which are reduced at more positive potentials than the free cation. The process occurs through a catalytic reaction (Hernandez et al., 1984).

3) A process similar to that described above (see remark 2) has been observed for the cobalt and nickel complexes of ampicillin.

4) Ternary complexes have also been detected with ampicillin and dopamine for each investigated metal ion.

5) The reduction of the 1:1 copper - bleomycin complex exhibits an irreversible one-electron wave on glassy carbon and mercury electrodes. The reduction at the mercury electrode leads to the dissociation of the complex since the Cu(0)/Cu(I) couple appears. The Cu(I) - bleomycin complex is stable on glassy carbon.

6) Both the neutral and the protonated forms of ionomycin can complex several divalent metal ions in 80 % methanol - water mixture.

7) The strongest complexes are formed in the order Ag(I) < C_6H_5Hg(I) << Hg(II). Alkaline degradation products and hydroxamic acid derivatives of penicillin and cephalosporins are very similar to penicillamine in terms of complexation behavior due to the predominant role of the thiol group.

8) Penicillins G and V interact with copper (II) ions but these weak complexes undergo a rapid hydrolysis reaction leading to the respective penicilloïc acid - copper (II) complexes which are more stable (log K = 2.6 to 4.5 for G and 2.2 to 4.2 for V).

9) Cu^{2+} is slowly reduced to Cu^+ giving rise to a Cu(I)-MP complex which can be oxidized at a glassy carbon electrode into the chemically less inert Cu(II)-MP species. The

Table 1

Chelating Properties of Drugs

Drug (ligand)	Metal ion	Stoich. (M:L)	log β or log K	Method	Supporting electrolyte	Reference	Rem
Amoxicillin	Ni(II)	1:1	3.52	Amp.titn	0.2 M KNO_3	Kapetanovic and Veselinovic (1988)	
Amoxicillin	Ni(II)	1:1	-	DPP	0.2 M $NaNO_3$	Erceg et al. (1991)	1
Ampicillin	Ni(II) Co(II)	1:1 1:1	5.16 5.01	Amp.titn Amp.titn	0.2 M KNO_3 0.2 M KNO_3	Veselinovic and Kapetanovic (1988)	3
Ampicillin	Cu(II) Fe(III) Zr(IV)	1:2 1:2 1:2	1.77 6.13 20.62	Pot.titn Pot.titn Pot.titn	0.1 M KCl 0.1 M KCl 0.1 M KCl	Abd El Wahed et al. (1984)	4
Bleomycin	Cu(II)	1:1	-	CV + pol.	BR or HEPES buffer	Fultz and Durst (1991)	5
Cephalexin	Ni(II) Co(II) Cu(II)	1:1 1:1 1:2	3.16 3.09 9.44	Amp.titn Amp.titn Amp.titn + DPP	0.2 M KNO_3 0.2 M KNO_3 0.2 M $NaNO_3$	Kapetanovic et al. (1990a) Kapetanovic et al. (1990b)	2
Ionomycin	Sr(II) Ni(II)	1:1 1:1	5.30 10.25	Pot.titn Pot.titn	MeOH-H_2O (80:20), 0.05 M TEAP	Stiles et al. (1991)	6
Ionomycin, protonated	Sr(II) Ni(II)	1:1 1:1	8.4 5.9	Pot.titn Pot.titn			
Penicillin derivatives	Ag(I) Hg(II)	- -	- -	Pot.titn Pot.titn	Acetate buffer Acetate buffer	Van Krimpen et al. (1988)	7
Penicillin G Penicillin V	Cu(II) Cu(II)	- -	2.61 2.24	Pot.titn Pot.titn	- -	Cressman et al. (1966)	8
Penicilloïc acid G	Cu(II)	-	4.50	Pot.titn	-	Cressman et al. (1966).	8
Penicilloïc acid V	Cu(II)	-	4.20	Pot.titn	-		
6-mercapto-purine	Cu(II)	1:1	-	C V	0.1 M TEAP in DMF	Bilewicz and Muszalska (1991)	9
Aclacynomicin	Cu(II)	1:1	-	Pot.titn	HEPES buffer	Fiallo and Garnier-Suillerot (1987)	10
Carminomycin	Cu(II)	1:2	-	Pot.titn	HEPES buffer	Fiallo and Garnier-Suillerot (1987)	10
Azathioprine	Co(II) Cu(II) Ni(II)	1:2 1:2 1:2	8.42 10.22 8.47	Pot.titn Pot.titn Pot.titn	$\mu = 0.005$ $\mu = 0.005$ $\mu = 0.005$	Singh et al. (1991)	

cont'd

Table 1 (continued)

Drug (ligand)	Metal ion	Stoich. (M:L)	log β or log K	Method	Supporting electrolyte	Reference	Rem
α-methyl-L-dopa	Cu(II)	1:2	3.66	Pot. titn	0.1 M KCl	Abd El Wahed and Ayad (1984)	
	Fe(III)	1:2	16.25				
	Zr(IV)	1:2	9.31				
Folic acid	Cd(II)	1:1	5.56	DPP	0.15 M $NaClO_4$	Abo El Maali et al. (1989a)	11
		1:2	4.71				
		1:3	6.76				
	Pb(II)	1:1	5.78	DPP	0.15 M $NaClO_4$	Abo El Maali et al. (1989b)	11
		1:2	4.52				
		1:3	6.99				
	Cu(II)	1:2	4.97	DPP	0.15 M $NaClO_4$, pH = 4.5	Abo El Maali et al. (1989c)	11
	Cu(II)	1:1	3.23	DPASV	0.15 M $NaClO_4$	Abo El Maali et al. (1989c)	11
		1:3	11.57	pH = 7.4			
	Cu(II)	1:1	4.78	DPASV	0.15 M $NaClO_4$ + 2 mM phosphate buffer, pH = 7.4	Abo El Maali et al. (1989c)	11
		1:2	3.62				
		1:3	5.06				
Menadione	Cu(II)	1:1	3.65	DPP	0.15 M $NaClO_4$	Abo El Maali et al. (1990)	12
		1:2	8.67	DPP	pH = 7.4		
Naftazone	Cu(II)	1:1	7.18	DPP	0.15 M $NaClO_4$ pH = 5.0	Abo El Maali et al. (1990)	12
	Cu(II)	1:1	5.12	DPP	0.15 M $NaClO_4$ pH = 7.4		
Pilocarpine	Ni(II)	-	-	DPP	Veronal Na HNO_3-$NaNO_3$	Lopez and Fojon (1990)	13
Cimetidine	Ni(II)	-	-	DPP	Veronal Na HNO_3-$NaNO_3$	Lopez and Fojon (1992)	13
Piroxicam	Cu(II)	1:2	11.9	DPP	0.15 M $NaClO_4$ + 1 mM phosphate buffer, pH = 7.4	Abo El Maali et al. (1989d)	
		1:4	20.7	DPP	idem, pH = 4.5		
		1:2	10.1	DPP	idem, pH = 2.1		
	Pb(II)	1:1	6.13	DPP	idem, pH = 7.4		
Tenoxicam	Cu(II)	1:4	18.7	DPP	0.15 M $NaClO_4$ + 1 mM phosphate buffer, pH = 7.4	Abo El Maali et al. (1989d)	
	Cu(II)	1:4	24.4	DPP	idem, pH = 4.5		
	Cu(II)	1:1	4.1	DPP	idem, pH = 2.1		
		1:2	6.6				
		1:3	4.2				
	Pb(II)	1:1	3.8	DPP	idem, pH = 7.4		
		1:2	6.1				

Cu(I)-MP complex can be used as a mediator allowing an easier oxidation of ascorbic acid and for the determination of MP by cathodic stripping voltammetry at a copper amalgam electrode. The Cu(I)-MP complex is 2.9 x 10^8 times more stable than the Cu(II)-MP form.

10) Copper is complexed by aclacynomycin through the phenolate oxygen on C^6 and the carbonyl oxygen on C^5, forming a 6-membered ring. The hydroxy group is not involved in the carminomycin chelate. A 1:2 stoichiometry is postulated.

11) Studies performed on folic acid derivatives show that the active chelating site is the pterin moiety, plus the carboxylic group of the glutamate side chain. Pterin alone has a very weak complexing capacity. Complexation of lead and cadmium is inhibited in acidic solutions, due to the protonation of the nitrogen atoms of the pterin ring and only copper can be complexed at pH 4.5. The strength of the complexes increases according to the sequence Cd(II) < Pb(II) < Cu(II) at pH 7.4. The presence of phosphate ions hinders (Cu(II)) or completely inhibits (Cd(II) and Pb(II)) complex formation.

12) Naphthoquinonic compounds exhibit relatively weak complexing properties towards metal ions. This tendency is influenced by the presence of a substituent on the quinonic ring at the 2- or 3-position (increased by an electron donor substituent, decreased by a substituent with withdrawal effect). The use of a semicarbazone derivative of 1,2-naphthoquinone (naftazone) gives rise to one stable 1:1 complex; this is attributed to the participation of a nitrogen atom of the side chain to the complexation reaction, forming a stable six-membered ring.

13) The complex occurring between nickel(II), veronal (used as a buffer with HNO_3 and $NaNO_3$) and pilocarpine allows to determine this alkaloïd through the polarographic catalytic prewave of nickel (Lopez Fonseca and Fojon, 1990). The same behavior has been described for cimetidine (Lopez Fonseca and Fojon, 1992). Simpler complexes of nickel with pilocarpine have been studied by Jaber (1986) from an analytical point of view. Similarly, Clark et al. (1978) have developed a polarographic method for the determination of pilocarpine based on the potential shift of the copper peak.

B. Coulometric titrations

Amperostatic precipitation titrations are usually performed by stripping silver(I) ions from a silver rod in acidic medium or mercury(I) or (II) ions from a mercury pool in both acidic or alkaline solutions. Complexation reactions involve electrogeneration of mercury(II) ions or of chelating agents such as aminopolycarboxylic acids or thiol derivatives. These ligands are generated in most cases through the reduction of their mercury complexes (Bishop, 1975). Table 2 demonstrates the usefulness of amperostatic coulometry for the determination of drugs, often down to 0.2 mg.

C. Polarographic and voltammetric studies

Polarographic and voltammetric techniques have been applied to mechanistic and analytical studies involving metal ion - drug interactions. Organometallic compounds, mainly the platinum antitumor drugs, have been investigated using cyclic voltammetry. It has been demonstrated that their redox properties highly depend on the presence of chloride ligands but also on the free chloride concentration. Moreover, platinum particles deposited during the reduction step catalyze the redox processes of the compounds (Mebsout et al., 1987; 1988; 1989a; 1989b). Comparative studies of several analogues have been carried out (Elferink et al., 1987) and quantitative analysis has been performed using HPLC (Treskes et al., 1990; Dewaal et al., 1990; Digua et al., 1993). The polarographic analysis of the antirheumatic drug auranofin has permitted its determination at the micromolar level (Hernandez et al., 1989). Another widely investigated field involves the oxidation of the mercury electrode and the formation of an anodic polarographic wave. This subject has been recently reviewed (Budnikov et al., 1989) and recent examples include N-acetylcysteine (Belal et al., 1991a), ethinamate and methylpentynol carbamate (Belal, 1991b), carbimazole and methimazole (Fijalek and Zuman, 1990), orotic acid (Calvo et al., 1988a; 1989; 1990) and thioproline (Karpinski and Radomski, 1989; Karpinski et al., 1990). The method has proved to be useful for detection in flowing streams as examplified for allopurinol (Cataldi et al., 1989), amethopterine (Guerrieri and Palmisano, 1987) and 5-fluorouracil (Cataldi et al., 1988a).

Table 2

Coulometric Studies

Compound	Supporting electrolyte + precursor	Generated species	Electrodes Work	Aux	Detection system	Reference
Thiamazole, BAL, methylthiouracil pentothal, disulfiram.	Borate buffer pH 9.3 + 0.5 M KNO_3	Hg^{2+}	Hg	Pt	Potentio. Hg - SCE	Mairesse et al. (1973)
Aminopolycarboxylic acids.	0.1 M Na acetate, 85 % methanol	Hg^{2+}	Hg	Pt	Potentio. Hg - SCE	Vandenbalck et al. (1975)
Cryogenine, isoniazid, paracetamol, phenacetine.	0.05 M $Ce_2(SO_4)_3$ or 0.3 M $MnSO_4$ in 3 M H_2SO_4	Ce^{4+} or Mn^{3+}	Pt Pt	Pt Pt	Potentio. Pt - SCE or Ampero. Pt - Pt	Chateau and Patriarche (1978) Chateau and Patriarche (1977)
Isoniazid, nialamid, procarbazine.	0.1 M $AgNO_3$ in 5 M HNO_3 at -10°C	Ag^{2+}	Au	Pt	Potentio. Pt - SCE	Chateau and Patriarche (1978).
Phenelzine, cryogenine, ascorbic acid.	0.02 M HCl	Au^{3+}	Au plated Pt	Pt	Ampero. Pt - Pt	Chateau et al. (1979)
Levopromazine, chlorpromazine, promethazine, diethazine.	0.05 M $Ce_2(SO_4)_3$ in 3 M H_2SO_4	Ce^{4+}	Pt	Pt	Ampero. Pt - Pt	Patriarche and Lingane (1970)
Menadione, menadione bisulfite, methylene blue.	0.4 M $TiCl_4$ in 8 M H_2SO_4	Ti^{3+}	Pt	Cd	Potentio. Pt - SCE	Viré et al. (1977)

D. Cathodic stripping voltammetry
Sensitivity has been considerably improved by accumulating the mercury derivative at the electrode surface and several relevant examples are listed in Table 3.

E. Chemically modified electrodes
Numerous modification procedures have been developed to quantify various compounds through a catalytic process but few drugs have been investigated.

Ascorbic acid has been amperometrically determined using a platinum electrode modified by Prussian Blue (Li and Dong, 1987) or a graphite electrode modified by ferrocene or dimethylferrocene (Kulys and Drungiliene, 1991) resulting in a marked increase of the recorded current. Ruthenium analogues have been successfully employed as modifiers of carbon electrodes for the determination of insuline down to 0.1 μM (Cox and Gray, 1989), and streptomycin, novobiocin and neomycin (Leech et al., 1990).

More widely investigated is the carbon paste electrode (CPE) which is easily modified by direct mixing of the modifier (Kalcher, 1990). For example, cobalt phthalocyanine has often been used to catalyze the oxidation of several sulfur-containing drugs such as cysteine, lipoïc acid, disulfiram (Linders et al., 1986a), thiamazol and mercaptopurine (Linders et al., 1986b).

It should be noted that most of the developed modified electrodes have been used as detectors in flowing systems as reviewed elsewhere (Wang, E., et al., 1991; Cox and Gray, 1990; Wallace et al., 1990; Kalcher, 1990).

Table 3

Cathodic Stripping Voltammetry

Compound	Drug family	Supporting electrolyte	Linear range	Reference
Amethopterine	antineoplastic	0.01 M phosphate buffer, pH 7	$0.2 - 6 \times 10^{-9}$ M	Cataldi et al. (1988b)
Captopril	cardiotonic	6.3×10^{-3} M $HClO_4$ pH 2.2	$9 \times 10^{-10} - 3 \times 10^{-6}$ M	Passamonti et al. (1987)
Fluorouracil	antineoplastic	borate buffer, pH 7.8	$0.2 - 12 \times 10^{-7}$ M	Miranda Ordieres et al. (1987)
Orotic acid	vitamin	NaOH, pH 11.2	d.l.: 5×10^{-10} M	Calvo et al. (1988b)
Thioproline	antitumor	acetate buffer, pH 5	$10^{-9} - 10^{-7}$ M	Costa Garcia and Tunon Blanco (1988)
Citiolone	hepatoprotectant	carbonate, pH 10.9	$1 \times 10^{-9} - 3.5 \times 10^{-7}$ M	Casassas et al. (1990)
Thiamine	vitamin	0.05 M phosphate buffer, pH 7.4	$10^{-6} - 10^{-4}$ M	Ciszewski and Wang (1992)
Thiopronine	hepatoprotectant	-	$10^{-9} - 10^{-7}$ M	Casassas et al. (1991).
Uracil derivatives	18 derivatives	borate, pH 7.6	$0.5 - 5 \times 10^{-7}$ M	Bouzid and Mc Donald (1988a)
Uracil deriv. by FIA	18 derivatives	borate / KNO_3, pH 7.6	$10^{-8} - 10^{-7}$ M	Bouzid and Mc Donald (1988b)

IV. CONCLUSION

Metal ion - drug interactions represent a wide field which is only evoked in this brief overview. In addition to chemical reactions, mainly redox and complexation interactions, which are of prime importance for clinical and pharmacological studies, it can be seen that many aspects of this subject are also concerned with analytical procedures belonging to pharmaceutical chemistry and that electrochemical methods are particularly suited to investigate these systems.

REFERENCES

Abd El Wahed, M.G., and Ayad, M. (1984). *Anal. Lett.*, *17*: 205.

Abo El Maali, N., Viré, J.-C., Patriarche, G.J., and Ghandour, M.A. (1989a). *Electroanalysis*, *1*: 87.

Abo El Maali, N., Viré, J.-C., Patriarche, G.J., and Ghandour, M.A. (1989b). *Anal. Lett.*, *22*: 1251.

Abo El Maali, N., Viré, J.-C., Patriarche, G.J., and Ghandour, M.A. (1989c). *Electroanalysis*, *1*: 341.

Abo El Maali, N., Viré, J.-C., Patriarche, G.J., and Ghandour, M.A. (1989d). *Anal. Lett.*, *22*: 3025.

Abo El Maali, N., Viré, J.-C., Patriarche, G.J., and Ghandour, M.A. (1990). *Anal. Lett.*, *23*: 529.

Amatore, C. (1990). *J. Organomet. Chem.*, *22 (Organomet. Radical Processes)*: 1.

Baumann, E.W., and Wallace, R.M. (1969). *Anal. Chem.*, *41*: 2072.

Belal, F. (1991b). *J. Assoc. Off. Anal. Chem.*, *74*: 592.

Belal, F., Metwally, M.E., and Moustafa, M.A. (1991a). *J. Pharm. Belg.*, *46*: 320.

Bilewicz, R., and Muszalska, E. (1991). *J. Electroanal. Chem.*, *300*: 147.

Bishop, E. (1975). *Comprehensive Analytical Chemistry*, Vol. IID (C. Wilson and D. Wilson, eds.), Elsevier, Amsterdam.

Bouzid, B., and Mc Donald, A.M.G. (1988a). *Anal. Chim. Acta*, *211*: 155.

Bouzid, B., and Mc Donald, A.M.G. (1988b). *Anal. Chim. Acta*, *211*: 175.

Budnikov, G.K., Ulakhovich, N.A., Maistrenko, V.N., and Medyantseva, E.P. (1989). *Fresenius Z. Anal. Chem.*, *335*: 81.

Calvo, L., Rodriguez, J., Vinagre, F., and Sanchez, A. (1988a). *Analyst*, *113*: 321.

Calvo, L., Rodriguez, J., Vinagre, F., and Sanchez, A. (1988b). *Fresenius Z. Anal. Chem.*, *330*: 146.

Calvo, L., Rodriguez, J., Vinagre, F., and Sanchez, A. (1989). *Anal. Lett.*, *22*: 177.

Calvo, L., Rodriguez, J., Vinagre, F., and Sanchez, A. (1990). *Fresenius Z. Anal. Chem.*, *338*: 80.

Casassas, E., Arino, C., Esteban, M., and Redondo, A. (1990). *Anal. Lett.*, *23*: 981.

Casassas, E., Arino, C., Esteban, M., and Redondo, A. (1991). *Anal. Lett.*, *24*: 1183.

Cataldi, T., Guerrieri, A., Palmisano, F., and Zambonin, P.G. (1988a). *Anal. Chim. Acta*, *207*: 183.

Cataldi, T., Guerrieri, A., Palmisano, F., and Zambonin, P.G. (1988b). *Analyst*, *113*: 869.

Cataldi, T., Palmisano, F., and Zambonin, P.G. (1989). *Analyst*, *114*: 1449.

Chappuis, P. (1991). *Les Oligoéléments en Médecine et Biologie*, Lavoisier, Paris.

Chateau, M., and Patriarche, G.J. (1977). *Mikrochim. Acta*, *I*: 265.

Chateau, M., and Patriarche, G.J. (1978). *Anal. Chim. Acta*, *102*: 215.

Chateau, M., Christian, G.D., and Patriarche, G.J. (1979). *Mikrochim. Acta*, *I*: 415.

Ciszewski, A., and Wang, J. (1992). *Analyst*, *117*: 985.

Clark, C.G., Moody, G.J., and Thomas, J.D.R. (1978). *Anal. Chim. Acta*, *98*: 215.

Cosofret, V.V. (1980). *Ion-Selective Electrode Rev.*, *2*: 159.

Costa Garcia, A., and Tunon Blanco, P. (1988). *J. Electroanal. Chem.*, *245*: 157.

Cox, J.A., and Gray, T.J. (1989). *Anal. Chem.*, *61*: 2462.

Cox, J.A., and Gray, T.J. (1990). *Contemporary Electroanalytical Chemistry*, (A. Ivaska, A. Lewenstam and R. Sara, eds.), Plenum, New York, p. 267.

Cressman, W.A., Sugita, E.T., Doluisio, J.T., and Niebergall, P.J. (1966). *J. Pharm. Pharmacol.*, *18*: 801.

Crow, D.R. (1969). *Polarography of Metal Complexes*, Academic Press, London.

Davis, D.G. (1964). *Comprehensive Analytical Chemistry*, Vol. IIA (C. Wilson and D. Wilson, eds.), Elsevier, Amsterdam, p. 65.

De Robertis, A., De Stefano, C., and Sammartano, S. (1987). *Talanta*, *34*: 933.

Dewaal, W.A., Maessen, F.J., and Kraak, J.C. (1990). *J. Pharm. Biomed. Anal.*, *8*: 1.

Digua, K., Kauffmann, J.-M., Ghanem, G., and Patriarche, G.J. (1993). *J. Liq. Chromatogr.*, *15*: 3295.

Elferink, F., Leeuwenkamp, O.R., Pinedo, H.M., and Van der Vijgh W.J. (1987). *J. Electroanal. Chem.*, *238*: 297.

Erceg, M.R., Kapetanovic, V.P., and Suznjevic, D.Z. (1991). *Coll. Czechosl. Chem. Commun.*, *56*: 152.

Eyal, E., and Rechnitz, G.A. (1971). *Anal. Chem.*, *43*: 1090.

Fiallo, M.M.L., and Garnier-Suillerot, A. (1987). *J. Inorg. Biochem.*, *31*: 43.

Fijalek, Z., and Zuman, P. (1990). *Anal. Lett.*, *23*: 1213.

Forster, R.J., and Vos, J.G. (1992). *Comprehensive Analytical Chemistry*, Vol. XXVII (Svehla, G., ed.), Elsevier, Amsterdam, p. 465.

Foye, W.O. (1961). *J. Pharm. Sci.*, *50*: 93.

Fultz, M.L., and Durst, R. (1991). *Electroanalysis, 3*: 519.

Guerrieri, A., and Palmisano, F. (1987). *Anal. Chem., 59*: 2127.

Hernandez, J., Sanchez, A., and Delgado, M. (1984). *Anal. Chim. Acta, 160*: 335.

Hernandez, J., Sanchez A., and Delgado, M. (1989). *J. Pharm. Sci., 78*: 589.

Jaber, A.M.Y. (1986). *Anal. Lett., 19*: 2039.

Jenkins, C.L., Knevel, A.M., and Digangi, F.E. (1967). *Quantitative Pharmaceutical Chemistry*, Mc Graw Hill, New York, p. 128.

Kalcher, K. (1990). *Electroanalysis, 2*: 419.

Kapetanovic, V., and Veselinovic, D. (1988). *Arch. Pharm., 321*: 559.

Kapetanovic, V., Suznjevic, D., and Veselinovic, D. (1990a). *Electroanalysis, 2*: 481.

Kapetanovic, V., Veselinovic, D., and Suznjevic, D. (1990b). *Anal. Lett., 23*: 1857.

Karpinski, Z., and Radomski, J. (1989). *Anal. Lett., 22*: 2291.

Karpinski, Z., Karny, M., and Kublik, Z. (1990). *J. Electroanal. Chem., 288*: 129.

Kuan, S.S., and Guilbault, G.G. (1987). *Biosensors. Fundamentals and Applications* (A. Turner, I. Karube and G.S. Wilson, eds.), Oxford University Press, Oxford, p. 135.

Kulys, J., and Drungiliene, A. (1991). *Electroanalysis, 3*: 209.

Leech, D., Wang, J., and Smyth, M.R. (1990). *Analyst, 115*: 1447.

Li, F., and Dong, S. (1987). *Electrochim. Acta, 32*: 1511.

Linders, C.R., Patriarche, G.J., and Kauffmann, J.-M. (1986a). *Anal. Lett., 19*: 193.

Linders, C.R., Kauffmann, J.-M., and Patriarche, G.J. (1986b). *J. Pharm. Belg., 41*: 373.

Lopez Fonseca, J.M., and Fojon, D. (1990). *Anal. Chim. Acta, 230*: 97.

Lopez Fonseca, J.M., and Fojon, D. (1992). *Anal. Lett., 25*: 1095.

Luque de Castro, M.D., and Izquierdo, A. (1991). *Electroanalysis, 3*: 457.

Mairesse, C.A., Vandenbalck, J.L., and Patriarche, G.J. (1973). *J. Pharm. Belg., 28*: 300.

Mebsout, F., Kauffmann, J.-M., and Patriarche, G.J. (1987). *J. Pharm. Biomed. Anal., 5*: 223; (1988). *ibid., 6*: 441.

Mebsout, F., Kauffmann, J.-M., and Patriarche, G.J. (1989a). *Electroanalysis, 1*: 161; (1989b). *ibid., 1*: 257.

Midorikawa, T., Tanoue, E., and Sugimura, Y. (1990). *Anal. Chem., 62*: 1737.

Milner, G.W.C., and Phillips, G. (1967). *Coulometry in Analytical Chemistry*, Pergamon, Oxford.

Miranda Ordieres, A.J., Garcia Guttierez, M.J., Costa Garcia, A., Tunon Blanco, P., and Smyth, W.F. (1987). *Analyst, 112*: 243.

Murray, R.W. (1984). *Electroanalytical Chemistry*, Vol. XIII (A.J. Bard, ed.), Dekker, New York, p. 191.

Passamonti, P., Bartocci, V., and Pucciarelli, F. (1987). *J. Electroanal. Chem., 230*: 99.

Patriarche, G.J., and Lingane, J.J. (1970). *Anal. Chim. Acta, 49*: 25.

Pavlovich, G.Z., and Luthy, R.G. (1988). *Water Res., 22*: 327.

Ross, J.W., Jr., and Frant, M.S. (1969). *Anal. Chem., 41*: 1900.

Sigel, H., Zuberbuehler, A., and Yamauchi, O. (1991). *Anal. Chim. Acta, 255*: 63.

Singh, S., Gulati, M., and Gupta, R.L. (1991). *Int. J. Pharm., 68*: 105.

Stiles, M.K., Craig, M.E., Gunnell, S.L., Pfeiffer, D.R., and Taylor, R.W. (1991). *J. Biol. Chem., 266*: 8336.

Suarez, M., Calvo, A., and Leon, L.E. (1986). *Anal. Lett., 19*: 1219.

Treskes, M., Dejong, J., Leeuwenkamp., O.R., and Van der Vijgh W.J. (1990). *J. Liq. Chromatogr., 13*: 1321.

Vandenbalck, J.L., Mairesse, C.A., and Patriarche, G.J. (1975). *Analusis, 3*: 473.

Vander Linden, W. (1981). *Comprehensive Analytical Chemistry*, Vol. XI (G. Svehla, ed.), Elsevier, Amsterdam, p. 275.

Van Krimpen, P.C., Van Bennekom, W.P., and Bult, A. (1988). *Pharm. Weekbl. (Scient. Ed.), 10*: 259; and references therein.

Veselinovic, D., and Kapetanovic, V. (1988). *J. Pharm. Belg., 43*: 163.

Viré, J.-C., Patriarche, G.J., and Christian, G.D. (1977). *Mikrochim. Acta, I*: 17.

Vydra, F., Stulik, K., and Julakova, E. (1976). *Electrochemical Stripping Analysis*, Ellis Horwood Ltd, Chichester.

Wallace, G.G., Meaney, M., and Smyth, M.R. (1990). *Contemporary Electroanalytical Chemistry* (A. Ivaska, A. Lewenstam and R. Sara, eds.), Plenum, New York, p. 283.

Wang, E., Ji, H., and Hou, W. (1991). *Electroanalysis, 3*: 1.

Wang, J. (1985). *Stripping Analysis: Principles, Instrumentation and Applications*, Verlag Chemie, Deerfield Beach, Florida.

Wilson, C.O., Gisvold, O., and Doerge, R.F. (1977). *Textbook of Organic Medicinal and Pharmaceutical Chemistry*, 7th ed., J.B. Lippincott Co., Philadelphia, p. 50.

Zapotocky, J.A. (1966). *Pharmaceutical Chemistry*, Vol. I (L.G. Chatten, ed.), Marcel Dekker, New York, p. 129.

OXIDANT SCAVENGING AND METAL BINDING BY ANTI-INFLAMMATORY DRUGS

by

Okezie I. Aruoma

The Pharmacology Group,
University of London King's College,
Manresa Road
London SW3 6LX
U.K.

I. INTRODUCTION

Oxidants such as the superoxide radical (O_2^-), hydroxyl radical ($^{\cdot}OH$), hydrogen peroxide (H_2O_2) and hypochlorous acid (HOCl), collectively referred to as reactive oxygen species (ROS), are generally accepted to be formed at sites of inflammation (Fig. 1) and appear to contribute to the tissue damage in some acute and chronic inflammatory diseases (such as rheumatoid arthritis), and some forms of adult respiratory distress syndrome (Weiss, 1989; Dowling et al., 1986; Halliwell et al., 1985; 1992; Allan et al., 1987). Some of the drugs used in the treatment of rheumathoid arthritis are shown in Table 1 (for further account, see Brooke et al., 1986). It has been suggested that many anti-inflammatory drugs might exert part of their action by scavenging oxidants (Bragt, 1984; Biemond et al., 1986; Vapaatalo, 1986; Aruoma et al., 1987a; Wasil et al., 1987; Aruoma and Halliwell, 1988; Ahnfelt-Ronne and Nielsen, 1987) and by decreasing the formation of O_2^- and H_2O_2 by activated phagocytes (Oyanagui, 1978).

Hiller et al. (1983) found that several anti-inflammatory drugs could scavenge hydroxyl radical generated in free solution by pulse radiolysis at an almost diffusion-controlled rate.

One factor influencing tissue damage during oxidant stress is the concentration and location of metal ions that can promote $^{\cdot}OH$ formation (Halliwell and Gutteridge, 1986). The hydroxyl radical is very reactive and cannot travel any significant distance when formed before it reacts with a target. Reactions of hydrogen peroxide with haemoglobin, liberated as a result of traumatic micro-bleeding in the joint (Blake et al., 1981) can generate oxo-haem species and associated globin radicals. These radicals are capable of oxidizing many biological molecules (Puppo and Halliwell, 1986). The pannus overgrowing the cartilage contain many macrophages presumably secreting O_2^- and H_2O_2, eiconosoids, interleukin 1 and the free radical nitric oxide ($NO^{\cdot}$). Under certain circumstances, the reaction between $NO^{\cdot}$ and O_2^- might also produce $^{\cdot}OH$ (Beckman et al., 1990; Stamler et al., 1992). A number of factors combine to determine how compounds could efficiently scavenge reactive oxygen species. The effective concentration of the scavenger is important. Their ability to chelate iron and present the iron - scavenger complex in the form that is unable to favor oxidant formation is equally important.

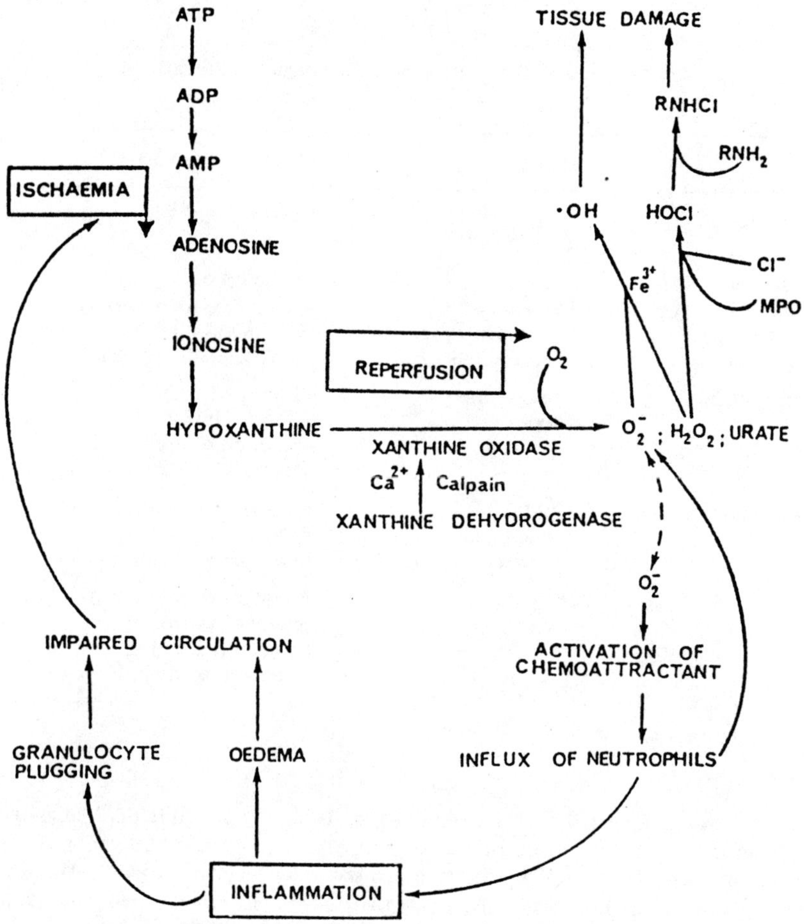

Figure 1.
Proposed relationship between inflammation, ischaemia-reperfusion and xanthine oxidase generated superoxide radical.
The scheme implicates xanthine oxidase generated O_2^-, neutrophil infiltration and derived oxidants (H_2O_2, •OH, HOCl, and RNHCl Chloramines) as being involved in mediating tissue damage. The neutrophil enzyme myeloperoxidase MPO catalyses the production of HOCl. This may then constitute the basis of postischaemic tissue injury and inflammatory disorders (for example, those of the bowel).

II. ANTI-INFLAMMATORY DRUGS AND REACTIVE OXYGEN SPECIES

A. Reaction with hypochlorous acid

Neutrophils contain myeloperoxidase which catalyzes the oxidation of chloride ions (Cl^-) by H_2O_2 to a highly reactive oxidant (HOCl). HOCl is a selective oxidant and reacts quickly with methionine and with the thiol functions in proteins (Travis and Salvesen, 1983; Weiss, 1986; 1989). Studies of the reaction between anti-inflammatory drugs and HOCl show that most of the drugs are able to react with HOCl (Pekoe et al., 1982; Wasil et al., 1987). For example, pre-incubation of the HOCl with varying concentrations of the drugs for 5 min decreased inactivation of α_1-antiproteinase (Table 2) in a concentration-dependent manner. α_1-Antiproteinase is a major circulating inhibitor of serine proteases such as elastase in human plásma (Travis and Salvesen, 1983). Phenylbutazone and penicillamine were

Table 1

Drugs Used in the Treatment of Rheumatoid Arthritis

Class	Examples	Mechanisms of action
Analgesics	Paracetamol Codeine	Central or peripheral pain relief
Non-steroidal anti-inflammatory drugs	Aspirin Indomethacin Phenylbutazone Flurbiprofen	Inhibition of cyclo-oxygenase activity resulting in inhibition of prostanoid synthesis
Corticosteroids	Prednisolone Dexamethasone	Wide spectrum of activities including inhibition of excosanoid synthesis, cytokine synthesis, etc
Disease-modifying drugs	Gold Penicillamine (hydroxy) chloroquine	Mechanism unclear, may act as immunosuppressants
Immunomodulators	Azathioprine Cyclophosphamide Chlorambucil Levamisole	Modulation of immune system by cytotoxic action or by stimulation

effective in scavenging HOCl on a 1:1 molar basis. Thiomalic acid, the thiol component of gold sodium thiomalate could also protect α_1-antiproteinase against inactivation by HOCl. Scavenging of HOCl by anti-inflammatory drugs will only be physiologically significant if at the concentration of drug that can be achieved in vivo its reaction with HOCl is fast enough to protect the biological target from attack (Wasil et al., 1987).

B. Reaction with hydroxyl radicals

The antiproteinases are also susceptible to attack by hydroxyl radicals. α_1-Antiproteinase can be inactivated by exposure to $^{\cdot}OH$ generated by pulse radiolysis and/or by an Fe^{3+}-EDTA/H_2O_2/ascorbic acid system (Aruoma et al., 1989). The protection of α_1-antiproteinase by anti-inflammatory drugs is shown in Table 3 (see Aruoma et al., 1989). Thus the inactivation of the antiproteinase by $^{\cdot}OH$ generated under anoxic condition was decreased by adding a range of anti-inflammatory drugs to the reaction mixture. The thiol compound penicillamine promoted damage to the protein under conditions that favor the formation of oxysulphur radicals (Aruoma et al., 1989; Hiller et al., 1983). This observation has raised the possibility that certain sulphur-containing compounds might react in such a way as to produce secondary radicals that could contribute to the side effects of the drug (Aruoma et al., 1989).

The drug phenylbutazone was one of the first non-steroidal anti-inflammatory drugs (NSAID) to be used in the treatment of rheumatoid arthritis (see Table 1). The drug is effective against ankylosing spondylitis but it has a number of side effects. As such, it has not been recommended for use in the treatment of osteoarthritis or soft-tissue rheumatism. Reaction of phenylbutazone with ROS may not necessarily be a good thing. A drug like penicillamine reacts with ROS to produce drug-derived radicals (Aruoma et al., 1989; Evans et al., 1992). Interestingly, both compounds react with $^{\cdot}OH$ with rate constant ($M^{-1}s^{-1}$) of 4.4-6.5 x 10^9 (see Table 4 column A) for penicillamine and 1.3-1.7 x 10^{10} (by the deoxyribose method, see Table 4) and 1.1-1.5 x 10^{10} (by pulse radiolysis) for phenylbutazone.

Table 2

Effect of Anti-inflammatory Drugs on the Inactivation of α_1-Antiproteinase by HOCl

Drug present	Concentration in reaction system 1 (μM)	Elastase activity % maximum rate
None	-	99
Omit HOCl	-	2
Phenylbutazone	50	52
	100	9
	200	1
Piroxicam	50	99
	400	80
	800	38
	1000	38
Indomethacin	200	98
	600	68
	800	56
	1000	51
Primaquine	50	98
	100	78
	200	48
	400	10
	600	3
Penicillamine disulphide	50	92
	100	21
	200	13
	400	2
Thiomalic acid	200	44
	800	21
	1000	0

α_1-antiproteinase (final concentration 1.0 mg/ml) was mixed with the final concentration of drug stated and then HOCl was added to a final concentration of 75 μM. The final reaction volume was 35.1 μl, and also contained phosphate-buffered saline at pH 7.4. This reaction mixture (reaction mixture I) was incubated at 25° for 1 hr. 3 ml of buffer was then added, followed by pancreatic elastase. After further incubation for 30 min, elastase substrate was added and the elastase activity remaining was assayed as a rise in absorbance at 410nm. Results are expressed as a % of maximum elastase activity (100% is an A_{410} of 0.060/min), being the mean values of duplicate experiments that agree to within 5%.

From Wasil et al. (1987)

Table 3

Protection of α_1-Antiproteinase Against Hydroxyl Radical Attack by Anti-inflammatory Drugs

Drug added to reaction mixture	Elastase activity ($\Delta A \times 10^{-3}$/sec)	% protection of α_1AP by drug (as % decrease in elastase activity)
None	13.1	-
None (unirradiated α_1-antiproteinase)	0	-
5-Aminosalicylate 120 μM	11.9	9
240 μM	8.7	34
Piroxicam 120 μM	10.8	18
240 μM	7.6	42
DL-Penicillamine disulphide 120 μM	12.2	7
240 μM	8.8	33
DL-Penicillamine 120 μM	11.7	11
240 μM	9.2	30
Diclofenac sodium 120 μM	11.0	16
240 μM	7.8	40
Chloroquine 120 μM	10.3	21
240 μM	7.6	42
Sulphapyridine 120 μM	10.4	21
240 μM	7.6	42
Hydroxychloroquine 120 μM	8.7	34
240 μM	7.5	43
Indomethacin 120 μM	6.9	47
240 μM	6.6	50

α_1-Antiproteinase (l mg/ml) was pulse-irradiated in N_2O-saturated buffer over a period of approximately 6 min, to generate 180 μmol ·OH/dm^3. Drugs were included in the reaction mixture at the final concentration stated. Elastase alone gave a ΔA_{410} of 1.84×10^{-2}/sec.

From Aruoma et al. (1989)

III. IRON BINDING AND ANTIOXIDANT CHARACTERIZATION

Some ·OH scavengers have a degree of metal-binding capacity, and can pull metal ions away from sensitive targets. If the metal ion - scavenger complex is still redox active, some or all of the ·OH radical generated by it will be directed towards the scavenger, and the target will be protected. However, would anti-inflammatory drugs be able to interfere with site-specific ·OH radical formation caused by iron ions bound to critical sites? Most of the damage done to living organisms by exposing them to high-energy ionizing radiation is due to the radiation-dependent splitting of water to form ·OH. In the case of Fenton-type chemistry, the site of hydroxyl radical generation is often determined by the availability and location of the metal ion catalyst. The damage caused by the ·OH depends on where it is formed (Halliwell and Gutteridge, 1989). If anti-inflammatory drugs were to protect against this type of damage, such action would depend on the effective concentration of the drugs at the site of inflammation, the rate constant for reaction with ROS and the ability of the drugs to chelate iron (in the case of iron-dependent reaction) ions.

Anti-inflammatory drugs are able to inhibit the oxidation of deoxyribose in the presence of Fe-EDTA (Aruoma and Halliwell, 1988), indicating their ability to scavenge ·OH. It is also possible that EDTA keeps the Fe bound to itself and does not let it bind to

Table 4

Scavenging of Hydroxyl Radicals by Anti-inflammatory Drugs in the Deoxyribose Assay

Drugs	Second order rate constant (+EDTA) (x $10^{10}M^{-1}s^{-1}$)	Apparent second order rate constant (-EDTA) (x $10^{10}M^{-1}s^{-1}$)
Penicillamine	0.44-0.65	0.33-0.39
Diclofenac sodium	1.00-2.40	2.20-4.60
Penicillamine disulphide	0.40-0.82	0.41-0.54
Phenobarbital	0.92-1.56	1.60-1.70
Piroxicam	0.54-0.67	1.00-4.00
Azathioprine	0.60-1.25	4.60-5.00
Phenylbutazone	1.30-1.70	1.70-3.20
Ibuprofen	1.58-1.80	1.30-2.40
Paracetamol	1.20-1.40	1.60-4.00
Mefenamic acid	2.10-2.20	2.40-3.00
Indomethacin	0.98-1.20	1.20-1.30
Chloroquine	0.75-1.01	2.60-3.80
Naproxen	1.20-2.20	1.10-1.80

Use of the deoxyribose method for the detection of hydroxyl radicals has been reviewed in Aruoma (1994). The value in the Table were abstracted from Aruoma (1988) and Aruoma and Halliwell (1988).

the drugs. Figure 2 shows a selection of plots obtained for different anti-inflammatory drugs (Aruoma and Halliwell, 1988). Competition plots (Fig. 2, line Y of graphs A, B, C, and D) were linear in all cases and the rate constants (Halliwell et al., 1987; Aruoma, 1994) calculated from their slopes (Table 4) indicate that the drugs react with ·OH radical at essentially a diffusion-controlled rate. The application of the deoxyribose assay as a tool for the detection of hydroxyl radicals has been described (Aruoma, 1994). Anti-inflammatory drugs also inhibit oxidation of deoxyribose in reaction mixtures not containing EDTA (Fig. 2, line X) suggesting that they have some degree of iron-binding capacity. The initial phase of effective inhibition is due to a transfer of iron from the deoxyribose to the added scavenger (Aruoma et al., 1987b), giving an iron - scavenger complex that is poorly effective (for reasons of redox potential or of self scavenging) in generating ·OH radical that can attack deoxyribose. Once sufficient scavenger has been added to convert all available iron into an iron-scavenger complex, the remaining deoxyribose degradation is caused by ·OH generated in free solution by that iron-scavenger complex (Gutteridge, 1984; 1987; Aruoma et al., 1987b; Aruoma and Halliwell, 1988). Any further additions of scavenger will inhibit the remaining deoxyribose degradation simply by competing for that ·OH, and hence the slopes of the competition plots become similar to those for inhibition in the presence of EDTA (as is clearly shown in Fig. 2: line Y when iron is in the form of Fe-EDTA, line X when iron is present in the reaction mixture as just $FeCl_3$).

IV. METAL COMPLEXES OF ANTI-INFLAMMATORY DRUGS

Although it has been suggested that chelation of iron and other metals may contribute to anti-inflammatory action, the use of iron chelator desferrioxamine (DFO) in humans has highlighted the toxicity problems (Blake et al., 1985). Desferrioxamine causes reversible ocular abnormalities which may be related to its high iron-binding capacity. Recent report on anti-inflammatory effects of D-myo-inositol-1.2.6-triphosphate, an isomer of inositol

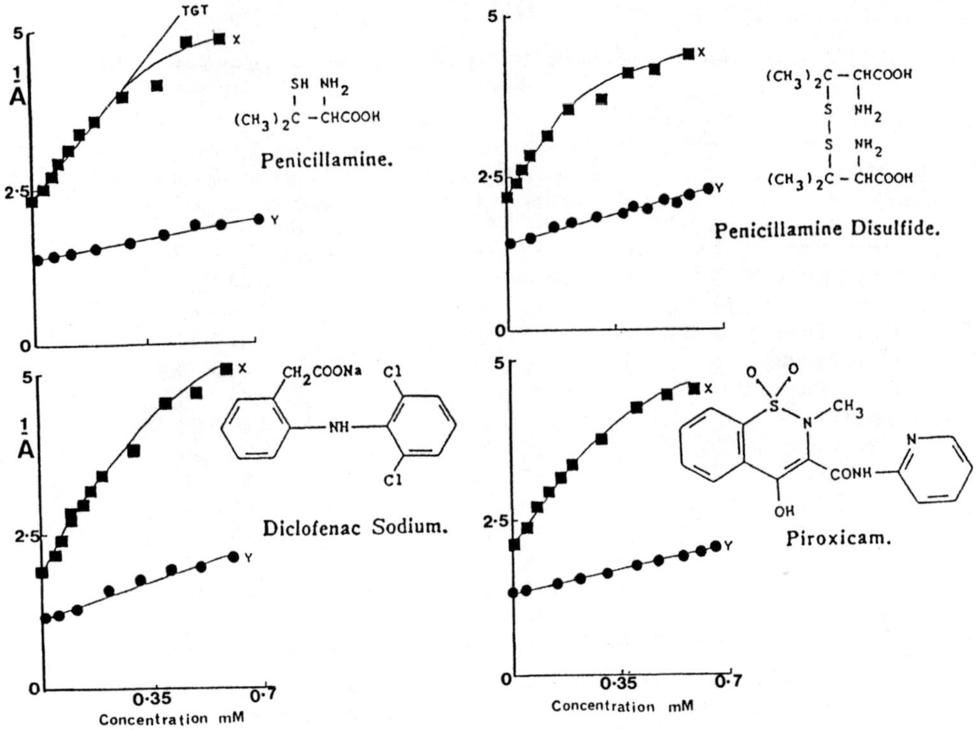

Figure 2.
Scavenging of hydroxyl radicals in the deoxyribose assay. This is illustrated with the drugs penicillamine, diclofenac sodium, penicillamine disulphide and piroxicam. The rate constants (k_s) for reaction with •OH (shown in Table 4) were calculated form the equation (in the presence of EDTA and in the absence of ETDA for the apparent rate constants):

$$k_s = \text{slope} \times k_{DR} \times [DR] \times A°$$

where k_{DR} is 3.1 x 10^9 $M^{-1}s^{-1}$. A° is the absorbance in the absence of added drug. [DR] is the concentration of deoxyribose in the reaction mixture.
Line X, EDTA absent; Line Y, EDTA present.
(From Aruoma and Halliwell, 1988).

triphosphate (Claxson et al., 1990) suggests that the drug can chelate iron, decrease metal-induced lipid peroxidation and afford protection against DFO ocular toxicity (Claxson et al., 1990; Good et al., 1989).

Certain metal complexes of anti-inflammatory drugs have been suggested as effective agents against ROS (Bury et al., 1987; Harrison et al., 1985a; 1985b; Underhill et al., 1986; Grootveld et al., 1984; Brown et al., 1980). For example, copper complexes of a number of non-steroidal anti-inflammatory drugs are more potent anti-inflammatory agents and less ulcerogenic than their parent anti-inflammatory drugs (Brown et al., 1980; Sorenson, 1989). This was demonstrated for the copper complexes of penicillamine, flufenamic acid, levamisole, aspirin and 2-amino-2-thiazoline (Brown et al., 1980). Brown et al. (1980) reported that there was no obvious difference between the activities of copper(I) and copper(II) complexes, suggesting the possibility of a common mechanism.

Upon administration, it is necessary for the copper complexes to remain bound to minimize the possibility of free copper ions becoming available at the site of inflammation. Metal ions, especially those of copper and iron, are catalysts for free radical formation in biological systems. Recent studies show that mixtures of copper ions with H_2O_2 and O_2^- or

Table 5

Oxidative Damage to Isolated DNA: a Comparison or Iron and Copper Ions

System used	nmole modified bases/mg DNA produced	
	total bases	8-OH-Gua
DNA alone	2.32 ± 0.19	1.02 ± 0.09
Plus Cu^{2+}/H_2O_2	14.30 ± 0.39	9.02 ± 0.12
Plus Fe^{3+}/H_2O_2	3.63 ± 0.24	1.29 ± 0.09
Plus Cu^{2+}/H_2O_2/ascorbate	83.20 ± 8.84	48.20 ± 6.80
Plus Fe^{3+}/H_2O_2/ascorbate	8.15 ± 0.84	2.14 ± 0.24

Calf thymus DNA was exposed to H_2O_2 alone, or H_2O_2 plus ascorbate, in the presence of added Cu^{2+} or Fe^{3+} ions at pH 7.4. No extra DNA damage over the control (DNA alone) was seen if metal ions were not added.
Results are mean ± S.D. from three experiments each performed in triplicate. For further details, see Aruoma et al. (1991).

ascorbate can do considerable damage to DNA and proteins (see for example, Aruoma et al., 1991; Halliwell and Aruoma, 1993). Copper ions mediate much more damage to DNA bases by H_2O_2 in the presence or absence of ascorbate than do iron ions in comparable reaction mixtures (Table 5). It is possible that the copper ions facilitated more damage seen in Table 4 because they bind to DNA in the vicinity of guanine and "site direct" ·OH formation to the base (Halliwell and Aruoma, 1993). Under normal conditions, much less copper is present in the human body than iron. For example, an adult body has on average 4.5 g of iron but only 0.08 g of copper. Interestingly, proteins such as albumin are able to inhibit formation of free radicals from free copper ions in free solution. Thus much care is needed in evaluating the efficacy of the metal complexes of the anti-inflammatory drugs. The integrity of the complexes in vivo should be considered in view of the ability of free copper ions to form free radicals.

CONCLUSION

In the event that free radicals and other reactive oxygen species are produced in free solution at sites of inflammation, the ability of anti-inflammatory drugs to scavenge them would depend on the concentrations of the drugs at the site, the iron-binding properties of the drugs and on the various rate constants of reaction of the drugs with ROS. One complication is that some of the anti-inflammatory drugs might form secondary radicals that could themselves cause damage (Hiller et al., 1983; Wasil et al., 1987; Aruoma and Halliwell, 1988; Aruoma et al., 1989; Evans et al., 1992).

ACKNOWLEDGEMENT

I thank the UK Medical Research Council, Nestec Research Centre, Switzerland, and the UK Ministry of Agriculture, Fisheries and Food for research support.

REFERENCES

Ahnfelt-Ronne, I., and Nielsen, O.H. (1987). *Agents Actions*, *21*: 191.

Allan, R.E., Outhwaite, J., Morris, C.J., and Blake, D.R. (1987). *Ann. Rheum. Dis.*, *46*: 843.

Aruoma, O.I. (1988). *PhD Thesis*, University of London.

Aruoma, O.I. (1994). *Methods Enzymol.*, *233*: 57.

Aruoma, O.I., and Halliwell, B. (1988). *Xenobiotica*, *18*: 459.

Aruoma, O.I., Wasil, M., Halliwell, B., Hoey, B.M., and Butler, J. (1987a). *Biochem. Pharmacol.*, *36*: 3739.

Aruoma, O.I., Grootveld, M., and Halliwell, B. (1987b). *J. Inorg. Biochem.*, *29*: 289.

Aruoma, O.I., Halliwell, B., Butler, and Hoey, B.M. (1989). *Biochem. Pharmacol.*, *38*: 4353.

Aruoma, O.I., Halliwell, B., Gajewski, E., and Dizdaroglu, M. (1991). *Biochem. J.*, *273*: 601.

Beckman, J.S., Beckman, T.W., Chen, J., Marshall, P.A., and Freeman, B.A. (1990). *Proc. Nat. Acad. Sci. USA*, *87*: 1620.

Biemond, P., Swaak, A.J.G., Panders, J.M.A., Biendorff, C.M., and Koster, J.F. (1986). *Ann. Rheum. Dis.*, *45*: 249.

Blake, D.R., Hall, N.D., Bacon, P.A., Dieppe, P.A., Halliwell, B., and Gutteridge, J.M.C. (1981). *Lancet*, *ii*: 1142.

Blake, D.R., Winyard, P., and Lunec, J. (1985). *Quart. J. Med.*, *56*: 345.

Bragt, P.C. (1984). *Agents Actions*, *15*: 54.

Brooke, P.M., Kean, W.F., and Buchanan, W.W. (1986). *The Clinical Pharmacology of Anti-inflammatory Agents*, Taylor & Francis, London, p. 81.

Brown, D.H., Smith, W.E., Teape, J.W., and Lewis, A.J. (1980). *J. Med. Chem.*, *23*: 729.

Bury, A., Underhill, A.E., Kemp, D.R., O'Shea, N.J., Smith, J.P., and Gomm, P.S. (1987). *Inorg. Chim. Acta*, *138*: 85.

Claxson, A., Morris, C., Blake, D.R., Siren, M., Halliwell, B., Gustafsson, T., and Lofkvist, B. (1990). *Agents Actions*, *29*: 68.

Dowling, E.J., Symons, A.M., and Parke, D.V. (1986). *Agents Actions*, *19*: 203.

Evans, P.J., Cecchini, R., and Halliwell, B. (1992). *Biochem. Pharmacol.*, *44*: 981.

Good, P.A., Claxson, A., Morris, C.J., and Blake, D.R. (1989). *Free Radicals, Disease States, and Anti-radical Intervention* (C. Rice-Evans, ed.), Richelieu Press, London.

Grootveld, M., Razi, M.T., and Sadler, P.J. (1984). *Clin. Rheumatol.*, *3*: 5.

Gutteridge, J.M.C. (1984). *Biochem. J.*, *224*: 761.

Gutteridge, J.M.C. (1987). *Biochem. J.*, *243*: 709.

Halliwell, B., and Aruoma, O.I., eds. (1993). *DNA and Free Radicals*, Ellis Horwood, London, New York.

Halliwell, B., and Gutteridge, J.M.C. (1986). *Arch. Biochem. Biophys.*, *246*: 501.

Halliwell, B., and Gutteridge, J.M.C. (1989). *Free Radicals in Biology and Medicine*, 2nd ed., Clarendon Press, Oxford.

Halliwell, B., Gutteridge, J.M.C., and Blake, D.R. (1985). *Phil. Trans. Roy. Soc. London B*, *311*: 659.

Halliwell, B., Gutteridge, J.M.C., and Aruoma, O.I. (1987). *Anal. Biochem.*, *165*: 215.

Halliwell, B., Gutteridge, J.M.C., and Cross, C.E. (1992). *J. Lab. Clin. Med.*, *6*: 598.

Harrison, D.O., Thomas, R., Underhill, A.E., Everall, F.A., Gomm, P.S., and Low, G.D. (1985a). *J. Coord. Chem.*, *14*: 107.

Harrison, D.O., Thomas, R., Underhill, A.E., Fletcher, J.K., Gomm, P.S., and Hallway, F. (1985b). *Polyhedron*, *4*: 681.

Hiller, K.O., Hodd, P.L., and Willson, R.L. (1983). *Chem.-Biol. Interact.*, *47*: 293.

Oyanagui, Y. (1978). *Biochem. Pharmacol.*, *27*: 777.

Pekoe, G., van Dyke, K., Mengoli, H., Peden, D., and English, D. (1982). *Agents Actions*, *12*: 232.

Puppo, A., and Halliwell, B. (1986). *Biochem. J.*, *249*: 185.

Sorenson, J.R.J. (1989). *Chem. Brit.*, *25*: 169

Stamler, J.S., Single, D.J., and Loscalzo, J. (1992). *Science*, *258*: 1898.

Travis, J., and Salvesen, G.S. (1983). *Ann. Rev. Biochem.*, *52*: 655.

Underhill, A.E., Blundell, R.P., Gomm, P.S., and Jacks, M.E. (1986). *Inorg. Chim. Acta*, *124*: 133.

Vapaatalo, H. (1986). *Med. Biol.*, *64*: 1

Wasil, M., Halliwell, B., Moorhouse, C.P., Hutchison, D.C.S., and Baum, H. (1987). *Biochem. Pharmacol.*, *36*: 3847.

Weiss, S.J. (1986). *Acta Physiol. Scand.*, *Suppl. 548*: 9.

Weiss, S.J. (1989). *N. Engl. J. Med.*, *320*: 365.

METAL-ANTIBIOTIC INTERACTIONS

by

P.C. L'Herminez - van Krimpen

Mariaplaats 14,
2713 EC Zoetermeer
The Netherlands

I. INTRODUCTION

Antibiotics are used to prevent and control infections, both for human and veterinary applications. In most cases the mechanism of the antibacterial action includes changing of the local environment, resulting in acceleration or inhibition of bacterial growth. In general, these changes do not have a large effect on mammal tissue, resulting in low toxicities.

Normally, the applied dose of the antibiotic depends on the sensitivity of the organism involved, which is determined by dose-effect relationships in vitro, in combination with the results of studies on the biological availability of the drug in the applied dosage form. In most cases no attention is paid to interactions which may occur within the body, with the exception of interactions with food or drinks which may occur in the gastrointestinal tract (a study of these effects is often included in bioavailability studies).

However, important classes of antibiotics are known to have affinities towards normal constituents of biological tissues and fluids, such as proteins, salts and minèral ions. For instance, penicillins are known to form inactive complexes with serum proteins. Fortunately, the magnitude of the binding strength is too small to induce major inhibitory effects.

Other antibiotic drugs are known to have very strong affinities in vitro towards biological constituents such as metal ions, but the magnitude and the effect of these interactions in vivo have not been investigated.

In some cases, interactions with metal ions in vivo have been predicted from the affinity between drug and metal ion of interest in vitro, e.g. by a description of the complex formation behavior of the antibiotic. From this behavior, combined with a knowledge of the local conditions pertinent to the tissue or fluid of interest (concentration of the metal ion, pH, ionic strength, protein concentration, etc), interactions are foreseen. However, the actual biological importance of these interactions has to be investigated through in vivo experiments.

Especially β-lactam antibiotics (penicillin, cephalosporin and their derivatives), and tetracycline and its derivatives, which are very important classes of antibiotics, have a number of potential coordinating sites and are known to effectively interact with several metal ions.

II. MECHANISMS OF INTERACTIONS

A. Complexation

Coordination between metal ions and ligand donor atoms or groups is generally explained using the HSAB principle (see Part One, Chapter 3), even though application of this principle has been severely criticized (see Part One, Chapter 3-A).

Classifications of metal ions and ligands derived from HSAB considerations, and which may be extended to metal ion - antibiotic interactions, may be found elsewhere in Volume 1 (see Part Three, Chapters 1-C1 and 2-A1).

Metal ion complexation by antibiotics may occur in large volumes of biological fluids (blood, gastric juice) or microfluids (interstitial fluids), resulting in more or less stable *free* complexes. It is also possible that the (free) antibiotic forms complexes with metal ions already *bound* to mammal or bacterial structures, or that free (hydrated) metal ions form complexes with a previously *bound* antibiotic molecule. All these types of complexes may lead to inactivation of the antibiotic, either by enhanced degradation of the drug, or by irreversible binding to specific sites. These mechanisms will be discussed in detail where appropriate.

B. Oxidation/reduction

Complex formation between metal ions and antibiotics can be regarded as a 'non-completed' redox reaction in which the intermediate formed is stable. However, 'true' redox reactions, resulting in transfer of charge, may occur.

C. Effect on physiological transport of metal ions

For physiologically occurring metal ions, such as calcium, potassium, and sodium, special channels are present in mammal cell walls. The influx and efflux of these metal ions are often related to activation or deactivation of the cell.

Several antibiotics are known to block those channels, resulting either in loss of their antibacterial activity, or in an impaired cell function, or both. Other antibiotics are known to form channels in the mammal cell walls themselves, resulting in an increased flux of the physiological metal ions. These effects cannot be regarded as a 'true' interaction between the antibiotic and the metal ion itself. They have been discussed in detail elsewhere (see Chapter 1-A4 of this Part Four) and will not be developed here.

III. METAL INTERACTIONS WITH SPECIFIC ANTIBIOTICS

A. Penicillins and cephalosporins

Penicillins and cephalosporins are the most frequently prescribed antibacterial drugs. Their use has not been limited to human medicine: especially penicillins are often used for veterinary purposes. Together with their developed analogues, e.g. penems and monobactams, these compounds are referred to as β-lactam antibiotics, due to their common β-lactam structure. The general structures of these compounds are given in Figure 1.

The β-lactam ring is relatively unstable, a feature which is in fact essential for the antibacterial activity. The ring is opened by a bacterial enzyme, forming an intermediate whose stability is such that formation of the bacterial cell wall cannot proceed further. The

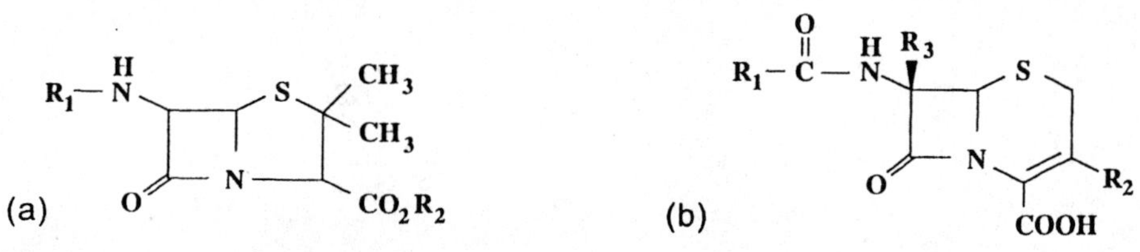

Figure 1.
General structures of penicillins (a) and cephalosporins (b).

result is a rather badly cross-linked bacterial cell wall structure, which is not strong enough to prevent the contents of the cell from bursting out.

Each of the β-lactam antibiotic derivatives contains various (potential) donor sites for metal complexation (N, O, and S). However, the function, position and accessibility of these donor atoms differ strongly for the parent molecule and the degradation products, resulting in different complexation behaviors.

In neutral media such as most of the biological fluids, the β-lactam ring can be opened by nucleophiles (hydroxyl ions, primary and secondary amine, amino acids, etc), resulting in the formation of (inactive) penicilloic and cephalosporoic acids, respectively. In these compounds, the opened β-lactam ring results in a carboxylic acid and a secondary amine function (Fig. 2a). The penicilloic acids are relatively stable, except in the presence of soft metal ions: in that case the degradation proceeds further to a penilloaldehyde and penicillamine (Figs. 2b and 2c, respectively). All these compounds have a number of (potential) sites for metal complexation: especially penicillamine is known for its complexing behavior.

1. *Mercury(II) and silver(I)*

Complex formation between intact molecules of penicillins or cephalosporins and soft or borderline metal ions in the HSAB classification does not play an important role in vivo. This is due to the relatively low stability of the resulting species, which contrasts with that of the complexes formed by the degradation products.

The high coordination capacity of the inactive compounds, compared to that of the parent molecules, can be illustrated by the mercurimetric titration of penicillins. Before titration of the solution with mercury(II) ions, the compounds are degraded to their penicilloic acids. A blank titration of the non-degraded compound is used for the determination of the penicilloic acid already present in the sample (thus indicating that intact penicillin does not yield stable complexes with the very soft metal ion Hg^{2+}).

However, penicillins and cephalosporins degrade quite easily, even under physiological conditions, and complex formation with the degradation products further enhances the degradation of the parent molecules.

Especially the very soft metal ions Hg^{2+} and Ag^{+} are known to form complexes with penicilloic acids and penicillamine (Van Krimpen et al., 1988). In the case of cephalosporins, it is assumed that besides complex formation, oxidation or complete degradation of the molecule may also occur, resulting in a mixture of insoluble compounds.

2. *Zinc(II) and copper(II)*

Compared to the very soft metal ions Hg^{2+} and Ag^{+}, Zn^{2+} and Cu^{2+} ions are classified as borderline. Although the formation constants of their complexes with β-lactam

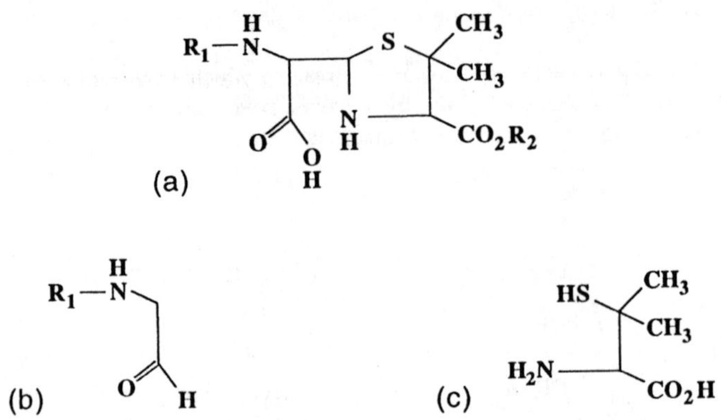

Figure 2.
Structure of penicilloic acid (a), penilloaldehyde (b) and penicillamine (c).

antibiotics are not very large, several in vitro interactions have been described. For instance, for ampicillin (Fig. 1: R_1 = C_6H_5-$CHNH_2$-CO; R_2 = H) a complex with a stoichiometric stability constant (log K_1) of 1.6 has been reported (Abd el Wahed and Ayad, 1984). For fenoxymethylpenicillin (Fig. 1: R_1 = C_6H_5-O-CH_2-CO; R_2 = H) a log K_1 of 2.2 was found by Cressman et al. (1966), while for benzylpenicillin (Fig. 1: R_1 = C_6H_5-CH_2-CO; R_2 = H) a value of 2.6 was reported by Cressman et al. (1966) and by Fazakerley et al. (1976). For all these compounds, the β-lactam nitrogen or amine function and the carboxyl function are indicated as the sites of complexation.

No physiological effect is to be expected from such complexes.

3. *Iron(III)*

Interaction of methicillin, a semisynthetic penicillin derivative, has been described by Aldea et al. (1984) both in vitro and in rats. The effect of ampicillin and carbenicillin was reduced by iron(III) ions (Miles et al., 1986). The molar ratio of the chelate varied between 0.2 and 2.4.

Penicillins form complexes with iron ions (Abd el Wahed and Ayad, 1984). The sites of complexation are reported to be the β-lactam nitrogen (amine) function and the carboxylic acid group, similar to the complexation with zinc and copper.

The formation of complexes with Fe^{3+} ions seems to be responsible for the development of an iron-deficiency anaemia which often results from penicillin therapy (L'Herminez, 1991). This iron deficiency is not very frequent, which is consistent with the (relatively small) stoichiometric stability constant log K_1 of approximately 6 for penicillins (a strong complex, such as the sulphur-mercury(II) complex, has a log K_1 of approximately 12) (Van Krimpen et al., 1988).

B. Tetracyclines

Tetracyclines are, after the penicillins and cephalosporins, the most prescribed antibiotics. The structure of tetracycline is given as an example in Figure 3.

From this structure, it can be seen that tetracyclines are considerably complex ligands. Three electron donor sites can be distinguished in the molecule, namely the C4 dimethyl amino group, the C10-C11-C12 diketophenolic moiety and the C1-C2-C3 tricarbonyl methane moiety, by order of increasing acidity (Leeson et al., 1963). The formation of metal complexes is very well-known for these type of antibiotics, especially with calcium and magnesium ions.

1. *Calcium(II) and magnesium(II)*

Complex formation between all tetracyclines and calcium and magnesium ions has been extensively studied. The gastrointestinal absorption of tetracyclines is impaired by milk and milk products and calcium and magnesium salts, although not only by the formation of complexes, but also by the increase in gastric pH.

The stability constants of the complexes formed by calcium or magnesium ions with a number of tetracyclines have been reported, derived from in vitro potentiometric data (Brion et al., 1981; Berthon et al., 1983). The stoichiometry of the predominant

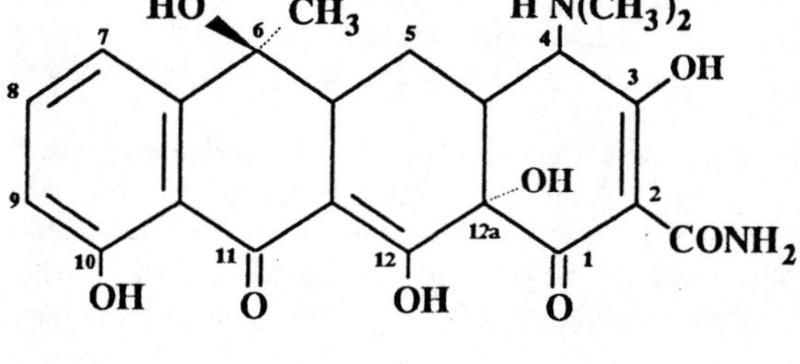

Figure 3.
Structure of tetracycline.

complexed species at physiological pH is not the same for the two metal ions: while the predominant calcium complex of tetracycline at pH 7 is ML_2H_2 (Brion et al., 1981), the predominant complex of magnesium at the same pH is MLH, a form which was not characterized in the case of calcium (Berthon et al., 1983). With doxycycline (6-deoxy-5β-hydroxy tetracycline), complex formation is similar for both metals: at pH 7 the predominant complex form is MLH. These distinct coordination behaviors among tetracycline derivatives have later been interpreted (Lambs et al., 1988a; 1988b).

In biological fluids, the concentrations of both calcium and magnesium ions are high compared with that of the tetracyclines during treatment. This favors the formation of mixed-metal complexes, including both calcium *and* magnesium ions. The distribution of mixed-metal complexes of four tetracyclines has been simulated, revealing that especially oxytetracycline (= 5-OH tetracycline) and tetracycline itself do form mixed complexes with Ca^{2+} and Mg^{2+} (metal ratio: 1:1). For tetracycline, approximately 18% of the plasma concentration is a mixed-metal complex, while 62% is a calcium complex and 11% a magnesium complex. For oxytetracycline the mixed-metal complex concentration is even almost 31%, while almost 34% are different calcium complexes and 28% exist in the form of different magnesium complexes (Lambs et al., 1984). Surprisingly, for doxycycline the simulated concentration of the mixed-metal complex was zero. For details on these speciation studies, see Part Five, Chapter 3-F.

2. *Iron(III)*

Iron(III) ions are known to enhance infection. In an in vivo study on the effect of the administration or Fe^{3+} ions, separate or together, with tetracycline, it was found that the inhibition of the tetracycline action by the metal was an unlikely event (3 out of 16 cases) (Miles and Maskell, 1985). Instead, the inhibition of the infection-enhancing effect of Fe^{3+} by tetracycline is much more likely (12 out of 16). Although Fe^{3+} ions had a large effect on the action of tetracycline on *Klebsiellae* in vitro (Miles and Maskell, 1986), another study by the same group (Miles et al., 1986) revealed that the Fe^{3+}: tetracycline molar ratio at the point of reversal of the antibiotic action in vivo was approximately 46:1, and that this molar ratio increased with increasing concentrations of tetracycline.

3. *Miscellaneous metals*

Besides their well-known and well-documented complexation with Ca^{2+} and Mg^{2+} ions, tetracyclines also interact with many other metal ions. With aluminum(III) ions, this interaction is in most cases seen during the oral administration of the drugs, and is caused partly by chelate formation, and partly by increase in the gastric pH, both resulting in decreased absorption.

Several tetracyclines have also been shown to form complexes with zinc and copper ions in vitro (Brion et al., 1985). Using a simulation program, it was concluded that the influence of zinc complexes on the bioavailability of the drugs in vivo is in general insignificant. Only the absorption of oxytetracycline is expected to be impaired. Using the same technique, no influence was expected from copper on the bioavailability in blood plasma, and vice versa. Copper is also not likely to interfere with the gastrointestinal absorption of the drugs (Brion et al., 1986). Details on these effects are available in Part Five, Chapter 3-F.

Recently it has been suggested that copper complexes of tetracyclines catalyze the formation of free radicals whereby nucleic acids are damaged (Dizdaroglu et al., 1990). Details on the metal requirements for the action of antibiotics are available in *Bioinorganic Medicine*, Part Five, Chapter 4-C.

Oxytetracycline, which is known to be electrochemically active, has been proven to yield in vitro an anodic shift of the corrosion potential of stainless steel, vitallium and titanium (Von Fraunhofer et al., 1989). This means that corrosion of the surfaces of those metals, which are often used for surgical implants, are inhibited. The mechanism of this inhibition (either complexation or some form of film deposition) was not elucidated.

C. Aminoglycosides

The aminoglycosides are cations, and therefore are less likely to form complexes with metal ions. Although most aminoglycosides are known to have neurotoxic effects related to calcium concentration, this is more likely due to blockade of calcium channels than to direct interaction with calcium ions.

REFERENCES

Abd el Wahed, M.G., and Ayad, M. (1984). *Anal. Lett.*, *17*: 205.

Aldea, V., Cristea, A., and Boros, D. (1984). *Farmacia (Bucharest)*, *32*: 245.

Berthon, G., Brion, M., and Lambs, L. (1983). *J. Inorg. Biochem.*, *19*: 1.

Brion, M., Berthon, G., and Fourtillan, J.-B. (1981). *Inorg. Chim. Acta*, *55*: 47.

Brion, M., Lambs, M., and Berthon, G. (1985). *Agents Actions*, *17*: 229.

Brion, M., Lambs, M., and Berthon, G. (1986). *Inorg. Chim. Acta*, *123*: 61.

Cressman, W.E., Sugita, E.T., Doluisio, J.T., and Niebergall, P.J. (1966). *J. Pharm. Pharmacol.*, *18*: 801.

Dizdaroglu, M., Aruoma, O.I., and Halliwell, B. (1990). *Biochemistry*, *29*: 8447.

Fazakerley, G.V., Jackson, G.E., and Linder, P.W. (1976). *J. Inorg. Nucl. Chem.*, *38*: 1397.

Lambs, L., Brion, M., and Berthon, G. (1984). *Agents Actions*, *14*: 743.

Lambs, L., Venturini, M., Decock-le Révérend, B., Kozlowski, H., and Berthon, G. (1988a). *J. Inorg. Biochem.*, *33*: 193.

Lambs, L., Venturini, M., Decock-le Révérend, B., Kozlowski, H., and Berthon, G. (1988b). *Inorg. Chem.*, *27*: 3001.

Leeson, L.J., Krueger, J.E., and Nash, R.A. (1963). *Tetrahedron Lett.*, *18*: 1155.

L'Herminez, P.C. (1991). Unpublished results.

Miles, A.A., and Maskell, J.P. (1985). *J. Med. Microbiol.*, *20*: 17.

Miles, A.A., and Maskell, J.P. (1986). *J. Antimicrob. Chemother.*, *17*: 481.

Miles, A.A., Maskell, J.P., and Evans, S.J.W. (1986). *J. Antimicrob. Chemother.*, *18*: 185.

Van Krimpen, P.C., Van Bennekom, W.P., and Bult, A. (1988). *Pharmaceut. Weekbl. Sci. Ed.*, *10*: 259.

Von Fraunhofer, J.A., Berberich, N., and Seligson, D. (1989). *Biomaterials*, *10*: 136.

METAL ION INTERACTION WITH ANTITUMOR ANTIBIOTICS

by

Masami Otsuka and Yukio Sugiura

Institute for Chemical Research,
Kyoto University,
Uji, Kyoto 611
Japan

I. INTRODUCTION

Antitumor antibiotics exert their activity according to diverse mechanisms, i.e. inhibition of nucleic acid biosynthesis, DNA cleavage, and tubulin binding, etc. Among them, bleomycin and streptonigrin are unique because they bind metal ions to attack DNA.

Since metal ion plays a crucial role in the bleomycin action, the biological, biochemical, physicochemical and chemical characteristics of the metal-bleomycin complexes have been extensively investigated. Furthermore, understanding on the mechanism of action of bleomycin associated with metal ion was greatly facilitated by the synthetic model approaches. On the other hand, metal ion is also significant for the antitumor activity of streptonigrin. Metal complexation and the mechanistic aspects of the biological activity of bleomycin and streptonigrin will be described in this chapter.

II. BLEOMYCIN

A. Characteristics of the metal-bleomycin complexes

Bleomycins are metalloglycopeptide antibiotics used for the clinical treatment of Hodgkin's lymphoma, carcinomas of the skin, head, and neck, and tumors of the testis mostly in combination with radiation or other chemotherapeutic agents (Umezawa, 1977). The bleomycin complexes radiolabeled with ^{55}Co, ^{99m}Tc, and ^{111}In are also useful in tumor diagnosis (Nieweg et al., 1982). The drug was isolated from *Streptomyces verticillus* as a copper chelate and the structure was shown to be a glycopeptide comprising unusual components (Fig. 1) (Takita et al., 1978a). Naturally occurring bleomycins consist of many congeners with amine constituent of varied structure at the C-terminus. Initially, a mixture of metal-free bleomycin A2 and B2 was used for the clinical treatment. However, afterwards, second-generation bleomycins have been developed. Thus, naturally occurring bleomycins were displaced by semi-synthetic bleomycins, i.e. peplomycin and liblomycin which are less toxic and have improved antitumor activity (Tanaka, 1977; Umezawa et al., 1985).

In addition to its clinical usefulness, bleomycin attracts current interest because of its unique biochemical functions. The Fe(II)-bleomycin-O_2 complex system cleaves isolated DNA specifically at G-C (5'→3') and G-T (5'→3') sequences (Takeshita et al., 1978). It is believed that the bithiazole and the terminal amine residues of bleomycin contribute to the specific binding to DNA, and the diaminopropionamide-pyrimidine-β-hydroxyhistidine

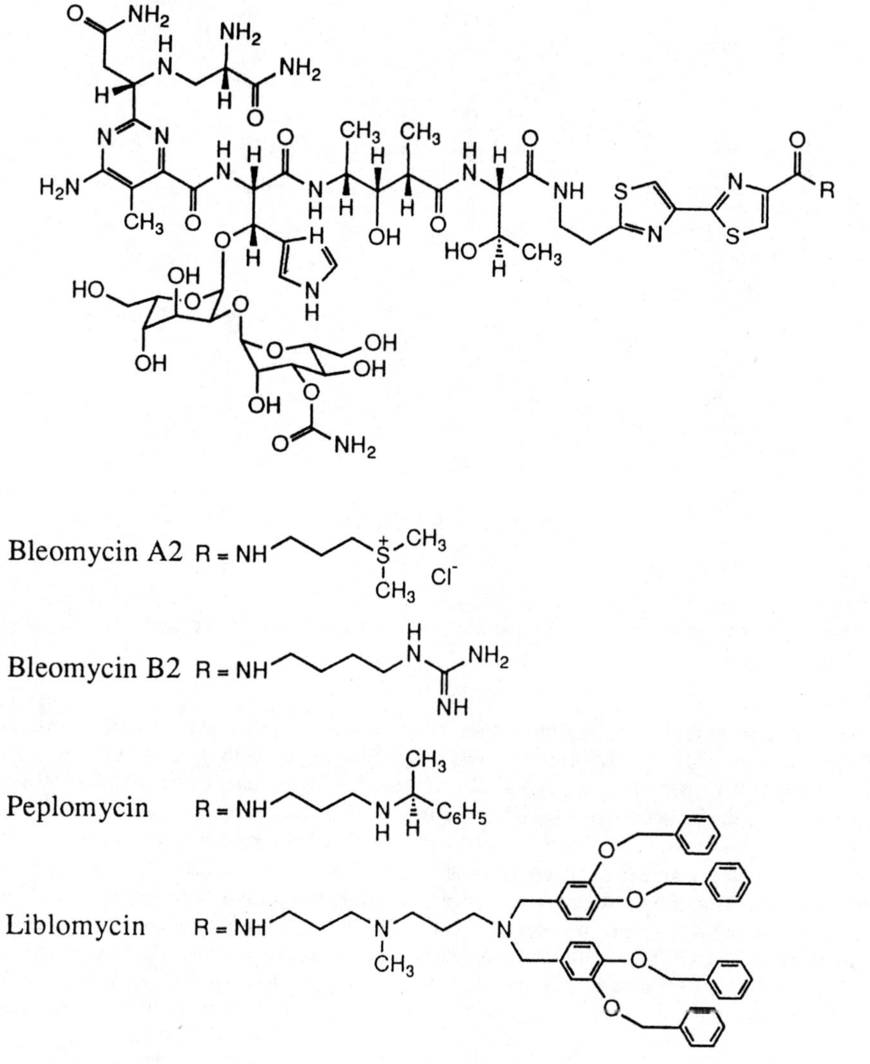

Figure 1.
Structure of bleomycins.

region is capable of dioxygen activation by the formation of iron chelate (Fig. 2) (Takita et al., 1978b; Sugiura et al., 1985; Hecht, 1986; Stubbe and Kozarich, 1987). Thus, bleomycin oxidatively cleaves cellular DNA to exert the antitumor activity.

Bleomycin forms chelates with various transition metals such as Fe(II), Cu(II), Co(II), Mn(II), Ni(II), and Zn(II) (Dabrowiak, 1980; Sugiura et al., 1985). Potentiometric studies revealed that the divalent metal complexes of bleomycin have substantially similar coordination core, although the stability of each complex is not the same, i.e. in the order Fe(II) < Co(II) < Ni(II) < Cu(II) > Zn(II) (Sugiura et al., 1979a).

The X-ray crystallographic analysis of the Cu(II) complex of P-3A, a biosynthetic intermediate of bleomycin, demonstrated that the Cu(II) site has a distorted square-pyramidal structure with 5-5-5-6 chelate ring members coordinated by the α-amino group, the secondary amino group, the pyrimidine, the deprotonated peptide and the histidine imidazole, and that the Cu(II) ion is displaced about 0.20Å from the basal plane in the direction of the α-amino nitrogen (Fig. 3) (Iitaka et al., 1978). The 1:1 Cu(II)-bleomycin

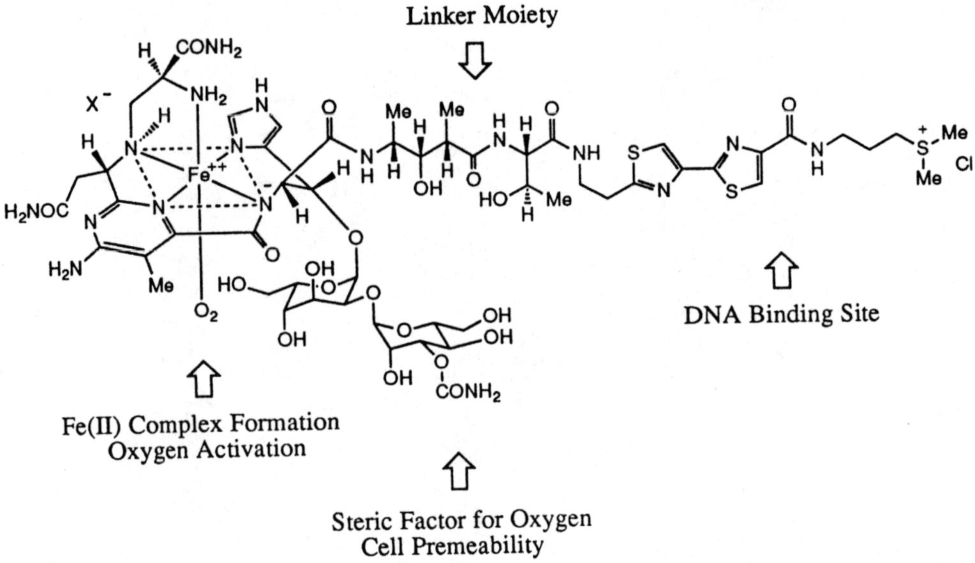

Figure 2.
Proposed structure of the Fe(II)-bleomycin-O_2 complex and the presumed role of each domain.

complex showed an absorption maximum at 595 nm (ε 120), circular dichroism extrema at 555 nm (Δε +1.21) and 665 nm (Δε -0.61), and ESR signal with $g_{\parallel}$ = 2.211, $g_{\perp}$ = 2.055, and $A_{\parallel}$ = 178 x 10^{-4} cm^{-1} (Sugiura et al., 1979a). These indicate that the Cu(II)-bleomycin complex also has a square-pyramidal configuration with the secondary amino nitrogen, the pyrimidine, the deprotonated peptide and the histidine imidazole nitrogen as planar ligand donors, and the α-amino nitrogen as axial donor. The X-ray structure of the P-3A-Cu(II) complex provided the most direct evidence to support the proposed structure of the Fe(II)-bleomycin complex shown in Figure 2. The cyclic voltammogram of the Cu(II)-bleomycin complex for Cu(II)/Cu(I) exhibited a redox wave with an $E_{1/2}$ value of -319 mV vs NHE (Ishizu et al., 1981). The Cu(II)-bleomycin complex was shown to be reduced by intracellular reducing agents and the resulting Cu(I) ion is transported to cellular protein (Takahashi et al, 1977). Based on ^{1}H NMR and potentiometric pH titration data, the Cu(I)-bleomycin complex was shown to have a geometry very different from that of the corresponding Cu(II) complex (Oppenheimer et al., 1981). In addition, if anaerobic reduction of the Cu(II)-bleomycin was carried out in the presence of Fe(II), the chelated copper was replaced by iron (Freedman et al., 1982).

The low-spin Co(II)-bleomycin complex and its dioxygen adduct were studied by ESR (Sugiura, 1980a). The Co(II)-bleomycin complex showed the ESR features ($g_{\perp}$ = 2.272, $g_{\parallel}$ = 2.025, $A_{\parallel}^{Co}$ = 92.5 G, and $A_{\parallel}^{N}$ = 13 G), typical of a low-spin square-pyramidal structure with $(d_{xy})^2$ $(d_{yz})^2$ $(d_{zx})^2$ $(d_{z}2)^1$ electronic configuration. The ESR spectrum of the Co(II)-bleomycin complex undergoes a dramatic change upon oxygenation. The effective g values ($g_{\perp}$ = 2.007 and $g_{\parallel}$ = 2.098), the relationship of $g_{\parallel} > g_{\perp} \sim 2.00$ and the relatively smaller A_{iso}^{Co} value (15 G) suggested a considerable delocalization of the unpaired spin density from the Co(II) ion. These ESR characteristics closely resemble those of various monooxygenated Co(II) complexes, e.g. oxy-Co(II)-myoglobin and oxy-Co(II)-peroxidase. The inactivity of Co(II)-bleomycin complex against DNA cleavage may be due to the irreversibility of Co(II)-Co(III) redox reaction. Imidazole was identified as a ligand in Co(II)-bleomycin by electron spin echo envelope spectroscopy (Burger et al., 1981a).

Aerobic oxidation of the Co(II)-bleomycin gave the Co(III)-bleomycin complex which has been known to exist in three different forms, i.e. brown, green, and orange. The brown complex was found to be a short-lived mononuclear superoxo Co(III)-complex (Vos et al.,

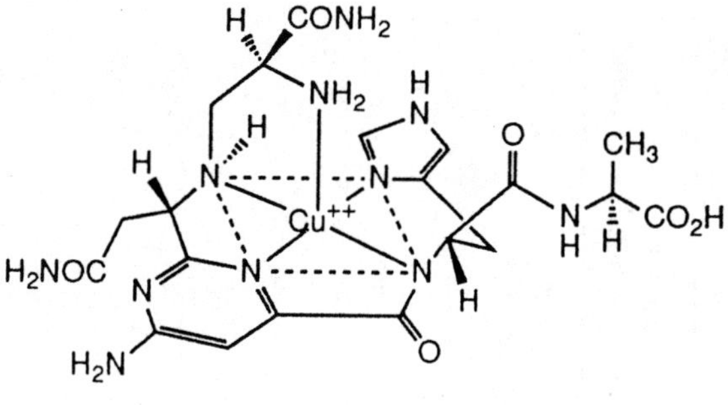

Figure 3.
Structure of Cu(II)-P-3A complex based on the X-ray crystallographic analysis.

1980a; 1980b; 1981). The reaction of two molecules of the brown complex, accompanied by the loss of oxygen, yielded the green complex, namely the dinuclear μ-peroxo Co(III) complex. The dimerization followed a second order rate low with $k = 200 \pm 50\ M^{-1}\ sec^{-1}$ at 25°C (Garnier-Suillerot et al., 1981; Albertini and Garnier-Suillerot, 1982). Upon heating, the green Co(III)-bleomycin complex was converted into the orange Co(III)-bleomycin complex which does not contain active form of oxygen (DeRiemer et al., 1979).

The iron-bleomycin complex forms adducts with CO, NO, C_2H_5NC, OH^-, N_3^-, CN^-, and CH_3NH_2. These adduct complexes were characterized by electronic, ESR, ^1H NMR, and Mössbauer spectroscopies (Table 1) (Sugiura, 1980b; Sugiura et al., 1979b; 1980; 1982a; 1982b). Mössbauer and ^1H NMR spectra showed that the Fe(II)-bleomycin complex has a high-spin ferrous iron. The visible absorption spectra of the Fe(II)-bleomycin-CO, Fe(II)-bleomycin-C_2H_5NC, and Fe(II)-bleomycin-NO complexes differ considerably from that of the 1:1 Fe(II)-bleomycin complex, and the absorption maxima are shifted to a longer wavelength in the order C_2N_5NC > NO > CO. A short-lived Fe(II)-bleomycin-O_2 complex was reported to have an absorption maximum at 385 nm (ε 3,000) (Burger et al., 1979a). The Mössbauer features of CO and C_2H_5NC adducts of the Fe(II)-bleomycin complex were consistent with an S = 0 ferrous assignment.

The ESR, ^1H NMR, and Mössbauer spectra of the OH^-, CH_3NH_2, and N_3^- adducts of the Fe(III)-bleomycin complex were typical of a low spin ferric iron (S = 1/2). The magnitude of the proton chemical shifts (over ± 50 ppm) of the Fe(III)-bleomycin-CN^- complex indicated the presence of a high-spin ferric iron in the cyanide adduct complex (Sugiura et al., 1980), although CN adducts of ferric hemoproteins are usually of low-spin type. The unusual behavior of the CN^- ion toward the iron-bleomycin complex is consistent with the observation that cyanide enhances the activity of bleomycin against DNA cleavage (Burger et al., 1979b).

Several controversial structures of the metal-bleomycin complex have been proposed. The first proposal for the ferrous complex was based on the X-ray structure of the Cu(II)-P-3A complex (Takita et al., 1978b). This chelate involves five nitrogens of the α-amino group, the secondary amino group, the pyrimidine, the deprotonated peptide and the imidazole, which was supported on the basis of extensive physico-chemical studies (Sugiura et al., 1985a).

On the other hand, there have been many reports which appeared to be inconsistent with the above proposal. The paramagnetically shifted ^1H NMR resonances of the Fe(II)-bleomycin complex, assigned by the saturation transfer method, suggested that the α-amino group, the pyrimidine and/or the secondary amine, the imidazole, the pentanoic acid and the carbamoyl group of the mannose may be coordinated to the iron (Pillai et al., 1980). A ^1H NMR study on the Fe(II)-bleomycin-CO complex indicated that the six coordination sites of the Fe(II) ion are occupied by the α-amino group, the secondary amino group, the pyrimidine, the imidazole, the mannose carbamoyl group, and the carbon monoxide (Oppenheimer et al., 1979a). The ^1H NMR spectrum of the Zn(II) complex of bleomycinic acid, a bleomycin congener devoid of terminal amine, showed that Zn(II) binding induced

Table 1

Electronic, ESR, ^{1}H NMR and Mössbauer Parameters of the Iron-bleomycin Complexes with Oxygen-analogous Ligands

Complex	Electronic spectra (77 K) λ_{max}, nm(ε)		ESR Spectra (77 K)			^{1}H NMR spectra (25 °C) (ppm from TSP)	Mössbauer spectra (110 K)		Iron spin state
			g_1	g_2	g_3		δ (mms^{-1})	ΔE_Q (mms^{-1})	
Fe(II)-bleomycin	476 (380)		no ESR signals			47.4, 45.1, 42.9, 34.7, 23.8, 20.6, 13.9, 9.9, -4.1, -5.2, -6.7, -11.8, -15.1	1.10	2.99	S = 2
Fe(II)-bleomycin-NO	470 (2300)		2.041	1.976	2.008	broad (48.0, 33.1, 29.8, -17.1, -28.8)	0.36	0.78	S = 1/2 (˙NO)
Fe(II)-bleomycin-CO	380 (3000)		no ESR signals			no paramagnetic signals	0.19	0.66	S = 0
Fe(II)-bleomycin-C_2H_5NC	495 (2700)		no ESR signals			no paramagnetic signals	0.18	0.62	S = 0
Fe(III)-bleomycin-OH^-	365 (2000)	384 (1900)	1.893	2.185	2.431	11.6	0.37	2.53	S = 1/2
Fe(III)-bleomycin-CN^-	390 (2000)	465 (800)	g = 4 (broad, 4 K)			64.5, 13.8, 12.6, -6.2, -10.8, -14.9. -19.4, -27.3, -41.9, -46.5	0.48	0.89	S = 5/2
Fe(III)-bleomycin-N_3^-	355 (2400)	375 (2200)	1.857	2.223	2.492	17.2	0.32	2.95	S = 1/2
Fe(III)-bleomycin-CH_3NH_2	350 (sh, 2000)		1.847	2.179	2.540	25.1	0.30	2.36	S = 1/2

large chemical shift changes for the diaminopropionamide, the β-aminopropionamide, the pyrimidinyl methyl group, the β-hydroxyhistidine, 3' of the mannose and the pentanoic acid (Oppenheimer et al., 1979b). The two-dimensional NMR study on the Fe(II)-bleomycin-CO complex suggested the involvement of the following residues in the iron binding: the secondary amino group, the pyrimidine, the deprotonated peptide, the imidazole, the mannose carbamoyl group and the carbon monoxide (Akkerman et al., 1990). A study on the solution structure of the Zn(II)-bleomycin complex by two-dimensional NMR and distance geometry calculations indicated that the complex is formed involving the α-amino group, the secondary amino group, the deprotonated peptide, imidazole and the mannose carbamoyl group (Akkerman et al., 1988). A ^1H NMR relaxation study on the Mn(II)-bleomycin complex suggested the involvement of the bithiazole in the coordination (Sheridan and Gupta, 1981).

The major product of acid hydrolysis of both orange and green Co(III)-bleomycin complexes was isolated and examined by ^1H NMR, absorption, and CD spectroscopies (Tsukayama et al., 1981; Dabrowiak et al., 1981). The hydrolysis product was shown to contain pseudotetrapeptide A (Fig. 4) bound to the cobalt in a manner identical to that of Cu(II)-P-3A complex.

Thus, there have been many controversial proposals for the structures of the metal-bleomycin chelates. On the basis of physicochemical measurements of bleomycin itself, biosynthetic intermediates, and synthetic analogues, which will be described in the following sections, five-nitrogen coordination involving the α-amino group, the secondary amino group, the pyrimidine, the deprotonated peptide and the imidazole groups is thought to be most reasonable for the iron-bleomycin complex.

B. Oxygen activation by the iron-bleomycin complex

Oxygen activating capability of bleomycin was unambiguously demonstrated by the ESR spin trapping experiments (Sugiura and Kikuchi, 1978; Sugiura, 1980b). Hydroxyl or superoxide radicals generated from the Fe(II)-bleomycin-oxygen system was trapped with *N-tert*-butyl-α-phenylnitrone (BPN) and the resulting stable spin adducts were observed by ESR at room temperature.

The catalytic process of the oxygen activation of the Fe(II)-bleomycin complex was observed by ESR spectroscopy (Sugiura and Kikuchi, 1978; Sugiura, 1980b). The pink-yellow colored Fe(II)-bleomycin complex is ESR silent at 77 K. However, when this complex was exposed to oxygen, several ESR-active species were generated. Immediately after the contact with oxygen, the transient low-spin Fe(III) complex species with small rhombic splitting of the g values (g_1 = 1.937, g_2 = 2.171, and g_3 = 2.254) was observed together with a small free radical species at g = 2.0046 (a = 28.6 G) which may be attributed to a nitrogen radical in the bleomycin molecule. This transient Fe(III) species disappeared in the course of reaction, and, instead, a stable low-spin Fe(III) species was

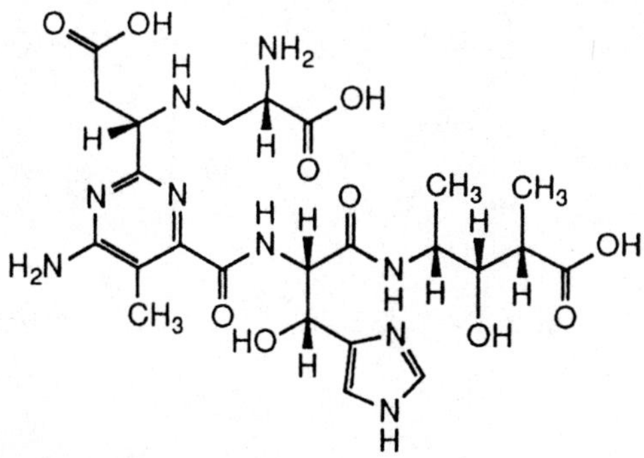

Figure 4.
Pseudotetrapeptide A.

formed. This was characterized by the large rhombic splitting (g_1 = 1.893, g_2 = 2.185, and g_3 = 2.431) and easily underwent reduction by a reducing agent. During this reversible redox reaction, iron-bleomycin complex generates oxygen radical species which were detected by the ESR spin trapping experiment.

Thus, the catalytic cycle of the oxygen activation involves two low-spin species, i.e. the transient Fe(III) complex and the stable Fe(III) complex. The transient Fe(III) complex seems to be the active species responsible for the DNA degradation by bleomycin. While the latter stable Fe(III) complex is inactive, addition of H_2O_2 to this inactive Fe(III) complex resulted in the formation of the active Fe(III) species (Kuramochi et al, 1981; Burger et al., 1981b). Therefore, the actual active species of bleomycin appears to be the Fe(III)-bleomycin-O_2H^- or a species closely related to it (Sugiura et al., 1982c). The low-spin ferric character of the transient complex was confirmed by the Mössbauer spectrum ($\Delta E_Q = 3 \pm 0.2$ and $\delta = 0.10 \pm 0.07$ mm/sec) (Burger et al., 1983).

Bleomycin appears to activate molecular oxygen via the catalytic cycle depicted in Figure 5 (Sugiura et al., 1982c; 1985a). The reaction of the Fe(II)-bleomycin complex with oxygen yields the Fe(II)-bleomycin-O_2 complex. Then, the active, transient species, Fe(III)-bleomycin-O_2H^-, is produced via the formation of the intermediate Fe(III)-bleomycin-O_2^- complex. The transient species induces the DNA cleavage, generating the stable Fe(III)-bleomycin-OH^- complex, which is reduced to the initial Fe(II)-bleomycin complex by a cellular reducing agent. This mechanism is in accordance with the above mentioned ESR observation of the catalytic process.

C. Sequence-specific DNA cleavage by the metal-bleomycin complexes

Many experimental observations support that the bithiazole - terminal amine region contributes to the DNA binding of bleomycin. Bleomycin exhibits fluorescence bands at 353 and 405 nm which are assigned to the bithiazole and 4-aminopyrimidine, respectively. The bithiazole fluorescence was shown to be quenched by DNA and the apparent equilibrium constant for the bleomycin-DNA complex, determined by the fluorescence quenching method, was $1.2 \times 10^5\ M^{-1}$ (Chien et al., 1977). Moreover, in the ^1H NMR spectrum of bleomycin, the signals for the bithiazole - terminal amine regions were selectively broadened by addition of DNA, suggesting the participation of the bithiazole moiety in DNA binding (Chien et al., 1977; Chen et al., 1980). Fluorescence quenching experiments using the copper complex of bleomycin revealed that poly (dG-dC) and poly dG • poly dC more strongly interact with bleomycin than poly (dA-dT) or poly dA • poly dT (Kasai et al., 1978). This is inconsistent with G-C (5'→3') and G-T (5'→3') specific cleavage of DNA with bleomycin.

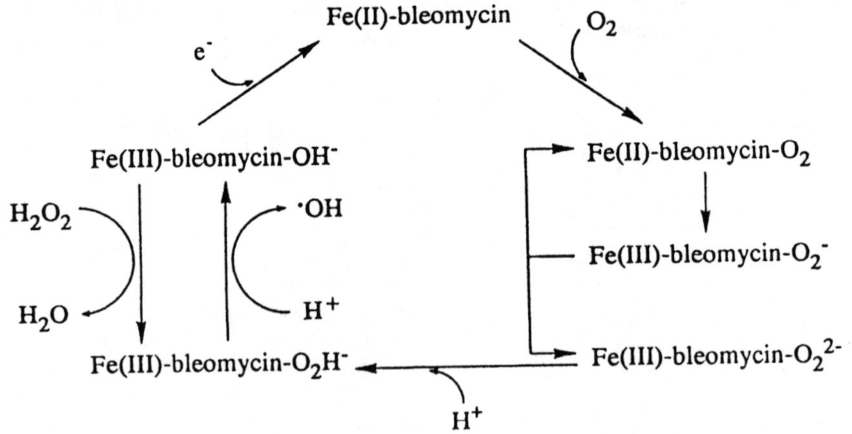

Figure 5.
Hypothetical mechanism for the catalytic activation of oxygen by the iron-bleomycin complex.

When calf thymus DNA was added to the Fe(II)-bleomycin-NO complex, the ESR spectrum of the ferrous-NO complex clearly changed and more rhombic signals were obtained (Sugiura and Ishizu, 1979b; Sugiura et al., 1981). The Co(II)-bleomycin-O_2 complex was shown to be stable for long periods when bound to DNA (Chikira et al., 1989). The ESR study of this Co(II)-bleomycin-O_2 associated with DNA fibers revealed that the localization of the O-O bond is restricted to a plane perpendicular to the fiber axis.

Whereas the finding of the DNA cleavage by bleomycin was reported shortly after the discovery of the antibiotic (Suzuki et al., 1969), the detailed mode of DNA cleavage could not be elucidated until the development of a new methodology for DNA sequencing (Maxam and Gilbert, 1980). The Maxam-Gilbert method allowed us to determine which base sequences in DNA are preferentially cleaved by bleomycin. The iron-bleomycin system cleaves DNA preferentially at G-C (5'→3') and G-T (5'→3') sequences, as shown in the cleavage of a pBR 322 DNA fragment by the Fe(II)-bleomycin and Fe(II)-*deglyco*-bleomycin complex systems (Sugiura et al., 1983). The DNA cleavage experiments using lactose operon pL J3 or bacteriophage ϕX 174 DNA fragments also exhibited the guanine-pyrimidine (5'→3') sequence preference (D'Andrea and Haseltine, 1978; Takeshita et al., 1978). It should be noted that the Fe(II)-bleomycin and Fe(II)-*deglyco*-bleomycin complexes showed almost identical nucleotide sequence specificity in the DNA cleavage (Sugiura et al., 1983). This suggests that the disaccharide moiety of bleomycin produces virtually no noticeable effect on the specificity for the binding and cleavage of DNA.

The Co(III)-bleomycin complex was shown to cause DNA breakage in the presence of light (Chang and Meares, 1982). The Fe(II)-bleomycin complex and the Co(III)-bleomycin-light system cleaved DNA at the same sites, i.e. pyrimidine residues at the 3' side of guanine were preferentially cleaved by the Co(III)-bleomycin-light system (Chang and Meares, 1984; Saito et al., 1989a). DNA damage mediated by the green Co(III)-bleomycin complex was more efficient than that caused by the brown Co(III) complex (Saito et al., 1989a). The light-induced cleavage of DNA by the green Co(III)-bleomycin complex appears to be independent of oxygen.

The UV light irradiation toward the inactive Fe(III)-bleomycin-OH^- complex clearly induced the DNA cleavage activity (Sugiura et al., 1982c). ESR spin trapping experiment showed that the Fe(III)-bleomycin complex system resulted in the generation of strong ESR signals of the spin adduct with the irradiation of UV light. The DNA cleavages of Mn-bleomycin complex system were induced by 2-mercaptoethanol, hydrogen peroxide, or UV light, and these three induction systems gave nucleotide sequence cleavage modes remarkably similar, but not identical, to those of the corresponding Fe-bleomycin complex systems (Suzuki et al., 1985).

The light irradiation of the 1:1 Cu(II)-peplomycin complex induced isomerization of the 2,4'-bithiazole ring system to 4,4'-bithiazole to afford *photo*-peplomycin (Fig. 6) in high yield, together with a small amount of byproduct (Morii et al., 1986). On the other hand, the irradiation of the 5:1 Cu(II)-peplomycin complex system enhanced the formation of the minor product, which was now named *lumi*-peplomycin (Fig. 6) (Morii et al., 1987). *Lumi*-peplomycin was shown to have thiazolyl-isothiazole structure. Fe(II)-O_2-2-mercaptoethanol and Co(III)-UV light complex systems of *photo*-peplomycin and *lumi*-peplomycin showed the nucleotide sequence cleavage mode to be essentially the same as that of the parent bleomycin.

Treatment of metal-free bleomycin with triethylamine in aqueous alcohol at room temperature induced migration of the carbamoyl group of 3-*O*-position of the mannose to 2-*O*-position to give *iso*-bleomycin (Fig. 7) (Nakayama et al., 1973). Treatment of the copper complex of bleomycin under the same conditions resulted in the epimerization at the α-position of the pyrimidine ring to afford *epi*-bleomycin (Muraoka et al., 1976). *Iso*-bleomycin and *epi*-bleomycin exhibited remarkably lowered antimicrobial activity.

Bleomycin was found to be inactivated by an enzyme named bleomycin hydrolase which hydrolyzes the carbamoyl group of the diaminopropionamide moiety, resulting in the formation of *deamido*-bleomycin (Fig. 7) (Umezawa et al., 1974). The inactivity of *deamido*-bleomycin is presumably due to the less efficient oxygen activation by the Fe(II)-*deamido*-bleomycin complex in which the fifth coordination site is occupied by the carboxyl

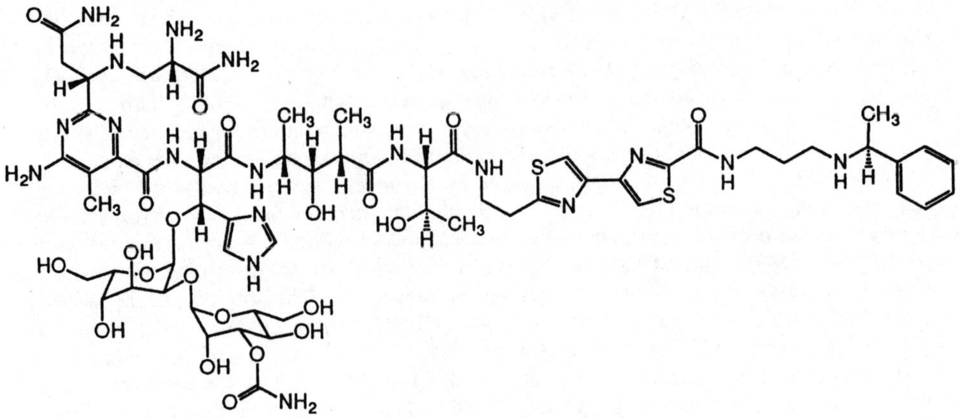

Photo-peplomycin

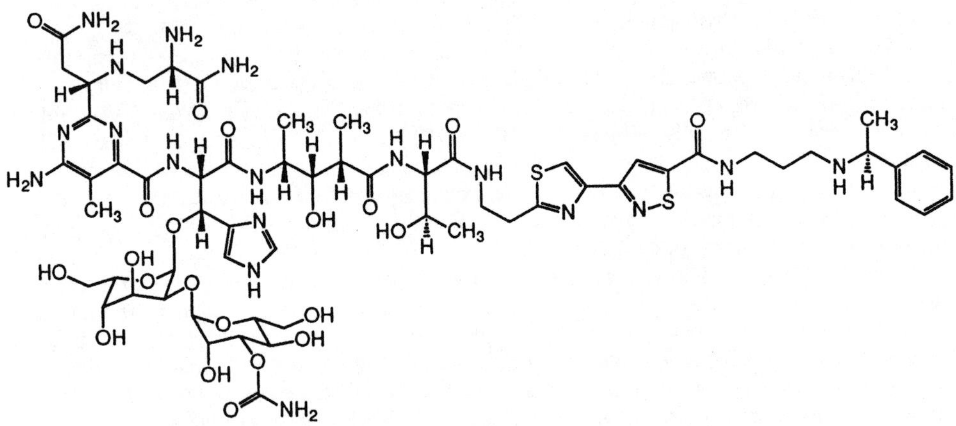

Lumi-peplomycin

Figure 6.
Photo-peplomycin and *lumi*-peplomycin.

group at physiological pH (Fig. 8) (Sugiura, 1980b). Therefore, it can be said that bleomycin possesses a switch device in the diaminopropionamide moiety to regulate its in vivo activity depending on the abundance of bleomycin hydrolase in the tissue.

D. Synthetic models for the metal-bleomycin complexes

The understanding of the unprecedented mode of bleomycin action was greatly facilitated by investigations on synthetic models. In fact, there has been many important reports on the model compounds of bleomycin which were designed and synthesized in order to elucudate the chelation chemistry of bleomycin and to obtain a basis for the rational design of molecules which can reproduce the biochemical function of bleomycin.

Ohno et al. reported a series of pyridine-based models of bleomycin, namely PYMLs, which were designed systematically (Fig. 9) (Ohno and Otsuka, 1990). The first synthetic model ligand, PYML-1, has a pyridine ring and simplified amino acid residues (Otsuka et al., 1981). The Cu(II), Co(II), and Fe(III) complexes of PYML-1 exhibited ESR signals, visible absorption, CD spectra, and redox characteristics well in accordance with those of bleomycin (Sugiura et al., 1983). The oxygen-activating capability of the Fe(II)-PYML-1

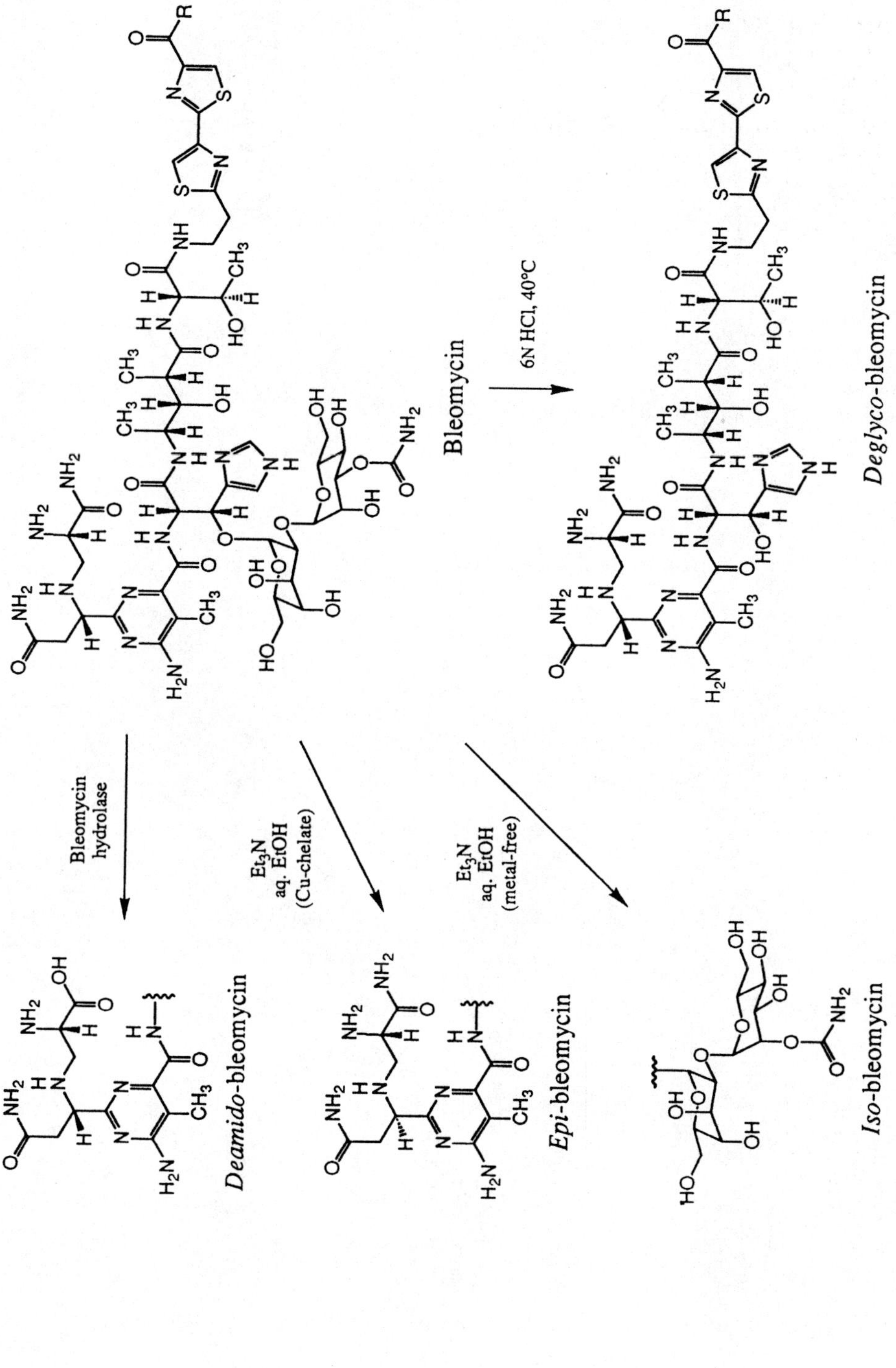

Figure 7.
Transformation of bleomycin.

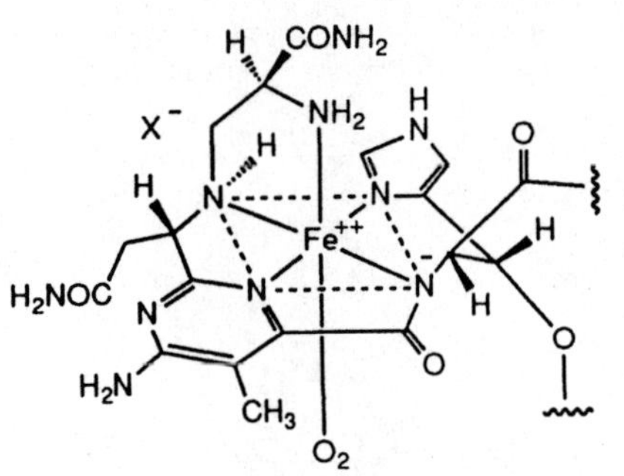

Bleomycin

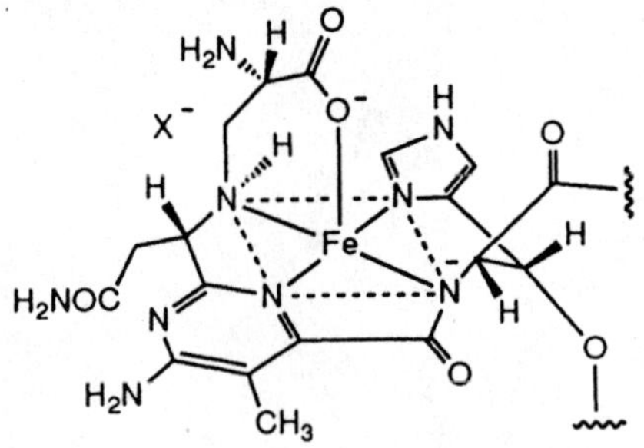

Deamido-bleomycin

Figure 8.
Change of axial donor in the iron complex of *deamido*-bleomycin.

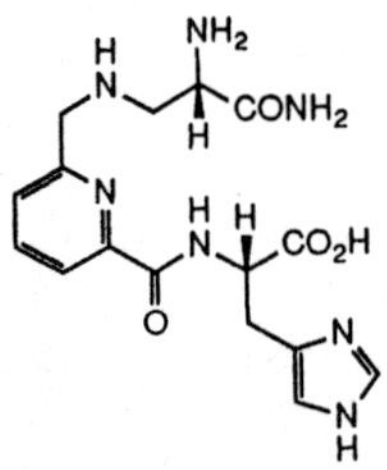

PYML-1

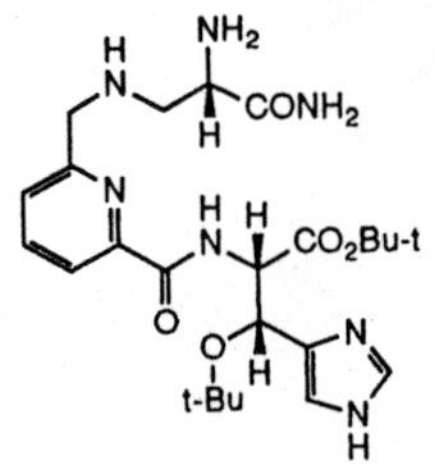

PYML-4

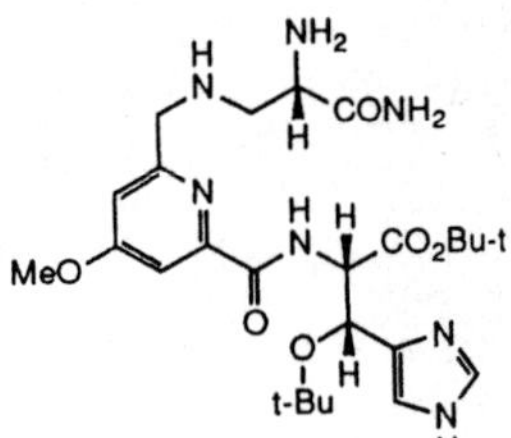

PYML-6

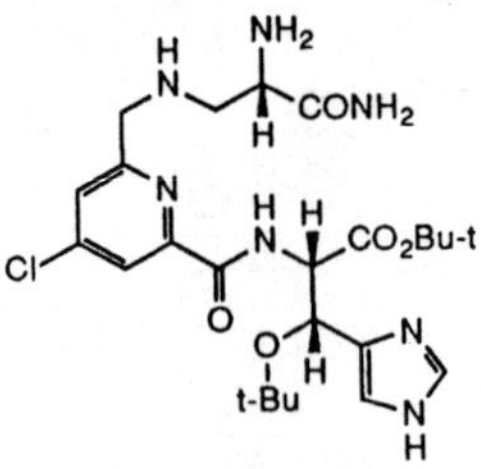

PYML-7

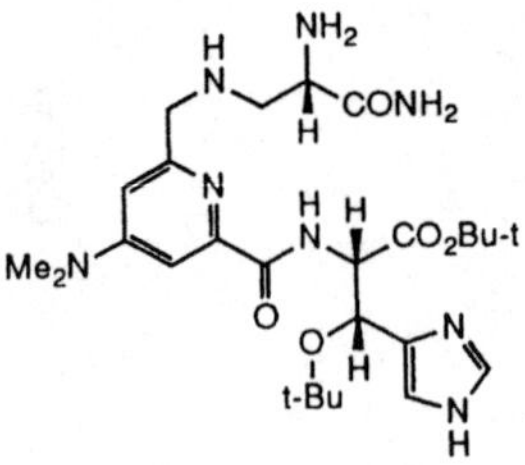

PYML-8

Figure 9.
Synthetic models for the metal-binding site of bleomycin (1).

complex was demonstrated by ESR spin trapping experiments and the concentration of hydroxyl radicals generated from the Fe(II)-PYML-1-O_2 complex was estimated to be 18% of that of the Fe(II)-bleomycin-O_2 complex (Table 2).

Various steric and electronic factors of the PYML skeleton were examined. In particular, the disaccharide moiety of bleomycin appeared to be significant as a major steric factor. Therefore, PYML-4 was designed by replacing the disaccharide moiety by a *tert*-butyl group (Kittaka et al., 1986; 1988a). ESR spin trapping experiments revealed that the dioxygen-activating capability of the Fe(II)-PYML-4 complex increased up to 71% of that of the Fe(II)-bleomycin complex (Table 2). Thus, the *tert*-butyl group was shown to exert a profound steric effect in improving the interaction with oxygen.

On the other hand, it was considered that the efficiency of the oxygen activation is dependent upon the electronic property of the ligand around the iron. Therefore, models having an electron-donating methoxyl substituent (PYML-6) and an electron-withdrawing chloro group (PYML-7) were synthesized and compared (Fig. 9) (Sugano et al., 1986; Kittaka et al., 1988b). PYML-6 showed oxygen-activating power virtually equivalent (97%) to that of bleomycin whereas PYML-7 was less active (55%) than unsubstituted PYML-4 (Table 2). PYML-8, equipped with a strongly electron-donating 4-dimethylaminopyridine moiety, was more potent than bleomycin in oxygen-activating capability (Table 2, Fig. 9) (Suga et al., 1989; 1991).

Hénichart et al. reported a model compound consisting the ethylenediamine, pyridine, and histidine methyl ester, namely AMPHIS (Fig. 10) (Hénichart et al., 1982; 1985). The Cu(II)-AMPHIS complex exhibited ESR parameters well in accordance with those of the Cu(II)-bleomycin complex. Oxygen-activating capability of the Fe(II)-AMPHIS complex was demonstrated by ESR spin trapping experiments.

Hecht et al. prepared the (pyrimidoblamic acid)-(β-hydroxyhistidine)-(4-amino-3-hydroxy-2-methylpentanoic acid) region of bleomycin, designated PBA-HIS-AHM in Figure 10, in the course of their total synthesis of bleomycin and *deglyco*-bleomycin (Kilkuskie et al., 1985). Oxidation of *cis*-stilbene with the PBA-HIS-AHM-Fe(III)-iodosobenzene system afforded the corresponding epoxide in 150% yield based on the PBA-HIS-AHM used. Similar yields of *trans*-epoxide were obtained when the PBA-HIS-AHM-Fe(III) was incubated in the presence of *cis*-stilbene, oxygen, and ascorbate.

Boger and Dang (1992a) reported a concise synthesis of desacetamido P-3A (Fig. 10) based on the inverse electron demand [4+2] cycloaddition reaction. This methodology led to the synthesis of advanced analogues of bleomycin having a pyrimidine ring as will be described later.

Mascharak et al. (1983) designed and prepared model compounds, PypepH, PmpepH, and PMAH (Fig. 10), aiming at the X-ray crystal analysis of the metal complexes. As PypepH and PmpepH lack the side chain corresponding to the diaminopropionamide part of bleomycin, the X-ray structure of their metal complexes were different from that of the

Table 2

Spin Concentrations of Hydroxyl Radicals from Fe(II) Complexes of Bleomycin and the Synthetic Models

Fe(II) complex	Relative spin concentration
Bleomycin	100
PYML-1	18
PYML-4	71
PYML-6	97
PYML-7	55
PYML-8	125

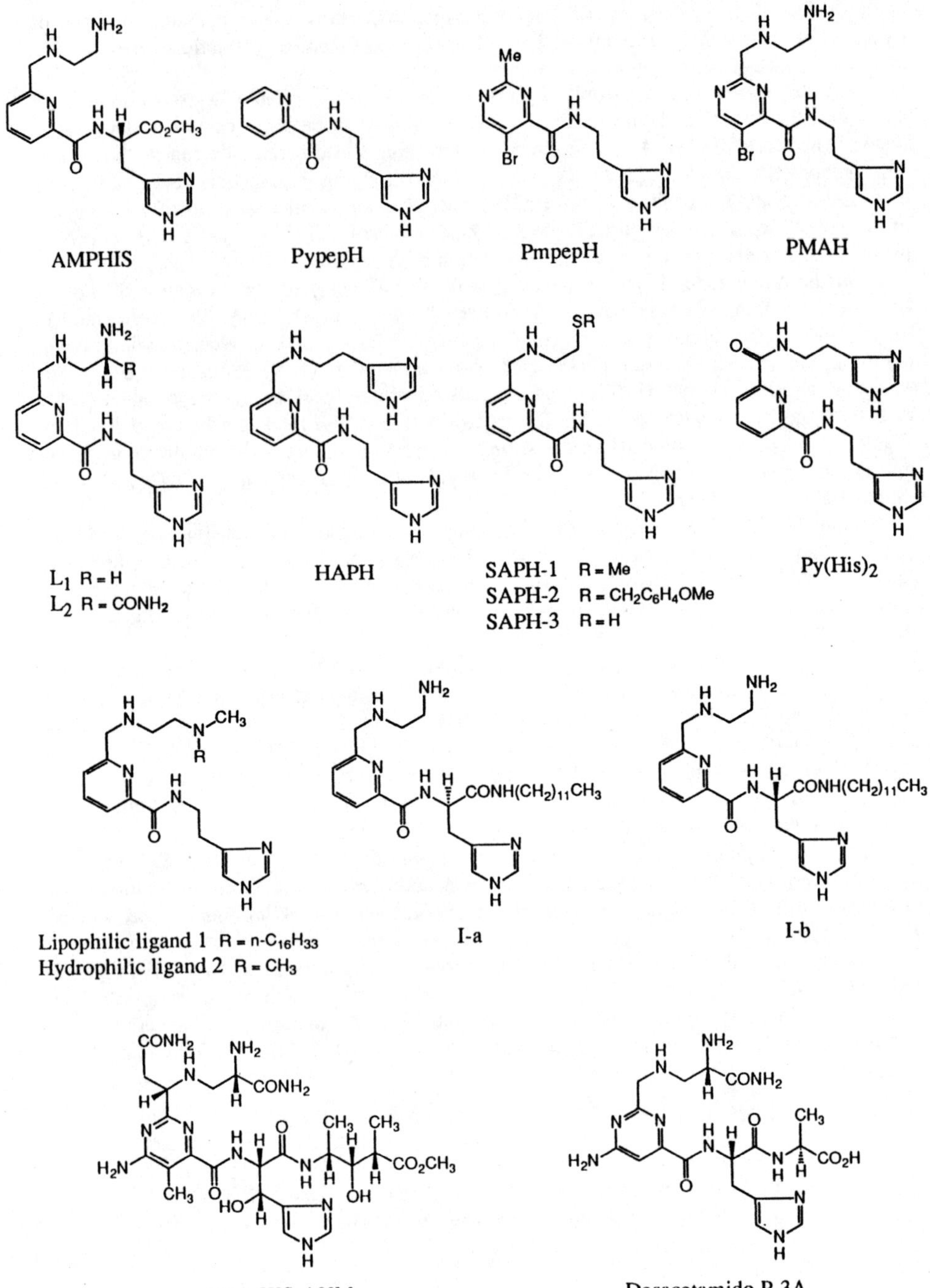

Figure 10.
Synthetic models for the metal-binding site of bleomycin (2).

Cu(II)-P-3A complex (Brown et al., 1986; 1988a; Tao et al., 1987; Delany et al., 1988). In contrast, X-ray analysis of the Cu(II)-PMAH complex revealed that the coordination geometry around the copper is distorted square-pyramidal with five nitrogen atoms of the pyrimidine, the imidazole, the secondary amine, the deprotonated peptide and the primary amino group (Brown et al., 1988b; 1989a). Interestingly, the X-ray analysis of the Co(III)-PMAH complex indicated that only a change in the sixth ligand, e.g. H_2O, *N*-methylimidazole, and Cl^-, gives rise to the formation of the brown, orange, and green products, respectively (Brown et al., 1989b; Tan et al., 1992).

Kimura et al. prepared models L_1 and L_2 (Fig. 10) and reported detailed study on the chelation chemistry (Kurosaki et al., 1988; Kimura et al., 1992). Visible spectra, metal redox properties, and ESR spectra showed that L_1 and L_2 are similar to bleomycin as ligands. L_1 and L_2 exhibited pH-dependent coordination with Cu(II), Fe(II), and Zn(II). Noteworthy is the successful X-ray analysis of the Cu(II)-L_2 complex, demonstrating the distorted square-pyramidal coordination geometry similar to the Cu(II)-PMAH complex.

Wu et al. (1992) carried out a molecular dynamics calculation on the metal complexes of PYML-4 and PMAH. The results of the olefin epoxidation by the iron complexes of PYML were reasonably explained. Significance of the intramolecular hydrogen bonding between the secondary amino group of the ligand and the coordinated dioxygen in facilitating the formation of the active complex species was pointed out.

Shepherd et al. reported model compounds having a modified axial donor, namely, HAPH, SAPH-1, SAPH-2, and SAPH-3 (Fig. 10) (Lomis et al., 1988; 1989a; 1989b). ESR spin trapping experiments indicated that the Fe(II)-HAPH-O_2 complex and the Fe(II)-HAPH-H_2O_2 complex generated the hydroxyl radicals in 53% and 63% efficiency, respectively, compared with that of the Fe(II)-bleomycin-O_2 complex. No oxygen radical was trapped for the Fe(II)-SAPH-3-O_2 complex system.

Hirao et al. designed a ligand consisting of pyridine and histamine, designated $Py(His)_2$ in Figure 10 (Hirao et al., 1990). Oxidation of α-methylstyrene or styrene catalyzed by $Mn(OAc)_2$ or $VO(OEt)Cl_2$ was enhanced by $Py(His)_2$.

Cristini et al. (1989) reported unique lipophilic ligand 1 and hydrophilic ligand 2 (Fig. 10). The Fe(III) complex of lipophilic ligand 1 formed a micellar aggregation in water and catalyzed the H_2O_2 oxidation of *p*-nitrophenyl methyl sulfide. Sawai et al. prepared enantiomeric lauryl derivatives I-a and I-b (Fig. 10), which also form micellar aggregates in water (Shinozuka et al., 1991). The Fe(II) complexes of I-a and I-b were shown to be capable of converting closed circular pBR322 DNA (form I) into nicked (form II) and linear (form III) DNA. The compound I-b exhibited greater activity compared with the antipode I-a.

Ohno et al. reported a series of DNA-cleaving molecules based on the PYML-6 metal-binding domain, namely "man-designed bleomycins" (Fig. 11) (Ohno and Otsuka, 1990). DNA cleavage by bleomycin-like compounds required not only the dioxygen-activating moiety but also the DNA-binding moiety; whereas a 5 μM concentration of bleomycin cleaved DNA, a 500 ~ 1000μM concentration of PYML-6 was required to induce efficient DNA cleavage despite the comparable oxygen-activating power of PYML-6 (Owa et al., 1992). The bithiazole moiety, the DNA-binding site of bleomycin, was coupled with the PYML-6 skeleton and a DNA cleaving molecule PYML(6)-bleomycin was synthesized (Fig. 11) (Otsuka et al., 1986; Ohno et al., 1988; Kittaka et al., 1988b). The nucleotide sequence cleavage mode of PYML(6)-bleomycin was remarkably similar to that of bleomycin. Therefore, Ohno et al. designed PYML(6)-(4*R*-APA)-distamycin where the PYML-6 metal core and the A,T-specific distamycin moiety were combined through (*R*)-4-aminopentanoic acid (4*R*-APA) linker (Fig. 11) (Otsuka et al., 1989). PYML(6)-(4*R*-APA)-distamycin predominantly cleaved A,T-rich regions which were not cleaved by bleomycin. Thus, the base specificity in the DNA cleavage by man-designed bleomycins was shown to depend on the base specificity of the DNA-binding site.

Structure-function relationships of the linker moiety of man-designed bleomycins were studied. PYML(6)-(AHM)-distamycin, having (2*S*, 3*S*, 4*R*)-4-amino-3-hydroxy-2-methylpentanoic acid (abbreviated as AHM) as the linker, and PYML(6)-(4*S*-APA)-distamycin, containing (*S*)-4-aminopentanoic acid (4*S*-APA), were prepared (Owa et al.,

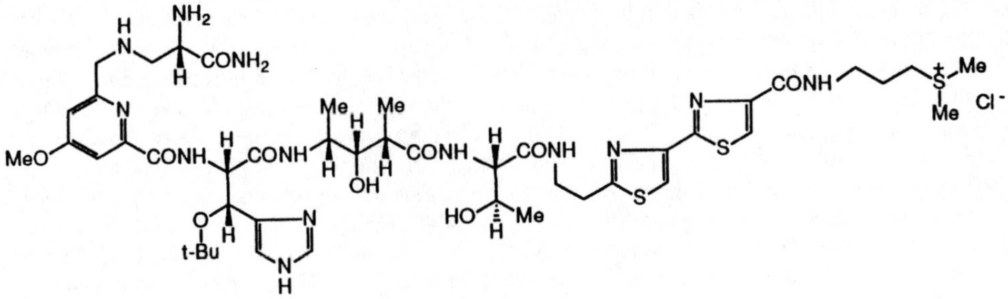

PYML(6)-bleomycin

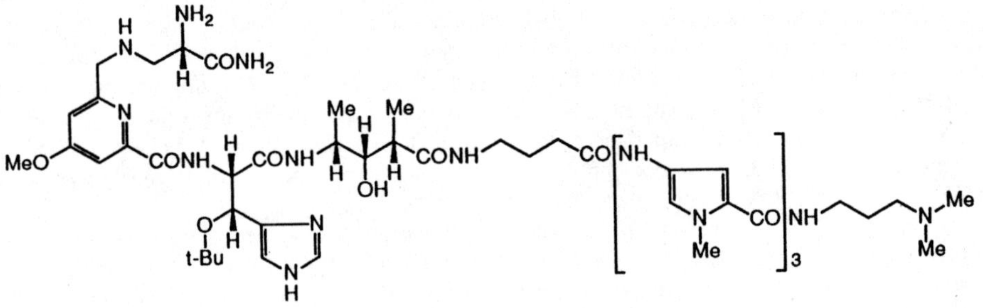

PYML(6)-(AHM)-distamycin

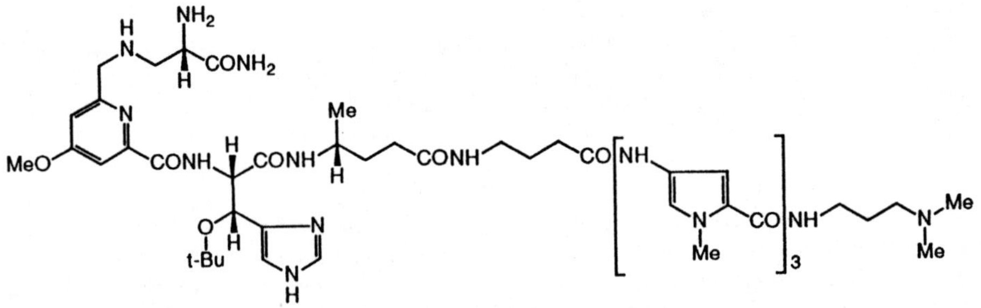

PYML(6)-(4*R*-APA)-distamycin

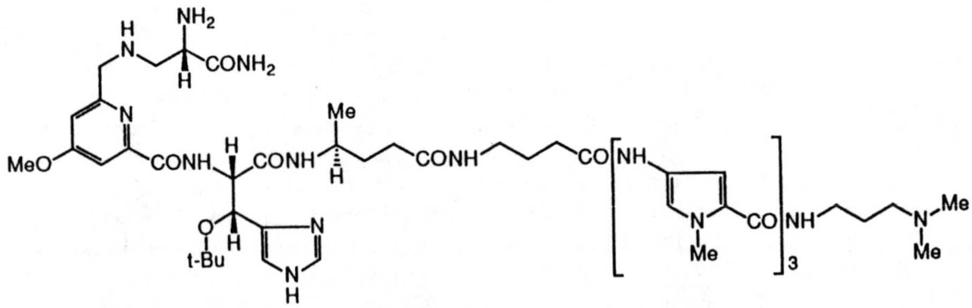

PYML(6)-(4*S*-APA)-distamycin

Figure 11.
Man-designed bleomycins.

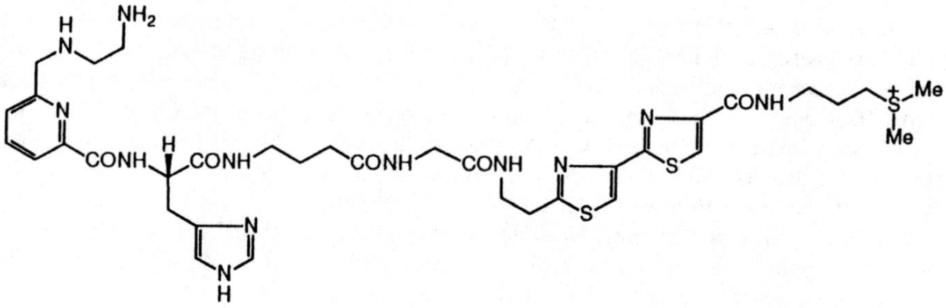

AMBI-A_2

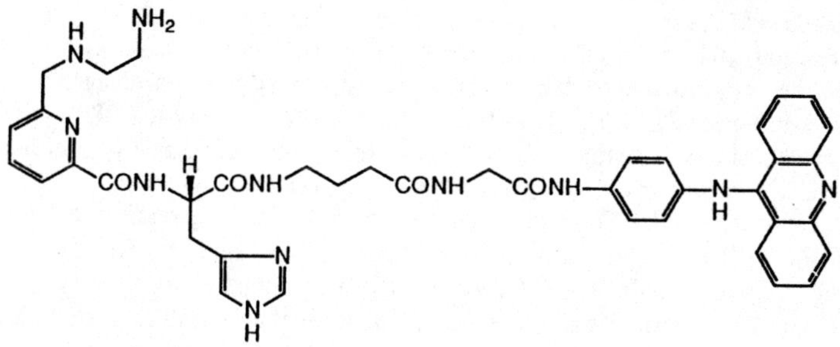

AGGA

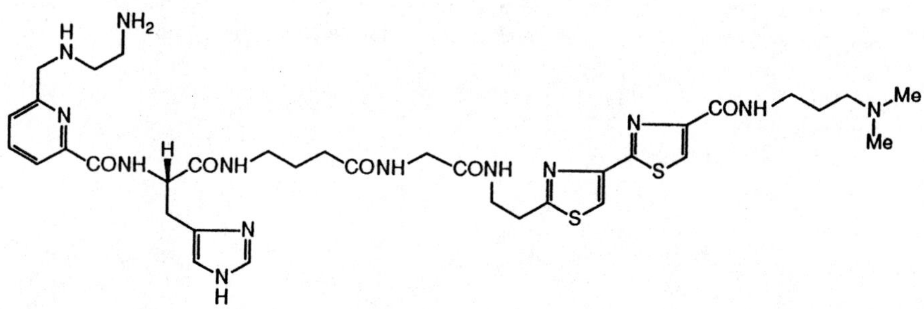

SBLM

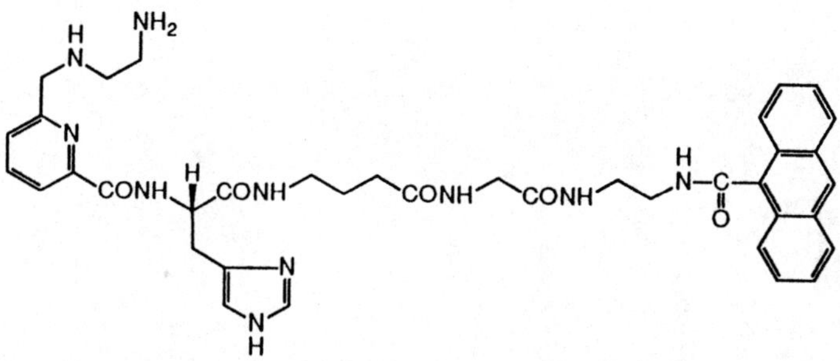

ABLM

Figure 12.
Synthetic models of bleomycin (1).

1992). Whereas DNA cleavage by PYML(6)-(AHM)-distamycin was found to be mostly the same as that caused by PYML(6)-(4*R*-APA)-distamycin, the cleavage activity of PYML(6)-(4*S*-APA)-distamycin was remarkably less than that of PYML(6)-(4*R*-APA)-distamycin by a factor of 10. Computer simulation suggested a steric repulsion between the C5-methyl group of the 4*S*-APA linker and the nucleic acid, whereas there seemed to be no such interaction in the case of the 4*R*-APA linker. It was noteworthy that PYML(6)-(AHM)-distamycin showed potent cytotoxicity against L1210 (Owa et al., 1992).

Hénichart et al. reported DNA-cleaving molecule based on AMPHIS (Fig. 12). AMBI-A_2 is a bleomycin analogue having AMPHIS metal-binding region and bithiazole DNA-binding region connected by a simplified linker (Kenani et al., 1987). The Cu(II)-AMBI-A_2 complex exhibited ESR parameters similar to that of the bleomycin complex. Oxygen-activating capability of the Fe(II)-AMBI-A_2 complex was demonstrated by ESR spin trapping. The Fe(II)-AMBI-A_2 complex converted supercoiled pVM216 DNA (form I) into relaxed closed circular DNA (form II) and open linear DNA (form III). AGGA is another bleomycin analogue, reported by Hénichart et al., possessing 9-aminoacridine as an intercalative DNA-binding site (Fig. 12) (Bailly et al., 1987a; 1987b; 1988). Although AGGA reproduced the metal-binding and the oxygen-activating properties of bleomycin as demonstrated by CD and ESR spectroscopies, it did not cleave DNA.

Saito et al. reported similar DNA-cleaving molecules based on pyridine metal-binding site and a bithiazile (SBLM) or anthracene (ABLM) (Fig. 12) (Saito et al., 1989b). Upon photoirradiation, supercoiled ϕX174 RFI DNA (form I) was almost quantitatively cleaved to give nicked circular DNA (form II) in the presence of the cobalt complexes of SBLM and ABLM. Photocleavage of DNA by the cobalt complex of SBLM occurred uniformly at all bases.

Boger et al. reported pyrimidine-based analogues of bleomycin, namely deglyco GABA,Gly-desacetamidobleomycin A_2 and deglyco desacetamidobleomycin A_2 (Fig. 13)

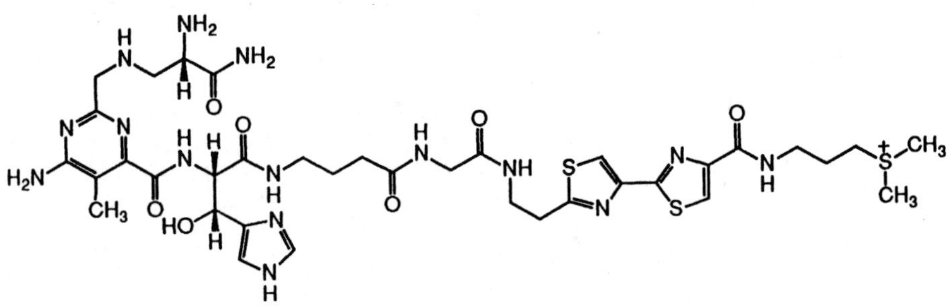

Deglyco GABA,Gly-desaceatamidobleomycin A_2

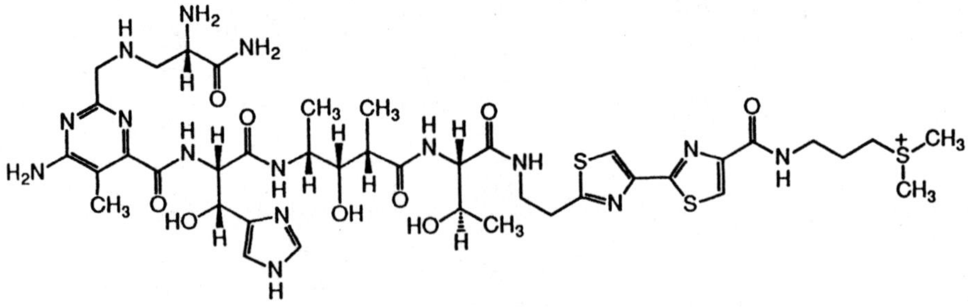

Deglyco desaceatamidobleomycin A_2

Figure 13.
Synthetic models of bleomycin (2).

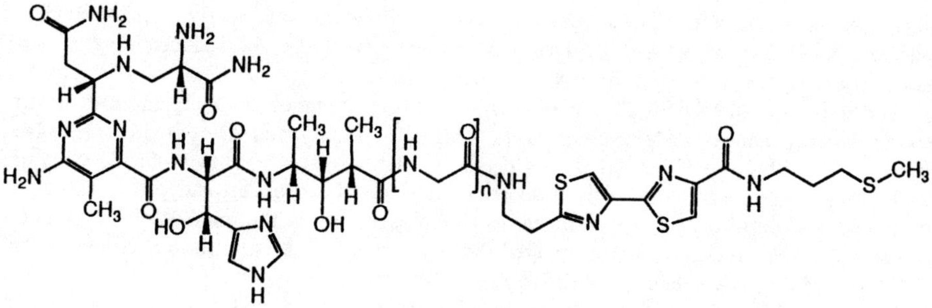

Gly_0-BLM n = 0
Gly_1-BLM n = 1
Gly_2-BLM n = 2
Gly_4-BLM n = 4

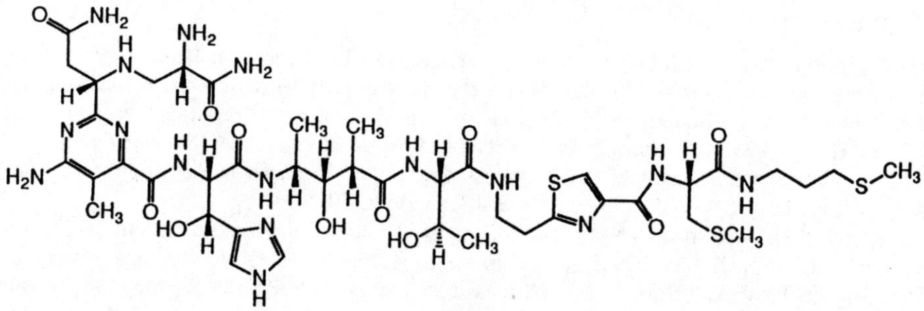

Bleomycin monothiazole A

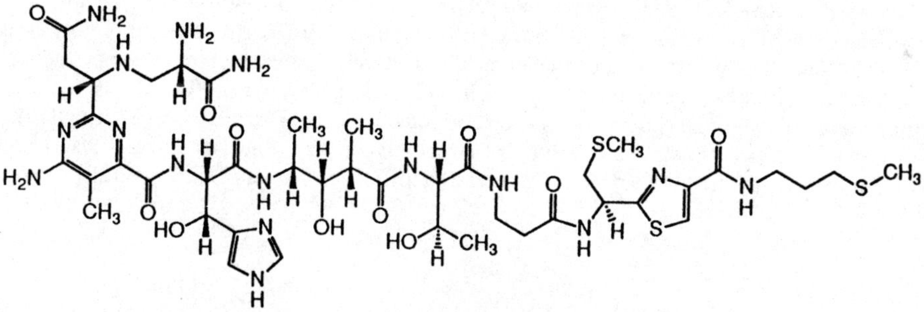

Bleomycin monothiazole B

Figure 14.
Synthetic models of bleomycin (3).

(Boger et al., 1992b; 1992c). The Fe(II) complexes of deglyco GABA,Gly-desacetamidobleomycin A_2 and deglyco desacetamidobleomycin A_2 induced single- and double-strand cleavage of supercoiled ϕX174 RFI DNA.

Hecht et al. reported *deglyco*-bleomycin analogues equipped with glycine spacers of varying length, namely Gly_0-BLM, Gly_1-BLM, Gly_2-BLM, and Gly_4-BLM (Fig. 14) (Carter et al., 1990). All four analogues produced cleavage at the same preferred sites of SV40 DNA fragment although the spacers altered the distance between the pyrimidine moiety and the bithiazole moiety as much as 14 Å depending on the number of the glycine unit. The identical cleavage pattern invoked the significance of the pyrimidine moiety as the determinant of the base sequence specificity in DNA cleavage.

Hecht et al. synthesized bleomycin monothiazole A and bleomycin monothiazole B (Fig. 14) in order to examine the role of individual thiazoles in oxygen activation and DNA cleavage (Hamamichi et al., 1992). The Fe(III)-H_2O_2 complexes of bleomycin monothiazole A and B were shown to be excellent catalysts for the oxygenation of naphthalene and styrene and also they mediated the demethylation of *N,N*-dimethylaniline. However, bleomycin monothiazole A and B were much less active in DNA cleavage compared with bleomycin and the DNA degradation occurred without discernible sequence selectivity.

III. STREPTONIGRIN

Streptonigrin is an aminoquinone antibiotic active against lymphoma, melanoma, carcinoma, and cancers of breast, cervix, head and neck (Davis et al., 1978). However, the clinical use of streptomycin was limited due to its toxicity, especially severe bone marrow depression (Wilson et al., 1961). Streptonigrin was isolated from *Streptomyces flocculus* (Rao and Cullen, 1959) and the chemical structure was determined by degradation study and X-ray analysis (Fig. 15) (Rao et al., 1963; Chiu and Lipscomb, 1975).

Metal binding seemed to be crucial for the expression of the antitumor activity because the addition of Cu(II) and Fe(II) were shown to enhance the DNA cleavage caused by streptonigrin (Lown and Sim, 1976). Whereas heteroatoms and functional groups in the ring B - ring C region constitute an excellent environment for the metal binding, many alternative metal-binding modes may be possible due to the free rotation of the carbon-carbon bond between the rings B and C.

The formation of Zn(II) and Cu(II) complexes of streptonigrin was demonstrated by potentiometric titrations and absorption spectra (Gould and Weinreb, 1982; Hajdu and Armstrong, 1981). Formation of the ternary complex, Zn(II)-streptonigrin-DNA, was reported (Rao, 1979). It was postulated that reductively activated streptonigrin induces DNA damage in the presence of metal ions and molecular oxygen (Lown, 1977). The semiquinone radical of streptonigrin has been detected by ESR (Bachur et al., 1978; 1979).

Despite the potent antitumor activity of streptonigrin, its clinical application had to be given up because of the high toxicity. Structural modification and analogue synthesis did not

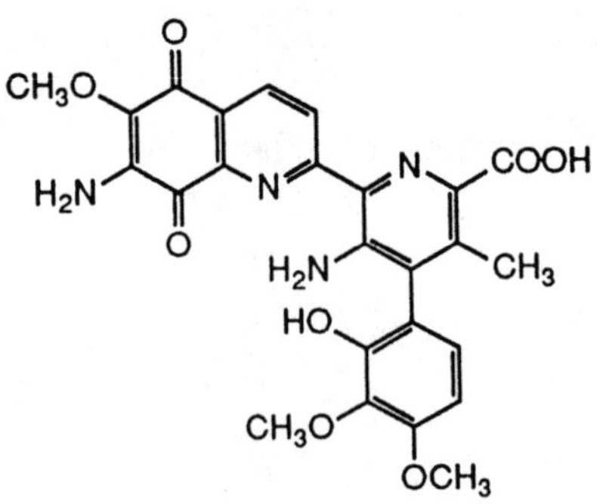

Figure 15.
Streptonigrin.

lead to development of streptonigrin derivatives with low toxicity. However, the novel structure and metal-binding property of streptonigrin provided a new insight into chelating agents having antitumor activity.

IV. CONCLUSION

Bleomycin, an array of unusual hexapeptide and a disaccharide, cleaves DNA preferentially at G-C (5'→3') and G-T (5'→3') in the presence of ferrous ion and oxygen. The molecule can be regarded as an integrated entity of components endowed with the following functions: (i) iron binding and dioxygen activation by the amine-pyrimidine-imidazole region, (ii) stabilization of the iron-oxygen complex and contribution to the membrane transport by the disaccharide moiety, (iii) interaction with certain sequence of nucleotide by the bithiazole - terminal amine region, and (iv) assemblage of each part of the molecule at an appropriate distance and in a good orientation by the linker moiety. The metal-binding region embedded in the aminoquinone structure of streptonigrin is a novel oxygen-activating system. Thus, it is concluded that bleomycin and streptonigrin provide a prototype for the rational design of new metal complexes aiming at efficient antitumor activity. Moreover, the binding and cleavage of DNA by bleomycin and streptonigrin could be regarded as examples of a specific interaction of metal complex and biological receptor.

REFERENCES

Akkerman, M.A.J., Haasnoot, C.A.G., and Hilbers, C.W. (1988). *Eur. J. Biochem., 173*: 211.

Akkerman, M.A.J., Neijman, E.W.J.F., Wijmenga, S.S., Hilbers, C.W., and Bermel, W. (1990). *J. Amer. Chem. Soc., 112*: 7462.

Albertini, J.P., and Garnier-Suillerot, A. (1982). *Biochemistry, 21*: 6777.

Bachur, N.R., Gordon, S.L., and Gee, M.V. (1978). *Cancer Res., 38*: 1745.

Bachur, N.R., Gordon, S.L., Gee, M.V., and Kon, H. (1979). *Proc. Nat. Acad. Sci. USA, 76*: 914.

Bailly, C., Bernier, J.L., Houssin, R., Helbecque, N., and Hénichart, J.P. (1987a). *Anti-Cancer Drug Design, 1*: 303.

Bailly, C., Catteau, J.-P., Helbeque, N., Bernier, J.-L., Houssin, R., Denis, C., and Hénichart, J.-P., (1987b). *J. Inorg. Biochem., 31*: 211.

Bailly, C., Pommery, N., and Hénichart, J.-P. (1988). *Cancer Lett., 38*: 321.

Boger, D.L., and Dang, Q. (1992a). *J. Org. Chem., 57*: 1631.

Boger, D.L., Menzes, R.F., Dang, Q., and Yang, W. (1992b). *Bioorg. Med. Chem. Lett., 2*: 261.

Boger, D.L., Menzes, R.F., and Dang, Q. (1992c). *J. Org. Chem., 57*: 4333.

Brown, S.J., Tao, X., Stephan, D.W., and Mascharak, P.K. (1986). *Inorg. Chem., 25*: 3377.

Brown, S.J., Tao, X., Wark, T.A., Stephan, D.W., and Mascharak, P.K. (1988a). *Inorg. Chem., 27*: 1581.

Brown, S.J., Mascharak, P.K., and Stephan, D.W. (1988b). *J. Amer. Chem. Soc., 110*: 1996.

Brown, S.J., Hudson, S.E., Stephan, D.W., and Mascharak, P.K. (1989a). *Inorg. Chem., 28*: 468.

Brown, S.J., Hudson, S.E., Mascharak, P.K., and Olmsted, M.M. (1989b). *J. Amer. Chem. Soc., 111*: 6446.

Burger, R.M., Horwitz, S.B., Peisach, J., and Wittenberg, J.B. (1979a). *J. Biol. Chem., 254*: 12299.

Burger, R.M., Peisach, J., Blumberg, W.E., and Horwitz, S.B. (1979b). *J. Biol. Chem., 254*: 10906.

Burger, R.M., Adler, A.D., Horwitz, S.B., Mims, W.B., and Peisach, J. (1981a). *Biochemistry, 20*: 1701.

Burger, R.M., Peisach, J., and Horwitz, S.B. (1981b). *J. Biol. Chem., 256*: 11636.

Burger, R.M., Kent, T.A., Horwitz, S.B., Munck, E., and Peisach, J. (1983). *J. Biol. Chem.*, *258*: 1559.

Carter, B.J., Murty, V.S., Reddy, K.S., Wang, S.-N., and Hecht, S.M. (1990). *J. Biol. Chem.*, *256*: 4193.

Chang, C.-H., and Meares, C.F. (1982). *Biochemistry*, *21*: 6332.

Chang, C.-H., and Meares, C.F. (1984). *Biochemistry*, *23*: 2268.

Chen, D.M., Sakai, T.T., Glickson, J.D., and Patel, D.J. (1980). *Biochem. Biophys. Res. Commun.*, *92*: 197.

Chien, M., Grollman, A.P., and Horwitz, S.B. (1977). *Biochemistry*, *16*: 3641.

Chikira, M., Antholine, W.E., and Petering, D.H. (1989). *J. Biol. Chem.*, *246*: 21478.

Chiu, Y.H., and Lipscomb, W.N. (1975). *J. Amer. Chem. Soc.*, *97*: 2525.

Cristini, M., Scrimin, P., and Tonellato, U. (1989). *Tetrahedron Lett.*, *30*: 2987.

Dabrowiak, J.C. (1980). *J. Inorg. Biochem.*, *13*: 317.

Dabrowiak, J.C., and Tsukayama, M. (1981). *J. Amer. Chem. Soc.*, *103*; 7543.

D'Andrea, A.D., and Haseltine, W.A. (1978). *Proc. Nat. Acad. Sci. USA*, *75*: 3608.

Davis, H.L., HonHoff, D.D., Henney, J.T., and Rozencweig, M. (1978). *Cancer Chemother. Pharmacol.*, *1*: 83.

Delany, K., Arora, S.K., and Mascharak, P.K. (1988). *Inorg. Chem.*, *27*: 705.

DeRiemer, L.H., Meares, C.F., Goodwin, D.A., and Diamanti, C.I. (1979). *J. Med. Chem.*, *22*,: 1019.

Freedman, J.H., Horwitz, S.B., and Peisach, J. (1982). *Biochemistry*, *21*: 2203.

Garnier-Suillerot, A., Albertini, J.P., and Tosi, L. (1981). *Biochem. Biophys. Res. Commun.*, *102*: 499.

Gould, S.J., and Weinreb, S.M. (1982). *Fortschr. Chem. Org. Natur.*, *41*: 77.

Hajdu, J., and Armstrong, E.C. (1981). *J. Amer. Chem. Soc.*, *103*: 232.

Hamamichi, N., Natrajan, A., and Hecht, S.M. (1992). *J. Amer. Chem. Soc.*, *114*: 6278.

Hecht, S.M. (1986). *Accounts Chem. Res.*, *19*: 383.

Hénichart, J.-P., Houssin, R., Bernier, J.-L., and Catteau, J.P. (1982). *J. Chem. Soc., Chem. Commun.*, 1295.

Hénichart, J.-P., Bernier, J.-L., Houssin, R., Lohez, M., Kenani, A., and Catteau, J.P. (1985). *Biochem. Biophys. Res. Commun*, *126*: 1036.

Hirao, T., Mikami, S., and Oshiro, Y. (1990). *Synlett*, 541.

Iitaka, Y., Nakamura, H., Nakatani, T., Murata, Y., Fujii, A., Takita, T., and Umezawa, H. (1978). *J. Antibiot.*, *31*: 1070.

Ishizu, K., Murata, S., Miyoshi, K., Sugiura, Y., Takita, T., and Umezawa, H. (1981). *J. Antibiot.*, *34*: 994.

Kasai, H., Naganawa, H., Takita, T., and Umezawa, H. (1978). *J. Antibiot.*, *31*: 1316.

Kenani, A., Lohez, M., Houssin, R., Helbecque, N., Bernier, J.-L., Lemay, P., and Hénichart, J.-P. (1987). *Anti-Cancer Drug Design*, *2*: 47.

Kilkuskie, R.E., Suguna, H., Yellin, B., Murugesan, N., and Hecht, S.M. (1985). *J. Amer. Chem. Soc.*, *107*: 260.

Kimura, E., Kurosaki, H., Kurogi, Y., Shionoya, M., and Shiro, M. (1992). *Inorg. Chem.*, *31*: 4314.

Kittaka, A., Sugano, Y., Otsuka, M., Ohno, M., Sugiura, Y., and Umezawa, H. (1986). *Tetrahedron Lett.*, *27*: 3631.

Kittaka, A., Sugano, Y., Otsuka, M., and Ohno, M. (1988a). *Tetrahedron*, *44*: 2811.

Kittaka, A., Sugano, Y., Otsuka, M., and Ohno, M. (1988b). *Tetrahedron*, *44*: 2821.

Kuramochi, H., Takahashi, K., Takita, T., and Umezawa, H. (1981). *J. Antibiot.*, *34*: 576.

Kurosaki, H., Anan, H., and Kimura, E. (1988). *Nippon Kagaku Kaishi*, 691.

Lomis, T.J., Siuda, J.F., and Shepherd, R.E. (1988). *J. Chem. Soc., Chem. Commun.*, 290.

Lomis, T.J., Elliot, M.G., Siddiqui, S., Moyer, M., Koespel, R.R., and Shepherd, R.E. (1989a). *Inorg. Chem.*, *28*: 2367.

Lomis, T.J., Martin, J., McCloskey, B., Siddiqui, S., and Shepherd, R.E. (1989b). *Inorg. Chim. Acta*, *157*: 99.

Lown, J.W. (1977). *Bioorganic Chemistry* (E.E. Van Tamelen, ed.), Academic Press, New York, p. 95.

Lown, J.W., and Sim, S.K. (1976). *Can. J. Chem.*, *54*: 2563.

Mascharak, P.K., Sugiura, Y., Kuwahara, J., Suzuki, T., and Lippard, S.J. (1983). *Proc. Nat. Acad. Sci. USA*, *80*: 6095.

Maxam, A.M., and Gilbert, W. (1980). *Methods Enzymol.*, *65*: 499.

Morii, T., Matsuura, T., Saito, I., Suzuki, T., Kuwahara, J., and Sugiura, Y. (1986). *J. Amer. Chem. Soc.*, *108*: 7089.

Morii, T., Saito, I., Matsuura, T., Kuwahara, J., and Sugiura, Y. (1987). *J. Amer. Chem. Soc.*, *109*: 938.

Muraoka, Y., Kobayashi, H., Fujii, A., Kunishima, M., Fijii, Y., Nakayama, Y., Takita, T., and Umezawa, H. (1976). *J. Antibiot.*, *29*: 853.

Nakayama, Y., Kunishima, M., Omoto, S., Takita, T., and Umezawa, H. (1973). *J. Antibiot.*, *26*: 400.

Nieweg, O.E., Beekhuis, H., Paans, A.M.J., Piers, D.A., Vaalburg, W., Welleweers, J., Wiegman, T., and Woldring, M.G. (1982). *Eur. J. Nucl. Med.*, *7*: 104.

Ohno, M., and Otsuka, M. (1990). *Recent Progress in the Chemical Synthesis of Antibiotics* (G. Lukacs and M. Ohno, eds.), Springer-Verlag, Berlin Heidelberg, p. 387.

Ohno, M., Otsuka, M., Kittaka, A., Sugano, Y., Sugiura, Y., Suzuki, T., Kuwahara, J., Umezawa, K., and Umezawa, H. (1988). *Int. J. Exp. Clin. Chemother.*, *1*: 12.

Oppenheimer, N.J., Rodriguez, L.O., and Hecht, S.M. (1979a). *Proc. Nat. Acad. Sci. USA*, *76*: 5616.

Oppenheimer, N.J., Rodriguez, L.O., and Hecht, S.M. (1979b). *Biochemistry*, *18*: 3439.

Oppenheimer, N.J., Chang, C., Rodriguez, L.O., and Hecht, S.M. (1981). *J. Biol. Chem.*, *256*: 1514.

Otsuka, M., Yoshida, M., Kobayashi, S., Ohno, M., Sugiura, Y., Takita, T., and Umezawa, H. (1981). *J. Amer. Chem. Soc.*, *103*: 6986.

Otsuka, M., Kittaka, A., Ohno, M., Suzuki, T., Kuwahara, J., Sugiura, Y., and Umezawa, H. (1986). *Tetrahedron Lett.*, *27*: 3639.

Otsuka, M., Masuda, T., Haupt, A., Ohno, M., Shiraki, T., Sugiura, Y., and Maeda, K. (1989). *J. Amer. Chem. Soc.*, *112*: 838.

Owa, T., Haupt, A., Otsuka, M., Kobayashi, S., Tomioka, N., Itai, A., Ohno, M., Shiraki, T., Uesugi, M., Sugiura, Y., and Maeda, K. (1992). *Tetrahedron*, *48*: 1193.

Pillai, R.P., Lenkinski, R.E., Sakai, T.T., Geckle, J.M., Rama Krishna, N., and Glickson, J.D. (1980). *Biochem. Biophys. Res. Commun.*, *96*: 341.

Rao, K.V. (1979). *J. Pharm. Sci.*, *68*: 853.

Rao, K.V., and Cullen, W.P. (1959). *Antibiot. Ann.*, 950.

Rao, K.V., Biemann, K., and Woodward, R.B., (1963). *J. Amer. Chem. Soc.*, *85*: 2532.

Saito, I., Morii, T., Sugiyama, H., Matsuura, T., Meares, C.F., and Hecht, S.M. (1989a). *J. Amer. Chem. Soc.*, *111*: 2307.

Saito, I., Morii, T., Obayashi, T., Sera, T., Sugiyama, H., and Matsuura, T. (1989b). *J. Chem. Soc., Chem. Commun.*, 360.

Sheridan, R.P., and Gupta, R.K. (1981). *J. Biol. Chem.*, *256*: 1242.

Shinozuka, K., Morishita, H., Yamazaki, T., Sugiura, Y., and Sawai, H. (1991). *Tetrahedron Lett.*, *32*: 6869.

Stubbe, J., and Kozarich, J.W. (1987). *Chem. Rev.*, *87*: 1107.

Suga, A., Sugiyama, T., Sugano, Y., Kittaka, A., Otsuka, M., Ohno, M., Sugiura, Y., and Maeda, K. (1989). *Synlett*, 70.

Suga, A., Sugiyama, T., Otsuka, M., Ohno, M., Sugiura, Y., and Maeda, K. (1991). *Tetrahedron*, *47*: 1191.

Sugano, Y., Kittaka, A., Otsuka, M., Ohno, M., Sugiura, Y., and Umezawa, H. (1986). *Tetrahedron Lett.*, *27*: 3635.

Sugiura, Y. (1980a). *J. Amer. Chem. Soc.*, *102*: 5216.

Sugiura, Y. (1980b). *J. Amer. Chem. Soc.*, *102*: 5208.

Sugiura, Y., and Ishizu, K. (1979b). *J. Inorg. Biochem.*, *11*: 171.

Sugiura, Y., and Kikuchi, T. (1978). *J. Antibiot.*, *31*: 1310.

Sugiura, Y., Ishizu, K., and Miyoshi, K. (1979a). *J. Antibiot.*, *32*: 453.

Sugiura, Y., Ogawa, S., and Morishima, I. (1980). *J. Amer. Chem. Soc.*, *102*: 7944.

Sugiura, Y., Takita, T., and Umezawa, H. (1981). *J. Antibiot. 34*: 249.

Sugiura, Y., Suzuki, T., Kawabe, H., Tanaka, H., and Watanabe, K. (1982a). *Biochim. Biophys. Acta, 716*: 38.

Sugiura, Y., Kuwahara, J., and Suzuki, T. (1982b). *J. Chem. Soc., Chem. Commun.*, 908.

Sugiura, Y., Suzuki, T., Kuwahara, J., and Tanaka, H. (1982c). *Biochem. Biophys. Res. Commun.*, *105*: 1511.

Sugiura, Y., Suzuki, T., Otsuka, M., Kobayashi, S., Ohno, M., Takita, T., and Umezawa, H. (1983). *J. Biol. Chem.*, *258*: 1328.

Sugiura, Y., Takita, T., and Umezawa, H. (1985). *Metal Ions in Biological Systems*, Vol. 19 (H. Sigel, ed.), Marcel Dekker, New York, p. 81.

Suzuki, H., Nagai, K., Yamaki, H., Tanaka, N., and Umezawa, H. (1969). *J. Antibiot.*, *22*: 446.

Suzuki, T., Kuwahara, J., Goto, M., and Sugiura, Y. (1985). *Biochim. Biophys. Acta, 824*: 330.

Takahashi, K., Yoshioka, O., Matsuda, A., and Umezawa, H. (1977). *J. Antibiot.*, *30*: 861.

Takeshita, M., Grollman, A.P., Ohtsubo, E., and Ohtsubo, H. (1978). *Proc. Nat. Acad. Sci. USA, 75*: 5983.

Takita, T., Muraoka, Y., Nakatani, T., Fujii, A., Umezawa, Y., Naganawa, H., and Umezawa, H. (1978a). *J. Antibiot.*, *31*: 801.

Takita, T., Muraoka, Y., Nakatani, T., Fujii, A., Iitaka, Y., and Umezawa, H. (1978b). *J. Antibiot.*, *31*: 1073.

Tan, J.D., Hudson, S.E., Brown, S.J., Olmsted, M.M., and Mascharak, P.K. (1992). *J. Amer. Chem. Soc.*, *114*: 3841.

Tanaka, W. (1977). *Jpn. J. Antibiot.*, *30 (Suppl.)*: S-41.

Tao, X., Stephan, D.W., and Mascharak, P.K. (1987). *Inorg. Chem.*, *26*: 754.

Tsukayama, M., Randall, M., Santillo, F.S., and Dabrowiak, J.C. (1981). *J. Amer. Chem. Soc.*, *103*: 458.

Umezawa, H. (1977). *Lloydia*, *40*: 67.

Umezawa, H., Hori, S., Sawa, T., Yoshioka, T., and Takeuchi, T. (1974). *J. Antibiot.*, *27*: 419.

Umezawa, H., Takita, T., Saito, S., Muraoka, Y., Takahashi, K., Ekimoto, H., Minamide, S., Nishikawa, K., Fukuoka, T., Nakatani, T., Fujii, A., and Masuda, A. (1985). *Bleomycin Chemotherapy* (B.I. Sikic, M. Rozenweig and S.K. Carter, eds.), Academic Press, Orlando, p. 289.

Vos, C.M., and Westra, G. (1981). *J. Inorg. Biochem.*, *15*: 253.

Vos, C.M., Westra, G., and van Zanten, B. (1980a). *J. Inorg. Biochem.*, *12*: 45.

Vos, C.M., Westra, G., and Schipper, D. (1980b). *J. Inorg. Biochem.*, *13*: 165.

Wilson, W.L., Labra, C., and Barrist, E. (1961). *Antibiot. Chemother. (Basel)*, *11*: 147.

Wu, Y.-D., Houk, K.N., Valentine, J.S., and Nam, W. (1992). *Inorg. Chem.*, *31*: 720.

METAL COMPLEXES OF BENZODIAZEPINES

by

Carlo Preti and Giuseppe Tosi

Università di Modena,
Departimento di Chimica,
Via G.Campi 183
41100 Modena
Italy

I. INTRODUCTION

Benzodiazepines are quite peculiar ligands capable of binding a metal center through cyclic as well as exocyclic heteroatoms.

A few papers appeared in 1971-73 dealing with 1,5-benzodiazepines complexes. However, systematic studies regarding the behavior of 1,4-benzodiazepines as ligands started in 1976 only.

II. SYNTHESIS AND CHARACTERIZATION OF METAL COMPLEXES

A. 1,5-Benzodiazepine complexes

The first coordination compound of 1,5-benzodiazepines was an iron(II) sulphate of 2,4-dimethyl-3H-1,5-benzodiazepine of the type $C_{11}H_{12}N_2$, $FeSO_4$, 1.5 H_2O, obtained as dark blue crystals as early as 1933 (Emmert and Gsottschneider, 1933).

The same ligand has been the object of other studies in a more recent paper (Ouchi et al., 1971); the authors took into account the reactions with different transition metals of the deep violet 2,4-dimethyl-3H-1,5-benzodiazepine hydrochloride which contains the corresponding benzodiazepinium cation, L. The crystalline products obtained are of the type ML_2X_4 with M= Co(II), Mn(II), Zn(II) and X= Cl, Br, with M= Cd(II) and X= Br, with M= Cu(II) and X= Cl; MLX_2 with M= Cu(I) and X= Br; MLX_3 with M= Cd(II) and X= Cl; and MLX_4 with M= Sb(III) and X= Cl. The authors concluded that these derivatives were likely to be ionic compounds consisting of the benzodiazepinium cation and the halogenometallato anion.

One year later, a second paper appeared (Hunter and Webb, 1972) dealing with the same compounds. In addition, the authors reported metal sulphates of the type $L_2M(H_2O)_6(SO_4)_2$, where M= Mn(II), Fe(II), Co(II), Ni(II), Cu(II), Zn(II). On the basis of IR and electronic spectroscopic studies as well as magnetic properties, they concluded that the diazepine is present as the delocalized cation in which neither nitrogen atom is available for coordination. They hypothesized, however, that other diazepines containing electron-withdrawing substituents may be capable of producing metal complexes. In fact, they reported in a second paper (Hunter and Webb, 1973) the Ni(II) and Cu(II) halides and perchlorate complexes of a series of 2,3,4,5-tetrahydro-1H-1,5-benzodiazepines.

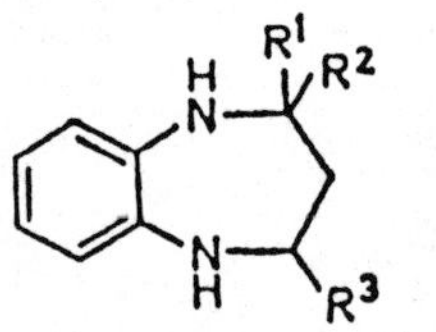

R^1= Me, R^2= R^3= H
R^1= R^3= Me, R^2= H
R^1= R^2= R^3= Me

scheme 1

From their magnetic properties and from the results of electronic and vibrational spectra, the nickel perchlorate complexes seem to be square-planar; the copper complex of the trimethylbenzodiazepine is also square planar, but a decrease in the number of the methyl substituents in the diazepine ring allows the approach of a further ligand, consequently forming five-coordinate complexes.

The nickel halide complexes are in three different stereochemistries; the paramagnetic derivatives NiL_2Cl_2 and $NiLCl_2$ are five-coordinate square-pyramidal species, the former through the axial coordination of one of the two chlorine atoms, while the second appeared dimeric through chlorine-bridging atoms. The bromide and iodide complexes, NiL_2X_2, are square planar.

In the study of oxidative transformations using dioxygen and metal complexes (Szeverenyi et al., 1988), it was suggested that Mn(II) chloride promotes the conversion of 1H-2,3-dihydro-2,2,4-trimethyl-1,5-benzodiazepine to a new compound, as reported in the scheme:

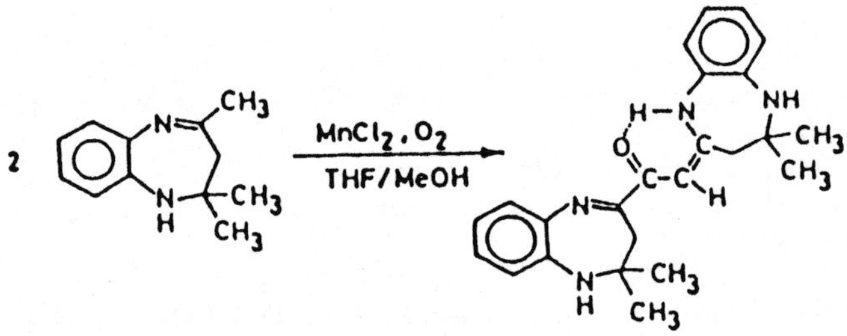

scheme 2

It has been observed that the complexation of manganese with the reaction product, giving the $MnL_2Cl_2,2H_2O$ derivative, may eliminate the ability of the metal to promote the oxidative reaction of the 1,5-benzodiazepine.

The structure of $MnL_2Cl_2,2H_2O$ has been determined by X-ray diffraction, the crystals being monoclinic, space group P21/n; the donor atoms are in the C=O and C=N groups, resulting in a relatively weak metal-ligand interaction. The six donor atoms form a distorted octahedron around the Mn(II) ion.

B. 1,4-Benzodiazepine complexes

The 1,4-benzodiazepines have been the object of more numerous and more detailed studies including in many cases crystal and molecular structure determinations using X-ray techniques.

The research started with the studies of Preti and Tosi (1976; 1978; 1979a; 1979b) who considered the interaction of 7-chloro-2-methylamino-5-phenyl-3H-1,4-benzodiazepin-4-oxide (chlordiazepoxide), 1,3-dihydro-7-nitro-5-phenyl-2H-1,4-benzodiazepin-2-one (nitrazepam) and 7-chloro-1,3-dihydro-1-methyl-5-phenyl-2H-1,4-benzodiazepin-2-one (diazepam) with Co(II), Ni(II), Cu(II), Zn(II), Cd(II), Hg(II), Pd(II) and Pt(II). The complexes obtained were true coordination compounds and not salts of the diazepinium cations and tetrahalomatallate anions, as demonstrated by conductivity measurements and detailed IR spectrophotometric studies. The new complexes were in 1:1, 1:1.5, 1:2

metal:ligand molar ratios; they were studied by means of magnetic susceptibility measurements, IR and far IR spectra, electronic spectra and conductivity measurements. The evidence suggested that the Co(II) and Ni(II) complexes have a pseudotetrahedral symmetry. The 1:1 and 1:2 Zn(II), Cd(II) and Hg(II) derivatives can be considered tetrahedral, with terminal bonded ligands and terminal and bridging halides in the 1:1, and terminal bonded ligands and halides in the 1:2 derivatives. In the case of nitrazepam, the hexacoordinate trans-octahedral D_{4h} $Zn(L)_4Br_2$ complex has been obtained in addition to the 1:2 derivatives.

A coordination of the ligands through a nitrogen of the diazepine ring has been suggested for all the above complexes based on the IR results. This hypothesis has been fully confirmed by the works of Mosset et al. (1980) and of Miyamae et al. (1982) who independently published the structure of the square planar $Cu(diazepam)_2Cl_2$ complex. The boat conformation of the seven-membered diazepine ring is maintained in the complex. Mosset et al. (1980) have shown from EPR and ^{13}C NMR studies that a square planar coordination of the copper(II) ion is maintained in solution, similar to that found in the solid state.

With Pd(II) and Pt(II) and the above three ligands, complexes in 1:1 and 1:2 metal:ligand molar ratios have been synthesized (Preti and Tosi, 1979b). The 1:2 derivatives are trans-square planar of D_{2h} stereochemistry with the ligands N-bonded and terminal halide atoms, while the question of the 1:1 derivatives is still a matter of discussion. Many possibilities can be suggested always considering square planar geometries: (i) terminal N-bonded ligands and terminal and bridging halide atoms; (ii) terminal halide atoms and bridging N-bonded ligands; (iii) bridging halide atoms and the benzodiazepine ligand acting as N-bonded through the N4 atom, the ortho carbon atom of the phenyl ring giving a five-membered cyclometallated system. The first or/and second hypothesis has been suggested by Preti and Tosi (1979b) based on IR spectra results, while the last one has been suggested by Cinellu et al. (1988) working with diazepam and 7-chloro-1-(cyclopropylmethyl)-1,3-dihydro-5-phenyl-2H-1,4-benzodiazepin-2-one (prazepam) as ligands.

New diamagnetic halide complexes of Rh(III), Ir(III), Pd(II) and Pt(II) with 7-chloro-1,3-dihydro-3-hydroxy-5-phenyl-2H-1,4-benzodiazepin-2-one (oxazepam) have been reported by Benedetti et al. (1984). The complexes are of the ML_3X_3 type (M= Rh, Ir) and ML_2X_2 (M= Pd, Pt) with the exception of the rhodium and palladium iododerivatives which have 1:6 and 1:4 metal:ligand molar ratios, respectively. The electronic spectral parameters suggest an octahedral configuration for Rh(III) and Ir(III) with the presence of MN_3X_3 chromophores, and of a RhN_6 chromophore as regards the iodo-derivative in strong covalently bonded complexes. The electronic spectra of the diamagnetic Pd(II) and Pt(II) derivatives are indicative of square planar geometries. The IR spectra allow to propose a coordination through the nitrogen atom in the 4-position of the diazepine ring.

The free ligand oxazepam and a representative complex, $[Pd(oxazepam)_2Cl_2]$, H_2O, have been extensively studied through ^1H and ^{13}C NMR spectroscopy. The ^1H and ^{13}C NMR spectra obtained from freshly prepared DMSO d_6 solutions of the ligand and of the complex suggest that the presence of a metal ion induces a systematic perturbation of the resonance conditions of the Ci nuclei of the ligand, characterized by a moderate decrease in intensity and broadening of all the peaks. This behavior, in agreement with other literature data (Mosset et al., 1980) is, as a whole, consistent with coordination of the metal ion via the N4 nitrogen of the diazepine ring and excludes the possibility of different coordinations.

Gold(III) 1:1 chloride complexes have been studied by Minghetti et al. (1982; 1984) using the ligands nitrazepam, diazepam, prazepam and, furthermore, 1,3-dihydro-1-methyl-7-nitro-5-phenyl-2H-1,4-benzodiazepin-2-one (nimetazepam) and 7-chloro-5-(o-chlorophenyl)-1,3-dihydro-3-hydroxy-2H-1,4-benzodiazepin-2-one (lorazepam). The results of the IR, Raman, ^1H NMR spectroscopies and of the X-ray structure determination of $Au(prazepam)Cl_3$ suggest a coordination of the ligands through the 4-nitrogen atom. In presence of alkali in methanol solution, $(PPh_3)AuCl$ reacts with 1-unsubstituted 1,4-benzodiazepines such as nitrazepam and lorazepam (Cinellu at al., 1990) to give neutral gold(I) species $(L-H)Au(PPh_3)$ where the benzodiazepine anion (L-H) is bonded to the metal via the N1 atom, as shown by X-ray studies in the case of the nitrazepam derivative. The complex is monoclinic, space group P21/c. Metallation of the ligand through the C3 atom is ruled out by the ^{1}H NMR spectra. The complex (L-H)Au(PPh_3) reacts with [(Ph_3P)Au(S)][BF_4] (S= solvent), to give the dinuclear cationic species [(Ph_3P)Au{m(L-H)}Au(PPh_3)][BF_4] where the deprotonated nitrazepam bridges two (Ph_3P)Au groups through the N1 and N4 atoms.

By reaction with Pd(II), the ligands diazepam and prazepam (Cinellu et al., 1988) give the PdL_2Cl_2 and [(L-H)PdCl]$_2$ products. In the 1:2 complexes, the benzodiazepine acts as monodentate ligand through the N4 atom, as shown by the structure determined by X-ray diffraction for two different crystalline modifications, i.e. Pd(prazepam)$_2Cl_2$, $CHCl_3$ and Pd(prazepam)$_2Cl_2$, CH_2Cl_2. Both complexes were trans-square planar and monoclinic, space group P21/n the first and P21/c the latter. Due to the different quality of the diffracting crystals, the structure determination of the first molecule is markedly more precise than that of the second (Figs. 1 and 2).

The only substantial difference lies in the conformation of the cyclopropyl moiety, in particular as regards C20 and C21. In the first compound, the conformation of the cyclopropyl ring is similar to that of the free ligand molecule (Brachtel and Jansen, 1981), i.e. the C20 and C21 atoms are displaced from the plane N1-C18-C19 in the same direction as the fragment C2-C3-N4; in the second, on the contrary, the atoms C20 and C21 are on the side of the plane N1-C18-C19 where the C6-C11 phenyl ring is placed. Thus, the torsion angle relative to the N1, C18, C19 atoms and the midpoint of the vector C20-C21 is negative in the case of the first molecule and positive for the second.

As for the [(L-H)PdCl]$_2$ derivatives, the authors suggest cyclometallated complexes for diazepam and prazepam, involving both C- and N- intramolecular coordination of the deprotonated ligands, with bridging chlorides. This hypothesis was checked by reaction with neutral and anionic ligands such as triphenylphosphine and thallium(I)acetylacetonate and using FAB mass spectrometry techniques (Cinellu et al., 1989). At the moment, no structural data are available on this type of compounds and none of the above cyclometallated species has been clearly ascertained.

Wide and complete X-ray structural studies on different metals and different ligands have recently been carried out (Antolini et al., 1986; Benedetti et al., 1987; 1989; Corradini et al., to be published), in particular on the complexes Hg(temezepam)Cl_2,

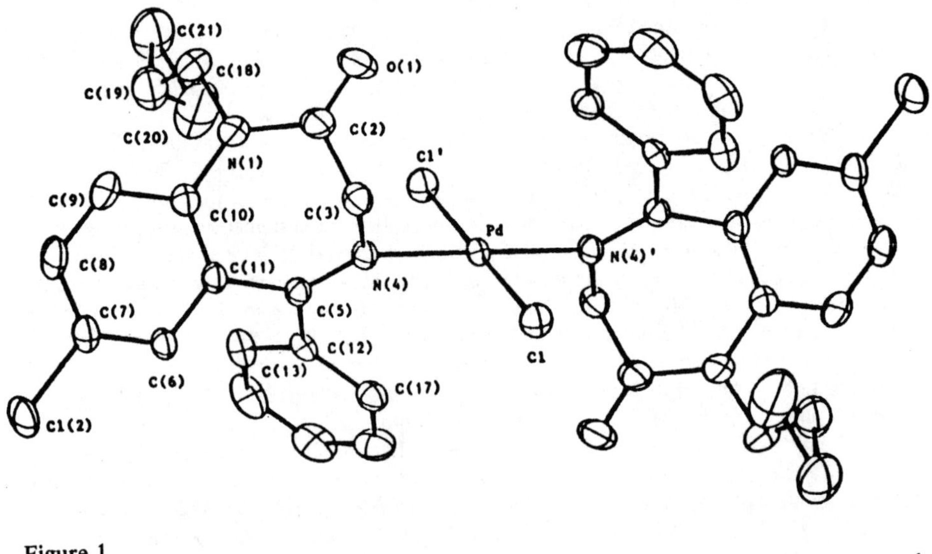

Figure 1.
ORTEP view of the Pd(prazepam)$_2Cl_2$ molecule in crystals of $CHCl_3$ containing compound. (Courtesy of Cinellu et al., 1988).

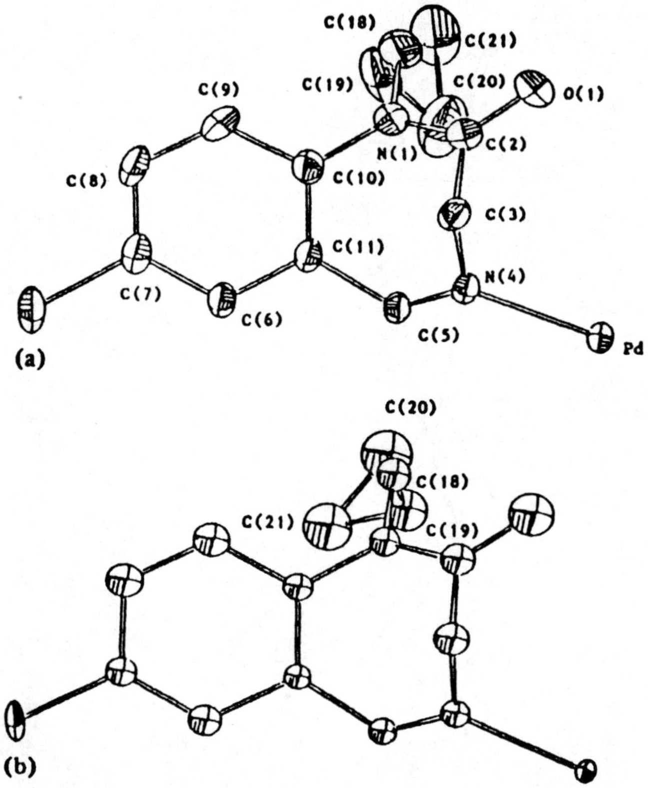

Figure 2.
Projection of a fragment of the prazepam molecule onto the best plane of the C6-C11 phenyl ring in the compound Pd(prazepam)$_2$Cl$_2$,CHCl$_3$ (a) and Pd(prazepam)$_2$Cl$_2$,CH$_2$Cl$_2$ (b). (Courtesy of Cinellu et al., 1988).

Pt(bromazepam)Cl$_2$, Cd(triazolam)$_2$Cl$_2$, Cu(nimetazepam)$_4$Cl$_2$ where
temazepam=7-chloro-1,3-dihydro-5-phenyl-1-methyl-3-hydroxy-2H-1,4-benzodiazepin-2-one;
bromazepam= 7-bromo-1,3-dihydro-5-(2-pyridyl)-2H-1,4-benzodiazepin-2-one;
triazolam= 8-chloro-6-(o-chlorophenyl)-1-methyl-4H-s-triazolo[4,3-a]1,4-benzodiazepine;
nimetazepam= 7-nitro-1,3-dihydro-5-phenyl-1-methyl-2H-1,4-benzodiazepin-2-one.
The mercury-temazepam complex belongs to the monoclinic space group P21/a (Fig. 3).

The general conformation of the ligand molecule in this mercury complex is very similar to that of the free ligand, and the mercury atom was found in an almost undistorted mercuric chloride geometry; the HgCl$_2$ molecule dimerizes by means of bridging chlorine atoms to form a moiety of imposed planarity. The dimer units Hg$_2$Cl$_4$ are bonded by bridging temazepam ligands, each of which is in contact with two mercury atoms of the same dimer unit through the keto oxygen and the N4 nitrogen atoms, as confirmed by far-IR spectra; however, no Hg-O and Hg-N stretching modes have been observed in the far-IR region, this supporting the fact that the suggested metal-ligand interactions are quite weak.

The platinum-bromazepam complex belongs to the monoclinic space group C2/c; the crystal consists of van der Waals packing units without significant intermolecular contacts. A stereoview of the molecule is shown in Figure 4.

As for the cadmium-triazolam complex, the crystals are monoclinic, space group C2/c. The structure consists of discrete molecules without hydrogen bonding interactions, in which the cadmium atom is coordinated in a distorted tetrahedral geometry by two chlorine

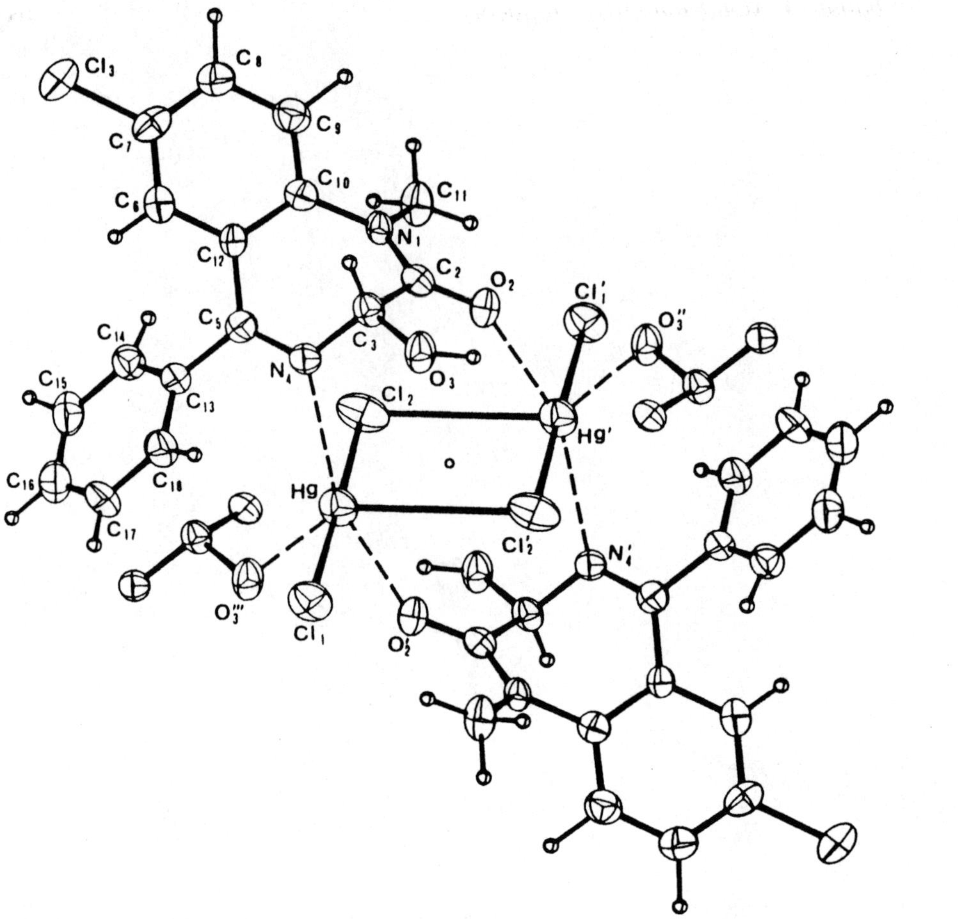

Figure 3.
Stereoview of the environment of two ligand molecules. (Courtesy of Antolini et al., 1986).

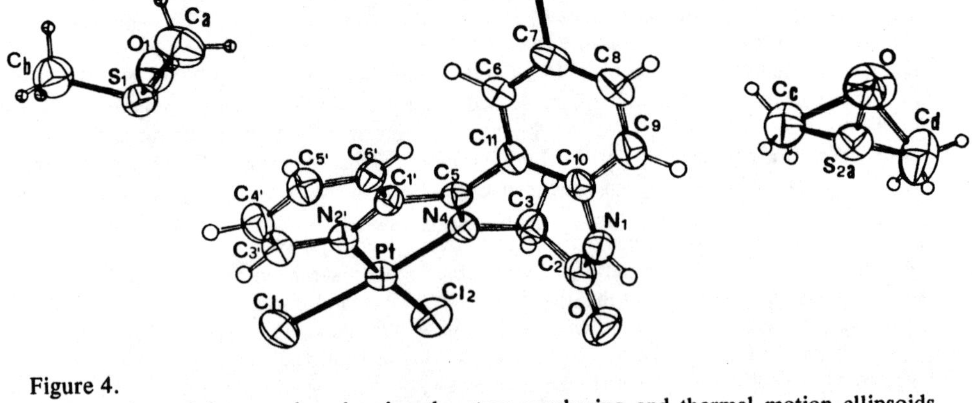

Figure 4.
ORTEP view of the complex showing the atom numbering and thermal motion ellipsoids (50%) for the nonhydrogen atoms. The hydrogen atoms are represented as spheres of arbitrary radius. (Courtesy of Benedetti et al., 1987).

atoms and two nitrogens of two triazolic rings of triazolam molecules. The intermolecular van der Waals' interactions are probably responsible for the crystal packing of the molecular units.

The copper-nimetazepam complex belongs to the space group P1. The structure consists in two ligand molecules linked to the copper ion through the N4 atom of the diazepine ring (Fig. 5).

The copper is in a planar environment by conditions of crystallographic symmetry. This unusual perfect planarity could be explained by taking into account the perturbation that may be induced by the proximity of the hydrogen atoms on C15 and its centrosymmetric which are close enough to warrant special attention: Cu-H= 2.64(2) Å. The crystal packing consists in π-π interactions involving also the two other non-bonded molecules of ligands present in the lattice.

Real and Borras have reported bromazepam complexes of Mn(II), Fe(II), Co(II), Ni(II), Cu(II) and Zn(II) chloride, perchlorate, sulphate and thiocyanate (Real and Borras, 1984a; 1984b; 1986). Three papers by Real et al. (1987a; 1987b; 1987c) report the structures of copper(II) complexes with bromazepam as ligand and oxalate and sulphate as counterions. In the complex [Cu(bromazepam)(C_2O_4)(H_2O)], H_2O (Real et al., 1987a), monoclinic, space group C2/c, the Cu(II) ion displays a square pyramidal coordination with

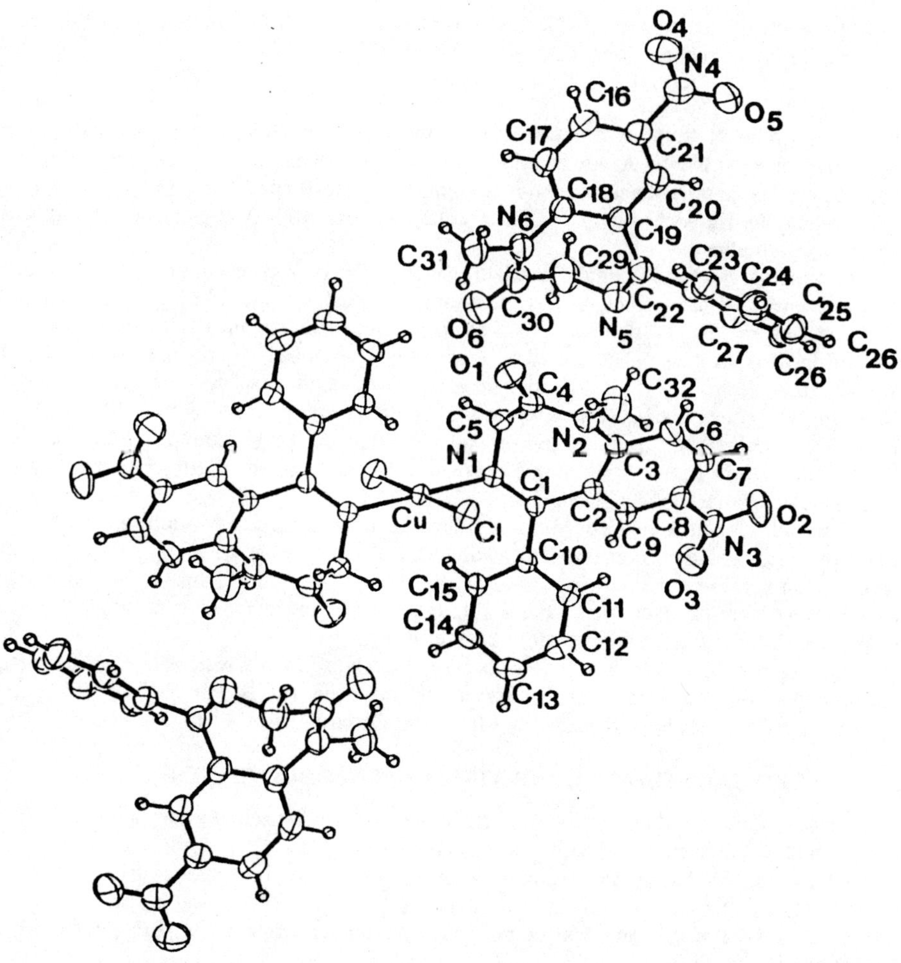

Figure 5.
ORTEP view of the [Cu(nimetazepam)$_2$Cl$_2$], 2nimetazepam complex.

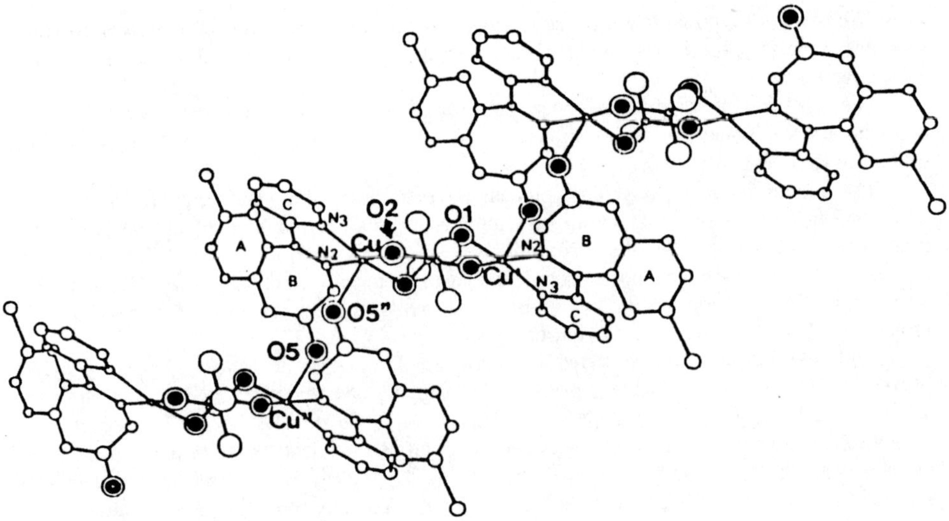

Figure 6.
Geometry of the $[Cu(bromazepam)SO_4]_n$ chains. The black circles correspond to the oxygen atoms exhibiting the connections. (Courtesy of Real et al., 1987c).

the two α-diimine N-atoms of bromazepam and two oxalate O-atoms in the basal plane, the copper ion being 0.257(5) Å out of this plane and a water molecule occupying the apical site. The electronic reflectance and the room temperature ESR spectra are compatible with a square pyramidal stereochemistry in CuO_2N_2 chromophores with strong planar bonding and weaker axial bonding.

A dimeric square pyramidal oxalato Cu(II) bromazepam complex has been also obtained whose X-ray structure has been solved (Real et al., 1987b). The complex $Cu_2(bromazepam)_2(PF_6)_{1.5}(ClO_4)_{0.5}$, 1.5 H_2O, triclinic, space group P1, consists of dimer units bridged by oxalate ligands; each copper ion is surrounded by two nitrogen atoms of bromazepam, N_{imine} and $N_{pyridine}$, two oxygen atoms of the oxalate ligand and a water molecule. The compound $Cu(bromazepam)SO_4$ is orthorombic, space group Pbcn (Real et al., 1987c); the crystal structure consists of a packing of $[Cu(bromazepam)SO_4]_n$ chains held together via a rather weak van der Waals network between the two noncoordinated oxygen atoms of the sulphate group (Fig. 6).

The copper atom is pentacoordinated to three oxygen atoms and two nitrogen atoms, in a square pyramidal environment. The sixth octahedron position is occupied by the O4 oxygen atom of the sulphate group weakly interacting with the copper. The ligand is tricoordinated towards two metal atoms through the N2 and N3 nitrogen atoms for one copper and through the O5 oxygen atom for the second.

A thermogravimetric study on some chromium(III) derivatives of diazepam, chlordiazepoxide, nitrazepam and oxazepam of the type $[LH]_3$ $[Cr(NCS)_6]$ and [LH] $[Cr(NH_3)_2(NCS)_4]$ has been reported by Marcu and Grecu (1979).

III. ANALYTICAL STUDIES INVOLVING 1,4-BENZODIAZEPINES

Analytical methods for the quantitative determination of 1,4-benzodiazepines based on metal complexation have been proposed by various authors.

Grecu and Barbu (1966) reported a gravimetric and photocolorimetric dosage of chlordiazepoxide based on the formation of [LH] $[Cr(NH_3)_2(NCS)_4]$, and a gravimetric and volumetric dosage of the same benzodiazepine based on the formation of $[LH]_3$ $[Cr(NCS)_6]$ (Grecu and Barbu, 1968).

A new gravimetric and photometric method is reported (Grecu and Marcu, 1973) based on the formation of diazepam tetraiodobismutate [LH] $[BiI_4]$. The error in the proposed method is of ± 0.30% in the case of gravimetric determinations and of ± 1.22% for the photometric ones.

Ferrous ions form purple complexes with bromazepam and its 1-(3-dimethylaminopropyl)-derivative. Sabatino et al. (1969) suggested the use of these complexes in the spectrophotometric determination of iron and studied the pH dependence of the complex formation. Many years later, Fukumoto (1980) proposed the use of the same iron(II) chelate compound for the spectrophotometric determination of bromazepam.

An indirect analytical method, based on complexation of bromazepam with copper(II) ions, has been proposed using differential pulse anodic stripping voltammetry in a series of papers dealing with the investigation of the complexation of Fe(II), Cu(II), and Co(II) ions with bromazepam by means of UV spectrophotometric, polarographic and voltammetric techniques (Smyth et al., 1977; 1982). The behavior of the Cu(II)-bromazepam complex was studied by anodic stripping voltammetry (Smyth et al., 1977). The addition of Co(II) ions does not change the UV spectral behavior of bromazepam, but changes its polarographic behavior markedly.

Compounds involving Fe(II), Cu(II), and Co(II) bromazepam perchlorates can be separated by HPLC techniques (Smyth et al., 1982). Copper(II) and iron(II) chelates have also been used for selective determinations of trace concentrations of the above 1,4-benzodiazepine either by differential pulse voltammetry at the hanging mercury drop electrode, or by spectrometric procedures.

REFERENCES

Antolini, L., Preti, C., Tosi G., and Zannini, P. (1986). *J. Crystallogr. Spectrosc. Res.*, *16*: 115.

Benedetti, A., Preti, C., and Tosi, G. (1984). *J. Mol. Struct.*, *116*: 397.

Benedetti, A., Fabretti, A.C., Preti, C., Tosi, G., and Zannini, P. (1987). *J. Crystallogr. Spectrosc. Res.*, *17*: 771.

Benedetti, A., Fabretti, A.C., and Preti, C. (1989). *J. Crystallogr. Spectrosc. Res.*, *19*: 651.

Brachtel, G., and Jansen, M. (1981). *Cryst. Struc. Commun.*, *10*: 669.

Cinellu, M.A., Ganadu, M.L., Minghetti, G., Cariati, F., Demartin, F., and Manassero, M. (1988). *Inorg. Chim. Acta*, *143*: 197.

Cinellu, M.A., Minghetti, G., Banditelli, G., Bandini, A.L., Pelli, B., and Traldi, P. (1989). *Inorg. Chim. Acta*, *161*: 57.

Cinellu, M.A., Stoccoro, S., Minghetti, G., Bandini, A.L., and Demartin, F. (1990). *Inorg. Chim. Acta*, *168*: 33.

Corradini, F., Franchini, G.C., Preti, C., Tagliazucchi, M., and Tosi, G., unpublished results.

Emmert, B., and Gsottschneider, H. (1933). *Ber.*, *66*: 1871.

Fukumoto, M. (1980). *Chem. Pharm. Bull.*, *28*: 3678.

Grecu, I., and Barbu, S. (1966). *Société de Pharmacie* (Galati, Romania).

Grecu, I., and Barbu, S. (1968). *Ann. Pharmac. Franç.*, *26*:405.

Grecu, I., and Marcu, P. (1973). *Farmacia Bucarest*, *21*: 339.

Hunter, P.W.W., and Webb, G.A. (1972). *J. Inorg. Nucl. Chem.*, *34*: 1511.

Hunter, P.W.W., and Webb, G.A. (1973). *J. Chem. Soc. Dalton Trans.*, 26.

Marcu, P., and Grecu, I. (1979). *Pharmazie*, *34*: 432.

Minghetti, G., Foddai, C., Cariati, F., Ganadu, M.L., and Manassero, M. (1982). *Inorg. Chim. Acta*, *64*: L235.

Minghetti, G., Ganadu, M.L., Foddai, C., Cinellu, M.A., Cariati, F., Demartin, F., and Manassero, M. (1984). *Inorg. Chim. Acta*, *86*: 93.

Miyamae, H., Obata, A., and Kawazura, H. (1982). *Acta Cryst.*, *B38*: 272.

Mosset, A., Tuchagues, J.P., Bonnet, J.J., Haran, R., and Sharrock, P. (1980). *Inorg. Chem.*, *19*: 290.

Ouchi, A., Takeuchi, T., Nakatani, M., and Takahashi, Y. (1971). *Bull. Chem. Soc. Japan*, *44*: 434.

Preti, C., and Tosi, G. (1976). *J. Coord. Chem.*, *6*: 81.

Preti, C., and Tosi, G. (1978). *Transition Met. Chem.*, *3*: 246.

Preti, C., and Tosi, G. (1979a). *J. Inorg. Nucl. Chem.*, *41*: 263.

Preti, C., and Tosi, G. (1979b). *J. Coord. Chem.*, *8*: 223.

Real, J.A., and Borras, J. (1984a). *Synth. React. Inorg. Met.-Org. Chem.*, *14*: 843.

Real, J.A., and Borras, J. (1984b). *Synth. React. Inorg. Met.-Org.Chem.*, *14*: 857.

Real, J.A., and Borras, J. (1986). *Synth. React. Inorg. Met.-Org. Chem.*, *16*: 13.

Real, J.A., Borras, J., Solans, X., and Font-Altaba, M. (1987a). *Transition Met. Chem.*, *12*: 79.

Real, J.A., Borras, J., Solans, X., and Font-Altaba, M. (1987b). *Transition Met. Chem.*, *12*: 254.

Real, J.A., Borras, J., Munoz, M.C., Mosset, A., and Galy, J. (1987c). *J. Inorg. Biochem.*, *31*: 221.

Sabatino, J.D., Weber, O.W., Padmanabhan, G.R., and Senkowski, B.Z. (1969). *Anal. Chem.*, *41*: 905.

Smyth, M.R., Beng, T.S., and Smyth, W.F. (1977). *Anal. Chim. Acta*, *92*: 129.

Smyth, W.F., Scannell, R., Goggin, T.K., and Lucas-Fernandez, D. (1982). *Anal. Chim. Acta*, *141*: 321.

Szeverenyi, Z., Fulop, V., Kalman, A., and Simandi, L.I. (1988). *Inorg. Chim. Acta*, *147*: 135.

METAL COMPLEXES OF ANTIVIRAL AGENTS

by

Peter A. Hunt and David P. Fairlie

The 3D Centre,
University of Queensland,
Brisbane, Qld 4072
Australia

I. INTRODUCTION

Certain organic molecules that possess antiviral activity can interact with metal ions to form metal complexes. In rare cases, these complexes are also antiviral agents and these are mentioned in *Bioinorganic Medicine* (Part Five, Chapter 1-K) along with other metal complexes *as* antiviral agents. This chapter deals only with possible interactions between endogenous metals and exogenous organic molecules used as antiviral agents.

Antiviral chemotherapy is usually aimed at preventing one or more steps (cf *Bioinorganic Medicine*, Part Five, Chapter 1-K) in viral replication (Wiltink, 1992), focussing on virally-encoded proteins. Examples of events (and inhibitors) are: (i) recognition by, and attachment of the virus to, the host cell (e.g. immunoglobulins, glycoproteins, anions like dextran sulfate and sulfonate dyes); (ii) penetration of the cell membrane (e.g. amantadine); (iii) uncoating of the virion by removing the nuclear capsid (e.g. TIMP); (iv) replication of viral components in the host cell (e.g. nucleoside analogues, protease inhibitors, thiosemicarbazones, phosphono-acetate); (v) protein assembly and encapsulation (e.g. isatin β-thiosemicarbazone); and (vi) expulsion ("budding") of new viral particles from the host. Inhibitors include ionic species that interfere with viral attachment, aromatic molecules that inhibit transcriptases, and peptides that compete with substrates for proteases. Some antiviral compounds from above can interact with metals.

II. METAL-CHELATOR INTERACTIONS

Exogenous antiviral agents may interact with endogenous metal ions that are either complexed in protein environments (section III), or "free" or at least readily exchangeable (e.g. in blood plasma). Only a tiny fraction of bioconstituent metal ions are "free" but they are in equilibrium with reversibly protein-complexed metals, so their continued availability is critical for supply to catalytic centers of metalloenzymes (e.g. oxidoreductases, polymerases, hydrolases) and to proteins where metals play a structural role maintaining protein conformation. Exogenous antiviral agents could compete for "free" metal ions and limit supply; those that *chelate* metals are more likely to reduce metabolic pools of metal ions, and strong chelators can extract metals from proteins. Few antiviral agents (Perrin and Stünzi, 1982; Hutchinson, 1985; Robins, 1986) that chelate metals are known to act via (or interact with) metals in vivo. Reported stability constants for relevant metal-ligand interactions (Sillén and Martell, 1964; 1971) must be recalculated for physiological pH (e.g. Zn^{2+}: Fairlie et al., 1992). Coordinating atoms of ligands below are asterisked.

A. Simple chelators

8-Hydroxyquinoline (**1**), which blocks replication of Rous sarcoma virus (Rhode et al., 1976; Levinson et al., 1977) in cells by inhibiting RNA-dependent DNA polymerase, and *isonicotinic acid hydrazides* (e.g. **2**, R = H), which also inhibits Rous sarcoma virus (Levinson et al., 1977), both form strong 5m ring N,O- chelates with Zn^{2+} and Cu^{2+} (Albert, 1965). Tri-, bi- (e.g. **3**) and mono- (e.g. **4**) dentate *amines* complex metal ions, especially Cu^{2+} and Zn^{2+}, and have antiviral activity. *Adamantanes* (e.g. amatadine, **4**) are active against influenza A. *EDTA* can chelate metals strongly enough to extract them from metalloproteins and this is the likely cause of inhibition of influenza neuraminidase (Rafelson et al., 1963) and tobacco rattle virus (Perrin and Stünzi, 1982), where essential Ca^{2+} may be removed (Montelius et al., 1990). In vitro EDTA, 1,10-phenanthroline and derivatives inhibit poliovirus polymerases and influenza virus RNA transcriptases which require Zn^{2+} (Hutchinson, 1985).

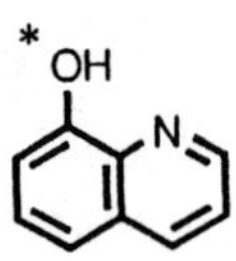

(1)

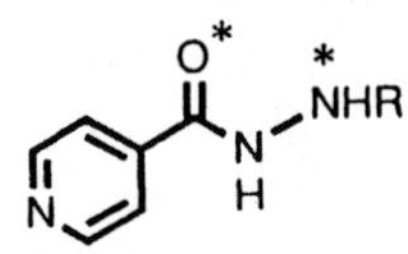

(2)

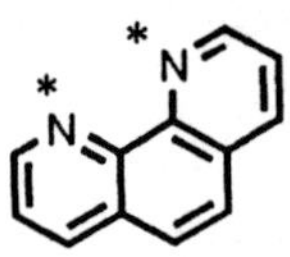

(3)

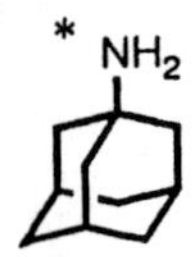

(4)

B. Thiosemicarbazones, thioureas, thiazolidinones

Thiosemicarbazones are active against both DNA (e.g. Herpes and Rous sarcoma) and RNA (e.g. Mengo, influenza) viruses in vivo, especially pox viruses, but they are also quite toxic (Perrin and Stünzi, 1982). Isatin β-thiosemicarbazones (**5**) are deactivitated if sulfur is replaced by oxygen or NH, and inhibit assembly and maturation of vaccinia viruses. They were used against smallpox in humans (Bauer et al., 1963), and can act as chelating agents (via S∩N∩O), forming 5m rings, and/or reducing agents (e.g. Cu^{2+} to Cu^{+}) (Perrin and Stünzi, 1981). Metal-binding has not been established in vivo. In some cases thiosemicarbazones remove metals such as iron (Spector et al., 1991) from viral enzymes. *Thioureas* (e.g. **6**, R = OH), particularly diphenyl-thioureas, inhibit picornaviruses in vitro, especially if complexed strongly to Zn^{2+} or Cu^{2+} (Galabov et al., 1980). *Thiazolidinones* (e.g. **7**) inhibit uncoating of echovirus-12 (Eggers, 1977), perhaps via the known coordination of the N and exocyclic S to Zn^{2+} (Fabretti and Peyronel, 1978).

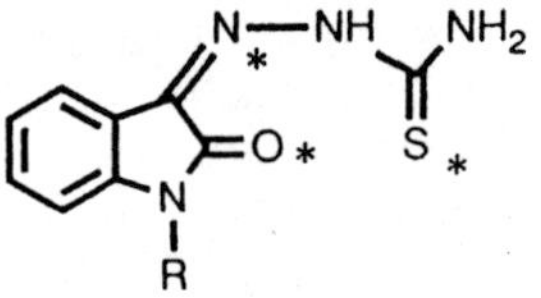

(5)

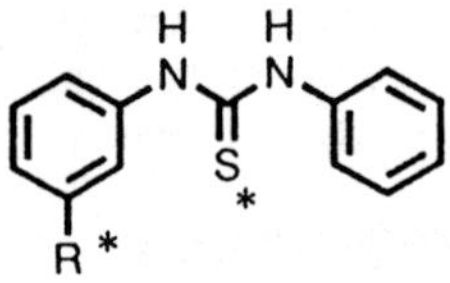

(6)

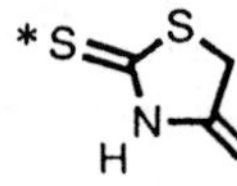

(7)

C. Phosphonocarboxylates, polyethers and β-diketones

Phosphonocarboxylates, especially phosphonoacetate (**8**) and phosphonoformate, are known metal-chelators (Stünzi and Perrin, 1979) that have general "pyrophosphate-like" activity against herpes viruses (Hay et al., 1977) by blocking DNA polymerases (known to contain active site Zn^{2+}) and DNA synthesis (Perrin and Stünzi, 1982; Hutchinson, 1985). They also inhibit influenza virus RNA polymerase and this is enhanced by Mn^{2+} or Mg^{2+}. They form 5-membered rings with metal ions, possibly chelating Zn^{2+} in DNA polymerase. *Polyether* macrocycles (e.g. **9**), resembling the ionophoric antibiotics, can sequester metal ions and are μM inhibitors of for example picorna virus RNA polymerase (Kauer, 1976/1977; Korant et al., 1984). *β-Diketones*, especially 4-substituted 3,5-heptanediones (e.g. **10**) are active in vitro against RNA and DNA viruses (e.g. rhino, influenza, picorna, herpes) and in vivo against herpes viruses (Diana et al., 1978a; 1978b). The antiviral action in cell culture of a lipophilic analogue, arildone (**10**, R = $-(CH_2)_{16}O-C_6H_3-2-Cl-4-OMe$), which inhibits herpes simplex virus 2 (by blocking DNA polymerase) and poliovirus (by preventing uncoating of virion), may be due to metal chelation (Perrin and Stünzi, 1982).

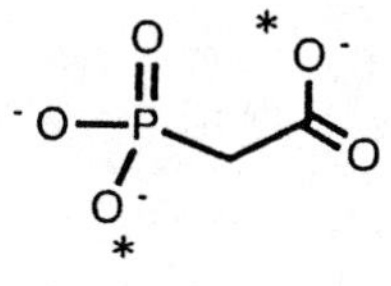

(8)

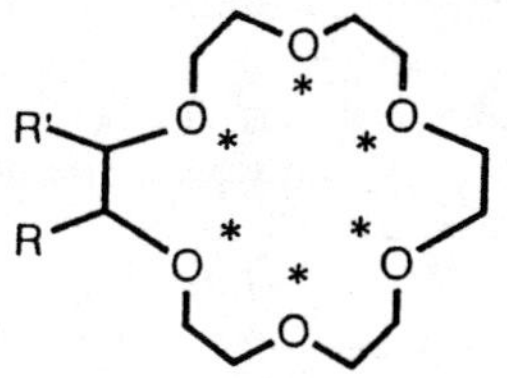

(9)

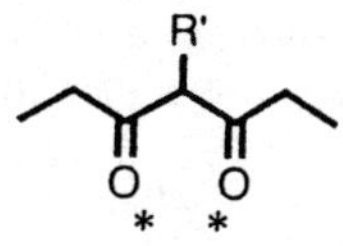

(10)

D. Flavones, quinones and polyphenols

Simple *flavones* (e.g. **11**) are antiviral (Bélädi et al., 1977; Veckenstedt et al., 1978; Hitano et al., 1988), particularly against enveloped viruses. They are known to chelate metals, especially Fe^{3+}, Al^{3+}, Zn^{2+} (Tomas et al., 1972), via the carbonyl oxygen and adjacent (deprotonated) hydroxyl substituent producing complexes of comparable or superior stability to His complexes of these metals. However flavones could be antiviral due to non-specific interactions with proteins important in viral reproduction. For example they inhibit in vitro HIV-1 proteinase (Brinkworth et al., 1992), a processing enzyme required for replication of HIV-1, without binding to metals (Woon et al., 1992). They inhibit in vitro rhino- and coxsackie viruses by blockading events prior to RNA synthesis (Ishitsuka et al., 1982). *Quinones* (e.g. adriamycin, daunomycin) inhibit DNA polymerases of retroviruses by interfering with nucleic acid base pairs rather than metal ions. Naphthaquinones (**12**) and anthraquinones (**13**) exhibit antiviral activity probably through binding to DNA by intercalation (Sill et al., 1974). However these and other antiviral *polyphenols* (Okuda et al., 1992) can also bind metal ions, usually via a phenolate.

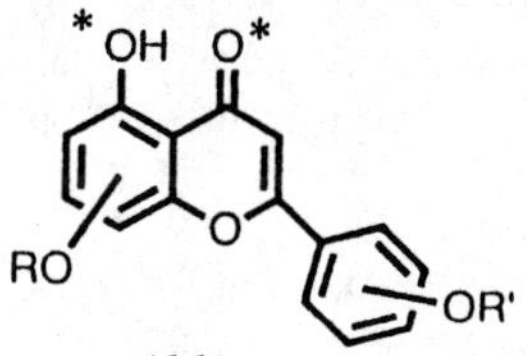

(11)

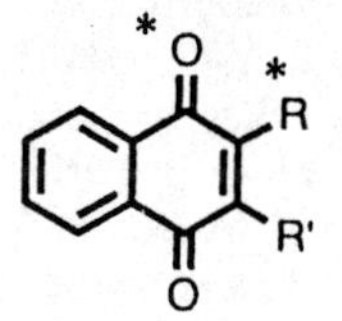

(12)

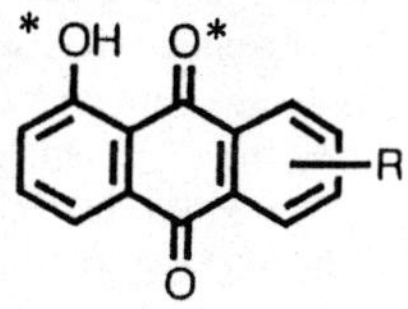

(13)

E. Nucleoside analogues

Many antiviral drugs contain purine or pyrimidine bases, which are known to bind metals. For example iodoxuridine (**14**) was the forerunner (Robins, 1986) of a host of pyrimidine nucleoside antiviral drugs with activity in humans against DNA viruses (e.g. herpes). Adenosine nucleoside analogues like Ara-A (**15**) are used to treat herpes and vaccinia viruses. Guanosine analogues like acyclovir (**16**) effectively inhibit herpes viruses by blocking viral but not cellular DNA polymerases (Robins, 1986). Their mechanism of action involves phosphorylation of the pendant alcohol by herpes thymidine kinase. 3'-Azidothymidine (**17**) is selectively active against certain RNA viruses called retroviruses, including those involved in AIDS and certain leukemias. Activity against HIV is a result of inhibition of reverse transcriptase and thence viral replication (Darby, 1992). Some

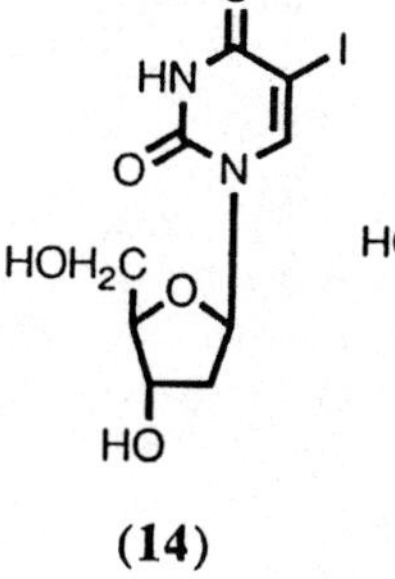

(14)

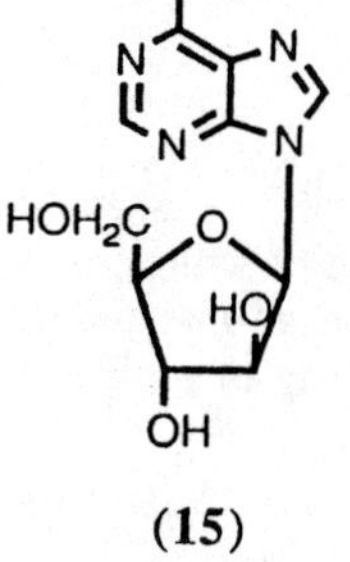

(15)

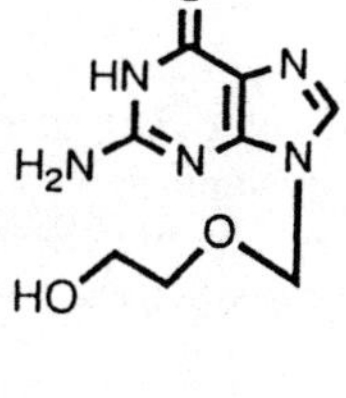

(16)

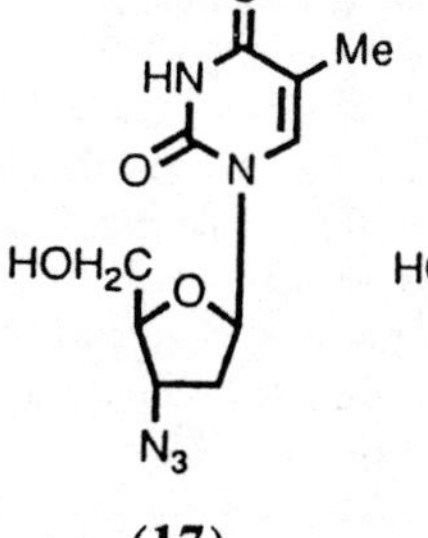

(17)

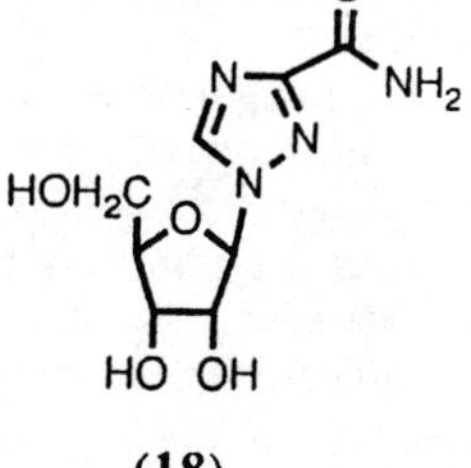

(18)

guanosine and adenosine analogues inhibit both DNA and RNA viruses; among these is the guanine nucleoside analogue ribavirin (**18**), which inhibits IMP dehydrogenase and RNA synthesis (Browne, 1979). Like other nucleosides and nucleotides (De Clercq, 1991), ribavirin has broad spectrum antiviral activity.
Such antiviral drugs can interact with metal ions via their heterocyclic ring. Metals rarely complex the endocyclic base nitrogens, generally preferring instead to bind adenosines (**19**) at N7 or N1, guanosines (**20**)/inosines via N7 or N1, cytosines (**21**) at N3 or O2, and thymidines (**22**)/uridines complex via N3 or O4 (Martin and Mariam, 1979). Ribavirin (**18**) does coordinate metal ions via a ring N4 and exocyclic oxygen (Bukovec et al., 1981).

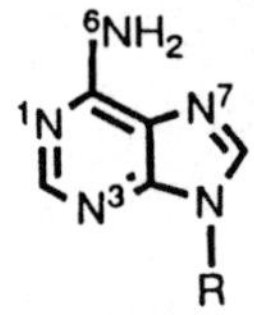

(19)

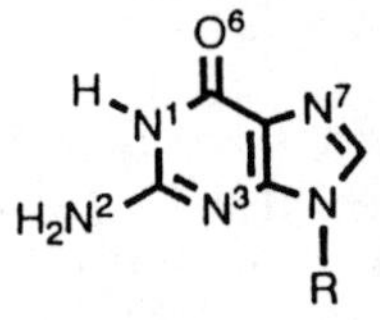

(20)

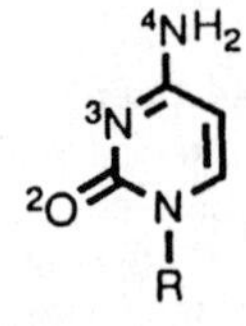

(21)

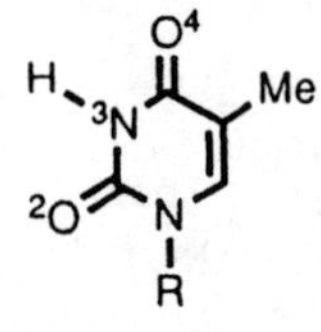

(22)

III. METALLOPROTEINS AS ANTIVIRAL DRUG TARGETS

A. Metalloproteins and viruses

One future for metal-antiviral interactions lies in the structural identification of more viral metalloproteins that can be intercepted during viral replication. The above investigations of metal-ligand interactions may have little relevance to future design of metal-targetted antiviral agents because the metal is usually tied up in protein environments. By virtue of their non-selectivity, all the above chelating agents may also be toxic due to indiscriminate complexing with metal ions wherever they encounter them. The challenge is to develop antiviral agents which *selectively* complex a particular (metallo)protein and avoid drug-induced toxicity arising from complexing to cellular (metallo)proteins.

Only about 10% of the ~1000 protein structures presently in the Brookhaven Database (Bernstein et al., 1977) are metalloproteins. The discovery and structural identification of new proteins is rapidly increasing and one can expect to find new metalloproteins that are critical for the replication and infectivity of viruses. Knowledge of the 3-dimensional structure of these metalloproteins can then be used to rationally design selective antiviral agents (e.g. Rossmann and McKinlay, 1992; Tomasselli et al., 1991). There are at present very few known examples of viral metalloproteins and none are well established yet as antiviral drug targets.

Zinc is a key metal in a number of DNA and RNA polymerase and integrase enzymes (Riordan and Vallee, 1976; Perrin, 1977), as well as nucleotidyl transferases, that are crucial to viral replication. All are inhibited by competing zinc chelators (Perrin, 1977). Papillomavirus proteins E6 and E7 have zinc-finger binding domains and have been proposed to mediate zinc binding through a Cys-X-X-Cys sequence (Barbosa et al., 1989). Also in Abelson leukemia virus-infected pre-B cells it has been demonstrated that expression of high levels of the cell surface metalloproteinase 6C3/BP-1Ag correlated with the ability of Abelson leukemia to invade the bloodstream (Weissman et al., 1990). Ca^{2+} is required for cell fusion mediated by the gp120-CD4 interaction between HIV-1 and T-cells (Dimitrov et al., 1993).

B. Drug design: metalloprotein-inhibitor optimization

The design of a specific and potent inhibitor of a viral metalloprotein is, in theory, identical to that for any enzyme and involves the development of a set of hydrogen, van der Waals and ionic bonding partners that complement the protein residues that line the active site. However, the presence of the metal atom does present a variety of problems to computer-aided design programs.

1. *Metals and computer simulations*

Computer-aided drug design has taken giant steps forward in recent years with more sophisticated algorithms, hardware and software advances. However, these advances relate to organic- and bio-, rather than bioinorganic-, chemistry where *ab initio* atomic parameter calculations are easier to apply. Changes in the ligand environment, oxidation state and stereochemistry of metals dramatically alter their properties. Creating parameters for all possible metal-ligand coordinations is impractical. Variations of the metal and its surrounding ligands affect charge distributions and bond strengths. Various force field parameter sets which have been produced do solve some problems of metal parameterization, but these are either system dependent, rely upon extrapolation from a small trial data set, or concentrate on a limited set of ligands, metals or coordination types. Relatively few publications have dealt with metals in proteins.

A parameter set specifically for iron-porphyrin systems (Lopez and Kollman, 1989) was produced from the original AMBER force field package. It correctly reproduced structural characteristics and partition coefficients for four known iron-heme complexes. These modifications worked reasonably well but geometry disparities remained that tended to be caused by differences in high-spin versus low-spin Fe. Models of carbonic anhydrase (Hoops et al., 1991) for both high (Zn-OH) and low (Zn-OH_2) pH dealt only with fixed coordination geometries rather than allowing variations about the metal. The SHAPES empirical force field (Allured et al., 1991) reproduces coordination geometries when applied to square-planar complexes of Rh but has not been applied to metalloproteins or other metals.

One molecular mechanics force field (Bernhardt and Comba, 1992) has been applied to a variety of metals (Cu(II), Ni(II), Co(II), Fe(III), Cr(III), Zn(II), and Rh(III)) bonded to hard amino- or soft thio-ligands. Only two variable parameters were required to simulate the appropriate crystal data but, although this was a significant improvement in the application of a single force field, some case-by-case modifications were needed. Also this work only addressed small metal complexes, not proteins where long range non-bonded effects of the metal can impact on enzyme conformation and dynamics.

A potential function has been developed for metal centers within proteins (Vedani, 1986) based on small zinc complexes reported in the Cambridge Crystallographic Database. This method overcame distortions from the varying coordination geometries but did not model metal-bound anions or a "ligand-sphere transition" mechanism (Dutler and Ambar, 1982) well. The work was extended (Vedani and Huhta, 1990) to Zn^{2+} and Co^{2+} where incorporation into the molecular mechanics program YETI allowed somewhat for directionality in hydrogen-bonding and some estimate of charge transfer between the metal and ligand.

2. *Designing inhibitors for metalloproteins*

The principles to consider when designing inhibitors for structurally known proteins (including metalloenzymes) are those of maximizing (i) surface contact between inhibitor and protein, and (ii) hydrogen bonding and electrostatic contacts in the protein-inhibitor complex. To illustrate these two maxims (with reference to metalloenzymes) we have chosen two inhibitors of thermolysin which have been co-crystallized with the enzyme. The enzyme and inhibitor residues are shown in bold, the zinc atom by a star, the hydrogen bonds by dashed lines and the enzyme's solvent accessible surface by dots.

The active site of thermolysin is an open ended groove in which the essential zinc atom sits close to the middle and floor. Zn^{2+} is bound by side chains of Glu-166, His-146 and His-142, leaving a site vacant for water or a substrate/inhibitor. In Figure 1, both inhibitors completely fill a depression or "pocket" (to the left of the zinc) in the enzyme surface with the side chain of Leu, thus making the greatest surface contact with it. If the side chain was smaller (e.g. methyl of Ala), the pocket would not be filled to the same extent, resulting in a weaker interaction. A larger substituent (e.g. Trp) would be too big and the inhibitor would not fit into the active site cleft. Similarly, the extension of one of the inhibitors with a carboxybenzyl ester (to the right of the zinc atom) fills more of the active site cleft (seen more clearly in Figure 2) and increases binding selectivity.

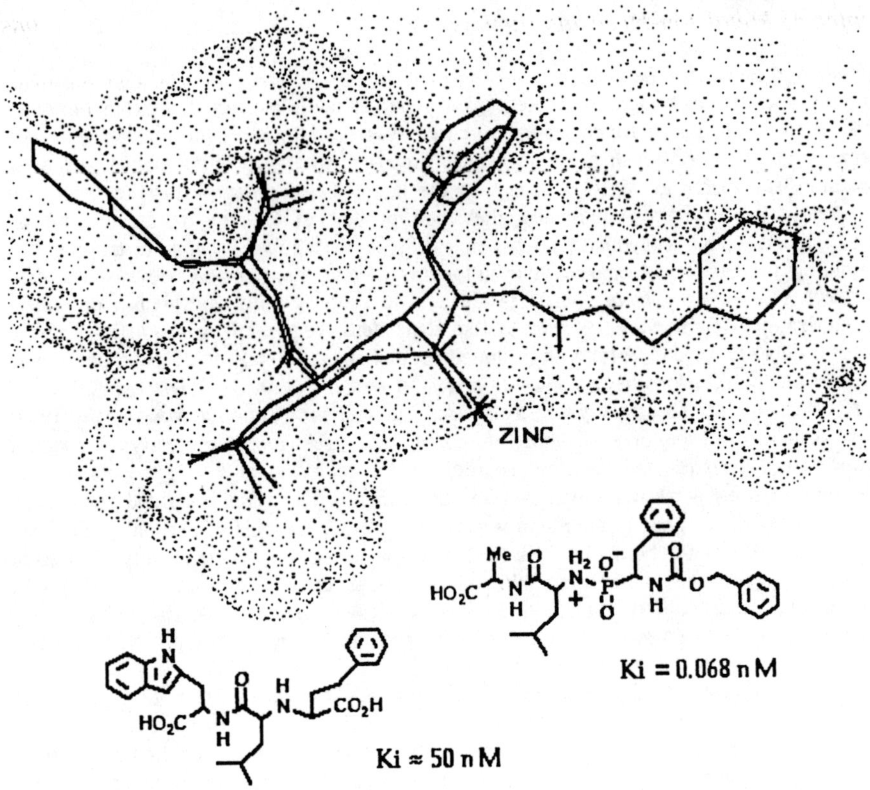

Figure 1.

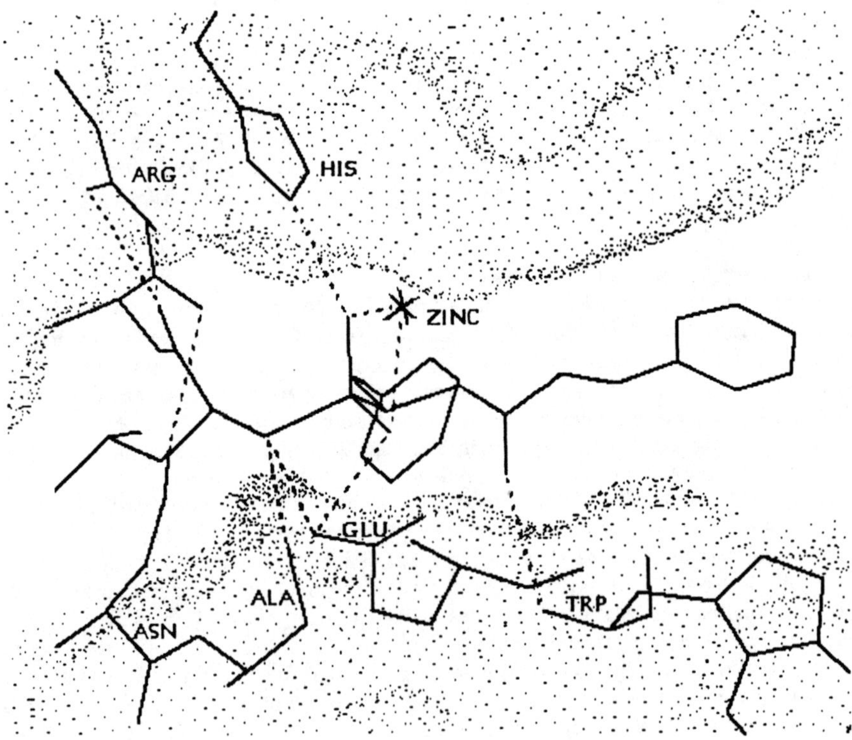

Figure 2.

The hydrogen bonding between the better inhibitor and the enzyme is shown on Figure 2. The dipolar or charged residues which line the cleft (Arg-203, His-231, Asn-112, Glu-143, Trp-115, and Ala-113 backbone) all interact with complementary groups on the inhibitor. The water molecule which binds in the vacant zinc site on the native enzyme is replaced by phosphinate (or carboxylate in the other inhibitor).

The binding of the active site metal is achieved by a charged group which, with the charged amino group, combine to make a mimic of the transition state formed during peptide cleavage. The extra hydrogen bond of the inhibitor carboxybenzyl ester matches well with the backbone NH of the active cleft Trp and it is this complementarity in matching hydrogen-bonding donors with acceptors and pairing opposite electrostatic charges/dipoles that determines the specificity of an inhibitor for a particular enzyme. Even greater specificity for thermolysin might be attained by extending the inhibitor in the cleft or increasing the hydrogen-bonding partners (e.g. Tyr-193, Leu-202, Tyr-211 one way and Asp-150, Asn-165, and Tyr-110 the other). Of course other factors such as bioavailibility and drug delivery must also be considered in drug design.

To design an enzyme inhibitor, ideally one needs to create a 3-dimensional representation of the (metallo)enzyme's active site either from structure-activity, NMR or X-ray data. This allows determination of size, shape and electrostatic fields associated with the site, mobility and complimentarity of the putative enzyme-inhibitor complex. For the latter, the choice of metal forcefield parameters can have dramatic effects upon the derived inhibitor structure. The use of computers to aid design of metalloenzyme inhibitors is possible with the present state of metal parameterization and force field algorithms, but *general* application of these parameters and fields is not yet feasible. Such problems must be overcome if molecular modelling is to be usefully applied to metalloproteins.

REFERENCES

Albert, A. (1965). *Selective Toxicity*, Wiley, New York, p. 222.

Allured, V.S., Kelly, C.M., and Landis, C.R. (1991). *J. Amer. Chem. Soc.*, *113*: 1.

Barbosa, M.S., Lowry, D.R., and Schiller, J.T. (1989). *J. Virol.*, *63*: 1404.

Bauer, D.J., St. Vincent, L., Kempe, L.H., and Downie, A.M. (1963). *Lancet*, *2*: 494.

Béládi, I., Pusztai, R., Mucsi, I., Bekay, M., and Gabor, M. (1977). *Ann. N. Y. Acad. Sci.*, *284*: 358.

Bernhardt, P.V., and Comba, P. (1992). *Inorg. Chem.* *31*: 2638.

Bernstein, F.C., Koetzle, T.F., Williams, G.J.B., Meyer , E.F., Jr., Brice, M.D., Rodgers, J.R., Kennard, O., Shimanouchi, T., and Tasmuni, M. (1977). *J. Mol. Biol.*, *112*: 535.

Brinkworth, R.I., Stoermer, M.J., and Fairlie, D.P. (1992). *Biochim. Biophys. Res. Comm.* *188*: 631.

Browne, M.J. (1979). *Antimicrob. Agents Chemother.*, *15*: 747.

Bukovec, P., Golic, L., Orel, B., and Kobe, J. (1981). *J. Carbohyd. Nucleos. Nucleot.*, *8*: 1.

Darby, G. (1992). *Biochem. Soc. Trans.*, *20*: 505.

De Clercq, E. (1991). *Nucleos. Nucleot.*, *10*: 167.

Diana, G.D., Carabateas, U.J., Johnson, R.E., Williams, G.L., Pancic, F., and Collins, J.C. (1978a). *J. Med. Chem.*, *21*: 889.

Diana, G.D., Carabateas, U.J., Salvador, U.J., Williams, G.L., Zalay, E.S., Pancic, F., Steinberg, B.A., and Collins, J.C. (1978b). *J. Med. Chem.*, *21*: 689.

Dimitrov, D.S, Broder, C.C., Berger, E.A, and Blumenthal, R. (1993). *J. Virol.*, *67*: 1647.

Dutler, H., and Ambar, A., (1982). *The Coordination Chemistry of Metalloenzymes* (I. Bertini, S.R. Drago and C. Luchinat, eds.), NATO Adv. Study Inst. Series, Reidel, Dordrecht, p. 135.

Eggers, H.J. (1977). *Virology*, *78*: 241.

Fabretti, A.C., and Peyronel, G. (1978). *Spectrochim. Acta*, *34A*: 667.

Fairlie, D.P., Whitehouse, M.W., and Taylor, R.M. (1992). *Agents Actions*, *36*: 152.

Galabov, A.S., Galabov, B.S., and Neykova, N.A. (1980). *J. Med. Chem.*, *23*: 1048.

Hay, J., Brown, S.M., Jamieson, A.T., Rixon, F.J., Moss, H., Dargan, D.A., and Subak-Sharpe, J.H. (1977). *J. Antimicrob. Chemother. Suppl. A*, *3*: 63.

Hitano, T., Yasuhara, T., Miyamoto, K., and Okuda, T. (1988). *Chem. Pharm. Bull.*, *36*: 2286.

Hoops, S.C., Anderson, K.W., and Merz, K.M., Jr. (1991). *J. Amer. Chem. Soc.*, *113*: 8262.

Hutchinson, D.W. (1985). *Antiviral Res.*, *5*: 193.

Ishitsuka, H., Ohsawa, C., Ohiwa, T., Umeda, I., and Suhara, Y. (1982). *Antimicrob. Agents Chemother.*, *22*: 611.

Kauer, J.C. (1976). *U. S. Patent* 3,997,565; (1977). *U. S. Patent* 4,024,158.

Korant, B.D., Lonberg-Holm, K., and La Colla, P. (1984). *Targets for the Design of Antivirals* (E. De Clercq and R.T. Walker, eds.), Plenum, New York, p. 61.

Levinson, W., Rhode, W., Mikelens, P., Antony, A., and Ramakushna, T. (1977). *Ann. N.Y. Acad. Sci.*, *284*: 525.

Lopez, M.A., and Kollman, P.A. (1989). *J. Amer. Chem. Soc.*, *111*: 6212.

Martin, R.B., and Mariam, Y.H. (1979). *Metal Ions in Biological Systems*, Vol. 8 (H. Sigel, ed.), Marcel Dekker, New York, p. 57.

Montelius, I., Liljas, L., and Unge, T. (1990). *J. Mol. Biol.*, *212*: 331.

Okuda, T., Yoshida, T., and Hatano, T. (1992). *ACS Symp. Ser.*, *507*: 160.

Perrin, D.D. (1977). *Chemoprophylaxis of Virus Infection in the Respiratory Tract*, Vol. 2 (J.S. Oxford, ed.), CRC, West Palm Beach, FL, p. 201.

Perrin, D.D., and Stünzi, H. (1981). *Pharm. Ther.*, *12*: 255.

Perrin, D.D., and Stünzi, H. (1982). *Metal Ions in Biological Systems*, Vol. 14 (H. Sigel, ed.), Marcel Dekker, New York, p. 207.

Rafelson, M.E., Schneir, M., and Wilson, V.W. (1963). *Arch. Biochem. Biophys.*, *103*: 424.

Riordan, J.F., and Vallee, B.L. (1976). *Trace Elements in Human Health and Disease*, Vol. 1 (A.S. Prasad and D. Oberleas, eds.), Academic Press, New York, p. 227.

Robins, R.K. (1986). *Chem. Eng. News, January issue*: 28.

Rohde, W., Mikelens, P., Jackson, J., Blackman, J., Whitcher, J., and Levinson, W. (1976). *Antimicrob. Agents Chemother.*, *10*: 234.

Rossmann, M.G., and McKinlay, M.A. (1992). *Infect. Agents Dis.*, *1*: 310.

Sill, A.D., Andrews, E.R., Sweet, F.W., Hoffman, J.W., Tiernan, P.L., Grisar, J.M., Fleming, R.W., and Mayer, G.D. (1974). *J. Med. Chem.*, *17*: 965.

Sillén, L.G., and Martell, A.E. (1964, 1971). *Stability Constants of Metal-Ion Complexes*, Special publications No. 17 and No. 25, The Chemical Society, London.

Spector, T., Harrington, J.A., and Porter, D.J.T. (1991). *Biochem. Pharmacol.*, *42*: 91.

Stünzi, H., and Perrin, D.D. (1979). *J. Inorg. Biochem.*, *10*: 309.

Tomas, F., Carpena, O., and Mataix, J. (1972). *An. Quim.*, *68*: 115.

Tomasselli, A.G., Howe, W.J., Sawyer, T.K., Wlodawer, A., and Heinrickson, R.L. (1991). *Chimicaoggi*, *9*: 6.

Veckenstedt, A., Béládi, I., and Mucsi, I. (1978). *Arch. Virol.*, *57*: 255.

Vedani, A., Dobler, M., and Dunitz, J.D. (1986). *J. Comput. Chem.*, *7*: 701.

Vedani, A., and Huhta, D.W. (1990). *J. Amer.Chem. Soc.*, *112*: 4759.

Weissman, I.L., Shih, C.C., and Sherwood, P. (1990). *Origins of Human Cancer* (J. Brugge, ed.), 2nd Meeting, Cold Spring Harbor, New York, p. 463.

Wiltink, E.H.H (1992). *Pharmaceut. Weekbl.-Sci. Ed.*, *14*: 268.

Woon, T.C., Brinkworth, R.I., and Fairlie, D.P. (1992). *Int. J. Biochem.*, *24*: 911.

PLATINUM ANTI-TUMOR COMPOUNDS: THEIR CHEMISTRY IN RELATION TO THE MECHANISM OF ACTION

by

Jan Reedijk

Leiden Institute of Chemistry,
Gorlaeus Laboratories,
Leiden University,
P.O. Box 9502
2300 RA Leiden
The Netherlands

I. HISTORIC OVERVIEW

The first antitumor activity for inorganic compounds was reported more than 25 years ago (Kirschner et al., 1966). Only a few years later, the classical compound *cis*-$PtCl_2(NH_3)_2$ (synthesized by Peyrone as early as 1844) was found to be able to kill tumors in animals and first clinical trials on human tumors were reported to be successful (Rosenberg, 1971).

As a physicist, Rosenberg was impressed by the similarity between the mitotic spindle figures and the field lines of a dipole. This led him to investigate - as early as 1962(!) - how electric fields may interfere with cell division of cultured bacterial cells. The field generated between platinum electrodes apparently seemed to stop cell division without hampering cell growth, in this way inducing filamentous growth (Rosenberg et al., 1965).

Subsequent experiments by the same group made clear that this phenomenon was caused by traces of compounds like *cis*-$PtCl_2(NH_3)_2$ and *cis*-$PtCl_4(NH_3)_2$ formed during electrolysis by interaction of the electrolyte used (NH_4Cl) and the supposedly "inert" Pt electrodes (Rosenberg, 1971). Quite soon after this important discovery, several other Pt(II) and Pt(IV) compounds were found to show similar effects on bacterial growth (Rosenberg et al., 1969). Surprisingly, only the *cis*-isomers - and not the *trans*-isomers - appeared to be effective in these experiments. In a next set of experiments on a worldwide scale, the antitumor activity of these and other platinum compounds have been studied. In particular, *cis*-$PtCl_2(NH_3)_2$ turned out to be very active against a variety of tumors induced in animals, such as Sarcoma 180 and Leukemia L1210 in mice (Rosenberg et al., 1967a; 1967b; 1969) (for an early review, see Roberts and Thompson (1979)). In many cases a total regression of the tumors was observed. Clinical trials of cis-Pt started already in 1971 (Wagstaff et al., 1989).

Inorganic chemistry-oriented papers dealing with the mechanism of the reactions in this exciting research area were published as soon as 1973 (Hadjiliadis et al., 1973). Since then, it has become well known that *cis*-$PtCl_2(NH_3)_2$ (often abbreviated as cis-Pt, cisplatin, cisDDP, CDDP) reacts with cellular components such as DNA and proteins (see reviews by Lempers and Reedijk (1991), Reedijk (1987; 1992), Sundquist and Lippard (1990)).

Details of these interactions at the molecular level remain scarce, however, despite many years of research in this area. After the discovery of DNA binding almost two decades ago (Roberts and Thompson, 1979), most attention has been focussed on studies of binding between cis-Pt and nucleic acids (Reedijk, 1992). It is now generally accepted that the antitumor properties of cis-Pt are based upon a selective interaction with DNA inside the cells.

The chemistry of Pt-antitumor compounds is a timely research field indeed and even textbooks nowadays present a brief description of the history and mechanism of such compounds (Fraústo da Silva and Williams, 1991).

Due to its successful curing results, cisplatin has developed to a very important antitumor drug; the major targets of cisplatin and derivatives are urogenital tumors, tumors of the head and neck, whereas more recently also treatment of small-cell lung cancer, bladder and squamous cell carcinoma have been reported (Wagstaff et al., 1989; Barnard, 1989; see *Medicine*, Part Five. Chapter 2-C). Nowadays, cis-Pt is routinely used and has been particulary successful in the treatment of testicular and ovarian cancers (Barnard, 1989). Its severe renal toxicity has been largely circumvented by the use of diuretics and pre-hydration. In 1978 cis-Pt was officially approved as a drug in the USA, and within a decade this first metal-containing antitumor compound had become the biggest selling antitumor drug. Moreover, the use of cis-Pt led to major improvements in response rates for head, neck and lung tumors, even though responses to chemotherapy are often of limited duration for these tumors (Loehrer and Einhorn, 1984). Cis-Pt is also being used in combination therapy with antitumor drugs such as vinblastine, bleomycin and adriamycin, or in combination with radiotherapy (Bajetta et al., 1982; Prestayko et al., 1979; Douple, 1985).

Progress in the curing of testicular and ovarian cancers by use of cis-Pt has been quite spectacular (Barnard, 1989), especially for early recognized testicular cancer in which results are approaching 100%.

Despite this success, the therapeutic use of cis-Pt still suffers from serious drawbacks and a continuous search for new Pt compounds is required and encouraged. In particular, important side effects of cis-Pt need to be overcome in new generations of Pt compounds: (i) its poor solubility in saline; (ii) its severe toxicity effects; (iii) the development of resistance (see e.g. Barnard, 1989; Van der Vijgh, 1991). Also, the applicability of cis-Pt is still limited to a narrow range of tumors. During the last decade, this search for new compounds has been stimulated significantly by the results of mechanistic studies, in particular DNA-binding studies.

A descriptive summary of the history of *cis*-$PtCl_2(NH_3)_2$ would require too much space; only the following major and generally accepted highlights will be discussed here (details have been reviewed; see e.g. Reedijk (1992) and references therein):

1966: Growth of bacteria blocked in an electric field is caused by dissolved Pt compounds from electrodes;
1969: Such Pt compounds, like *cis*-$PtCl_2(NH_3)_2$ and *cis*-$PtCl_4(NH_3)_2$ appear active against tumors;
1971: First clinical trials for *cis*-$PtCl_2(NH_3)_2$ successful;
1974: First indications that DNA is the major target for the Pt compounds;
1976: Improvement of cis-Pt by use of combination therapy, with compounds like vinblastine, adriamycin and bleomycin;
1978: First indications that a cisPt-GG adduct plays a role;
1979: Routine use of cis-Pt in curing of certain tumors; especially successful treatment of urogenitaric tumors;
1981: Clinical trials with analogs and tests with rescue agents, like NaDDTC or $Na_2S_2O_3$;
1983: Cis-Pt can form bis-N7 chelate at -GG- sequences in ss and ds oligonucleotides similar to those in DNA;
1985: Distortion of DNA after cis-Pt chelation appears to be small; a kink in the DNA appears to be the major structural change in ds DNA;
1987: X-ray structures prove chelation at GG sites in ss DNA;
1989: Discovery of a Damage Recognition Protein binding at Pt-GG, later known as SSRP1 protein;

1991: Introduction of "non-classical" Pt drugs and Pt drugs that can be administered orally;
1992: Clinical effectiveness of cis-Pt can be explained by inefficient repair of the major drug-DNA adduct.

II. PLATINUM COMPOUNDS

A. Synthesis and structure-activity relationships

Already at an early stage it became clear that only the *cis* isomer of $PtCl_2(NH_3)_2$ showed antitumor activity. Both compounds are easily synthesized according to the old recepy (Peyrone, 1844), although more recently advanced recepies have been reported (Dhara, 1970; Morita and Bailar, 1983). Not unexpectedly, the initial success of cis-Pt had resulted in the emergence of many derivative compounds, among which the so-called CBDCA derivatives are clinically used on a still increasing scale (Barnard et al., 1986; Barnard, 1989; Kelland et al., 1992a).

A next generation of active compounds including the Pt(IV) compounds that show promising activity upon oral administration (Kelland et al., 1992b) again have variation in the amine part and contain two coordinating anions among halides, carboxylates and phosphates. Kinetic studies over the years have shown that comparable Pt compounds always react a factor of 10^3 slower than the corresponding Pd compounds. This explains why comparable Pd(II) compounds are not active (they decompose and isomerize on the time scale of a few hours or less).

Apart from the now classical cis-Pt, a variety of other platinum compounds has been synthesized and subsequently tested for antitumor activity (Barnard et al., 1986; Douple, 1985; Howell, 1991), with the aim of obtaining higher solubility, better antitumor activity and lower toxicity. The synthesis of most of these new platinum compounds is largely governed by two classical, quite empirical, rules, concerning the trans-labilizing power of the coordinated ligands and the difference in platinum-ligand bond strength.

A variety of studies on these derivatives during the last two decades has made clear that there are a number of structural requirements which have to be fulfilled for a complex of this type to show antitumor activity. These requirements are accepted to be:

(i) The compounds should have a *cis* geometry, with general formulae [*cis*-$PtX_2(Am)_2$] for Pt(II) and [*cis*-$PtY_2X_2(Am)_2$] for Pt(IV) compounds (Cleare and Hoeschele, 1973; reviews: Reedijk, 1987; Reedijk, 1992). It is evident that this geometric restriction is automatically valid for didentate amines such as ethylenediamine and derivatives.

(ii) The group of active Pt(IV) compounds is octahedrally coordinated and possess axially bound chloride ligands (or - to improve the solubility - hydroxo ligands), i.e. two Y ligands in the trans orientation. These compounds are kinetically far more inert than the corresponding Pt(II) compounds that lack these axial ligands. Most likely the Pt(IV) complexes are reduced in vivo to the corresponding Pt(II) complexes, which are in fact the active species (Chaney et al., 1991; Van der Veer et al., 1986; Roat and Reedijk, 1993). They can therefore be considered as a type of "pro-drug", which first need in vivo activation (ligand substitution and reduction) to the square planar Pt(II) compounds before exhibiting anti-neoplastic activity. This hypothesis is supported by the observation that platinum(IV) compounds hardly - if at all - react with DNA under ambient conditions (Chaney et al., 1991; Roat and Reedijk, 1993), and that appreciable amounts of Pt(II) derivatives can be detected in the urine of Pt(IV)-treated patients (Chaney et al., 1991).

(iii) The ligand X (i.e. the leaving group) must be an anion with an intermediate binding strength to Pt(II). Complexes with labile anions such as NO_3^- or ClO_4^-, are often highly toxic (Cleare, 1974; Barnard, 1989). Complexes with strongly bound anions are quite inert, although in some cases the dissociation of these anions appears to be activated in vivo (Cleare, 1974). Examples of effective anions in antitumor compounds are Cl^-, SO_4^{2-}, citrate(3-), glycolate, oxalate(2-) and other (chelating) carboxylic acid residues.

(iv) At least one of the two *cis* amine ligands should have an N-H group (Cleare and Hoeschele, 1973; van Kralingen et al., 1980). Compounds of the category described above and which lack this property have been found inactive. This N-H group is likely required for a hydrogen-bond donor function (vide infra); however, steric effects cannot be ruled out a priori.

It cannot be excluded, however, that other compounds lacking the N-H group but acting through a different mechanism, will show antitumor activity; (vide infra).

B. Promising new Pt drugs

As indicated above, a growing group of very promising "second and third generation" platinum antitumor drugs that meet all criteria for structural requirements and that answer the above-mentioned characteristics is now known, and several of these drugs are ready for use in many clinics all over the world (Kelland et al., 1992a). A few important examples of these compounds, all of which fulfil the above requirements are shown in Figure 1. These compounds cannot be discussed all in detail, but as is seen from their structures, they all have common characteristics that are likely to be related to the mechanism of action. Many other new compounds, e.g. with phosphates, phosphonates or sulfonates are under study (Keppler, 1990; Valsecchi et al., 1991).

C. Unusual other Pt compounds with cytostatic activity

It should again be mentioned that all the above- and below-mentioned studies with classical Pt compounds do not exclude the possibility that other structurally quite different Pt compounds may be found that are also active, and perhaps even more active.

In fact, more recently, some very active Pt (non-classical) compounds have been found, including dinuclear ones, that do show promising in vitro activities (Farrell and Qu, 1989; Farrell et al., 1988; 1990a; 1990b; Hollis et al., 1989; 1991; Alul et al., 1992). Some representative examples are given in Figure 2. These compounds will be excluded from further discussions in this overview, because their mechanisms of action are hardly known and are likely to differ from that of the classical compound.

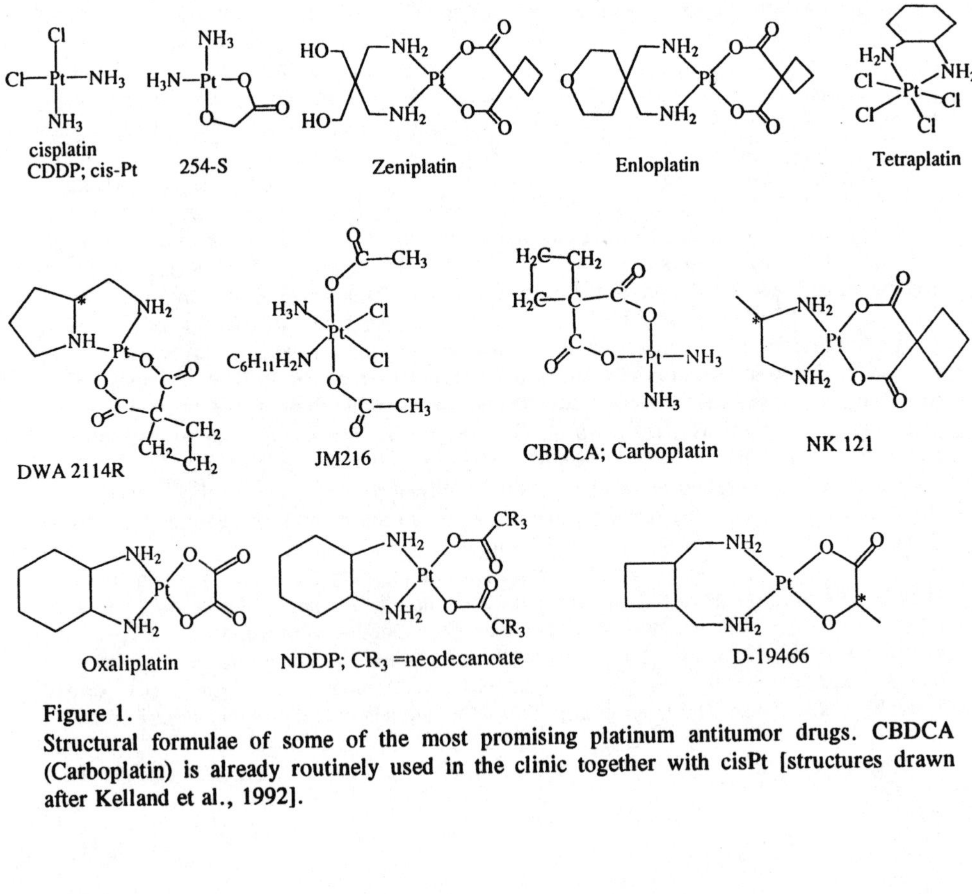

Figure 1.
Structural formulae of some of the most promising platinum antitumor drugs. CBDCA (Carboplatin) is already routinely used in the clinic together with cisPt [structures drawn after Kelland et al., 1992].

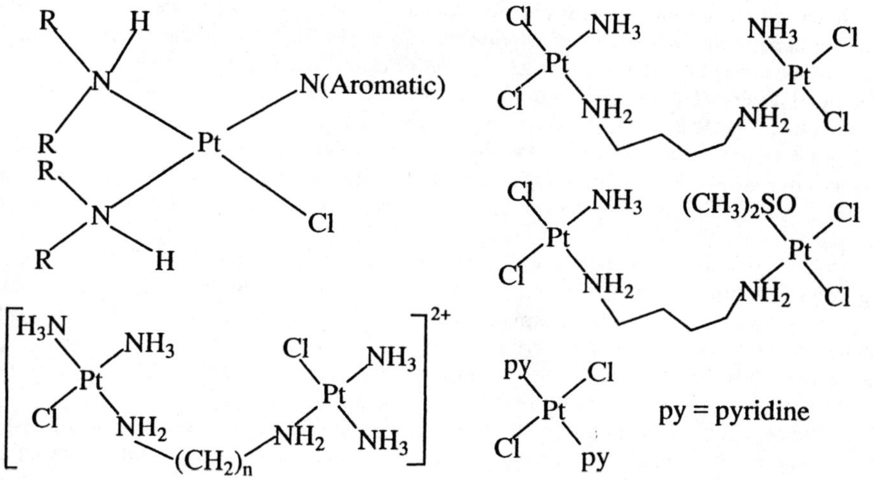

Figure 2.
Structures of a selection of unconventional Pt drugs with antitumor activity.

III. REACTIONS OF PLATINUM COMPOUNDS

A. Hydrolysis reactions of cis-Pt in vivo and in vitro

Upon dissolution of cis-Pt in water, the labile chloride ions are slowly replaced by water molecules (hydrolysis) in a stepwise manner, as shown in eq. 1-7 (Reishus and Martin, 1961; Lee and Martin, 1976; Lim and Martin, 1976; Fagianni et al., 1978; Appleton et al., 1989a; 1989b; Miller and House, 1989; 1990). The process of hydrolysis reactions takes a few hours at 37°C and millimolar concentrations.

$$[cis\text{-}Pt(NH_3)_2Cl_2] + H_2O \rightarrow [cis\text{-}Pt(NH_3)_2Cl(H_2O)]^+ + Cl^- \quad (1)$$
$$[cis\text{-}Pt(NH_3)_2Cl(H_2O)]^+ + H_2O \rightarrow [cis\text{-}Pt(NH_3)_2(H_2O)_2]^{2+} + Cl^- \quad (2)$$
$$[cis\text{-}Pt(NH_3)_2Cl_2] + OH^- \rightarrow [cis\text{-}Pt(NH_3)_2Cl(OH)] + Cl^- \quad (3)$$
$$[cis\text{-}Pt(NH_3)_2Cl(OH)] + OH^- \rightarrow [cis\text{-}Pt(NH_3)_2(OH)_2] + Cl^- \quad (4)$$
$$[cis\text{-}Pt(NH_3)_2Cl(H_2O)]^+ \rightarrow [cis\text{-}Pt(NH_3)_2Cl(OH)] + H^+ \quad (pK_a = 6.85) \quad (5)$$
$$[cis\text{-}Pt(NH_3)_2(H_2O)_2]^{2+} \rightarrow [cis\text{-}Pt(NH_3)_2(OH)(H_2O)]^+ + H^+ \quad (pK_a = 5.93) \quad (6)$$
$$[cis\text{-}Pt(NH_3)_2(OH)(H_2O)]^+ \rightarrow [cis\text{-}Pt(NH_3)_2(OH)_2] + H^+ \quad (pK_a = 7.87) \quad (7)$$

The relative amounts of all these Pt species vary as a function of pH and (externally added) chloride concentration. As a matter of fact, only platinum species with a coordinated water molecule are regarded to be reactive, since - in contrast to coordinated chloride or hydroxide - this ligand can be easily substituted by other donor molecules. Hydroxo species are formed as indicated in eq. 3-7, with *cis*-$Pt(NH_3)_2(OH)_2$ as the stable end product in alkaline solution. However, this species very easily dimerizes and trimerizes at higher concentrations, producing ions like $[cis\text{-}Pt(NH_3)_2]_2(\mu\text{-}OH)_2$ and $[cis\text{-}Pt(NH_3)_2]_3(\mu\text{-}OH)_3$, as has been proven with e.g. ^{195}Pt NMR spectroscopy (Faggianni et al., 1978). Recently, accurate pK_a values have been presented (Appleton et al., 1989b) for the (de)hydronation equilibria; these pK_a values have been added above to equations (5-7).

The kinetics of hydrolysis of cis-Pt has been investigated in great detail by Miller and House (1989; 1990) for the several hydrolysis reactions. They concluded that acid hydrolysis of cis-Pt in vivo is unlikely to proceed beyond $[cis\text{-}Pt(NH_3)_2Cl(H_2O)]^+$.

When entered in blood plasma, the chloride ion concentration is sufficiently large (about 100 mM) to prevent cis-Pt hydrolysis and the neutral platinum species most likely crosses the cell membrane. Inside the cell, the chloride ion concentration is much lower (about 4 mM), and now the hydrolysis reactions will occur. Since water is a far better leaving group than chloride or hydroxide (Basolo et al., 1960; Gray and Olcott, 1962), the aqua species are most likely the reactive form of cis-Pt in vivo, and therefore hydrolysis is the rate-limiting step in the reaction of cis-Pt with biomolecules like proteins and DNA (Johnson et al., 1980; Know et al., 1986; Bancroft et al., 1990).

As a matter of fact, recent ^{195}Pt NMR results of Bancroft et al. (1990) indicate that, in agreement with the kinetic hydrolysis studies of Miller and House (1990), most likely [*cis*-$Pt(NH_3)_2Cl(H_2O)]^+$ is the predominant species that reacts with biomolecules (at least with DNA).

Other Pt amine compounds which are antitumor active have different kinetics of the hydrolysis reactions, and usually react much slower. The hydrolytic species might even be slightly (or perhaps completely) different for other Pt compounds; in fact very few studies on derivatives have been studied so far. The second-generation drug CBDCA (see Fig. 1 for the structure) is known to hydrolyze (in a 1 mM solution) with a half-life of a few days at 37 °C (compared to only one hour for cis-Pt, vide supra).

The requirement for both for Pt(II) and Pt(IV) compounds to have amine ligands with a N-H group on the amine (Van Kralingen et al., 1980) has been a lead toward the design and synthesis of new derivative Pt compounds with biological activity.

B. CisPlatin under in vivo conditions

Concerning the in vivo mechanism of action, only a brief summary will be given here of the reaction pathway leading to the ultimate cell-killing event: binding to the DNA in the cellular nucleus.

After injection or infusion in the blood, the cis-Pt drug is transported through the body, almost unchanged (clearly due to the relative high concentration of chloride ions in blood, as described above for the equilibrium reactions), entering both normal and tumor cells. Either inside the cell, or during the relatively slow passage of the cell wall ($t_{1/2}$ of a few hrs at 37 °C, vide supra), hydrolysis occurs followed by binding to nucleic acids and - likely - many other possible targets. The transport of cis-Pt (and other compounds) through the body is schematically depicted in Figure 3. Many molecular aspects of this transport are still far from being understood. The section will deal in more detail with some of these reactions in the body.

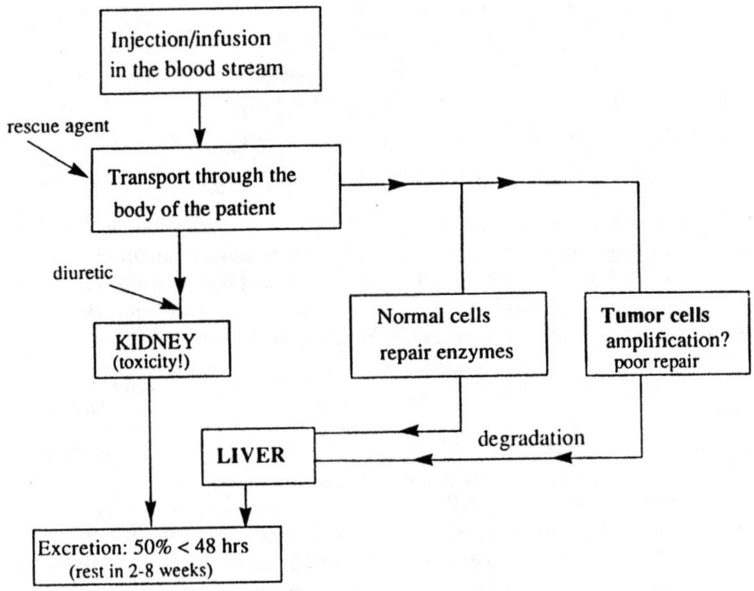

Figure 3.
Schematic representation of the pathway of cis-Pt in the body.

IV. TARGETS IN THE CELL

A. General aspects

Although the major target for cis-Pt in the cells appears to be DNA (see reviews listed by Reedijk (1992) and others cited there), many low-molar-mass competitors for cisplatin-DNA reactions in the cell are known. To be mentioned are: (i) simple ions and molecules like Cl^-, $(HPO_4)^{2-}$, OH^-, and H_2O (concentration 55 M!), (ii) amino acids, peptides and polyphosphates like His, Met, Cys; glutathione, metallothionein; ATP.

From a general coordination point of view, metal compounds (including those of Pt) have several possible binding sites in the cell. Among the possible cell targets are the nucleobases of DNA, but others are also possible such as the S-donor ligand glutathione. The sites at proteins are believed to be the most likely origin of the severe toxic side effects of cis-Pt and its derivatives (Corden, 1987; Lempers and Reedijk, 1990a). It has been known for some time that significant amounts of the cis-Pt administered are lost as a result of binding to proteins; some of these bonds can be "rescued" by certain agents, like thiourea. Possibly, cis-Pt derivatives with different kinetics in DNA (and protein) binding may result in better protocols (Bodenner et al., 1986a; Borch and Markman, 1989).

To explore this type of protein-binding reactions, a variety of Pt amine compounds (including the inactive trans isomer of cis-Pt and also the reference compound [PtCl(dien)]Cl) have been reacted with synthetic peptides and with proteins. Competition between proteins and nucleic acids has been studied with in vitro reactions for Pt compounds and nucleopeptides. It has been found that the degree of hydrolysis of cis-Pt determines the rate of binding to DNA (at guanine-N7) and to S-donor atoms (most rapidly at thioethers) in proteins and peptides (Lempers and Reedijk, 1990b; Djuran et al., 1991).

Using advanced NMR techniques in combination with spectroscopy and X-ray diffraction studies, the structures and conformations of the Pt-DNA and Pt-peptide adducts obtained have been determined. The importance of hydrogen-bonding in the kinetics of the approach and the stability of the formed adducts is shown to be of great importance for the understanding of the mechanism of action for cis-Pt and related compounds. The results of these studies are being used to make predictions for possible third-generation Pt compounds.

Although the low-molar approach did receive much attention in most studies (i.e. binding to DNA fragments and peptides that can be studied and the molecular level), it should be mentioned that recently fascinating results have been obtained in the more biological part of this field. To be mentioned are the results of Toney et al. (1989), Chu and Chang (1990), Visse et al. (1991), Pil and Lippard (1992) and Szymkowski et al. (1992), about proteins that bind at the Pt-DNA site.

B. Competition between proteins and DNA

It is quite clear that in the cells, after the relatively slow hydrolysis, cis-Pt must eventually have a preference for DNA over proteins (kinetics of approach); from protein side chains the following ligands are in principle available:

(i) the amine side groups, but these always occur in the hydronated state (and form $-NH_3^+$) under ambient conditions;
(ii) the carboxylates, although negatively charged, have a small Pt binding affinity;
(iii) the imidazole groups of histidine side chains; these are likely candidates for binding;
(iv) the S-donor atoms in amino acids (like cysteine and methionine), small peptides (such as glutathione) and proteins are strong competitors under certain conditions (Djuran et al., 1991).

On the other hand, DNA contains a highly negatively charged phosphate chain and, as a result, at least an electrostatic interaction can take place with the positively charged Pt ions and the amine N-H groups, thereby bringing the platinated species close to the N-donor atoms of the bases (this is likely to be important in the kinetics of the process).

Recent experiments from our laboratories (Lempers and Reedijk, 1990a; 1990b; Van Boom and Reedijk, 1993) have suggested that at least one other possibility should be considered, namely transfer of the Pt species from the S-bound site to the N-bound position in the nucleic acid base. Detailed studies on S-adenosylhomocysteine (Fig. 4) and S-

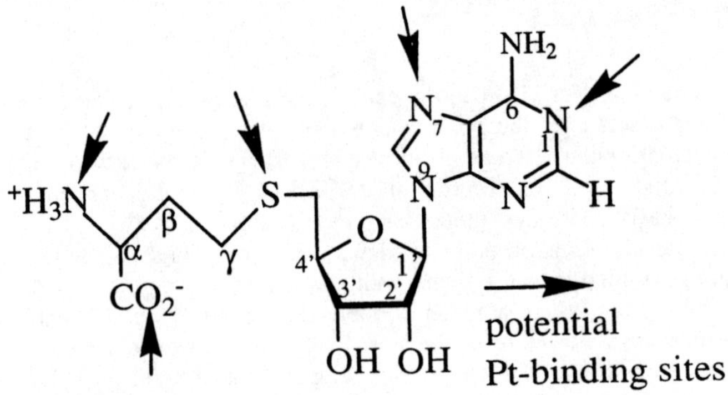

Figure 4.
Structure of *S* adenosyl-L-homocysteine; SAH.

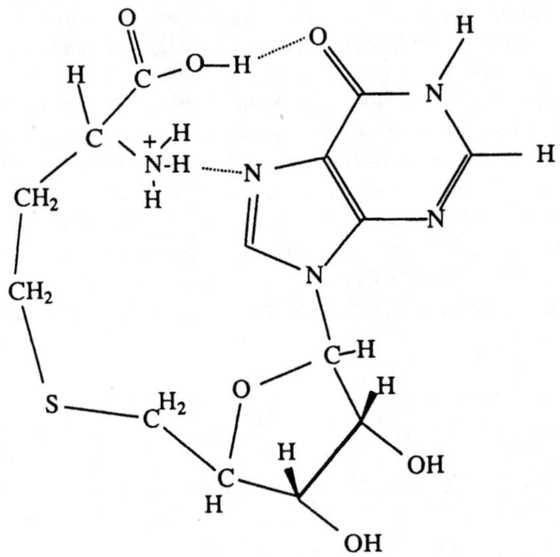

Figure 5.
Structure of *S* guanosyl-L-homocysteine (SGH) and its possible folding at acidic and neutral pH.

guanosylhomocysteine (Fig. 5), have made clear that a rapid binding of platinum - studied so far as the monofunctional DNA-binding species $Pt(dien)^{2+}$ - takes place at the thioether atom. In subsequent reactions, depending on pH, transfer of the Pt species to an N-donor site may take place (vide infra).

C. Binding to and structure of Pt-DNA adducts

The preferred binding sites on DNA for Pt compounds are now well known (Reedijk, 1992 and references therein). The first binding step almost invariably occurs at the guanine-N7 site, but binding at adenine-N7 may also occur. It appears that the amines and the leaving groups coordinated to Pt play a very important role in the kinetics of the binding to this site, and also have an important effect of the resulting DNA structure after the Pt binding.

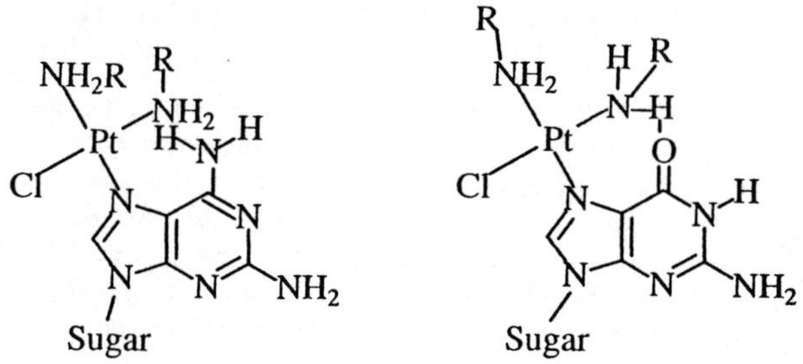

Figure 6.
Schematic structure of the cis-Pt monochloro adducts with G and A (illustrating H bonding differences).

The preference of G-N7 over A-N7, at least on the nucleobase level, is easily seen from Figure 6. An attractive H-bonding with O6 must indeed result in both kinetic and thermodynamic preferences over the situation with repulsion at the NH_2 group at position 6.

The available conclusions with respect to DNA ds binding of Pt compounds drawn from the work of others as well as from our own work, are briefly summarized below (details are presented in the following Chapter 5-A2):

(i) In the cis-Pt adducts at GG (and AG) sequences, the two bases are coordinated via N7 in a "head-to-head" orientation; the deoxyribose moiety of the 5' guanine has changed to an almost pure N-type conformation (the S-type conformation is present in free d(GpG)). Other conformational characteristics of the sugar hardly change upon platination.

(ii) Detailed analysis of proton and phosphorus NMR spectra have resulted in the conclusion that the double-helix is somewhat destabilized after the platination. The decrease of the "melting" (fraying) temperature of the duplex by 10-20 °C at NMR concentrations (3 mM) is a strong indication for such a destabilization. Base-pair formation by hydrogen bonding, however, remains possible after the platination, as reflected by the appearance of all imino-proton resonances. For the central GG sequence, these signals are only observed at the lower temperature, albeit shifted to lower field and broadened.

(iii) Detailed analysis of chemical shift and coupling constant from the NMR spectra in combination with conformational analysis suggests a rather small but significant distortion of the double helix of the larger oligonucleotides upon platination, which has been described as a "kink" of about 40(10)° in the helical axis at or around the GG lesion. In addition, a helical twist (unwinding) of about 13° has been proposed (Keck and Lippard, 1992).

(iv) Comparison of CD spectra and ^{31}P NMR spectra of platinated DNA from several sources with the platinated ds decanucleotide strongly suggests similar distortions for oligonucleotides and ds DNA.

(v) The interactions with the 5'-phosphate group in the DNA appears to be important, i.e. it is involved in a hydrogen bond with an NH_3 (or other amine) ligand of platinum. Such H bonding is quite likely to be the origin for the observation that active "classic" platinum antitumor drugs require an acid N-H group (vide supra). The phosphate-ammonia interaction could induce and/or stabilize (additional) DNA distortions, thereby interfering with the replication process. Details of this H bonding in the ds species have most recently been modelled by Herman et al. (1990).

For newly prepared Pt compounds, similar - but not identical - types of DNA binding and distortions can be expected. Knowing these distortions in detail is important for understanding the mechanism and, subsequently, for developing advanced drugs. Unfortunately, until now no X-ray structures of platinated ds DNA are available to confirm that the solution structures, as determined with NMR and biochemical methods also exist in the solid state.

V. REACTIONS OF Pt COMPOUNDS WITH S-DONOR LIGANDS

A. Introduction

It has long been known that sulfur-containing molecules have a high affinity for platinum (Basolo and Pearson, 1967). Possible reactive biomolecules are cysteine, methionine, S-adenosyl-L-homocysteine, S-adenosyl-L-methionine, glutathione, metallothionein, and other proteins. Only in the last decade some information about (bio)chemically and medically interesting Pt-sulfur interactions has become available. Usually, interactions of platinum antitumor compounds with sulfur-containing biomolecules are considered to have an overall negative effect on the antitumor activity. Such interactions can be responsible for the inactivation of Pt(II) species, development of resistance and toxic side effects such as nephrotoxicity. This type of interactions and their relevance in the mechanism of action of cis-Pt as an antitumor drug will be discussed briefly in this section. Also, some thoughts about the mechanism of action of agents reducing nephrotoxicity, i.e. those with sulfur like sodium diethyldithiocarbamate (NaDDTC) and sodium thiosulfate (STS), will be discussed and reviewed.

B. Inactivation of platinum amine compounds

Sulfur-containing biomolecules may react in vivo with platinum antitumor compounds, thereby hampering, retarding or even preventing Pt binding to the critical DNA target. Such reactions are possible mechanisms of inactivation. Glutathione (GSH, a cysteine-containing tripeptide, see Fig. 7 for its structure) which is the predominant intracellular thiol with concentrations varying from 0.5 up to 10 mM does inhibit the reaction of DNA with $[Pt(en)Cl_2]$ (Eastman, 1987a, 1987b) and also with cis-Pt (Andrews et al., 1986; Dedon and Borch, 1987). Many years ago (Burchenall et al., 1978; Newman et al., 1979) the antitumor activity of cis-Pt was proven to be inhibited by co-administered methionine. A bis-adduct formed from cis-Pt and methionine has been isolated from the urine of patients (Sternson et al., 1984). Such products have been characterized in great detail (in vitro) more recently by Norman et al. (1992).

The reactions of cis- and trans-Pt in intact cells have been investigated in some detail with the use of ^1H NMR by Berners-Price and Kuchel (1990a, 1990b). As expected they observed that both Pt compounds react with GSH, leading to binding of the Pt species. Trans-Pt reacts more rapidly with GSH, which would suggest that in the case of cis-Pt higher percentages of the drug dose may reach the cell nucleus before inactivation by cellular GSH would take place. The more rapid reaction of trans-Pt compared to cis-Pt can be explained by the more labile chloride atoms in trans-Pt due to the trans effect; i.e. the chloride in trans-Pt is a better trans labilizer than is the amino group in cis-Pt (Basolo and Pearson, 1967).

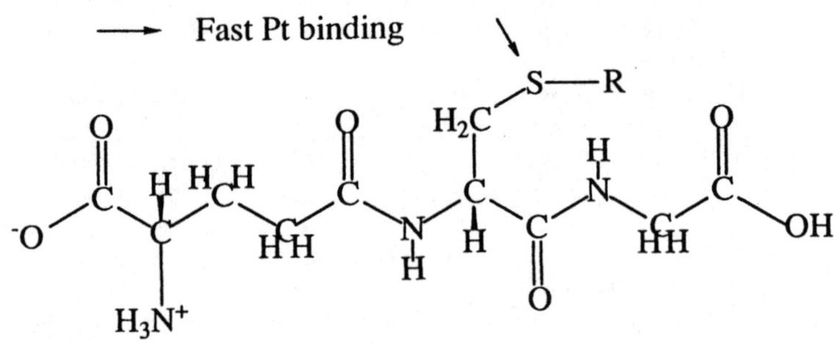

Glutathione (R=H) and S-methyl glutathione (R=Me)

Figure 7.
Schematic structure of glutathione (GSH).

Another mechanism of inactivation for Pt drugs and in particular cis-Pt might be the reaction of sulfur containing biomolecules with the cis-Pt-DNA monoadducts, which prevents these from rearranging to toxic bifunctional adducts. Supportive for such a mechanism is the observation that GSH can be crosslinked to DNA by cis-Pt [41b; Know et al., 1986) and [Pt(en)Cl_2] (Eastman and Barry, 1987).

It is also known that cis-Pt-DNA monoadducts can be experimentally quenched with thiourea, which reduces drug toxicity (Zwelling et al., 1979; Micetich et al., 1983). Trans-Pt also yields monofunctional adducts after reaction with DNA, and these rearrange somewhat slower than cis-Pt into bifunctional adducts (Eastman and Barry, 1987), clearly for sterical reasons. The relatively long-lived monofunctional adducts react efficiently with GSH and proteins (Eastmand and Barry, 1987; Lippard and Hoeschele, 1979; Olinski et al., 1987).

It now appears that almost all S-containing biomolecules can inactivate Pt amine compounds by the two mechanisms and it is probable that the inactivation is not limited to the above described binding by GSH, methionine and cysteine. The high percentages of cis-Pt reported by Kraker et al. (1985) as bound to metallothionein (MT, a protein with a molar mass of 6-7 kg mol^{-1} and containing 33% cysteine; see Part Two, Chapter 1-D5) in Ehrlich tumor cells (30%) is an indication for this. Such MT binding is also known in cis-Pt resistant cells (70%) and in rat tissues (25-40%) (Bakka et al., 1981; Sharma and Edwards, 1983; Zezalowski et al., 1984).

In a comparison of cis-Pt and trans-Pt it was shown that both inactivation mechanisms (i.e. the reaction with free Pt(II) species and with Pt(II)-DNA monoadducts) are more pronounced for trans-Pt, which could - at least partly - explain the reduced toxicity of this antitumor-inactive compound. This statement agrees with studies in which especially the cytotoxicity of trans-Pt, but not of cis-Pt (Andrews et al., 1985; 1986), to a human ovarian carcinoma cell line was found to be increased markedly by reduced GSH levels.

C. Nephrotoxicity of Pt compounds and application of rescue agents

A limitation of cis-Pt in its use as an antitumor drug is its concentration dependent nephrotoxicity (Krakoff, 1979; Fillastre and Raguenez-Viotte, 1989; Daugaard and Abildgaard, 1989), besides a variety of other side effects (Von Hoff et al., 1979). Nowadays, the nephrotoxicity effects can be reduced by mannitol-induced diuresis (Hayes et al., 1979) and hypersalination (Ozols et al., 1984); see *Medicine*, Part Six, Chapter 1-B.

The well-known affinity of sulfur for platinum complexes has led to investigations of numerous sulfur nucleophiles as inhibitors of cis-Pt nephrotoxicity, including NaDDTC, STS, WR-2721, mesna, methionine, thiourea, cysteine, N acetylcysteine, penicillamine, and GSH. More recently, biotin, sulfathiazole and others have been added (Jones et al., 1992). Of these, NaDDTC, STS and WR-2721 are undergoing preclinical and/or clinical evaluation; for a review see Van der Vijgh (1991). Some of the more promising compounds (see Fig. 8) are briefly discussed below.

The well-known salt NaDDTC has proven to be a very effective inhibitor of nephrotoxicity; it is also known to be effective against bone-marrow toxicity (Bodenner et al., 1986b; Gringeri et al., 1988). It should be administered 1-4 h after cis-Pt to avoid interfering with the antitumor properties of cis-Pt (Borch et al., 1980; Gale et al., 1982; Borch et al., 1984). In fact NaDDTC does not interfere with the antitumor properties of cis-Pt, which is in agreement with the observation that NaDDTC could not reverse Pt-DNA crosslinks (Borch et al., 1984), except for the cis-Pt-adenosine 1:2 complex and the cis-Pt-guanosine 1:1 complex (Bodenner et al., 1986a) which are considered unimportant for the antitumor activity of cis-Pt. As a matter of fact, NaDDTC and thiourea (vide infra) are the only rescue agents that provide protection from nephrotoxicity when administered after cis-Pt at a time when most of the reactive platinum species have been taken up by cells, or are being already excreted in the urine of the patient. Pretreatment with NaDDTC appears to be inefficient in inhibiting nephrotoxicity, probably because it inactivates cis-Pt (vide supra), although one study is known in which treatment with NaDDTC 12 h prior to cis-Pt was effective in protecting kidney damage (Jones and Basinger, 1989). Since NaDDTC is a powerful inducer of MT synthesis (Sunderman and Fraser, 1983), metallothionein could be

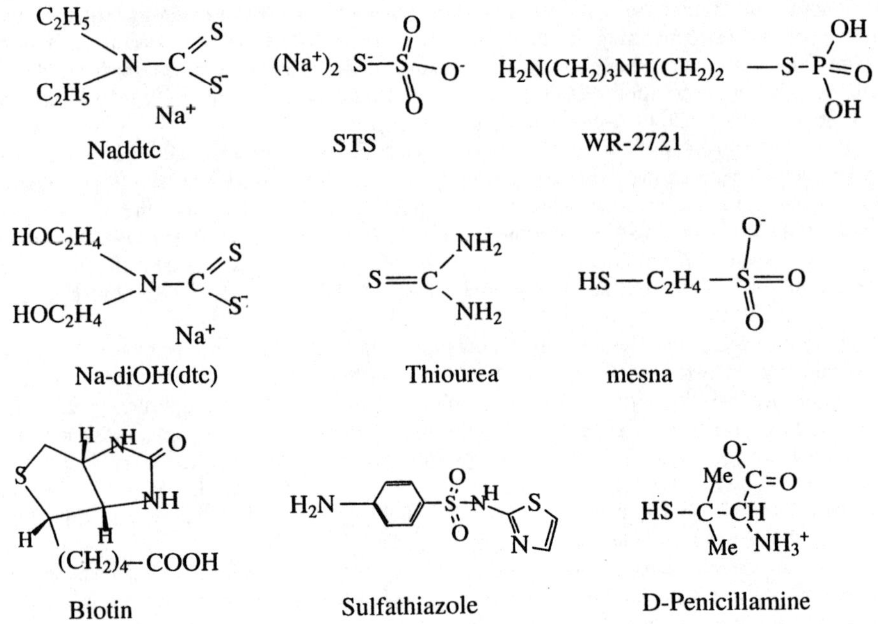

Figure 8.
Structure of some well-known nephrotoxicity inhibitors. (Abbreviations: see Table).

the actual inhibitor of the nephrotoxicity (i.e. by reacting with free cis-Pt in the kidney). The time of administration of NaDDTC, i.e. after cis-Pt treatment, agrees with the hypothesis that NaDDTC acts as a real rescue agent. So, it reduces the nephrotoxicity by removing the platinum from certain proteins, thereby restoring the original structure of the protein (Borch and Markman, 1989).

To investigate what kind of Pt-enzyme interactions could be reversed, a study of the exchange reactions of model compounds with NaDDTC was undertaken. With well-identified model compounds like GSMe (Fig. 7) and $Pt(dien)^{2+}$ (Lempers et al., 1988) it was aimed to mimick Pt-cysteine and Pt-methionine adducts within a protein (Lempers and Reedijk, 1990a). It was shown in particular that the Pt-methionine type bond in $[Pt(dien)GS\text{-}Me]^{2+}$ reacts fast with NaDDTC, thereby restoring the original structure of the thioether linkage. The Pt-cysteine type bond in $[Pt(dien)GS]^{+}$ appears to be inert toward NaDDTC treatment. These observations are in agreement with the results of exchange reactions of NaDDTC with the not well characterized but biologically more relevant cis-Pt-methionine and cis-Pt-GSH systems, which also points to the lability of Pt-methionine type interactions only (Andrews et al., 1984). The inert Pt-S bond in Pt-cysteine complexes under neutral conditions is probably the result of a very strong negatively charged $Pt\text{-}S^-$ binding compared to the rather weak neutral Pt-S binding in Pt-methionine complexes. Extrapolating the results obtained with the model adducts to a protein leads to the following hypothesis for the mechanism of reducing the nephrotoxicity by NaDDTC.

NaDDTC would only reduce the nephrotoxicity when the Pt-binding to proteins would be caused by Pt coordination to methionine residues and not to cysteine residues. This differs from earlier ideas which suggest that NaDDTC would revert the Pt-cysteine type bonds in proteins (Borch and Markman, 1989). It is known that NaDDTC does not react with all the platinated proteins (Hegedus et al., 1987). Therefore, the nephrotoxicity of cis-Pt is likely to be the result of inactivation of certain enzymes through binding at cysteines and/or methionines; NaDDTC can reduce the nephrotoxicity only by reverting the Pt-S bonds of the methionine type.

So, NaDDTC would be a very effective rescue agent in the clinic; however, it is known to cause severe side effects, including burning of the mouth, chest tightness, and extreme anxiety (Hacker and Roberts, 1987). Fortunately, it has been demonstrated recently that the use of other dithiocarbamates, like sodium di(hydroxyethyl)dithiocarbamate (see Fig. 8), possess certain promising advantages (Jones et al., 1986a; 1986b). The polar groups on these NaDDTC derivatives can be expected to result in cis-Pt complexes with a different character than those formed with NaDDTC itself. Also the solubility of DDTC and other derivatives will be much higher, which may induce lower toxic side effects. Borch and Markman (1989) have reported on the modulation of cis-Pt toxicity by using e.g. a combination of NaDDTC and hypertonic saline.

Thiourea probably acts in a way comparable to NaDDTC and should also be administered after cis-Pt treatment. Like NaDDTC, thiourea is able to remove platinum from platinated enzymes such as leucine aminopeptidase and γ-glutamyltranspeptidase (Dedon et al., 1986), and fumarase (Boelrijk et al., 1991). However, thiourea appears to be less useful as an inhibitor of nephrotoxicity; it also reacts quite rapidly with Pt-DNA crosslinks (Filipski et al., 1979).

The rescue agent STS (Fig. 8), which as such is non-toxic, is injected during the period 1 h before and ½ h after cis-Pt treatment (Howell and Taetle, 1980). The optimal protocol in fact uses STS intravenously (iv) in conjunction with intraperitoneally (ip) cis-Pt (Goel et al., 1989). Concurrent injection of STS and cis-Pt, ip or iv (Aamdahl et al., 1988), partly reduces the antitumor activity, probably by inactivation of cis-Pt. A likely explanation for the nephrotoxicity-protecting effect is that STS extensively concentrates in the kidney where it is known to react rapidly with cis-Pt (Elferink et al., 1986).

While NaDDTC is acting as a "true" rescue agent, i.e. by reversing Pt-biomolecule interactions, STS most likely acts by locally inactivating cis-Pt and lowering its concentration in the kidney. This may also explain why STS is not effective when administered after cis-Pt treatment.

The mechanism by which the more recently used WR-2721 (Fig. 8) reduces nephrotoxicity is not yet very well understood (Treskes et al., 1992). WR-2721 protects against nephrotoxicity when administered just prior to cis-Pt (Glover et al., 1984; 1988) and it is known that WR-2721 is preferentially taken up by normal cells and not by tumor cells (Yuhas, 1980). Recently, it was concluded by Calabro-Jones et al. (1988) that the uncharged form of the dephosphorylated WR-2721 (known as WR-1065) is the actual species taken up by both normal and tumor cells. They proposed that the conversion of WR-2721 into WR-1065 is slower in tumors compared with normal tissues, possibly because tumors generally have lower levels of alkaline phosphatase. These authors also proposed that once formed WR-1065 will have a decreased uptake rate in tumors, probably as a consequence of the lower pH of these (Kennedy et al., 1981) as compared with normal tissues; i.e. the neutral form of WR-1065 will only constitute 0.1% of the total drug present at pH 7 and 1% of the total at pH 8. The reactive species WR-1065 is therefore likely to bind directly to cis-Pt, thereby preventing side reactions of cis-Pt.

Recent work of Jones et al. (1992) has shown that another interesting approach is the use of S-containing well-known drugs clinically used for other diseases and purposes. They have studied a variety of compounds such as biotin, captopril, sulfathiazole and cephalexin as potential antagonists for the nephrotoxicity of cis-Pt. Both biotin and sulfathiazole (see Fig. 8) were found to be the most promising candidates for further application.

Finally, it should be mentioned that a new interesting method has been developed to protect cells from cis-Pt nephrotoxicity. Pretreatment of mice by administration of bismuth salts induces MT synthesis, especially in the kidney, but not in the tumor (Naganuma et al., 1987). This induction of MT decreased the renal toxicity of cis-Pt while the antitumor activity was not affected (Satoh et al., 1988); these observations can be rationalized by assuming that cis-Pt binds to MT in the kidney as has been previously observed by other investigators (Sharma and Edwards, 1983; Zelazowski et al., 1984).

D. Competition between Pt-N and Pt-S binding

It has now been generally accepted that the hydrolysis of the chloro ligand is the rate-determining step in the reaction of cis-Pt with DNA (vide supra). Concerning the rate-determining step of platinum amine compounds with sulfur-containing biomolecules, the available data are in many cases ambiguous. It has been reported that the chloride hydrolysis is the rate-determining step in reactions of cis-Pt with leucine aminopeptidase (Dedon and Borch, 1987), γ-glutamyl transpeptidase (Bodenner et al., 1986a) and also with albumin (LeRoy and Thompson, 1989). On the other hand, it has been suggested earlier that a direct binding to proteins may occur without prior aquation (Repta and Long, 1980), and this reaction has in fact been observed with cysteine (Corden, 1987), GSH (Andrews et al., 1986; Corden, 1987), adenosine triphosphatase (Dedon and Borch, 1987) and with metallothionein (Otvos et al., 1989).

To investigate the kinetics in more detail, the reaction rates of a simple pair of model compounds, [Pt(dien)X] (X = Cl^-, H_2O), with GSH, GS-Me and 5'-GMP have been investigated and compared (Djuran et al., 1991). These products are the first well identified complexes between S-containing biomolecules and platinum amine compounds (Lempers and Reedijk, 1990a; 1990b) and therefore are ideally suited as model compounds for kinetic studies. The results are in agreement with the above-mentioned hypothesis that chloride hydrolysis is the rate-determining step in the reaction of $[Pt(dien)Cl]^+$ with 5'-GMP, forming [Pt(dien)(5'-GMP-N7)]. This reaction can be completely inhibited in saturated NaCl.

On the contrary, the reaction of Pt(dien)X with GSH and GS-Me is second order in Pt and S, and the rate constant is nearly independent of the Cl^- concentration. The slightly slower reaction at higher [Cl^-] indicates in fact that only a small fraction proceeds through the hydrolysis pathway. The results (Djuran et al., 1991) provide clear evidence that sulfur primarily reacts directly with Pt amine compounds, substituting for Cl^- without prior aquation. The hydrolyzed species $[Pt(dien)(H_2O)]^{2+}$ will almost selectively react with 5'-GMP, whereas the chloro species $[Pt(dien)Cl]^+$ will only react with sulfur, under competition conditions. This information is relevant to the development of new Pt drugs. In case one could develop a compound with structural properties such that the direct attack by sulfur is inhibited, but with a similar rate of chloro hydrolysis compared to cis-Pt, this would lead to compounds with improved antitumor properties and lower toxicity.

E. Pt(II)-S bonds as intermediates towards DNA binding?

The high affinity of many platinum compounds for sulfur and the availability of many sulfur-containing biomolecules have raised the question whether Pt-sulfur biomolecule interactions could serve as a drug reservoir for platination at DNA, necessary for the antitumor activity of cis-Pt. Two reaction paths are possible, i.e. spontaneous release of platinum from the sulfur, or nucleophilic displacement of platinum from sulfur by e.g. guanine (at N7). As yet there is no real evidence for the existence of such (re)activation mechanisms. In fact, it has been reported that Pt-protein interactions in the plasma (albumin) are not reversible under normal conditions (Repta and Long, 1980; Cole and Wolf, 1980). The fact that a mixture of cis-Pt methionine products does not show antitumor properties (Daley-Yates and McBrien, 1984) indicates no methionine-induced platination of DNA. Much more research is required to investigate the existence of a reactivation mechanism. From our earlier investigations, it is predicted that if such a reactivation phenomenon were operational, the most likely candidate would be the labile Pt-methionine bond as has been shown by its rapid reaction with NaDDTC, STS and thiourea (vide supra).

In recent years there has also been a growing interest in Pt-S interactions from a chemical point of view, mainly to understand the exact role (vide supra) of such interactions in the mechanism of the antitumor activity of cis-Pt. Below, some reactions between Pt-amine compounds and sulfur-containing biomolecules will be evaluated, with special attention for cis-Pt and the related, antitumor-inactive complexes trans-Pt and [Pt(dien)Cl]Cl. [Pt(dien)Cl]Cl is monofunctional in nature and is often used to model the first binding step of cis-Pt (Admiraal et al., 1992). In addition, stable complexes can be expected with $Pt(dien)^{2+}$, the chelate effect of dien rendering all Pt-N bonds thermodynamically more stable and kinetically more inert, whereas cis-Pt often gives complicated mixtures containing all kinds of degradation products (vide infra).

It is well known from in vitro studies that both trans-Pt and cis-Pt bind to enzymes and proteins (see e.g. Dedon and Borch, 1987; Bodenner et al., 1986a; Gonias et al., 1988;

Pizzo et al., 1988). In all these and many other studies it has been suggested that cysteines and methionines are the binding sites and that - where reported - in general trans-Pt reacts more readily (see also the reaction with glutathione; Lempers and Reedijk, 1991) than cis-Pt. Usually, high ratios of Pt : protein are required (at least > 100) to observe detectable amounts of binding within a reasonable time. The presence of such interactions in vivo is not proven, although cis-Pt and $[Pt(en)Cl_2]$ are known to react in vivo with plasma proteins (Cole and Wolff, 1980; Robbins, 1982) which makes it reasonable that reaction takes place, at least in those cases with albumin, α_2-macroglobulin and/or α_1-antitrypsin.

Bongers et al. (1988) have described the interaction of MT with an excess of K_2PtCl_4. A product is formed containing as much as 7 mol Pt/mol MT (at neutral pH) with coordination to sulfur only. Upon reaction of [*cis*-$Pt(Am)_2Cl_2$] with MT, release of the amine ligands has been observed (Otvos et al., 1989; Bongers et al., 1988). Therefore, similar products as formed with K_2PtCl_4 can be expected. Such a release of amine ligands upon coordination of [*cis*-$Pt(Am)_2Cl_2$] to methionine (Appleton et al., 1988; Norman and Sadler, 1988; Norman et al., 1992a), cysteine (Appleton et al., 1989a, 1989b) and GSH (Dedon and Borch, 1987; Appleton et al., 1989a, 1989b; Berners-Price and Kuchel, 1990b) have already been observed and is rationalized by the large trans effect of a coordinated sulfur atom, which labilizes the amine ligand trans to the sulfur. For this reason, amine release upon coordination of sulfur to [*trans*-$Pt(Am)_2Cl_2$] is not expected. It has even been speculated that the different biological activities of cis-Pt and trans-Pt might be related to S-induced amine release only from cis-Pt (Ismail and Sadler, 1983).

Many investigations of the reaction of cis-Pt with GSH are known (Appleton et al., 1988; 1989a; 1989b; Dedon and Borch, 1987), but due to the complexicity of the system, i.e. because of amine release, not all reaction products could be characterized unambiguously. It is quite likely that eventually a polymeric structure is formed with formula $[Pt(GS)_2]_n$ (Appleton et al, 1988; Berners-Price and Kuchel, 1990a; 1990b)), involving loss of NH_3. Initially, intermediate species like [*cis*-$Pt(NH_3)_2(GS)Cl$] and $[Pt_2(NH_3)_4(GS)_2]$ can be present which lose NH_3 upon standing, eventually forming the polymeric $[Pt(GS)_2]_n$ with coordination exclusively via the S atom but with several different Pt-S and Pt-S-Pt environments.

The reaction of trans-Pt with GSH (1:2) results in a monomeric complex in which two GS^- units are coordinated via the cys-S atom; (equation 8). The unstable intermediate species, [*trans*-$Pt(NH_3)_2(GS)Cl$] and [*trans*-$\{Pt(NH_3)_2Cl\}_2GS$] could also be detected by Berners Price and Kuchel (1990a, 1990b).

$$[\textit{trans}\text{-}Pt(NH_3)_2Cl_2] \xrightarrow[-Cl^-]{+GS^-} [\textit{trans}\text{-}Pt(NH_3)_2(GS)Cl] \xrightarrow[-Cl^-]{+GS^-} [\textit{trans}\text{-}Pt(NH_3)_2(GS)_2] \quad (10)$$

Cysteine reacts with cis-Pt yielding species like [*cis*-$Pt(NH_3)_2$(Cys-*S*)$_2$] and [*cis*-$Pt(NH_3)_2$(Cys-*N,S*)] (Appleton et al., 1989a).

A number of complexes has been reported as products from the reaction of cis-Pt and methionine; i.e. [*cis*-$Pt(NH_3)_2$(Met-*S*)$_2$], [*cis*-$Pt(NH_3)_2$(Met-*N,S*)], [*cis*-$PtCl(NH_3)$(Met-*N,S*)] (some examples are given in fig. 9), [*trans*-$Pt(NH_3)$(Met-*S*)(Met-*N,S*)] and [*trans*-Pt(Met-*N,S*)$_2$] (Norman et al., 1992b). From these species [*cis*-$Pt(NH_3)_2$(Met-*N,S*)] and [*cis*-$Pt(NH_3)_2$(Met-*S*)$_2$] lose NH_3 on standing (Appleton et al., 1989b). Methionine yields a similar complex after reaction with trans-Pt, as formed between GSH and trans-Pt (i.e. [*trans*-$Pt(NH_3)_2$(Met-*S*)$_2$] (Volshtein et al., 1965).

The reaction of [Pt(dien)Cl]Cl with S-adenosyl-L-homocysteine (SAH, a biologically relevant molecule, as it is the co-product of the methyl-transfer reaction by S-adenosyl-L-methionine; see Fig. 4) results in a mixture of complexes (Lempers and Reedijk, 1990b); i.e. the mononuclear complex $[Pt(dien)(SAH\text{-}S)]^{2+}$ (1), with platination of SAH at the sulfur atom, the mononuclear complex $[Pt(dien)(SAH\text{-}N)]^{+}$, which has a $Pt(dien)^{2+}$ unit coordinated to the amino group of the homocysteine unit and the dinuclear complex $[\{Pt(dien)\}_2(SAH\text{-}S,N)]^{3+}$, which has a $Pt(dien)^{2+}$ unit coordinated to the sulfur atom as well as a $Pt(dien)^{2+}$ coordinated to the amino group of the cysteine unit. The Pt-methionine type bond in the first adduct formed is labile in the presence of the dehydronated amino group, resulting in the formation of a species with coordination at the NH_2 group. Such

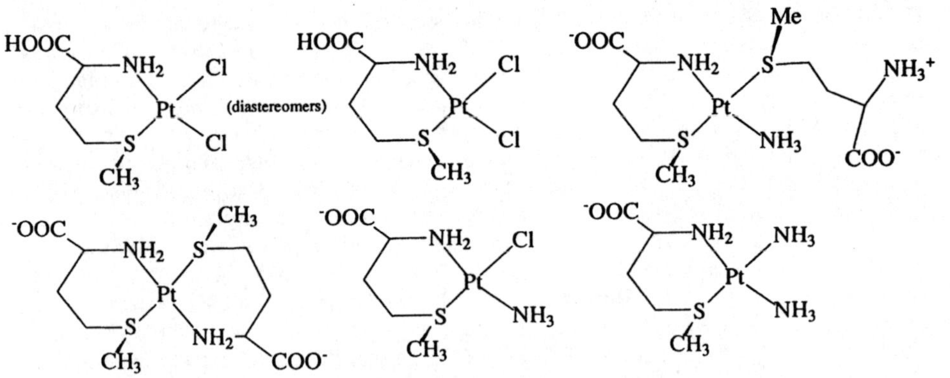

Figure 9.
Structure of some Pt-methionine adducts (after Norman et al.; 1992).

migrations of platinum are quite uncommon and this was the first case in which a Pt was found to migrate from a sulfur to a nitrogen and vice versa.

In a subsequent study with SAH-Boc (where the NH_2 group is blocked by a *t*-butyl group), we found that the Pt unit, after initial binding to the S atom, migrates slowly to the adenine-N7 and adenine-N1 position (Van Boom et al., 1994).

When a similar reaction is carried out with S-guanosyl-L-homocysteine (Van Boom and Reedijk, 1993), again the initial binding (pH=7) takes place at the thioether sulfur, but within a few hours, migration is taking place to the guanine-N7 atom!

In all Pt complexes described above, coordination is observed to the sulfur atom (i.e. with SAH, cysteine, methionine, GSH, MT and other proteins) which is consistent with a high reactivity of sulfur. It is therefore likely to expect that significant amounts of platinum antitumor drugs will bind in vivo to sulfur-containing biomolecules. Although the kinetic reactivity of sulfur is high, the Pt-S bond of the methionine type is labile in the presence of other nucleophiles (vide supra for reactions with NaDDTC and the interconversion of the Pt-SAH adducts). The relatively labile Pt-S methionine bond, but not the Pt-S cysteine bond, now appears to be involved in:
(i) The mechanism suppressing the nephrotoxicity; i.e. NaDDTC reduces the nephrotoxicity by reversing Pt-S bonds of the methionine type only.
(ii) The development of permanent nephrotoxicity. Nephrotoxicity cannot be cured by NaDDTC and thiourea when administered more than 4 h after cis-Pt treatment. A possible mechanism for the development of permanent nephrotoxicity might be that the platinum migrates from a methionine to a cysteine within a protein, resulting in irreversible binding.
(iii) The storage of drug for platination at DNA. Although there is no evidence for the existence of such a phenomenon, in principle Pt-methionine interactions are labile and therefore Pt-methionine complexes are likely candidates to serve as a drug reservoir for platination at DNA; i.e. through spontaneous release and binding to DNA or nucleophilic displacement of platinum by e.g. guanine.

F. Reduction of Pt(IV) compounds exhibiting antitumor activity

As already indicated in previous sections, Pt(IV) complexes are most likely reduced to the corresponding active Pt(II) complexes in vivo. Evidence for such a mechanism comes from the fact that Pt(IV) compounds only bind to 5'-GMP in the presence of ascorbic acid (Van der Veer et al., 1986). Other possible reducing agents are SH-containing biomolecules, of which GSH is the most abundant compound in a cell. Indeed, it has been shown that in the presence of GSH, tetraplatin [Pt(dach)Cl_4] reacts readily with DNA (Eastman, 1987b). At high concentrations of GSH, inhibition of DNA binding was observed due to a "simple" inactivation reaction between GSH and the Pt(II) species formed. In accordance with these observations, [*cis*-Pt(i-PrNH_2)$_2$$Cl_2$] and Pt(dach)$_2$$Cl_2$ have been identified as metabolites of the respective Pt(IV) complexes (Chaney et al., 1991).

VI. CONCLUDING REMARKS AND OUTLOOK TO THE FUTURE

A. General observations

The intensive research efforts spent during the last two decades to elucidate the detailed working mechanism of cis-Pt have not yet resulted in a complete picture.

A first important question deals with the similarity in cis-Pt and trans-Pt; they both bind to DNA with comparable kinetics (Bancroft et al., 1990), but their antitumor properties are totally different. The difference in antitumor properties for these two isomers may be related to:

(i) Differences in platinum-DNA adduct induction. As intrastrand chelation of neighboring purines is possible with cis-Pt, but - for steric reasons - not possible for trans-Pt, the AG- and the GG-chelate are likely candidates for the crucial lesion. A reasonable hypothesis states that these lesions, contrary to the GNG-chelate and the interstrand crosslinks - which can be formed by trans-Pt too - are less or not at all recognized by cellular repair mechanisms, but on the other hand do interfere with DNA replication.

(ii) Difference in binding to proteins (competition; inactivation) and cell-wall transport (kinetics).

A second important question deals with the role of rescue agents, added either before, or at the same time, or after cis-Pt administration. It is not yet clear whether they mainly inactivate the Pt drugs, or prevent it from binding to lethal sites (such as enzymes), or perhaps have a role as a transport agent (intermediate) towards the ultimate DNA target.

Other relevant questions concerning the uptake and transport of cis-Pt into cells (are certain cell walls selective for uptake of cis-Pt?) and the cellular transport and degradation are far from being answered yet. These and related questions for cis-Pt derivatives promise a very interesting future for interdisciplinary research by chemists, biologists, pharmacologists and physicians.

The very interesting aspects of resistance and cross-resistance to cis-Pt and related compounds could not be discussed in this chapter. For details the reader is referred to work of Kelley and Rosencweig (1989), to Canon et al. (1990), or to Scanlon et al. (1991), dealing with suggestions to circumvent the resistance problems. A review is given by Eastman (1991).

The results discussed above have clearly shown that there is an increasing evidence for the importance of Pt-sulfur interactions, not only from a chemical point of view, but also from biological, pharmacological and medical points of view. Although such interactions are not directly responsible for the antitumor activity of cis-Pt, they definitely contribute to some of the mentioned overall negative effects (inactivation, resistance, nephrotoxicity). Much more research is needed to understand these negative effects in detail, which could ultimately lead to compounds with better antitumor properties and reduced side effects.

B. Towards new drugs

Given the clinical success of cis-Pt and CBDCA, one could ask whether or not there is still a future for new platinum drugs. On the other hand, the results described above on the chemistry of the side effects (which may differ significantly depending on the amine and the leaving groups) and cell-wall transport (hardly known, but likely to vary from compound to compound), justify a serious search for derivatives of cis-Pt. Morover, since the repair proteins are now known to bind at the platinated sites in the DNA, the size, shape and electronic properties (H-bonding; hydrophobic) of the amines will play a major role in developing drugs that are cross-resistant to cis-Pt.

Criteria of relevance for future "classical" Pt drugs are accepted to be:

(i) good water solubility and stability in 0.15 M NaCl;
(ii) in blood, reactions with ligands containing S donor atoms should be suppressed;
(iii) in cells, hydrolysis reactions should be optimized;
(iv) reaction kinetics with DNA should be optimal; resulting distortion of the DNA should be limited;
(v) undesired reactions in cells should be suppressed as well;
(vi) cross-resistance properties to cis-Pt and CBDCA.

These properties are to be controlled by variations in: (i) the chosen amine ligand at platinum (chelating or not; H-bond donor; steric effects), and (ii) the chosen leaving groups at platinum (non toxic; optimal kinetics; acting as a pro-drug).

Many interesting new approaches in the design of antitumor drugs have appeared during recent years and a variety of platinum complexes are being designed, having the following characteristics:
(i) containing carrier molecules as ligands for achieving higher drug concentrations, or slower release in tumor tissues (see e.g. Schönenberger et al., 1984);
(ii) containing other chemotherapeutic agents, like intercalators (Bowler et al., 1989; Sundquist et al., 1988; 1990) and phosphono carboxylates (Hollis et al., 1990) as co-ligands in the hope of obtaining some sort of synergistic effect;
(iii) containing more than one platinum atom (Farrell and Qu, 1989; Farrell et al., 1988; 1990a; 1990b; Bloemink et al., 1992);
(iv) containing radiosensitizers as ligands (Douple, 1985; Farrell and Skov, 1987) for use in radiation therapy.

C. Mechanistic studies

Mechanistic studies of cis-Pt and derivatives in the future are expected to concentrate on:
(i) Refinements in the structure-activity relationships. Synthesis of classical Pt amine compounds with at least a N-H donor (asymmetry; steric effects; protein recognition); orally administered drugs;
(ii) Binding of Pt compounds (active ones and inactive ones) to and removal from proteins;
(iii) Binding of Pt compounds to and removal from nucleic acids;
(iv) Binding of Pt compounds to cell-wall components;
(v) Exchange reactions of platinated sulfurs of the methionine type with mononucleotides and longer DNA fragments, with the aim to investigate the possibility that such Pt-S methionine bonds could act as drug reservoir for platination at DNA;
(vi) Exchange reactions of rescue agents with platinated proteins, which are known to form Pt S cysteine bonds (molecular basis for nephrotoxicity);
(vii) Study of the classical compounds: details of distorted DNA structures after Pt binding (cis-Pt; other Pt compounds);
(viii) The understanding of the first binding step of cis-Pt (and derivatives) to DNA; is there a preference for (certain?) -GG- sequences?
(ix) A very exciting development concerns the interplay of chromosomal proteins and platinated DNA. *E. coli* produces UvrABC excision repair proteins, and the UvrAB complex binds to the convex side of a cisplatin-induced kink in DNA (Visse et al., 1991). Details of this binding (structure, stability) and consequences need to be investigated, also for the human equivalents (Szymkowski et al., 1992).

Finally, the study of the rapidly growing class of non-classical compounds needs (and will receive) attention. There appear to be almost unlimited variation possibilities for such compounds. What are the cellular targets for these exciting compounds? Is it (also) the DNA, or perhaps protein? Exploring such new structural classes of Pt antitumor drugs may allow the discovery of new agents that possess chemical and biological properties quite different from those of cisplatin. The many recently discovered examples of new classes may provide the basis for expanding the spectrum of activity of Pt-based drugs. Extension to drugs containing other metals (Keppler, 1990; see also following Chapter 5-B) and not necessarily binding to DNA is also very important for the future.

ACKNOWLEDGEMENTS

The research described has been sponsored by the Netherlands Organisation for Chemical Research (SON), with financial aid of the Netherlands Organisation for the Advancement of Research (NWO).

The authors acknowledges EEC support (grant nr ST2J-0462-C) allowing regular scientific exchange with the group of Prof. Dr. J.C. Chottard (Paris). Mrs. Dr. Annemarie J. Fichtinger-Schepman is thanked for continuous collaboration and many useful suggestions.

The author also wishes to thank Johnson & Matthey (Reading, UK) for their generous loan of K_2PtCl_4. The several colleagues and co-workers from our laboratory - their names are listed as our co-authors - are thanked for their many useful contributions.

List of important abbreviations:

cis-Pt	[*cis*-$PtCl_2(NH_3)_2$]
trans-Pt	[*trans*-$PtCl_2(NH_3)_2$]
CBDCA	[*cis*-$Pt(C_6H_6O_4)(NH_3)_2$]
DNA	deoxyribonucleic acid
RNA	ribonucleic acid
Am	amine
diamo	didentate amine
en	ethylenediamine
dien	diethylenetriamine
dach	1,2-diaminocyclohexane
N het	heterocyclic amine
mepy	methylpyridine
Me_2SO	dimethyl sulfoxide
MeBzSO	methyl benzyl sulfoxide
GSH	γ-L-glutamyl-L-cysteinylglycine; glutathione
GS^-	GSH, dehydronated at the thiol group
GS-Me	S methyl glutathione
MT	metallothionein
cys	cysteine
met	methionine
SAH	S-adenosyl-L-homocysteine
SGH	S-guanosyl-L-homocysteine
NaDDTC*	sodium diethyldithiocarbamate
STS	sodium thiosulfate
WR-2721	[*S*-2-(3-aminopropylamino)ethyl phosphorothioic acid]
WR-1065	dephosphorylated WR-2721
mesna	sodium 2-mercaptoethane sulfonate
d	deoxyribo
A	adenine
G	guanine
Guo	guanosine
GMP	guanosine monophosphate
p	phosphate group (irrespective of charge)
r_b	number of Pt atoms bound per nucleobase
py	pyridine

*now recommended as Naddtc by IUPAC

REFERENCES

Aamdal, S., Fodstad, O., and Pihl, A. (1988). *Cancer Chemother. Pharmacol.*, *21*: 129.

Admiraal, G., Alink, M., Altona, C., Dijt, F.J., van Garderen, C.J., de Graaff, R.A.G., and Reedijk, J. (1992). *J. Amer. Chem. Soc.*, *114*: 930-938

Alul, R., Cleaver, M.B., and Taylor, J.S. (1992). *Inorg. Chem.*, *31*: 3636.

Andrews, P.A., Wung, W.E., and Howell, S.B. (1984). *Anal. Biochem.*, *143*: 46.

Andrews, P.A., Murphy, M.P., and Howell, S.B. (1985). *Cancer Res.*, *45*: 6250.

Andrews, P.A., Murphy, M.P., and Howell, S.B. (1986). *Mol. Pharmacol.*, *30*: 643.

Appleton, T.G., Connor, J.W., and Hall, J.R. (1988). *Inorg. Chem.*, *27*: 130.

Appleton, T.G., Hall, J.R., Ralph, S.F., and Thompson, C.S.M. (1989a). *Inorg. Chem.*, *28*: 1989.

Appleton, T.G., Connor, J.W., Hall, J.R., and Prenzler, P.D. (1989b). *Inorg. Chem.*, *28*: 2030.

Bajetta, E., Rovej, R., Buzzoni, R., Vaglini, M., and Bonadonna, G. (1982). *Cancer Treatm. Rep.*, *66*: 1299.

Bakka, A., Endresen, l., Johnsen, A.B.S., Edminson, P.D., and Rugstad, H.E. (1981). *Toxicol. Appl. Pharmacol.*, *61*: 215.

Bancroft, D.P., Lepre, C.A., and Lippard, S.J. (1990). *J. Amer. Chem. Soc.*, *112*: 6860.

Barnard, C.F.J. (1989). *Plat. Met. Rev.*, *33*: 162;

Barnard, C.F.J., Cleare, M.J., and Hydes, P.C. (1986). *Chem. Brit.*, 22: 1001.

Basolo, F., Gray, H.B., and Pearson, R.G. (1960). *J. Amer. Chem. Soc.*, *82*: 4200.

Basolo, F., and Pearson, R.G. (1967). *Mechanisms of Inorganic Reactions*, J. Wiley and Sons, New York, pp. 351, 453.

Berners Price, S.J., and Kuchel, P.W. J. (1990a). *J. Inorg. Biochem.*, *38*: 327.

Berners Price, S.J., and Kuchel, P.W. J. (1990b). *J. Inorg. Biochem.*, *38*: 305.

Bloemink, M.J., Reedijk, J., Farrell, N., Qu, Y., and Stetsenko, A.I. (1992). *J. Chem. Soc., Chem. Commun.*, 1002.

Bodenner, D.L., Dedon, P.C., Keng, P.C., Katz, J.C., and Borch, R.F. (1986a). *Cancer Res.*, *46*: 2745.

Bodenner, D.L., Dedon, P.C., Keng, P.C., and Borch, R.F. (1986b). *Cancer Res.*, *46*: 2751.

Boelrijk, A.E.M., Boogaard, P.J., Lempers, E.L.M., and Reedijk, J. (1991). *J. Inorg. Biochem.*, *41*: 17.

Bongers, J., Bell, J.U., and Richardson, D.E. (1988). *J. Inorg. Biochem.*, *34*: 55.

Borch, R.F., and Markman, M. (1989). *Pharm. Ther.*, *41*: 371.

Borch, R.F., Katz, J.C., Lieder, P.H., and Pleasants, M.E. (1980). *Proc. Nat. Acad. Sci. USA*, *77*: 5441.

Borch, R.F., Bodenner, D.L., and Katz, J.C. (1984). *Platinum Coordination Complexes in Cancer Chemotherapy* (M.P. Hacker, E.B. Douple and I.H. Krakoff, eds.), Martinus Nijhoff, Boston, p. 154.

Bowler, B.E., Ahmed, K.J., Sundquist, W.I., Hollis, L.S., Whang, E.E., and Lippard, S.J. (1989). *J. Amer. Chem. Soc.*, *111*: 1299.

Burchenal, J.H., Kalaher, K., Dew, K., Lokys, L., and Gale, G. (1978). *Biochimie*, *60*: 961.

Calabro Jones, P.M., Aguilera, J.A., Ward, J.F., Smoluk, G.D., Fahey, R.C. (1988). *Cancer Res.*, *48*: 3634.

Canon, J.L., Humblet, Y., and Symann, M. (1990). *Eur. J. Cancer*, *26*: 1.

Chaney, S.G., Gibbons, G.R., Wyrick, S.D., and Podhasky, P. (1991). *Cancer Res.*, *51*: 969.

Chu, G., and Chang, E. (1990). *Proc. Nat. Acad. Sci. USA*, *87*: 3324.

Cleare, M.J. (1974). *Coord. Chem. Rev.*, *12*: 349.

Cleare, M.J., and Hoeschele, J.D. (1973). *Bioinorg. Chem.*, *2*: 187.

Cole, W.C., and Wolf, W. (1980). *Chem.-Biol. Interact.*, *30*: 223.

Corden, B.J. (1987). *Inorg. Chim. Acta*, *137*: 125.

Daley Yates, P.T., and McBrien, D.C.H. (1984). *Biochem. Pharmacol.*, *33*: 3063.

Daugaard, G., and Abildgaard, U. (1989). *Cancer Chemother. Pharmacol.*, *25*: 1.

Dedon, P.C., and Borch, R.F. (1987). *Biochem. Pharmacol.*, *36*: 1955.

Dedon, P.C., Qazi, R., and Borch, R.F. (1986). *Biochemical Mechanisms of Platinum Antitumor Drugs* (D.C.H. McBrien and T.F. Slater, eds.), IRL Press Limited, Oxford, p. 199.

Dhara, S.C. (1970). *Ind. J. Chem.*, *8*: 193.

Djuran, M., Lempers, E.L.M., and Reedijk, J. (1991). *Inorg. Chem.*, *30*: 2648.

Douple, E.B. (1985). *Plat. Met. Rev.*, *29*: 118.

Eastman, A. (1987a). *Chem.-Biol. Interact.*, *61*: 241.

Eastman, A. (1987b). *Biochem. Pharmacol.*, *36*: 4177.

Eastman, A. (1991). *Molecular and Clinical Advances in Anticancer Drugs* (R.F. Ozols, ed.), Kluwer, Boston.

Eastman, A., and Barry, M.A. (1987). *Biochemistry, 26*: 3303.

Elferink, F., van der Vijgh, W.J.F., and Klein, I. (1986). *Clin. Chem., 32*: 641.

Faggianni, R., Lippert, B., Lock, C.J.L., and Rosenberg, B. (1978). *Inorg. Chem., 17*: 1941.

Farrell, N., and Qu, Y. (1989). *Inorg. Chem., 28*: 3416.

Farrell, N., and Skov, K.A. (1987). *J. Chem. Soc., Chem. Commun.*, 1043.

Farrell, N., de Almeida, S.G., and Skov, K.A. (1988). *J. Amer. Chem. Soc., 110*: 5018.

Farrell, N., Qu, Y., and Hacker, M.P. (1990a). *J. Med. Chem., 33*: 2179.

Farrell, N., Kiley, D.M., Schmidt, W., and Hacker, M.P. (1990b). *Inorg. Chem., 29*: 397.

Filipski, J., Kohn, K.W., Prather, R., and Bonner, W.M. (1979). *Science, 204*: 181.

Fillastre, J.P., and Raguenez-Viotte, G. (1989). *Toxicol. Lett., 46*: 163.

Fraústo da Silva, J.J.R., and Williams, R.J.P. (1991). *The Biological Chemistry of the Elements*, Oxford.

Gale, G.R., Atkins, L.M., and Walker, E.M. (1982). *Ann. Clin. Lab. Sci., 12*: 345.

Glover, D., Glick, J.H., Weiler, C., Yuhas, J., and Kligerman, M.M. (1984). *Int. J. Radiat. Oncol. Biol. Phys., 10*: 1781.

Glover, D., Fox, K.R., Weiler, C., Kligerman, M.M., Turrisi, A., and Glick, J.H. (1988). *Pharm. Ther., 39*: 3.

Goel, R., Cleary, S.M., Horton, C., Kirmani, S., Abramson, I., Kelly, C., and Howell, S.B. (1989). *J. Nat. Cancer Inst., 81*: 1552.

Gonias, S.L., Swaim, M.W., Massey, M.F., and Pizzo, S.V. (1988). *J. Biol. Chem., 263*: 393.

Gray, H.B., and Olcott, R.J. (1962). *Inorg. Chem., 1*: 481.

Gringeri, A., Keng, P.C., and Borch, R.F. (1988). *Cancer Res., 48*: 5708.

Hacker, M.P., and Roberts, J.D. (1987). *Platinum and Other Metal Coordination Compounds in Cancer Chemotherapy* (M. Nicolini, ed.), Martinus Nijhoff, Boston, p. 163.

Hadjiliadis, N., Kourounakis, P., and Theophanides, T. (1973). *Inorg. Chim. Acta, 7*: 226.

Hayes, D.M., Cvitkovic, E., Golbey, R.B., Scheiner, E., Helson, L., and Krakoff, I.H. (1979). *Cancer, 39*: 1372.

Hegedus, L., van der Vijgh, W.J.F., Klein, I., Kerpel Fronius, S., and Pinedo, H.M. (1987). *Cancer Chemother. Pharmacol., 20*: 211.

Herman, F., Kozelka, J., Stoven, V., Guittet, E., Girault, J.-P., Huynh-Dinh, T., Igolen, J., Lallemand, J.-Y., and Chottard, J.-C. (1990). *Eur. J. Biochem., 194*: 119-133.

Hollis, S.L., Amundsen, A.R., and Stern, E.W. (1989). *J. Med. Chem., 32*: 128.

Hollis, L.S., Miller, A.V., Amundsen, A.R., Schurig, J.E., and Stern, E.W. (1990). *J. Med. Chem., 33*: 105.

Hollis, S.L., Sundquist, W.I., Burstyn, J.N., Heiger-Bernays, W.J., Bellon, S.F., Ahmed, K.J., Amundsen, A.R., Stern, E.W., and Lippard, S.J. (1991). *Cancer Res., 51*: 1866.

Howell, S.B., ed. (1991). *Platinum and Other Metal Coordination Compounds in Cancer Chemotherapy*, Plenum Press, New York.

Howell, S.B., and Taetle, R. (1980). *Cancer Treat. Rep., 64*: 611.

Ismail, I.M., and Sadler, P.J. (1983). *Platinum, Gold and other Metal Chemotherapeutic Agents* (S.J. Lippard, ed.), American Chemical Society, Washington, DC, p. 171.

Johnson, N.P., Hoeschele, J.D., and Rahn, R.O. (1980). *Chem.-Biol. Interact., 30*: 151.

Jones, M.M., and Basinger, M.A. (1989). *J. Appl. Toxicol., 9*: 229.

Jones, M.M., Basinger, M.A., Mitchell, W.M., and Bradley, C.A. (1986a). *Cancer Chemother. Pharmacol., 17*: 38.

Jones, M.M., Basinger, M.A., Craft, W.D., Domingo, J.L., and Llobet, J.M. (1986b). *Arch. Toxicol., 59*: 167.

Jones, M.M., Basinger, M.A., and Holscher, MA. (1992). *Fund. Appl. Toxicol. 18:181*

Keck, M.V., and Lippard, S.J. (1992). *J. Amer. Chem. Soc.*, *114*: 3386.

Kelland, L.R., Clarke, S.J., and McKeage, M.J. (1992a). *Plat. Met. Rev.*, *36*: 178

Kelland, L.R., Murrer, B.A., Abel, G., Giandomenico, C.M., Mistrym, P., and Harrap, K.R. (1992b). *Cancer Res.*, *52*: 822.

Kelley, S.L., and Roxencweig, M. (1989). *Eur. J. Cancer*, *25*: 1135.

Kennedy, K.A., Teicher, B.A., Rockwell, S., and Sartorelli, J.R. (1981). *Molecular Targets and Actions for Cancer Chemotherapeutic Agents* (A.C. Sartorelli, J.R. Bertino and J.S. Lazo, eds.), New York, Academic Press, p. 85.

Keppler, B.K. (1990). *New J. Chem.*, *14*: 389.

Kirschner, S., Wei, Y.K., Francis, D., and Bergman, J.G. (1966). *J. Med. Chem.*, *9*: 369

Know, R.J., Friedlos, F., Lydall, D.A., and Roberts, J.J. (1986). *Cancer Res.*, *46*: 1972.

Kraker, A., Schmidt, J., Krezoski, S., and Petering, D.H. (1985). *Biochem. Biophys. Res. Commun.*, *130*: 786.

Krakoff, I.H. (1979). *Cancer Treat. Rep.*, *63*: 1523.

Lee, K.W., and Martin, D.S., Jr. (1976). *Inorg. Chim. Acta*, *17*: 105.

Lempers, E.L.M., and Reedijk, J. (1990a). *Inorg. Chem.*, *29*: 217-220 .

Lempers, E.L.M., and Reedijk, J. (1990b). *Inorg. Chem.*, *29*: 1880.

Lempers, E.L.M., and Reedijk, J. (1991). *Adv. Inorg. Chem.*, *37*: 175.

Lempers, E.L.M., Inagaki, K., and Reedijk, J. (1988). *Inorg. Chim. Acta*, *152*: 201.

LeRoy, A.F., and Thompson, W.C. (1989). *J. Nat. Cancer Inst.*, *81*: 427.

Lim, M.S., and Martin, R.B. (1976). *J. Inorg. Nucl. Chem.*, *38*: 1911.

Lippard, S.J., and Hoeschele, J.D. (1979). *Proc. Nat. Acad. Sci. USA*, *76*: 6091.

Loehrer, P.J., and Einhorn, L.H. (1984). *Ann. Intern. Med.*, *100*: 704.

Micetich, K., Zwelling, L.A., and Kohn, K.W. (1983). *Cancer Res.*, *43*: 3609.

Miller, S.A., and House, D.A. (1989). *Inorg. Chim. Acta*, *166*: 189.

Miller, S.A., and House, D.A. (1990). *Inorg. Chim. Acta*, *173*: 53.

Morita, H., and Bailar, J.C., Jr. (1983). *Inorg. Synth.*, 22: 239.

Naganuma, A., Satoh, M., and Imura, N. (1987). *Cancer Res.*, *47*: 983.

Newman, A.D., Ridgeway, H., Speer, R.J., and Hill, J.M. (1979). *J. Clin. Hematol. Oncol.*, *9*: 208.

Norman, R.E., and Sadler, P.J. (1988). *Inorg. Chem.*, *27*: 3583.

Norman, R.E., Ranford, J.D., and Sadler, P.J. (1992). *Inorg. Chem.*, *31*: 877.

Olinski, R., Wedrychowski, A., Schmidt, W.N., Briggs, R.C., and Hnilica, L.S. (1987). *Cancer Res.*, *47*: 201.

Otvos, J.D., Petering, D.H., and Shaw, C.F. (1989). *Comments Inorg. Chem.*, *9*: 1.

Ozols, R.F., Corden, B.J., Collins, J., and Young, R.C. (1984). *Platinum Coordination Complexes in Cancer Chemotherapy* (M.P. Hacker, E.B. Douple and I.H. Krakoff, eds.), Martinus Nijhoff, Boston, p. 321.

Peyrone, M. (1844). *Ann. Chemie und Pharmacie*, *51*: 1-29.

Pil, P.M., and Lippard, S.J. (1992). *Science*, *256*: 234.

Pizzo, S.V., Swaim, M.W., Roche, P.A., and Gonias, S.L. (1988). *J. Inorg. Biochem.*, *33*: 67.

Prestayko, A.W., D'Aoust, J.C., Issell, B.F., and Crooke, S.T. (1979). *Cancer Treat. Rev.*, *6*: 17.

Reedijk, J. (1987). *Pure Appl. Chem.*, *59*: 181.

Reedijk, J. (1992). *Inorg. Chim. Acta*, *198-200*: 873.

Reishus, J.W., and Martin, D.S., Jr. (1961). *J. Amer. Chem. Soc.*, *83*: 2457.

Repta, A.J., and Long, D.F. (1980). *Cisplatin Current Status and New Developments* (A.W. Prestayko, S.T. Crooke and S.K. Carter, eds.), Academic Press, New York, p. 285.

Roat, R., and Reedijk, J. (1993). *J. Inorg. Biochem.*, *32*: 263.

Robbins, A.B. (1982). *Chem.-Biol. Interact.*, *38*: 349.

Roberts, J.J., and Thomson, A.J. (1979). *Progr. Nucl. Acid Res. Mol. Biol.*, *22*: 71.

Rosenberg, B. (1971). *Plat. Met. Rev.*, *15*: 42.

Rosenberg, B., Van Camp, L., and Krigas, T. (1965). *Nature*, *205*: 698.

Rosenberg, B., Renshaw, E., Van Camp, L., Hartwick, J., and Drobnik, J. (1967a). *J. Bacteriol.*, *93*: 716.

Rosenberg, B., Van Camp, L., Grimley, E.B., and Thomson, A.J. (1967b). *J. Biol. Chem.*, *242*: 1347.

Rosenberg, B., Van Camp, L., Trosko, J.E., and Mansour, V.H. (1969). *Nature*, *222*: 385.

Satoh, M., Naganuma, A., and Imura, N. (1988). *Cancer Chemother. Pharmacol.*, *21*: 176.

Scanlon, K.J., Kashani-Sabet, M., Tone, T., and Funato, T. (1991). *Pharm. Ther.*, *52*: 385.

Schönenberger, H., Wappes, B., Jennerwein, M., and Berger, M. (1984). *Cancer Treat. Rev.*, *11*: 125.

Sharma, R.P., and Edwards, I.R. (1983). *Biochem. Pharmacol.*, *32*: 2665.

Sugiyama, S., Hayakawa, M., Kato, T., Hanaki, Y., Shimizu, K., and Ozawa, T. (1989). *Biochem. Biophys. Res. Commun.*, *159*: 1121.

Sunderman, F.W., Jr., and Fraser, C.B. (1983). *Ann. Clin. Lab. Sci.*, *13*: 489.

Sundquist, W.I., and Lippard, S.J. (1990). *Coord. Chem. Rev.*, *100*: 293.

Sundquist, W.I., Bancroft, D.P., Chassot, L., and Lippard, S.J. (1988). *J. Amer. Chem. Soc.*, *110*: 8559.

Sundquist, W.I., Bancroft, D.P., and Lippard, S.J. (1990). *J. Amer. Chem. Soc.*, *112*: 1590.

Sternson, L.A., Repta, A.J., Shih, H., Himmelstein, K.J., and Patton, T.F. (1984). *Platinum Coordination Complexes in Cancer Chemotherapy* (M.P. Hacker, E.B. Douple and I.H. Krakoff, eds.), Martinus Nijhoff, Boston, p. 126.

Szymkowski, D.E., Yarema, K., Essigmann, J.M., Lippard, S.J., and Wood, R.D. (1992). *Proc. Nat. Acad. Sci. USA*, *89*: 10772.

Toney, J.H., Donahue, B.A., Kellett, P.J., Bruhn, S.L., Essigmann, J.M., Lippard, S.J. (1989). *Proc. Nat. Acad. Sci. USA*, *86*: 8328.

Treskes, M., Nijtmans, L.G.J., Fichtinger-Schepman, A.M.J., and Van der Vijgh, W.J.F. (1992). *Biochem. Pharmacol.*, *43*: 1013.

Valsecchi, M., Spinelli, S., Alfonso, G.D., Pasini, A., Moret, M., and Farrell, N. (1991). *Platinum and Other Metal Coordination Compounds in Cancer Chemotherapy* S.B. Howell, ed.), Plenum Press, New York.

Van Boom, S.S.G.E., and Reedijk, J. (1993). *J. Chem. Soc. Chem. Commun.*, 1397.

Van Boom, S.S.G.E., Hensen, A.M.H., and Reedijk, J. (1994). in preparation.

Van der Veer, J.L., Peters, A.R., and Reedijk, J. (1986). *J. Inorg. Biochem.*, *26*: 137.

Van der Vijgh, W.J.F. (1991). *Clin. Pharmacokin.*, *21*: 242.

Van Kralingen, C.G., Reedijk, J., and Spek, A.L. (1980). *Inorg. Chem.*, *19*: 1481.

Visse, R., de Ruijter, M., Brouwer, J., Brandsma, J. A., and van de Putte, P. (1991). *J. Biol. Chem.*, *266*: 7609.

Volshtein, L.M., Krylova, L.F., and Mogilevkina, M.F. (1965). *Russ. J. Inorg. Chem.*, *10*: 1077.

Von Hoff, D.D., Schilsky, R., Reichert, C.M., Reddick, R.L., Rozencwerg, M., Young, R.C., and Muggie, F.M. (1979). *Cancer Treat. Rep.*, *63*: 1527.

Wagstaff, A.J., Ward, A.P., Benfield, A., and Heel, R.C. (1989). *Drugs*, *37*: 162

Yuhas, J.M. (1980). *Cancer Res.*, *40*: 1519.

Zelazowski, A.J., Garvey, J.S., and Hoeschele, J.D. (1984). *Arch. Biochem. Biophys.*, *229*: 246.

Zwelling, L.A., Filipski, J., and Kohn, K.W. (1979). *Cancer Res.*, *39*: 4989.

OLIGONUCLEOTIDE MODELS OF PLATINUM-DNA INTERACTIONS

by

Jean Claude Chottard

Université René Descartes (Paris V),
Laboratoire de Chimie et Biochimie Pharmacologiques et Toxicologiques,
Unité de Recherche Associée au CNRS n°400
45 rue des Saints-Pères
75270 Paris Cedex 06
France

I. QUESTIONS ADDRESSED BY OLIGONUCLEOTIDE MODEL STUDIES

Modelling platinum-DNA interactions with oligonucleotides can lead to a detailed understanding of these interactions, thanks to the power of analytical tools such as HPLC, NMR spectroscopy and X-ray crystallography. This understanding, up to the atomic level, is of primary importance for the design of new drugs targeted at specific DNA sequences and for the study of the recognition of the various platinum-DNA adducts by selective binding and repair proteins, which are involved in the biological processing of the lesions.

Oligonucleotide models are well suited to address the following questions: (i) what controls the kinetics and selectivity of the first platination step on DNA? (ii) what controls the intrastrand chelation selectivity on single and double-stranded DNA and the competition between intrastrand and interstrand chelations? (iii) what are the deformations of DNA due to the platinum adducts? and (iv) can they be related to the biological consequences of the lesions?

Many reviews covering these topics have appeared in the recent years (Sherman and Lippard, 1987; Reedijk, 1987; Reedijk et al., 1987; Lippert, 1989; Sundquist and Lippard, 1990; Lepre and Lippard, 1990; Bruhn et al., 1990; Lempers and Reedijk, 1991) and the purpose of this contribution is to present the most significant answers to the above questions and to point to those which are still lacking.

II. KINETICS AND SELECTIVITY OF DNA PLATINATION

The major platinum adducts from the reaction of *cis*-$[PtCl_2(NH_3)_2]$ (cisplatin or *cis*-DDP) with DNA, both in vitro and in vivo, are the so-called GG- and AG-adducts which are intrastrand N7,N7 chelates of the *cis*-$Pt(NH_3)_2^{2+}$ moiety. (Fichtinger-Schepman et al., 1985, 1987; Eastman, 1986; Plooy et al., 1985). In vitro, the GG- and AG-chelates account respectively for 60-65% and 20%, of the platinum bound; 4-10% correspond to chelates involving two non-neighboring guanines, either from the same strand or from the two strands. No GA-adduct has been detected. However, T4 DNA polymerase mapping of platinated-DNA restriction fragments showed that d(GpA) can react with *cis*-DDP, within unusual structures (Rahmouni et al., 1988); more recent results suggest a platination ratio

GA:AG = 1:10 (Decoville et al., 1993). Also, the arrest of *Taq* DNA polymerase extension of an oligonucleotide primer by *cis*-DDP adducts, followed by linear amplification, indicated $(G)_n$ ($n \geq 2$), and GA, AG and GC sequences as the most intense damage sites in the pUC8 plasmid (Murray et al.,1992). The percentage of GG-adducts (65%) exceeds the 37% value to be expected from an equal reactivity of all the guanines. Platination of DNA by *trans*-DDP was found to give the following adducts after digestion: dG-Pt-dC 50%, dG-Pt-dG 40%, dG-Pt-dA 10% (Eastman et al., 1988). As expected, *trans*-DDP does not form intrastrand chelates between adjacent nucleotides. Knowing that the dinucleotide adducts are formed by a two-step process: platination of a purine followed by chelation on the monoadducts, these results raise several questions which can be addressed with oligonucleotide models.

A. Kinetics and selectivity of the first platination step

DNA platination is kinetically controlled for both *cis*- and *trans*-DDP (Johnson et al., 1980) and this is also the case in the conditions of oligonucleotide studies. *Cis*- and *trans*-DDP react with nucleotides after one or two aquation steps. For $[PtCl_2(en)]$ at pH 7.4, 35°C and 4mM chloride (intracellular concentration), the calculated proportions of the hydrolysis species at equilibrium are the following: dichloro complex 25.3%, $[PtCl(H_2O)(en)]^+$ 17.5%, [PtCl(OH)(en)] 17.5%, $[Pt(H_2O)_2(en)]^{2+}$ 0.6%, $[Pt(OH)(H_2O)(en)]^+$ 24.1%, $[Pt(OH)_2(en)]$ 15% (Lim and Martin, 1976). For *cis*- and *trans*-DDP, the species likely to react with anionic DNA are the cationic electrophilic complexes with a labile aqua ligand. A kinetic study of the platination of the dinucleotide Ap_4A by *cis*-DDP has clearly shown that the first aquation is rate limiting, and has revealed the influence of weakly binding ligands like phosphates on the platination process (Segal and Le Pecq, 1985). In vivo, the aquation equilibrium may not be reached, the major platinating species for *cis*- and *trans*-DDP being the monoaqua derivative $[PtCl(NH_3)_2(H_2O)]^+$ (Johnson et al., 1980). Reactions between $[^{195}Pt]$*cis*-$[PtCl_2(NH_3)_2]$ or *trans*-$[PtCl_2(NH_3)_2]$ and DNA (20 to 60 base pair fragments) in 3mM chloride, 1mM sodium phosphate at pH 6.5, at final respective platinum and nucleotide concentrations of 16 and 227mM, showed that in each case the rate constant of platination was the same as that of the first aquation step of the platinum complex. The pseudo-first-order rate constants determined at 37°C for the *cis*- and *trans*-isomers were: $10.2 \pm 0.7 \times 10^{-5} s^{-1}$ ($t_{1/2} = 1.9 \pm 0.1h$) and $9.6 \pm 0.4 \times 10^{-5} s^{-1}$ ($t_{1/2} = 2.0 \pm 0.1h$). This shows that the monoaqua complex is the actual platinating species in these conditions (Bancroft et al., 1990). However, it should be noted that a nucleotide : platinum ratio of ca 14 corresponds to a ratio guanine : platinum of ca 3.5 which can hardly be considered as pseudo-first-order conditions for the first platination reaction.

For DNA platination monitored by circular dichroism and ethidium bromide fluorescence, the following second-order rate constants were determined: 2000-3000 $M^{-1}min^{-1}$ for *cis*-$[Pt(NH_3)_2(H_2O)_2]^{2+}$ and 280-400 $M^{-1}min^{-1}$ for *cis*- and *trans*-$[PtCl(NH_3)_2(H_2O)]^+$ (Schaller et al., 1987).

Platination reactions of oligonucleotide models have been run either with equilibrated aqueous solutions of cisplatin, or with its diaqua derivative (Scovell and O'Connor, 1977), or with solutions enriched in $[PtCl(H_2O)(NH_3)_2]^+$ (Hollis et al., 1989; Bancroft et al., 1990).

It has been early shown that the order of nucleophilicity towards *cis*- and *trans*-DDP is Guo > Ado > > Cyd (Robins, 1973) and GMP > AMP > > CMP > > UMP (Mansy et al., 1978) and that the Pt-binding sites are the purine N7 (GN7 and AN7) and cytosine N3 (CN3). Only the former two are potential ligands in the major groove of a duplex oligonucleotide (see Reedijk et al., 1987 for a discussion). Buoyant density measurements showed that both *cis*- and *trans*-DDP bind to DNAs in proportion to their (G+C) contents but that only *cis*-DDP is selective for d(GpG) sequences (Stone et al., 1976). Inhibition of DNA polymerase I by platinum adducts on a M13mp8 viral DNA template indicated preferential binding of *cis*-DDP to (dG)n sequences and of *trans*-DDP to d(GpNpG) sequences (N being any intervening nucleotide) (Pinto and Lippard, 1985).

The preference for *cis*-DDP binding to GG sequences has been used to prepare single-stranded oligonucleotides containing one GG-*cis*Pt adduct from oligonucleotides having no other Gs. These platinated oligonucleotides were used (i) for structural studies as single strands e.g. d(CCG*G*) (Marcelis et al., 1981), d(ATG*G*) and d(CCATG*G*) (Girault et al., 1984a), d(G*G*CC) (Inagaki and Kidani, 1984), d(TG*G*CCA) (Girault et al., 1982b), d(AG*G*CCT) (Caradonna et al., 1982), or in duplex structures d(TCTCG*G*TCTC) (den Hartog et al., 1984a) d(CTTCTCTTCTG*G*TCTTCTCT) (Schwartz et al., 1989); (ii) for the construction of site specifically modified genomes for biological studies (Pinto et al., 1986; Naser et al., 1988); (iii) for the construction of polymers, e.g. $[d(TCTCCTTCTG^*G^*TCTCTTCTC)$ - $d(GAAGAGACCAGAAGGAGAGA)]_n$ used to determine bending and unwinding of the duplex from gel electrophoresis mobility (Bellon and Lippard, 1990; Bellon et al., 1991).

Single strands containing only pyrimidines and an AG sequence were also used to prepare AG-*cis*Pt chelates, e.g. d(CTTCTCTTCTA*G*TCTTCTCT) (Marrot and Leng, 1989), d(CTCA*G*CCTC) (Fouchet et al., 1992). In these cases it is better to run the platination reaction with *cis*-$[Pt(NH_3)_2(H_2O)_2]^{2+}$ at low pH, ca 4.2-4.4, in order to avoid competitive cytosine chelation on the first formed guanine monoadduct.

The same strategy was used to prepare the GNG-*cis*Pt adducts d(TCTCG*TG*TCTC) (den Hartog, 1985b) and d(CTCTCTTCTG*TG* TCTTCTCT) (Marrot and Leng, 1989).

It was also applied to the preparation of the 1,3-intrastrand *trans*-$Pt(NH_3)_2$ adducts, d(G*TG*) (van der Veer et al., 1986a), d(G*CG*) (van der Veer et al., 1986b; Gibson and Lippard, 1987), d(A*GG*CCT) (Lepre et al., 1987), d(TCTCG*TG*TCTC) (den Hartog et al., 1985b), d(TTTTCG*CG*TTTT) (Ptak et al., 1989), d(TCTACG*CG*TTCT) (Comess et al., 1990) and d(CCTCG*AG*TCTCC) (Lepre et al., 1990). Whereas the latter adduct is stable, d(TCTACG*CG*TTCT) rearranges ($t_{1/2}$ 47h at 37°C) to give an equilibrium mixture containing the more stable 1,4-intrastrand N3(5), N7(8) chelate (Comess et al., 1990).

It should be stressed that the selectivity for GG or GNG platination on single strands is only observed in the absence of other Gs in the sequence (Chottard, 1988).

The influence of the adjacent bases on G platination has been studied with short oligonucleotides. Competition experiments between several dinucleotides reacting with [PtCl(dien)]Cl have given the following order of reaction rates: d(CpG):d(TpG):d(ApG) 4:3:2 (van der Veer et al., 1987). The reaction of d(GAG) with $[PtCl(dien)]^+$ revealed 65% of platinum binding at G(3) and 35% at G(1) (van der Veer et al., 1986b). However, in this case G(1) lacks both a 5'-base and a 5'-phosphodiester, the latter being often invoked for a platination directing effect. Accordingly, a study with d(pGGG) and *cis*-DDP showed that the major adduct was the G(1)G(2) chelate, suggesting a directing influence of the free 5'-phosphate on the first platination step (van der Veer, 1986). Such an effect agrees with the increase in the platination rate observed for 5'-GMP compared to 3'-GMP (Marcelis et al., 1984) and also found for d(pGpG) compared to d(GpG) (Girault et al., 1982a). One should be aware of the very small differences in activation energy which correspond to the differences in percentage of the various adducts formed on di- and trinucleotides. It is noteworthy that $[PtCl_2(en)]$, $[PtCl(en)(Me_2SO)]Cl$ and *cis*-$[PtCl(NH_3)_2$(4-methylpyridine)]Cl reacting with d(GpG) respectively give 60% and 40%, 70% and 30%, and 65% and 35% of the 5'- and 3'-adducts; a similar selectivity is observed with [PtCl(dien)]Cl and $[PtCl(NH_3)_3]Cl$ (Lempers et al., 1990; 1991). To get an insight into the competition between various Gs, the reaction of *cis*-$[Pt(NH_3)_2(H_2O)_2]^{2+}$ was investigated with the decanucleotides d(GCCGGATCGC) and d(GCGATCCGGC), and with their duplex form. The adducts were isolated by HPLC and characterized by enzymatic digestion (Gauthier, 1990; Chottard, 1988). On the single strands, all the possible platinum chelates between adjacent and non-adjacent guanines were formed. For the first decanucleotide, they account respectively for 38% and 45% of the platinum bound. The identified adducts were the following:

GCCG*GATCG*C	10%	G*CCGGATCG*C	8%
GCCGG*ATCG*C	11%	G*CCGG*ATCGC	8%
G*CCG*GATCGC	8%	GCC*G*GATCGC	8%
G*CCG*G*ATCG*C	8%	GCCG*G*ATCGC	30%

The proportions are identical for the reaction run in water and in $NaNO_3$ O.5M. As far as the first platination step is concerned, the least possible amount of reaction which has not occurred on G(4) or G(5) is 16% corresponding to the G(1)-G(9) chelate and the G(4)-G(5), G(1)-G(9) bis-chelate. However, one cannot exclude a first platination on G(1) or G(9) followed by chelation by G(4) or G(5). The platination of the second decanucleotide gave 7 adducts with G*CGATCCG*GC and GCGATCCG*G*C accounting respectively for 34% and 17% of the platinum bound, the proportion of the second adduct increasing with ionic strength. The reaction of the *cis*-DDP diaqua derivative with the duplex d(GCCGGATCGC)-d(GCGATCCGGC) either in water (Tm = 48°C) or in $NaNO_3$ 0.5M (Tm = 58°C) at 25°C or 4°C, gave 66% of the G(4)-G(5) chelate on the first strand and 33% of the G(18)-G(19) chelate on the second strand. This clearly shows a complete selectivity of the first platination step for the GG sequences within the duplex structure (Gauthier, 1990; Chottard, 1988). However, to get a full understanding of these results, one needs a determination of the rate constants for the platination of each guanine within a GG sequence as a function of the adjacent bases.

It is not yet understood which factors orient the first platination by *cis*-DDP towards $(dG)_n$ sequences ($n \geq 2$) and which guanines are platinated first. Calculations on a duplex have shown that the negative electrostatic potential of GN7 is (i) higher than that of AN7, (ii) higher in the central part of an oligonucleotide than at the ends, and (iii) increased for adjacent guanines (Pullman and Pullman, 1981).

Work is in progress in our group, based on kinetic studies with single- and double-stranded oligonucleotides reacting with *cis*-$[Pt(NH_3)_2(H_2O)_2]^{2+}$, $[Pt(NH_3)_3(H_2O)]^{2+}$ and *cis*-$[PtCl(NH_3)_2(H_2O)]^+$ (Gonnet et al., 1992), coupled with a molecular modelling analysis of the first platination transition state (Laoui et al., 1988).

B. Kinetics and selectivity of the chelation step

It was soon recognized that the monofunctional adducts formed on double-stranded DNA by *cis*- and *trans*-DDP had life-times of several hours (Eastman, 1983; Eastman, 1986; Butour and Johnson, 1986). However, the chelation rate measurements were based on various trapping techniques of the monoadducts which introduced some discrepancies between the data (for a discussion see Lepre and Lippard, 1990). Trapping experiments with ^{14}C-labeled thiourea gave half-lives of 1.1 and 2.4h for the chelation of the monoadducts formed respectively with *cis*-DDP and *trans*-1,2-diaminocyclohexanedichloroplatinum (Eastman, 1986; Page et al., 1990). For *trans*-DDP, the monoadducts were found to have a longer half-life on double-stranded than single-stranded DNA (Eastman and Barry, 1987).

Experiments using *cis*-DDP monofunctional chloro-adducts on DNA, generated from $(NH_3)_2$ (ethidium) monoadducts by removal of ethidium, gave a chelation half-time of about 4h, corresponding to the hydrolysis rate of the second chloride ligand (Malinge and Leng, 1988).

Using ^{195} Pt NMR to monitor the chelation of the monoadducts to diadducts of *cis*- and *trans*-DDP in 3mM chloride at pH 6.5 and 37°C, the following pseudo-first-order rate constants were determined: $9.2 \pm 1.4 \times 10^{-5} s^{-1}$ ($t_{1/2} = 2.1 \pm 0.3h$) and $6.3 \pm 0.1 \times 10^{-5} s^{-1}$ ($t_{1/2} = 3.1 \pm 0.1h$), which are identical to those determined for chelation on single-stranded DNA (Bancroft et al., 1990). These results confirm that hydrolysis of the chloride ligand of the monoadducts is rate-determining for the chelation step in this type of conditions.

In order to get actual rate constants for the chelation of the monoaqua-monoadduct species on DNA, the latter (0.05-1mM in nucleotide) was reacted with cis-$[Pt(NH_3)_2(H_2O)_2]^{2+}$ at pH 5.5 (MES buffer) and 37°C, at a ratio of 0.1 mole platinum per mole of DNA nucleotide. Quenching of the reactions at pH 8.3 with chloride followed by endonucleolytic digestion and HPLC analysis of the adducts, revealed two waves of cross-linking with rate constants 0.2-0.3 min^{-1} ($t_{1/2}$ = 3.5min) and 0.015-0.025 min^{-1}($t_{1/2}$ = 35 min), correlated with changes in optical properties. The second wave of cross-linking, assigned to a local "melting" of DNA would probably not occur under in vivo conditions with a much lower platinum-to-DNA molar ratio ($< 10^{-5}$). The results also suggest that the d(GpG), dGuo-Pt-dGuo, d(GpNpG), d(ApG) and d(ApNpG) chelates are formed at very similar rates (Bernges and Holler, 1991).

Concerning chelation selectivity on DNA, it has been early found that no d(GpA) adduct was formed with *cis*-DDP (Fichtinger-Schepman et al., 1985; Eastman, 1986; Johnson et al., 1985). An interpretation has been proposed based on the distances within the AGA sequence in B-DNA, between the platinum atom of the monoadduct bound to G-N7 and the two potential chelating sites: the 5'-AN7 at about 3Å and the 3'-AN7 at about 5Å (Dewan, 1984). As mentioned above, several results now claim the platination of GA and GC sequences (Rahmouni et al., 1988; Murray et al., 1992; Decoville et al., 1993). However, the corresponding platinum chelates have not yet been characterized from a platinated DNA digest.

At the oligonucleotide level, no study has yet given the chelation rate constants for 5'-G and 3'-G chelation on the adjacent 3'- and 5'-monoadducts. Such a work is going on in our laboratory on single-stranded and double-stranded oligonucleotides (Gonnet et al., 1992).

It is noteworthy that GA chelation does occur in single-stranded oligonucleotides. While ApG gives only the N7, N7 *cis*-$Pt(NH_3)_2$ chelate, d(ApG) and d(GpA) both give a mixture of N1, N7 and N7, N7 adducts. For d(AGA) and d(TAGAT), N1 coordination is not observed, but GA chelation predominates over AG chelation (67% and 75% for the tri- and tetranucleotides, respectively); an equal amount of each chelate is formed with d(pApGpA) (Dijt et al., 1989).

The G*AG*/A*G*/G*A* competition for the chelation step has been studied by comparing the reactions of d(GpApG) with [PtCl(dien)]Cl and *cis*-DDP. The former gave 65% of 3'-G and 35% of 5'-G monoadducts. Assuming the same selectivity for the first platination step by *cis*-DDP, one deduced that the 3'-G monoadduct was chelated both by the 5'-G and the 5'-A (45 and 20%, respectively) whereas the 5'-G was only chelated by the 3'-G (35%) giving an overall 80% of G*AG* and 20% of GA*G* chelates (van der Veer et al., 1986b).

For single-stranded oligonucleotides, the results of these two series of experiments suggest that the formation of a GA chelate, which results from the adenine chelation on its neighboring 5'-guanine-monoadduct, can only be observed in the absence of a competing 3'-guanine, either adjacent or separated by a small number of bases (see GCCG<u>GA</u>TC<u>G</u>C, vide supra).

Another competition for chelation of platinum monoadducts is that between the formation of intra- vs interstrand diadducts. *cis*-DDP gives up to 10% of G-Pt-G interstrand cross-links on DNA in vitro (Fichtinger-Schepman et al., 1985), but this proportion can decrease to 1% in mammalian cells (Plooy et al., 1984). It has been demonstrated, using polymerase inhibition and selective resistance of the adducts to cyanide removal, that interstrand cross-links are preferentially formed at d(GC) sites (Lemaire et al., 1991). This reaction has been studied in more detail with the 22-mer oligonucleotides: d(CTTCTCCT<u>TGCT</u>CTCCTTCTTC) and d(TCTCTCCTCT<u>AGCT</u>CTCCTTCT) associated to their complementary strands. The duplex G-$PtCl(NH_3)_2$ monoadducts have been prepared by 1M NaCl reaction with the [$Pt(NH_3)_2$(*N7*-N-methyl-2-diazapyrenium)-(*N7*-dG)] adduct. Incubation of the chloroadducts in 60mM $NaClO_4$ at 37°C led to interstrand cross-linking at the GC sites, with chelation-$t_{1/2}$ of 14 and 3h for the central sequences TGCT and AGCT, respectively. In the same conditions, the reaction of *cis*-$[Pt(NH_3)_2(H_2O)_2]^{2+}$ with the duplex of the first 22-mer (TGCT) was complete in 15 minutes. This suggests that the difference observed between the two cross-linking reactions might reflect a difference in the rates of aquation of the chloride ligand (Payet et al., 1993).

Trans-DDP was thought to be a better candidate than the *cis*-isomer for interstrand cross-linking due to the larger distance between the two leaving ligands. A complete analysis of the adducts formed by *trans*-DDP on DNA is not available. *Trans*-DDP is less selective than *cis*-DDP and the major adducts are the 1,3 chelates: d(G*pNpG*) and d(A*pNpG*) (Pinto and Lippard, 1985). The former represents 40% of the adducts found in vitro on double-stranded DNA and 60% on single-stranded DNA (Eastman et al., 1988). This suggests that such 1,3-chelation is not hindered by the duplex structure. As noted previously, the slow rearrangement of the single-stranded *trans*-[$Pt(NH_3)_2$ d(TCTACG*CG*TTCT)] chelate, compared to the stability of the *trans*-[$Pt(NH_3)_2$

Table 1

Rate Constants for the First Platination Step (k_1) and the Chelation Step (k_2) of Single-Stranded Oligonucleotides Reacting with *cis*-$[Pt(NH_3)_2(H_2O)_2]^{2+}$ in Water at pH 5.2 and 20°C [a]

Nucleotide	$k_1(M^{-1}.s^{-1})$	$k_2 \times 10^3 (s^{-1})$
d(TGGCTA)	139±5	2.4±0.2
d(TAGCCA)[b]	47±1	0.52±0.04

[a]Laoui, 1989a; Pseudo- first-order platination conditions: 0.125mM Pt/0.01mM hexanucleotide, UV monitoring at 272nm (AG) and 280nm (GG).
[b]Some *cis*-$[Pt(NH_3)_2\{d(TAG^*CCA)\}_2]$ bis-adduct is formed even at a 10^{-5}M concentration.

d(CCTCG*AG*TCTCC)] isomer, suggests a sequence-dependent assistance to ligand exchange on the inert Pt(II) (Lepre et al., 1990).

C. Discussion of the results of oligonucleotide studies relevant to DNA platination

The present situation of the modelling studies is the following:

- In a single-stranded oligonucleotide, a GG sequence, in the absence of other Gs, will selectively react with the aquated forms of *cis*-DDP to give the GN7,GN7 chelate of the *cis*-$Pt(NH_3)_2^{2+}$ moiety. The difference between the reactivities of the GG and AG sequences is illustrated by the overall rate constants for the platination and chelation steps of GG- and AG-containing hexanucleotides reacting with *cis*-$[Pt(NH_3)_2(H_2O)_2]^{2+}$ (Table 1).
- In the presence of other Gs, the selectivity for GG platination is only observed in duplex structures.
- In a single-stranded oligonucleotide, a GNG sequence, in the absence of other Gs, will selectively react with the aquated forms of *cis*- or *trans*-DDP to give the GN7,GN7 chelates of the *cis*- and *trans*-$Pt(NH_3)_2^{2+}$ moieties. A study of these reactions has not yet been reported for a duplex oligonucleotide.
- Concerning the kinetic data for the DNA platination and chelation steps, the relevant results have been obtained with 20 to 60 base-pair DNA fragments and various types of DNAs (for a review see Lepre and Lippard, 1990; Bernges and Holler, 1991). They do not differentiate between the various kinds of guanines according to their neighboring bases and to the sequence they belong to. They do not give an insight into the factors which control Pt-binding and chelation selectivities. To tackle these questions, oligonucleotide models seem best suited to give quantitative data susceptible of a structural interpretation based on molecular modelling. Several groups are actively working in this direction.

III. DNA DEFORMATION DUE TO PLATINUM ADDUCTS

Oligonucleotide studies have played the major role in the determination of the structures of the major *cis*- and *trans*-platinum adducts and in the prediction of the corresponding DNA deformations.

A. Di- and trinucleotide adducts of *cis*- and *trans*-$Pt(NH_3)_2^{2+}$ and of $Pt(triamine)^{2+}$

1. *Cis-$Pt(NH_3)_2$ chelates*

Soon after DNA was designated as the likely cellular target of *cis*-DDP, at the origin of its antitumor activity, several results pointed to intrastrand chelates as the most probable adducts (for a review see Sherman and Lippard, 1987). The first attempt to prove such a chelation was made by a CD study, as a function of pH, of *cis*-DDP adducts of ApA and ApC (Roos et al., 1974).

The first clear-cut evidence for the easy formation of dinucleotide-*cis*$Pt(NH_3)_2$ chelates, was obtained by ^1H NMR characterization of the single product formed upon reaction of *cis*-$[Pt(NH_3)_2(H_2O)_2](NO_3)_2$ with IpI (Chottard et al., 1978), GpG, d(GpG) and d(pGpG) (Chottard et al., 1980; Girault et al., 1982a). These complexes are N7,N7 chelates of the *cis*-$Pt(NH_3)_2^{2+}$ moiety, with a "head-to-head" arrangement of the purines and a pure N conformation for the sugar of the 5'-nucleotide instead of pure S in the free dinucleotide. A more detailed NMR study including computer simulations gave a complete picture of the most likely conformer of *cis*-$[Pt(NH_3)_2\{d(GpG\text{-}N7,N7)\}]^+$ in solution. Among the added features were: the dihedral angle of about 53° between the bases, the χ values of 110 and 115° confirming the *anti* conformation of the two Gs and the sugar pseudorotation characteristics concluding to 100%N and 70%S conformations for the 5'- and 3'-Gs (den Hartog et al., 1982) (Fig. 1). All these data were confirmed by the X-ray crystal structure of *cis*-$[Pt(NH_3)_2$ d(pGpG)] (Fig. 2) which revealed a hydrogen bond between an ammine ligand and the 5'-terminal phosphate (Sherman et al., 1985; 1988). This hydrogen bond might also be present with phosphodiester groups in longer oligonucleotides, as well as the weak hydrogen bonding present between an ammine ligand and the guanine 06 atoms, which were found in the crystal structure of *cis*-$[Pt(NH_3)_2\{d(CpGpG)\text{-}N7(2)N7(3)\}]$ (den Hartog et al., 1985c; Admiraal et al., 1987; Reedijk, 1992).

Dinucleotide adducts of the *cis*-$Pt(NH_3)_2^{2+}$ moiety, containing only one G, have been prepared and characterized by ^1H NMR in the cases of CpG and pCpG (*N3*(1), *N7*(2)) (Girault et al., 1984b), ApG and GpA (*N7*(1), *N7*(2)) (van Hemelryck et al., 1987), d(ApG), d(GpA)(Dijt et al., 1989). In all these adducts, the sugar of the 5'-nucleoside adopts a pure N conformation which appears as a characteristic of the Pt chelates between adjacent bases. In addition, they also exhibit a rotation of the 5'-base between the *anti* and *syn* conformations, not observed for the GG adducts.

Cis-DDP can also cross-link the two GN7 of d(GCG) (den Hartog et al., 1983) and d(GTG) (van der Veer, 1986a) as seen above. A detailed ^1H NMR study of the d(G^*CG^*) adduct concluded to a "bulged-out" structure for the intervening cytidine, accompanied for the two guanosines either by a rapid *syn* $\rightleftharpoons$ *anti* equilibrium on the NMR time scale or by a locked conformation between the *syn* and *anti* positions. The guanosine sugars (about 60% S) exhibit much more conformational mobility than the cytidine one (80% S).

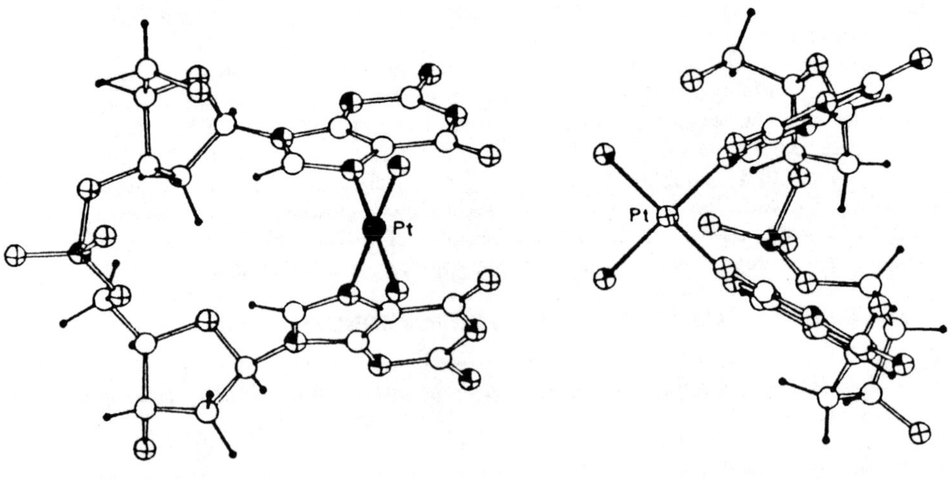

Figure 1.
cis-$[Pt(NH_3)_2$ d(GpG)] structure as deduced from NMR data.
ORTEP drawing of a possible conformer in solution, C = O, N = ⊕, O = ⊕, P = ●
(Reproduced with permission from den Hartog et al., 1982).

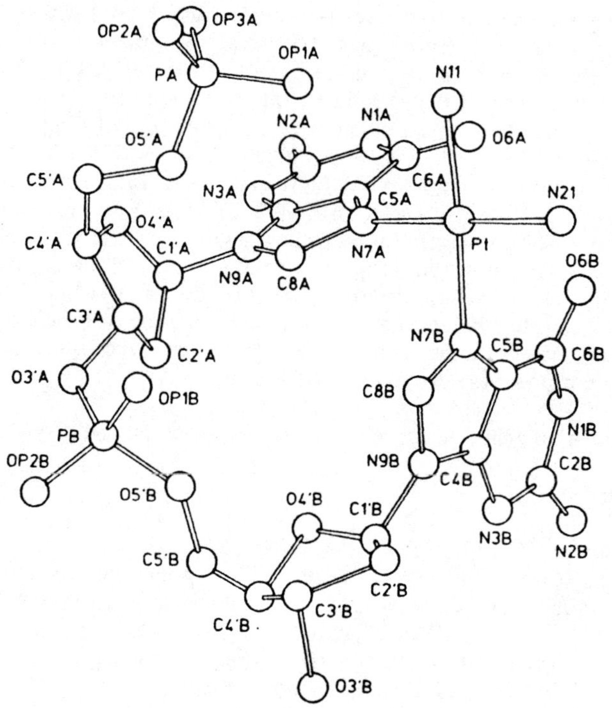

Figure 2.
cis-[Pt$(NH_3)_2$ d(pGpG)] crystal structure.
Molecular structure of one of the four crystallographically independent molecules in the unit cell. (Reproduced with permission from Sherman et al., 1985).

2. ***Trans*-Pt$(NH_3)_2$ *chelates***
Trans-DDP being devoid of antitumor properties, modelling of its binding to DNA started later than for the *cis*-isomer. At the oligonucleotide level, *trans*-DDP, like *cis*-DDP, binds to a guanine in the first step. It cannot cross-link an adjacent base to give a 1,2 chelate, but it readily forms 1,3 chelates of the G*NG* and A*NG* types (N being any intervening nucleotide). The trinucleotide adducts d(G*TG*) (van der Veer et al., 1986a) and d(G*CG*) (Gibson and Lippard, 1987) have been studied by NMR. As for the corresponding *cis*-Pt$(NH_3)_2$ chelates, platinum cross-linking of the two GN7 destacks the central pyrimidine. This destacking might be larger than for the *cis*-Pt$(NH_3)_2$ adducts and the perturbation of the sugar-phosphate backbone is accompanied by an N conformation for the deoxyribose of the 5'-guanosine.

3. *Pt(triamine) monoadducts*
The CD spectra of ApG, [Pt(dien){ApG-*N7*(2)}]$^+$, [Pt$(NH_3)_3${ApG-*N7*(2)}]$^+$, *cis*-[PtCl$(NH_3)_2${ApG-*N7*(2)}] and *cis*-[Pt$(NH_3)_2${ApG-*N7*(1),*N7*(2)}] clearly showed larger and similar signals for the platinum triamine monoadducts compared to ApG, a much larger signal for the N7,N7 chelate, and a very small signal for the diamminechlorocomplex. This suggested a perturbation of the base stacking due to platinum binding, associated with hydrogen bonding between amine NH and the phosphodiester group (Laoui, 1989b). This perturbation was also revealed by NMR studies of the G-Pt(dien) adducts of CGT (van Garderen et al., 1987) and AGA (Admiraal et al., 1992).

B. Single-stranded oligonucleotide adducts of *cis*- and *trans*-Pt$(NH_3)_2^{2+}$
We have seen di- and trinucleotide models of *cis*- and *trans*-DDP DNA adducts which represent the basic 1,2 and 1,3 purine chelates of the *cis*- and *trans*-Pt$(NH_3)_2$ moieties.

Several longer oligonucleotide adducts have been studied which are listed in Table 2 with the corresponding references. They gave ^1H NMR confirmation of the structural characteristics of the GG-*cis*Pt$(NH_3)_2$ and GNG *cis*-and *trans*-Pt$(NH_3)_2$ chelates. It was already mentioned in Section II.B that whereas all the oligonucleotide platinum chelates studied so far were found to be stable in aqueous solution, including *trans*-[Pt$(NH_3)_2$ d(A*GG*CCT)] (Lepre et al., 1987) and *trans*-[Pt$(NH_3)_2$ d(CCTCG*AG*TCTCC)] (Comess et al., 1990), the complex *trans*-[Pt$(NH_3)_2${d(TCTACG*CG*TTCT)}] slowly rearranges into the 1,4 chelate *trans*-[Pt$(NH_3)_2$ {d(TCTAC*GCG*TTCT)}] (Lepre et al., 1990).

Recently a ^1H NMR and molecular modelling study of several *cis*-Pt$(NH_3)_2$ adducts revealed that the GG-chelates can adopt two conformations differing by the helical arrangement of the purines. It is noteworthy that in the crystal unit cell of the *cis*-[Pt$(NH_3)_2$(dpGpG)] complex discussed above, one can distinguish two chelates with a right-handed helicity and two with the opposite one. In duplex structures, the right-handed chelate is the only one present (Kozelka et al., 1992).

C. Double-stranded oligonucleotide adducts of *cis*- and *trans*-Pt$(NH_3)_2$ $^{2+}$ and of Pt(dien)$^{2+}$

They are also listed in Table 2.

1. *Structures of duplex GG-cisPt$(NH_3)_2$ and AG-cisPt$(NH_3)_2$ adducts*

The following duplexes have been studied, which were prepared by adding the complementary strand to the purified single-stranded GG-*cis*Pt$(NH_3)_2$ adduct. d(TCTCG*G*TCTC) - d(GAGACCGAGA) (I) (den Hartog et al., 1984a), d(GATCCG*G*C)- d(GCCGGATCGC) (II) (van Hemelryck et al., 1984), d(GCCG*G*ATCGC) - d(GCGATCCGGC) (III) (van Hemelryck et al., 1986). In the three cases, platinum coordination lowers the melting temperature of the duplex, respectively by 15, 27 and 9°C. This destabilization depends on the GC content of the duplex. The CD spectra revealed that the three platinated duplexes keep a B-DNA type structure. ^1H and ^{31}P NMR studies of (I) and (II) (den Hartog et al., 1984b; 1985a; Herman et al., 1990) led to the following conclusions:

- The GG-*cis*Pt$(NH_3)_2$ unit within the duplex is in every respect identical to the dinucleotide chelate seen above.
- All the imino proton resonances were observed at low temperatures with some shifts and with some broadening and/or modification of the melting profile for the three CG*G*-CCG pairs, indicating that base-pairing was still present but mainly perturbed for the 5'-G*C pair.
- Analysis of the ^1H chemical shifts and coupling constants, and of the ^{31}P data (Fouts et al., 1987) concluded to a local distortion of the double helix, mainly affecting the CG*G*-CCG sequence, which suggested the presence of a kink (den Hartog et al., 1985a).

Another insight into the structure of the platinated duplex (I) was obtained from molecular mechanics calculations (Kozelka et al., 1985; 1986; 1987). The calculations predicted two structures of similar energies, without and with a kink of the helix axis. In both cases, base destacking was observed, mainly affecting the 5'-G* and its pairing with the opposite C. The kink angle was evaluated to 40-60°. The models revealed an ammine hydrogen bond to the phosphodiester group in 5' position to the G*G* chelate, as found in the crystal structure of *cis*-[Pt$(NH_3)_2$(dpGpG)] (Sherman et al., 1988). Other modelling studies proposed helix bends of 40 and 70° (Miller et al., 1985).

A deeper investigation of the structure of the platinated duplex (III) was conducted using sugar coupling constants and 2D-NMR data, to drive molecular mechanics calculations (Herman et al., 1990). This study clearly concluded to a kinked structure. It was characterized by a kink angle of 55-62° towards the major groove and an unwinding angle of 12-19°. The calculations also suggested the existence of an equilibrium between two types of conformations characterized by two stacking positions of the cytosine opposite the 5'-G*, either on its 5'-C or on its 3'-G-neighbor (Kozelka and Chottard, 1990) (Fig. 3).

A similar study of the duplex AG-*cis*Pt$(NH_3)_2$ adduct d(CTCA*G*CCTC)-d(GAGGCTGAG) (IV) (Fouchet, 1992) concluded to a kinked structure with respective

Table 2

Oligonucleotide Adducts of the *cis*-$Pt(NH_3)_2^{2+}$, *trans*-$Pt(NH_3)_2^{2+}$ and $Pt(triamine)^{2+}$ Moieties, Studied by NMR Spectroscopy, X-ray Crystallography, Molecular Modelling (MM) and Chemical Probe (CP) Techniques[a].

(a) *cis*-$Pt(NH_3)_2^{2+}$ chelates

Type of adducts	Technique	Reference
GG *cis*-$Pt(NH_3)_2$		
- single-stranded		
G^*pG^*	NMR,MM	Chottard et al. (1980) Girault et al. (1982a) Kozelka et al. (1992)
$d(G^*pG^*)$	NMR,MM	Girault et al. (1982a) den Hartog et al. (1982) Kozelka et al. (1992)
$d(pG^*pG^*)$	NMR,X-ray	Girault et al. (1982a) Sherman et al. (1985; 1988)
$d(CpG^*pG^*)$	NMR-X-ray	den Hartog et al. (1985c) Admiraal et al. (1987)
$d(CpCpG^*pG^*)$	NMR	Marcelis et al. (1981; 1983)
$d(ApTpG^*pG^*)$	NMR	Girault et al. (1984a) Neumann et al. (1984)
$d(CpCpApTpG^*pG^*)$	NMR	Girault et al. (1984a)
$d(TpG^*pG^*pCpCpA)$	NMR	Girault et al. (1982b)
$d(ApG^*pG^*pCpCpT)$	NMR	Caradonna et al. (1982)
- double-stranded[b]		
$[(GAG^*G^*) - (CCTC)],[(G^*G^*AG) - (CTCC)]$	MM	Hambley (1991)
$[(GATCCG^*G^*C) - (GCCGGATCGC)]$	NMR	van Hemelryck et al. (1984)
$[(TCTCG^*G^*TCTC) - (GAGACCGAGA)]$	NMR	den Hartog et al. (1984a; 1985a)
$[(GCCG^*G^*ATCGC) - (GCGATCCGGC)]$	NMR,MM	van Hemelryck et al. (1986) Herman et al. (1990) Kozelka and Chottard (1990)
$[(CTTCTCTTCTG^*G^*TCTTCTCT) - (GAGAGAAGACCAGAAGAGAA)]$	CP	Schwartz et al. (1989)

Table 2 (continued)

(a) *cis*-$Pt(NH_3)_2^{2+}$ chelates

Type of adducts	Technique	Reference
CG, AG, GA.*cis*-$Pt(NH_3)_2$ adducts		
- single-stranded		
C*pG*, pC*pG*	NMR	Girault et al. (1984b)
A*pG*, G*pA*	NMR	van Hemelryck et al. (1987)
d(A*pG*),d(pA*pG*)d(G*pA*),d(pG*pA*)	NMR	Dijt et al. (1989)
d(GA*G*)	NMR	van der Veer et al. (1986b)
d(A*G*A),d(AG*A*),d(pA*pG*pA), d(pApG*A*),d(TA*G*AT),d(TAG*A*T)	NMR	Dijt et al. (1989)
- double-stranded		
[(CTCA*G*CCTC) - (GAGGCTGAG)]	NMR,MM	Fouchet et al. (1992) Fouchet (1992)
[(CTTCTCTTCTA*G*TCTTCTCT) - (GAGAGAAGACTAGAAGAGAA)]	CP	Schwartz et al. (1989)
GNG.*cis*-$Pt(NH_3)_2$ adducts		
- single-stranded		
d(G*pCpG*)	NMR	den Hartog et al. (1983)
d(G*pApG*)	NMR	van der Veer et al. (1986b)
d(G*pTpG*)	NMR	van der Veer et al. (1986a)
d(CpG*pCpG*),d(pCpG*pCpG*) d(pG*pCpG*pC)	NMR	Marcelis et al. (1983)
- double-stranded		
[(TCTCG*TG*TCTC) - (GAGACACGAGA)]	NMR	den Hartog et al. (1985b)
[(TTTTCG*CG*TTTT) - (AAAACGCGAAAA)]	NMR	Ptak et al. (1989) Mazeau et al. (1989)
[(CTCTCTTCTG*TG*TCTTCTCT) - {AGAGAGAAGACACAGAAGAG)]	CP	Marrot and Leng (1989)

Table 2 (continued)

(b) *trans*-$Pt(NH_3)_2^{2+}$ chelates

Type of adducts	Technique	Reference
GNG.*trans*-$Pt(NH_3)_2^{2+}$ adducts		
- single-stranded		
d(G*pTpG*)	NMR	van der Veer et al. (1986a)
d(G*pCpG*)	NMR	Gibson and Lippard (1987)
d(TCTACG*CG*TTCT)	NMR	Comess et al. (1990)
- double-stranded		
[(CCTCG*AG*TCTCC) - (GGAGACTCGAGG)]	NMR,MM	Lepre et al. (1990)
ANG and CNN'G.*trans*-$Pt(NH_3)_2$ adducts		
- single-stranded		
d(A*GG*CCT)	NMR	Lepre et al. (1987)
d(TCTAC*GCG*TTCT)	NMR	Comess et al. (1990)

(c) Pt(triamine)$^{2+}$ monoadducts

G.Pt(dien) monoadducts		
- single-stranded		
d(CpG*pT)	NMR	van Garderen et al. (1987)
d(ApG*pA)	NMR	Admiraal et al. (1992)
- double-stranded		
[CGCG*AATTCGCG]$_2$[c]	X-ray	Wing et al. (1984)
[(TCTCG*TCTC) - (GAGACGAGA)]	NMR	van Garderen et al. (1989)
[(TCTCTCCTCTAG*CTCTCCTTCT) - (AGAGAAGGAGAGCTAGAGGAGA)] and similar duplexes with AG*T, CG*A, TG*A, TG*C and TG*T sequences.	CP	Brabec et al. (1992)

[a]The asterisks indicate the platination binding sites; [b]All the platinated duplexes belong to the deoxyribose series, therefore the d is omitted in the formulas as well as the phosphodiester symbol; [c]The platinum moiety is $PtCl(NH_3)_2$.

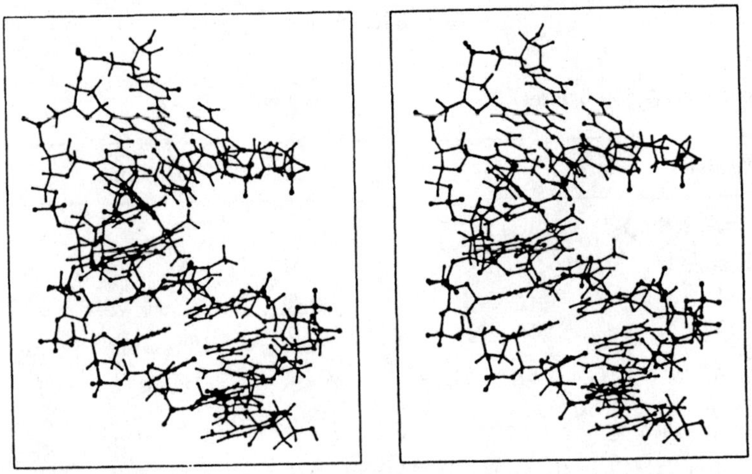

Figure 3.
Stereoview of one molecular model for *cis*-[Pt$(NH_3)_2${d(GCCG*G*ATCGC)-d(GCGATCCGGC)}].
This model corresponds to the conformation in which the cytosine (C17) opposite the platinated 5'-G*(G*4) is stacked on the cytosine (C16) on its 5'-side (see text). (Reproduced with permission from Herman et al., 1990).

kink and unwinding angles of 46° and 17°. Molecular mechanics and dynamics suggested stacking of the T opposite A* on its 5'-C neighbor and pointed to a conformational mobility of this T related to two non Watson-Crick hydrogen bonding to A* (Fouchet, 1992; Fouchet et al., to be published).

Both the GG- and AG-*cis*Pt$(NH_3)_2$ adducts give comparable kinked structures.

2. *Structures of duplex GNG-**cis**Pt$(NH_3)_2$ and **trans**Pt$(NH_3)_2$ adducts*

a. *cis*Pt adducts

The following *cis*Pt$(NH_3)_2$ 1,3-intrastrand cross-linked duplexes have been prepared, by adding the complementary strand to the purified G*NG* single-stranded adducts: d(TCTCG*TG*TCTC) - d(GAGACACGAGA) (V) (den Hartog et al., 1985b) and d(TTTTCG*CG*TTTT) - d(AAAACGCGAAAA) (VI) (Ptak et al., 1989). The decrease of the melting temperature was 26°C for both duplexes, that is more than 10°C larger than that observed for the comparable GG-adduct (I). NMR studies revealed a complete exchange of the N3H or N1H imino protons of the central AT or GC base pairs and some exchange of those of the two flanking G*C pairs. Together with other ^1H and ^{31}P NMR data, including the characteristic ones already presented for trinucleotide adducts, it was concluded that the central base pair of the chelate was disrupted and that the base between the chelating G*s was bulged out. Molecular mechanics calculations related to the NMR study of the adduct (VI) concluded to two possible structures for the platinated dodecamer, either with no kink of the helix axis or with a 60° kink. In both cases, a *syn* conformation was found for the 5'-G*, the sugars of the two chelating G*s exhibiting S conformations. A hydrogen bond was also found between an ammine ligand and the phosphodiester group in 5' position to the G*CG* chelate (Mazeau et al., 1989). The use of chemical probes to detect unpaired bases in a 20 mer G*TG*-*cis*Pt$(NH_3)_2$ duplex concluded to the unpairing of the 5'-G* and of the central T (Marrot and Leng, 1989).

b. *trans*-Pt adducts

The structure of only one *trans*-Pt$(NH_3)_2$ 1,3-intrastrand cross-linked duplex has been reported yet: d(CCTCG*AG*TCTCC) - d(GGAGACTCGAGG) (VII). The melting temperature of the duplex was not affected by the platinum chelate. No downfield shifted

^{31}P NMR signal was observed, excluding a large distortion of the phosphodiester backbone. However, ^{1}H NMR revealed that the imino protons of the three central base pairs were accessible to the solvent. A molecular dynamics model gave a slightly bent structure, by an angle of 18°, with a little affected sugar phosphate backbone and an important destacking of the three central base pairs (Fig. 4). The destabilization due to the corresponding base pair disruptions is compensated by several hydrogen bonds which bridge the two strands via the ammine ligands (Lepre et al., 1990).

3. *Distortion of a duplex structure by G-Pt(triamine) monoadducts*

CD and UV data did not reveal any perturbation of base stacking due to Pt(dien)-monoadducts on DNA (Macquet and Butour, 1978a; 1978b). By soaking crystals of $[d(CGCGAATTCGCG)]_2$ in a solution containing *cis*-DDP, monochlorodiammine platinum monoadducts were formed at G-N7 sites and found not to disrupt base stacking or pairing (Wing et al., 1984).

However, $Pt(dien)^{2+}$ monofunctional binding to poly(dG-dC)-poly(dG-dC) was shown to favor B to Z transition (Malfoy et al., 1981; Ushay et al., 1982). Recently, it has been shown that the monoadduct [Pt(dien)(dTCTCG*TCTC)]-d(GAGACGAGA)] led to a 16° decrease of the melting temperature of the duplex, comparable to that due to platinum chelates on duplex oligonucleotides of similar sizes and compositions, suggesting a structural perturbation confirmed by NMR analysis (van Garderen et al., 1989). Very recently, the study with chemical probes (vide infra) of eleven duplex Pt(dien)-monoadducts (9 to 20 base pairs) with the AGC, AGT, CGA, TGA, TGC and TGT sequences has concluded to a destabilization that is larger for the pyrimidine-G-pyrimidine sequences. However, electrophoretic mobility of multimers (vide infra) of the 20 base pair oligonucleotides showed that no bending was induced by the monoadducts (Brabec et al., 1992).

4. *Distortions of a duplex structure by interstrand cis- and trans-$Pt(NH_3)_2$ adducts*

Gel electrophoresis, chemical probes and molecular mechanics modeling have been used to characterize the deformation due to the interstrand cross-linking of 19-22 base-pair double-stranded oligonucleotides, by *cis*- and *trans*-$Pt(NH_3)_2^{2+}$.

In the case of *cis*-DDP, the cross-linked sequences d(TG*CT)-d(AG*CA) lead to a 55° bending of the duplex towards the major groove, with no perturbation of the twist angle and a distortion localized to the d(G*C/G*C) sequence (Sip et al., 1992). In the case of *trans*-DDP, the cross-linked d(TG*CT)-d(AGC*A) sequences exhibit a *syn* conformation of dG* and the duplex appears to be unwound and bent towards the major groove with respective angles of 12 and 26° (Brabec et al., 1993).

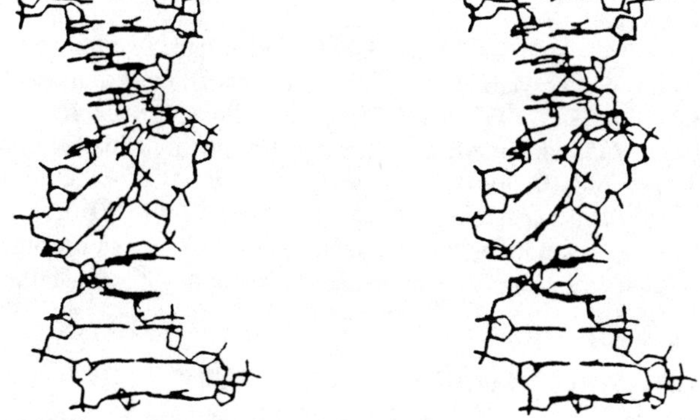

Figure 4.
Stereoview of a molecular model for *trans*$[Pt(NH_3)_2${d(CCTCG*AG*TCTCC)-d(GGAGACTCGGAGG)}]. (Reproduced with permission from Lepre et al., 1990).

D. Discussion of the results of oligonucleotide studies relevant to the deformation of DNA structure by platinum adducts

The first data concerning *cis*- and *trans*-DDP DNA deformations suggested large base destacking, disruption of base-pairing, and helix unwinding (Macquet and Butour, 1978a). They were often observed for relatively high Pt : base ratios, up to 0.1.

With the aim of completely describing the structures of the major Pt-DNA adducts, oligonucleotide model studies led to the following propositions for DNA deformation:

- *cis*-DDP chelation by two adjacent guanines leads to a localized distortion of the helix, affecting essentially the XG*G*-CCY sequence; the stacking is suppressed between the two G*s and perturbed between the opposite Cs; the base-pairing is perturbed, particularly for the 5'-G*C; the helix is kinked by 40-60° towards the major groove and unwound by 12-19°. The A*G* chelate induces a 46° kink and 17° unwinding of the helix.
- *cis*-DDP chelation by a G*NG* sequence leads to a 60° kink of the helix towards the major groove, with a bulged-out N base; the 5'-G* and N bases are unpaired; the 5'-G* is *syn*.
- *trans*-DDP chelation by a G*NG* sequence leads to a large perturbation of the stacking of these bases and to a disruption of their base-pairing that is compensated by hydrogen-bond bridging of the two strands by the ammine ligands; there is a slight 18° bend of the helix.
- Pt(triamine) monodentate binding to a guanine alters base stacking but does not bend the helix.
- *cis*- and *trans*-DDP interstrand cross-linking leads to a 55° bending of the helix for the *cis*-$Pt(NH_3)_2$ d(G*C/CG*) chelate and a 26° bending - 12° unwinding for the *trans*-$Pt(NH_3)_2$ d(G*C/C*G) chelate.

The previous structural studies were all conducted with the spectroscopic tools currently applicable to short oligonucleotides up to dodecanucleotides. The use of chemical probes (chloroacetaldehyde, osmium tetraoxide, diethyl pyrocarbonate) and artificial nuclease (copper-1,10-phenanthroline) allowed the study of longer duplexes (20 mers). It revealed a similar overall deformation for the G*G* and A*G*-*cis*$Pt(NH_3)_2$ adducts, but different distortions at the kink site (Marrot and Leng, 1989; Schwartz et al., 1989).

Gel electrophoretic mobility studies gave a more direct access to the bending of platinated DNA and to the corresponding angle value. The results are the following: (i) the kink angles for DNA bearing the *cis*-$Pt(NH_3)_2$-G*G*, A*G* and G*TG* cross-links are respectively 32,34 and 35° (Bellon and Lippard, 1990; Marrot and Leng, 1989 for G*TG*), (ii) the *trans*-$Pt(NH_3)_2$-G*TG* cross-link allows bending of the DNA in a nondirected fashion (Bellon and Lippard, 1990), and (iii) the $[Pt(dien)]^{2+}$ and *cis*-$[Pt(NH_3)_2$ {*N3*-cytosine)$]^{2+}$ monoadducts induce no bend of the DNA (Marrot and Leng, 1989; Bellon and Lippard, 1990; Brabec et al., 1992). Clearly, molecular mechanics calculations on short oligonucleotides without explicitly including solvent interactions overestimated the values of the kink angles. However, they correctly predicted, with the experimental support of NMR data, a helix bending towards the major groove due to the intrastrand G*G* and G*NG* adducts.

Another characteristic of platinated DNA and oligonucleotide models is helix unwinding. Here again, polyacrylamide gel electrophoresis allowed a quantitative evaluation for the *cis*-$Pt(NH_3)_2$ - G*G*, A*G* and G*TG* cross-links, giving respective unwinding angles of 13, 13 and 23°, and confirmed the hypothesis of a hinge joint at the *trans*-$Pt(NH_3)_2$-G*TG* cross-link (Bellon et al., 1991). For the G*G*- and A*G*-*cis*$Pt(NH_3)_2$ adducts, molecular mechanics calculations had predicted 12-20°, (Kozelka and Chottard, 1990; Fouchet, 1992). A gel electrophoretic mobility shift assay, run on supercoiled pUC19 plasmid DNA, revealed that triamine monoadducts induce a 6-8° unwinding of the duplex (Keck and Lippard, 1992).

IV. CONCLUSION AND OUTLOOK FOR THE FUTURE

Oligonucleotide model studies have played a major role in the determination of the DNA distortions induced by the various intrastrand platinum cross-links. X-ray crystal structures of the major duplex adducts would give valuable complementary informations and remain a challenging goal in the field. Improvement of molecular modelling techniques with better force fields and an explicit involvement of water is required to better account for the hydrogen bonds which seem to play a major role in the interactions between the amine ligands and the DNA bases and phosphates.

Several results already show that the different types of distortions will be differently recognized and processed by the various cellular systems (Bruhn et al., 1990). Therefore, from a biological and pharmacological point of view, it will be desirable to produce a chosen type of adduct. To attain such a goal, one requires a knowledge of the factors controlling the DNA platination steps. Here again, oligonucleotide studies should provide the necessary rate constants of the first platination and chelation steps, as a function of typical DNA sequences. This will then allow the design of targeted platinum drugs.

Several results also show that the repair enzymes and damage recognition proteins are sensitive to the distortions of the DNA structure and particularly to helix unwinding. One can try to favor one type of distortion, or to induce new ones, by designing complexes susceptible to give new adducts. One can also try to induce a specific recognition by first identifying the groups at the origin of tight protein binding and second by tuning their binding properties. To attain such goals, one requires a detailed knowledge of platinated DNA-protein interactions. This should be available from complexes formed between completely characterized platinated oligonucleotides and recognition proteins.

SUMMARY

DNA appears as the major target of the platinum drugs, at the origin of their antitumor activity. This interaction raises the following questions: (i) what controls the kinetics and selectivity of the first platinum binding? (ii) what controls the competition between the various intrastrand and interstrand chelations of the platinum monoadducts? (iii) what are the deformations of DNA due to the platinum adducts? (iv) can these deformations be related to the biological properties of the lesions? This overview will show how studies with oligonucleotide models contribute to answer these questions.

ACKNOWLEDGEMENTS

The contribution of our group to this research has been supported by the "Centre National de la Recherche Scientifique", the "Association pour la Recherche sur le Cancer" (ARC), the "Ligue Nationale Française contre le Cancer" and a contract with "Rhône-Poulenc Rorer" company. Their support is gratefully acknowledged. Stimulating discussions with the group of Prof. J. Reedijk were made possible thanks to support by the European Economic Community (Grant ST2J-0462-C). I am indebted to my coworkers and colleagues, whose names are mentioned in the references, for their determining contributions and their team spirit.

Abbreviations

cis-DDP	*cis*-$[PtCl_2(NH_3)_2]$
trans-DDP	*trans*-$[PtCl_2(NH_3)_2]$
*cis*Pt	*cis*-$Pt(NH_3)_2^{2+}$
*trans*Pt	*trans*-$Pt(NH_3)_2^{2+}$
DNA	deoxyribonucleic acid
en	ethylenediamine
dien	diethylenetriamine
G	guanosine and nucleotide in duplex formulas
d	deoxyribo
P	phosphate (when terminal) or phosphodiester group
rb	number of Pt atoms bound per nucleotide
Tm	melting temperature (from UV melting profile)
G^*, A^*, C^*	the asterisk indicates nucleotides bound to platinum
MES	2-(N-morpholino)ethanesulfonic acid buffer

REFERENCES

Admiraal, G., van der Veer, J.L., de Graaf, R.A.G., den Hartog, J.H.J., and Reedijk, J. (1987). *J. Amer. Chem. Soc.*, *109*: 592.

Admiraal, G., Alink, M., Altona, C., Dijt, F.J.,van Garderen, C.J., de Graaff, R.A.G., and Reedijk, J. (1992). *J. Amer. Chem. Soc.*, *114*: 930.

Bancroft, D.P., Lepre, C.A., and Lippard, S.J. (1990). *J. Amer. Chem. Soc.*, *112*: 6860.

Bellon, S.F., and Lippard, S.J. (1990). *Biophys. Chem.*, *35*: 179.

Bellon, S.F., Coleman, J.H., and Lippard, S.J. (1991). *Biochemistry*, *30*: 8026.

Bernges, F., and Holler, E. (1991). *Nucleic Acids Res.*, *19*: 1483.

Brabec, V., Reedijk, J., and Leng., M. (1992). *Biochemistry*, *31*: 12397.

Brabec, V., Sip, M., and Leng., M. (1993). *Biochemistry*, *32*: 11676.

Bruhn, S.L., Toney, J.H., and Lippard, S.J. (1990). *Progress in Inorganic Chemistry*, Vol. 38 (S.J. Lippard, ed.), John Wiley & Sons, p. 477.

Butour, J.L., and Johnson, N.P. (1986). *Biochemistry*, *25*: 4543.

Caradonna, J.P., Lippard, S.J., Gait, M.J., and Singh, M. (1982). *J. Amer. Chem. Soc.*, *104*: 5793.

Chottard, J.C. (1988). *Platinum and Other Metal Coordination Compounds in Cancer Chemotherapy* (M. Nicolini, ed.), Martinus Nijhoff, Boston, p. 47.

Chottard, J.C., Girault, J.P., Chottard, G., Lallemand, J.Y., and Mansuy, D. (1978). *Nouv. J. Chim.*, *2*: 551.

Chottard, J.C., Girault, J.P., Chottard, G., Lallemand, J.Y., and Mansuy, D. (1980). *J. Amer. Chem. Soc.*, *102*: 5565.

Comess, K.M., Costello, C.E., and Lippard, S.J. (1990). *Biochemistry*, *29*: 2102.

Decoville, M., Schwartz, A., Locker, D., and Leng, M. (1993). *FEBS Lett.*, *323*: 55.

den Hartog, J.H.J., Altona, C., Chottard, J.C., Girault, J.P., Lallemand, J.Y., de Leeuw, F., Marcelis, A.T.M., and Reedijk, J. (1982). *Nucleic Acids Res.*, *10*: 4715.

den Hartog, J.H.J., Altona, C., van Boom, J.H.J., Marcelis, A.T.M., van der Marel, G.A., Rinkel, L.J., Wille, G., and Reedijk, J. (1983). *Eur. J. Biochem.*, *134*: 485.

den Hartog, J.H.J., Altona, C., van Boom, J.H., van der Marel, G.A., Haasnoot, C.A.G., and Reedijk, J. (1984a). *J. Amer. Chem. Soc.*, *106*: 1528.

den Hartog, J.H.J., Altona, C., van Boom, J.H., and Reedijk, J. (1984b). *FEBS Lett.*, *176*: 393.

den Hartog, J.H.J., Altona, C., van Boom, J.H., van der Marel, G.A., Haasnoot, C.A.G., and Reedijk, J. (1985a). *J. Biomol. Struct. Dyn.*, *2:* 1137.

den Hartog, J.H.J., Altona, C., van den Elst, H., van der Marel, G.A., and Reedijk, J. (1985b). *Inorg. Chem.*, *24*: 983.

den Hartog, J.H.J., Altona, C., van der Marel, G., and Reedijk, J. (1985c). *Eur. J. Biochem.*, *147*, 371.

Dewan, J.C. (1984). *J. Amer. Chem. Soc.*, *106*: 7239.

Dijt, F.J., Chottard, J.C., Girault, J.P., and Reedijk, J. (1989). *Eur. J. Biochem.*, *179*: 333.

Eastman, A. (1983). *Biochemistry*, *22*: 3927.

Eastman, A. (1986). *Biochemistry*, *25*: 3912.

Eastman, A., and Barry, M.A. (1987). *Biochemistry*, *26*: 3303.

Eastman, A., Jennerwein, M.M., and Nagel, D.L. (1988). *Chem.-Biol. Interact.*, *67*: 71.

Fichtinger-Schepman, A.M.J., van der Veer, J.L., den Hartog, J.H.J., Lohman, P.H.M., and Reedjik, J. (1985). *Biochemistry*, *24*: 707.

Fichtinger-Schepman, A.M.J., Van Oosterom, A.T., Lohman, P.H.M., and Berends, F. (1987). *Cancer Res.*, *47*: 3000.

Fouchet, M.H. (1992). PhD Thesis, Université Pierre et Marie Curie (Paris VI), p. 65.

Fouchet, M.H., Gauthier, C., Guittet, E., Girault, J.P., Igolen, J., and Chottard, J.C. (1992). *Biochem. Biophys. Res. Commun.*, *182*: 555.

Fouchet, M.H., Guittet, E., Cognet, J., Kozelka, J., Igolen, J., and Chottard, J.C., to be published.

Fouts, C.S., Reily, M.D., Marzilli, L.G., and Zon, G. (1987). *Inorg. Chim. Acta, 137*: 1.

Gauthier, C. (1990). PhD Thesis, University Pierre et Marie Curie (Paris VI), p. 28.

Gibson, D., and Lippard, S.J. (1987). *Inorg. Chem., 26*: 2275.

Girault, J.P., Chottard, G., Lallemand, J.Y., and Chottard, J.C. (1982a). *Biochemistry, 21*: 1352.

Girault, J.P., Chottard, J.C., Guittet, E.R., Lallemand, J.Y., Huynh-Dinh, T., and Igolen, J. (1982b). *Biochem. Biophys. Res. Commun., 109*: 1157.

Girault, J.P., Chottard, J.C., Neumann, J.M., Tran-Dinh, S., Huynh-Dinh, T., and Igolen, J. (1984a). *Nouv. J. Chim., 8*: 7.

Girault, J.P., Chottard, G., Lallemand, J.Y., Huguenin, F., and Chottard, J.C. (1984b). *J. Amer. Chem. Soc., 106*: 7227.

Gonnet, F., Kozelka, J., and Chottard, J.C. (1992). *Angew. Chem., Int. Ed. Engl., 31*: 1483.

Hambley, T.W. (1991). *Inorg. Chem., 30*: 937.

Herman, F., Kozelka, J., Stoven, V., Guittet, E., Girault, J.P., Huynh-Dinh, T., Igolen, J., Lallemand, J.Y., and Chottard, J.C. (1990). *Eur. J. Biochem., 194*: 119.

Hollis, L.S., Amundsen, A.R., and Stern, E.W. (1989). *J. Med. Chem., 32*: 128.

Inagaki, K., and Kidani, Y. (1984). *Inorg. Chim. Acta, 92*: 19.

Johnson, N.P., Hoeschele, J.D., and Rahn, R.O. (1980). *Chem.-Biol. Interact., 30*: 151.

Johnson, N.P., Mazard, A.M., Escolier, J., and Macquet, J.P. (1985). *J. Amer. Chem. Soc., 107*: 6376.

Keck, M.V., and Lippard, S.J. (1992). *J. Amer. Chem. Soc., 114*: 3386.

Kozelka, J., and Chottard, J.C. (1990). *Biophys. Chem., 35*: 165.

Kozelka, J., Petsko, G.A., Quigley, G.J., and Lippard, S.J. (1985). *J. Amer. Chem. Soc., 107*: 4079.

Kozelka, J., Petsko, G.A., Lippard, S.J., and Quigley, G.J. (1986). *Inorg. Chem., 25*: 1075.

Kozelka, J., Archer, S., Petsko, G.A., Lippard, S.J., and Quigley, G.J. (1987). *Biopolymers, 26*: 1245.

Kozelka, J., Fouchet, M.H., and Chottard, J.C. (1992). *Eur. J. Biochem., 205*: 895.

Laoui, A. (1989a). PhD Thesis, Université Denis Diderot (Paris VII), p. 69.

Laoui, A. (1989b). PhD Thesis, Université Denis Diderot (Paris VII), p. 38.

Laoui, A., Kozelka, J., and Chottard, J.C. (1988). *Inorg. Chem., 27*: 2751.

Lemaire, M.A., Schwartz, A., Rahmouni, A., and Leng, M. (1991). *Proc. Nat. Acad. Sci. USA, 88*: 1982.

Lempers, E.L.M., and Reedjik, J. (1991). *Adv. Inorg. Chem., 37*: 175.

Lempers, E.L.M., Bloemink, M.J., Brouwer, J., Kidani, Y., and Reedijk, J. (1990). *J. Inorg. Biochem., 40*: 23.

Lempers, E.L.M., Bloemink, M.J., and Reedijk, J. (1991). *Inorg. Chem., 30*: 201.

Lepre, C.A., and Lippard, S.J. (1990). *Nucleic Acids and Molecular Biology*, Vol. 4 (F. Eckstein and D.M.J. Lilley, eds.), Springer Verlag, Berlin Heidelberg, p. 9.

Lepre, C.A., Strothkamp, K.G., and Lippard, S.J. (1987). *Biochemistry, 26*: 5651.

Lepre, C.A., Chassot, L., Costello, C., and Lippard, S.J. (1990). *Biochemistry, 29*: 811.

Lim, M.C., and Martin, R.B. (1976). *J. Inorg. Nucl. Chem., 38*: 1911.

Lippert, B. (1989). *Progress in Inorganic Chemistry*, Vol. 37 (S.J. Lippard, ed.), John Wiley & Sons, p. 1.

Macquet, J.P., and Butour, J.L. (1978a). *Biochimie, 60*: 901.

Macquet, J.P., and Butour, J.L. (1978b). *Eur. J. Biochem., 83*, 375.

Malfoy, B., Hartman, B., and Leng, M. (1981). *Nucleic Acids Res., 9*: 5659.

Malinge, J.M., and Leng, M. (1988). *Nucleic Acids Res.*, *16*: 7663.

Mansy, S., Chu, G.Y.H., Duncan, R.E., and Tobias, R.S. (1978). *J. Amer. Chem. Soc.*, *100*: 607.

Marcelis, A.T.M., Canters, G.W., and Reedijk, J. (1981). *Rec. Trav. Chim. Pays-Bas*, *100*: 391.

Marcelis, A.T.M., den Hartog, J.H.J., van den Marel, G.A., Wille, G., and Reedijk, J. (1983). *Eur. J. Biochem.*, *135*: 343.

Marcelis, A.T.M., Erkelens, C., and Reedijk, J. (1984). *Inorg. Chim. Acta*, *91*: 129.

Marrot, L., and Leng, M. (1989). *Biochemistry*, *28*: 1454.

Mazeau, K., Vovelle, F., Rahmouni, A., Leng, M., and Ptak, M. (1989). *Anti-Cancer Drug Design*, *4*: 63.

Miller, K.J., Taylor, E.R., Basch, H., Krauss, M., and Stevens, W.J. (1985). *J. Biomol. Struct. Dyn.*, *2*: 1157.

Murray, V., Motyka, H., England, P.R., Wickham, G., Lee, H.H., Denny, W.A., and McFadyen, W.D. (1992). *J. Biol. Chem.*, *267*: 18805.

Naser, L.J., Pinto, A.L., Lippard, S.J., and Essigman, J.M. (1988). *Biochemistry*, *27*:4357.

Page, J.D., Husain, I., Sancar, A., and Chaney, S.G. (1990). *Biochemistry*, *29*: 1016.

Payet, D., Gaucheron, F., and Leng, M. (1993). *Nucleic Acids Res.*, *21*: 5846.

Pinto, A.L., and Lippard, S.J. (1985). *Proc. Nat. Acad. Sci. USA*, *82*: 4616.

Pinto, A.L., Naser, L.J., Essigman, J.M., and Lippard, S.J. (1986). *J. Amer. Chem. Soc.*, *108*: 7405.

Plooy, A.C.M., Fichtinger-Schepman, A.M.J., Schutte, H.J., Van Dijk, M., and Lohman, P.H.M. (1985). *Carcinogenesis*, *6*: 561.

Plooy, A.C.M., van Dijk, M., and Lohman, P.H.M. (1984). *Cancer Res.*, *44*: 2043.

Ptak, M., Rahmouni, A., Mareau, K., Thuong, N.T., and Leng, M. (1989). *Anti-Cancer Drug Design*, *4*: 53.

Pullman, A., and Pullman, B. (1981). *Quart. Rev. Biophys.*, *14*: 289.

Rahmouni, A., Schwartz, A., and Leng, M. (1988). *Platinum and Other Metal Coordination Compounds in Cancer Chemotherapy* (M. Nicolini, ed.), Martinus Nijhoff, Boston, p. 127.

Reedijk, J. (1987). *Pure Appl. Chem.*, *59*: 181.

Reedijk, J. (1992). *Inorg. Chim. Acta*, *198-200*: 873.

Reedijk, J., Fichtinger-Schepman, A.M.J., van Oosterom, A.T., and van de Putte, P. (1987). *Struct. Bonding*, *67*: 53.

Robins, A.B. (1973). *Chem.-Biol. Interact.*, *6*: 35 and *7*: 11.

Roos, I.A.G., Thomson, A.J., and Mansy, S. (1974). *J. Amer. Chem. Soc.*, *96*: 6484.

Schaller, W., Reisner, H., and Holler, E. (1987). *Biochemistry*, *26*: 943.

Schwartz, A., Marrot, L., and Leng, M. (1989). *Biochemistry*, *28*, 7975.

Scovell, W.M., and O'Connor, T. (1977). *J. Amer. Chem. Soc.*, *99*: 120.

Segal, E., and Le Pecq, J.B. (1985). *Cancer Res.*, *45*:492.

Sherman, S.E., and Lippard, S.J. (1987). *Chem. Rev.*, *87*: 1153.

Sherman, S.E., Gibson, D., Wang, A.H.J., and Lippard, S.J. (1985). *Science*, *230*: 412.

Sherman, S.E., Gibson, D., Wang, A.H.J., and Lippard, S.J. (1988). *J. Amer. Chem. Soc.*, *110*: 7368.

Sip, M., Schwartz, A., Vovelle, F., Ptak, M., and Leng, M. (1992). *Biochemistry*, *31*: 2508.

Stone, P.J., Kelman, A.D., and Sinex, F.M. (1976). *J. Mol. Biol.*, *104*: 793.

Sundquist, W.I., and Lippard, S.J. (1990). *Coord. Chem. Rev.*, *100*: 293.

Ushay, H.M., Santella, R.M., Caradonna, J.P., Grunberger, D., and Lippard, S.J. (1982). *Nucleic Acids Res.*, *10*: 3573.

Van der Veer, J.L. (1986). PhD Thesis, Leiden University.

Van der Veer, J.L., Ligtvoet, G.J., van den Elst, H., and Reedijk, J. (1986a). *J. Amer. Chem. Soc.*, *108*: 3860.

Van der Veer, J.L., van den Elst, H., den Hartog, J.H.J., Fichtinger-Schepman, A.M.J., and Reedijk, J. (1986b). *J. Inorg. Chem.*, *25*: 4657.

Van der Veer, J.L., Noteborn, H.P., van den Elst, H., and Reedijk, J. (1987). *Inorg. Chim. Acta*, *131*: 221.

Van Garderen, C.J., Altona, C., and Reedijk, J. (1987). *Recl. Trav. Chim. Pays-Bas*, *106*: 196.

Van Garderen, C.J., van der Elst, H., van Boom, J.H., and Reedijk, J. (1989). *J. Amer. Chem. Soc.*, *111*: 4123.

Van Hemelryck, B., Guittet, E., Chottard, G., Girault, J.P., Huynh-Dinh, T., Lallemand, J.Y., Igolen, J., and Chottard, J.C. (1984). *J. Amer. Chem. Soc.*, *106*: 3037.

Van Hemelryck, B., Guittet, E., Chottard, G., Girault, J.P., Huynh-Dinh, T., Lallemand, J.Y., Igolen, J., and Chottard, J.C. (1986). *Biochem. Biophys. Res. Commun.*, *130*: 758.

Van Hemelryck, B., Girault, J.P., Chottard, G., Valadon, P., Laoui, A., and Chottard, J.C. (1987). *Inorg. Chem.*, *26*: 787.

Wing, R.M., Pjura, P., Drew, H.R., and Dickerson, R.E. (1984). *EMBO J.*, *3*: 1201.

METAL-CONTAINING ANTICANCER AGENTS

by

Michael J. Clarke and Stacey Limauro

Department of Chemistry,
Boston College,
Chestnut Hill, MA 02167
U.S.A.

I. INTRODUCTION

Interest in transition metal chemotherapeutic agents predates the clinically used platinum agents; indeed, Paracelsus is reputed to have used mixtures of heavy metals against cancer (Williams, 1974). Prior to Rosenberg's discovery of cisplatin, Dwyer had shown that phenanthroline complexes of Ru(II) exhibited anticancer activity, but had neurotoxic side effects (Dwyer et al., 1965). Studies on the effects of chelating agents on tumor growth led to later investigations of the anticancer effects of copper complexes (Crim and Petering, 1967; Petering, 1980). Cu-thiosemicarbazone complexes appear to bind cellular thiols, thereby interfering with DNA synthesis by inhibiting ribonucleotide reductase (Farrell, 1989). In general, however, the most successful chemotherapeutics target nucleic acids. This review covers non-platinum DNA-binding agents with anticancer activity through February 1993.

II. RUTHENIUM

A. Complexes with nitrogen ligands

Compared with corresponding square-planar Pt complexes, Ru complexes are usually significantly less toxic, but also often require a larger dose, so that therapeutic ratios are often not improved. The two additional coordination sites for ruthenium as opposed to Pt(II) provide for more modes of binding to nucleic acids as well as a third dimension in stereochemical interactions with the chiral DNA helix (Pyle and Barton, 1990). Ammine complexes of Ru(II) have a fairly strong affinity for heterocyclic nitrogen sites (Taube, 1973; Kuehn and Taube, 1976) and selectively label exposed imidazole nitrogens both on nucleic acids (Clarke et al., 1986; McNamara and Clarke, 1992) and proteins (Winkler and Gray, 1992). Inhibition of DNA replication (Kelman et al., 1977), mutagenic activity and induction of the SOS repair mechanism (Yasbin et al., 1980), binding to nuclear DNA (Marx et al., 1989a), and reduction of RNA synthesis (Marx et al., 1989b) strongly suggest DNA as the target molecule for ruthenium ammines.

Differences between the chemistry of Ru(II) and Ru(III) can be taken to advantage by utilizing Ru(III) complexes as prodrugs which are activated in vivo to yield a Ru(II) form which can rapidly bind to heterocyclic nitrogens. Low O_2 content in tumor cells should favor the formation and maintenance of the lower valent state, which binds much more rapidly to nucleic acids (Clarke, 1993). Since Ru(III) has a d^5 electronic configuration, it can accept π-electron density from π-donor ligands into its half-filled d_π-orbital. This, coupled with its higher charge, causes it to be relatively "hard" in character relative to Ru(II) and bind firmly to halides and anionic oxygen ligands. The lower charge and d^6 configuration of Ru(II) yields more extended, filled d_π-orbitals which can donate electron density to ligand π*-orbitals. As a result, Ru(II) is fairly "soft" and has little affinity for π-donor ligands, while often adhering strongly to π-acceptors (Wishart et al., 1986). The d^4 configuration of Ru(IV) causes it to strongly accept π-electrons so that it often exists as Ru=O, which is capable of hydride abstraction, so that Ru(IV)-polypyridyl complexes can cleave DNA through sugar oxidation (Gupta et al., 1993).

Prodrug complexes containing Ru(III) (or possibly Ru(IV)) with at least several nitrogen ligands and one or more acido ligands are often fairly stable to substitution. When such complexes have biologically accessible reduction potentials, in vivo reduction to Ru(II) is likely. Such reduction is most readily effected by electron-transfer proteins capable of donating a single electron following an interface with a more powerful 2e-reductant such as NADH. Reductants of this type exist in both the mitochondrial electron-transfer chain and in microsomal electron-transfer systems, with microsomal proteins being the more efficient (Clarke et al., 1980). Ammine-ruthenium complexes can also be reduced by a prevalent transmembrane electron transport system (Laliberté et al., 1987). Oxidation of ammine-ruthenium(II) complexes back to Ru(III) inside the cell is also possible, especially where oxygen is abundant (Stanbury et al., 1980). While whole mitochondria will not oxidize $[(NH_3)_6Ru]^{2+}$, cyt-b_{562} on the inner face of submitochondrial particles (Kunz and Konstantinov, 1984) and cytochrome oxidase will (Tsofina et al., 1986; Moroney et al., 1984).

Reduction of complexes of the general type $[N_{3-y}X_yRu]^{3-y}$, where N represents a nitrogen ligand that is not a strong π-acceptor and X is an acido ligand, generally leads to the rapid aquation of at least one of the acido ligands (Marchant et al., 1977). As long as the Ru(II) oxidation state is maintained, imidazole rings on histidine and purines are the most likely biological ligands. However, in the presence of oxygen or other oxidants, transient coordination of acido ligands favors re-oxidation back to Ru(III) (Stritar and Taube, 1969). Owing to the higher metabolism of tumor cells, oxygen is rapidly utilized and neovascularization often fails to keep pace with rapidly growing tumors (Palmer et al., 1990; Vaupel et al., 1991; Snyder, 1988; Schiffer et al., 1989). This results in glycolysis as the primary energy source, which tends to lower tumor pH and favor reduction of complexes with a pH-dependent reduction potential (Vaupel et al., 1989).

Tumor uptake of transition metal complexes may proceed by a number of different pathways: (i) nonspecific uptake due to increased blood flow and higher tumor cell permeability. Charged complexes may enter through endocytosis, while neutral ones may also diffuse through the cell membrane; (ii) specific uptake, such as that mediated by transferrin. Fe and Ru are sufficiently similar in their biological chemistry in that both are concentrated by the villi of the small intestines and are then widely distributed (Bruce et al., 1962; Clarke, 1980a). Both Fe^{3+} and Ru^{3+} have a high affinity for phenolate ligands (Pell et al., 1989), which are involved in the Fe^{3+} site of transferrin (Kratz and Messori, 1993). Since tracer studies with $^{103}RuCl_3$ injected into mice demonstrate substantial transferrin binding (Srivastava et al., 1981; Som et al., 1983; Srivastava et al., 1979), transport of at least some Ru(III) ions can be effected by transferrin. This mechanism may be favored for ruthenium complexes with a sufficient number of good leaving ligands, and transferrin binding may partially explain why some ruthenium complexes are retained by the blood (Anto et al., 1988).

Since small ions are excreted fairly readily by the kidneys, it is likely that the nonspecific uptake of ruthenium ions into tumors occurs within a few hours, while the transferrin-mediated uptake occurs on a somewhat longer time scale (Som et al., 1983). Once inside the cell, it is likely that some activation-by-reduction mechanism operates to cause Ru(II) to bind. Indeed, reduction is one way of releasing metal ions from transferrin and the lower pH of tumor tissue should also help dissociate Ru(III) from this protein (Kratz and Messori, 1993). Once Ru(II) binds to a nitrogen site on its target molecule, it will remain so even when oxidized to Ru(III), since nitrogen ligands are generally inert to substitution on either ion (Taube, 1973).

While not an anticancer agent, $[(H_2O)(NH_3)_5Ru^{II}]^{2+}$ serves as a simple model of monodentate coordination to DNA and reveals many of the main features of Ru-nucleic acid interactions. Owing to the attraction of the large DNA or RNA polyanions for cations, binding of $[(H_2O)(NH_3)_5Ru^{II}]^{2+}$ proceeds fairly rapidly and is strongly ionic-strength dependent. For tRNA's the rate for binding to guanine N7 sites (G^7) is rate = $k[Ru][P_{RNA}]$, where k = 5.96 M^{-1} s^{-1} at 25°C and μ = 0.045 (McNamara and Clarke, 1992). The initial reaction involves G^7 sites, which are relatively exposed in the major groove of B-DNA, while a second phase probably has to do with the Ru(II) attacking interior sites made available on separating the nucleic acid strands (Clarke et al., 1986; McNamara and Clarke, 1992). Ru-DNA binding constants are 5100 M^{-1} and 7800 M^{-1} for helical and single-stranded CT-DNA, respectively (Clarke et al., 1986). For a mixture of tRNA's, the binding constant is somewhat lower at 2900 M^{-1} (McNamara and Clarke, 1992).

Coordination of $[(NH_3)_5Ru^{III}]$ also occurs on deprotonated exocyclic amines of adenosine and cytidine (Clarke, 1980b; Clarke, 1978). Such complexes exhibit a remarkably short N-Ru bond arising from substantial π-d_π interactions (Graves and Hodgson, 1979). Hydrogen bonding between ammine ligands and the adjacent nitrogen on the pyrimidine ring is negated upon protonation of the ring site, which causes the metal to rotate about the N_{exo}-C bond. Both rotamers are evident by electrochemistry and NMR in the pH range around the pK_a's of the isomers (~3.3). A second motion of the metal occurs upon reduction to Ru(II), which rapidly linkage isomerizes to the adjacent π-acceptor ring nitrogen. Taken together the reversible pirouette and side-step movements, which can be choreographed by pH and electrochemical potential, constitute a true terpsichorean motion (Clarke, 1989a).

The relative selectivity of ammineruthenium ions for G^7 strongly suggest this as the initial site of attack on DNA in vivo. Possibilities for interfering with DNA metabolism include: (i) failure of replicating enzymes to recognize the metallated G, with subsequent halting of replication; (ii) protein-, intra-, or interstrand crosslinking by the metal; (iii) chemical reactions of the guanine residue induced by the presence of the metal ion; and, (iv) Fenton's chemistry taking place at the metal ion to generate radicals capable of strand cleavage. However, given the low oxygen concentration and the relative inefficiency of $[Guo(NH_3)_5Ru^{III}]$ in cutting DNA by Fenton's chemistry (Clarke et al., 1986), significant Fenton's chemistry is unlikely.

No studies have been done to survey for adjacent G^7-G^7 crosslinks with *cis*-ammineruthenium ions and these appear to be sterically difficult; however, there is evidence for a macrocyclic chelate of GpG with *trans*-Ru-DMSO complexes (Esposito et al., 1992). Local helix disruption might be generated by steric effects as well as a weakening of the hydrogen bonding and π-stacking abilities of the base through polarization of electron density toward the metal cation and O6 and/or phosphate hydrogen bonding to ammine protons. Additionally, the linkage isomerization reactions already discussed may allow the metal to migrate from a less obtrusive G^7 to a strongly helix disruptive binding site on the exocyclic amines of A and C.

The polarizing power of Ru(III) allows it to function as a general acid in promoting hydrolysis of the glycosidic bond in $[(dG^7)(NH_3)_5Ru^{III}]$, ($t_{1/2}$ = 1.5 days, 56 °C, pH 7) (Clarke and Morrissey, 1984). Consequently, hydrolysis of DNA might be induced by π-acceptor cation coordination at N7, but this has not been observed (Clarke et al., 1986). A reaction unique to Ru(III) is the base-catalyzed air oxidation of the nucleosides to 8-oxo-nucleosides when coordinated to $[(NH_3)_5Ru^{III}]$ at the N7 position (Gariepy et al., 1989). Following autoxidation, the glycosidic bond undergoes base-catalyzed cleavage. DNA cleavage has now been observed with *trans*-$[py(NH_3)_4Ru^{III}]$; however, this can proceed without O_2 (LaChance-Galang and Clarke, 1993).

Compounds such as *cis*-$[Cl_2(NH_3)_4Ru]Cl$ and *fac*-$[Cl_3(NH_3)_3Ru]$ have exhibited good to excellent antitumor activity in several tumor screens; however, the poor solubility of the latter renders it with poor pharmacological properties (Clarke 1980c; 1983; 1989b; 1993). A similar problem arises with *mer*-$[Cl_3(NH_3)_3Rh]$, which is active but insoluble (Cleare and

Hydes, 1980). Keppler has increased the solubility of ruthenium complexes by charging with fewer nitrogen ligands and more halides (Keppler et al., 1987; 1989; 1990; Seelig et al., 1992). *Trans*-complexes of the type $HB[RuB_2Cl_4]$ (where B = imidazole or indazole) have proved to be most active against colorectal tumors. Given the multiple leaving groups, these complexes may be more likely to be transported to the tumor site by transferrin. While steric effects with ammine ligands probably minimize intrastrand G^7-G^7 crosslinking, *trans* crosslinks to other than adjacent sites and crosslinks to exocyclic amines on A and C remain possibilities. In vitro crosslinking studies with the indazole compound revealed few intrastrand or DNA-protein crosslinks; however, this may be due to maintaining the complex as Ru(III), rather than providing for reduction to Ru(II) (Fruhauf and Zeller, 1991).

B. Dimethylsulfoxide complexes

Dimethylsulfoxide complexes of ruthenium are surprisingly nontoxic with LD_{50}'s up to ~1 g/kg. Unfortunately, correspondingly high therapeutic doses are needed to obtain a response in animals (Farrell, 1989). Both *cis* and *trans*-$[Cl_2(DMSO)_4Ru]$ show marginal activity against the primary tumor of the Lewis lung carcinoma, but are significantly effective against its metastases (Sava et al., 1989). The *cis* isomer is similarly effective against MCa mammary carcinoma (Sava et al., 1984), while the *trans* isomer markedly inhibits both primary tumor growth and metastases of B16 melanoma in mice.

The target of these complexes is thought to be DNA, since *cis*-$[Cl_2(DMSO)_4Ru]$ induces filamentous growth and λ-prophage activity as well as inhibiting the growth of bacteria with defective DNA repair systems (Monti-Bragadin et al., 1975; 1987). In the structure of *cis*-$[Cl_2(DMSO)_4Ru]$, three of the four DMSO ligands are S-bound while one is O-bound. Dissolution of the compound in water results in rapid loss of the O-bound ligand followed by reversible loss of a *cis*-chloride (Alessio et al., 1988a). Chloride loss is suppressed at $[Cl^-]$ = 150 mM, which is similar to that in blood, but is nearly complete at intracellular chloride concentrations (~3 mM). Maintenance of the neutral complex probably facilitates its crossing lipid membranes to enter the cell, whereas intracellular chloride should aid its cellular retention, since a cationic species, *mer*-$[(DMSO)_3Cl(H_2O)_2Ru]^+$, can be formed, which is capable of rapidly binding to DNA.

In *trans*-$[Cl_2(DMSO)_4Ru]$, all the DMSO's are S-bound and two are rapidly lost in water to yield the *cis*-diaqua species, which then undergoes the reversible aquation of a chloride to give *fac*-$[(H_2O)_3Cl(DMSO)_2Ru]^+$ (Alessio et al., 1988b). It may be significant that the rate of chloride loss for both *cis* and *trans*-$[Cl_2(DMSO)_4Ru]$ is similar to that of cisplatin. Comparing the expected intracellular cationic species for the two isomers, the greater degree of aquation and the lower steric hindrance of *trans*-derived product may account for the much lower LD_{50} (84 mg/kg) of *trans*-$[Cl_2(DMSO)_4Ru]$ compared to the *cis* isomer (Mestroni et al., 1989). When *cis*-ammonias are substituted opposite the DMSO's, the Ru-S bonds are shortened by 0.11 Å suggesting that intrastrand G^7-G^7 crosslinks would yield a stable species (Henn et al., 1991). While individual guanosines bind reversibly to *trans*-$[Cl_2(DMSO)_4Ru]$ (Cauci et al., 1991) and 5'dGMP forms an N7-PO_4 chelate rather than a bis-5'dGMP complex (Alessio et al., 1989), NMR evidence indicates a fairly stable chelate with GpG (Esposito et al., 1992).

Coordination of *cis*-$[Cl_2(DMSO)_4Ru]$ to DNA does not significantly affect the B-conformation and increases DNA thermal stability. On the other hand, binding to poly(dGdC) at low ionic strength induces a B to Z transition. *Trans*-$[Cl_2(DMSO)_4Ru]$ binds much more rapidly to DNA with greater changes in its CD spectra, which is attributed to disruption of the DNA structure due to the formation of bifunctional adducts along the chain (Mestroni et al., 1989). The *trans* isomer induces the B to Z transition in poly(dGdC) much more rapidly than the *cis*-complex.

The mutagenic activity of *cis*-$[Cl_2(DMSO)_4Ru]$ in bacteria is suppressed by inorganic phosphate, which is likely due to the formation of a phosphato chelate (Alessio et al., 1988). While exhibiting lower mutagenicity in bacteria, the *trans* isomer shows higher cytotoxicity and mutagenic activity in the eukaryotic V79 Chinese hamster lung cells in tissue culture. DNA extracted from cells separately treated with the isomers showed a fivefold higher content of Ru in the *trans* case. Given the greater potency of the *trans* isomer and its efficacy against tumor metastases, this drug may be useful as an adjuvant chemotherapeutic following surgery (Sava et al., 1989).

C. Ruthenium-red

Preparations of the mixed-valent complex, ruthenium red (Ru-red), $[(NH_3)_5RuORu(NH_3)_4ORu(NH_3)_5]^{6+}$, mixed with other μ-O-Ru complexes have been widely used as a cytological stain for almost a century (Mangin, 1893). These substances bind to polyanions such as DNA and the protective mucopolysaccharide (mainly hyaluronic acid) coat, which protects some tumor cells (Oberc-Greenwood et al., 1986). Such binding probably accounts for the radio-imaging of tumors with ^{103}Ru-red (Anghileri, 1975). Ru-red also inhibits tumor growth, which may arise from its ability to interfere with calcium metabolism (Anghileri et al., 1986).

Ruthenium red blocks Ca^{2+} transport in a number of biological systems (Hurley, 1988). An impurity in Ru-red preparations that is responsible for most of the inhibition of Ca^{2+} uptake in mitochondria has recently been identified as an oxygen-bridged, dimeric species, μ-O-$[X(NH_3)_4Ru]_2^{3+}$, where X = Cl^- or OH^- (Ying et al., 1991). The tripositive charge of these complexes probably allows them to ion-pair and hydrogen-bond with anionic, carboxylate sites on the surface of proteins. Substitution of the apical ligands may also result in covalent binding to carboxylate residues.

III. RHODIUM

The rhodium acetate dimer, μ-$(CH_3CO_2)_4Rh_2(H_2O)_2$, shows good antitumor activity against the Ehrlich ascites tumor, sarcoma 180 and P388, but little activity against L1210 and B16 (Bear, 1986). Activity increases in the series $[(RCO_2)_4Rh_2]$ with the lipophilicity of R and is independent of the reduction potential. The acetate complex reacts specifically with adenosine, and relatively little with guanosine, cytidine and uridine. Moreover, it reacts with single stranded poly(dA) and DNA, but not with poly(dG), poly(C) or double helical DNA (Erck et al., 1974). Crystallography of the complex with 1-methyladenosine shows the apical positions on each Rh(II) coordinated to the N7 of separate adenosines (Rubin et al., 1991). Equilibrium binding constants for several μ-carboxylato complexes varied from 1100 M^{-1} to 4500 M^{-1} (Rainen et al., 1975).

Complexes of this type inhibit DNA polymerase I and RNA polymerase from *E. Coli*. Treatment with the acetate complex also inhibits cellular DNA synthesis and arrests cellular development in the G_2 phase of the cell cycle; however, the butyrate complex inhibits DNA synthesis during S-phase development and is most toxic to cycling cells (Rao et al., 1980). The complexes also bind to a variety of proteins and irreversibly inhibit proteins with cysteines in the active site (Howard et al., 1976). Mammalian decomposition of the μ-acetate eventually results in oxidation to CO_2.

IV. CYCLOPENTADIENYLS

Neutral complexes of the type $(Cp)_2MX_2$, where Cp = cyclopentadienide and X = halide, have shown pronounced cytostatic activity against diverse animal tumors and xenografted human carcinomas (Kopf-Maier and Kopf, 1987). The antitumor activity is dependent upon the nature of the metal ion with M = Ti, V, Nb and Mo showing marked activity, Ta and W exhibiting marginal effectiveness, and Zr and Hf being inactive. Variations in X have substantially less effect, even when X is neutral (Kopf-Maier, 1989). Moreover, cationic complexes of the type $[(Cp)_2MCl_2]^+$, where M = Nb, Mo and Re, are active (Kopf-Maier and Klapotke, 1992a; 1992b).

Titanocene dichloride is the most potent of the neutral metallocenes. The liver and intestine serve as the main routes of its excretion. Since liver cells accumulate the drug, hepatotoxicity is dose-limiting. Blood clearance is biphasic with a rapid phase ($t_{1/2}$ = 5 h) and a slower phase ($t_{1/2}$ = ~3 d). Selective tumor uptake is not evident for 2-3 days, but thereafter Ti levels in the tumor increase relative to muscle and blood. The metal ion accumulates in cellular regions rich in nucleic acids with highest concentrations in the chromatin. In behavior analogous to that of cisplatin, treatment of Ehrlich ascites tumor cells revealed a premitotic G_2 block combined with marked mitotic depressions and additional accumulations at the G_1/S boundary. Addition of the metallocenes to DNA or RNA results in a change in the secondary structure of the nucleic acid (Kopf-Maier, 1989).

Of the Cp_2MCl_2 complexes containing Ti, V and Mo, hydrolysis of Mo proceeds the most quickly and is complete within 1 h with $[Cl^-]/[Mo] = 2$ at neutral pH. The Mo complex is also the least acidic ($pK_{a1} = 5.5$, $pK_{a2} = 8.5$), while the Ti complex is the most acidic ($pK_{a1} = 3.51$, $pK_{a2} = 4.35$) so that their aquated forms are different at neutral pH. In addition, the Ti-Cp is hydrolytically unstable. In aqueous solution, Cp_2VCl_2 binds fairly rapidly and selectively to nucleotide phosphates and hydrogen bonds strongly as $[Cp_2V(OH_2)]^{2+}$ to an anionic phosphodiester in the solid state; consequently, it has relatively little effect on base pairing (Toney et al., 1986). The "softer" Cp_2MoCl_2 has a hydrolytically stable Mo-Cp bond and forms 1:1 complexes with 5'NMP's (N = G, A, C and T) with little selectivity and substantial exchange between the ligands. Since it forms bonds with both the phosphates and heterocyclic bases, it generates major conformational changes with these nucleotides (Kuo et al., 1991).

Several crystallographic studies indicate that the Cp groups of the metallocenes sterically prevent crosslinking DNA. The structure of $[(Cp)_2Ti^{III}(1,3Me_2Xan^-)]$ exhibits chelation between N7 and O6, suggesting a similar mode for Ti(IV)-guanine binding (Cozak et al., 1986). On the other hand, the structure of $[(Cp)_2Mo(5'dGMP)]_2$ reveals a macrocycle with each Mo coordinated to the N7 of one 5'dGMP and a phosphate oxygen of the second 5'dGMP, while in solution this is in equilibrium with the monomeric N7/PO_4 chelate (Kuo et al., 1991).

The crystal structure of $[(Cp)_2Mo(9MeAde^-)]PF_6$ shows the metal to be chelated between N1 and an ionized N6 of 9-methyladenine, which in solution is in equilibrium with the N6/N7 chelate. Similarly, the structure of $[(Cp)_2Mo(1MeCyt^-)]PF_6$ exhibits chelation between N3 and the exocyclic N4 (Kuo et al., 1991). Both crystal structures are consistent with partial M-N π-bonding, with a concomitant weakening of the exocyclic-N π-bonding to the pyrimidine ring, similar to that observed for $[1MeCyt(NH_3)_5Ru^{III}]^{2+}$ (Clarke et al., 1980). In aqueous solution, the Mo is chelated by the N7 and phosphate oxygen of 5'dAMP and the N3/PO_4 of 5'dCMP, with the latter ligand also exhibiting monodentate phosphate binding. 5'dTMP also forms a N3/PO_4 chelate (Kuo et al., 1991). Since interaction with the phosphate oxygens in phosphodiesters occurs through H-bonding and is relatively unimportant, monodentate binding to bases or chelation involving the phosphate terminus of nucleic acids may be possible modes of interference with nucleic acid metabolism.

V. TITANIUM

Budotitane, *cis*-$[(CH_3CH_2O)_2(1\text{-phenylbutane-1,3-dionato})_2Ti^{IV}]$ is the first non-platinum transition metal anticancer agent to be tested in clinical trials (Keppler et al., 1991). Together with the metal, the 1,3-diketonate forms a six-membered planar structure, which exhibits both reactivity and NMR spectra suggestive of aromatic character. Depending on the arrangements of the methyl and phenyl groups on the diketonates, three geometric isomers are possible in the *cis* configuration shown in Figure 1, all of which exist as enantiomeric pairs. Activity is enhanced by unsubstituted aromatic ring systems on the periphery of the molecule. Hydrolysis of the ethoxy groups is complicated by the drugs ability to form insoluble particulates. In a soluble form, the half-life of the ethoxy groups is ~ 20 sec, while that of the diketonates is on the order of hours.

Formulation of the drug is done with a solubilizer, glycerinepolyethylene-glycolericinoleate (CremophorEL) and 1,2-propyleneglycol to form a coprecipitate in which the drug is enveloped between layers of the solubilizer. Dissolution of the formulated drug in water yields micelles in which the compound is protected from hydrolysis (Keppler et al., 1991). The compound is quite effective against a number of ascites tumors and induced colorectal tumors in animals (Heim et al., 1989). Early clinical trials indicate it to be fairly well tolerated by patients with the main side effects being kidney and liver toxicity. In vitro and in vivo experiments with budotitane have revealed no significant DNA damage, which has led to the suggestion that a DNA intercalative mode may be responsible for inhibition of tumor cell growth (Fruhauf and Zeller, 1991).

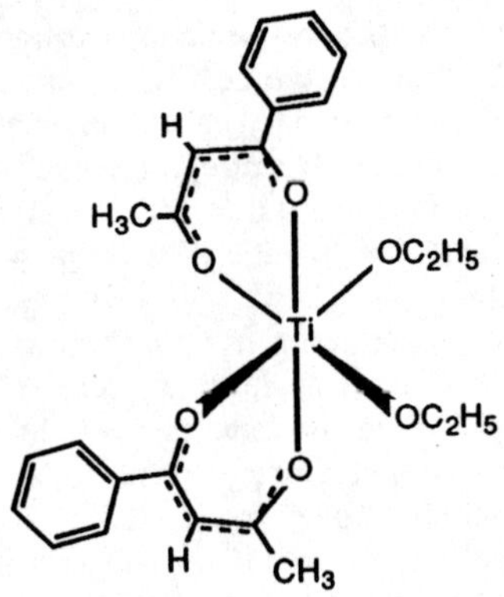

Figure 1.
Budotitane.

ACKNOWLEDGMENT

This work was supported by PHS Grant GM26390.

REFERENCES

Alessio, E., Mestroni, G., Nardin, G.A., Wahib, M., Calligaria, M., Sava, G., and Zorzet, S. (1988). *Inorg. Chem.*, *27*: 4099.

Alessio, E., Attia, W., Calligaris, M., Cauci, S., Dolzani, L., Mestroni, G., Monti-Bragadin, C., Nardin, G., Quadrifoglio, F., and Sava, G. (1988). *Develop. Oncol.*, *54*: 617.

Alessio, E., Xuu, Y., Cauci, S., Mestroni, G., Quadrifoglio, F., Viglino, P., and Marzilli, L.G. (1989). *J. Amer. Chem. Soc.*, *111*: 7068.

Anghileri, L.H. (1975). *Strahlentherapie*, *149*: 173.

Anghileri, L.J., Marchal, C., Matrat, M., and Crone-Escanya, M.C. (1986). *Neoplasma*, *33*: 603.

Anto, A., Anto, I., Hiraki, T., and Hisada, K. (1988). *Int. J. Rad. Appl.*, *15*: 133.

Bear, J.L. (1986). *Precious Metals 1985: Proceedings of the Ninth International Precious Metals Conference* (E.E. Zysk and J.A. Bonucci, eds.), Int. Precious Metals, Allentown, PA, p. 337.

Bruce, R.S., Carr, T.E.F., and Collins, M.E. (1962). *Health Phys.*, *8*: 397.

Cauci, S., Viglino, P., and Esposito, G. (1991). *J. Inorg. Biochem.*, *43*: 739.

Clarke, M.J. (1978). *J. Amer. Chem. Soc.*, *100*: 5068.

Clarke, M.J. (1980a). *Metal Ions in Biological Systems*, Vol. 11 (H. Sigel, ed.), Marcel Dekker, New York, p. 231.

Clarke, M.J. (1980b). *Inorg. Chem.*, *19*: 1103.

Clarke, M.J. (1980c). *Inorganic Chemistry in Biology and Medicine* (A.E. Martell, ed.), American Chemical Society, Washington, D.C., p. 157.

Clarke, M.J. (1983). *Chemistry and Biochemistry of Platinum, Gold and Other Chemotherapeutic Agents* (S.J. Lippard, ed.), American Chemical Society, Washington, D.C., p. 335.

Clarke, M.J. (1989a). *Prog. Clin. Biochem. Med.*, *10*: 25.

Clarke, M.J., ed. (1989b). *Ruthenium and Other Non-Platinum Metal Complexes in Cancer Chemotherapy*, Vol. 10, Springer-Verlag, Heidelberg.

Clarke, M.J. (1993). *Metal Complexes in Cancer Chemotherapy* (B.K. Keppler, ed.), VCH Publishing House, Heidelberg, p. 152.

Cleare, M.J., and Hydes, P.C. (1980). *Metal Ions in Biological Systems*, Vol. 11 (H. Sigel, ed.), Marcel Dekker, New York, p. 1.

Clarke, M.J., and Morrissey, P.E. (1984). *Inorg. Chim. Acta*, *80*: L69.

Clarke, M.J., Bitler, S., Rennert, D., Buchbinder, M., and Kelman, A.D. (1980). *J. Inorg. Biochem.*, *12*: 79.

Clarke, M.J., Jansen, B., Marx, K.A., and Kruger, R. (1986). *Inorg. Chim. Acta*, *124*: 13.

Cozak, D., Mardhy, A., Olivier, M.J., and Beauchamp, A.L. (1986). *Inorg. Chem.*, *25*: 2600.

Crim, J.A., and Petering, H.G. (1967). *Cancer Res.*, *27*: 1278.

Dwyer, F.P., Mayhew, E., Roe, E.M.F., and Shulman, A. (1965). *Brit. J. Cancer*, *19*: 195.

Erck, E., Rainen, L., Whileyman, J., and Chang, I.M. (1974). *Proc. Soc. Exp. Biol. Med.*, *145*: 1278.

Esposito, G., Cauci, S., Fogolari, F., Alessio, E., Scocchi, M., Quadrifoglio, F., and Viglino, P. (1992). *Biochem.*, *31*: 7094.

Farrell, N. (1989). *Transition Metal Complexes as Drugs and Chemotherapeutic Agents*, Kluwer Academic Publishers, Boston.

Fruhauf, S., and Zeller, W.J. (1991). *Cancer Res.*, *51*: 2943.

Gariepy, K.C., Curtin, M.A., and Clarke, M.J. (1989). *J. Amer. Chem. Soc.*, *111*: 4947.

Graves, D.J., and Hodgson, D.K. (1979). *J. Amer. Chem. Soc.*, *101*: 5608.

Gupta, N., Grover, N., Neyhart, G.A., Singh, P., and Thorp, H.H. (1993). *Inorg. Chem.*, *32*: 310.

Heim, M.E., Flechtner, H., and Keppler, B.K. (1989). *Ruthenium and Other Non-Platinum Metal Complexes in Cancer Chemotherapy* (M.J. Clarke, ed.), Springer-Verlag, Heidelberg, p. 217.

Henn, M., Alessio, E., and Mestroni, G. (1991). *Inorg. Chim. Acta*, *187*: 39.

Howard, R.A., Spring, T.G., and Bear, J.L. (1976). *Cancer Res.*, *36*: 4402.

Hurley, T.W. (1988). *Amer. J. Physiol.*, *23*: 621.

Kelman, A.D., Clarke, M.J., Edmonds, S.D., and Peresie, H.J. (1977). *J. Clin. Hemat. Oncol.*, *7*: 274.

Keppler, B.K., Rupp, W., Juhl, U.M., Endres, H., Niebel, R., and Balzer, W. (1987). *Inorg. Chem.*, *26*: 4366.

Keppler, B.K., Henn, M., Juhl, U.M., Berger, M.R., Niebl, R., and Wagner, F.E. (1989). *Ruthenium and Other Non-Platinum Metal Complexes in Cancer Chemotherapy* (M.J. Clarke, ed.), Springer-Verlag, Heidelberg, p. 41.

Keppler, B.K., Berger, M.R., and Heim, M.E. (1990). *Cancer Treat. Rev.*, *17*: 261.

Keppler, B.K., Friesen, C., Moritz, H.G., Vongerichten, H., and Vogel, E. (1991). *Struct. Bond.*, *78*: 97.

Kopf-Maier, P. (1989). *Ruthenium and Other Non-Platinum Metal Complexes in Cancer Chemotherapy* (M.J. Clarke, ed.), Springer-Verlag, Heidelberg, p. 151.

Kopf-Maier, P., and Klapotke, T. (1992a). *J. Cancer Res. Clin. Oncol.*, *118*: 216.

Kopf-Maier, P., and Klapotke, T. (1992b). *Cancer Chemother. Pharmacol.*, *29*: 361.

Kopf-Maier, P., and Kopf, H. (1987). *Chem. Rev.*, *87*: 1137.

Kratz, F., and Messori, L. (1993). *J. Inorg. Biochem.*, *49*: 79.

Kuehn, C.G., and Taube, H. (1976). *J. Amer. Chem. Soc.*, *98*: 689.

Kunz, W.S., and Konstantinov, A. (1984). *FEBS Lett.*, *175*: 100.

Kuo, L.Y., Kanatzdis, M.G., Sabat, M., Tipton, A.L., and Marks, T.J. (1991). *J. Amer. Chem. Soc.*, *113*: 9027.

LaChance-Galang, K., and Clarke, M.J. (1993). Unpublished results.

Laliberté, J.F., Sun, I.L., Crane, F.L., and Clarke, M.J. (1987). *J. Bioenerg. Biomembr.*, *19*: 69.

Mangin, L. (1893). *C.R. Acad. Sci. Paris*, *116*: 653.

Marchant, J.A., Matsubara, T., and Ford, P.C. (1977). *Inorg. Chem.*, *16*: 2160.

Marx, K.A., Kruger, R., and Clarke, M.J. (1989a). *Mol. Cell. Biochem.*, *86*: 155.

Marx, K.A., Seery, C., and Malloy, P. (1989b). *Mol. Cell. Biochem.*, *90*: 37.

McNamara, M., and Clarke, M.J. (1992). *Inorg. Chim. Acta, 195*: 175.

Mestroni, G., Alessio, E., Calligaris, M., Attia, W.M., Quadrifoglio, F., Cauci, S., Sava, G., Zorzet, S., Pacor, S., Monti-Bragadin, C., Tamaro, M., and Dolzani, L. (1989). *Ruthenium and Other Non-Platinum Metal Complexes in Cancer Chemotherapy* (M.J. Clarke, ed.), Springer-Verlag, Heidelberg, p. 74.

Monti-Bragadin, C., Ramani, L., Samer, L., Mestroni, G., and Zassinovich, G. (1975). *Antimicrob. Agents Chemother.*, *7*: 825.

Monti-Bragadin, C., Giacca, M., Dolzani, L., and Tamaro, M. (1987). *Inorg. Chim. Acta.*, *137*: 31.

Moroney, P.M., Scholes, T.A., and Hinkle, P.C. (1984). *Biochemistry, 23*: 4991.

Oberc-Greenwood, M.A., Muul, L.M., Gately, M.K., Kornblith, P.L., and Smith, B.W. (1986). *J. Neurooncol.*, *3*: 387.

Palmer, B.D., Wilson, W.R., and Pullen, S.M. (1990). *J. Med. Chem.*, *33*: 112.

Pell, S.D., Sherban, M.M., Tramontano, V., and Clarke, M.J. (1989). *Inorg. Synth.*, *26*: 65.

Petering, D.H. (1980). *Metal Ions in Biological Systems*, Vol. 11 (H. Sigel, ed.), Marcel Dekker, New York, p. 197.

Pyle, A.M., and Barton, J.K. (1990). *Bioinorganic Chemistry* (S.J. Lippard, ed.), Wiley, New York, 413.

Rainen, L., Howard, R.A., Kimball, A.P., and Bear, J.L. (1975). *Inorg. Chem.*, *14*: 2752.

Rao, P.N., Smith, M.L., Pathak, S., Howard, R.A., and Bear, J.L. (1980). *J. Nat. Cancer Inst.*, *64*: 905.

Rubin, J.R., Haromy, T.P., and Sundaralingam, M. (1991). *Acta Cryst. C.*, *47*: 1712.

Sava, G., Zorzet, S., Giraldi, T., Mestroni, G., and Zassinovich, G. (1984). *Eur. J. Cancer Clin. Oncol.*, *20*: 841.

Sava, G., Pacor, S., Zorzat, S., Alessio, E., and Mestroni, G. (1989). *Pharmacol. Res.*, *21*: 617.

Schiffer, D., Chio, A., Giordana, M.T., Mauro, A., Migheli, A., and Vigliani, M.C. (1989). *Acta Neuropathol. (Berlin)*, *77*: 369.

Seelig, M.H., Berger, M.R., and Keppler, B.K. (1992). *J. Cancer Res. Clin. Oncol.*, *118*: 195.

Snyder, G.K. (1988). *J. Appl. Physiol.*, *65*: 2332.

Som, P., Oster, Z.H., Matsui, K., Gugliemi, G., Persson, B., Pellettieri, M.L., Srivastava, S.C., Richards, P., Atkins, H.L., and Brill, A.B. (1983). *Eur. J. Nucl. Med.*, *8*: 491.

Srivastava, S.C., Richards, P., Meinken, G.E., Som, P., Atkins, H.L., Larson, S.M., Grunbaum, A., Rasey, J.S., Dowling, M.G., and Clarke, M.J. (1979). *Radiopharmaceuticals II* (J.A. Sorensen, ed.), Society of Nuclear Medicine, New York, p. 265.

Srivastava, S.C., Richards, P., Meinken, G.E., Larson, S.M., and Grunbaum, Z. (1981). *Radiopharmaceuticals* (R.P. Spencer, ed.), Grune & Stratton, Inc., New York, 207.

Stanbury, D.M., Haas, O., and Taube, H. (1980). *Inorg. Chem.*, *19*: 518.

Stritar, J.A., and Taube, H. (1969). *Inorg. Chem.*, *8*: 2281.

Taube, H. (1973). *Surv. Prog. Chem.*, *6*: 1.

Toney, J.H., Brock, C.P., and Marks, T.J. (1986). *J. Amer. Chem. Soc.*, *113*: 7263.

Tsofina, L., Liberman, E.A., Vygodina, T.V., and Konstantinov, A.A. (1986). *Biochem. Int.*, *12*: 103.

Vaupel, P., Okunleff, P., and Neuringer, L.J. (1989). *Adv. Exp. Med. Biol.*, *248*: 835.

Vaupel, P., Schlenger, K., and Knoop, C. (1991). *Cancer Res.*, *51*: 3316.

Williams, D.R. (1974). *Educ. Chem.*, *1974*: 124.

Winkler, J.R., and Gray, H.B. (1992). *Chem. Rev.*, *92*: 369.

Wishart, J.F., Bino, A., and Taube, H. (1986). *Inorg. Chem.*, *25*: 3318.

Yasbin, R.E., Matthews, C.R., and Clarke, M.J. (1980). *Chem.-Biol. Interact.*, *30*: 355.

Ying, W.-L., Emerson, J., Clarke, M.J., and Sanadi, D.R. (1991). *Biochemistry*, *30*: 4949.

METAL BINDING OF NUCLEIC ACIDS AND HELIX INVERSION

by

T. Theophanides

National Technical University of Athens,
Radiation Chemistry and Biospectroscopy,
Zografou Campus, Zografou
157 80, Athens
Greece

I. INTRODUCTION

Metal ions play an important role in biological systems in general, and in particular in nucleic acids, in the regulation of gene membrane structure and permeability, in enzymes, etc (Eichhorn, 1973). It is interesting, however, to know how these metal ions are bound to or interact with nucleic acids and change their structures and conformations. Metal-nucleic acid interactions may reveal, for instance, why some metal ions are cytotoxic or why they induce cell growth and generate abnormalities by attacking the genetic material or by provoking structural changes in the double helix of DNA. The studies of metal ion-interactions with nucleic acids and the genetic material in general could also help us understand the environmental effects and the influence of these on our lives.

Metal ions in fact can critically modify the structure and the hydration shell of natural DNA and nucleotides by changing, stabilizing or destabilizing the natural conformation of a nucleotide in its water environment (Theophanides, 1984). They may affect also their biological activity. These metal ions often interact as coordination complexes or simple salts which are hydrated. Hydrated metal ions could interact with the nucleotides of nucleic acids (i) through their hydration shell, as Van der Waals contacts of the water hydrogens (hydrogen-bonds), (ii) through their electrical charges as positive cations, and (iii) by forming chemical bonds with the most reactive sites of nucleotides (Theophanides and Anastassopoulou, 1988).

It is well known that the nucleotides possess several donor sites which can interact with metal ions or attract metal cations. These sites are the electronegative atoms of the bases and the phosphate groups (PO_2^-). Rarely the metal cations will interact with the sugar part, except in extreme cases of high pH values or to form hydrogen bonds through their hydroxyl groups with the hydrated metal cations $[M(H_2O)_6]^{n+}$, where n is the charge of the cation.

The most reactive sites of the purine bases (guanine and adenine) are the N7 sites of the imidazole ring and the N1 site of adenine; for the pyrimidine bases (cytosine and thymine or uracil), it is the N3 site (Barton and Lippard, 1980). In the case of guanine, the carboxyl group (C6=O) may also be involved in interactions with metal ions, either indirectly through hydrogen bonding or directly by forming a chemical bond (Marzilli et al., 1980). It is, therefore, clear that metal ion binding should be found either at the

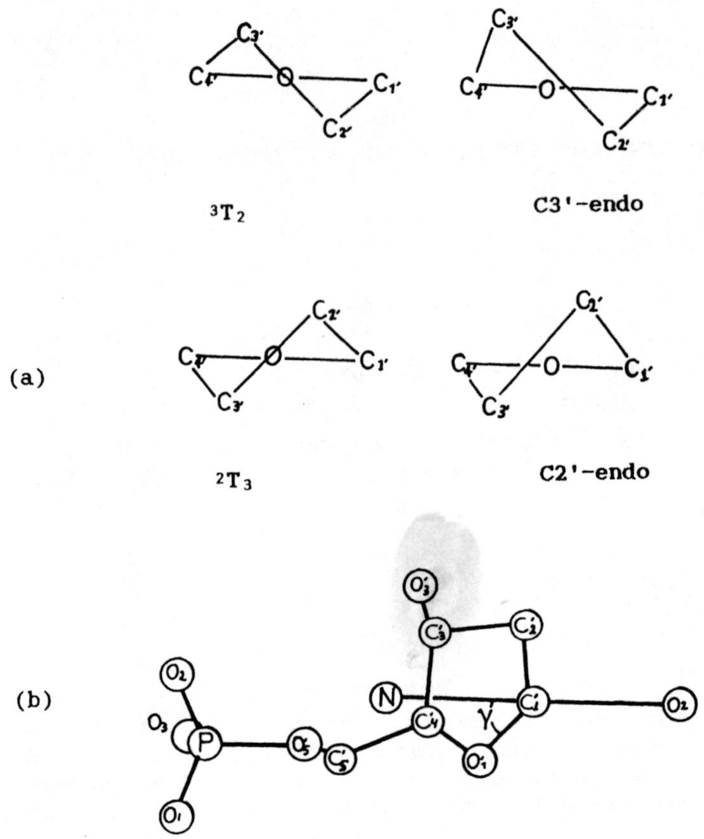

Figure 1.
(a) sugar puckering, (b) the conformation at the glycosyl bond of a nucleoside.

phosphate groups of DNA (PO_2^-) or at the bases, but rarely in the sugar part (Fig. 1). In some cases, chelation may also take place between the N7 site and the phosphate group (N7-PO2-) or at two N7 sites of two adjacent guanines (N7-N7). These types of metal binding should result in major destabilization of the double helix (Theophanides, 1989).

The biochemistry of metal ions should provide new insights into the molecular chemistry of DNA as well as membrane chemistry and should lead to a better understanding of diseases and physiopathology in general.

II. INTERACTION OF METAL IONS WITH NUCLEIC ACIDS

Metal ions play important roles in several parts of the body, and in particular in the cell (Theophanides, 1981). It is now well known that metal ions in the presence of biological molecules influence their conformations by stabilizing some and destabilizing others, altering in that way their biological activities.

In general, alkaline and alkaline-earth metal ions (Li^+, Na^+, K^+, Mg^{2+}, Ca^{2+}, Ba^{2+},...), as well as rare-earth or lanthanide ions (La^{3+}, Eu^{3+}, Tb^{3+},...), the so-called hard metals or hard Lewis acids, are linked or interact with the phosphate anionic groups of DNA (PO_2^-) and stabilize the double helix (Eichhorn and Shin, 1968). In the case of nucleotides, they stabilize the natural conformation of the nucleotide as sodium salts. On the other hand, soft metals or soft Lewis acids such as the transition metal cations with electronic configuration d^{7-8} (Co^{2+}, Ni^{2+}, Pd^{2+}, Pt^{2+}...) or the post-transition metal cations with electronic configuration d^{9-10} (Cu^{2+}, Zn^{2+}, Hg^{2+},...) interact with the soft basic sites, i.e. the heterocyclic nitrogen atoms, in particular the N7 atom of purines and the

N3 atom of pyrimidines. However, there is also a possibility of base-phosphate mixed coordination, chelation leading to stabilization of unusual (non-natural) structural geometries and conformations. Of course, all these donor sites compete for metal ions depending also on the reaction conditions (solvent, pH, concentration, etc).

III. DNA DOUBLE HELICES AND METAL IONS

The natural structures of DNA double helices are normally stabilized by sodium cations (Na^+_{aq}) which neutralize the phosphate anions (PO_2^-) of the backbone strands of sugar-phosphate. It is known that at concentrations of NaCl below 10^{-3} M the double helices separate into single strands and the hydrogen bonds of G...C and A...T which hold the bases of opposite strands together break up and the strands become single (Freifelder, 1978). By forming ionic bonds with the phosphates, hard metal ions thus help to keep the base-base hydrogen bonds and stabilize the double helices by forming a third helix of aquated sodium cations (Na^+_{aq}) which is at a distance of the ionic bond PO_2^- Na^+ from the backbone of DNA.

The conformation of the sugar ring and the sugar-base (glycosyl bond) depends very much on the relationship between the sugar ring pucker and allowed regions for the base position in nucleosides and nucleotides. The rotational potential about the glycosyl bond determines the overall stereochemistry of DNA. This relationship of course depends on a small balance of energy (rotational potentials) and thus the interaction of metal ions (hard or soft) will affect this balance of energy causing an interconversion of helices (Theophanides and Anastassopoulou, 1988).

The sugar ring conformations are known to be limiting conformations. For instance, the C2'-endo is close to $^2T_3$, while the C3'-endo conformation is $^3T_2$ (Fig. 1) (Theophanides, 1989). The C3'-endo conformation at a glycosyl torsion angle of about $\gamma = -40°$ (anti) and $+150°$ (syn) is a region of low energy. This is the notation by Donohue and Trueboold (1960). All the authors do not use identical definitions of the sign of γ.

Any interaction with the nucleotide, such as Van der Waals, hydrogen bonding, electrostatic (metal cations) and torsion potential should be taken into consideration in order to determine the helix inversion of a DNA.

X-ray structures of nucleosides and nucleotides in agreement with molecular mechanics calculations confirm the fact that electrostatic interactions are stabilizing factors for these molecules and are unimportant in determining the exact geometry of minimum energy, but may be quite important in determining conformational energies and rotational barriers. Covalent interactions, on the other hand, do influence the minimum energy. When a metal forms chemical bonds with a site of the base, then obviously we should expect major conformational changes in DNA's. This has been shown to be true in the trends observed in X-ray crystal structures of nucleosides and nucleotides (Hodgson, 1977).

A chemical bond with a metal, M-N7, where M is a soft cation such as Pt^{2+} with electronic configuration d8 creates geometrical structures near the N7 site of guanine in DNA, such as the square planar structure of platinum(II) with the other ligands occupying the four corners of the square plane (Theophanides, 1981). This of course will introduce geometrical changes in the double helix as it does in the interactions of cis-$Pt(NH_3)_2Cl_2$ (cis-platin) with nucleic acids (Gellert and Bau, 1979), and disrupt the helix considerably.

IV. HELIX CONFORMATIONAL INVERSION BY METAL IONS: INFRARED AND RAMAN "MARKER BANDS"

The first double-helical interconversion was observed in double helical poly(dG-dC) by Raman spectroscopy and optical rotatory dispersion measurements in aqueous solutions at low and high NaCl concentrations (Pohl and Jovin, 1972). The right-handed double helix form of poly(dG-dC) was converted at high concentrations of Na^+_{aq} cations in another double helix form, which was later identified with the left-handed helical form of DNA or Z-DNA.

These experiments (Pohl et al., 1973) were also tried with calf thymus DNA in the A,B and C forms, tRNA (yeast) and poly(G).poly(C). It was found in the Raman spectra

Table 1

FT-IR "Marker Bands" of Monovalent and Bivalent Metal-Nucleotide Complexes (Theophanides and Anastassopoulou, 1988)

Compound	Sugar conformation from X-rays	"Marker band" (cm^{-1}) sym. P-O str. sugar-phosphate	breathing of guanine ring
$Na_2GMP.7H_2O$	C2'-*endo*/*anti*	821	688
$CdGMP.8H_2O$	C3'-*endo*/*anti*	799	687
$MgGMP.6H_2O$	C3'-*endo*/*anti*	802	693
$Mg(Guo)_2Br_2.H_2O$	C3'-*endo*/*anti*	794	695
$NiGMP.8H_2O$	C3'-*endo*/*anti*	804	678
$CoGMP.8H_2O$	C3'-*endo*/*anti*	804	678
$Na_2IMP.7H_2O$	C2'-*endo*/*anti*	826	691
$NaHIMP.6H_2O$	C2'-*endo*/*anti*	824	693
$BaIMP.6H_2O$	C2'-*endo*/*anti*	827	695
$NiIMP.7H_2O$	C3'-*endo*/*anti*	804	700
$CoIMP.7H_2O$	C3'-*endo*/*anti*	804	700
RNA (A-DNA)	C3'-*endo*/*anti*	810	675
DNA (B-DNA)	C2'-*endo*/*anti*	825	690
CuRNA (A-DNA)	C3'-*endo*/*anti*	816	680
CuDNA (Z-DNA)	C3'-*endo*/*anti*	812	624

that the 832 cm^{-1} band of the symmetric stretching of C5'-O-P-O-C3' of the sugar-phosphate changed to 814 cm^{-1} at high NaCl concentrations (4.4 M NaCl). In addition, the ring stretching or ring deformation mode (breathing mode) changed from 682 cm^{-1} to 627 cm^{-1} upon addition of solid NaCl to 4.4 M. Using $MgCl_2$ salt produced the inversion at a concentration of only 0.8 M.

Regarding the X-ray atomic resolution of the crystal $d(CG)_3$ (hexamer) a left-handed helical form was found in the presence of $MgCl_2$. In this structure the sugar-phosphate backbone had an irregular zig-zag form and the structure was called Z-DNA (Thamann et al., 1981). The high salt (4M NaCl or $MgCl_2$) form of this hexamer was found to have changes in the Raman bands similar to the symmetric stretching of C5'-O-P-O-C3' and the breathing mode of the purine ring G (Thamann et al., 1981).

These ions affect the characteristic bands of the phosphodioxy group (PO_2^-) and the phosphodiester C5'-O-P-O-C3' linkages and also the C2'-endo, C3'-endo equilibrium, causing condensation of DNA and most likely interaction of the $[Mg(H_2O)_6]^{2+}$ sphere with additional sites of nucleotides, such as N7, O6 of guanine, etc.

The above characteristic ring mode of purine base arises from coupling with sugar vibrations involving the C2'-endo/anti and C3'-endo/anti or C3'-endo/syn conformations. The FT-IR spectra also reveal "marker bands" in this region for nucleotides and DNA's (Theophanides and Tajmir-Riahi, 1985). The data are given in Table 1.

The IR spectra of metal adducts of more nucleotides and DNA show an inversion from B-DNA into A-DNA conformation (Manfait and Theophanides, 1986). The transition from B-DNA to Z-DNA is induced by the metal ions Na^+, Mg^{2+}, Mn^{2+} and Co^{2+} at critical concentration.

REFERENCES

Barton, J.K., and Lippard, T.J. (1980). *Nucleic Acid-Metal Interactions* (T.G. Spiro, ed.), John Wiley & Sons, Inc., p. 180.

Donohue, J., and Trueblood, K.N. (1960). *J. Mol. Biol.*, 2: 363.

Eichhorn, G.L. (1973). *Inorganic Chemistry* (G.L. Eichhorn, ed.), Elsevier Scientific Publishing Company, Amsterdam, p. 1191.

Eichhorn, G.L., and Shin, Y.A. (1968). *J. Amer. Chem. Soc., 90*: 7323.

Freifelder, D.M. (1978). *The DNA Molecule* (D.M. Freifelder, ed.), W.H. Freeman and Company, San Francisco, p. 163.

Gellert, R.W., and Bau, R. (1979). *Metal Ions in Biological Systems,* Vol. 8 (H. Sigel, ed.), Marcel Dekker, New York, p. 1.

Hodgson, D.J. (1977). *Prog. Inorg. Chem., 23*: 211.

Manfait, M., and Theophanides, T. (1986). *Spectroscopy of Biological Systems* (R.J.H. Clark and R.E. Hester, eds.), John Wiley & Sons Inc., p. 31.

Marzilli, L.G., Kistermachter, T.J., and Eichhorn, G.L. (1980). *Nucleic Acid-Metal Ion Interactions* (T.G. Spiro, ed.), John Wiley & Sons, Inc., p. 180.

Pohl, F.M., and Jovin, T.M. (1972). *J. Mol. Biol., 67*: 375.

Pohl, F.M., Ranade, A., and Stockburger, M. (1973). *Biochim. Biophys. Acta, 335*, 85.

Thamann, T.J., Lord, R.C., Wang, A.H.J., and Rich, A. (1981). *Nucl. Acid. Res., 9*: 5443.

Theophanides, T. (1981). *Can. J. Spectrosc. 26*: 165.

Theophanides, T. (1984). *Intern. J. Quant. Chem., 24*: 933.

Theophanides, T. (1989). *Spectroscopy of Inorganic Bioactivators, Theory and Application-Chemistry, Physics, Biology, and Medicine* (T.Theophanides, ed.), Kluwer Academic Publishers, The Netherlands, p. 265.

Theophanides, T., and Anastassopoulou, J. (1988). *Metal-Based Anti-tumour Drugs* (M.F. Gielen, ed.), Freund Publishing House Ltd., England, p. 151.

Theophanides, T., and Tajmir-Riahi, H.A. (1985). *Structure and Motion: Membranes, Nucleic Acids and Proteins* (E. Clementi and G. Gorongiu, eds.), M.H. Sarma & R.H. Sarma, ISBN Adenine Press, p. 521.

COPPER COMPLEXES AS DRUGS

by

William E. Antholine

Department of Radiation Biology,
Medical College of Wisconsin,
8701 Watertown Plank Rd.
Milwaukee, WI 53226
U.S.A.

Robert Byrnes and David H. Petering*

Department of Chemistry,
University of Wisconsin-Milwaukee,
P.O. Box 413
Milwaukee, WI 53211
U.S.A.

I. INTRODUCTION

The past three decades have witnessed a renaissance in inorganic chemistry and with it the flowering of interest in metals in biological systems. Mostly, this has focused on essential metal metabolism and function. To a lesser extent the chemistry underlying heavy metal toxicity has been examined. In addition, a combination of fortuitous findings and directed research has begun to show that metal complexes have their place as therapeutic agents. For example, among the approximately 20 major drugs used in human cancer chemotherapy, two happen to require metals, cis-dichlorodiammine Pt(II) and bleomycin (Fe) (Petering et al., 1990; Radtke et al., 1991). A third, adriamycin, may also be activated by iron and utilize intracellular iron to support its cardiotoxic properties (Gianni et al., 1985; Myers, 1988; Olson and Mushlin, 1990). The discovery of the chemotherapeutic properties of these compounds did not involve deliberate efforts to isolate or synthesize metallodrug species.

Complexes of other metal ions have been made and tested for chemotherapeutic activity, for example, dicyclopentadienyl titanium dichloride (Köpf-Maier and Köpf, 1983). In fact, the earliest metal-binding ligand/metal complex demonstrated to have excellent, broad spectrum anticancer activity in rodent models of solid tumors was 3-ethoxy-2-oxobutyraldehyde bis(thiosemicarbazone) (Petering, 1980). Its Cu(II) complex as well as a number of analog structures are the active forms. Though never examined extensively in

*To whom correspondence should be addressed.

clinical trials, bis(thiosemicarbazones) and their copper complexes remain attractive candidates for antitumor agents (Regelson et al., 1967). In addition, they have stimulated continuing studies of other ligands and Cu-containing structures that are highly effective in causing cytotoxicity or other biological effects.

Several reviews have been written about copper complexes and their rich biological chemistry (Petering, 1980; Petering and Antholine, 1988; Petering et al., 1984; Sorenson, 1989). The present contribution builds upon them and focuses on newer results in this area, particularly as they illustrate the variety of interesting chemical questions that copper complexes present to researchers.

II. GENERAL MODEL FOR COPPER METABOLISM AND SPECIATION AMONG COPPER-BINDING LIGANDS

In developing a perspective on the potential biological activity of a copper-binding ligand or copper complex, several related topics need to be considered. These include the capacity and specificity of the ligand to compete for copper in the organism and the stability and inherent reactivity of copper complexes that either form in vivo or are directly administered to the organism (Petering and Antholine, 1988).

Relatively little is known about mammalian Cu metabolism. Apparently, Cu bound to histidine, albumin, or ceruloplasmin, the major pool of Cu in plasma, can be accumulated by various cell types (Linder and Goode, 1991) (Fig. 1). Once in cells, the pathways to its final binding sites are not established. One study has implicated metallothionein as an intermediate (Tereo and Owen, Jr., 1973).* Another has suggested that in Cu-resistant cells, Cu is first bound to glutathione and then transferred to metallothionein (Freedman et al., 1989). However, there is no consistent evidence that MT is an obligatory intermediate in copper metabolism. In fact, in vitro experiments suggest that copper bound to MT is not readily available for transfer to apoCu-binding proteins such as superoxide dismutase (Geller and Winge, 1982). Other protein-bound intermediates have been observed. Probably, some oxidation-reduction chemistry is involved to labilize the copper ion and facilitate ligand exchange, but this has not been established.

Only a few copper proteins are found in mammalian cells. Cytochrome oxidase, superoxide dismutase, lysyl oxidase, and dopamine-β-hydroxylase are common ones. In the first two proteins, which represent the major depots of Cu in cells, copper is not readily available to competing ligands nor is it easily extracted from lysyl oxidase or plasma ceruloplasmin (Gacheru et al., 1990; Hirose et al., 1982). So, apart from defined copper

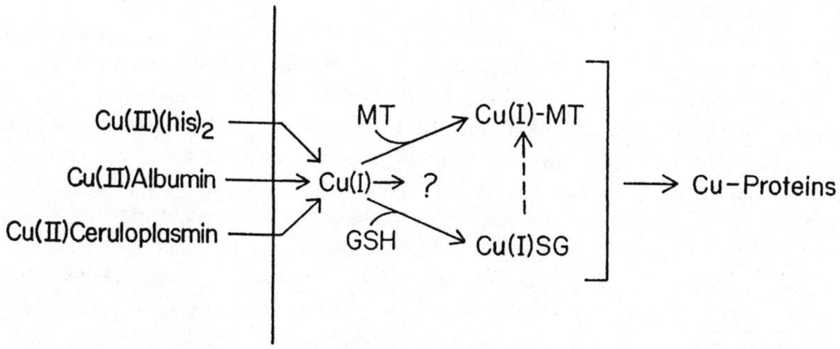

Figure 1.
Ligand exchange during copper metabolism.

---

*Abbreviations: MT, metallothionein; Blm, bleomycin; GSH, glutathione; H_2KTS, 3-ethoxy-2-oxobutyraldehyde bis(thiosemicarbazone); RDR, ribonucleotide reductase; NC, neocuproine or 2,9-dimethyl-1,10-phenanthroline; OP, orthophenanthroline or 1,10-phenanthroline.

Table 1

Apparent Stability Constants of Copper Complexes

Cu complex	log Stability constant[a]
Cu-albumin[b]	16.2
$Cu(II)(his)_2$[c]	15.2
Cu(I)-MT[d]	greater than 14
Cu(I)-SG[e]	13.2

[a]approximately pH 7.4 and 25°C
[b]Sarkar (1981)
[c]Antholine et al. (1977)
[d]Cu^+ stoichiometrically displaces Zn^{2+} and Cd^{2+} from metallothionein, which bind to the protein with average log stability constants of 11.2 and about 14 (Petering and Shaw, 1991).
[e]preliminary results

complexes such as $Cu(his)_2$ or Cu-albumin in the extracellular medium, there are few sites of competition for Cu by metal-binding ligands.

Another facet of the chelation of copper ions from the organism is the likelihood that metal ions such as Zn^{2+} or Fe^{2+} may also become bound to the ligand. Unless the issue of non-specificity of metal binding by ligands is explicitly considered, the underlying bioinorganic basis of their biological responses may be misinterpreted.

When copper complexes are added to cells or fluids, there are strong, identifiable sites of competition for the metal. For example, albumin and histidine have large apparent stability constants for Cu(II) and metallothionein and glutathione have high affinities for Cu(I) (Table 1). Because of the compartmentalization of biological forms of Cu and Cu-binding ligands in organisms, it is possible that Cu-drug species can be formed in one place and react in another, undergoing dissociation. In the sections that follow, these generalities will be examined with specific examples.

III. BIS(THIOSEMICARBAZONATO) COPPER COMPLEXES

A. Cancer chemotherapeutic complexes

One of the first efforts to design antitumor drugs that involved metal ions was initiated by French and coworkers more than three decades ago (Blanz, Jr. and French, 1965; French and Freedlander, 1958). They constructed a number of bis- and monothiosemicarbazone compounds that were intended to chelate essential metals in organisms and, thereby, disrupt cell proliferation. Several of these showed significant antineoplastic activity in early screening (French and Freedlander, 1960). One agent, 3-ethoxy-2-oxobutyraldehyde-bis(thiosemicarbazone), H_2KTS, was examined in detail by H.G. Petering and coworkers. They and others found that the free ligand administered orally to rodents bearing a variety of solid tumors, both new and established, had outstanding activity (Petering, 1980; Petering and Petering, 1975; Petering (H.G.) et al., 1964; Mihich and Nichol, 1965). The antitumor action was lost in copper-deficient animals, was enhanced when the copper complex (I) was substituted for the free ligand, and depended critically on the extent of N4 methyl substitution on the thiosemicarbazone group (Crim and Petering (H.G.), 1967; Petering (H.G.) et al., 1967; Van Giessen et al., 1973).

A number of subsequent mechanistic studies have contributed to the following model for the chemistry of reaction of H_2KTS/Cu(II)KTS with tumor cells (Fig. 2). (a) The free ligand has a very large apparent stability constant for Cu(II), which ensures that it can acquire Cu from plasma donors such as $Cu(his)_2$ (Petering, 1972a). Its affinity for Zn^{2+} is too low to compete with plasma ligands for this metal ion (Petering, 1974). (b) Once

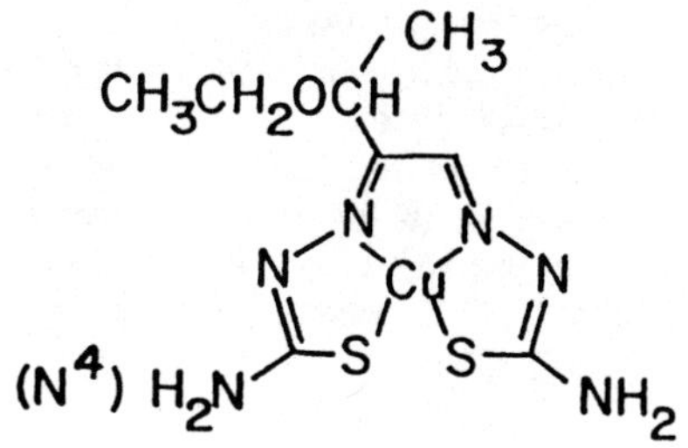

I

formed, the neutral Cu(II) complex, having substantial water and lipid solubility, can diffuse into cells. (c) There it rapidly reacts with reducing agents, probably thiols, and undergoes reductive dissociation (Petering, 1972b; Minkel and Petering, 1978). The released ligand, also a neutral molecule, partitions itself back into the extracellular medium, where it may acquire more copper. (d) Because of the cellular abundance of glutathione and its large affinity for Cu^+, it is likely that a Cu-glutathione complex forms, which continues to undergo cycles of oxidation-reduction, leading to the reduction of substantial amounts of O_2 (Chan-Stier et al., 1976). (e) If significant metallothionein is present, it readily binds the first equivalents of Cu which enter the cell, probably removing them from further oxidation-reduction reactions (Kraker et al., 1985). It is likely that the persistent presence of extra copper in these cells, unbound to its normal physiological ligands, causes oxidant stress resulting in DNA and membrane damage through the generation of hydroxyl radicals or more subtle effects following the oxidation of key thiol groups.

Corresponding chemical studies have emphasized the importance of N4 methylation on the physical and chemical properties of copper-bis(thiosemicarbazones). Moving from CuKTS and CuKTSMe to $CuKTSMe_2$ converts the structure from one with high biological activity to one with much lower activity (Van Giessen et al., 1973; Minkel et al., 1978). The reduction potential of Cu(II) decreases 100 mV in the transition from the monomethyl to dimethyl derivative, making $CuKTSMe_2$ much less reactive with thiols and more stable in cells (Winkelmann et al., 1974). Nevertheless, it is cytotoxic to cells at concentrations higher than those used for CuKTS (Antholine et al., 1984). The highly non-polar complex $CuKTSMe_2$ localizes in the interior of biological membranes, becoming oriented parallel to the lipid tails (Subczynski et al., 1987). A key tool in the cellular study of CuKTS and other copper complexes is electron spin resonance spectroscopy. It can detect the presence of parent and altered Cu(II) complexes, follow their reduction to diamagnetic Cu(I), and do so at ambient temperature (Antholine et al., 1985a; 1985b; Minkel and Petering, 1978).

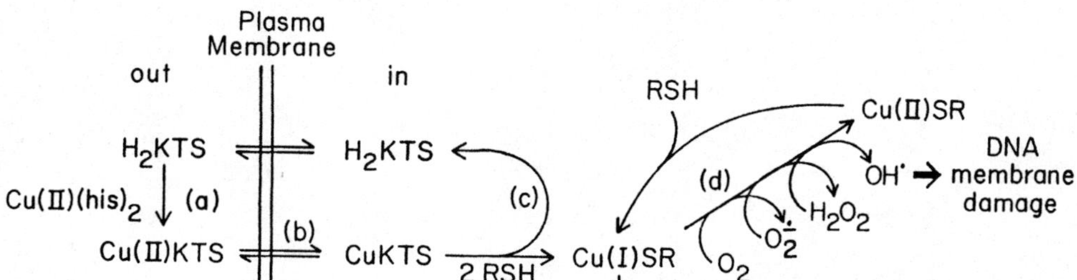

Figure 2.
Reaction of 3-ethoxy-2-oxobutyraldehyde bis(thiosemicarbazonato) Cu(II) with plasma and cells.

B. Complexes for positron emission tomography

^{62}Cu has a positron-emitting nucleus of use in positron emission tomography or PET scanning. Recently, Mark Green and others have turned to copper bis(thiosemicarbazones) as effective complexes to image the brain and heart (Shelton et al., 1989). Ideally, after injection of the emitter, all of the agent should be transported into the target tissue in a single pass of the blood through the tissue and react in some manner so that it is immobilized there. ^{62}CuPTSMe, 2-oxopropionaldehyde bis(N4 thiosemicarbazonato) ^{62}Cu(II), largely fulfills these requirements, but the related complex $^{62}CuPTSMe_2$, having N4 dimethylation, clears from the brain relatively soon after it is taken up and, therefore, is not a useful compound. The properties described above, favorable aqueous-lipid partitioning with substantial solubility in both media and reactivity with thiols leading to reduction and dissociation of Cu(I) and its subsequent binding to cellular sites, have satisfactorily explained the differential suitability of these compounds as imaging agents (Green et al., 1988; John and Green, 1990).

IV. α-N-HETEROCYCLIC CARBOXALDEHYDE THIOSEMICARBAZONATO COPPER COMPLEXES

A second class of thiosemicarbazones, α-N-heterocyclic carboxaldehyde thiosemicarbazones, also derived from the work of French as well as Brockman, have received extensive pharmacological and some human clinical study as antitumor agents (Agrawal and Sartorelli, 1975; Blanz, Jr. and French, 1965; Brockman et al., 1956; DeConti et al., 1972; French and Blanz, Jr., 1966). A principal target for their action is the enzyme ribonucleotide reductase (RDR) (Moore et al., 1970; Sartorelli et al., 1971). This is also probably the site for the activity of monothiosemicarbazones against herpes simplex virus type 1, which contains a large amount of ribonucleotide reductase (Spector et al., 1991). On the basis of experiments with crude preparations of RDR enzyme, it was originally thought that this tridentate ligand binds to Fe at the active site of the protein, forming a ternary complex (Moore et al., 1970). Probably, it simply chelates Fe(III) from the μ-oxo bridged metal center of the enzyme (Nordlund et al., 1990). However, the iron complexes of 2-formylpyridine thiosemicarbazone and 2-acetyl-pyridine-5-[(2-chloroaniline)-thiocarboxyl]-thiosemicarbazone also inhibit the enzyme and the copper complex of the first ligand is an even more potent inhibitor (Saryan et al., 1979; Spector et al., 1991).

When 5-hydroxy-2-formylpyridine thiosemicarbazone and related ligands are administered to organisms, they bind sizable amounts of Fe, which can be detected in the urine (DeConti et al., 1972; French et al., 1965). One site of reaction may be iron-transferrin, which undergoes reduction and ligand substitution in the presence of the ligand, ascorbate, and nitrilotriacetate (Ankel and Petering, 1980). So, the iron complex preferentially forms in vivo. Nevertheless, the preformed Cu(II) complex of this (**II**) and related monothiosemicarbazones have large apparent stability constants, which indicate that they will be stable in biological media such as plasma (Antholine et al., 1977).

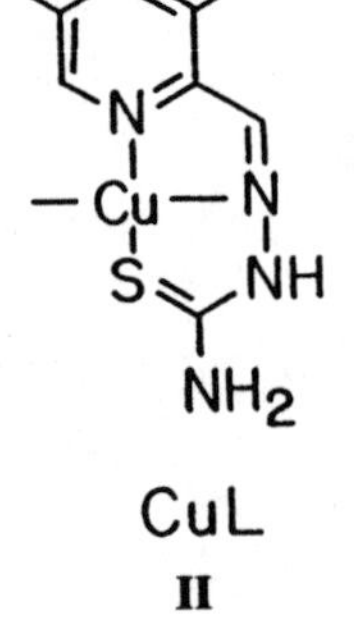

CuL

II

The tridentate nature of **II** (CuL) qualitatively changes the chemistry of these complexes in comparison with **I**. First, adducts with Lewis bases can be formed which have substantial stability (Antholine et al., 1977; Knight et al., 1979). Secondly, during

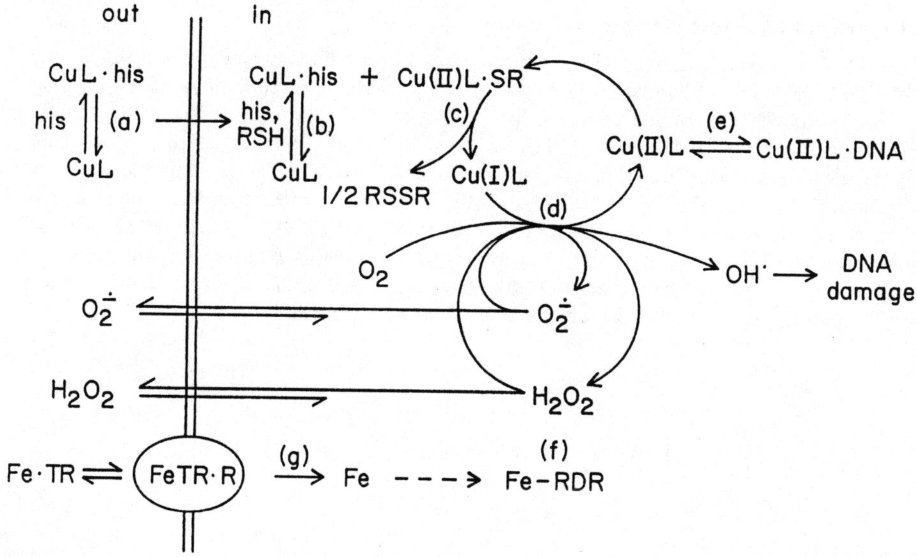

Figure 3.
Reactions of α-N-heterocyclic carboxaldehyde thiosemicarbazonato Cu(II) complexes with plasma and cells.

oxidation-reduction reactions of **II** with thiols (RSH), reoxidation of the complex (reaction 3) competes favorably with ligand substitution (reaction 2):

$$Cu(II)L + RSH \rightarrow Cu(I)L + RS^{\cdot}\ (½\ RSSR) \quad (1)$$

$$Cu(I)L + RSH \rightarrow Cu(I)SR + HL \quad (2)$$

$$Cu(I)L + O_2 \rightarrow Cu(II)L + O_2^{-} \quad (3)$$

A third property of this complex that may be biologically important is that it can interact with DNA, binding with its planar structure perpendicular to the helix axis (Chikira et al., 1991). A potential site-directing feature of CuL is its ability to form adducts (4):

$$Cu(II)L + B{:} \rightarrow B\text{-}Cu(II)L \quad (4)$$

Both histidine-like and thiol adducts have been detected in plasma, culture media, or cells (Saryan et al., 1981). Cu(II)L binds to the thiol site in cat hemoglobin, for example, and apparently does so even in red blood cells (Antholine and Taketa, 1982; 1984).

A model which summarizes how CuL and a related structure, pyridoxal thiosemicarbazonato Cu(II), react with tumor cells is shown in Figure 3 (Byrnes et al., 1990; Saryan et al., 1981). It incorporates formation of adducts outside and inside the cells, which have been demonstrated by ESR spectroscopy (a,b), reaction with intracellular thiols, probably involving glutathione (c), and redox cycling to catalyze the generation of reduced forms of dioxygen including superoxide and hydrogen peroxide (d), which can diffuse through membranes and react, respectively, with extracellular superoxide dismutase and catalase (Byrnes et al., 1990). This process culminates in the generation of highly reactive hydroxyl radicals, which can cause DNA strand breakage, possibly through a DNA-bound copper center (e). (f) CuL also induces the loss of the tyrosyl free radical signal in RDR in cells and in purified preparations (Narasimhan et al., 1991). The mechanism of this reaction is not yet understood. The effect may be indirect, for the transport of iron into cells from ^{59}Fe-transferrin is strongly inhibited after incubation with CuL (g). Thus, cells may be deprived of the iron needed to constitute the rather labile Fe site of RDR.

V. COPPER BLEOMYCIN

Bleomycin is commonly used in drug combinations to treat human cancer (Petering et al., 1990). Originally isolated from *Streptomyces verticilis* with about 10% of its metal binding sites occupied with Cu(II), the antitumor properties of this natural product were originally demonstrated on Cu(II)Blm (**III**). Later, because of its lower toxicity, the free ligand became the species used for chemotherapy. Blm has a very large log apparent stability constant (18.8) for Cu(II), Cu^{2+} displaces Fe(III) from Blm, and CuBlm has been detected in the urine of animals exposed to Blm (Antholine et al., 1982; Kanao et al., 1973; Solaiman et al., 1980). Nevertheless, it is iron not copper that is required for the cytotoxic action of Blm on cultured tumor cells (Radtke et al., 1991). Therefore, the speciation of the drug is complex and, conceivably, affects its antitumor properties.

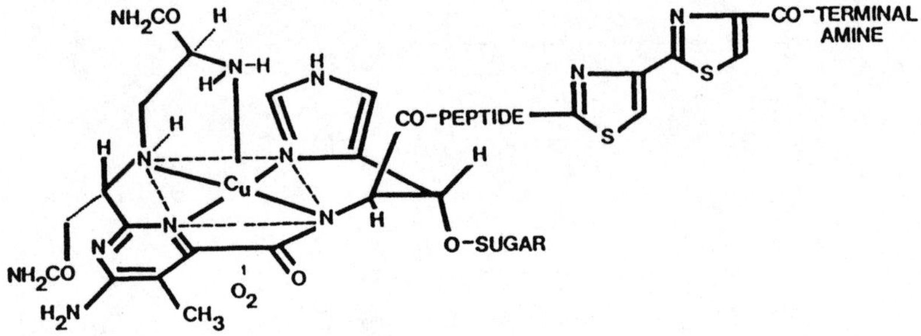

Cu-bleomycin

III

The following are reactions that may be involved in the formation and possible interconversion of Cu- and FeBlm (Fig. 4). (a) CuBlm can form in plasma, probably by exchange of Blm with amino acids in low-molar-mass complexes such as $Cu(his)_2$, Cu(his)(gln), etc. When taken up into cells, it is apparently converted to FeBlm (Radtke et al., 1991). The mechanism of this process may involve the reduction and dissociation of Cu(II)Blm (b), using a thiol as the reducing and Cu(I)-complexing agent (reactions 1 and 2) (Antholine et al., 1982; Freedman et al., 1981). Although cysteine is effective in this reaction, the common cellular thiol, glutathione (GSH), is not (Antholine et al., 1982).

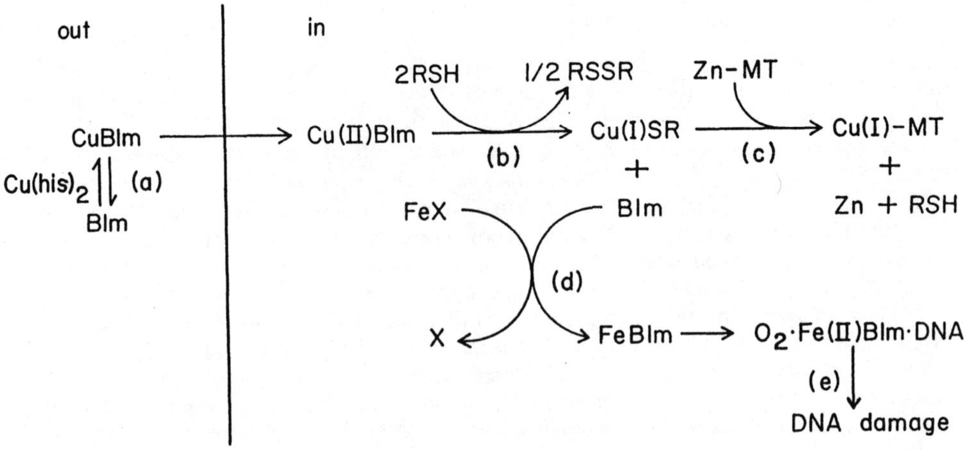

Figure 4.
Speciation and reactions of metallobleomycins in plasma and cells.

Possibly the bulky nature of both the metal-binding site of CuBlm and environment of the SH group of the GSH tripeptide retards the redox reaction. (c) One study showed that metallothionein with its high density of sulfhydryl groups was the final depot for Cu(I) when cysteine and Cd-MT were incubated with CuBlm (Takahashi et al., 1987).

Once Blm is present in cells, it must find a labile source of Fe to form FeBlm (d). What this may be is not known. One report suggests that ribonucleotide reductase may supply iron to the drug (McClarty et al., 1986). Once FeBlm forms, (e) it can bind to DNA and by an O_2-dependent redox reaction cause single- and double-strand cleavage (Petering et al., 1990).

VI. BIS(2,9-DIMETHYL-1,10-PHENANTHROLINE) COPPER(I)

Interest in the chemistry and biological actions of phenanthrolines is long standing. The basic ligand structure, 1,10-phenanthroline (OP), is highly cytotoxic as is $Cu(II)(OP)_2$ (**IVa**) (Byrnes et al., 1992a). The latter causes DNA damage by a mechanism involving reduction of O_2 to hydroxyl radical, in which Cu serves as the catalyst. Because of the planar-aromatic nature of the phenanthroline ligand, it can bind to DNA, providing a site for directed reaction of hydroxyl radical. Indeed, the properties of DNA strand cleavage induced by **IV** suggest that it occurs as the result of ˙OH produced by a copper phenanthroline species directly bound to the polymer (Sigman, 1986).

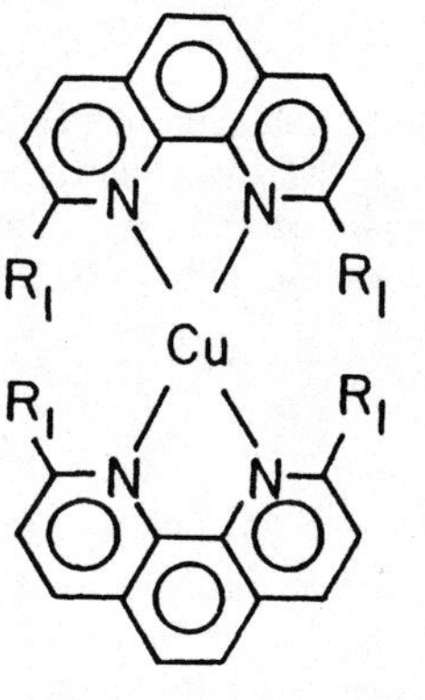

a R = H

b R = CH_3

IV

The surprise is that bis(2,9-dimethyl-1,10-phenanthroline)Cu(I) (**IVb**), Cu(I) neocuproine, is also highly cytotoxic to tumor cells and displays inhibitory effects on leukemic cells in mice (Mohindru et al., 1983; Byrnes et al., 1992b). This complex has a very large log apparent stability constant (19.9). Furthermore, it is oxidatively stable in the presence of O_2 and can not directly catalyze the reduction of O_2 to ˙OH. The key to its activity is two-fold (Fig. 5). First, **IVb** with its lipophilic character is readily accumulated by cells (a). Second, GSH has a large enough affinity for Cu(I) that it can kinetically and thermodynamically compete for Cu(I) bound to $Cu(I)(NC)_2$ (b). Once a copper glutathione complex forms, copper becomes redox active with dioxygen, leading to the generation of hydroxyl radical (c).

An additional facet of the chemistry of copper neocuproine is seen in the reaction of $Cu(II)(NC)_2$ with tumor cells (Fig. 5). The Cu(II) complex is not stable in growth media (d). Instead, a Cu-histidine-like species is formed in the extracellular medium, which is transported into the cells (Byrnes et al., 1992b). Once inside, $Cu(I)(NC)_2$ is generated as the ESR signal of the Cu-histidine-like species declines. The reduction potential of $Cu(II)(his)_2$

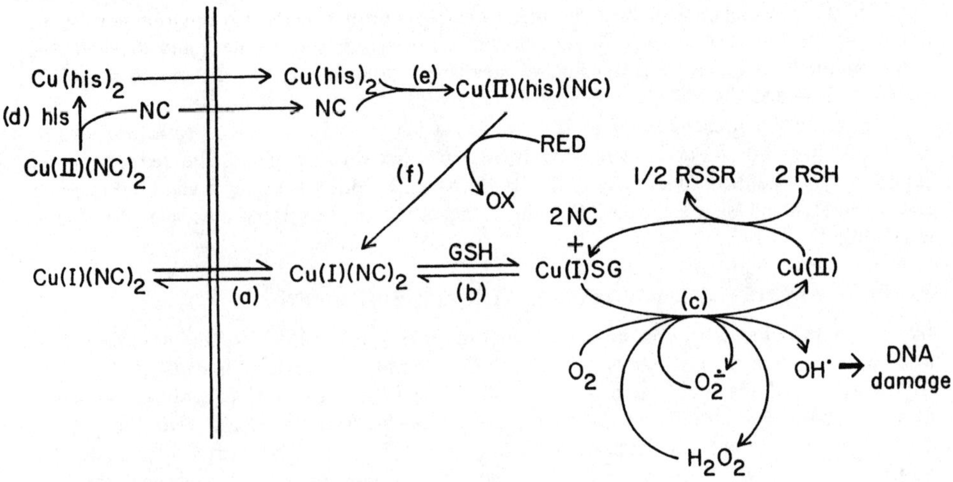

Figure 5.
Reaction of Cu-bis(2,9-dimethyl-1,10-phenanthroline) in cells.

may be too low for direct intracellular reduction of this complex. As an alternative, it has been shown in model reactions that NC can compete with histidine, forming a ternary complex (e), which readily undergoes reduction by reductants (red) because of its higher redox potential and because NC binds strongly to Cu(I) (reaction 5 and 6) (f):

$$Cu(his)_2 + NC \rightarrow Cu(his)(NC) + his \quad (5)$$
$$Cu(his)(NC) + red + NC \rightarrow Cu(I)(NC)_2 + ox + his \quad (6)$$

VII. COPPER ANTIOXIDANTS

A number of adverse physiological conditions have been postulated to involve the presence and biological action of elevated levels of forms of activated oxygen (Halliwell and Gutteridge, 1988). Collectively termed oxidative stress, they are initiated by the reduction of dioxygen as follows:

$$O_2 + e^- \rightarrow O_2^- \quad (7)$$
$$2O_2^- + 2H^+ \rightarrow H_2O_2 + O_2 \quad (8)$$
$$H_2O_2 + M^{(n-1)+} \rightarrow {}^{\bullet}OH + OH^- + M^{n+} \quad (9)$$

Hydroxyl radicals produced by these reactions can abstract hydrogen atoms from biomolecules, such as DNA or membrane lipids, positioned within a few angstroms of their site of generation. Once carbon-based radicals have formed, a variety of structurally damaging chemistry can ensue (Von Sonntag et al., 1981; Girotti, 1985).

The source of electrons in reaction 7 may be xenobiotic reducing agents or abnormal reductive conditions in cells resulting from, for example, anaerobiosis (Aust and White, 1985). Reaction 8 is normally catalyzed by superoxide dismutase in cells (Fridovich, 1991). The final conversion of hydrogen peroxide to hydroxyl radical requires metal-ion mediated electron transfer (Minotti and Aust, 1989). The metal center in hydroxyl radical generation acts as a catalyst by coupling reactions 9 and 10. Beyond the expectation that M^{n+} is either some species of cellular Fe or Cu, little is known about its precise form(s) (Minotti and Aust, 1987; Aruoma et al., 1989).

$$M^{n+} + e^- \rightarrow M^{(n-1)+} \quad (10)$$

To interrupt this pathway, superoxide dismutase mimics have been designed to accelerate reaction 8 (de Alvare et al., 1976; Nagano et al., 1989). These include copper complexes. Under normal conditions of oxidative metabolism, cellular mechanisms of protection, including the operation of superoxide dismutase and catalase, suffice to minimize the formation of hydroxyl radical. However, when reaction 7 and those which follow are enhanced, additional superoxide dismutation may be needed.

Two arguments have been advanced for targeting superoxide ion. The first is based on the hypothesis that O_2^- is the preferential reducing agent for reaction 10. This has been criticized by pointing out that there are other cellular reducing agents, such as thiols and ascorbate, that might work just as well as superoxide ion in this reaction (Fee, 1980). How can it be specific for just one form of reducing equivalents? The other argument is that O_2^-, itself, can damage cells without conversion to $^{\bullet}OH$ (Fridovich, 1991). There is controversy about the reactivity of superoxide ion with biomolecules (Fee, 1980; Sawyer and Valentine, 1981). Nevertheless, for example, DNA strand scission caused by exposure of cells to $Cu(II)(OP)_2$ is greatly diminished by addition of Cu,Zn-superoxide dismutase (SOD) which catalyzes the dismutation of O_2^- diffusing into the extracellular medium (Byrnes et al., 1992a).

An important feature of cellular reduction of O_2 is that both O_2^- and H_2O_2 can travel relatively long distances in the cell before undergoing reaction because of their limited reactivities in the absence of redox active metal ions. They can also readily diffuse through cell membranes. Thus, a key point is that mimics which do not diffuse into or across membranes may still be effective antioxidants. If they can be kept out of cells, potential toxic side effects may be minimized.

Two of the most widely used forms of antioxidant copper are bis(3,5-diisopropyl salicylato) Cu(II) (CuDIPS, **V**) and the enzyme, Cu,Zn-SOD, itself (Flohé, 1988). CuDIPS possesses little toxicity and has anti-inflammatory activity. It has been repeatedly hypothesized that this biological activity is due to its capacity to dismute superoxide ion (deAlvare et al., 1976; Sorenson, 1987 and references therein).

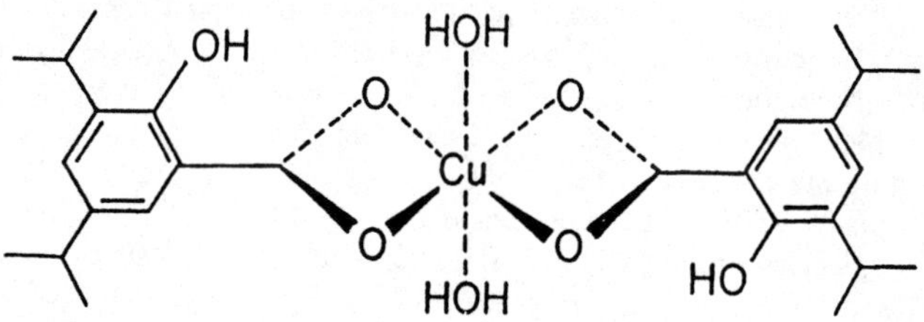

$Cu(DIPS)_2$

V

The mechanism by which CuDIPS serves as an antioxidant is uncertain. In nonpolar solvents, Cu(II) and DIPS are thought to form a species with a stoichiometry of $Cu_2(DIPS)_4$, which is reduced to $Cu(DIPS)_2$ in aqueous media (Greenaway et al., 1988). Because of its lipophilic character, the complex may partition preferentially into membrane. Nevertheless, in the presence of EDTA, thiols, histidine, or citrate, the complex is not stable and is dissociated into Cu and DIPS (Huber et al., 1987; Khan and Sorenson, 1991; Bligh et al., 1992). The complex probably does not have a large apparent stability constant and may readily undergo ligand substitution with competitors such as histidine (Pezeshk et al., 1987). Furthermore, reaction of Cu(II)DIPS with glutathione is known to produce Cu(I)-SG, which suggests another means similar to reactions 1 and 2 to accomplish the dissociation of Cu-DIPS (Khan and Sorenson, 1991).

If CuDIPS simply serves as a vehicle for the delivery of Cu to cells or other compartments such as plasma, then the released copper may be toxic as it is in the case of bis(thiosemicarbazonato) Cu complexes (**I**). Indeed, two groups have noted that CuDIPS can be cytotoxic, perhaps by utilizing pathways including reactions 7, 9, and 10 (Harrison et al., 1988b; Reiners and Colby, 1988). In this model, the antioxidant properties of CuDIPS must be related to how Cu is released from the complex, is selectively deposited, and then reacts in cells. While there are normal, unspecified pathways for intracellular Cu, it is not known to what extent Cu, entering the cell in non-physiological ways, has access to these paths. Therefore, before Cu enters into normal metabolic paths, it may interact with sensitive parts of the cell through a rapid succession of ligand substitution reactions, characteristic of Cu(II) and Cu(I) complexes.

An example of an in vitro reaction described for either CuDIPS or Cu^{2+} that could be hypothetically involved in the effects of Cu on oxidative stress involves inactivation of protein kinase C, a central protein involved in cell proliferation and tumor promotion (Nilsson, 1989). One may hypothesize that Cu competes with Zn bound to the zinc site of the protein and then oxidizes the thiolate ligands to inactivate the enzyme (Quest et al., 1992). The Zn-binding site would provide site direction for Cu-binding in the cell and also serve as a reactant for an oxidation-reduction reaction like reaction 1.

An alternative mechanism to explain the possible superoxide dismutase action of CuDIPS is that it remains intact in the extracellular medium bound to serum albumin and does not enter cells (Shuff et al., 1992). This complex might retain superoxide dismutase activity and scavenge O_2^- that equilibrates from inside cells.

Copper,zinc-superoxide dismutase follows a two-step "ping-pong" mechanism (Klug-Roth et al., 1973; Fielden et al., 1974):

$$Cu(II) + O_2^- \underset{k_{-1}}{\overset{k_1}{\rightleftharpoons}} Cu(I) + O_2 \qquad (11)$$

$$Cu(I) + O_2^- + 2H^+ \rightarrow H_2O_2 \qquad (12)$$

Many forms of low-molar-mass complexes of copper can catalyze reactions 11 and 12 at high rates (Goldstein and Czapski, 1986). For bovine Cu,Zn-SOD, the turnover rate is $2.3 \times 10^9\ M^{-1}\ s^{-1}$ (Fielden et al., 1974). For other forms of Cu, such as CuDIPS, the rate constants are still very large (10^4 to $10^6\ M^{-1}s^{-1}$) (Goldstein and Czapski, 1986). For this reason, it might be considered that an enzyme structure is not really necessary for intracellular O_2^- dismutation by Cu. However, a unique feature of the protein is its small value for k_{-1}, which can be 10^{-3} to 10^{-5} lower than that for simple copper complexes, thereby diminishing the chance that the enzyme can participate in superoxide ion generation (Czapski and Goldstein, 1988). As important is the fact that Cu(II),Zn-SOD is not reduced by ascorbate or glutathione, principal biological reductants, because the copper site is buried within the protein (Fee and DiCorleto, 1973). For this reason and because of the specificity of the binding site for O_2^-, Cu(I), Zn-SOD can not react with H_2O_2 in reaction 9 to produce hydroxyl radical, whereas copper complexes frequently can participate in reactions 7, 9, and 12.

VIII. COPPER COMPLEXES AND WOUND HEALING

A human plasma tripeptide, glycyl-L-histidyl-L-lysine, possesses substantial affinity for Cu(II) and its biological properties suggest a role for the complex in tissue regenerative phenomena (Pickart, 1987a). The tripeptide-copper complex, Cu-GHK (**VI**), stimulates

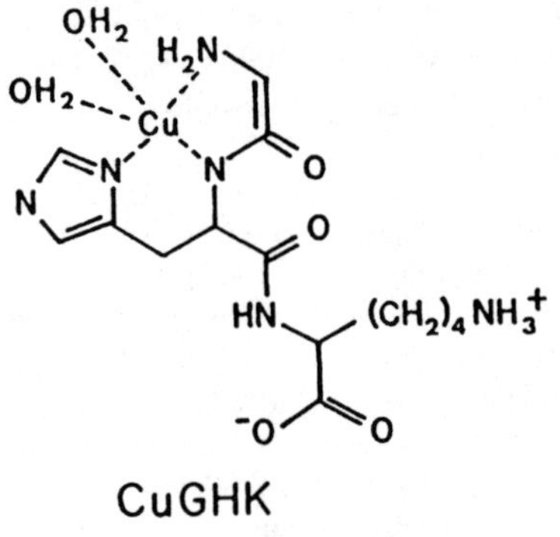

VI

nerve outgrowth and collagen synthesis in cell culture systems as well as angiogenesis in vivo (Marquart et al., 1988; Pickart, 1987a; 1987b). The latter may reflect the apparent requirement of new vessel formation on copper (Folkman and Shing, 1992). Cu-GHK and a number of close analogs stimulate the healing of superficial wounds, the rate of bone repair, and the healing of gastric and duodenal ulcers in model systems (Pickart, 1987b; 1991a; 1991b; 1992). How Cu-GHK stimulates cellular responses in these activities is not known.

IX. CONCLUSIONS

Most of the examples described in this chapter are variations on the theme of oxidation-reduction chemistry of copper. In each case copper serves as a catalyst for the reductive activation of O_2. It is evident that the nature of the ligand bound to copper controls the details of the redox chemistry as well as the sites of localization of the metal ion. One surprise observed in the reactions of bis(1,10-phenanthroline) Cu(II) and pyridoxalthiosemicarbazonato Cu(II) with cells is that the apparent suppression of hydroxyl-radical dependent DNA damage by extracellular catalase has no effect on the cytotoxicity of these complexes (Byrnes et al., 1990; 1992a). Thus, despite the prominence of this type of chemistry in cells treated with these agents, it may not be critical for their capacity to inhibit cell proliferation. An alternative is that the reduction process, itself, may involve the oxidation of key cellular thiols, which, in turn, is responsible for the growth inhibitory properties of these complexes. For example, it has been suggested that oxidation of sulfhydryl groups of a membrane Ca ATPase leads to the influx of Ca^{2+} which triggers the cell killing process (Nicotera et al., 1985). Future studies will need to examine the range of sites that undergo redox reaction with these copper complexes and define how such reactions cause cytotoxicity.

ACKNOWLEDGEMENTS

The authors appreciate the support of American Cancer Society grant DHP-31B and NIH grants CA-22184 and CA-58611.

REFERENCES

Agrawal, K.C., and Sartorelli, A.C. (1975). *Antineoplastic and Immunosuppressive Agents, Part II, Handbook of Experimental Pharmacology*, Vol. 38 (A.C. Sartorelli and D.G. Johns, eds.), Springer-Verlag, Berlin, p. 793.

Ankel, E., and Petering, D.H. (1980). *Biochem. Pharmacol.*, *29*: 1833.

Antholine, W.E., and Taketa, F. (1982). *J. Inorg. Biochem.*, *16*: 145.

Antholine, W.E., and Taketa, F. (1984). *J. Inorg. Biochem.*, *20*: 69.

Antholine, W.E., Knight, J.M., and Petering, D.H. (1977). *Inorg. Chem.*, *16*: 569.

Antholine, W.E., Solaiman, D., Saryan, L.A., and Petering, D.H. (1982). *J. Inorg. Biochem.*, *17*: 75.

Antholine, W.E., Basosi, R., Hyde, J.S., Lyman, S., and Petering, D.H. (1984). *Inorg. Chem.*, *23*: 3543.

Antholine, W.E., Kalyanaraman, B., and Petering, D.H. (1985a). *Environ. Health Perspect.*, *64*: 19.

Antholine, W.E., Lyman, S., Petering, D.H., and Pickart, L. (1985b). *Biochemical and Inorganic Aspects of Copper Coordination Chemistry* (K.D. Karlin and J. Zubieta, eds.), Adenine Press, p. 125.

Aruoma, O.I., Halliwell, B., Laughton, M.J., Quinlan, G.J., and Gutteridge, J.M.C. (1989). *Biochem. J.*, *258*: 617.

Aust, S.D., and White, B.C. (1985). *Free Rad. Biol. Med.*, *1*: 1.

Bligh, S.W.A., Boyle, H.A., McEwen, A.B., Sadler, P.J., and Woodham, R.H. (1992). *Biochem. Pharmacol.*, *43*: 137.

Blanz, Jr., E.J., and French, F.A. (1965). *Cancer Res.*, *28*: 2419.

Brockman, R.W., Thomson, J.R., Bell, M.J., and Skipper, H.E. (1956). *Cancer Res.*, *16*: 67.

Byrnes, R.W., Mohan, M., Antholine, W.E., Xu, R.X., and Petering, D.H. (1990). *Biochemistry*, *29*: 7046.

Byrnes, R.W., Antholine, W.E., and Petering, D.H. (1992a). *Free Rad. Biol. Med.*, *12*: 457.

Byrnes, R.W., Antholine, W.E., and Petering, D.H. (1992b). *Free Rad. Biol. Med.*, *13*: 469.

Chan-Stier, C.H., Minkel, D., and Petering, D.H. (1976). *Bioinorg. Chem.*, *6*: 203.

Chikira, M., Sato, T., Antholine, W.E., and Petering, D.H. (1991). *J. Biol. Chem.*, *266*: 2859.

Crim, J.A., and Petering, H.G. (1967). *Cancer Res.*, *27*: 1278.

Czapski, G., and Goldstein, S. (1988). *Free Rad. Res. Commun.*, *4*: 225.

de Alvare, L.R., Goda, K., and Kimura, T. (1976). *Biochem. Biophys. Res. Commun.*, *69*: 687.

DeConti, R.C., Toftness, B.R., Agrawal, K.C., Tomchick, Mead, J.A.R., Bertino, J.R., Sartorelli, A.C., and Creasey, W.A. (1972). *Cancer Res.*, *32*: 1455.

Fee, J.A. (1980). *Metal Ion Activation of Dioxygen* (T.G. Spiro, ed.), John Wiley and Sons, New York, p. 209.

Fee, J.A., and DiCorleto, P.E. (1973). *Biochemistry*, *12*: 4893.

Fielden, E.M., Roberts, P.B., Bray, R.C., Lowe, D.J., Mautner, G.N., Rotilio, G., and Calabrese, L. (1974). *Biochem. J.*, *139*: 49.

Flohé, L. (1988). *Mol. Cell. Biochem.*, *84*: 123.

Folkman, J., and Shing, Y. (1992). *J. Biol. Chem.*, *267*: 10931.

Freedman, J.H., Horwitz, S.B., and Peisach, J. (1981). *Biochemistry*, *21*: 2203.

Freedman, J.H., Ciriolo, M.R., and Peisach, J. (1989). *J. Biol. Chem.*, *264*: 5598.

French, F.A., and Blanz, Jr., E.J. (1966). *J. Med. Chem.*, *9*: 585.

French, F.A., and Freedlander, B.L. (1958). *Cancer Res.*, *18*: 1298.

French, F.A., and Freedlander, B.L. (1960). *Cancer Res.*, *21*: 505.

French, F.A., Lewis, A.E., Sheena, A.H., and Blanz, Jr., E.J. (1965). *Fed. Proc.*, *24*: 402.

Fridovich, I. (1991). *Oxidative Damage and Repair* (K.J.A. Davies, ed.), Pergamon Press, New York, p. 78.

Gacheru, S.N., Trackman, P.C., Shah, M.A., O'Gara, C.Y., Spacciapoli, P., Greenaway, F.T., and Kagan, H.M. (1990). *J. Biol. Chem.*, *265*: 19022.

Geller, B.L., and Winge, D.R. (1982). *Arch. Biochem. Biophys.*, *213*: 109.

Gianni, L., Zweier, J.L., Levy, A., and Myers, C.E. (1985). *J. Biol. Chem.*, *260*: 6820.

Girotti, A.W. (1985). *Free Rad. Biol. Med.*, *1*: 87.

Goldstein, S., and Czapski, G. (1986). *Free Rad. Biol. Med.*, *2*: 3.

Green, M.A., Klippenstein, D.L., and Tennison, J.R. (1988). *J. Nucl. Med.*, *29*: 1549.

Greenaway, F.T., Norris, L.J., and Sorenson, J.R.J. (1988). *Inorg. Chim. Acta*, *145*: 279.

Halliwell, B., and Gutteridge, J.M.C. (1988). *Free Radicals in Biology and Medicine*, Clarendon Press, Oxford, England, Chapter 8.

Harrison, J.R., Rillema, D.P., Ham, J.H., IV, and Sando, J.J. (1986). *Cancer Res.*, *46*: 5571.

Hirose, J., Ohhira, T., Hirata, H., and Kidani, Y. (1982). *Arch. Biochem. Biophys.*, *218*: 179.

Huber, K.R., Sridhar, R., Griffith, E.H., Amma, E.L., and Roberts, J. (1987). *Biochim. Biophys. Acta*, *915*: 267.

John, E.K., and Green, M.A. (1990). *J. Med. Chem.*, *33*: 1764.

Kanao, M., Tomita, S., Ishihara, S., Murakami, A., and Okada, H. (1973). *Chemotherapy (Tokyo)*, *21*: 1305.

Khan, M.F., and Sorenson, J.R.J. (1991). *J. Inorg. Biochem.*, *41*: 221.

Klug-Roth, D., Fridovich, I., and Rabani, J. (1973). *J. Amer. Chem. Soc., 95*: 2786.

Knight, J.M., Whelan, H., and Petering, D.H. (1979). *J. Inorg. Biochem., 11*: 327.

Köpf-Maier, P., and Köpf, H. (1983). *Platinum Coordination Complexes in Cancer Chemotherapy* (M.P. Hacker, E.B. Douple and I.H. Krakoff, eds.), Martinus Nijhoff Publishing, Boston, p. 279.

Kraker, A. Krezoski, S., Schneider, J., Minkel, D., and Petering, D.H. (1985). *J. Biol. Chem., 260*: 13710.

Linder, M.C., and Goode, C.A. (1991). *Biochemistry of Copper*, Plenum Press, New York, p. 43.

Marquart, F.X., Pickart, L., Laurent, M., Gillery, P., Monboisse, J.C., and Borel, J.P. (1988). *FEBS Lett., 238*: 343.

McClarty, G.A., Chan, A.K.M., and Wright, J.A. (1986). *Cancer Res., 46*: 4516.

Mihich, E., and Nichol, C.A. (1965). *Cancer Res., 25*: 1410.

Minkel, D.T., and Petering, D.H. (1978). *Cancer Res., 38*: 117.

Minkel, D.T., Saryan, L.A., and Petering, D.H. (1978). *Cancer Res., 38*: 124.

Minotti, G., and Aust, S.D. (1987). *J. Biol. Chem., 262*: 1098.

Minotti, G., and Aust, S.D. (1989). *Chem.-Biol. Interact., 71*: 1.

Mohindru, A., Fisher, J.M., and Rabinowitz, M. (1983). *Biochem. Pharmacol., 32*: 3627.

Moore, E.C., Zedeck, M.S., Agrawal, K.C., and Sartorelli, A.C. (1990). *Biochemistry, 9*: 4492.

Myers, C.E. (1988). *Organ Directed Toxicities of Anticancer Drugs* (M.P. Hacker, J.S. Lazo and T.R. Tritton, eds.), Martinus Nijhoff Publishing, Boston, p. 17.

Nagano, T., Hirano, T., and Hirobe, M. (1989). *J. Biol. Chem., 264*: 9243.

Narasimhan, J., Antholine, W.E., Chitambar, C.R., and Petering, D.H. (1991). *Arch. Biochem. Biophys., 289*: 393.

Nicotera, P., Moore, M., Mirabelli, F., Beellomo G., and Orrenius, S. (1985). *FEBS Lett., 181*: 149.

Nilsson, K. (1989). *Cancer Lett., 47*: 169.

Nordlund, P., Sjöberg, B.M., and Eklund, H. (1990). *Nature, 345*: 593.

Olson, R.D., and Mushlin, P.S. (1990). *FASEB J., 4*: 3076.

Petering, D.H. (1972a). *Bioinorg. Chem., 1*: 255.

Petering, D.H. (1972b). *Bioinorg. Chem., 1*: 273.

Petering, D.H. (1974). *Biochem. Pharmacol., 23*: 567.

Petering, D.H. (1980). *Metal Ions in Biological Systems*, Vol. 11 (H. Sigel, ed.), Marcel Dekker, New York, p. 197.

Petering, D.H., and Antholine, W.E. (1988). *Reviews in Biochemical Toxicology*, Vol. 9 (E. Hodgson, J.R. Bend and R.M. Philpot, eds.), Elsevier, New York, p. 225.

Petering, D.H., and Petering, H.G. (1975). *Antineoplastic and Immunosuppressive Agents, Vol 2. Handbook of Experimental Pharmacology*, Vol 38 (A.C. Sartorelli and D.G. Johns, eds.), Springer-Verlag, Berlin, p. 841.

Petering, D.H., and Shaw III, C.F. (1991). *Meth. Enzymol., 205*: 475.

Petering, D.H., Antholine, W.E., and Saryan, L.A. (1984). *Anticancer and Interferon Agents, Synthesis and Properties, Drugs and the Pharmaceutical Sciences*, Vol. 24, (R.M. Ottenbrite and G.B. Butler, eds.), Marcel Dekker, New York, p. 203.

Petering, D.H., Byrnes, R.W., and Antholine, W.E. (1990). *Chem.-Biol. Interact., 73*: 133.

Petering, H.G., Buskirk, H.H., and Underwood, G.E. (1964). *Cancer Res., 24*: 267.

Petering, H.G., Buskirk, H.H., and Crim, J.A. (1967). *Cancer Res., 27*: 1115.

Pezeshk, A., Antholine, W.E., and Petering, D.H. (1987). *J. Inorg. Biochem., 31*: 283.

Pickart, L. (1987a). *Biology of Copper Complexes* (J.R.J. Sorenson, ed.), Humana Press, Clifton, New Jersey, p. 273.

Pickart, L. (1987b). Chemical Derivatives of GHL-Cu, U.S. Patent Number 4,665,054.

Pickart, L. (1991a). Methods and Compositions for Healing Bone Using Gly-HIs-Lys: Copper, U.S. Patent Number 5,059,588.

Pickart, L. (1991b). Methods and Compositions for Healing Ulcers, U.S. Patent Number 5,023,237.

Pickart, L. (1992). Anti-oxidative and Anti-inflammatory Metal: Peptide Complexes and Uses Thereof, U.S. Patent Number 5,118,665.

Quest, A.F.G., Bloomenthal, J., Bardes, E.S.G., and Bell, R.M. (1992). *J. Biol. Chem.*, *267*: 10193.

Radtke, K., Byrnes, R., Lornitzo, F., Antholine, W.E., and Petering D.H. (1991). *J. Inorg. Biochem.*, *43*: J037.

Regelson, W., Holland, J.F., and Talley, R.W. (1967). *Cancer Chemother. Rep.*, *51*: 171.

Reiners, J.J., Jr., and Colby, A.B. (1988). *Carcinogenesis*, *9*: 629.

Sarkar, B. (1981). *Metal Ions in Biological Systems*, Vol. 12 (H. Sigel, ed.), Marcel Dekker, New York, p. 233.

Sartorelli, A.C., Agrawal, K.C., and Moore, E.C. (1971). *Biochem. Biophys. Res. Commun.*, *20*: 3119.

Saryan, L.A., Ankel, E., Krishnamurti, C., and Petering, D.H. (1979). *J. Med. Chem.*, *22*: 1218.

Saryan, L.A., Mailer, K., Krishnamurti, C., Antholine, W.E., and Petering, D.H. (1981). *Biochem. Pharmacol.*, *30*: 1595.

Sawyer, D.T., and Valentine, J.S. (1981). *Accounts Chem. Res.*, *14*: 393.

Shelton, M.E., Green, M.A., Mathias, C.J., Welch, M.J., and Bergmann, M.D. (1989). *J. Nucl. Med.*, *30*: 1843.

Shuff, S.T., Chowdhary, P., Khan, M.F., and Sorenson, J.R.J. (1992). *Biochem. J.*, *43*: 1601.

Sigman, D.S. (1986). *Accounts Chem. Res.*, *19*: 180.

Solaiman, D., Rao, E.A., Antholine, W., and Petering, D.H. (1980). *J. Inorg. Biochem.*, *12*: 201.

Sorenson, J.R.J., ed. (1987). *Biology of Copper Complexes*, Humana Press, Clifton, New Jersey.

Sorenson, J.R.J. (1989). *Progress in Medicinal Chemistry*, Vol. 26 (G.P. Ellis and G.B. West, eds.), Elsevier Science Publishers, New York, p. 437.

Spector, T., Harrington, J.A., and Porter, D.J.T. (1991). *Biochem. Pharmacol.*, *42*: 91.

Subczynski, W.K., Antholine, W.E., Hyde, J.S., and Petering, D.H. (1987). *J. Amer. Chem. Soc.*, *109*: 46.

Takahashi, K., Takita, T., and Umezawa, H. (1987). *J. Antibiot.*, *40*: 348.

Tereo, T., and Owens, Jr., C.A. (1973). *Amer. J. Physiol.*, *224*: 682.

Van Giessen, G.J., Crim, J.A., Petering, D.H., and Petering, H.G. (1973). *J. Nat. Cancer Inst.*, *51*: 139.

Von Sonntag, C.V., Hagen, U., Schon-Bopp, A., and Schulte-Frohlinde, D. (1981). *Adv. Radiat. Biol.*, *9*: 109.

Winkelmann, D.A., Bermke, Y., and Petering D.H. (1974). *Bioinorg. Chem.*, *3*: 261.

GOLD COMPLEXES AS DRUGS

by

Peter J. Sadler and Rodney E. Sue

Department of Chemistry,
Birkbeck College,
University of London,
Gordon House and Christopher Ingold Laboratories,
29 Gordon Square
London WC1H 0PP
U.K.

I. THE CHEMISTRY OF GOLD

As early as 2500 BC gold was used in Chinese and Arabic medicine. In the 8th century it was advocated as an elixir of youth, and in the Middle Ages gold mixtures were prescribed for a range of conditions (Sadler, 1976a). In 1890 Koch discovered that $[Au(CN)_2]^-$ had antitubercular activity, although this was later shown to have little effect on the disease (Sutton, 1986). In 1929, Forestier found that gold was effective for the treatment of rheumatoid arthritis; today this treatment involves the use of injectable gold(I) thiolates, supplemented in the clinic by a Au(I) phosphine complex, (auranofin). Auranofin is also being used in the clinic for psoriatic arthritis, juvenile rheumatoid arthritis and is on clinical trial as an antiasthmatic.

A. The element and its properties

Gold, atomic number 79, has an outer shell electronic configuration of $5d^{10}6s^1$. The yellow metal is malleable and ductile, resistant to chemical attack, and only agents such as cyanide and aqua regia will degrade it. Higher oxidation states of gold are readily reduced to Au(0) (Puddephatt, 1978).

B. Colloidal gold

"Potable gold" used in the past as a "cure-all" was obtained by reduction of a solution of gold in aqua regia and diluted with essential oils such as oil of rosemary, giving a colloidal solution of gold (Sadler, 1976a). The surfaces of colloidal gold particles carry a negative charge and so can adsorb strongly to proteins. This has led to the use of colloidal gold as a cytochemical marker in electron microscopy for protein binding studies (Beesley, 1989).

C. Physical methods for the study of gold compounds

There are few methods for the direct probing of gold in gold complexes. Techniques used in the study of gold complexes are described in Table 1. ^{197}Au (100% abundance) NMR has not been useful for examining gold complexes since the nucleus is quadrupolar ($I=3/2$) with

Table 1

Physical Methods for The Study of Gold Compounds

Technique	Information	Comments
X-ray crystallography	bond lengths and angles, 3-dimensional structure	need crystals
X-ray spectroscopy absorption - EXAFS, scattering - WAXS	bond lengths, distances to nearest neighbours, number and type of Au-bonded atoms, oxidation state	requires synchrotron radiation, can be used for solutions and amorphous solids, difficult to distinguish N from O, and P from S or Cl
Mössbauer spectroscopy	oxidation state, total electron density at Au (isomer shift) and symmetry (quadrupole splitting)	short source half-life (^{197}Pt, 18 h), low temp. required (4 K), for solids only
Nuclear magnetic resonance spectroscopy (NMR)	number and type of coordinated ligands, conformation, ligand exchange rates (dynamics)	probes ligand rather than Au, usually ^{1}H, ^{13}C, ^{31}P, ^{15}N, generally need solutions, although solid state spectra are possible, can be used to probe metabolism in biological media or intact organisms
Electronic absorption spectroscopy, and circular dichroism (CD)	electronic energy levels and symmetry (CD)	absorptions are often weak in the visible region, charge-transfer absorptions in the UV region
Infrared and Raman spectroscopy	ligand, and Au - ligand vibrational frequencies	ligand bands often do not change substantially on coordination, bands due to Au - ligand vibrations can be difficult to assign (Au - Cl, S ca 340 cm^{-1}, Au - Br ca 230 cm^{-1}, Au - I ca 180 cm^{-1})
Mass spectrometry	molar mass of parent and fragments	compound must be introduced into the ion source, fragmentation pattern can be obtained

a large quadrupole moment and low gyromagnetic ratio. Nuclear quadrupole resonance has also not found wide applications for gold. Electron paramagnetic resonance (EPR) has been used to study several Au(II) compounds (Puddephatt, 1978), but the biologically important oxidation states of Au(I) and Au(III) are not EPR active.

Total gold determinations are usually carried out by atomic absorption spectroscopy (AAS) where the detection limit is ca 0.5 ppm, or by inductively coupled plasma-mass spectrometry (ICP-MS) coupled with high performance liquid chromatography (HPLC) which has a lower detection limit than AAS (Elder et al., 1993). The use of AAS to monitor serum gold levels has greatly improved the safety of use of gold drugs since the 1970's.

D. Oxidation states

Au(I) is the most important form of gold biologically, and most gold drugs contain gold in this oxidation state. It is possible that Au(I) can be oxidized to Au(III) in vivo, e.g. via the myeloperoxidase system of white cells (Beverly and Couri, 1987). Au(III) may be responsible for some of the toxic side effects of Au(I) drugs and may involve oxidation of protein disulfides by Au(III) (Witkiewicz and Shaw, 1981). Au(II) could be an important intermediate in biological reactions, but has not yet been recognized as such, while Au(V) is not likely to be accessible in vivo.

Gold must be stabilized by π-acceptor ligands (those capable of accepting back-bonding from the metal) because there is a tendency to disproportionation to Au(III) and colloidal Au(0). An approximate order of thermodynamic stability (Sadler, 1982) for some Au(I) complexes with biologically-important ligands is:

$$CN^- \sim CysS^- \sim PR_3 >> Met\text{-}S\text{-}CH_3 \sim His(=N\text{-}) > Cl^- >> COO^-.$$

Generally the Au(I) complexes are prepared from Au(III) precursors and there is a variety of methods available, such as reduction with excess phosphine or thiodiglycol. Electrolytic methods for producing Au(I) often have the advantage of giving fewer by-products and allow a ready choice of counter-anion (Kissner et al., 1993).

E. Coordination geometries

Au(I) can adopt coordination numbers of two (linear), three (trigonal), and four (tetrahedral). Linear two-coordination is the most common, often with weak additional bonds in the solid state. A recent analysis of Au...Au interactions in the solid state shows a large number of contacts in the range 2.50 - 4.00 Å. For comparison, the Au...Au distance in metallic gold is 2.89 Å and the van der Waals distance is 3.60 Å, illustrating the tendency for both bonding and non-bonding Au-Au intermolecular contacts (Pathaneni and Desiraju, 1993). For Au(III), square-planar four coordination is the most common, but five- (square-pyramidal) and six-coordination (octahedral) are known.

II. ANTIARTHRITIC GOLD THIOLATE AND GOLD PHOSPHINE COMPLEXES

A. Structures

A number of Au(I)-thiolato complexes have been used in antiarthritic therapy; these have included aurothioglucose (Solganol), sodium aurothiosulfate (Sanochrysin), sodium aurothiomalate (Myocrisin), sodium aurothiopropanol sulfonate (Allocrysin) and the Au(I) complex of 4-amino-2-mercaptobenzoic acid (Krysolgan) (Fig. 1). The thiolate complexes are polymeric in the solid state and in solution, forming rings or chains.

Only Sanochrysin has been studied by X-ray crystallography. The $[Au(S_2O_3)_2]^-$ ion is linear and two-coordinate with a bond angle of 176° and Au-S distance of 2.28 Å while the nearest Au neighbors are 3.30 Å away (Ruben et al., 1974).

Both aurothiomalate and aurothioglucose have been studied by EXAFS since neither has been crystallized (Mazid et al., 1980; Elder et al., 1983). In both cases, the gold is bound to two S atoms at a distance of 2.37 Å, giving polymers with bridging thiolate sulfurs. WAXS measurements, which allow detection of Au...Au contacts (of 3.35, 5.8 and 8.1 Å), suggest a linear hexamer structure (Elder et al., 1985) as shown in Figure 1. Such structures require that the thiolate be present in stoichiometric excess over gold in these complexes, which is usually the case (Grootveld et al., 1984; ca 9% molar excess). At low

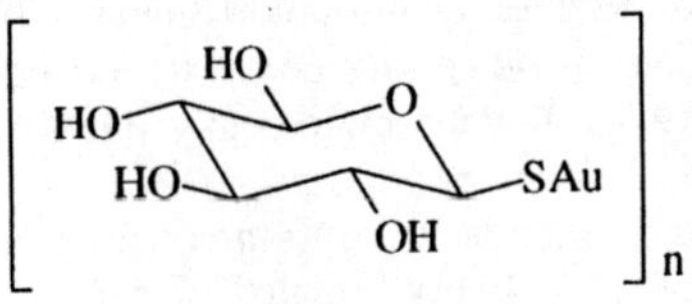

aurothioglucose
(Solganol)

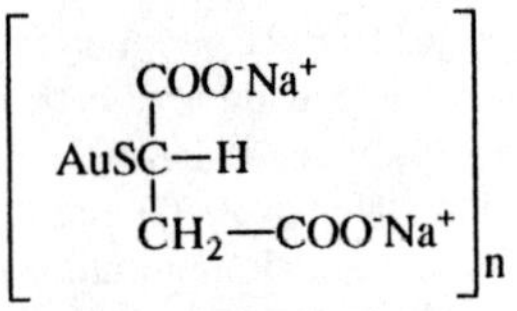

sodium aurothiomalate
(Myocrisin)

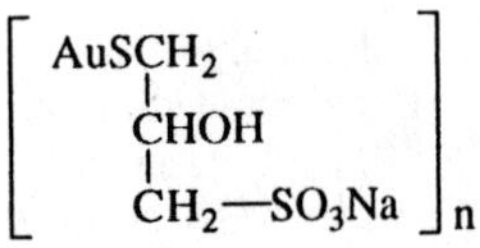

sodium aurothiopropanol sulfonate
(Allocrysin)

$[O_3S{-}S{-}Au{-}S{-}SO_3]Na_3$

sodium aurothiosulfate
(Sanochrysin)

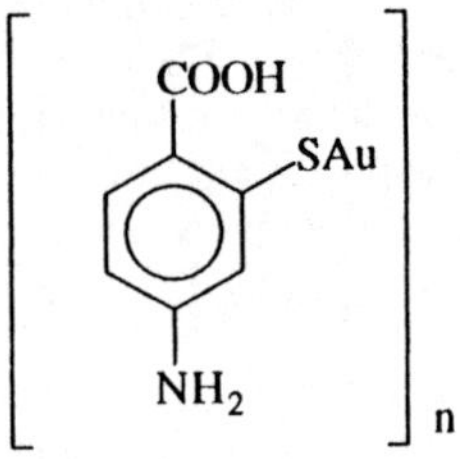

gold(I) complex of 4-amino-2-mercaptobenzoic acid
(Krysolgan)

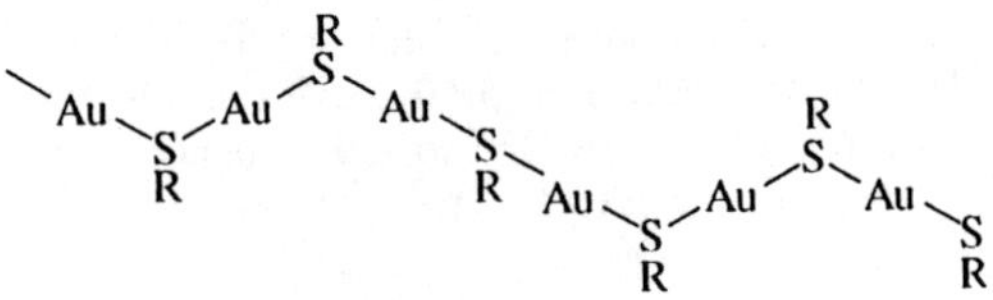

polymeric chains

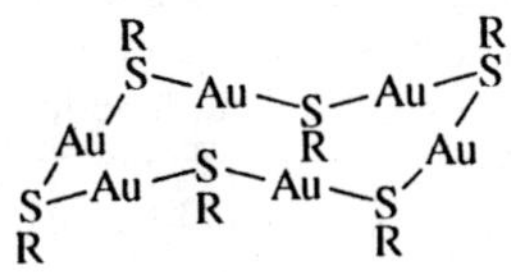

rings

Figure 1.
Injectable gold antiarthritic drugs and gold(I)-thiolate chains and rings.

ionic strength the conformation of the ligand in aurothiomalate is similar to that of free thiomalate, while at higher ionic strength the structures of a number of conformations can be detected by NMR (Isab and Sadler, 1981). EXAFS studies of aurothiomalate and aurothioglucose show that Au(I) is coordinated to 2 S atoms at distances of 2.29 Å and 2.30 Å, respectively (Elder et al., 1983).

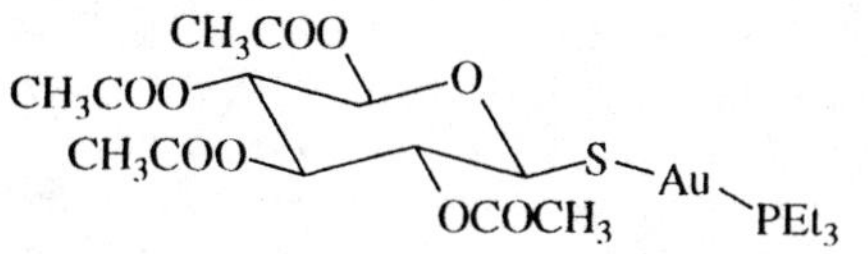

Auranofin ("Ridaura") (1)

Auranofin resulted from a search for an orally-active drug to overcome the problems associated with painful intramuscular injection of gold thiolates and their severe side effects such as exfoliative dermatitis, bone marrow suppression and nephrosis (Sutton, 1986). Et_3PAuCl and $(Et_3P)_2AuCl$ were found to have oral anti-inflammatory activity in animal models (Walz et al., 1972; Hill, 1978), but diarrhoea was a troublesome side effect. The tetraacetylthioglucose derivative, auranofin ("Ridaura", **1**) was finally selected for clinical use. X-ray crystallography shows that auranofin is monomeric with linear 2-coordinate Au(I) (Hill and Sutton, 1980), and ^1H, ^{13}C and ^{31}P NMR show that the structure is similar in solution (Razi et al., 1983a; 1983b).

Auranofin is a lipophilic complex and blood serum levels are maintained with a daily dose as opposed to the weekly or monthly injections of aurothiomalate. Whereas aurothiomalate is almost immediately taken into the blood stream and distributed to all parts of the body, auranofin is absorbed only slowly.

Auranofin has been shown to be an inhibitor of the incorporation of ^3H-thymidine and ^{14}C-amino acids in mitogen-stimulated human lymphocytes, i.e. the precursors to DNA and protein synthesis (Finkelstein et al., 1977). Both auranofin and Myocrisin have been implicated in the control of oxidative damage in rheumatoid arthritis (Grootveld et al., 1990). Au(I) thiolate drugs may be involved with suppressing the production of reactive oxygen species. Phagocytosis may result in the generation of superoxide ion which can be oxidized to singlet oxygen. This in turn is capable of peroxidation of unsaturated fatty acids, and may be involved in the inflammatory process. Auranofin deactivates singlet oxygen with a quenching constant of the order of 10^7 $M^{-1}s^{-1}$ (Corey et al., 1987).

B. Ligand exchange

Thiolate exchange reactions of aurothiomalate (AuTm) are facile via three-coordinate intermediates, and the polymers are eventually degraded to $[Au(SR)_2]^-$. The thiols with the lowest pK_a's form the most stable complexes; an approximate order of stability for complexes with some biologically important ligands is (Isab and Sadler, 1982):

cysteine methyl ester, D-penicillamine > β-D-thioglucose > N-acetylcysteine > glutathione, thiomalate, mercaptoacetate.

The high stability of gold complexes with thiols such as 2,3-dimercaptopropanol has made them useful as gold scavengers in cases of gold toxicity (Thompson and Whittaker, 1947). Such thiolate exchange reactions play a role in the transport and metabolism of gold drugs. Aurothiomalate is largely transported around the body on serum proteins such as albumin. Mössbauer and EXAFS studies show that the thiomalate is exchanged for cysteine-34 in albumin (Shaw et al., 1984):

$$\text{Alb-Cys-34-SH} + 1/n[\text{AuTm}]_n \rightarrow \text{Alb-Cys-34-S-Au-Tm}$$

Auranofin is readily deacetylated under acidic conditions, and triethylphosphine-Au(I)thioglucose has been detected after passage of the drug through the intestinal wall (Tepperman et al., 1984). Auranofin and related complexes are readily taken up into red cells where the binding sites are glutathione and hemoglobin (Cys-β-93). Et_3P-Au-Cl binds preferentially to the free Cys thiolates of hemoglobin and albumin, but when these are saturated, then weaker binding to histidine (His) residues occurs (Razi et al., 1983b; Malik and Sadler, 1979; Shaw et al., 1988; Kinsch and Stephan, 1984). Histidine binding is probably responsible for the rather unusual low-spin state to high-spin state change which this complex can induce in cytochrome *c* (Otiko and Sadler, 1980).

Studies of the metabolism of triply-radiolabelled auranofin (^{32}P, ^{195}Au, ^{35}S) have shown that excretion of the sulfur and phosphorus occurs faster than excretion of gold (Intoccia et al., 1982), consistent with the displacement of the (acetylated) thioglucose ligand by natural thiolate ligands in vivo, and subsequent displacement of triethylphosphine and its rapid oxidation to $OPEt_3$. The latter process may be accompanied by reduction of a protein disulfide bond (Malik et al., 1980; Razi et al., 1983b).

Smoking has been shown to increase the uptake of gold into red blood cells. This is attributable to inhalation of HCN in smoke (Graham et al., 1982). Cyanide reacts with aurothiomalate to form the mixed ligand complex $[Au(CN)(Tm)]^-$, and then $[Au(CN)_2]^-$ (Graham et al., 1985). $[Au(CN)_2]^-$ is readily taken into red cells whereas aurothiomalate is not. $[Au(CN)_2]^-$ has been detected as a metabolite in the urine of patients treated with both aurothiomalate and auranofin (Elder et al., 1993). $[Au(CN)_2]^-$ may therefore be a common metabolite for all gold antiarthritic complexes.

III. GOLD-PROTEIN AND GOLD-NUCLEOTIDE COMPLEXES

A. Gold-albumin

Over 80% of the gold in circulation in the body after administration of gold drugs is bound to albumin (M 66.5 kg mol^{-1}; Finkelstein et al., 1976). This protein contains 35 cysteine residues, all but one of which are oxidized as disulfide bridges. The remaining cysteine, Cys-34, is present in the reduced form (SH) in 60-70% of albumin molecules while 30 - 40% is a mixed disulfide with free cysteine or glutathione. The high affinity of Cys-34 for Au(I) is consistent with its reported low pK_{SH} (ca 5). EXAFS and Mössbauer spectra of albumin after modification with aurothiomalate suggest the presence of Alb-Cys-34-S-Au-Tm with Au-S distances of 2.28 Å (Shaw et al., 1984). Au-P and Au-S distances of 2.29 Å have been determined by EXAFS for Alb-Cys-34-S-Au-PEt_3 (Coffer et al., 1986) obtained by treatment of albumin with auranofin. Triple radiolabelling experiments (^3H, ^{195}Au, ^{14}C) confirm that auranofin binds to albumin with release of the tetraacetylthioglucose ligand (Ecker et al., 1986). In accordance with the pK_{SH} values, Au(I)PEt_3 readily transfers from hemoglobin Cys-β-93 to albumin Cys-34 (Shaw et al., 1988). ^1H NMR studies suggest that the binding of Et_3PAu^+ to Cys-34 of albumin influences the environment of His3 at the N-terminus. In vitro, reactions between Et_3PAuSR and albumin lead to the release of SR, which in turn can reduce the Cys-blocked albumin in the solution (ca 40% for bovine serum albumin) with concomitant production of disulfides Cys-SR (e.g. SR = thioglucose, tetraacetylthioglucose) and cystine (Ni Dhubhghaill et al., 1992). ^{31}P NMR studies show that treatment of albumin with Et_3PAuCl gives a Au(I) complex with phosphorus and sulfur ligands ($\delta^{31}P$ 42.0 ppm) as further evidence for complexation at Cys-34. Additional peaks at $\delta^{31}P$ 34.4 and 30.3 ppm are indicative of Au(I) binding to other sites such as histidine and methionine (Malik and Sadler, 1979). There is also evidence for the sulfur-bridged species Alb-Cys-S-$(AuPEt_3)_2^+$ ($\delta^{31}P$ 35.6 ppm, Xiao and Shaw, 1992). Once Et_3Au^+ is bound to Cys-34 of albumin, the phosphine can be displaced by other ligands such as CN^- with release of the phosphine. This may influence the rate of gold metabolism in smokers (Isab et al., 1988).

B. Gold-metallothionein

Metallothionein (MT) is a protein of 61 residues of which 20 are cysteine residues. It is thought to have a role in intracellular storage of Cu(I) and Zn(II) and may also play a part in the transport and detoxification of other metal ions (e.g. Hg(II) and Cd(II)) (see Part Two, Chapter 1-D5). When aurothiomalate reacts with metallothionein, Au(I) displaces Zn(II) and Cd(II) (Laib et al., 1985). If the protein is in excess then thiomalate is completely displaced, but if the drug is in excess, then thiomalate may be retained as MT-

Cys-Au-Tm. Up to 20 Au-Tm bonds can form, consistent with gold binding to each of the cysteine residues, with Au(I)-S distances of 2.30 Å as shown by EXAFS. In contrast, auranofin does not displace zinc or cadmium, although it will bind to the apoprotein. Et_3PAuCl binds strongly to metallothionein (Ecker et al., 1986; Shaw and Laib, 1986), indicating the ease of displacement of Cl compared to thiolate S.

Cultured human epithelial cells can be stimulated to MT synthesis by exposure to $[AuCl_4]^-$ (Glennas and Rugstadt, 1985). Auranofin also induces MT synthesis in Chinese hamster ovary cells via activation of gene transcription, and appears to provide a gold resistance mechanism (Butt et al., 1986).

C. Other gold proteins and peptides

Au(I) binds predominantly to cysteine sulfur in peptides, and much more weakly to methionine sulfur and histidine nitrogen. Au(III) on the other hand binds strongly to N and O side chains of peptides and deprotonated amide nitrogens. Au(III) also tends to oxidize Met and Cys sulfurs (Witkiewicz and Shaw, 1981).

In $[Au(Gly\text{-}L\text{-}His)Cl]^+$ the peptide is coordinated to square-planar Au(III) by the glycyl amino group, the deprotonated peptide N and the N3 of the His imidazole ring. The Au-N(peptide) bond (1.94 Å) is slightly shorter than the Au-N(amino) and Au-N(His) bonds (2.00 Å). This monomer (2) was crystallized at pH 1.5-2, while at pH 6 to 7 a tetramer $[Au(Gly\text{-}L\text{-}His)]_4$ (3) is formed. Here the fourth coordination site is occupied by a bridging N from the deprotonated imidazole ring. The four Au atoms form a distorted tetrahedron (Wienken et al., 1992).

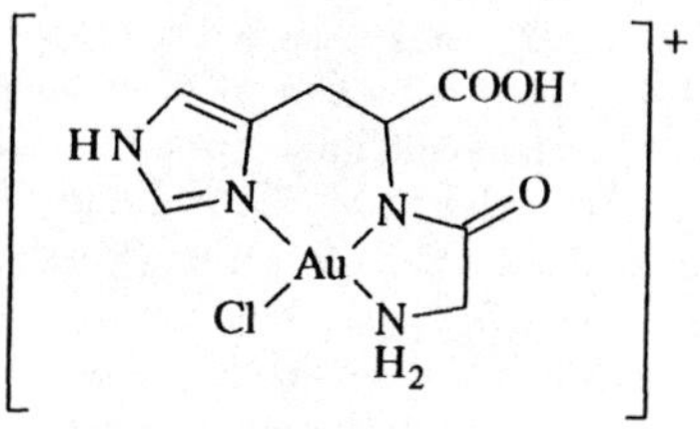

Monomer (2)

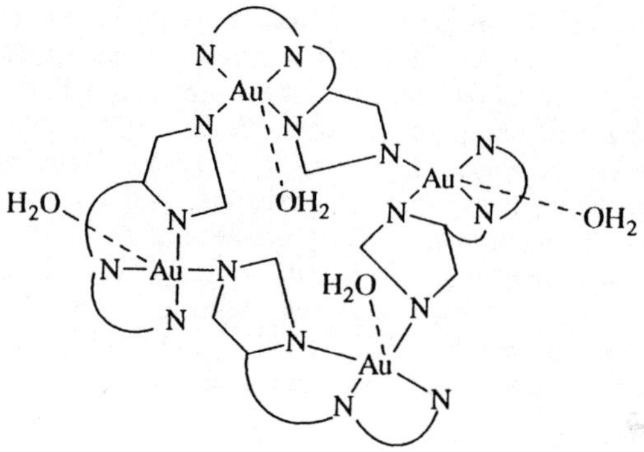

Tetramer (3)

Gold Complexes Of Gly-L-His

The only simple gold-amino acid complex which has been crystallized is triphenylphosphine-(N-benzoyl-L-alaninato)gold(I) (**4**). As expected, the gold is linearly coordinated by the phosphine (Au - P, 2.22 Å) and the amino acid carboxylate group (Au - O, 2.07 Å; Jones and Schelbach, 1991). This is one of the few examples of Au(I) - O bonds.

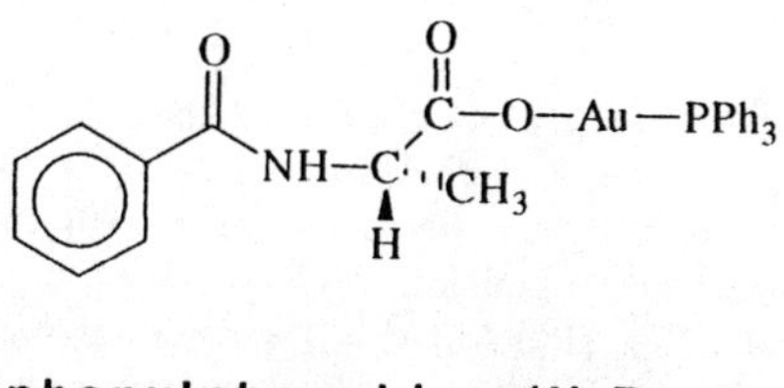

Triphenylphosphine-(N-Benzoyl-L-alaninato)gold(I) (4)

Gold-protein complexes are often prepared by X-ray crystallographers (by soaking crystals in solutions of gold complexes) to solve the phase problem (Sadler, 1976b). For example, the tetrachloroaurate salt of the antibiotic valinomycin (a cyclododecadepsipeptide) was solved by this method (Pinkerton et al., 1969).

Schuhmann et al. (1990) have reported that Au(III) induces antinucleolar autoantibodies in animals. They propose that Au(I) drugs produce toxic side effects via oxidation to Au(III), for example in phago-lysosomes (aurosomes). Self-proteins or peptides become modified by Au(III) and give rise to sensitized T-cells and adverse immunological reactions. However, Romagnoli et al. (1992) have found that lymphocytes of patients who develop toxic skin reactions to gold proliferate in vitro when challenged with different gold compounds. Unlike the findings of Schuhmann et al.(1990), they found that gold specific T-cells recognized Au(I) and not Au(III).

Zinc fingers are DNA-binding proteins, typically with two histidines and two cysteines as Zn(II) ligands (see Part Two, Chapter 2-C). These may be targets for gold binding (Ranford et al., 1991). Indeed, aurothiomalate has recently been found to inhibit binding of the progesterone receptor (PR) to its DNA response element (Handel et al., 1991). Displacement of Zn(II) by Au(I) is likely to disrupt DNA binding since Au(I) is likely to adopt a linear coordination in the flexible loops of this protein rather than the tetrahedral stereochemistry adopted by Zn(II).

D. Gold-nucleotide complexes

Nucleotides with only nitrogen or oxygen donors would be expected to bind only weakly to Au(I). This is consistent with the lack of mutagenicity and carcinogenicity of most gold complexes, and is an obvious advantage for gold drugs. Instead, the cytotoxicity of gold(I) phosphine complexes has been attributed to either the phosphine moiety or to interactions of the gold phosphine complexes with proteins such as DNA polymerase. Indeed it has been shown that auranofin does not interact with pBR322 DNA, although the analogous Et_3PAuX complexes (X = Cl, Br) do bind to DNA, under certain conditions (pH 9.5), with a preference for guanine and cytosine. This binding can be inhibited by the addition of a thiosugar, and illustrates the preference of Au(I) for softer donors such as sulfur (Mirabelli et al., 1986a; Blank and Dabrowiak, 1984). However, Au(III) does bind strongly to nitrogen and oxygen donors and a number of Au(III) interactions with nucleotides have been reported. For example, Et_3PAuBr_3 binds to *Hind III/NciI*, a 139 base-pair restriction fragment from pBR322 and this inhibits cleavage at guanine N(7) (Ward and Dabrowiak, 1987).

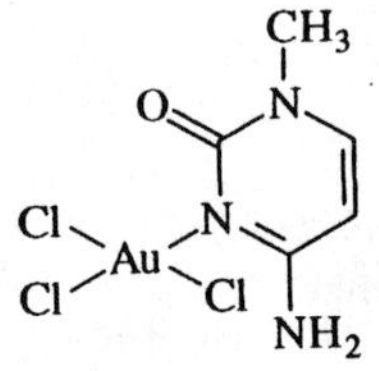

Trichloro(1-methylcytosinato)gold(III)

(5)

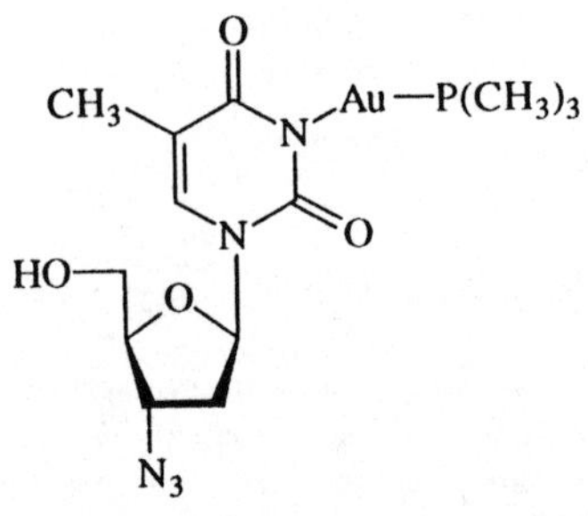

[(CH₃)₃PAu(AZT)]

(6)

Gold is bound to the nucleobase in trichloro(1-methylcytosinato)gold(III) (**5**). This complex is square-planar with the Au(III) bound to N(3) of 1-methylcytosine (Au-N(3) distance of 2.031Å); the angle between the planes of 1-methylcytosine and the $AuCl_3$ is 85°, which should minimize steric repulsion (Holowczak et al., 1985). Complexes of 3'-azido-3'-deoxythymidine (HAZT), $[R_3PAu(AZT)]$ exhibit anti-inflammatory activity for R = CH_3, and anti-HIV activity for R = CH_3, Ph. The trimethylphosphine complex (**6**) has been crystallized and contains linear 2-coordinate Au(I) (Pill et al., 1991).

Pyridinegold(I)chloride and pyridinegold(III)trichloride both appear to react with a number of different conformations of pBR322 DNA to produce inter-strand cross-links and single-strand breaks as does Et_3PAuCl_3. The amount of cross-linking can be reduced by the addition of thiols such as 2-mercaptoethanol, again illustrating the strength of Au(I)-P and Au(I)-S bonds (Mirabelli et al., 1986a; 1986b).

IV. GOLD ANTITUMOR COMPLEXES

A. Cytotoxic gold phosphines and other complexes

Auranofin, the antiarthritic drug, is potently cytotoxic to human cancer cells in culture (Simon et al., 1979). However, it is active only against intraperitoneal (ip) P388 leukemia in mice, and then only when administered ip. It does not alter cell cycle distribution but DNA, RNA and protein syntheses are inhibited at concentrations that are lethal to cells (Mirabelli et al., 1985).

A large number of gold monodentate phosphine-thiolate complexes have since been tested for antitumor activity, with variation of both phosphine and thiolate ligands (Mirabelli et al., 1986c). Comparison of Au(I) thiolates with phosphine Au(I) thiolates suggests that it is the phosphine that is the toxic agent and that the metal plays a role in delivery of the complex to the target site.

^{31}P NMR studies show that Au(I) complexes with monodentate phosphines undergo fast ligand exchange reactions with free phosphine at 213 K (Mays and Vergnano, 1979). Four-coordinate tetrahedral Au(I)-bisphosphine complexes undergo much slower ligand redistribution reactions. There is no evidence for 5-coordinate Au(I) complexes, hence all exchange reactions require ring-opening of the chelates. Four-coordinate complexes with 5- or 6-membered rings readily form, but bisphosphines with longer chains between phosphorus atoms form annular 3-coordinate Au(I) dimers or bridged digold species, with trigonal and linear coordination at gold, respectively.

Bridged linear digold(I)-bisphosphine compounds (Fig. 2) such as (thioglucose)Au(μ-dppe)Au(thioglucose) are also cytotoxic in vitro, and exhibit in vivo anticancer activity against a wider range of tumors compared to auranofin. A large number of analogues have been prepared and tested for their anticancer activity (Mirabelli et al., 1987). The related Au(III) complex, $[Cl_3Au(\mu\text{-dppe})AuCl_3]$, also shows activity (Snyder et al., 1986). Annular

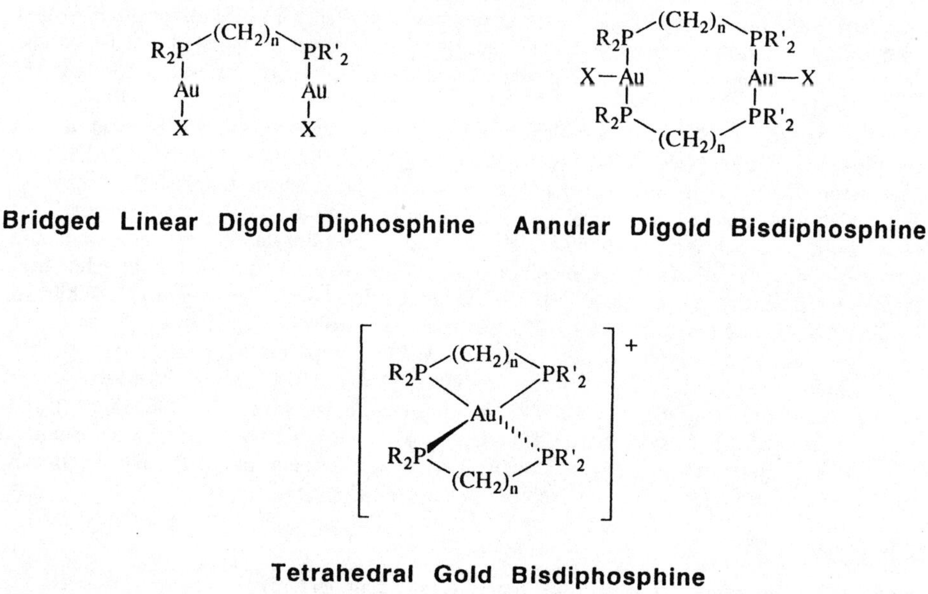

Figure 2.
General structures of cytotoxic gold(I)-diphosphine complexes.

Au(I) complexes such as $[Au_2(Ph_2PCH_2CH_2PEt_2)_2Cl_2]$ have 3-coordinate trigonal gold centers which are probably stabilized by short Au...Au contacts (Fig. 2; Berners-Price and Sadler, 1986). Replacement of P with As or S results in inactivity, and the activity is optimized for 2-carbon bridges between the phosphorus atoms; also replacement of the phenyl group can alter the activity (Snyder et al., 1986; Mirabelli et al., 1987). These complexes are readily converted into tetrahedral complexes (Fig. 2) in blood plasma, or on reaction with thiolates, sulfide, or further bisphosphine ligand (Berners-Price et al., 1987; Bates and Waters, 1984). In these complexes Au(I) has a flattened tetrahedral geometry with average Au - P bond lengths of ca 2.4 Å (Berners-Price et al., 1984; Bates and Waters, 1984). The tetrahedral complexes exhibit surprisingly high kinetic and thermodynamic stability, with sharp ^{31}P NMR signals for free ligand and complex at room temperature. Unlike auranofin, the tetrahedral complexes remain intact in human plasma and the lipophilic tetrahedral complexes such as $[Au(dppe)_2]^+$ readily partition into lipoproteins or bind within cell membranes (Berners-Price and Sadler, 1987).

$[Au(dppe)_2]^+$ is highly toxic to tumor cells in vitro, although cells can exhibit some resistance at low doses. The complex was found to be active to a P388 subline that was resistant to cisplatin. Moreover, administration of a combination of cisplatin and $[Au(dppe)_2]Cl$ was more effective against P388 leukemia than either agent alone (Berners-Price et al., 1986). The primary cytotoxic lesion appears to involve DNA-protein crosslinks. The free diphosphine ligand dppe is also active against P388 leukemia but is much less potent, suggesting that part of the action of Au(I) is to protect the ligand from oxidation reactions before it reaches the target site. In the series of diphosphine ligands $Ph_2P(CH_2)_nPPh_2$, activity is highest for $n = 2$ and 3 or where the bridge between the phosphorus atoms is *cis*-CH=CH, i.e. for those ligands which can form strong chelate rings. In the series of tetrahedral Au(I) complexes, $[Au(R_2PCH_2CH_2PR'_2)_2]^+$ antitumor activity is gradually lost on replacing R or R' = Ph with Et (Berners-Price et al., 1990). It is possible that chelation of the free ligand to Cu(I) inside cells is essential for activity. Cu(I)- and Ag(I)-dppe complexes are known to be active (Berners-Price et al., 1986; 1990; Snyder et al., 1986).

Arsenic analogues, such as the bridged digold and tetrahedral gold complexes of the ligand dadpe $(Ph_2PCH_2CH_2AsPh_2)$ have been tested against antimetabolite-sensitive leukemia cells, alkylating-agent-sensitive Walker tumor cells, and V.79 Chinese hamster lung cells. Both the bridged and tetrahedral complexes are cytotoxic although they are not as potent as the dppe complexes. The lower potency is probably a reflection of the increased kinetic lability and lower thermodynamic stability of Au-As bonds compared to Au-P bonds, and hence the ligand is more easily displaced and detoxified by oxidation (Ni Dhubhghaill et al., 1990).

In contrast to R_3PAuX complexes, the antitumor activity of the bridged digold complexes increases when X is a good leaving group. This is probably related to the conversion of the linear bridged complexes to the kinetically stable tetrahedral complexes. The major drawback to the clinical use of tetrahedral complexes such as $[Au(dppe)_2]^+$ is their ability to uncouple mitochondrial oxidative phosphorylation by increasing the permeability of the mitochondrial membrane to cations, with collapse of the mitochondrial membrane potential (Hoke et al., 1988). In preclinical trials, $[Au(dppe)_2]^+$ exhibited cardiac, hepatic and vascular toxicity (Rush et al., 1987; Hoke et al, 1989).

There are a few reports of non-phosphine Au(III) complexes with anticancer activity. These include $[Au(CN)_2]^-$, $[Au(III)Me_2Cl_2]AsPh_4$, $[Me_2Au(III)(SCN)_2Au(III)Me_2]$ (Sadler et al., 1984) and $[Au(DDC)Br_2]$, (DDC = diethyldithiocarbamate) (Ni Dhubhghaill and Sadler, unpublished). Despite these difficulties, gold(I)-phosphine complexes show promise as antitumor agents, particularly as complements to other metal drugs already in clinical use, and there remains much to be explored in their development.

REFERENCES

Bates, P.A., and Waters. J.M. (1984). *Inorg. Chim. Acta, 81*: 151.

Beesley, J.E. (1989). *Colloidal Gold: A New Perspective For Cytochemical Marking*, Oxford University Press, Oxford.

Berners-Price, S.J., and Sadler, P.J. (1986). *Inorg. Chem.*, *25*: 3822.

Berners-Price, S.J., and Sadler, P.J. (1987). *J. Inorg. Biochem.*, *31*: 267.

Berners-Price, S.J., Mazid, M.A., and Sadler, P.J. (1984). *J. Chem. Soc., Dalton Trans.*, 969.

Berners-Price, S.J., Mirabelli, C.K., Johnson, R.K., Mattern, M.R., McCabe, F.L., Faucette, L.F., Sung, C.M., Mong, S.-M., Sadler, P.J., and Crooke, S.T. (1986). *Cancer Res.*, *46*: 5486.

Berners-Price, S.J., Jarrett, P.S., and Sadler, P.J. (1987). *Inorg. Chem.*, *26*: 3074.

Berners-Price, S.J., Girard, G.R., Hill, D.T., Sutton, B.M., Jarrett, P.S., Faucette, L.F., Johnson, R.K., Mirabelli, C.K., and Sadler, P.J. (1990). *J. Med. Chem.*, *33*: 1386.

Beverly, B., and Couri, D. (1987). *Fed. Proc.*, *46*: 854.

Blank, C.E., and Dabrowiak, J.C. (1984). *J. Inorg. Biochem.*, *21*: 21.

Butt, T.R., Sternberg, E.J., Mirabelli, C.K., and Crooke, S.T. (1986). *Mol. Pharmacol.*, *29*: 204.

Coffer, M.T., Shaw, C.F., III, Eidsness, M.K., Watkins, J.W., II, and Elder, R.C. (1986). *Inorg. Chem.*, *25*: 333.

Corey, E.J., Mehrotra, M.M., and Khan, A.U. (1987). *Science*, *236*: 68.

Ecker, D.J., Hempel, J.C., Sutton, B.M., Kirsch, R., and Crooke, S.T. (1986). *Inorg. Chem.*, *25*: 3139.

Elder, R.C., Eidsness, M.K., Heeg, M.J., Tepperman, K.G., Shaw, C.F., III, and Schaeffer, N. (1983). *ACS Symposium Series*, *209*: 385.

Elder, R., Ludwig, K., Cooper, J.N., and Eidsness, M.K. (1985). *J. Amer. Chem. Soc.*, *107*: 5024.

Elder, R., Zhao, Z., Zhang, Y., Dorsey, J.G., Hess, E.V., and Tepperman, K. (1993). *J. Rheumatol.*, *20*: 268.

Finkelstein, A.E., Walz, D.T., Batista, V., Mizraji, M., Toisman, F., and Misher, A. (1976). *Ann. Rheum. Dis.*, *35*: 251.

Finkelstein, A.E., Burrone, O.R., Walz, D.T., and Misher, A. (1977). *J. Rheumatol.*, *4*: 245.

Forestier, J. (1929). *Bull. Mém. Soc. Méd. Hôp. (Paris)*, *53*: 3231.

Glennas, A., and Rugstadt, H.E. (1985). *J. Rheumatol.*, *14*: 25.

Graham, G.G., Haavisto, T,M., and McNaught, P.J. (1982). *J. Rheumatol.*, *9*: 527.

Graham, G.G., Bales, J.R., Grootveld, M.C., and Sadler, P.J. (1985). *J. Inorg. Biochem.*, *25*: 163.

Grootveld, M.C., Razi, M.T., and Sadler, P.J. (1984). *Clin. Rheumatol.*, *3 (Suppl. 1)*: 5.

Grootveld, M., Blake, D.R., Sahinoglu, T., Claxson, A.W.D., Mapp, P., Stevens, C., Allen, R.E., and Furst, A.(1990). *Free Rad. Res. Commun.*, *10*: 199.

Handel, M.L., deFazio, A., Watts, C.K.W., Day, R.O., and Sutherland, R.L. (1991). *Mol. Pharmacol.*, *40*: 613.

Hill, D.T. (1978). U.S. Patent: 4,057,630 - *Chem. Abs.*, *88*: 94838w.

Hill, D.T., and Sutton, B.M. (1980). *Cryst. Struct. Comm.*, *9*: 679.

Hoke, G.D., Rush, G.F., Bossard, G.E., McArdle, J.V., Jensen, B.D., and Mirabelli, C.K. (1988). *J. Biol. Chem.*, *263*: 11203.

Hoke, G.D., Macia, R.A., Meunier, P.C., Bugelski, P.J., Mirabelli, C.K., Rush, G.F., and Matthews, W.D. (1989). *Toxicol. Appl. Pharmacol.*, *100*: 293.

Holowczak, M.S., Stancl, M.D., and Wong, G.B. (1985). *J. Amer. Chem. Soc.*, *107*: 5789.

Intoccia, A.P., Flanagan, T.L., Walz, D.T., Gutzait, L., Swagzdis, J.E., Flagiello Jr., J., Hwang, Y.-H. B., Dewey, R.H., and Noguchi, H. (1982). *J. Rheumatol.*, *9* (suppl. 8*)*: 90.

Isab, A.A., and Sadler, P.J. (1981). *J. Chem. Soc., Dalton Trans.*, 1657.

Isab, A.A., and Sadler, P.J. (1982). *J. Chem. Soc., Dalton Trans.*, 135.

Isab, A.A., Hormann, A.L., Coffer, M.T., and Shaw, C.F., III (1988). *J. Amer. Chem. Soc.*, *110*: 3278.

Jones, P.G., and Schelbach, R. (1991). *Inorg. Chim. Acta, 182*: 239.

Kinsch, E.M., and Stephan, D.W. (1984). *Inorg. Chim. Acta, 91*: 263.

Kissner, R., Latal, P., and Geier, G. (1993). *J. Chem Soc. Chem. Commun.*, 136.

Laib, J.E., Shaw, C.F., III, Petering, D.H., Eidsness, M.K., Elder, R.C., and Garvey, J.S. (1985). *Biochemistry, 24*: 1977.

Malik, N.A., and Sadler, P.J. (1979). *Biochem. Soc. Trans., 7*: 731.

Malik, N.A., Otiko, G., and Sadler, P.J. (1980). *J. Inorg. Biochem., 12*: 317.

Mays, M.J., and Vergnano, P.A. (1979). *J. Chem. Soc. Chem. Commun.*, 1112.

Mazid, M.A., Razi, M.T., Sadler, P.J., Greaves, G.N., Hurman, S.J., Koch, M.H.J., and Phillips, J.C. (1980). *J. Chem. Soc. Chem. Commun.*, 1261.

Mirabelli, C.K., Johnson, R.K., Sung, C.M., Faucette, L.F., Muirhead, K., and Crooke, S.T. (1985). *Cancer Res., 45*: 32.

Mirabelli, C.K., Sung, C.M., Zimmerman, J.P., Hill, D.T., Mong, S., and Crooke, S.T. (1986a). *Biochem. Pharm., 35*: 1427.

Mirabelli, C.K., Zimmerman, J.P., Bartus, H.R., Sung, C.M., and Crooke, S.T. (1986b). *Biochem. Pharm., 35*: 1435.

Mirabelli, C.K., Johnson, R.K., Hill, D.T., Faucette, L.F., Girard, G.R., Kuo, G.Y., Sung, C.M., and Crooke, S.T. (1986c). *J. Med. Chem., 29*: 218.

Mirabelli, C.K., Hill, D.T., Faucette, L.F., McCabe, F.L., Girard, G.R., Bryan, D.B., Sutton, B.M., O'Leary-Bartus, J., Crooke, S.T., and Johnson, R.K. (1987). *J. Med. Chem., 30*: 2181.

Ni Dhubhghaill, O.M., Sadler, P.J., and Kuroda, R. (1990). *J. Chem Soc., Dalton Trans.*, 2913.

Ni Dhubhghaill, O.M., Sadler, P.J., and Tucker, A. (1992). *J. Amer. Chem. Soc., 114*: 1118.

Otiko, G., and Sadler, P.J. (1980). *FEBS Lett., 116*: 227.

Pathaneni, S.S., and Desiraju, G.R. (1993). *J. Chem. Soc., Dalton Trans.*, 319.

Pill, T., Polborn, K., Kleinschmidt, A., Erfle, V., Breu, W., Wagner, H., and Beck, W. (1991). *Chem. Ber., 124*: 1541.

Pinkerton, M., Steinrauf, L.K., and Dawkins, P. (1969). *Biochem. Biophys. Res. Commun., 35*: 512.

Puddephatt, R.J. (1978). *The Chemistry of Gold*, Elsevier, Amsterdam.

Ranford, J.D., Rhodes, M.D., and Sadler, P.J. (1991). *Metallodrugs in Metallothioneins: Synthesis, Structure and Properties of Metallothioneins, Phytochelatins and Metal Thiolate Complexes* (M.J. Stillman, C.F. Shaw III and K.T. Suzuki, eds.), VCH Publ., New York, p. 414.

Razi, M.T., Sadler, P.J., Hill, D.T., and Sutton, B.M. (1983a). *J. Chem. Soc., Dalton Trans.*, 1331.

Razi, M.T., Otiko, G., and Sadler, P.J. (1983b). *ACS Symposium Series, 209*: 371.

Romagnoli, P., Spinas, G.A., and Sinigaglia, F. (1992). *J. Clin. Invest., 89*: 254.

Ruben, H., Zalkin, A., Faltens, M.O., and Templeton, D.H. (1974). *Inorg. Chem., 13*: 1836.

Rush, G.F., Alberts, D.W., Meunier, P., Leffler, K., and Smith, P.F. (1987). *Toxicologist, 7*: 59.

Sadler, P.J. (1976a). *Gold Bull., 110*: 10.

Sadler, P.J. (1976b). *Struct. Bonding, 29*: 171.

Sadler, P.J., (1982). *J. Rheumatol., 9 (Suppl. 8)*: 71.

Sadler, P.J., Nasr, M., and Narayanan, V.L. (1984). *Platinum Coordination Complexes in Cancer Chemotherapy* (M.P. Hacker, E.B. Douple and I.H. Krakoff, eds.), Martinus Nijhoff Publishing, Boston, p. 290.

Schuhmann, D., Kubicka-Muranyi, M., Mirtschewa, J., Günther, J., and Gleichmann, E. (1990). *J. Immunol., 145*: 2132.

Shaw, C.F., III, and Laib, J.E. (1986). *Inorg. Chim. Acta, 123*: 197.

Shaw, C.F., III, Schaeffer, N.A., Elder, R.C., Eidsness, M.K. Trooster, J.M., and Calis, G.H.M. (1984). *J. Amer. Chem. Soc., 106*: 3511.

Shaw, C.F., III, Coffer, M.T., Klingbeil, J., and Mirabelli, C.K. (1988). *J. Amer. Chem. Soc., 110*: 729.

Simon, T.M., Kunishima, D.H., Vibert, G.J., and Lorber, A. (1979). *Cancer, 44*: 1965.

Snyder, R.M., Mirabelli, C.K., Johnson, R.K., Sung, C.M., Faucette, L.F., McCabe, F.L., Zimmerman, J.P., Whitman, M., Hempel, J.C., and Crooke, S.T. (1986). *Cancer Res., 46*: 5054.

Sutton, B.M. (1986). *Gold Bull., 19*: 15.

Tepperman, K., Finer, R., Donovan, S., Elder, R.C., Doi, J., Ratliff, D., and Ng, K. (1984). *Science, 225*: 430.

Thompson, R.H.S., and Whittaker, V.P. (1947). *Biochem. J., 41*: 342.

Walz, D.T., DiMartino, M.J., Sutton, B.M., and Misher, A. (1972). *J. Pharmacol. Exp. Ther., 181*: 292.

Ward, B., and Dabrowiak, J.C. (1987). *J. Amer. Chem. Soc., 109*: 3810.

Wienken, M., Lippert, B., Zangrando, E., and Randaccio, L. (1992). *Inorg. Chem., 31*: 1983.

Witkiewicz, P.L., and Shaw, C.F., III (1981). *J. Chem. Soc. Chem. Commun.*, 1111.

Xiao, J., and Shaw III, C.F. (1992). *Inorg. Chem., 31*: 3706.

RADIOACTIVE METAL COMPLEXES USED FOR TUMOR DIAGNOSIS

by

Kazuko Horiuchi S., Ph D., and Akira Yokoyama, Ph D.

Department of Radiopharmaceutical Chemistry,
Faculty of Pharmaceutical Sciences,
Kyoto University,
Kyoto 606
Japan

I. GENERALITIES ON RADIOACTIVE METAL COMPLEXES IN TUMOR DIAGNOSIS

The use of radioactive diagnostic agents in Medicine comprises a relatively new field called *Nuclear Medicine*. The clinical application of radioactivity centered upon diagnosis and therapy constitutes one of the greatest advances in Non-Invasive Medicine. In the field of tumor diagnosis, the recognition of favorable prognostic value of early cancer detection has resulted in a great effort directed toward the use of radioactive agents for tumor localization. A drug containing radioactive atoms within its structure can offer the most suitable character for tumor detection or therapy. Drugs containing radioactive nuclides are called *radiopharmaceuticals.*

Table 1

Metallic Radionuclides Used in Diagnostic Agents

Radionuclide	Decay mode	Half-life	Photon energy (keV)
^{67}Ga	EC	78.1 h	93, 185
^{68}Ga*	β^+	68.3 min	511
^{99m}Tc*	IT	6.02 h	140
^{111}In	EC	67.2 min	172, 247
^{113m}In*	IT	99.4 min	392
^{201}Tl	EC	73.5 h	69, 71, 80 (X-ray)
^{62}Cu	β^+	9.8 min	511

* Generator-produced, EC= electron capture, IT= isomeric transition, β^+= positron

Table 2

Generators of Interest in Nuclear Medicine

Parent	$t_{1/2}$	Daughter	$t_{1/2}$	γ-energy (keV)
^{68}Ge	275 d	^{68}Ga	68.3 min	511
^{99}Mo	67 h	^{99m}Tc	6.02 h	140 (98)
^{113}Sn	115 d	^{113m}In	1.7 h	393 (64)
^{62}Zn	9.1 h	^{62}Cu	9.8 min	511

Among clinically useful radionuclides, *radiometals* abound and they have received by far the most attention due not only to their nuclear physic characteristics (Table 1) but also to their inherent capacity to coordinate with a great variety of ligands (Bulman, 1987). In fact, radiometals and radiometal complexes such as $^{201}Tl^{+}$, ^{67}Ga-citrate and ^{99m}Tc-methylene diphosphonate have been extensively used in radiodiagnosis for tumor or metastasis localization (Coleman, 1991; Chilton et al., 1990b; Davis, 1990). These radioactive nuclides are *short-lived,* possess *low gamma energy emitting,* and when intravenously administered to the patient, their localization at the tumor or metastasis site can be visualized with external radiation detectors. In the past few years of Nuclear Medicine practice, the development of high resolution instrumentation has indicated the great importance of using radionuclides at very *high doses* and *high specific activity*; positron-emitting radionuclides with two opposing high energy 511 keV γ rays generated by electron-positron annihilation fulfill the required characteristics. In diagnostic or therapeutical practice, the advantages of dealing with *carrier-free radionuclides or very low masses* of the element are that chemical toxicity rarely arises and that the radionuclide can be used as a tracer without disturbing the system under study. A carrier-free radioactive solution most often has a very high specific activity, and the concentration may well be of the order of 10^{-9} mol dm^{-3}, a typical concentration found in generator-produced radionuclides (Table 2).

Conforming with the above, the use of radiometallic nuclides in tumor diagnosis constitutes, nowadays, a very challenging task. *High tumor radioactivity localization* is most desirable, but it is also important to take into account the regional blood or tissue biodistribution in order to reach a *high target/non-target ratio.* In other words, high radioactivity accumulation in the tumor is not the only parameter to consider for selecting a good diagnostic agent. The agent should also be cleared from surrounding non-target tissue within the patient diagnostic examination interval. At present, tumor imaging diagnosis is performed 1-3 hrs post-injection of ^{99m}Tc- or ^{201}Tl-labeled agents, but other radionuclides such as ^{67}Ga or ^{111}In require 24-48 hrs, an interval inconvenient to the patient. Thus, in the evaluation of radiodiagnostic agents, bioavailability, distribution and clearance are the major pharmacokinetic variables, but the *time* required to attain the objective differs from that of conventional pharmaceuticals.

II. DESCRIPTION OF THE PHYSICO-CHEMICAL PROPERTIES AND SELECTION CRITERIA OF RADIONUCLIDES

Radioactive metal complexes to be used as tumor diagnostic agents are meant to be administered to the patient for external detection of malignancy; so, an appropriate selection of radionuclides is required to minimize the radiation dose to the staff and patient, and maximize the information to be gained. Three main factors can be considered in the selection of electromagnetic radiation-emitting radionuclides: radionuclide physical characteristics, radionuclide availability and desirable chemistry.

A. Radionuclide physical characteristics

Electromagnetic radiations (gamma and X-ray) are the only suitable forms, since particulate radiation (β^-) is completely absorbed by tissue and occasions a high radiation burden. The main factors to be considered are:

(i) Decay mode. The most suitable decay modes for diagnostic imaging are electron capture (EC) and isomeric transition (IT) from metastable nuclides because there is no particulate radiation associated with these processes.

(ii) Photon energy - ideally 100-200 keV. The energy of gamma rays should be high enough to readily penetrate and escape from the body, yet enough to be efficiently stopped by the gamma camera. Although 500 keV is the upper limit, ideally about 200 keV is most desirable.

(iii) Half-life. In choosing an appropriate radionuclide physical half-life, T_p, the effective half-life, T_e, should be taken into consideration. T_e has been defined as the time required to remove half of the radioactivity from an organ by a combination of T_p and of the biologic elimination, T_b (i.e. $T_e = T_p \times T_b / (T_p + T_b)$), and equals 1 to 1.5 times the imaging time.

(iv) Abundance. The radiation should be of high abundance. It is also normally an advantage, but not essential, that the gamma source be monoenergetic rather than spread out over a range of energies as in the case of ^{67}Ga which has principal gamma rays at 92, 182, 300 and 390 keV.

B. Radionuclide availability

In medical practice, ready availability of the radionuclide whenever a diagnosis is required is one of the very important factors to have in mind, particularly if the radionuclide physical life is taken into consideration; a half-life close to the length of the diagnostic test is desirable. This implies the use of short-lived radionuclides; so, for hospitals not located close to a radionuclide production facility, the only way to get these is by the use of a *generator*, i.e. a system containing a long-lived parent radionuclide which decays to a short-lived daughter used for preparing the diagnostic agent (Table 2).

Among the various metallic radionuclides reported as useful for radiodiagnosis (Table 1), there is no single *ideal* radionuclide, but for many studies the generator-produced ^{99m}Tc physical and radiation characteristics closely approach the ideal requirements; it decays by isomeric transition emitting a 140 keV gamma ray in high abundance and its half-life of six hours is a convenient length for the diagnostic test.

In recent years, those radionuclides with gamma radiation like ^{99m}Tc are called Single Photon Emitting nuclides and the radioisotopic imaging methodology for studying regional function using the tomographic system is called *Single Photon Emission Computed Tomography* (*tomos; greek= cut, section*) or *SPECT* (Phelps, 1977). Although most of the radionuclides are gamma-emitters, a few are positron-emitters and the methodology using positron is called *Positron Emission Tomography or PET* (Ter-Pogossian, 1975; Coleman, 1991); the term "single photon" refers to the fact that the photons are detected one at the time, not in coincidence pairs like the photons in PET.

C. Desirable chemistry

The almost ideal physical property of the radiometal technetium has enormously contributed to its use in over 80% of Nuclear Medicine procedures. However, research and development on new technetium radiopharmaceuticals have been hindered by the intricate chemical character of the element and its lack of a stable isotope. Like Mn and Re, technetium belongs to group VIIB with valence states from -1 to +7, and can also display a large amount of coordination numbers. The most well known stable oxidation state is +7 represented by the pertechnetate ion, TcO_4^-, which resembles the permanganate ion, MnO_4^-, and the perrhenate ion, ReO_4^-; in addition, +1, +3, and +5 oxidation states have also been reported in Tc-radiopharmaceuticals. This versatile chemistry is the matter of a great challenge (Deutsch et al., 1983; 1989; Pikerton et al., 1985), but might at the same time constitute a field for new drug-design exploration.

The chemistry of the other radiometallic nuclides with properties for diagnostic use has been better explored. These belong mostly to the IIIA group, and include gallium (Ga), indium (In), and thallium (Tl) (Moerlein and Welch, 1981; Green and Welch, 1989). In aqueous solution, the trivalent state is important for Ga and In, but the monovalent state is a dominant feature for Tl. Variable coordination numbers from 4 to 6 and 8 have been reported. The polymerization of trivalent Ga and In to achieve coordination saturation is common to all group IIIA elements (Cotton and Wilkinson, 1988).

III. BASIC CONCEPTUAL APPROACHES IN RADIOACTIVE DIAGNOSTIC AGENTS

As illustrated in Table 1, only very few elements have radioisotopes suitable for radiodiagnostic studies. In the past, most empirical approach has often ensued. However, with the advent of the ultra-short-lived cyclotron-produced positron-emitting (β^+) radioisotopes of carbon (^{11}C, $t_{1/2}$ = 20.3 min), nitrogen (^{13}N; $t_{1/2}$= 9.96 min) and oxygen (^{15}O, $t_{1/2}$=123 s), the radiolabeling of organic molecules has initiated a new era in the development of radiopharmaceuticals. Various radiolabeled biomolecules (^{11}C-deoxyglucose, ^{18}F-fluorodeoxyglucose) (F-18, $t_{1/2}$ = 109.7 min) have made it possible to measure biochemical and physiological parameters in humans, and a rational interpretation has emerged from a knowledge of the fundamental basis of biochemistry and tracer kinetic modeling (Huang et al., 1983; Phelps et al., 1983; Barrios, 1986; Huang and Phelps, 1986; Chilton and Witcofski, 1990a; Balant et al., 1990). However, the widespread use of these on-site-cyclotron produced ultra-short lived radionuclides is still thwarted by their high cost, and radiometallic nuclides are better suited for clinical practice. The great versatility of inorganic compounds allows the rational design of radiometallic drugs mimicking the pharmacokinetics of radioorganic drugs; in recent years, radiometallic-containing agents for brain perfusion studies have been prepared whose modeling bases are introduced in the next paragraphs. Although this might at first appear irrelevant to tumor diagnostic agents, it is a "state of the art" example achieved with inorganic radiopharmaceuticals and a valuable conceptual guideline for future tumor-seeking agents.

A. Basis on brain radiopharmaceutical development

A good example of development of successful radiopharmaceuticals is the in vivo quantification of regional cerebral blood flow achieved with ^{15}O-labeled water (^{15}O-H_2O) (Huang et al., 1981) (Fig. 1). Great advances in the basic tracer kinetic mathematical model and the biochemical strategies employed have allowed the quantification of cerebral perfusion in different neurological disorders, stroke, dementia, epilepsia, Alzheimer's disease (Wagner, 1987; 1989; 1992), the tumor blood flow also being exemplary in this respect (Mathias et al., 1990). Immediately after, the clinical applicability of the principle was assessed by the use of radioiodinated ^{123}I-iodoamphetamine (^{123}I-IMP) and ^{123}I-propanediamine (^{123}I-HIPDM), respectively (Kung, 1990a; Stocklin, 1992).

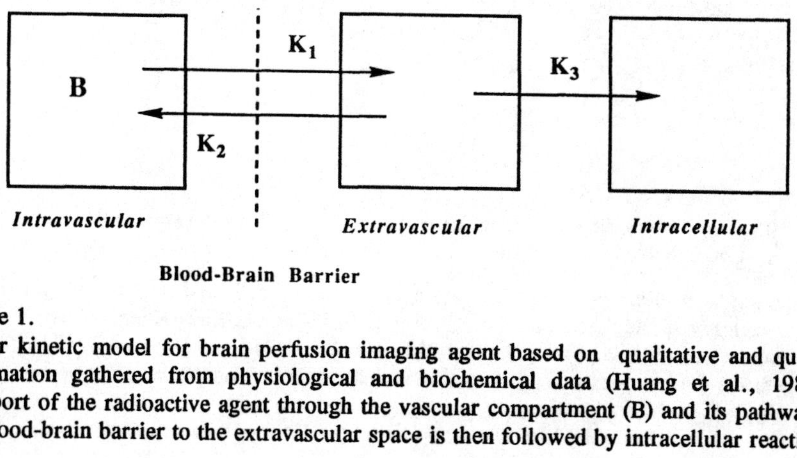

Figure 1.
Tracer kinetic model for brain perfusion imaging agent based on qualitative and quantitative information gathered from physiological and biochemical data (Huang et al., 1981). The transport of the radioactive agent through the vascular compartment (B) and its pathway across the blood-brain barrier to the extravascular space is then followed by intracellular reaction.

Meanwhile, small size, compactness, neutrality and lipid solubility had been pointed out as important criteria for the design of radiometallic complexes likely to cross the blood-brain-barrier (BBB) (Hosotani et al., 1986); however, such complexes were rapidly cleared from the brain, acting as freely diffusible tracers. The addition of methyl groups to the neutral lipid-soluble ^{99m}Tc-complex based on the N_4-ligand system of propyleneamine (PnAO) led to the development of ^{99m}Tc-D,L-hexamethyl propylene amine oxime (^{99m}Tc-HM-PAO) (Neirinckx et al., 1987) (Fig. 2) which undergoes intracellular conversion into a hydrophilic compound with poor diffusion. Later, a diester of a neutral ^{99m}Tc-complex of the N_2S_2 ligand series, the ^{99m}Tc-ethyl cysteinate dimer (^{99m}Tc-ECD), provided through hydrolysis the desired brain retention necessary to quantification (Walovitch et al., 1988a; 1988b). These radiometallic drugs (Fig. 2) fulfill the basic criteria drawn from the PET perfusion imaging agent, and are showing in initial clinical studies an efficacy parallel to those reported for the radioiodinated brain perfusion agent. So, a good knowledge of tracer kinetics has helped to build up a basis for the development of brain Tc-radiopharmaceuticals (Kung, 1990b).

B. Basis on metal-containing antitumor agents

Among metal-containing antitumor agents, cis-dichlorodiamine-platinum, *cis*-$[PtCl_2(NH_3)_2]$ (Cleare, 1974; Reedijk et al., 1987), *Cu-KTS* (Petering, 1980), the copper complex of 3-ethoxy-2-oxobutyraldehyde bis-(thiosemicarbazone) known as kethoxal bis(thiosemicarbazone) or *KTS* (Fig. 3), and various metal complexes of the antibiotic *bleomycin* can be mentioned (Dabrowiak, 1980; Niewig et al., 1982).

Attempts to radiolabel these antitumor agents with diagnostic and therapeutical radionuclides have been reported. The occurrence of gamma-emitting isotopes of Pt, ^{195m}Pt for example, has made the radiolabeling of cis-platinum possible but with a low target /non-target ratio; at present, the lack of research on the production of high specific activity ^{195m}Pt can be argumented to explain the sequential effect of cis-platinum administered at high concentration (Owens et al., 1985). As for Cu-KTS, the ^{64}Cu-radiolabeled complex has also shown a low target/non-target tissue distribution ratio (Pastakia et al., 1980). However, diffusibility through cellular membranes can be visualized from its lipophilic and compact chemical character (Fig. 3); in our laboratory, this KTS character has served as the drug-design basis for the development of the first brain Tc-radiopharmaceuticals (Hosotani et al., 1986), essentially the foundation of the above-mentioned Tc-HM-PAO or Tc-ECD.

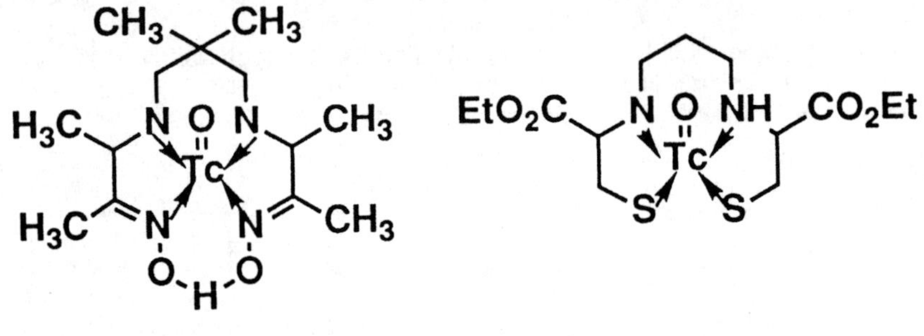

Tc-HM-PAO
(Hexamethyl propylene amine oxime)

Tc-ECD
(Ethylcystinate dimer)

Figure 2.
Technetium-99m brain perfusion imaging agents. Neutral and lipid soluble ^{99m}Tc complexes of clinical efficacy in the detection of cerebral vascular abnormalities (Kung, 1990). Tc-HM-PAO (Tc complex of hexamethyl propylene amine oxime) and Tc-ECD (Tc complex of ethylcystinate dimer) fulfill the basic criteria of extraction proportional to blood flow, BBB penetrability (lipid solubility), high brain uptake, high brain/blood ratio and brain retention related to the intracellular formation of hydrophilic compound (Tc-HM-PAO) or ester hydrolysis (Tc-ECD).

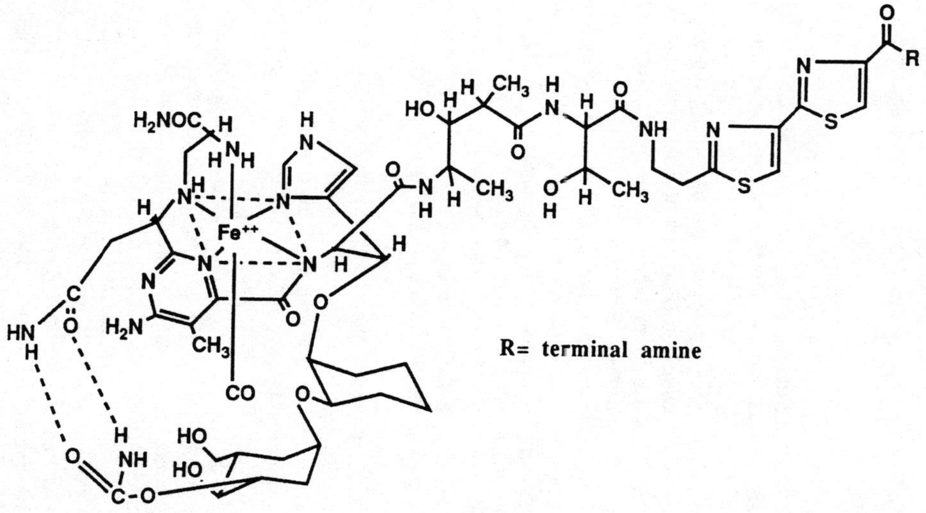

R= terminal amine

Fe-Bleomycin

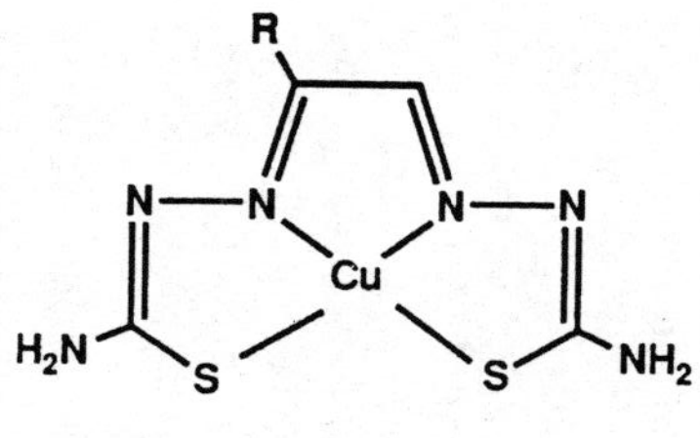

R= CH_3-CH-OC_2H_5

Cu- KTS

[Kethoxal bis(thiosemicarbazone)]

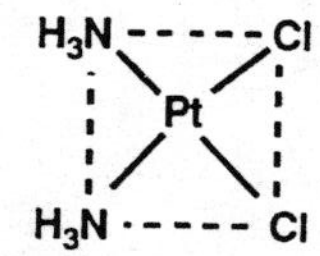

cis-[$PtCl_2(NH_3)_2$]

Figure 3.

Structure of metal-containing antitumor agents being radiolabeled with diagnostic and therapeutical radionuclides. Clinical use of radiolabeled bleomycin or cis-[$PtCl_2(NH_3)_2$] hampered by the lack of studies on high specific activity radiometallic nuclides (^{57}Co, ^{195m}Pt) production; while the structure of kethoxal bis(thiosemicarbazone) or KTS ligand enables the radiolabeling with the generator-produced ^{62}Cu.

In spite of the great number of tetradentate Tc-chelates described in the literature, no tumor-specific Tc-radiopharmaceutical has been reported. Recently, however, tumor blood flow has been shown to be measurable with a copper complex of a KTS derivative, pyruvaldehyde-di(4-methylthiosemicarbazone) (PTSM) (Mathias et al., 1990). The reduction of this complex from Cu(II) to Cu(I) in cell mitochondria is an exploitable phenomenon for the future development of tumor-related metal complexes (Fujibayashi et al., 1993).

As for the antibiotic bleomycin, an interaction with DNA has been reported; Sugiura et al. (1980) described the coordination of Fe^{2+} with the β-aminoalanine-pyrimidine-β-hydroxyl-histidine portion as being essential to the oxido-reduction reaction involved in the DNA cleavage. Also, while the dithiazole portion binds in the minor DNA groove (Kuwabara et al., 1987), the sugar portion seems related to the biological distribution. Bleomycin has been radiolabeled with ^{57}Co (Kono et al., 1972; Konikowski et al., 1975), ^{99m}Tc (Yokoyama et al., 1978) and ^{111}In (Niewig et al., 1982); although ^{57}Co-Bleomycin has shown higher tumor accumulation in patients (Rasker et al., 1975), ^{99m}Tc-Bleomycin has found clinical applicability in tumor diagnosis, particularly of sarcomas and adenocarcinomas in areas such as neck, thyroid and extremities (Odori, 1979a; 1979b). The use of the higher tumor-specificity of Co-Bleomycin is unfortunately hindered by the lack of research on the production of Co radionuclides for diagnostic or therapeutical purposes; while ^{99m}Tc is a better suited radionuclide, research has been limited to the preparation of a stable ^{99m}Tc-Bleomycin (Yokoyama et al., 1978).

Almost no progress in the development of radiodiagnostic agents based on metal-containing antitumor agents has been observed in the past years. Thus, there is a great need for a more rational drug-design involving a better knowledge not only of biochemistry and tracer kinetic modeling, but of metal coordination chemistry and its interaction in biologicals systems, so that a pharmacokinetical strategy for tumor-detecting agents can be finalized (see e.g. Part Five, Chapter 3).

C. Basis of Ga-citrate diagnostic agent

Largely stimulated by the observed preferential tumor uptake and clinical use of ^{67}Ga-citrate, considerable research has been reported on the localization mechanism of this complex (Hayes, 1978; 1983; Larson et al., 1980; Ando et al., 1985). Today, a clear unified hypothesis is still lacking. Most of the work carried out for optimizing the clinical use of ^{67}Ga-citrate refers to the analogy of Ga with iron (Fe), this probably originating in the finding that upon ^{67}Ga-citrate administration gallium rapidly binds to transferrin (Tf) (Bates et al., 1967; Vallabhajosula et al., 1980; Hoffer, 1980; Weiner, 1989).

Resembling aspects with iron on ionization potentials, ionic radii, coordination number and stability constants have often been reported, but differences have also been mentioned. Of particular importance is the difference in Tf conditional binding constants; as reported by Harris and Pecoraro (1983), Tf binds gallium about 300 times more weakly than the ferric ion. In our laboratory, a rational approach built on reaction rate differences in metal ion transfer from Ga-citrate to Tf (slow) and Fe-nitrilotriacetate (Fe-NTA) to Tf (fast) has permitted to shorten the tumor imaging time interval in vivo, increasing the target/non-target ratio (Saji et al., 1985).

In the biological utilization of metals, the stability constants of the involved chelates, the nature of the ligands and their steric configuration are undoubtedly important, but the solubility, size and electrical charge of the chelates can also have great effects. For example, the concept of polynuclear gallium complexes (Glickson et al., 1973; 1974) and their metal exchange kinetics have been reported to be affected by polynuclear-mononuclear equilibria (Bates et al., 1967; Spiro and Saltman, 1969). Major interest in this concept has been gained by the study of the bioavailability of iron in the gastrointestinal tract, the size of polynuclear iron species being claimed to provide a chemically attractive way of optimizing oral preparations (Schneider and Forth, 1987; Schneider, 1988). Another interesting notion that emerged with the thermodynamic study of the gallium-transferrin binding is the dissociation of about 75% of the gallium from Tf to form $Ga(OH)_4^-$, a thermodynamically favored phenomenon attributed to the hydrolytic tendency of the Ga^{3+} ion (Harris and Pecoraro, 1983). In the authors' laboratory, the scheme shown in Figure 4

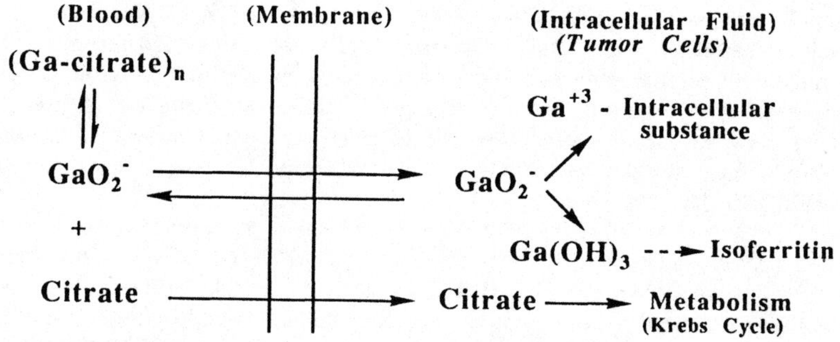

Figure 4.
Proposal for the ^{67}Ga tumor cell uptake. The extensive clinical use of ^{67}Ga-citrate in tumor diagnosis has stimulated considerable research on localization mechanism, serving as the premises for the concept of polynuclear gallium complexes and its role in the ^{67}Ga tumor localization.

has been drawn to account for the tumor localization mechanism of ^{67}Ga-citrate, and this constitutes the key basis for the drug design presented in the next paragraph.

IV. RADIOMETALLIC DRUG DESIGN BASIS FOR TUMOR DIAGNOSTIC AGENT

A. Pentavalent ^{99m}Tc-dimercaptosuccinate complex [Tc(V)DMS]

A judicious combination of the elements of knowledge described in the above paragraph gave rise to the drug design presented in Figure 5 and, after the screening of various candidates (Yokoyama and Saji, 1980), the selection of the ligand dimercaptosuccinic acid (DMS) led to the pentavalent technetium-99m chelate of dimercaptosuccinic acid *[Tc(V)-DMS]* (Yokoyama et al., 1985b). Tc(V)-DMS is a polynuclear technetium radiopharmaceutical formulated in a pentavalent state by the reduction of TcO_4^- with stannous chloride at a slightly alkaline pH. The pentavalent state and the polynuclear character of Tc(V)-DMS have been experimentally demonstrated via analytical methodology (Yokoyama et al., 1985b; Horiuchi et al., 1986).

As shown in Figure 5, the key concept borrowed from the Ga-citrate work is the formulation of a polynuclear Tc complex which can dissociate a $Tc(V)O_4^{3-}$ oxoanion capable of crossing the cell membrane and then being trapped within the cell while the remaining undissociated stable Tc-chelate portion can be systemically excreted. The clearest support to this approach has been gathered from the abundant clinical results reported in

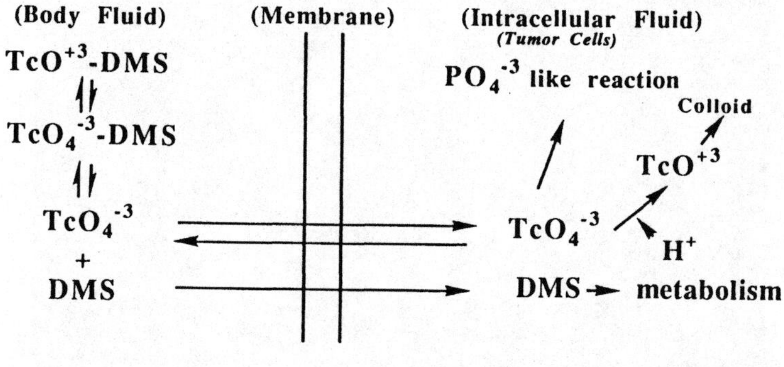

Figure 5.
Drug design basis of pentavalent Tc-dimercaptosuccinate complex. Upon injection of Tc(V)-DMS complex into the blood stream, Tc complex equilibria are altered and depolymerization or dissociation occurs, inducing the formation of active Tc species or dissociated Tc species (TcO_4^{3-}) capable of crossing the tumor cell membrane and being then retained within the cell.

numerous publications (Ohta et al., 1984; 1985; 1986; 1988; 1989; Miyauchi et al., 1986; Clarke et al., 1988; Patel et al., 1988; Watkinson et al., 1989; 1990; Adams et al., 1990); as an example, the scintigraphic imaging of a patient with medullary thyroid carcinoma (MTC) is presented in Figure 6. In an experimental animal model (nude mouse bearing a MTC tumor), a direct correlation between the radioactivity accumulation in this tumor of calcitonin-producing cells and the MTC-generated calcitonin level has been demonstrated (Horiuchi et al., 1991).

In our search for the building of a drug design basis, a correlation between tumor accumulation and dissociation of the polynuclear Tc(V)-DMS complex has been pursued using Ehrlich Ascites Tumor Cells (EATC) (Horiuchi et al., 1986). Most of the experimental data indicated that the simple dilution of the Tc(V)-DMS complex and the pH of the cell culture medium played a major role in inducing an augmented EATC uptake. In fact, the more diluted the Tc(V)-DMS complex solution, the higher percentage of dose radioactivity accumulated in the EATC (Horiuchi et al., 1986), further striking increases being observed as the pH of the incubation medium pH was lowered. This result cannot be merely attributed to the formation of a mononuclear complex; the generation of an ionic species might be predictable at low pH, but if this phenomenon of reactive species dissociation under slight pH variation can be associated with the well documented pH lowering in tumor tissues, then the Tc(V)-DMS drug design might be substantiated.

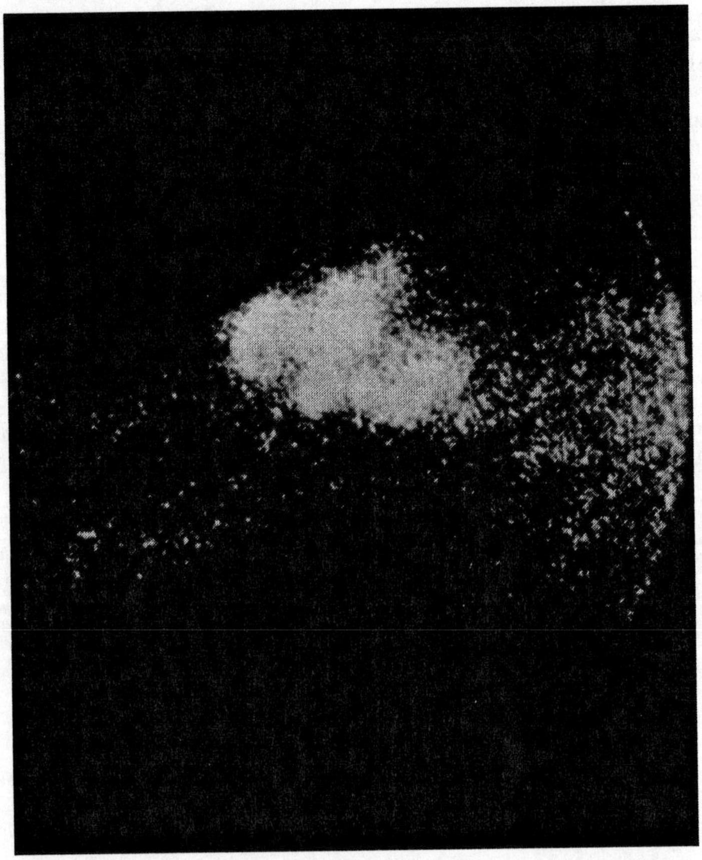

Figure 6.
Preoperative Tc(V)-DMS scintigram of a patient with medullary thyroid carcinoma. Please note the lack of Tc(V)-DMS accumulation in the normal part of right thyroid lobe. (From Ohta et al., 1985, with permission).

As regards the dissociated anionic species, the pentavalent oxoanion TcO_4^{3-} has been hypothesized (Fig. 5) and also predicted (Kissel and Feldberg, 1969); there is still no clear evidence of its existence but clinical data as well as the study with the above-mentioned MTC animal model suggest the participation of an oxoanion with a biological behavior similar to that of PO_4^{3-}. In an experimental model carried out using hydroxyapatite, a synthetic calcium phosphate, competitive binding of Tc(V)-DMS with hydroxyapatite was detected only in the presence of different concentrations of PO_4^{3-}. The Thin Layer Chromatography of Tc(V)-DMS has produced data supporting the dissociation of a polynuclear technetium complex; in fact, while the chromatographic development of the radiolabeled Tc(V)-DMS in acidic medium detected a single peak, that of the diluted sample showed the generation of a new radioactive peak with faster migration rate than the non-diluted sample; the new species increases with the degree of dilution (Horiuchi et al., 1986). Thus, although the study on the mechanism of Tc(V)-DMS tumor localization is still in progress, it is our belief that the already collected data offer valuable bases for the drug design scheme outlined in Figure 5. This research on Tc(V)-DMS has exploited the notion of hydrolytic polynucleation of metals, offering a new approach for the development of tumor diagnostic agents. Formulation of radiotherapeutical agents can also be envisaged with the technetium congener, rhenium, ^{186}Re or ^{188}Re, both β-emitting nuclides as the DMS complex (^{186}Re(V)-DMS) (Bisudanan et al., 1991).

B. Bifunctional radiopharmaceuticals

The demand for more biospecific radiopharmaceuticals has evolved into the generation of the so-called *Bifunctional Radiopharmaceuticals*, i.e. radiopharmaceuticals containing a biologically active molecule (glucose, fatty acids, amino acids) together with a metal chelating group with the ability to bind radiometallic nuclides (Yokoyama et al., 1985a). Research has been centered on the development of *Bifunctional Chelating Agents (BCA)* for the formulation of radiolabeled macromolecules such as albumin, enzymes, and antibodies against tumor-associated antigens (Sundberg et al., 1974; Meares et al., 1976; Koppel, 1990; Brinkley, 1992). BCAs carry both a chelating moiety and a functional group capable of forming a covalent bond with macromolecules (Fig. 7). Among the various BCAs reported, the cyclic DTPA anhydride (DTPA=diethylenetriaminepentaacetic acid) (Hnatowich et al., 1983) and deferoxamine (DFO) (Yokoyama et al., 1982) have been widely used for labeling different monoclonal antibodies (MoAb) and/or antibody fragments with ^{111}In and ^{67}Ga, respectively. Certainly, the hybridoma production of MoAbs and their radiolabeling has promoted the detection of tumors of various etiologies: ovarian, melanoma, lymphoma and neuroblastoma neoplasia, etc (Larson, 1991). Recently also, the small peptide Tyr-3-octreotide, a somatostatin analog, labeled via DTPA has been reported as suitable for localizing somastostatin receptor-positive tumors (Krenning, 1992). For the radiolabeling with Tc, a KTS-based ligand, CE-DTS, or p-carboxy-ethylphenylglyoxal-di(N-methylthiosemicarbazone) (Arano et al., 1986; 1987) or diamine dithiolate derivatives DADS (Fritzberg et al., 1988) have been conjugated to the protein; meanwhile, another approach has been to bind the radionuclide Tc directly to reduced disulfide sites on the protein itself (Schwarz and Steinstrasser, 1987).

Developing tumor diagnostic radiopharmaceuticals has become a challenging field; in the past few years, a great deal of efforts has gone into incorporating various radiometals into MoAbs. Moreover, as stated above, the translation of PET advances into clinical practice is occurring via SPECT at a fast pace. Thus, it is our belief that not only "molecular medicine" (Wagner, 1992) as a translation of molecular biology, but also a better translation of bioinorganic chemistry because our life depends on bioinorganic principles, might help to develop diagnostic radiopharmaceuticals. The tumor-imaging properties of Tc(V)-DMS derived from Ga-citrate data might constitute an example of rationally designed drug (Yokoyama et al., 1985; Abrams, 1991); a better knowledge of tumor pathophysiology and a more rational use of bioinorganic chemistry principles along with tracer kinetic modelings will contribute to develop new tumor diagnostic agents radiolabeled with metallic radionuclides.

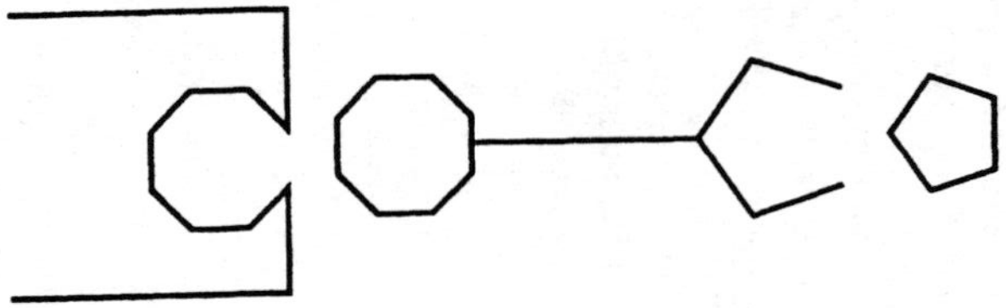

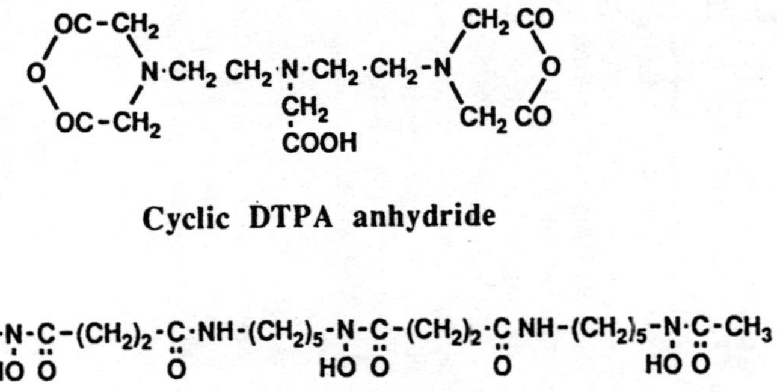

Cyclic DTPA anhydride

Deferoxamine

Figure 7.
Bifunctional radiopharmaceuticals. A schematic representation showing a *biomolecule* with affinity for the *receptor site*; the biomolecule is bound to a molecule containing a radiometal *chelating site*. A molecule holding two reactive sites for binding the biomolecule and chelating the radiometal is called bifunctional chelating agent (BCA). Two examples of BCAs are presented, the *cyclic DTPA anhydride* (based on the amide bond formation between the anhydride moiety and an amino group of the biomolecule, the resulting diethylenetriaminetetraacetic acid residue being the radiometal chelator) and *deferoxamine* (based on the free terminal amino group coupling reaction, via the glutaraldehyde with the amine group of the biomolecule; the three hydroxamate groups form a stable radiometal chelate).

REFERENCES

Abrams, M.J. (1991). *J. Nucl. Med.*, *32*: 849.

Adams, B.K., Fataar, A., Byrne, M.J., Levitt, N.S., and Matley, P.J. (1990). *J. Nucl. Med.*, *31*: 106.

Ando, A., Ando, I., Sanada, S., Hiraki, T., and Hisada, K. (1985). *Eur. J. Nucl. Med.*, *10*: 262.

Anghileri, L.J., Crone, M.C., Thouvenot, P., Brunotte, F., Marchal, C., and Robert, J. (1983). *Nucl. Med.*, *22*: 152.

Arano, Y., Yokoyama, A., Magata, Y., Horiuchi, K., Saji, H., and Torizuka, K. (1986). *Appl. Radiat. Isot.*, *37*: 587.

Arano, Y., Yokoyama, A., Furukawa, T., Horiuchi, K., Yahata, T., Saji, H., Sakahara, H., Nakashima, T., Koizumi, M., Endo, K., and Torizuka, K. (1987). *J. Nucl. Med.*, *28*: 1027.

Bates, G.W., Billups, C., and Saltman, P. (1967). *J. Biol. Chem.*, *242*: 2810.

Barrios, J.R. (1986). *Positron Emission Tomography and Autoradiography* (M.E. Phelps, J.C. Mazziotta and H.R. Schelbert, eds.), Raven Press, New York, p. 451.

Balant, L.P., Roseboom, H., and Gundert-Remy, U.M. (1990). *Advances in Drug Research* (B. Testa, ed.), Academic Press, London, p. 1.

Bisunadan, M.M., Blower, P.J., Clarke, S.E.M., Singh, J., and Went, M.J. (1991). *Appl. Radiat. Isot.*, *42*: 167.

Brinkley, M. (1992). *Bioconjugate Chemistry*, *3*: 2.

Bulman, R.A. (1987). *Structure and Bonding*, Vol. 67 (M.J. Clarke, J.B. Goodenough, J.A. Ibers, C.K. Jorgensen, D.M.P. Mingos, J.B. Neilands, G.A. Palmer, D. Reinen, P.J. Sadler, R. Weiss and R.J.P. Williams, eds.), Springer-Verlag, New York, p. 93.

Clarke, S.E.M., Lazarus, C.R., Wraight, P., Sampson, C., and Maisey, M. (1988). *J. Nucl. Med.*, *29*: 33.

Cleare, M.J. (1974). *Coord. Chem. Rev.*, *12*: 349.

Chilton, H.M., and Witcofski, R.L. (1990a). *Pharmaceuticals in Medical Imaging* (D.P. Swanson, H.M. Chilton and J.H. Thrall, eds.), MacMillan Publishing, New York, p. 279.

Chilton, H.M., Burchiel, S.W., and Watson, N.E., Jr. (1990b). *Pharmaceuticals in Medical Imaging* (D.P. Swanson, H.M. Chilton and J.H. Thrall, eds.), MacMillan Publishing, New York, p. 564.

Coleman, R.E. (1991). *Cancer*, *67*: 1261.

Cotton, F.A., and Wilkinson, G. (1988). *Advances in Inorganic Chemistry*, John Wiley & Sons Inc., New York, p. 208.

Dabrowiak, J.C. (1980). *Metal Ions in Biological Systems*, Vol. 11 (H. Sigel, ed.), Marcel Dekker Inc., New York, p. 305.

Davis, M. (1990). *J. Nucl. Med.*, *31*: 1658.

Deutsch, E., Libson, K., and Jurisson, S. (1983). *Progress in Inorganic Chemistry* (S.J. Lippard, ed.), John Wiley & Sons, New York, Vol 30, p. 75.

Deutsch, E., Libson, K., and Vanderheyden, J.-L. (1989). *Technetium and Rhenium in Chemistry and Nuclear Medicine* (M. Nicolini, G. Bandoli and U. Mazzi, eds.), Raven Press, New York, p. 13.

Fritzberg, A.R., Abrams, P.G., Beaumier, P.C., Kasina, S., Morgan, A.C., Rao, T.N., Reno, J.M., Sanderson, J.A., Srinivasan, A., Wilbur, D.S., and Vanderheyden, J.-L., (1988). *Proc. Nat. Acad. Sci. USA*, *85*: 4025.

Fujibayashi, Y., Wada, K., Taniuchi, H., Yonekura, Y., Konishi, J. and Yokoyama, A. (1993). *Biol. Pharm. Bull.*, *16*: 146.

Glickson, J.D., Ryel, R.B., Bordenca, M.M., Kim, K.H., and Gams, R.A. (1973). *Cancer Res.*, *33*: 2706.

Glickson, J.D., Webb, J., and Gams, R. (1974). *Cancer Res.*, *34*: 2957.

Green, M.A., and Welch, M.J. (1989). *Nucl. Med. Biol.*, *16*: 435.

Grifiths, G.L., Goldenberg, D.M., Jones, A.L., and Hansen, H.J. (1992). *Bioconjugate Chemistry*, *3*: 91.

Harris, W.R., and Pecoraro, V.L. (1983). *Biochemistry*, *22*: 292.

Hayes, R.L. (1978). *Semin. Nucl. Med.*, *8*: 183.

Hayes, R.L. (1983). *Int. J. Nucl. Med. Biol.*, *10*: 257.

Hnatowich, D.J., Childs, R.L., Lanteine, D., and Najafi, A.J. (1983). *J. Immunol. Method.*, *65*: 147.

Hoffer, P. (1980). *J. Nucl. Med.*, *21*: 282.

Horiuchi, K., Yomoda, I., Yokoyama, A., Endo, K., and Torizuka, K. (1986). *Technetium in Chemistry and Nuclear Medicine* (M. Nicolini, G. Bandoli and U. Mazzi, eds.), Raven Press, New York, p. 155.

Horiuchi, K., Yomoda, I., Ota, H., Endo, K., and Yokoyama, A. (1991). *Eur. J. Nucl. Med.*, *18*: 796.

Hosotani, T., Yokoyama, A., Arano, Y., Horiuchi, K., and Torizuka, K. (1986). *Int. J. Nucl. Med. Biol.*, *12*: 431.

Huang, S.C., Phelps, M.E., and Carson, R.E. (1981). *J. Cerebr. Blood Flow Metabol.*, *1*: S-31.

Huang, S., and Phelps, M E. (1986). *Positron Emission Tomography and Autoradiography* (M.E. Phelps, J.C. Mazziotta and H.R. Schelbert, eds.), Raven Press, New York, p. 287.

Huang, S.C., Carson, R.E., and Phelps, M.E. (1983). *Tracer Kinetics and Physiologic Modeling* (R.M. Lambrecht and A. Rescigno, eds.), Springer-Verlag, New York, p. 298.

Kissel, G., and Feldberg, S.W. (1969). *J. Physiol. Chem.*, *73*: 3082.

Konikowski, T., Haynie, T.P., and Glenn, H.J. (1975). *J. Nucl. Med.*, *16*: 738.

Kono, A., Kojima, M., and Maeda, T. (1972). *Radioisotopes*, *21*: 118.

Koppel, G.A. (1990). *Bioconjugate Chemistry*, *1*: 13.

Krenning, E.P., Bakker, W.H., Kooij, P.P.M., Breeman, W.A.P., Oei, H.Y., De Jong, M., Reubi, J.C., Visser, T.J., Bruns, C., Kwekkeboom, D.J., Reijs, A.E.M., Van Hagen, P.M., Koper, J.W., and Lambert, S.W.J. (1992). *J. Nucl. Med.*, *33*: 652.

Kung, H.F. (1990a). *Nucl. Med. Biol.*, *17*: 85.

Kung, H.F. (1990b). *Semin. Nucl. Med.*, *20*: 150.

Kuwabara, J., and Sugiura, Y. (1987). *Proc. Nat. Acad. Sci. USA*, *85*: 2459.

Larson, S.M. (1991). *Cancer*, *67 (Suppl.)*: 1253.

Larson, S.M., Rasey, J.S., Allen, D.R., Nelson, N.J., Grumbaun, Z., Harp, G.D., and Williams, D.L. (1980). *J. Nat. Cancer Inst.*, *64*: 41.

Mathias, C.J., Welch, M.J., Green, M.A., Perry, D.J., McGuire, A.H., Zhu, X., and Connett, J.M. (1990). *J. Nucl. Med.*, *31*: 909 (Abst.).

Meares, C.F., Goodwin, D.A., Leung, C.S.-H., Girgis, A.Y., Silvester, D.J., Nunn, A.D., and Lavender, P.J. (1976). *Proc. Nat. Acad. Sci. USA*, *73*: 3803.

Miyauchi, A., Endo, K., Ohta, H., Torizuka, K., Matsuzuka, F., Kobayashi, A., Kuma, K., and Maeda, M. (1986). *World J. Surg.*, *10*: 640.

Moerlein, S.M., and Welch, M.J. (1981). *Int. J. Nucl. Med. Biol.*, *8*: 277.

Neirinckx, R.D., Canning, L.R., Piper, L.M., Nowotnik, D.P., Pickett, R.D., Holmes, R.A., Volkert, W.A., Forster, A.M., Weisner, P.S., Marriot, J.A., and Chaplin, S.B. (1987). *J. Nucl. Med.*, *28*: 191.

Niewig, O.E., Beekhuis, H., Paans, A.M.J., Piers, D.A., Vaalburg, W., Welleweers, J., Wiegman, T., and Woldring, M.G. (1982). *Eur. J. Nucl. Med.*, *7*: 104.

Ohta, H., Yamamoto, K., Endo, K., Mori, T., Konishi, J., Morita, R., Hata, N., Ikekubo, K., Makimoto, K., Iida, Y., Konishi, J., Morita, R., Hata, N., Horiuchi, K., Yokoyama, A., Torizuka, K., and Kuma, K. (1984). *J. Nucl. Med.*, *25*: 323.

Ohta, H., Endo, K., Koizumi, M., Fujita, T., Sakahara, H., Nakashima, T., Torizuka, K., Hata, N., Masuda, H., Yomoda, I., Horiuchi, K., Yokoyama, A., Shimizu, Y., Ishii, M., and Kasahara, K. (1985). *Jpn. J. Nucl. Med.*, 22: 1653.

Ohta, H., Shane, F., Endo, K., Torizuka, K., Horiuchi, K., Yokoyama, A., and Ishii, M. (1986). *Clin. Nucl. Med.*, *11*: 842.

Ohta, H., Endo, K., Fujita, T., Koishi, J., Torizuka, K., Horiuchi, K., and Yokoyama, A. (1988). *Nucl. Med. Commun.*, *9*: 105.

Ohta, H., Endo, K., Kanoh, T., Konishi, J., and Kotoura, H. (1989). *J. Nucl. Med.*, *30*: 2049.

Odori, T. (1979a). *Jpn. J. Nucl. Med.*, *16*: 721.

Odori, T. (1979b). *Jpn. J. Nucl. Med.*, *16*: 829.

Owens, S.E., Thatcher, N., Sharma, H., Adam, R., Harrison, A., Smith, A., Zaki, A., Baer, J.C., McAuliffe, C.A., Crowther, D., and Fox, B.W. (1985). *Cancer Chemother. Pharmacol.*, *14*: 253.

Pastakia, B., Lieberman, L.M., Gatley, S.J., Young, D., Petering, D.H., and Minkel, D. (1980). *J. Nucl. Med.*, *21*: 67.

Patel, M.C., Patel, R.B., Ramathan, P., Ramamoorthy, N., Krishna, B.A., and Sharma, S.M. (1988). *Eur. J. Nucl. Med.*, *13*: 507.

Petering, D.H. (1980). *Metal Ions in Biological Systems*, Vol. 11 (H. Sigel, ed.), Marcel Dekker Inc., New York, p. 197.

Phelps, M.E. (1977). *Semin. Nucl. Med.*, *7*: 337.

Phelps, M.E., Mazziotta, J.C., and Huang, E.J. (1983). *Ann. Intern. Med.*, *98*: 339.

Pikerton, T.C., Desilets, C.P., Hoch, D.J., Mikelsons, M.V., and Wilson, G.M. (1985). *J. Chem. Educ.*, *62*: 965.

Rasker, J.J., Van de Poll, M.A.P.C., Beekhuis, H., Woldring, M.G., and Nieweg, H.O. (1975). *J. Nucl. Med.*, *16*: 1058.

Reedijk, J., Fichtinger-Schepman, A.J., Van Oosterom, A.L., and Van de Putte, P. (1987). *Structure and Bonding*, Vol. 67 (M.J. Clarke, J.B. Goodenough, J.A. Ibers, C.K. Jorgensen, D.M.P. Mingos, J.B. Neilands, G.A. Palmer, D. Reinen, P.J. Sadler, R. Weiss and R.J.P. Williams, eds.), Springer-Verlag, New York, p. 53.

Saji, H., Hata, N., Horiuchi, K., Yokoyama, A., Yamamoto, K., and Torizuka, K. (1985). *Int. J. Nucl. Med.*, *12*: 281.

Schneider, W. (1988). *Chimia*, *42*: 9.

Schneider, W., and Forth, W. (1987). *Arzneim.-Forsch.-Drug Res.*, *37*: 89.

Schwarz, A., and Steinstrasser, A. (1987). *J. Nucl. Med.*, *28*: 721.

Spiro, T.G., and Saltman, P. (1969). *Struct. Bonding*, *6*: 116.

Stocklin, G. (1992). *Eur. J. Nucl. Med.*, *19*: 527.

Sugiura, Y., Akita, T., and Umezawa, H. (1980). *Metal Ions in Biological Systems*, Vol. 19 (H. Sigel, ed.), Marcel Dekker Inc., New York, p. 81.

Sundberg, M.W., Meares, C.F., Goodwin, D.A., and Diamanti, C.I. (1974). *J. Med. Chem.*, *17*: 1304.

Ter-Pogossian, M.M., Phelps, M.E., and Hoffman, E.J. (1975). *Radiology*, *114*: 89.

Vallabhajosula, S.R., Harwig, J.F., Siemsen, J.K., and Wolf, W. (1980). *J. Nucl. Med.*, *21*: 650.

Wagner, H.N. (1987). *J. Nucl. Med.*, *28*: 1235.

Wagner, H.N. (1989). *J. Nucl. Med.*, *30*: 1283.

Wagner, H.N. (1992). *J. Nucl. Med.*, *33*: 10N.

Walovitch, R., Makuch, J., Garrity, S., Burgess, B., Morgan, R., Williams, S., Leveille, J., Douesnard, J., Rigo, P., Demonceau, G., Cantineau, R., and De Roo, M. (1988a). *Neurology*, *38*: 363.

Walovitch, R., Makuch, J., Knapik, G., Watson, A.D., and Williams, S.J. (1988b). *J. Nucl. Med.*, *29*: 747.

Watkinson, J.C., Lazarus, C.R., Mistry, R., Shaheen, O.H., Maisey, M.N., and Clarke, S.E. (1989). *J. Nucl. Med.*, *30*: 174.

Watkinson, J.C., Allen, S., Lazarus, C.R., Sinclair, J., Blake, G.M., and Clarke, S.E. (1990). *Nucl. Med. Commun.*, *11*: 343.

Weiner, R.E. (1989). *J. Nucl. Med.*, *30*: 70.

Yokoyama, A. (1985a). "Recent Progress in the Development of Bifunctional Radiopharmaceuticals", Proceedings of IAEA '84 on Radiopharmaceuticals and Labelled Compounds, Tokyo, pp. 251-265.

Yokoyama, A., and Saji, H. (1980). *Metal Ions in Biological Systems*, Vol. 10 (H. Sigel, ed.), Marcel Dekker, New York, p. 313.

Yokoyama, A., Terauchi, Y., Horiuchi, K., Okumura, S., Saito, Y.,Tanaka, H., Odori, T., Morita, R., Mori, T., and Torizuka, K. (1978). *Int. J. Appl. Rad. Isot.*, *29*: 549.

Yokoyama, A., Ohmomo, Y., Horiuchi, K., Saji, H., Tanaka, H., Yamamoto, K., Ishii, Y., and Torizuka, K. (1982). *J. Nucl. Med.*, *23*: 909.

Yokoyama, A., Hata, N., Horiuchi, K., Masuda, H., Saji, H., Ohta, H., Yamamoto, K., Endo, K., and Torizuka, K. (1985b). *Int. J. Nucl. Med. Biol.*, *12*: 273.

GALLIUM AND INDIUM COMPLEXES AS RADIOPHARMACEUTICALS

by

Arthur E. Martell

Department of Chemistry,
Texas A&M University,
College Station, TX 77843-3255
U.S.A.

and

Michael J. Welch

Division of Radiation Sciences,
The Edward Mallinckrodt Institute of Radiology,
Washington University School of Medicine,
St. Louis, Missouri 63110
U.S.A.

I. INTRODUCTION

There is great interest in the nuclear medicine application of the radionuclides ^{67}Ga, ^{68}Ga and ^{111}In. Gallium-67 and ^{111}In with half-lives of 77.9 hours and 67.9 hours respectively are currently utilized for such applications as tumor scanning (Edward and Hayes, 1969; Thakur et al., 1973), abscess localization (Lavender et al., 1971; Thakur et al., 1976), and thrombus detection (Thakur et al., 1975). Indium-111 is the metal radionuclide of choice for the labeling of monoclonal antibodies for diagnosis (Zalutsky, 1989).

Radiopharmaceuticals labeled with positron-emitting radionuclides are of particular importance because the distribution in humans can be quantitated in vivo utilizing positron-emission tomography (PET) (Burnham et al., 1970; Ter-Pogossian et al., 1975; Harper, 1977). The majority of positron-emitting radiopharmaceuticals in use today required an in-house cyclotron for their production. The $^{68}Ge/^{68}Ga$ generator system with a parent half-life of 288 days and a daughter half-life of 68 minutes is one of a very limited number of generator systems that will produce a short-lived positron-emitting radionuclide (Green and Tucker, 1961; Green and Welch, 1989). This generator is commercially available (DuPont Medical Products), and a series of gallium-68 radiopharmaceuticals have been described in the literature. Investigators in pulmonary research at Washington University Medical Center are at present evaluating ^{68}Ga-labeled proteins for the measurement of pulmonary transcapillary exchange rate (Mintun et al., 1987; Calandrino et al., 1988) and ^{68}Ga-labeled macroaggregrated albumin for the quantitation of pulmonary blood flow (Mintun et al., 1986). Previously ^{68}Ga-1,1,1-tris(5-methoxysalicylaldiminomethyl)ethane [(5-MeOSal)$_3$ TAME] has been developed as an agent which can be used to quantitate myocardial blood flow (Green et al., 1985). This agent, however, suffers from the fact that clearance from the myocardium is rapid and a blood pool subtraction technique has to be used. The development of an agent where clearance from the blood is rapid would again greatly enhance the application of ^{68}Ga-labeled radiopharmaceuticals. The application of ^{68}Ga would be greatly increased if agents could be developed which would readily penetrate the blood brain barrier (BBB) and quantitate cerebral blood flow (CBF). There has been great interest in the development of agents labeled with metal radionuclides which cross the BBB and are trapped in the brain. Agents which have been developed include Cu(II) pyruvaldehyde bis(N^4-methylthiosemicarbazone) [CuPTSM], (Green, 1987; Mathias et al., 1990c) as well as a series of technetium agents: ^{99m}Tc-hexamethylpropyleneamine oxime (HMPAO), (Nowotnik et al., 1985) ^{99m}Tc-L,L-ethyl cysteinate dimer (ECD), (Vallabhajosula et al., 1989) ^{99m}Tc-[Bis[2,3-butanedionedioximateo(1-)0][2,3-butanedionedioximato(2-)-0]2-methyl, 1-propylborato-(2-)N'N"N'"N""N'""]-chlorotechnetium (SQ32,097) (Linder et al., 1987). The cerebral extraction and retention of all of these agents has been measured in the primate brain using a technique described in the literature (Raichle et al., 1976; Dischino et al., 1983). It has been found that although they all do cross the BBB they underestimate CBF at high flow rates (Mathias et al., 1988b).

II. SYNTHESIS AND EVALUATION OF NEW LOW-MOLAR-MASS LIGANDS

In this area, work has been carried out on the development of ligands with two types of structures. One type is based upon the HBED/PLED series and the second based on substituted triazacylononanes. Following the synthesis and evaluation of substituted PLED and HBED complexes (Mathias et al., 1988a), three derivatives were synthesized where the charge was anticipated to be significantly different (Sun et al., 1991b). The formal charge on these ligands varies from -1 to 0 to +1. It was anticipated that with the varying charges, different biodistributions would be observed. Although the molecules have different formal charges, the biological behavior of all three were found to be similar (Sun et al., 1991b). Even though the formal charge on these complexes is different, various specific areas of the molecule have positive charges and others negative charges. Also at physiological pH, species with different degrees of protonation exist in aqueous solution. Thus biological distribution appears to depend on several factors in addition to the overall charge.

As a logical extension, an attempt was made to reduce the overall charge on small complexes utilizing different coordination groups. The first approach was to prepare a hexachelating ligand with a $N_3O_3^{3-}$ core which would produce neutral species with trivalent

Table 1

Biodistribution of Complexes with TACN-TS expressed as %ID/g ($\bar{x}$ ± s.d.)*

	In-111			Ga-68		
	1 min	5 min	1 hr	1 min	5 min	1 hr
Blood	1.76 ± 0.01	1.12 ± 0.07	0.19 ± 0.02	4.34 ± 1.23	1.98 ± 0.48	0.68 ± 0.28
Liver	2.08 ± 0.17	2.61 ± 0.22	0.77 ± 0.16	1.60 ± 0.21	2.44 ± 0.24	1.14 ± 0.15
Kidney	5.23 ± 1.16	6.86 ± 0.55	1.48 ± 0.18	5.95 ± 1.77	9.73 ± 2.18	3.19 ± 0.62
Muscle	0.25 ± 0.02	0.26 ± 0.03	0.24 ± 0.04	0.40 ± 0.07	0.38 ± 0.04	0.22 ± 0.03
Heart	0.77 ± 0.09	0.60 ± 0.09	0.24 ± 0.02	1.38 ± 0.28	0.79 ± 0.06	0.30 ± 0.07
Brain	0.07 ± 0.02	0.04 ± 0.01	0.01 ± 0.03	0.11 ± 0.04	0.07 ± 0.01	0.02 ± 0.01

* Performed in adult Sprague Dawley rats (n=5).

metals, (Moore et al., 1989) resulting in the synthesis of N,N',N"-tris(3,5-dimethyl-2-hydroxybenzyl)-1,4,7-triazacyclononane (TACN-TX).

Also synthesized and characterized was N,N',N"-tris(2-mercaptoethyl)-1,4,7-triazacyclononane (TACN-TS) (Moore et al., 1990). This complex has a $N_3S_3^{3-}$ core, avoiding unwanted size, charge and hydrophilic properties. Biological data (Table 1), shows that even this complex does not behave in vivo as a lipophilic complex. In vivo behavior of the indium and gallium complexes is likely due to the presence of the facial arrangement of thiolates about the metal center, an arrangement which may create a substantial dipole moment for the complex and account for the observed large hydrophilicity. The affinities of TACN-TS for indium(III) and gallium(III) were determined by potentiometric techniques. The determined values were used to calculate pM values for the complexes (where pM is -log [free M] in equilibrium with 10% excess ligand at physiological pH = 7.4, and at a total concentration of 1 x 10^{-6} M). The pM values of Ga^{3+} and In^{3+} are 25.2 and 26.9, respectively. These compare with Ga^{3+} values of 26.8 and 24.5 and In^{3+} values of 18.7 and 18.8 for the pentachelate ethylenediamine-based ligands, SHBED and PLED, respectively. The potentiometric data show that the complexes exist as neutral species at physiological pH, where the ligand is completely deprotonated and fully chelating the metal. The amounts of various species are shown in Figures 1 and 2 at equimolar concentration of metal and ligand (10^{-3}M).

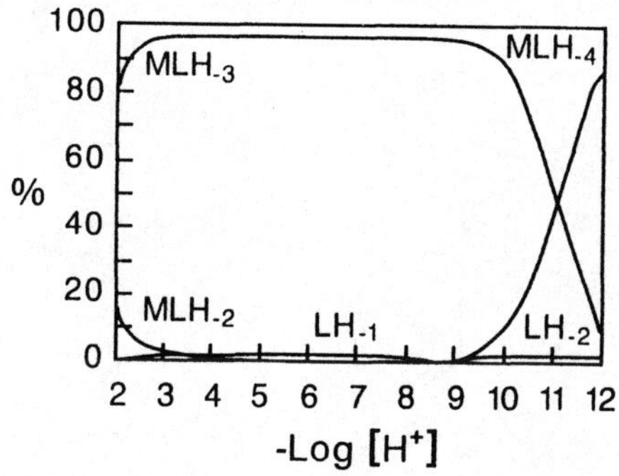

Figure 1.
The species anticipated to be present in an equimolar mixture of indium and ligand TS-TACN.

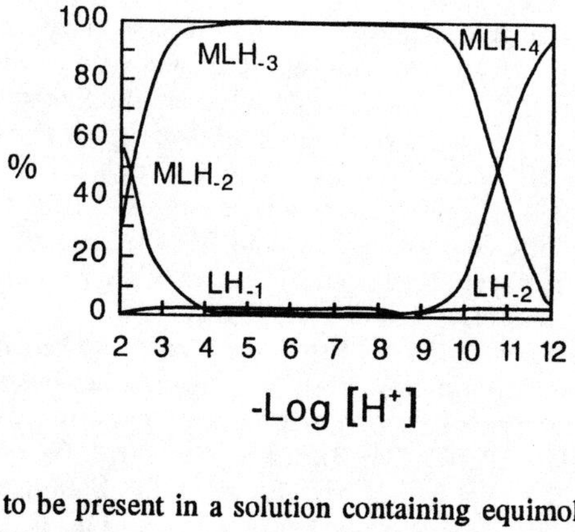

Figure 2.
The species anticipated to be present in a solution containing equimolar mixture of gallium and TS-TACN (10^{-3}M).

Table 2

Biodistribution of THM_2BED Complexes at 1 Hour Post Injection. Expressed as % ID.g ($\overline{x}$ ± s.d.:n = 5)

	Fe(III)	In(III)	Ga(III)
Blood	0.53 ± 0.08	3.39 ± 1.27	0.30 ± 0.02
Lung	1.91 ± 0.29	1.72 ± 0.40	1.74 ± 0.37
Liver	4.83 ± 0.52	0.94 ± 0.19	6.53 ± 0.66
Spleen	1.51 ± 0.18	0.68 ± 0.25	2.84 ± 1.02
Kidney	1.11 ± 0.13	3.99 ± 1.52	0.83 ± 0.10
Heart	0.72 ± 0.1	1.03 ± 0.03	0.71 ± 0.20
Brain	0.05 ± 0.005	0.10 ± 0.04	0.023 ± 0.003

Also synthesized were the ligands tetrakis-(2-hydroxy-3,5-methylbenzyl)ethylenediamine (THM_2BED) and tetrakis-(2-hydroxy-5-methylbenzyl)ethylenediamine (THMBED) as more lipophilic tracers (Scheme 1) (Madsen et al., 1992). These agents do have interesting biological behavior; complexes of the trivalent metals indium, gallium and iron with these two ligands show different biodistributions (Table 2).

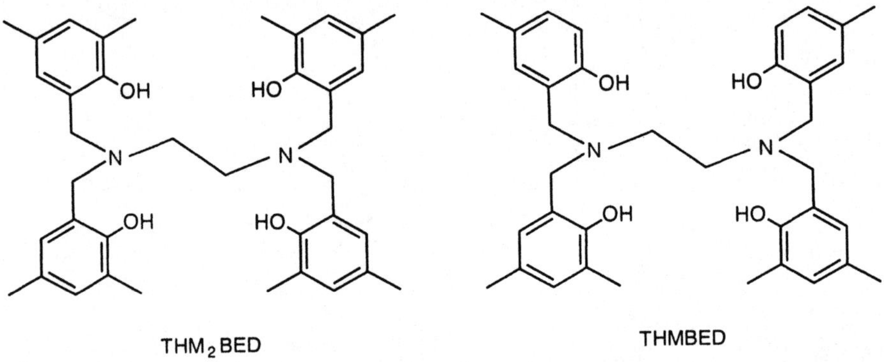

Scheme 1

These data show: (i) significant differences between the in vivo behavior of Fe(III), In(III) and Ga(III); (ii) with ^{67}Ga-THM_2BED a heart to blood ratio of ca 2.8 is observed 30 minutes after administration and the myocardial activity remains high over long-time frames. The myocardial uptake and retention of the compound indicates that this agent shows promise as a myocardial perfusion agent. Other agents with potential for use as myocardial perfusion agents have also been described in the literature (Green, 1990; Kung et al., 1990). At times >48 hours, significant brain retention of this compound is observed. Differences between Fe(III), In(III), and Ga(III) complexes were also seen in studies of the EHPG complexes (Kung et al., 1990); these agents can be utilized both for imaging agents (with In and Ga) and NMR contrast agents (with Fe). An example of these differences is given (Kung et al., 1990) where significant differences in liver uptake are observed between the three metal complexes of *meso* and *racemic* EHPG.

As an example of ligand design based on feedback from previous work in both thermodynamic and biodistribution studies, our laboratories have reported a series of ligands containing a new type of donor group having high affinity for trivalent metal ions (Ga(III), In(III), Fe(III), Al(III), Gd(III), etc), the 3-hydroxy-6-methyl-2-pyridylmethyl donor group. Thus far this new hydroxypyridyl group has been used to produce variations on PLED and on N,N',N"-tris(hydroxybenzyl)triazanonane (TACN-HB) described by

Weighardt and coworkers (Auerback et al., 1990). The new compounds, EDDA-HP, HPMPT, and TACN-HP are illustrated in Scheme 2.

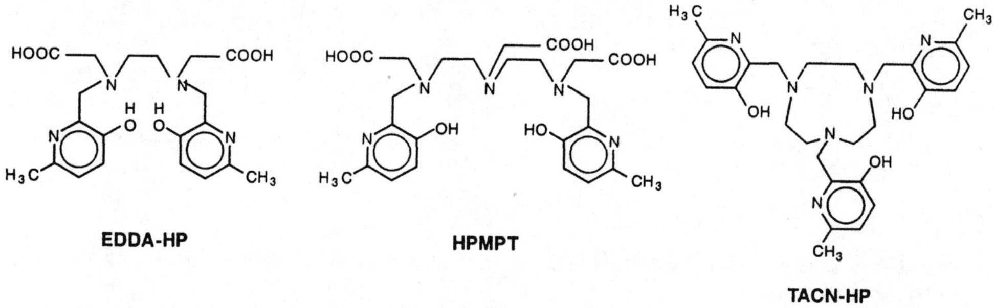

Scheme 2

The development of these new ligands represents an important breakthrough in the achievement of both exceptionally high stability constants and very high metal-binding efficiency. In fact, it has just been discovered that TACN-HP binds the ferric ion more strongly than any other known ligand, including enterobactin, at physiological pH. This unprecedented coordinating effectiveness can be explained by several favorable factors: (i) the coordinating hydroxyl groups have very low pK's because of the competing protonation of the pyridine nitrogens; (ii) there is no coordination inhibition of the hydroxyl groups by the sterically-demanding methyl groups present in analogous pyridoxyl derivatives such as PLED; (iii) the ring nitrogens (in TACN-HP) are part of the tightly-fitting coordinating framework and greatly assist metal coordination, (Sun et al., 1991a; Martell et al., 1990) in contrast to the macrocyclic rings in enterobactin models, and enterobactin itself, which serve merely as frameworks on which to attach pendant coordinating groups.

III. NEW CHELATING LIGANDS

The design of new ligands for improving the performance of metal chelate diagnostic agents involved the following guidelines: (i) the charge on the complex should be very small or zero, (ii) thermodynamic stability should be maximized, and (iii) hydrophilicity should be decreased. Also, with the working hypothesis that carboxyl groups are probably metabolized and destroyed in vivo, non-carboxylic donors such as phenolic and sulfhydryl groups will be preferred, as well as negative oxygen donors based on hydroxypyridines, hydroxamates, and hydroxypyridinones.

A. New ligands for Ga(III) and In(III)

The recent publication of the first cryptand with endocyclic hydroxamate groups provides a new ligand, **1**, (Sun and Martell, 1989) which meets the general specifications listed above. Preliminary molecular mechanics calculations on the Fe(III) chelate show that the fit of this ligand somewhat better around the metal center is not perfect, with the bridging groups on the left too short, and those on the right too long. While the macrobicyclic (cryptand) ligand may require readjustment, this is not as true for the open-chain hydroxamates, **2** and **3**, (Motekaitis et al., 1991) which may be viewed as essentially the left and right halves of the cryptand. Their greater flexibility will allow a closer fit of the donor groups to the metal ion, making up somewhat for the lower degree of pre-organization. Here again, steric effects due to less than optimum positioning of the hydroxamate oxygens can be fine-tuned by changing the lengths of the connecting chains, as is indicated by a, b, c in formula **2**, and l, m, n, formula **3**. It should be noted that ligands **1**, **2**, and **3** can be derivatized to produce bifunctional ligands by substituting the covalent linkage groups in the position labeled X.

In(III), with considerably higher effective radius than Ga(III), requires a larger cavity in **1**, and slightly different chain lengths in **2** and **3**. It is expected that the variations in

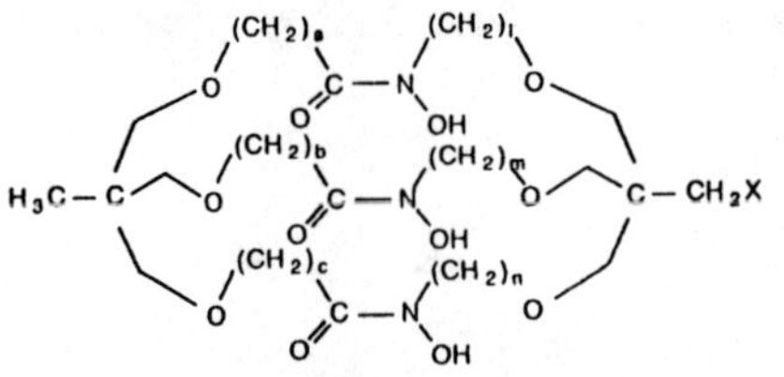

1 Trishydroxamate cryptand
a = b = c = 2, l = m = n = 3, X = H

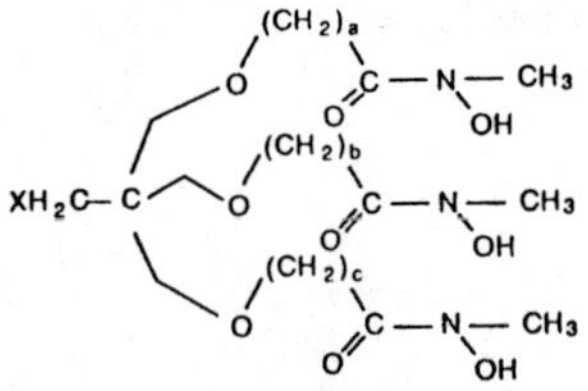

2 Trishydroxamate
a = b = c = 2, X = H

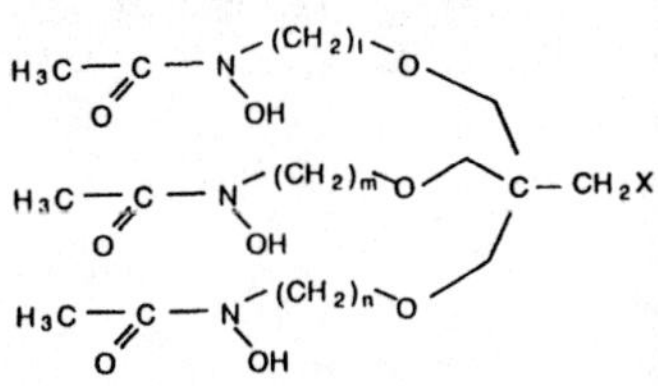

3 Trishydroxamate, l = m = n = 3, X = H

Formulas **1, 2, 3**

ligand structure described above should be able to satisfy the "best-fit" requirements for In(III) as well as Ga(III).

In a recent paper describing some new synthetic open-chain trishydroxamates, Ng et al. (1989) pointed out that the nature of the bridgehead is important (a flexible group is required) in maximizing affinity for the Fe(III) ion. The ligands indicated by **1, 2, 3,** above meet this suggested requirement quite effectively.

B. Phenolic groups

Many of the new ligands developed contain phenolic and carboxylate groups. The ligand structures can be modified by using other donor groups: *o*-hydroxybenzyl(phenolic), mercaptoethyl, and 3-hydroxy-2-pyridyl. The last replaces, and is superior to, the pyridoxyl donor groups previously employed. Formulas **4, 5,** and **6** are EDTA and NTA analogs in which all carboxylate donor groups are replaced by *o*-hydroxybenzyl and 3-hydroxy-6-methyl-2-pyridyl groups. The ligand THM$_2$BED has been found in preliminary studies to form a Ga(III) chelate which has some BBB permeability. It is not known, however, if all of the phenolic groups in **4** are coordinated to the metal ion, and potentiometric and spectrophotometric studies are being initiated to settle this important question, which has a bearing on whether **5** could be highly effective for Ga(III). It has the advantage of forming a neutral complex of fairly low molar mass (0.335 kg mol^{-1}). While its four donor groups would probably have tetrahedral geometry, instead of octahedral, it should be kept in mind that one of the most stable Ga(III) complexes of all, Ga(OH)$_4^-$, has tetrahedral geometry. For application as a blood flow tracer, only limited in vivo stability is needed.

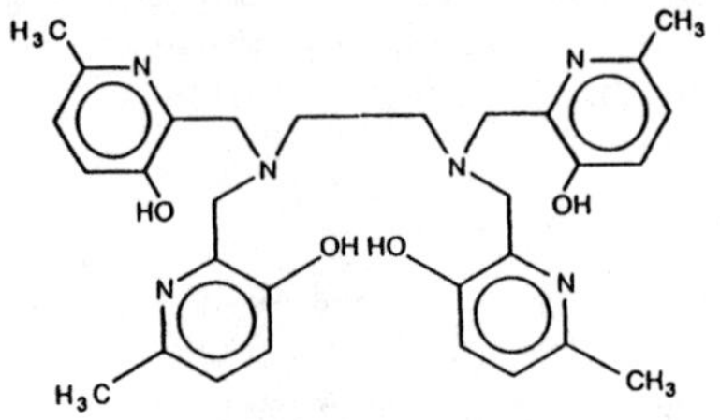

4 N,N,N',N'-tetrakis
(3-hydrosy-6-methyl
2-pyridylmethyl) ethylenediamine

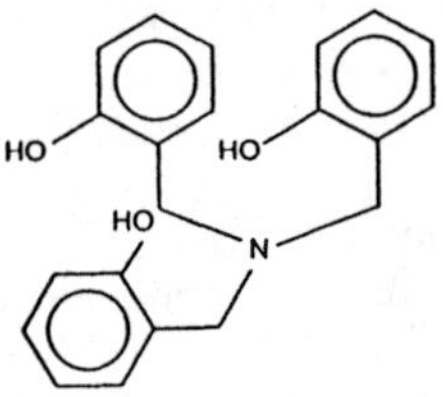

5 tris(o-hydroxybenzyl)amine

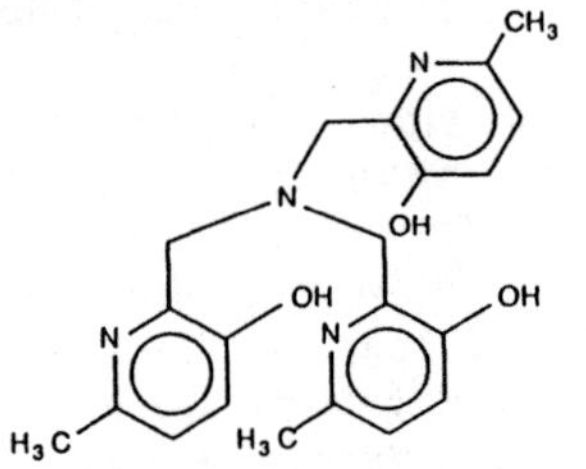

6 tris(3-hydroxy-6-methyl-2-pyridylmethyl)amine

Formulas **4, 5, 6**

C. Hydroxypyridinones

The seemingly successful application of 1,2-dimethyl-3-hydroxy-4-pyridinone for the treatment of iron(III) overload (Kontoghiorghes, 1985; Taylor and Kontoghiorghes, 1986; Kontoghiorghes and Hoffbrand, 1988) has resulted in the development of considerable interest in the study of 3-hydroxy-2- and 4-pyridinones, **7** and **8**, as well as the cyclic hydroxamic acid 1-hydroxy-2-pyridinones, **9**. In spite of this interest, and incorporation of these functional groups into multidentate structures, the binding strengths of their complexes with trivalent metal ions are not yet known (with the exception of reports on Fe(III) and In(III) binding by **9** (Kontoghiorghes, 1982; Nelson et al., 1988a; 1988b).

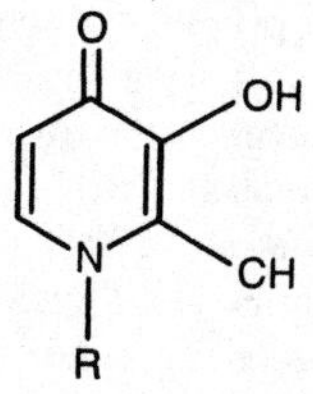

7 hydroxy-4-pyridinones 1,2-dimethyl-3-hydroxy-4-pyridinone, R-CH_3

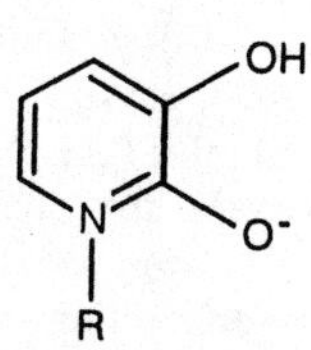

8 hydroxy-2-pyridinones 1-methyl-3-hydroxy-2-pyridinone, R=CH_3

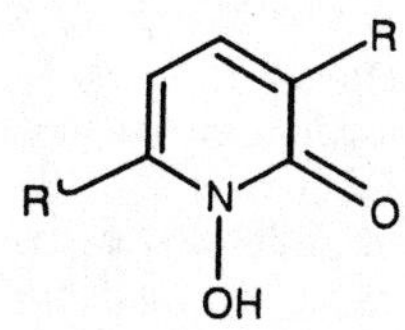

9 Substituted 1-hydroxy 2-pyridinone

Formulas **7, 8, 9**

The ability of these bidentate ligands to form stable tris complexes of Fe(III) (Scarrow and Raymond, 1988), Ga(III) and In(III) (Matsuba et al., 1988) suggests that they should be seriously considered for the formation of complexes of Ga(III) and In(III) as radiopharmaceuticals. The main problem with the lack of stability constant data seems to be the complete formation of the 1:1 (bidentate) Fe(III) chelate in acid solution, making determination of its formation constant by direct measurement impossible. In our hands, EDTA strongly competes with the 1:1 complex in acid solution, so that stability determination by a competition method is feasible.

It has been shown (Martell, 1981) that bidentate ligands with very high intrinsic metal ion affinities form 3:1 (completely coordinated) metal complexes that tend to dissociate in dilute solution because their complex formation equilibria are highly concentration dependent.

This problem was first pointed out by Martell (Nelson et al., 1988a; Martell and Motekaitis, 1989) for the complexation of Fe(III) by **7**, and the incorporation of three molecules of the bidentate ligand into a single hexadentate 6-coordinate ligand was suggested. A similar observation was made by Porter and coworkers, (Streater et al., 1990) who reported the synthesis of **10a** and **10b**.

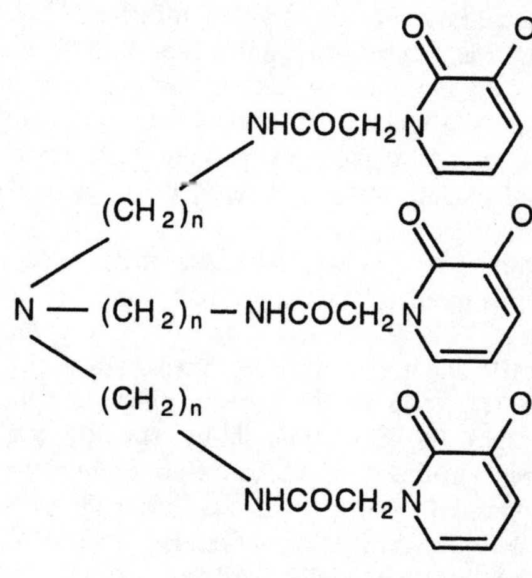

10a, n = 2; 10b, n = 3

Formulas **10a** and **10b**

D. Bifunctional chelates

One of the major applications of indium-111 has been of macromolecules labeled with bifunctional chelates. Because of the high stability constant of the transferrin complexes (Harris and Pecoraro, 1983; Welch and Welch, 1975) and the large amount of transferrin in human plasma (0.2 mg/100ml), indium and gallium complexes must have very high stability constants in order to survive in the plasma. Sundberg et al. (1974) prepared the initial bifunctional chelate 1-(ρ-benzenediazonium)ethylenediamine-N,N',N'-tetraacetic acid. This compound forms a bond between the metal-ligand complex and a protein via the diazo group. Many other approaches to labeling proteins and antibodies have been investigated. However, the majority of these require the attachment of either diethylenetriaminepentaacetic acid (DTPA) or ethylenediaminetetraacetic acid (EDTA) (Meares et al., 1976; 1984; Krejcarek and Tucker, 1977; Eckelman et al., 1975; Najafi et al., 1984). Desferrioxamine has also been used to attach indium to antibodies (Yokoyama et al., 1982). Although radiolabeled antibodies are being used extensively for tumor imaging and studying myocardial and vascular disorders, one of the major problems is that a significant fraction of the radioactivity is taken up and retained in the liver (DeNardo et al., 1985; Otsuka and Welch, 1985). This is in contrast to ^{125}I-labeled antibodies where the radioactivity is rapidly excreted. The exact mechanism of this liver "trapping" is the subject of some debate; several mechanisms have been postulated including:

(i) The complex undergoes exchange with an intracellular binding site which retains the metal;

(ii) The bond with which the chelate is attached to the antibody is resistant to enzymatic degradation and so the chelate remains attached to a relatively large peptide fragment which cannot escape from the liver cell;

(iii) Portions of the chelate (possibly the acetic acid) are themselves enzymatically degraded in the liver and the strongly bonding chelate ceases to exist;

(iv) Low-molar-mass metal-labeled chelates themselves are not retained in the liver, which suggests, that the "trapping" phenomenon is a function of the antibody and/or the technique used in labeling.

There is experimental evidence to support each of the above explanations. Meares and coworkers have carried out elegant studies utilizing EDTA bound to antibodies which suggest that (ii) is the correct explanation (Deshpande et al., 1988). Studies delivering double-labeled In-DTPA (^{111}In and ^{14}C-DTPA) to the liver in liposomes (MLV) have been carried out. The ^{14}C from the DTPA is found in the urine in forms other than DTPA and the ^{111}In and ^{14}C have different biodistributions (Mathias and Welch, 1987), which suggests either mechanism (i) or (iii). Other workers have carried out in vivo studies where In(III) and antibodies have very similar biodistributions and, therefore, claim mechanisms (iv) (Brechbiel et al., 1986; Carney et al., 1989). One of the major goals of current research is to attempt to determine the chemical constitutional factors which influence liver retention by one or more of the above mechanisms. This will enable the logical design and development of new bifunctional chelates for the enhanced non-target organ clearance and an overall reduction of radiation exposure. The work completed to date suggests that more lipophilic ligands may enhance the liver clearance of the metal and that high thermodynamic stability may be essential for small metal chelate molecules, used directly, or formed by breaking away from the antibody.

An important objective was to determine the stability constant of the ligands of interest and to determine their structure-biologication relationship. Motekaitis et al. (1990) represents some of our work in this area. Features observed from this work are: (i) the order in stability constants between Ga(III) and In(III) chelates varies when one switches from carboxylic ligands of the NTA, EDTA, and DTPA type to phenolic ligands based upon HBED or PLED; (ii) depending upon the magnitude of the stability constant, the ligand either clears through or is "trapped" in the liver of rats. New bifunctional chelates for protein and antibody labeling based upon HBED can be directed clear through the liver, and liver clearances is a major problem with indium-labeled antibodies. Several bifunctional chelates with potential applications for labeling antibodies and proteins have been synthesized. These chelates are listed in Scheme 3 and have been evaluated in animal models (Mathias et al., 1990a; 1990b; Schwarz et al., 1993).

Several in vivo models are used in our laboratories for the evaluation of radiolabeled antibodies. The first of these used a human parathyroid antibody (BB5) which is evaluated in nude mice (Schwarz et al., 1993). In this model, enlarged human parathyroid glands are minced into small pieces and implanted in intramuscular pockets in the flanks of nude mice. In some experiments, ca 10^9 cells of a nonspecific breast tumor carcinoma are implanted in the contralateral flank. Three to six weeks are allowed for basularity to develop before the

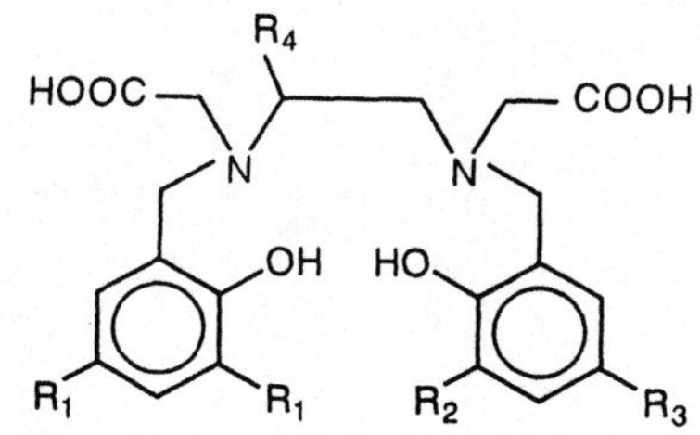

	R_1	R_2	R_3	R_4
HBED	H	H	H	
Me_4HBED	CH_3	CH_3	CH_3	
$BrMe_2$HBED	CH_3	H	$NHCOCH_2Br$	
BrHBED	H	H	H	$-CH_2$=$NHCOCH_2Br$
$BrMe_4$+HBED	CH_3	CH_3	CH_3	$-CH_2$=$NHCOCH_2Br$

Scheme 3

in vivo studies are carried out. Radiolabeled BB5 (or antibody fragment) were injected into the mouse tail vein and the animal sacrificed various times after injection.

The second animal model involves human colorectal carcinoma cells (GW39) in hamsters (Mathias et al., 1990a; 1990b). In this case, an in vitro assay is available for evaluating the immunoreactivity of the antibody (1A3). An in vitro immunoreactivity assay was developed so each radiolabeled antibody preparation can be tested prior to in vivo use. The direct binding assay uses GW39 cell suspensions as the target, and conditions of antigen excess.

The antibodies labeled with the bifunctional chelates $BrMe_2$HBED and BrdHBED appeared to have significant advantages over bifunctional chelates currently in use (Mathias et al., 1990a; 1990b; Schwarz et al., 1993). Disappointingly these chelates still do not enable the In(III) to be cleared from the liver. MAbs utilizing these chelates have been evaluated in humans (Philpott et al., 1994).

E. Bifunctional chelate linkages

The present authors have developed three new bifunctional chelates which are derived from HBED. In all cases the attachment to the antibody has utilized the bromoacetamide linkage initially developed by Meares (Meares et al., 1984). The bifunctional chelates have been attached to intact antibodies, presumably through free amino groups (the reactivity of the bromoacetamide group is much greater with sulfhydryl groups) (Mathies et al., 1990b). Different means of attaching the currently developed agents to antibodies will be investigated. Meares and coworkers have increased the labeling yield obtained with bromoacetamide-derivatized EDTA by reacting the antibody with 2-iminothiolane (Trauts' Reagent), which introduces a charged linker with a terminal SH group (Moi et al., 1985) (Scheme 4). Antibodies labeled in this manner will be investigated. The amino precursors that have been used to prepare the bromoacetamide derivatives will also be converted into isothiocyanate groups which react preferentially with free amino groups (Meares et al., 1984). Even though the chelates developed to date have some advantages over the ones in general use, the primary goal of developing an agent that clears the radioactivity from the body has not been achieved. To date this goal has not been achieved by others. Deshpande et al. (1989) have compared four different conjugation methods, and showed that a disulfide linker had the lowest liver uptake of activity in the kidney and the lowest immunoreactivity. A research group from the University of Durham, England, has been working independently on agents of this type (Craig et al., 1989; Cox et al., 1989; Murphy et al., 1988).

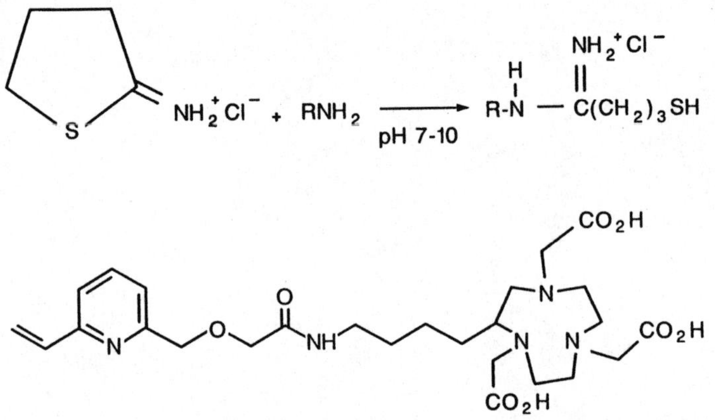

Schemes 4 and 5

Craig et al. (1989) have synthesized an activated bifunctional derivative (Scheme 5) of a triazacyclononane precursor by a ten-step synthesis. Rodwell and coworkers (Rodwell et al., 1986; Alvarez et al., 1986) have described a technique whereby the oligosaccharide moiety is oxidized to aldehydes which are then coupled to chelating groups utilizing sodium cyanoborohydride reduction. This technique reduces the amount of liver uptake of the agent.

ACKNOWLEDGEMENTS

The authors' work in this area is supported by NIH grant CA42925.

REFERENCES

Alvarez, V.L., Wen, M-L., and Lee, C., Lopez, A.D., Rodwell, J.D., and McKearn, T.J. (1986). *Nucl. Med. Biol.*, *13*: 347.

Auerback, U., Eckert, U., Wieghardt, K., Nuber, B., and Weiss, J. (1990). *Inorg. Chem.*, *29*: 938.

Brechbiel, M.W., Gansow, O.A., Atcher, R.W., Schlom, J., Esteban, J., and Simpson de Colcher, D. (1986). *Inorg. Chem.*, *25*: 2772.

Burnham, C.A., Aronow, S., and Brownell, G.L. (1970). *Phys. Med. Biol.*, *15*: 517.

Calandrino, F.S., Jr., Anderson, D.J., Mintun, M.A., and Schuster, D.P. (1988). *Amer. Rev. Resp. Dis.*, *138*: 421.

Carney, P.L., Rogers, P.E., and Johnson, K.D. (1989). *J. Nucl. Med.*, *30*: 374.

Cox, J.P.L., Jankowski, K.J., Kataky, R., Parker, D., Beeley, N.R.A., Boyce, B.A., Eaton, M.A.W., Miller, K., Millican, A.T., Harrison, A., and Walker, C. (1989). *J. Chem. Soc. Chem. Commun.*, 797.

Craig, A.S., Helps, I.M., Jankowski, K.J., Parker, D., Beeley, N.R.A., Boyce, B.A., Eaton, M.A.W., Millican, A.T., Millar, K., Phipps, A., Rhind, S.K., Harrison, A., and Walker, C. (1989). *J. Chem. Soc. Chem. Commun.*, 794.

DeNardo, G.L., DeNardo, S.J., Peng, J.S., O'Grady, L.F., Mills, S.L., Epstein, A.L., and Cardiff, R.D. (1985). *J. Nucl. Med.*, *26*: P67.

Deshpande, S.V., Denardo, S.J., Meares, C.F., McCall, M.J., and DeNardo, G.L. (1988) *J. Nucl. Med.*, *29*: P824.

Deshpande, S.V., Denardo, S.J., Meares, C.F., McCall, M.J., Adams, G.P., and Denardo, G.L. (1989). *Nucl. Med. Biol.*, *16*: 587.

Dischino, D.D., Welch, M.J., Kilbourn, M.R., and Raichle, M.E. (1983). *J. Nucl. Med.*, *24*: 1030.

Eckelman, W.C., Karesh, S.M., and Reba, R.C. (1975). *J. Pharm. Sci.*, *64*: 704.

Edward, C.L., and Hayes, R.L. (1969). *J. Nucl. Med.*, *10*: 103.

Green, M.A. (1987). *Int. J. Radiat. Appl. Instrum., Part B, 14*: 59.

Green, M.A. (1990). *J. Label Compound Radiopharm., 23*: 1227.

Green, M.A., and Welch, M.J. (1989). *J. Nucl. Med. Biol., 16*: 435.

Green, M.A., Welch, M.J., Mathias, C.J., Fox, K.A.A., Knabb, R.M., and Hoffman, J.C. (1985). *J. Nucl. Med., 26*: 170.

Green, M.W., and Tucker, W.D. (1961). *Int. J. Appl. Radiat. Isotop., 12*: 62.

Harper, P.V. (1977). *Int. J. Appl. Radiat. Isotop., 28*: 5.

Harris, W.R., and Pecoraro, V.L. (1983). *Biochemistry, 22*: 292.

Kontoghiorghes, G.J. (1982). *PhD Dissertation*, University of Essex.

Kontoghiorghes, G.J. (1985). *Lancet, 1*: 817.

Kontoghiorghes, G.J., and Hoffbrand, A.V. (1988). *Recent Adv. Haematol., 5*: 75.

Krejcarek, G.E., and Tucker, K.L. (1977). *Biochem. Biophys. Res. Commun., 77*: 581.

Kung, H.F., Liu, B., Mankoff, D., Kung, M.P., and Billings, J. (1990). *J. Nucl. Med., 31*: 727.

Lavender, J.P., Lowe, J., Barker, J.R., Burn, M.B., and Chaudhri, M.A. (1971). *Brit. J. Radiol., 44*: 361.

Linder, K., Feld, T., Juri, P.N., Nunn, A.D., and Treher, E. (1987). *J. Nucl. Med., 28*: 592.

Madsen, S.L., Welch, M.J., Motekaitis, R.J., and Martell, A.E. (1992). *Nucl. Med. Biol., 19*: 431.

Martell, A.E. (1981). *Development of Iron Chelators for Clinical Use* (A.E. Martell, W.F. Anderson and D.G. Badman, eds.), Elsevier, New York, p. 67.

Martell, A.E., and Motekaitis, R.J. (1989). Proceedings of the International Conference on Oral Chelation in the Treatment of Thalossaemia and Other Diseases, London.

Martell, A.E., Motekaitis, R.J., and Welch, M.J. (1990). *J. Chem. Soc. Chem. Commun.*, 1748.

Mathias, C.J., and Welch, M.J. (1987). *J. Nucl. Med., 28*: 657.

Mathias, C.J., Sun, Y., Welch, M.J., Green, M.A., Thomas, J.A., Wade, K.R., and Martell, A.E. (1988a). *Nucl. Med. Biol., 15*: 69.

Mathias, C.J., Welch, M.J., Lich, L., Raichle, M.E., Volkert, W.A., and Hung, J.C. (1988b). *J. Nucl. Med., 29*: 747.

Mathias, C.J., Sun, Y., Welch, M.J., Connett, J.M., Philpott, G.W., and Martell, A.E. (1990a). *Bioconjugate Chem., 1*: 204.

Mathias, C.J., Sun, Y., Welch, M.J., Connett, J.M., Philpott, G.W., and Martell, A.E. (1990b). *Inorg. Chem., 29*: 1475.

Mathias, C.J., Welch, M.J., Raichle, M.E., Mintun, M.A., Lich, L.L., McGuire, A.H., Zinn, K.R., John, E.K., and Green, M.A. (1990c). *J. Nucl. Med., 31*: 351.

Matsuba, C.A., Nelson, W.O., Rettign, S.J., and Orvig, C. (1988). *Inorg. Chem., 27*: 3935.

Meares, C.F., Goodwin, D.A., Leung, C.S.H., Girgis, A.Y., Silvester, D.J., Nunn, A.D., and Lavender, P.J. (1976). *Proc. Nat. Acad. Sci. USA, 73*: 3803.

Meares, C.F., McCall, M.J., Reardan, D.T., Goodwin, D.A., Diamanti, C.I., and McTigue, M. (1984). *Anal. Biochem., 142*: 68.

Mintun, M.A., Ter-Pogossian, M.M., Green, M.A., Lich, L.L., and Schuster, D.P. (1986). *J. Appl. Physiol., 60*: 317.

Mintun, M.A., Dennis, D.R., Welch, M.J., Mathias, C.J., and Schuster, D.P. (1987). *J. Nucl. Med., 28*: 1704.

Moi, M.K., Meares, C.F., McCall, M.J., Cole, W.C., and DeNardo, S.J. (1985). *Anal. Biochem., 148*: 249.

Moore, D.A., Fanwick, P.E., and Welch, M.J. (1989). *Inorg. Chem., 28*: 1504.

Moore, D.A., Fanwick, P.E., and Welch, M.J. (1990). *Inorg. Chem., 29*: 672.

Motekaitis, R.J., Martell, A.E., and Welch, M.J. (1990). *Inorg. Chem., 29*: 1463.

Motekaitis, R.J., Sun, Y., and Martell, A.E. (1991). *Inorg. Chem.*, *30*: 1554.

Murphy, J.R., Parker, D., Alexander, R.J., Amarjit, B., Carne, A.F., Eaton, M.A.W., Harrison, A., Millican A., Phipps, A., Rhind, S.K., Titmas, R., and Weatherby, D. (1988). *J. Chem. Soc. Chem. Commun.*, 156.

Najafi, A., Childs, R.L., and Hnatowich, D.J. (1984). *Int. J. Appl. Radiat. Isotop.*, *35*: 554.

Nelson, W.O., Karpishi, T.B., Rettig, S.J., and Orvig, C. (1988a). *Can. J. Chem.*, *66*: 123.

Nelson, W.O., Karpishin, T.B., Rettig, S.J., and Orvig, C. (1988b). *Inorg. Chem.*, *27*: 1045.

Ng, C.Y., Rodgers, S.J., and Raymond, K.N. (1989). *Inorg. Chem.*, *28*: 2062.

Nowotnik, D.P., Canning, L.R., Cumming, S.A., Harrison, R.C., Higley, B., Nechvatal, G., Pickett, R.D., Piper, I.M., Bayne, V.J., Forster, A.M., Weisner, P.S., Neirinckx, R.D., Volkert, W.A., Troutner, D.E., and Holmes, R.A. (1985). *Nucl. Med. Commun.*, *6*: 499.

Otsuka, F.L., and Welch, M.J. (1985). *Int. J. Nucl. Med. Biol.*, *12*: 331.

Philpott, G.W., Siegel, B.A., Schwarz, S.W., Connett, J.M., Rocque, P.A., Fleshman, J.W., Wallis, J.W., Baumann, M., Sun, Y., Martell, A.E., and Welch, M.J. (1994). *Dis. Col. Rect.*, in press.

Raichle, M.E., Eichling, J.O., Straatman, M.G., Welch, M.J., Larson, K.B., and Ter-Pogossian, M.M. (1976). *Amer. J. Physiol.*, *230*: 543.

Rodwell, J.D., Alvarez, V.L., Lee, C., Lopes, A.D., Goers, J.W.F., King, H.D., Powsner, H.J., and McKearn, T.J. (1986). *Proc. Nat. Acad. Sci. USA*, *83*: 2632.

Scarrow, R.C., and Raymond, K.N. (1988). *Inorg. Chem.*, *27*: 4140, and references therein.

Schwarz, S.W., Mathias, C.J., Sun, J.Y., Dilley, W.G., Wells, S.A., Martell, A.E., and Welch, M.J. (1993). *Nucl. Med. Biol.*, *18*: 477.

Streater, M., Taylor, P.D., Glider, H.C., and Porter, J. (1990). *J. Med. Chem.*, *33*: 1749.

Sun, Y., and Martell, A.E. (1989). *J. Amer. Chem. Soc.*, *111*: 8023.

Sun, Y., Martell, A.E., Reibenspies, J.H., and Welch, M.J. (1991a). *Tetrahedron*, *47*: 357.

Sun, Y., Mathias, C.J., Welch, M.J., Madsen, S.L., and Martell, A.E. (1991b). *Nucl. Med. Biol.*, *18*: 323.

Sundberg, M.W., Meares, C.F., Goodwin, D.A., and Diamanti, C.I. (1974). *Nature*, *250*: 587.

Taylor, D.M., and Kontoghiorges, G.J. (1986). *Inorg. Chim. Acta*, *125*: L35.

Ter-Pogossian, M.M., Phelps, M.E., Hoffman, E.J., and Mullani, N.A. (1975). *Radiology*, *114*: 89.

Thakur, M.L., Merrick, M.V., and Ganasekera, S.W. (1973). *Radiopharm. Label Compounds*, *2*: 183.

Thakur, M.L., Welch, M.J., Joist, J.H., and Coleman, R.E. (1975). *Thromb. Res.*, *9*: 345.

Thakur, M.L., Coleman, R.E., Mayhall, C.G., and Welch, M.J. (1976). *Radiology*, *119*: 731.

Vallabhajosula, S., Zimmerman, R.E., Picard, M., Stritzke, P., Mena, I., Hellman, R.S., Tikofsky, R.S., Stabin, M.G., Morgan, R.A., and Goldsmith, S.J. (1989). *J. Nucl. Med.*, *30*: 599.

Welch, M.J., and Welch, T.J. (1975). *Radiopharmaceuticals* (G. Subramanian, B. Rhodes, J. Cooper and V. Sodd, eds.), New York Society of Nuclear Medicine, p. 73.

Yokoyama, A., Ohmomo, Y., Horiuchi, K., Saji, H., Tanaka, H., Yamamoto, K., Ishii, Y., and Torizuka, K. (1982). *J. Nucl. Med.*, *23*: 909.

Zalutsky, M.R., ed. (1989). *Antibodies in Radiodiagnosis and Therapy*, CRC Press, Boca Raton, Florida.

TECHNETIUM COMPLEXES AS RADIOPHARMACEUTICALS

by

Helmut R. Maecke*

Institute of Nuclear Medicine,
University Hospital Basel,
4031 Basel
Switzerland

and

Michael Eisenhut

Universität Heidelberg,
Radiologische Universitäts-Klinik,
Abteilung Nuklearmedizin,
Radiopharmazeutische Chemie,
69120 Heidelberg
Germany

I. INTRODUCTION

Technetium-99m is the most commonly used radionuclide in diagnostic nuclear medicine. More than 80% of all scans in the hospital are being performed with ^{99m}Tc in complexed form. The main reason for this preeminence is nearly ideal nuclear properties, i.e. (i) the half-life of 6.02 h is perfectly compatible with the pharmacokinetics of most radiopharmaceuticals and allows for the injection of relatively high doses, (ii) ^{99m}Tc has no particulate emissions and the daughter nuclide ^{99}Tc is long lived ($T_{1/2}$ = 2.13·10^5 y) which limits the radiation dose to the patients, (iii) ^{99m}Tc emits a single 140 keV photon which is perfectly suited for high efficiency for the currently used Anger-gamma-cameras whose detection system is based on sodium iodide crystals. Moreover, the position in the periodic system (group VII element, $5s^2 4d^5$ electron configuration) allows for a rich coordination chemistry, a prerequisite for the synthesis of a large group of radiopharmaceuticals (compounds with a structure designed for tissue specific distribution and containing a radionuclide detectable with radiometric means).

* Author to whom correspondence should be addressed

A last important fact for the widespread use of Tc is its easy availability at low cost, which is due to current generator technology. The generator is based on the following decay scheme and chemistry involved (1).

$$^{99}MoO_4^{2-} \xrightarrow[67h]{\beta} {}^{99m}TcO_4^- \xrightarrow[6h]{\gamma} {}^{99}TcO_4^- \quad (1)$$

^{99}Mo is obtained from the fission of uranium in a nuclear reactor. The molybdate ion (as MoO_4^{2-} or even as the polyacid anion $Mo_7O_{24}^{6-}$) is loaded onto a chromatography column containing Al_2O_3 which strongly binds the double negatively charged anion (or polyanion) whereas the decay product TcO_4^- is easily eluted with physiological saline. A constant supply of $^{99m}TcO_4^-$ as a 10^{-7} - 10^{-8} M solution in 0.15 M sodium chloride is thus obtained which can be used for the synthesis of ^{99m}Tc-based radiopharmaceuticals, mainly Tc coordination compounds. Modifications in the first and second coordination sphere as well as in the oxidation state of the central metal allow for the synthesis of complexes with quite different 'organ specificity', biodistribution, pharmacodynamics, etc.

Because of the ideal radiophysical properties of ^{99m}Tc, the inorganic coordination chemistry of Tc has raised the interest of radiopharmaceutical and inorganic coordination chemists. Several reviews provide a background to the interrelation between structural and radiopharmaceutical aspects of technetium coordination compounds (Deutsch et al., 1983; Deutsch and Libson, 1984; Seiler, 1983; Jurisson et al., 1993; Clarke and Podbielski, 1987). Structural data were obtained on a macroscopic level by use of the long-lived radioisotope ^{99}Tc and single crystal X-ray structural analysis (Bandoli et al., 1982). As expected from the position in the periodic table, a diverse chemistry and a wide variety of oxidation states, coordination numbers and geometries have been found. Oxidation states range from +7 to -1; the most important in Tc-based radiopharmaceuticals are Tc(I), Tc(III) and Tc(V), the last being the most common. Coordination numbers of Tc(V) complexes range from 5 to 8 and the Tc(V) core can have one or two oxo groups to effectively reduce the high formal negative charge on Tc(V) (Davison and Jones, 1982).

Burns et al. (1978) classified technetium radiopharmaceuticals into two classes: (i) Tc essential radiopharmaceuticals where the biodistribution is determined by the first and second coordination sphere of the radiometal, (ii) Tc-tagged radiopharmaceuticals, which are usually biomolecules such as monoclonal antibodies (MoAb) or receptor binding molecules. Their biodistribution is mainly determined by the carrier molecule and ^{99m}Tc is the signal producing tag which can be detected by a gamma camera.

We present a short discussion of both groups of radiopharmaceuticals concentrating on those compounds used in daily nuclear medicine practice.

II. ^{99m}Tc LABELING OF MONOCLONAL ANTIBODIES

The labeling of monoclonal antibodies (MoAb) or their fragments with metallic radionuclides has been one of the major research areas in radiopharmaceutical chemistry. By comparison with the introduction of radiohalogens via the electrophilic aromatic substitution at tyrosine sites in proteins, metallic radionuclides need a chelating moiety for their binding to proteins. Various strategies and techniques have been developed to form stable binding sites for ^{99m}Tc.

A. Direct labeling

Intact IgG does not allow stable complexation of 'reduced ^{99m}Tc'. This designation is often used if the oxidation state is not known with certainty. Most likely the Tc is present in the +5 oxidation state. The unmodified amino acids of the protein do not provide stable coordination for 'reduced ^{99m}Tc'. A successful approach to create stable binding sites for 'reduced ^{99m}Tc' (Tc(V)) was achieved by formation of SH groups obtained from reductive cleavage of disulfide bridges (Fig. 1).

Due to the thiophilicity of Tc(V), the rate of complex formation is very fast, the labeling yields are around 98% or better and the stability of the labeled MoAb in vivo is high. This strategy was first introduced by Rhodes et al. (1980) by use of stannous ions for the reduction of the disulfide bridges. More reliable and MoAb less compromising

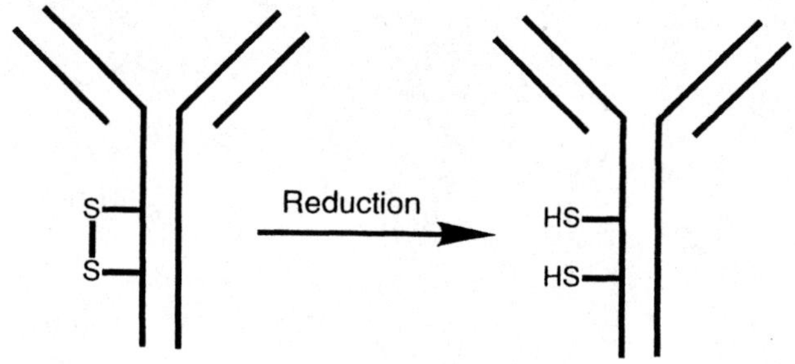

Figure 1.
Reductive cleavage of antibody disulfide groups to provide stable Tc(V) binding sites.

preparations have been obtained by use of mercaptoethanol (Schwarz and Steinsträsser, 1987; Mather and Ellison, 1990), Vitamin C, Dithiothreitol (Thakur et al., 1991) and water-soluble phosphines (Koprivova et al., 1993). The nature of the coordination sphere about Tc is not known, but one can assume that besides the thiol groups side-chain amino acid ligating groups complete the coordination sphere of Tc.

B. Indirect labeling

The introduction of bifunctional ligands into the antibody, providing a stable chelating moiety for 'reduced ^{99m}Tc' is called the indirect labeling method. Two main strategies have been followed.

1. *Preconjugation labeling approach*

a. Conjugation of precomplexed ^{99m}Tc

This approach uses a 'Tc-selective' bifunctional ligand which is labeled with 'reduced Tc', separated from excess ligand and conjugated to the antibody after complexation. This strategy affords highly stable immunoconjugates with retained immunoreactivity. Antibodies and their fragments have been labeled successfully with use of preformed ^{99m}Tc-4,5-bis(thioacetamido)pentanoate active ester by Fritzberg et al. (1988) (Fig. 2).

b. The avidin biotin approach

A completely different precomplexation route uses the high affinity of biotin (vitamin H, M = 0.244 kg mol^{-1} to avidin (K_D = 10^{-15} M^{-1}). Avidin is a 68 kg mol^{-1} glycoprotein which is found in egg white. The approach is a so-called pretargeting approach where

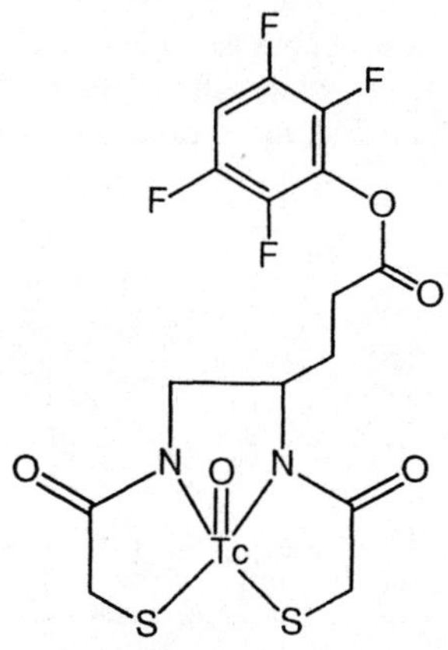

Figure 2.
Tetrafluorophenyl active ester of ^{99m}Tc-4,5-bis(thioacetamido)pentanoate.

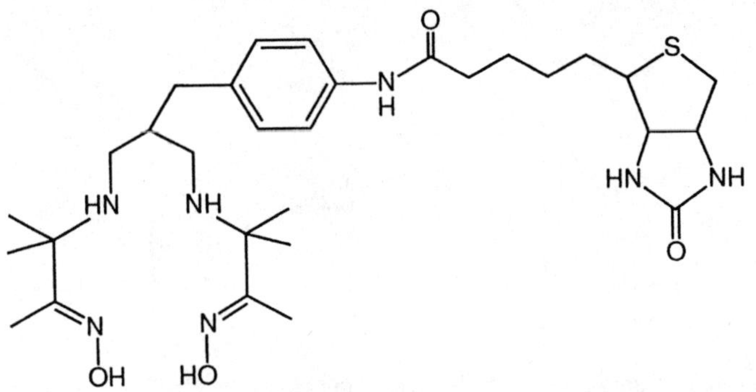

Figure 3.
Structure of biotinylated [^{99m}TcO] (6-p-aminobenzyl-3,3,9,9-tetramethyl-4,8-diazaundecane-2,10-dionedioxime).

biotinylated antibody is pretargeting the tumor followed by the injection of avidin. This second step avidinates the tumor; localization of the tumor is then performed by injection of a ^{99m}Tc labeled chelator-biotin conjugate (Fig. 3) (Koch and Mäcke, 1992). This approach was very successful in a trial with glioma patients (Paganelli et al., 1994).

2. *The postconjugation labeling approach*
The preconjugation approach is rather time consuming and labor intensive. Therefore, attempts have been made to conjugate a bifunctional ligand to the antibody first, followed by purification of the conjugate and lyophilization resulting in a kit formulation. To this 'reduced Tc' complexed to a weak transfer ligand like citrate, tartrate or glucoheptonate is added, followed by transchelation to the 'Tc receptor' bound to the MoAb. Two bifunctional ligands were especially successfully applied in preclinical and clinical studies. A bidentate ligand, 6-hydrazinopyridine-3-carboxylate, was introduced via the N-hydroxysuccinimide ester method and labeled with Tc(V) glucoheptonate. Specific activity and stability of the conjugates are high (Abrams et al., 1990). The mode of bonding to Tc is not fully understood, but formation of a Tc=N-N-R bond is assumed (Abrams et al., 1991). Additionally, a variety of tetradentate ligands have been described in the literature which react with 'reduced Tc' to form thermodynamically stable and kinetically inert ^{99m}Tc(V)O-complexes and have a functionality for antibody coupling. However, only two were finally brought to clinical application. Pak et al. (1989) use a mercaptoacetyltriglycine (MAG_3) type chelator (Fig. 4) which they have coupled to a Fab' fragment of an antibody via thiol groups to a maleimide bound to the antibody fragment as an anchoring group.

A similar approach leading to a lyophilized kit formulation was followed by Eisenhut et al. (1991). These authors used the NHS ester of 6-(4'-(4"-carboxyphenoxy)butyl)-2,10-dimercapto-2,10-dimethyl-4,8-diazaundecane (Fig. 5) for the introduction of a

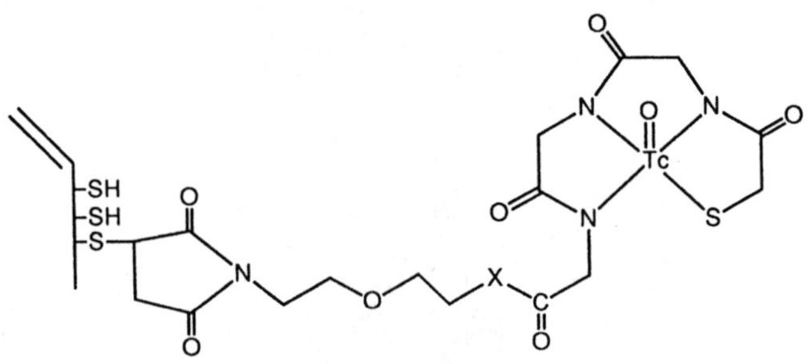

Figure 4.
Indirect labeling of MoAb with ^{99m}Tc via a mercaptoacetyltriglycine-type conjugate.

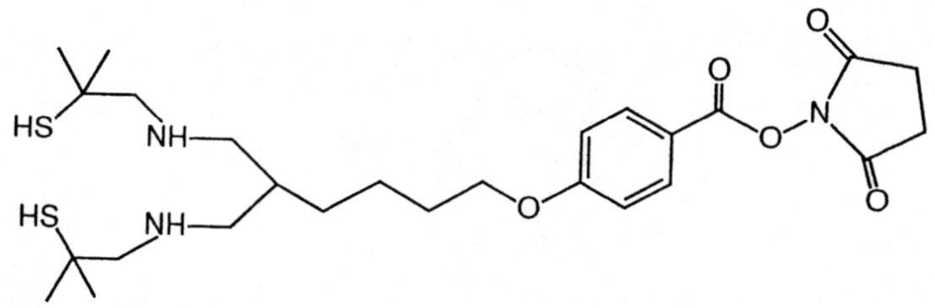

Figure 5.
N-hydroxisuccinimide ester of 6-(4'-(4"-carboxyphenoxy)butyl)-2,10-dimercapto-2,10-dimethyl-4,8-diazaundecane.

bis(aminoethanethiol)-type (BAT or N_2S_2) ligand to MoAb's. This ester conjugates rapidly to MoAb in the pH range 8 - 8.5 and complexes Tc(V) at ambient temperature.

III. ^{99m}Tc LABELED MYOCARDIAL PERFUSION AGENTS

Current efforts in the synthesis of new ^{99m}Tc complexes for myocardial perfusion studies in man reflect the medical demand to replace $^{201}Tl^+$, the most widely used radionuclide for this purpose. This agent has unfavorable nuclear physical characteristics. Its half-life of 73 h and its γ-rays of 69 - 80 keV are far from optimal and the radiation dosimetry is very unfavorable because of its long biological half-life of around 10 days. The uptake mechanism of Tl^+ into the myocytes is via the Na^+/K^+-ATPase pump (Sands et al., 1985). On the probably false assumption that monocationic Tc complexes are extracted by the same mechanism, Tc-based radiopharmaceuticals were developed to mimic Tl^+. Initial studies using Tc(III) complexes with bidentate π-acceptor ligands like o-phenylene bis(dimethylarsine) (DIARS) (Deutsch et al., 1981a) or bis(dimethylphosphino)ethane (DMPE) showed good myocardial uptake in dogs (Deutsch et al., 1981b), but disappointing myocardial images were found in humans (Gerson et al., 1983). An explanation was then given for this species difference. The uptake of cationic Tc(III) complexes by the human heart is transient because of their reduction to neutral Tc(II) species which are then tranferred to the liver (Deutsch et al., 1989). Finally, E. Deutsch's group (Rosetti et al., 1991) succeeded in making nonreducible monocationic Tc complexes which localize in the heart in proportion to the myocardial blood flow. These are based on a tetradentate Schiff base ligand occupying 4 binding sites of the equatorial plane and two monodentate tertiary phosphine ligands in the apical sites. The Schiff base ligand is 1,2-bis-[(dihydro-2,2,5,5-tetramethyl-3(2H)-furanonato)methylene]amino]ethane and the phosphine ligands are tris(3-methoxy-1-propyl)phosphine (De Rosch et al., 1992).

A big advance in the field of myocardial perfusion imaging agents has been based on substitution and redox inert Tc(I) complexes (d^6 electron configuration) with isonitrile ligands. These complexes are positively charged, extremely stable because of their low spin d^6 electron configuration, and nonreducible (Jones et al., 1984; Abrams et al., 1983). The most successful in this series is hexakismethoxy-isobutyl-isonitrile (MIBI) (Fig. 6) which has been registered in most countries and has replaced ^{201}Tl in most institutions.
This compound exhibits high myocardial uptake, is retained in the heart with a long half-life and shows rapid clearance from blood, lung and liver (Narahara et al., 1990).

A new group of compounds, developed only recently, is based on Tc(V) with modified bidentate bis-phosphino-ethane as ligands. The complex, optimized for good clinical performance, is shown in Figure 7 (Higley et al., 1993; Kelly et al., 1993).

A rather different approach uses neutral complexes with high extraction fraction but also rapid washout of the heart. These complexes are boronic acid adducts of Tc dioximes (BATO's) and form by template synthesis during addition of pertechnetate to 3 equivalents of 1,2-cyclohexanedioxime and methylboronic acid in HCl under reducing conditions (Treher et al., 1989). The most clinically useful of these complexes is shown in Figure 8.

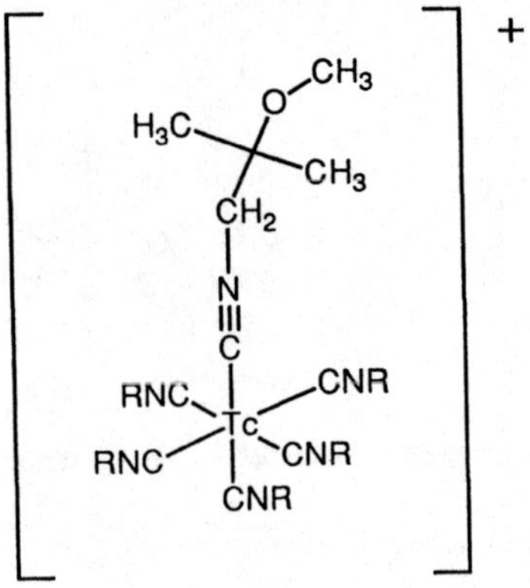

Figure 6.
Structure of Tc(Methoxyisobutylisonitrile)$_6$$^+$.

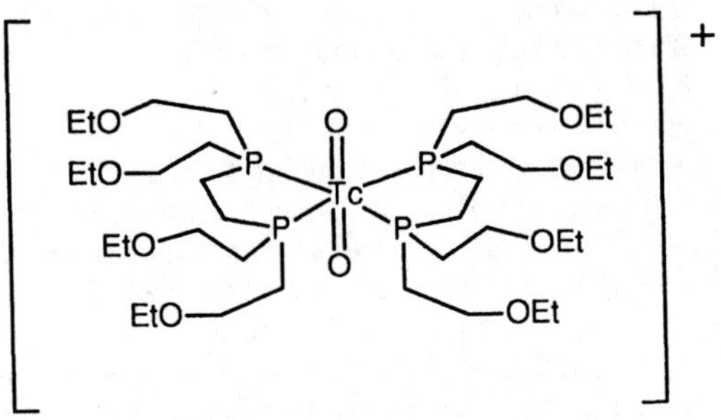

Figure 7.
Structure of TcO_2 (1,2-bis[bis(2-ethoxyethyl)phosphino]ethane).

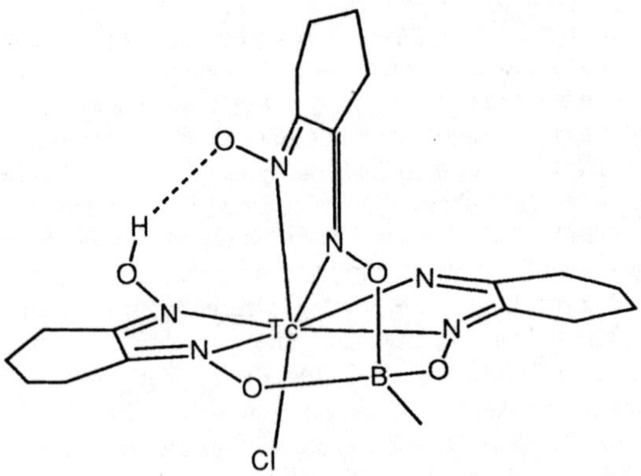

Figure 8.
Structure of CDO-MeB.

It is a sevencoordinate Tc(III) complex which is capped on one end of the molecule by the boronic acid derivative: six nitrogens are bound to Tc(III) in a trigonal configuration; the seventh ligand is a chloride which is labile at pH 7.4, 37°C, and undergoes chloro-hydroxo exchange with $T_{1/2}$ = 13 min (Jurisson et al., 1991), which may lead to a mixture of neutral and cationic forms in vivo. This compound, abbreviated as CDO-MeB, has been tested in several animal species in which it was shown to behave essentially in the same manner (Narra et al., 1989a; 1989b). Its blood clearance is fast, and its myocardial uptake (~4% of the initial dose) is rapid. The behavior in humans is similar. The fast washout of the heart is somewhat disadvantageous and necessitates a rapid imaging protocol. However, good SPECT and planar images have been obtained in humans (Seldin et al., 1989) and the radiopharmaceutical has been registered in Europe and the US.

IV. BRAIN IMAGING AGENTS

Technetium-99m based radiopharmaceuticals have played an important role in the assessment of regional cerebral blood flow. Although computed tomography (CT), magnetic resonance imaging (MRI) and ultrasound are well established diagnostic techniques for determining structural changes in the brain, brain function disturbances that cause ischemia but only small structural changes are very difficult to detect with these techniques. The development of new tracers was therefore of great importance. Such a tracer must have the following characteristics: (i) its extraction must be high and proportional to the blood flow over a wide range; (ii) it must cross the blood brain barrier easily by passive diffusion, i.e. it must be small ($M < 0.5$ kg mol^{-1}), lipophilic and neutral; (iii) it should possibly be trapped in the brain and retained in a fixed regional distribution during data acquisition; (iv) it should have little or no in vivo metabolism (Brodie and Hogben, 1957).

Several new tracers based on ^{99m}Tc have the required chemical characteristics. Most interesting is the development of a series of propylenediaminedioxime complexes (PnAO) (Fig. 9).

This development was based on the discovery by Troutner et al. (1983) that the prototype PnAO (3,3,9,9-tetramethyl-4,8-diazaundecane-2,10-dione dioxime) forms a neutral and lipophilic complex with Tc that is able to cross the blood brain barrier in rats but is not trapped. A number of Tc(V) complexes with modified PnAO structure have been synthesized and characterized by their visible, UV spectra, IR spectra, ^{1}H and ^{13}C NMR spectra as well as X-ray diffraction studies. Four X-ray crystal structures show that these complexes are of the monooxoTc(V) type with the oxo group at the apex of a square pyramid and the Tc sitting about 0.68 Å above the plane defined by the N_4 ligand. The overall zero charge can be understood by the fact that the two amino nitrogens and one of the oxime oxygens are deprotonated. The second oxime oxygen forms a stable intramolecular hydrogen bond resulting in a pseudomacrocyclic ring structure (Jurisson et al., 1986). The objective to make a brain flow tracer with the above mentioned properties was finally met with (RR,SS)-4,8-diaza-3,6,6,9-tetra-methyl-undecane-2,10-dione dioxime (d,l-HM-PAO). The ^{99m}Tc complex is again neutral, lipophilic, and is also trapped in the

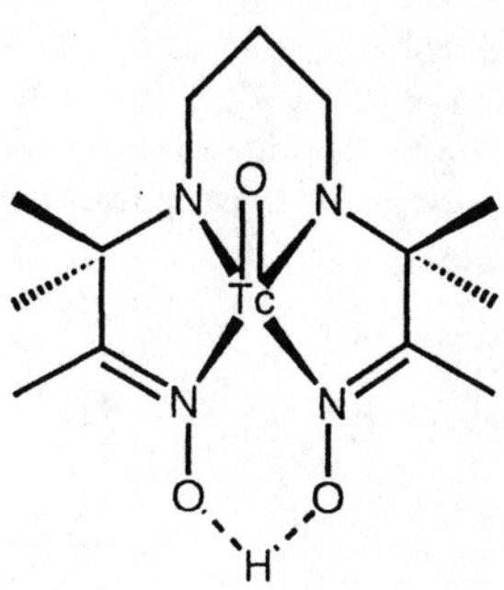

Figure 9.
Structure of Tc(V)O(PnAO).

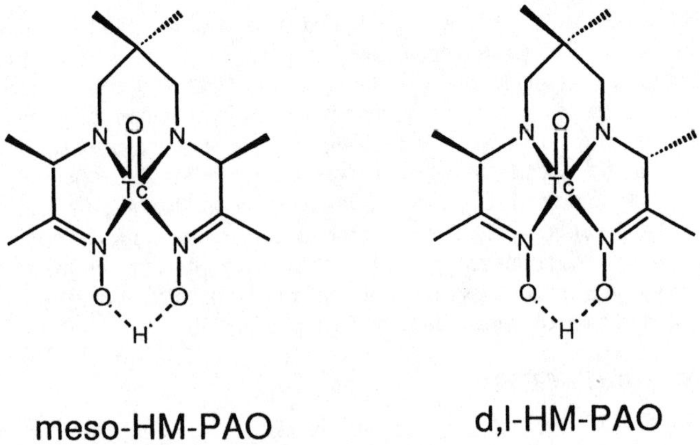

Figure 10.
Meso- and D,L-form of TcO(HM-PAO).

brain and shows blood flow dependent brain uptake in animals (Neirinckx et al., 1987; Holmes et al., 1985) and in humans (Ell et al., 1985; Leonard et al., 1986). The uptake mechanism in the brain is not well understood. It is assumed that in the cell the complex decomposes to a hydrophilic complex that cannot diffuse back. The HM-PAO ligand has two chiral centers and gives rise to stereoisomers (Fig. 10).
Interestingly [meso-HM-PAO]Tc(V) shows brain uptake but little retention. It is also more stable in vitro than the D,L-form that decays to hydrophilic species with $k = 0.17\ h^{-1}$, which limits its use to the first 30 min after reconstitution. It may form inside the cell a hydrophilic compound which precludes back diffusion, or it may bind to intracellular glutathione. Despite its limitations in terms of in vitro stability and the necessity to use the prepared compound within half an hour, ^{99m}Tc-HM-PAO is the radiopharmaceutical of choice for brain perfusion measurements and of high clinical utility. Its example shows that small stereochemical differences determine in vivo biodistribution and pharmacokinetics.

Another compound, ethylenecysteine dimer (ECD), which is based on a symmetrical diamino dithiol (DADT) ligand, has successfully entered clinical trials. The ligand N,N'-1,2-ethanediylbis-L-cysteine contains two ester functions and has therefore two chiral centers. The rationale behind the design of this compound was the assumption that, again, a neutral monooxo Tc(V) complex is formed by dissociation of the two thiol and one amino protons, and that the introduction of the two ester functions provides for a trapping mechanism. The complex indeed is trapped in the brain, this being most likely due to the hydrolysis of one or both of its ester groups to the acid (Fig. 11) (Walowitch et al., 1988a).

This hydrolysis is remarkably stereospecific, as only the L,L-isomer is trapped significantly whereas the D,D-isomer crosses the blood brain barrier but is not hydrolyzed and can diffuse back (Walowitch et al., 1988b). The L,L-complex shows concordance with the blood flow standards, radiolabeled microspheres and iodoantipyrine in terms of intercerebral distribution in monkeys (Walowitch et al., 1989), and was shown to behave in the expected way in patients. The metabolites, the mono- and diacid, are rapidly cleared from body tissues other than the brain. It thus appears that the brain to background facial

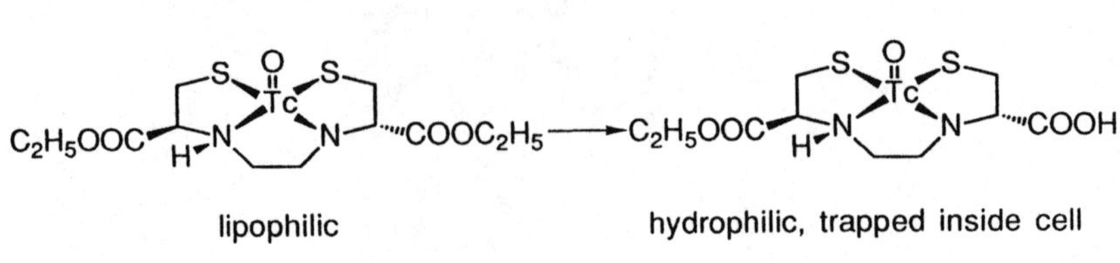

Figure 11.
Proposed trapping mechanism for ^{99m}Tc(ECD).

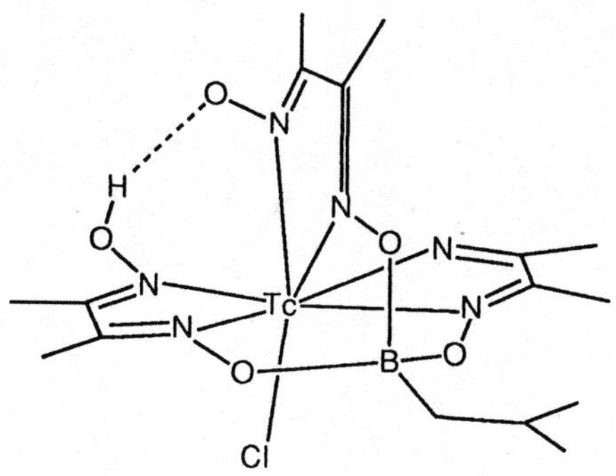

Figure 12.
Structure of Tc(III)Cl(DMG)$_3$2MP.

areas are improved compared to ^{99m}Tc-HM-PAO (Leveille et al., 1992; Holman and Devous, 1992). The stereospecific hydrolysis mechanism (probably due to an unspecified esterase) again is a first example of this kind of stereospecificity. The future will show if the scintigraphic images obtained depend not only on perfusion but also on activity or concentration of the enzyme in different disease conditions.

A third class of compounds extensively studied in the clinic belongs to the group of boronic acid adducts of Tc oximes (BATO). The compound chosen for clinical trials is TcCl(DMG)$_3$2MP and shown in Fig.12.

Unlike the other compounds, it shows slow but significant washout of the brain (Narra et al., 1989a; 1989b). A possible weak trapping mechanism may be given by TcCl $\rightarrow$ TcH_2O^+ exchange and a $TcH_2O^+ \rightleftharpoons TcOH^0$ equilibrium (Jurisson et al., 1991), but the very high lipophilicity of the complex (log P = 3.8) may cause a long residence time within the hydrophobic environment of the cell membrane bilayer.

V. ^{99m}Tc-BASED RADIOPHARMACEUTICALS FOR RENAL IMAGING

Nuclear medicine is a classical field for the assessment of renal function noninvasively. There are three parameters whose measurement with radiopharmaceuticals can help to diagnose and understand kidney problems and monitor changes in renal function: (i) the glomerular filtration rate (GFR), (ii) the effective renal plasma flow (ERPF), and (iii) the active renal mass. The suitability of an agent to be used in one of the three cases is determined by its molecular structure. The renal function can be obtained by either an indirect measurement taking plasma samples to determine plasma clearance by agents solely excreted by the kidneys, or, more importantly, by the use of the gamma camera which gives very similar functional results but also gives access to the split renal function.

The measurement of GFR uses a compound which is cleared solely by free filtration through the renal glomeruli. The compound should be inert in vivo, only eliminated by the kidneys and neither secreted nor reabsorbed by the renal tubules. Of the many agents cleared by GFR, only one is used in the clinic: ^{99m}Tc-DTPA (DTPA = diethylenetriaminepentaacidic acid). It provides a clearance measurement with only 3 - 5% below the true GFR obtained with the gold standard for GFR, inulin, which is a fructose polymer of molar mass of 5.2 kg mol^{-1}. Unfortunately, there are no clear structural data available on this compound, nor is the Tc oxidation state known with certainty. On the macroscopic level, oxidation states +3, +4 and +5 have been found or postulated (Seifert et al., 1982; Kempf et al., 1980; Bandoli et al., 1982). Under conditions as in a commercial kit formulation of ^{99m}Tc-DTPA, Eckelman et al. (1972) found at pH 1.5 - 2 with $SnCl_2$ as reductant an oxidation state of +4 without any structural characterization. The advantage of ^{99m}Tc-DTPA in renal imaging is its low price and a low radiation dose to the patient. Imaging with this agent is somewhat hampered by the fact that the extraction efficiency of a solely glomerularly filtered agent is only 20%, which results in a relatively high background due to the high activity in blood.

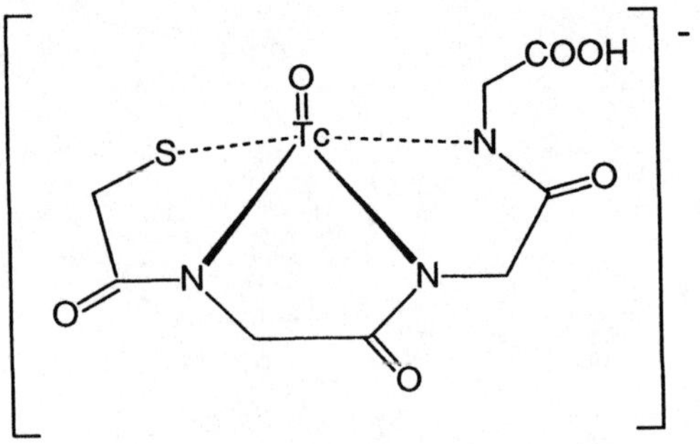

Figure 13.
Structure of $Tc(V)O(MAG)_3$.

A more efficient way of a compound to be cleared from the circulation is via secretion by the renal tubules. The characteristics of such an agent is similar to GFR agents but excretion now has to be by an active transport process. This process is 3 - 4 times more efficient and results in lower background and better image quality. The result of this kind of clinical monitoring of renal function is the effective renal plasma flow (ERPF). The gold standard in this group of compounds is p-aminohippuric acid and in renal functional imaging the radioiodine labeled analogue o-iodohippuran (OIH) was used. Technetium-99m labeled compounds have also been evaluated (Fritzberg et al., 1981; 1984; Costello et al., 1983; Jones and Davison, 1983) and the most suitable ligand is mercaptoacetyltriglycine which forms a negatively charged Tc oxo complex (Fritzberg et al., 1986). The $Tc(V)O^{3+}$ center affords deprotonation of the thiolate sulfur and all three amide nitrogens. The complex has square-pyramidal geometry with the oxo group in the apex and the Tc atom situated above the plane of the N_3S donor set (Nosco et al., 1990) (Fig. 13). This agent is now in clinical use and has replaced OIH in most hospitals (Taylor et al., 1986; Taylor and Eshima, 1988; Bubeck et al., 1990).

The third group of radiopharmaceuticals in kidney imaging is the one which measures active renal mass and provides information about morphology of the kidney. The two registered and marketed agents are the ^{99m}Tc complexes of 2,3-dimercaptosuccinic acid (DMSA) and glucoheptonate (GH). They are handled by the kidney quite differently than the other agents and for neither of them the mechanism of excretion is known (Blaufox, 1987). Also, no clear structural data are available on the two complexes. The composition of the DMSA complex is most likely $[TcO(DMSA)_2]^-$ with a square pyramidal geometry and the +5 oxidation state. If meso-DMSA is coordinated to Tc(V), three isomers are expected with syn-endo, syn-exo and anti orientation of COOH with respect to the Tc oxo group. All isomers have been shown by ^1H-NMR to be in a rapid equilibrium (Spies and Johanssen, 1981; Spies and Scheller, 1986). The presence of these isomers have also been shown for the analogues $[^{186}ReO(DMSA)_2]^-$ by HPLC and comparison to isolated $(Bu_4N)ReO(DMSA)_2$ (Bisunadan et al., 1991). Even less data are available on the structure of Tc(GH). Results by De Kieviet (1981) suggest that the oxidation state is +5 with a monooxo Tc(V) core. Both agents are only seldom used in the clinic because ultrasound, computer tomography and magnetic resonance imaging give more precise information on kidney morphology.

VI. BONE IMAGING AGENTS

Technetium-99m based bone imaging agents are extensively used for the detection of osteomyelitis, detection and staging of metastatic disease to the bone, bone infarction and infection. These diseases have in common the occurrence of immature new actively growing bone with easily accessible unsaturated coordination sphere of Ca^{2+} in the basic bone structure hydroxyapatite ($Ca_{10}(PO_4)_6(OH)_2$). Bone seeking radiopharmaceuticals are actively adsorbed on the surface of newly formed hydroxyapatite crystals and are analogues

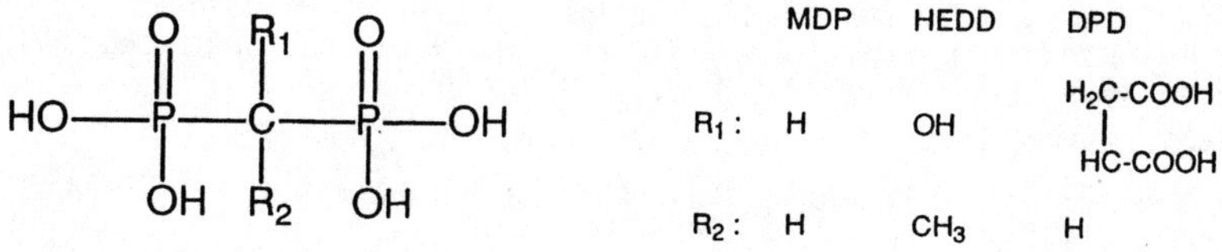

Figure 14.
Structure of diphosphonate ligands MDP, HEDP, DPD.

of Ca^{2+} like the strontium isotopes or are ligands for Ca^{2+}. Diphosphonates are known to be bone seekers and form stable complexes with Ca^{2+}. Several diphosphonates have been developed for labeling with ^{99m}Tc (Fig. 14) and are in use in nuclear medicine departments.

They have different substituents on the bridging C-atom. Their exact composition is not known and depends on the concentrations of TcO_4^- and diphosphonate. The rationale behind their development was based on the potential bifunctionality of these ligands: (i) they can form chelates with metals or bridge metal ions like Tc, (ii) they preserve high binding affinity for Ca^{2+} even if bound to another metal. Very elegant studies by Libson et al. (1980) and Jurisson et al. (1983) corroborated these two assumptions. An X-ray crystal structure of $[Tc(MDP)(OH)]_n$ showed this complex to be a polymeric species with one Tc per diphosphonate. Each Tc is bridging two diphosphonates and each diphosphonate is bound to two Tc atoms. Because of the polymeric nature of the complex and unknown degree of protonation of the OH or oxo group bridging two Tc, the oxidation state of Tc has not been determined with certainty, but it is assumed to be +4.

Moreover, the synthesis of different ternary Co(III)(ethylenediamine)$_2$(diphosphonate) complexes allowed to determine stability constants with Ca^{2+} in homogeneous aqueous solution. The stability constants range from $0.2 - 250 \cdot 10^4\ M^{-1}$ depending on the nature of the diphosphonate ligand. The high affinity confirms the potential of these phosphonates to bind two metals simultaneously. Furthermore, it was shown that the efficacy to localize bone lesions parallels the affinity for Ca^{2+}.

The composition of the ^{99m}Tc-diphosphonate radiopharmaceuticals was shown to be a mixture of different oligomers. It was studied on a carrier (addition of ^{99}Tc) and no-carrier added level, and different compounds were shown by HPLC, depending on Tc-concentration (Pinkerton et al., 1982). Even in daily used preparations in the clinic, the composition of injected agents may vary depending on the time point of last TcO_4^--generator elution. Technetium concentrations may vary between $10^{-6} - 10^{-8}$ M in such eluants (Wilson and Pinkerton, 1985; De Ligny et al., 1990). As all oligomeric species appear to show high bone affinity, it seems that only little difference in clinical performance can be expected in different preparations.

REFERENCES

Abrams, M.J., Davison, A., Jones, A.G., Costello, C.E., and Pang, H. (1983). *Inorg. Chem.*, *22*: 2798.

Abrams, M.J., Juweid, M., Tenkate, C.I., Schwartz, D.A., Hauser, M.M., Gaul, F.E., Fucello, A.J., Rubin, R.H., Strauss, H.W., and Fischman, A.J. (1990). *J. Nucl. Med.*, *31*: 2022.

Abrams, M.J., Larsen, S.K., and Zubieta, J. (1991). *Inorg. Chem.*, *30*: 2031.

Bandoli, G., Mazzi, U., Roncari, E., and Deutsch, E. (1982). *Coord. Chem. Rev.*, *44*: 191.

Bisunadan, M.M., Blower, P.J., and Clarke, S.E. (1991). *J. Chem. Soc. Chem. Commun.*, 1115.

Blaufox, M.D. (1987). *Contr. Nephrol.*, *56*: 31.

Brodie, B.B., and Hogben, C.A.M. (1957). *J. Pharm. Pharmacol.*, *9*: 345.

Bubeck, B., Brandou, W., Weber, E., Kölble, T., Parekh, N., and Georgi, P. (1990). *J. Nucl. Med.*, *31*: 1285.

Burns, H.D., Vorley, P., Wagner, H.N., Marzilli, L.G., and Risch, V. (1978). *The Chemistry of Radiopharmaceuticals* (N.D. Heindel, H.D. Burns, T. Honda, and L.W. Brady, eds.), Masson Publishing USA, Inc., New York, p. 269.

Clarke, M.J., and Podbielski, L. (1987). *Coord. Chem. Rev.*, *78*: 253.

Costello, C.E., Brodack, J.W., Jones, A.G., Davison, A., Johnson, D.L., Kasina, S., and Fritzberg, A.R. (1983). *J. Nucl. Med.*, *24*: 353.

Davison, A., and Jones, A. (1982). *Int. J. Appl. Radiat. Isot.*, *33*: 875.

De Kieviet, W.J. (1981). *J. Nucl. Med.*, *22*: 703.

De Ligny, C.L., Gelsema, W.J., Tji, T.G., Huigen, Y.M., and Vink, H.A. (1990). *Nucl. Med. Biol.*, *17*, 161.

De Rosch, M.A., Brodack, J.W., Grummon, G.D., Marmion, M., Nosco, D.L., Deutsch, K.F., and Deutsch, E. (1992). *J. Nucl. Med.*, *33*: 850.

Deutsch, E., and Libson, K. (1984). *Comments Inorg. Chem.*, *3*: 83.

Deutsch, E., Glavan, K.A., Sodd, V.J., Hishiyama, H., Ferguson, D.L., and Lukes, S.J. (1981a). *J. Nucl. Med.*, *22*: 897.

Deutsch, E., Bushong, W., Glavan, K.A., Elder, R.C., Sodd, V.J., Scholz, K.L., Fortman, D.L., and Lukes, S.J. (1981b). *Science*, *214*: 85.

Deutsch, E., Libson, K., Jurisson, S., and Lindoy, L.F. (1983). *Progress in Inorganic Chemistry*, Vol. 30 (S.J. Lippard, ed.), John Wiley & Sons New York, p. 75.

Deutsch, E., Ketring, A.R., Libson, K., Vanderheyden, J.-L., and Hirth, W.W. (1989). *Nucl. Med. Biol.*, *16*: 191.

Eckelman, W.C., Meinken, G., and Richards, P. (1972). *J. Nucl. Med.*, *13*: 577.

Eisenhut, M., Missfeldt, M., Lehmann, M.D., and Karas, M. (1991). *J. Lab. Comp. Radiopharm.*, *29*: 1283.

Ell, P.J., Hocknell, J.M.L., and Jarritt, P.H. (1985). *Lancet*, *ii*: 50.

Fritzberg, A.R., Klingensmith, W.C., III, Whitney, W.P., and Kuni, C.C. (1981). *J. Nucl. Med.*, *22*: 258.

Fritzberg, A.R., Kasina, S., Eshima, D., Johnson, D.L., Jones, A.G., Lister-James, J., Davison, A., and Brodack, J.W. (1984). *J. Nucl. Med.*, *25*: 16.

Fritzberg, A.R., Kasina, S., Eshima, D., and Johnson, D.L. (1986). *J. Nucl. Med.*, *27*: 111.

Fritzberg, A.R., Abrams, P.G., Beaumier, P.L., Kasina, S., Morgan, Ch.A., Rao, T.N., Reno, J.M., Sanderson, J.A., Srinivasan, A., Wilbur, D.S., and Vanderheyden, J. (1988). *Proc. Nat. Acad. Sci. USA*, *85*: 4025.

Gerson, M.C., Deutsch, E., and Nishiyama, H. (1983). *Eur. J. Nucl. Med.*, *8*: 371.

Higley, B., Smith, F.W., Smith, T., Gemmell, H.G., Das Gupta, P., Dragonten, V., Gvozdanovic, D.V., Graham, D., Hinge, D., Davidson, J., and Lahiri, A. (1993). *J. Nucl. Med.*, *34*: 30.

Holman, B.L., and Devous, M.D. (1992). *J. Nucl. Med.*, *33*: 1888.

Holmes, R.A., Chaplin, S.B., and Royston, K.G. (1985). *Nucl. Med. Commun.*, *6*: 443.

Jones, A.G., and Davison, A. (1983). *Technetium in Chemistry and Nuclear Medicine* (E. Deutsch, M. Nicolini and H.N. Wagner Jr., eds), Cortina International, Verona, p. 25.

Jones, A.G., Abrams, M.J., Davison, A., Brodack, J.W., Toothaker, A.K., Adelstein, S.J., and Kassis, A.I. (1984). *Nucl. Med. Biol.*, *11*: 225.

Jurisson, S.S., Benedict, J.J., Elder, R.C., Whittle, R., and Deutsch, E. (1983). *Inorg. Chem.*, *22*: 1332.

Jurisson, S., Schlemper, E.O., Troutner, D.E., Canning, L.R., Nowotnik, D.P., and Neirinckx, R.D. (1986). *J. Nucl. Med.*, *25*: 543.

Jurisson, S.S., Hirth, W., Linder, K.E., Di Rocco, R.J., Narra, R.K., Nowotnik, D.P., and Nunn, A.D. (1991). *Nucl. Med. Biol.*, *18*: 735.

Jurisson, S., Berning, D., Jia, W., and Ma, D. (1993). *Chem. Rev.*, *93*: 1137.

Kelly, J.D., Forster, A.M., Higley, B., Archer, C.M., Booker, F.S., Canning, L.R., Edwards, K.W.C.B., Gill, H.K., McPartlin, M., Nagle, K.R., Latham, I.A., Pickett, R.D., Storey, A.E., and Webbon, P.M. (1993). *J. Nucl. Med.*, *34*: 222.

Kempf, C.R., Clark, H.M., Aikens, D.A., Bailey, R.A., Bunce, S.C., and Applebaum, G. (1980). *J. Nucl. Med.*, *21*: 402.

Koch, P., and Mäcke, H.R. (1992). *Angew. Chem. Int. Ed.*, *31*: 1507.

Koprivova, H., Pèlegrin, A., Mach, J.P., and Mäcke, H.R. (1993). *Eur. J. Nucl. Med.*, *20*: 999.

Leonard, J.P., Nowotnik, D.P., and Neirinckx, R.D. (1986). *J. Nucl. Med.*, *27:* 1819.

Leveille, J., Demonceau, G., and Walovitch, R.C. (1992). *J. Nucl. Med.*, *33*: 480.

Libson, K., Deutsch, E., and Barnett, B.L. (1980). *J. Amer. Chem. Soc.*, *102:* 2476.

Mather, S.J., and Ellison, D. (1990). *J. Nucl. Med.*, *31*: 692.

Narahara, K.A., Villanuevameyer, J., Thompson, C.J., Brizendine, M., and Mena, I. (1990). *Amer. J. Cardiol.*, *66*: 1438.

Narra, R.K., Eckelman, W.C., and Kuczynski, B.L. (1989a). *J. Lab. Comp. Radiopharm.*, *26*: 418.

Narra, R.K., Nunn, A.D., Kuczinsky, B.L., Feld, T., Wedeking, P., and Eckelman, W.C. (1989b). *J. Nucl. Med.*, *30*: 1830.

Neirinckx, R.D., Canning, L.R., Piper, I.M., Nowotnik, D.P., Pickett, R.D., Holmes, R.A., Volkert, W.A., Forster, A.M., Weisner, P.S., Marriott, J.A., and Chaplin, S.B. (1987). *J. Nucl. Med.*, *28*: 191.

Nosco, D.L., Tofe, A.J., Dunn, T.J., Lyle, L.R., Wolfangel, R.G., Bushman, M.J., Grummon, G.D., Helling, D.E., Marmion, M.E., Miller, K.M., Pipes, D.W., Strubel, T.W., and Wester, D.W. (1990). *Technetium and Rhenium in Chemistry and Nuclear Medicine 3* (M. Nicolini, G. Bandoli and U. Mazzi, eds), Cortina International, Verona, p. 381.

Paganelli, G., Magnani, P., Zito, F., Lucignani, G., Sudati, F., Truci, G., Motti, E., Terreni, M., Pollo, B., Giovanelli, M., Canal, N., Scotti, G., Comi, G., Koch, P., Maecke, H.R., and Fazio, F. (1994). *Eur. J. Nucl. Med.*, *21:* 314.

Pak, K.Y., Nedelmann, M.A., Stewart, R., and Dean, R.T. (1989). *J. Nucl. Med.*, *30*: 793.

Pinkerton, T.C., Ferguson, D.L., Deutsch, E., Heinemann, W.R., and Libson, K. (1982). *Int. J. Appl. Radiat. Isot.*, *33*: 907.

Rhodes, B.A., Torvestad, D.A., Burchiel, S.W., and Austin, R.K. (1980). *J. Nucl. Med.*, *21*: 54.

Rosetti, C., Vanoli, G., Paganelli, G., Kwiatkowski, M., Zito, F., Colombo, F., Deutsch, E., and Fazio, F. (1991). *Nucl. Med.*, *32*: 1007.

Sands, H., Delano, M.L., Camin, L.L., and Gallagher, B.M. (1985). *Biochim. Biophys. Acta*, *812*: 665.

Schwarz, A., and Steinsträsser, A. (1987). *J. Nucl. Med.*, *28*: 721.

Seifert, S., Noll, B., and Münze, R. (1982). *Int. J. Appl. Radiat. Isot.*, *33*: 1391.

Seiler, H.G. (1983). *Metal Ions in Biological Systems*, Vol. 16 (H. Sigel, ed.), Marcel Dekker, New York, p. 317.

Seldin, D.W., Johnson, L.L., Blood, D.K., Muschel, M.J., Smith, K.F., Wall, R.M., and Cannon, P.J. (1989). *J. Nucl. Med.*, *30*: 312.

Spies, H., and Johannsen, B. (1981). *Inorg. Chim. Acta*, *48*: 255.

Spies, H., and Scheller, D. (1986). *Technetium and Rhenium in Chemistry and Nuclear Medicine 2* (M. Nicolini, G. Bandoli and U. Mazzi, eds.), Cortina International, Verona, p. 141.

Taylor, A., Jr., and Eshima, D. (1988). *J. Nucl. Med.*, *29*: 669.

Taylor, A., Jr., Eshima, D., Fritzberg, A.R., Christian, P.E., and Kasina, S. (1986). *J. Nucl. Med.*, *27:* 795.

Thakur, M.L., DeFulvio, J., Richard, M.D., and Park, C.H. (1991). *Nucl. Med. Biol.*, *18*: 227.

Treher, E.H., Francesconi, L.F., Malley, M.F., Gougoutas, J.Z., and Nunn, A.D. (1989). *Inorg. Chem.*, *28*: 341.

Troutner, D.E., Volkert, W.A., Hoffman, T.J., and Holmes, R.A. (1983). *J. Nucl. Med.*, *24*: 10.

Walowitch, R.C., Makuch, J., Knapik, J., Watson, A.D., and Williams, S. (1988a). *J. Nucl. Med.*, *29:* 747.

Walowitch, R.C., Makuch, J., Garrity, S., Burgess, B., Morgan, R., Williams, S., Leveille, J., Donesuard, J., Rigo, P., Demonceau, G., Cantineau, R., and DeRoo, M. (1988b). *Neurology*, *38*: 363.

Walowitch, R.C., Hill, T.C., Garrity, S.T., Cheesman, E.H., Burgess, B.A., O'Leary, D.H., Watson, A.D., Ganey, M.V., Morgan, R.A., and Williams, S.J. (1989). *J. Nucl. Med.*, *30*: 1892.

Wilson, G.M., and Pinkerton, T.C. (1985). *Anal. Chem.*, *57:* 246.

RELAXIVITY OF MRI MAGNETIC CONTRAST AGENTS. CONCEPTS AND PRINCIPLES

by

Seymour H. Koenig

IBM T.J. Watson Research Center,
Yorktown Heights, NY 10598
U.S.A.
(Present address: Relaxometry Inc., P.O. Box 760
Mahopac, NY 10541, U.S.A.)

and

Rodney D. Brown

Field Cycling Systems, Inc.,
297-101 Kinderkamak Rd. # 145
Oradell, NJ 17649
U.S.A.

I. INTRODUCTION

Contrast in magnetic resonance imaging (MRI) depends, in the main, on tissue-dependent values of $1/T_1$ and $1/T_2$ (the longitudinal and transverse relaxation rates, respectively) of protons of mobile water molecules of tissue (Koenig et al., 1984a), given that water density tends to vary little between adjacent soft tissues (except in brain). These intrinsic rates may be altered considerably by the presence of magnetic complexes; the fact that about one-third of present-day clinical MRI involves the administration of intravenous paramagnetic chelate complexes of Gd^{3+} is evidence of the utility of exogenous magnetic agents. The majority of such agents are typified by Gd(DTPA), used mainly to affect $1/T_1$. The octadentate ligand DTPA (diethylenetriaminepentaacidic acid) forms a long-lived complex with the Gd^{3+} ion, with a net double-negative charge, a complex that may be regarded as chemically inert - and therefore minimally toxic - during the 30 minutes or so required for clearance from the circulation through the kidney. In addition, Gd(DTPA) does not interact with any components of blood (Koenig et al., 1986); its relaxation properties in vivo may be predicted directly from those measured in vitro. Other complexes of Gd^{3+}, particularly nonionic (to minimize osmotic effects) and lipophilic derivatives, are in more limited use or in trials, as are complexes of other paramagnetic ions. One goal of the present work is to clarify why Gd(DTPA) is the "gold standard", and then develop general chemical-physics principles that can guide the development and evaluation of new agents. For example, it is not a clear advantage that Gd(DTPA) does not interact with the components of blood. It is possible, for example, that toxic effects of such an interaction may be offset by advantageous changes in $1/T_1$ and $1/T_2$ at imaging fields such that much lower dosages could be used.

The question of the magnetic field dependence of all relevant relaxation parameters is an important one (Koenig and Brown, 1985; 1988). MRI fields have drifted up as technology has improved, mainly because of the intrinsically higher signal-to-noise ratio, and thus better image resolution, at higher fields. On the other hand, since considerations of cost (which increases with MRI field) are always a factor, one finds MRI instrumentation available for clinical use spanning the approximate range 0.05 to 4 T, corresponding to 2 to 180 MHz proton Larmor frequency. The relaxation parameters $1/T_1$ and $1/T_2$, and the magnetization transfer rate (Koenig et al., 1978) are all field dependent, as are the incremental contributions of contrast agents to these rates. Thus, this entire presentation is dominated by considerations of the magnetic field dependence (called a nuclear magnetic relaxation dispersion, or NMRD, profile (Koenig and Schillinger, 1969)) of the relaxation parameters of nuclei (generally protons) of mobile water molecules in solution and in tissue. For magnetic agents generally, not only are the NMRD profiles that they generate a function of the dynamic interactions of water molecules with the agent (this determines the basic functional forms of the profiles), but the magnetic parameters of contrast agents depend on the MRI field B_0, thereby adding more complexity to the phenomena to be considered (cf Koenig and Brown, 1990 for a review). It should be stressed that even if all MRI were done at a single field, the field dependence of $1/T_1$ (at least) must be known to make useful predictions about the behavior of contrast agents in MRI applications.

The study of NMRD profiles, part of "Relaxometry," requires specialized instrumentation - called a "field-cycling relaxometer" (FCR) (Koenig and Brown, 1987) - available in a limited number of laboratories in the US and Europe. The present discussion - essentially a monograph - is based in the main on data taken in our laboratory in association with many collaborators. One important point is that little is understood about the magnetic field dependence of the magnetic parameters of the agents themselves (cf Koenig and Brown, 1990), parameters that are inherently the realm of electron spin resonance (ESR) measurements and analyses. However, ESR has traditionally emphasized solid-state problems, generally at low temperatures; measurements in (liquid) water near physiological temperatures are intrinsically more difficult, as are measurements at more than one value of magnetic field. In fact, major understanding of the magnetic parameters of paramagnetic complexes, large and small, in solution has been inferred from NMRD measurements (Koenig and Brown, 1990) along with improvements in the theory of electron-nuclear interactions (cf Bertini and Luchinat, 1986). It is now time to reverse this order since the relaxivity of contrast agents can be better understood by improved ESR experiments used in combination with NMRD profiles. A second goal of this monograph is to indicate where these measurements can be helpful, in the interest of stimulating such research.

We first summarize what is known from the NMRD profiles of Gd(DTPA), and their explication by current theory (cf Koenig and Brown, 1990), with emphasis on what parameters of the system can be estimated quite well from the size, shape, and chemistry of Gd(DTPA), and what parameters must be assigned phenomenological values. It becomes necessary to distinguish between "inner-sphere" and "outer-sphere" contributions to relaxation, the first from waters exchanging from the inner coordination sphere of the Gd^{3+} ion of the complex, the second from waters diffusing exterior to the complex. The theory of both contributions is reviewed. We then summarize the behavior of a variety of small agents, indicating what could have been predicted, and with what accuracy. Next, we consider the qualitative changes that occur when small paramagnetic complexes are rotationally immobilized, e.g. by attachment to polymers, protein, and membranes by covalent, ionic, or hydrophobic bonding.

We then develop the concepts of relaxation by relatively large solute magnetic particles by extrapolating from the results for the small complexes. A single Gd(DTPA) molecule (or ion) is in reality a fully magnetized complex; it has the full magnetic moment of its Gd^{3+} ion, although the spatial orientation of the magnetic moment may be changing rapidly. Clustering several Gd^{3+} ions in a larger complex rarely, if ever, induces magnetic interactions that make the individual moments tend to be parallel. This is unfortunate, since the effect of a magnetic particle on the proton $1/T_1$ and $1/T_2$ of solvent is proportional to the *square* of the magnetic moment of the ion, for both inner- and outer-sphere contributions. Thus, two aligned Gd^{3+} ions would have four times the effect of a single ion, i.e. twice the effect of the two ions when separated. On the other hand, there are compounds of iron, typified by the mixed-valence oxide *magnetite* (Fe_3O_4 or, more properly, $FeO.Fe_2O_3$), in which the mutual interaction of the magnetic ions generates a "permanent" macroscopic magnetization. Since the larger the size, the larger the magnetic moment of the particulate, and thus the larger the effect on $1/T_1$ and $1/T_2$ for a fixed amount of magnetic material, these agents tend to be administered as virus-size particles, which are ultimately sequestered in the RE (reticulo-endothelial) system. Even this limited tissue specificity has already made agents of this type clinically attractive. We show how the qualitative behavior of the large agents can be understood from their NMRD profiles, and how their profiles relate to those of smaller agents. In addition, we indicate where ancillary information would be useful in predicting, ultimately, the in vivo behavior of magnetic particulates.

For both classes of contrast agent, i.e. small and large complexes, the same physical principles apply regarding the influence of their magnetism on $1/T_1$ and $1/T_2$ of tissue water. Water protons are relaxed mainly when "near" the magnetic moments, rather than when far away. The concepts, particularly when considering outer-sphere relaxation, come down to what is meant by "near," and how rapidly a mobile water gets "far" from a magnetic center. Any system can be characterized by scales both in space and in time. The answers are very different for the two classes of agents, small and large. For all agents, "near" means within a particulate radius of the surface. For small chelates, this means ~4 Å and escape times ~100 ps ($\sim 10^{-10}$s). For large particulates, the escape time can become micro- or even milli-seconds and, in fact, water molecules may never get far from a relaxation center on a NMRD time scale. It is these quantitative differences in the spatial and temporal parameters that lead to qualitative differences in the apparent properties of small and large complexes when used as contrast agents in MRI; another goal of the present work is to demonstrate how to estimate these parameters and their influence on $1/T_1$ and $1/T_2$, first in vitro and then in vivo, again with emphasis on the need for ESR and other magnetic measurements.

II. Gd(DTPA)

A. $1/T_1$ NMRD profiles

This paradigm of the many small solute paramagnetic chelate complexes considered for use in MRI allows solvent water to be "near" in two ways. First, a single water molecule is complexed with each Gd^{3+} ion in its inner coordination environment, with the oxygen 2.6 Å from the center of the Gd^{3+} ion and the protons a distance (r) about 3.1 Å away. The residence lifetime of these water molecules (τ_M) is generally in the 1 to 10 ns range for Gd^{3+} complexes (Koenig and Epstein, 1975), corresponding to an exchange rate sufficiently rapid to allow the relaxation that occurs in the complex to be thoroughly averaged over all mobile water protons and with little influence of the value of τ_M on the observed phenomena.* (Indeed, it is just this property that makes τ_M difficult to measure. On the other hand, τ_M can have a large influence on the relaxivity of rotationally immobilized Gd^{3+} complexes, as discussed below.) Second, water molecules can freely diffuse in the exterior neighborhood of the solute complex, with a distance of nearest approach (a) of 4.6 Å. The theory of both "inner-sphere" and "outer-sphere" relaxivity is well developed, quantitatively, and quite reliable for small agents such as Gd(DTPA) (cf Koenig and Brown, 1990).

**Note added in proof*: Recent work (González et al., 1994) shows that $\tau_M \sim 0.25\ \mu s$ for Gd(DTPA). However, few of the conclusions herein have been altered significantly.

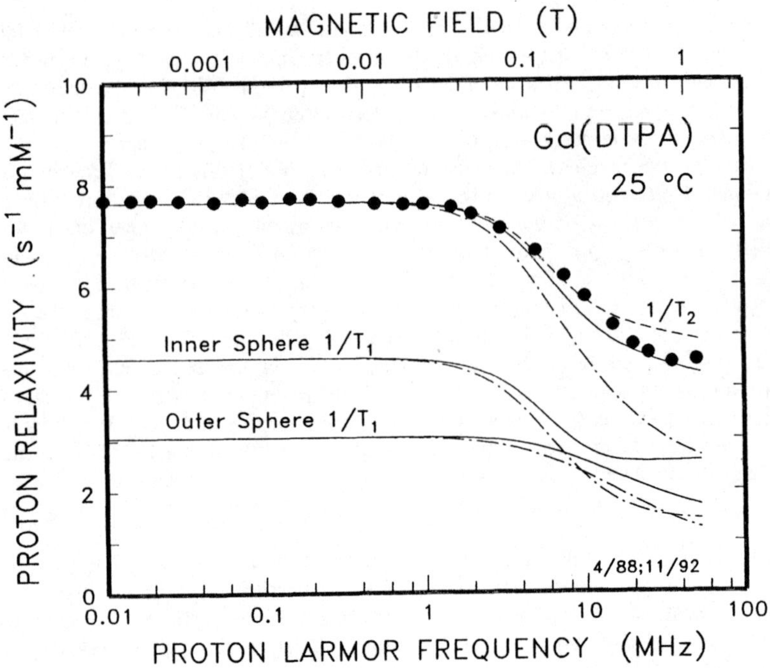

Figure 1.
The data points (●) are $1/T_1$, values for a water solution of Gd(DTPA) at 25°C, expressed as relaxivity, the incremental increase in $1/T_1$, per mM, over the solvent background. The lower two solid curves are the computed contributions of inner- and outer-sphere relaxation, as described in the text. The upper solid curve is the sum of the lower two. The lower two broken curves are the contributions expected if (cf eq. 9) $\blacklozenge_v = 0$; the upper broken curve is their sum. The upper dashed curve is the prediction of theory for $1/T_2$, based on the values of parameters derived from the fit to the measured $1/T_1$ values. Data after Koenig et al. (1984b).

The data points (Koenig et al., 1984b) (Fig. 1) show a longitudinal relaxivity profile of Gd(DTPA), here, $1/T_1$ of water protons, with the buffer rate (0.43 s^{-1}) subtracted, per mM solution of Gd(DTPA) at 25° C and near neutral pH. The solid curve through the data is a best fit of theory to data (Koenig and Brown, 1990), and the dashed line associated with the data is the expectation for $1/T_2$, seen to differ little from $1/T_1$ (this holds generally for small complexes over the field range shown). The solid curves of each of the lower pairs of curves show the inner- and outer-sphere contributions to the relaxivity, obtained by an iterative adjustment of the parameters that contribute to both (see below). The associated broken curves show what the contributions would be if the magnetic properties of Gd^{3+} in Gd(DTPA) were independent of field; their sum is the upper broken curve associated with the data points. The latter emphasizes that this field dependence can double the observed $1/T_1$ at high fields, i.e. at MRI fields.

B. Inner-sphere theory

One generality, valid for small complexes, is seen in Figure 1: the inner-sphere contribution of a single coordinated water is comparable, at all fields, to the combined contributions of all the outer-sphere waters (as will be seen, outer-sphere contributions decrease slowly with distance; one must include a volume of solvent about 1000-fold the solute size to obtain 90 % of the total outer-sphere contribution). The mathematical expressions that describe the relaxivity profiles (Fig. 1), although they have been given several times along with the various approximations and caveats stressed, bear repeating. The inner-sphere relaxivity depends, in general, on four parameters and a few reasonable assumptions. First, if there is more than one inner-coordinated water, it is assumed that they are chemically, physically, and sterically equivalent. Second, if there is but one, it is assumed that its two protons are equally, and maximally, distant from the paramagnetic ion. Gd^{3+} has the advantage, from the theoretical view, that additional model assumptions made to simplify the theory - e.g. treating the magnetic moment of the ion as localized at its nucleus, although it arises from orbital electrons, and assuming that all relaxation results arise only from magnetic dipolar (e.g. no contact) interactions - are most applicable to Gd^{3+}. This relates, of course, to the facts that not only are the electronic 4f levels that produce the electronic moment located deep within the Gd^{3+} electronic distribution and therefore relatively insensitive to ligands, but the f-shell is half-filled, leading to an S-ground state which is spherically symmetric and therefore only weakly coupled to thermal fluctuations in its environment; the more symmetric this environment, the smaller the coupling.

Equation 1 relates $1/T_{1is}$, the paramagnetic contribution to the observed longitudinal rates, to T_{1M}, the relaxation rate of protons of water molecules bound to the metal ions and in rapid exchange with solvent

$$\frac{1}{T_{1is}} = \frac{[C]q}{55.6}\frac{1}{T_{1M}+\tau_M} \quad . \tag{1}$$

Here, [C] is the molar concentration of paramagnetic ions (all assumed to be in identical chemical environments) and q the number of water ligands per ion; τ_M is assumed the same for each.

(As and aside, there are other examples of inner-sphere relaxation in which a water is held by hydrogen bonding of one of its protons to an anion coordinated directly to the metal ion. The first example of this is fluoro-methemoglobin, for which the relaxivity increases dramatically upon replacement of the long-lived inner-sphere water molecule of the Fe^{3+} ions of methemoglobin by F^-, which is hydrogen bonded to a water in rapid exchange with solvent (Koenig et al., 1981). An interesting aspect of this "fluoromet effect" is that it can lead to (approximate) half-integer values of q.)

The related expression for the paramagnetic contribution to $1/T_{2is}$ is more involved because of the possibility that a liganded water proton experiences a strong "chemical shift" field $\delta\omega$. This can introduce a random phase shift in the precession of the bound proton about the direction of B_0 when τ_M is appropriately short (which contributes to $1/T_2$, but not $1/T_1$). If the chemical shift is $\delta\omega$, then

$$\frac{1}{T_{2is}} = \frac{[C]q}{55.6}\frac{1}{T_{2M}+\tau_M}\times C_s \tag{2}$$

where

$$C_s = \frac{\left(1+\dfrac{((\delta\omega)T_{2M}\tau_M)^2}{\tau_M(T_{2M}+\tau_M)}\right)}{\left(1+\dfrac{((\delta\omega)T_{2M}\tau_M)^2}{(T_{2M}+\tau_M)^2}\right)} \quad . \tag{3}$$

In the limit of very rapid exchange, i.e. $\tau_M << T_{2M}$, $C_S \rightarrow (\delta\omega)^2 T_{2M}\tau_M$, and

$$\frac{1}{T_{2is}} = \frac{[C]q}{55.6}(\delta\omega)^2\tau_M \quad . \tag{4}$$

This inner-sphere limiting form has an analog in outer-sphere relaxation (eq. 14) to be considered below.

It should be noted that, for a chemical shift contribution, a water proton experiences an incremental change in the *longitudinal* component of magnetic field upon binding (generally linear in the applied field), which produces precessional phase shifts in the inner sphere of the paramagnetic complex. All outer-sphere relaxation is formally analogous to this term and, when the complex is sufficiently large, the contribution to relaxation is often called a "susceptibility effect". We will return to this point, noting for now that all such contributions can in principle be reduced by a Carr-Purcell pulse sequence. In the present case, the pulse separations would have to be comparable to or shorter than τ_M.

The expressions for $1/T_{1M}$ and $1/T_{2M}$ are

$$\frac{1}{T_{1M}}=\left(\frac{2}{15}\right)\gamma_I^2\gamma_S^2\hbar^2 S(S+1)\left(\frac{\tau_E}{r^6}\right)\{3j_{is}(\omega_I)+7j_{is}(\omega_S)\} \tag{5}$$

$$\frac{1}{T_{2M}}=\left(\frac{2}{15}\right)\gamma_I^2\gamma_S^2\hbar^2 S(S+1)\left(\frac{\tau_E}{r^6}\right)\{2+1.5j_{is}(\omega_I)+6.5j_{is}(\omega_S)\}\ . \tag{6}$$

Here, r (as above) is the distance of a liganded water proton from the center of the Gd^{3+} ion; S the electronic spin of the paramagnetic ion (= 3.5 for Gd^{3+}); and $\omega_I = \gamma_I B_0$ and $\omega_S = \gamma_S B_0$, the proton and electronic Larmor precession frequencies. For idealized conditions, the spectral density function, $j_{is}(\omega)$, is given by

$$j_{is}(\omega)=\left(\frac{1}{1+(\omega\tau_E)^2}\right), \tag{7}$$

a Lorentzian, where τ_E is the correlation time for the electron-proton dipolar interaction. Typically, $\omega_S \sim 660\ \omega_I$, so that the dispersions (eqs. 5 and 6) associated with ω_S occur at much lower fields than those associating with ω_I.

τ_E depends on the rate at which the dipolar interaction is modulated by electronic relaxation processes (i.e. mechanisms that influence the paramagnetic moment of the Gd^{3+} ions), characterized by τ_S; rotation of the complex, τ_R ; and dissociation of the complex, τ_M. The relation is

$$\frac{1}{\tau_E}=\frac{1}{\tau_S}+\frac{1}{\tau_R}+\frac{1}{\tau_M}\ . \tag{8}$$

In principle, there are both longitudinal and transverse relaxation processes for electronic relaxation and thus two components of τ_S (eq. 8) and, therefore, of τ_E; in practice this distinction is generally ignored, with few problems (Koenig, 1972). What is of greater import is that τ_S has its own dependence on field, analogous to nuclear relaxation times, and thus its own dispersion profile. Experience shows that adopting a theory for τ_S formally analogous to that for proton relaxation affords an adequate description of the field dependence of τ_S of Gd^{3+} (and its analog in the transition metal series, Mn^{2+}) in liquids near room temperature (Koenig and Epstein, 1975). As a result, the field dependence of τ_S is generally represented by

$$\frac{1}{\tau_S}=(\omega_{EL})^2\tau_V\times\left[\frac{0.2}{1+(\omega_S\tau_V)^2}+\frac{0.8}{1+(2\omega_S\tau_V)^2}\right];\ 1/\tau_{S_0}\equiv(\omega_{EL})^2\tau_V, \tag{9}$$

where ω_{EL} is an empirical electron-lattice interaction (arising from spin-orbit interactions), τ_V is a correlation time for the fluctuations of this interaction, and τ_{S_0} is the low field limit of τ_S. The two Lorentzians (eq. 9) result from assuming that the perturbation which produces electronic relaxation has the lowest symmetry needed to modulate the spin-orbit interaction by thermally induced molecular distortions.

Before considering outer-sphere effects, comments are in order about the values of the several parameters that determine $1/T_1$ and $1/T_2$. For water coordinated to Gd^{3+}, the value of r should vary little among complexes; that found here (3.06 Å, see below) is in good agreement with values inferred from X-ray data (which do not show proton locations explicitly). Note, however, that from equations 5 and 6, the sixth-power dependence of rates on r makes a 10% uncertainty (or change) in r equivalent to a change in q from one to two. This is a major limitation in trying to determine q from NMRD profiles (beyond the assumption that all waters are chemically equivalent).

The value of τ_R can be computed absolutely, using Stokes's law, from the known geometry of the complex and the solvent viscosity, with perhaps a 50% uncertainty because of the small size of the solute. However, this uncertainty can be reduced by comparisons of data for Gd(DTPA) with those of other Gd^{3+} agents. As noted, τ_M values are difficult to determine. The best values come from measurements of $1/T_1$ or $1/T_2$ of either ^2H or ^{17}O of solvent water, and ESR methods can be informative; there are few such data pertinent to those small paramagnetic complexes useful in MRI (González et al., 1994).

Little is known about ω_{EL} for Gd^{3+} complexes in liquids and even less ahout τ_V. Indeed, one of the best ways of determining their values is still through analyses of NMRD profiles generated by inner-sphere proton relaxation. The importance of τ_V in the present context is indicated by the difference between any pair of associated solid and broken curves (Fig. 1), but particularly for inner-sphere relaxivity; the broken curves were computed by assuming $\tau_V = 0$. As seen, the effect is almost a factor two in $1/T_1$ at 50 MHz. Once a successful resolution of the observed NMRD profiles into inner- and outer-sphere contributions is made (for small Gd^{3+} complexes), the high field $1/T_1$ values are determined by q, τ_R, and r, since the field dependence of τ_S (eq. 9) has removed its contribution to τ_E (eq. 8) and the values of the several parameters in the prefactor (eq. 5) are known.

Once values for τ_R and r are obtained, a comparison with the low-field inner-sphere results gives a value for τ_{S0}. The value of τ_V is then obtained from a fit to the details of the field dependence of $1/T_1$ of the inner-sphere relaxation in the intermediate field region. The accuracy of the results depends on the handling of the outer-sphere contribution.

C. Outer-sphere theory

Protons of solvent molecules diffusing exterior to a rigid, uniformly magnetized, solute sphere (a valid view for outer-sphere computations for Gd(DTPA)) experience fluctuations in the local magnetic field that are of two types: those that arise from a time dependence of the magnetic properties of the solute, characterized by τ_S, and those that arise from relative motion of solute and solvent. The first includes fluctuations in the orientation of the solute magnetic moment, which can arise from a variety of causes: the usual paramagnetic relaxation τ_S; thermally-induced jumping of the direction of the moment if the particle is small, magnetically ordered, and crystalline; and physical rotation of the solute if the paramagnetic dipole moment is coupled to the geometry of the solute particle (e.g. through spin-orbit interactions, or physically as in the iron storage protein ferritin). The second are fluctuations seen by solvent nuclei as waters diffuse relative to the solute particles, through the nonuniform fields near these particles. The contributions of each are a function of solute-solvent molecular separation; the diffusional contribution tends to dominate when this separation is small, since a diffusive jump then produces the greatest fractional change in local field. The contribution farther out tends to be dominated by what one would compute for stationary molecules. Mathematical complexities - mainly algebraic - arise when integrating the combined effects of the two contributions over all space. It turns out that accounting for the volume excluded by the particles from the start, both when considering the diffusive dynamics and in the final integration over all relaxing protons, actually simplifies the computations (Freed, 1978). The formal result, in a form particularly useful for small paramagnetic solutes with a single magnetic ion, is

$$\frac{1}{T_{1_{os}}} = \left(\frac{32\pi}{405}\right)\gamma_I^2\gamma_S^2\hbar^2 S(S+1)\frac{N_A}{1000}\left(\frac{[C]}{aD}\right)\{3j_{os}(\omega_I)+7j_{os}(\omega_S)\} \tag{10}$$

$$\frac{1}{T_{2_{os}}} = \left(\frac{32\pi}{405}\right)\gamma_I^2\gamma_S^2\hbar^2 S(S+1)\frac{N_A}{1000}\left(\frac{[C]}{aD}\right)\{2+1.5j_{os}(\omega_I)+6.5j_{os}(\omega_S)\} \tag{11}$$

where N_A is Avogadro's number; a is the distance of closest approach of solute and solvent; and D is the sum of solvent and solute diffusion constants. The spectral density function for diffusive fluctuations is more complex than for inner-sphere interactions (eq. 6)

$$j_{os}(\omega) = \mathrm{Re}\left(\frac{1+\frac{1}{4}(i\omega\tau+(\tau/\tau_S))^{1/2}}{1+(i\omega\tau+(\tau/\tau_S))^{1/2}+\frac{4}{9}(i\omega\tau+(\tau/\tau_S))+\frac{1}{9}(i\omega\tau+(\tau/\tau_S))^{3/2}}\right), \tag{12}$$

where Re means "the real part of", and $\tau = a^2/D$ is the correlation time for diffusion (the approximate time for solute and solvent to diffuse apart). When $\tau_S >> \tau$, i.e. when diffusion dominates the fluctuations in local field, then $j_{os} \rightarrow 1$ as $\omega \rightarrow 0$.

Equations 10 and 11 clarify the meaning of "near" in the outer-sphere case, since both $1/T_1$ and $1/T_2$ are inversely proportional to the distance of closest approach. Thus, excluding water from a volume with radius twice that of the solute molecules only lowers the relaxivity by a factor two. Put another way, one-half the contribution to outer-sphere relaxation comes from the region of space beyond a distance a from the solute surface. A general rule for outer-sphere relaxation, including "susceptibility" effects (see below) follows: only about one-half the relaxation occurs near the magnetic center, if "near" means within a linear dimension that characterizes a region of altered magnetic properties. As a result, other particles or structures "nearby," can significantly alter the relaxivity because of overlapping effects.

The pair of curves (Fig. 1) labeled "outer sphere" result from a fit to the difference between the data and the computed inner-sphere values. At high fields, because of the dispersion of τ_S, this is a one-parameter fit (eq. 10) which yields a value for a. One then computes the low-field outer-sphere contribution, using the same value for τ_{S_0} as used for the inner-sphere fit. If this result, added to the inner-sphere rate, does not match the data, another value for τ_{S_0} is chosen, and the process iterated. The final values for the pertinent parameters, for Gd(DTPA) at 25°C, are (Koenig and Brown, 1990) τ_R = 80 ps, τ_{S_0} = 85 ps, τ_V = 20 ps, and a = 4.59 Å; τ_M = 1 ns was assumed. The solid curve gives the computed outer-sphere contribution to $1/T_1$ and, again, the broken curve is for $\tau_V = 0$, meant to indicate the influence of the dispersion of the magnetic properties of the Gd^{3+} ions - in this particular compound - on the $1/T_1$ NMRD profiles of a water solution of Gd(DTPA).

III. OTHER SMALL Gd^{3+}-COMPLEXES

Gd(DOTA) is a small complex based on the polyaza macrocyclic 1,7,7,10-tetraazacyclododecanetetraacetic acid. The structural rigidity of the ligand makes the complex more kinetically inert than Gd(DTPA), which nonetheless has a higher thermodynamic stability. Nonionic derivatives of Gd(DOTA) are available for MRI use. The question here is whether the parameters that determine the relaxivities of DOTA-based and DTPA-based agents differ in a fundamental way. The answer is yes: τ_{S_0} for Gd(DOTA) is ten-fold longer than for Gd(DTPA). The reason for the longer τ_{S_0} is understood in principle: the higher symmetry of the macrocycle produces a more symmetric electric field, and weaker spin-orbit coupling between Gd^{3+} and complex. However, this has not been quantitated, and should be made part of a program of ESR studies. The longer τ_{S_0} as yet has not been of particular advantage since, at imaging fields, the dispersion due to τ_V removes the influence of τ_S, *but only for small, highly mobile complexes* (see below).

Figure 2A shows $1/T_1$ relaxivity at 50 MHz, plotted against an estimated value of integral q (eq. 1) for a range of small Gd^{3+} complexes, mostly based on NOTA, the triaza macrocyclic analog of DOTA (Geraldes et al., 1992). (The chemical structures are in Figure 3.) The outer-sphere value for Gd(DTPA) (Fig. 1) is taken as the q = 0 point. To the extent that the outer-sphere contribution is the same and the value of τ_R is appropriate for all complexes, the data points should all fall on a single line. They clearly do not (however, note the equivalence of Gd(DTPA) and Gd(DOTA)), although the general rule, indicated by the slope of the broken line - that each inner-sphere water contributes to $1/T_1$ an amount roughly equal to the total outer-sphere contribution - holds fairly well. The spread of the data about the line indicates the present range of ignorance, not about theory (in our view), but about the proper model assumptions, e.g. are inner-sphere waters bound symmetrically and are they chemically equivalent; how are the carboxylates configured; is there but a single solute conformer; must q be an integer; etc? Answers to these questions are beyond the scope of NMRD measurements; ESR and high resolution NMR studies on these complexes and their lanthanide analogs are suggested.

The data (Fig. 2B) are as for Figure 2A, but at 0.01 MHz, where the contribution of τ_{S_0} to the correlation time τ_E is important. The (imperfect) correlation between the

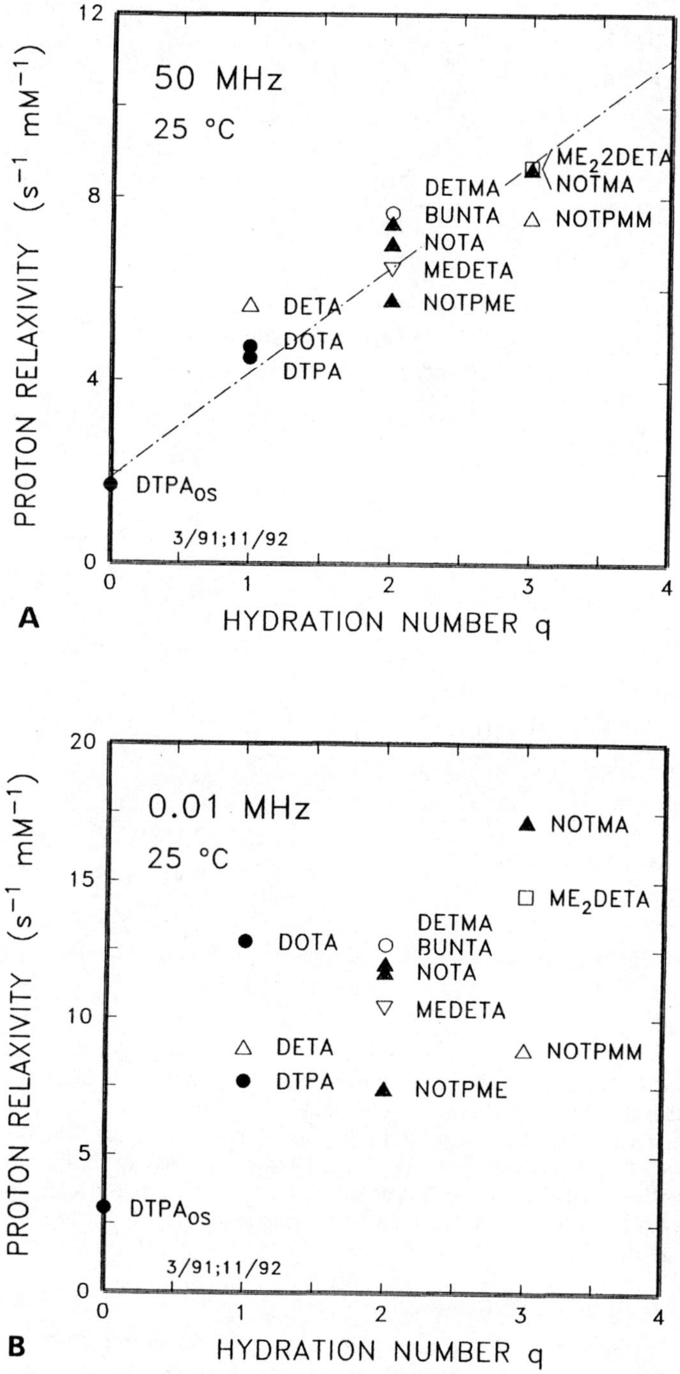

Figure 2.
A) The 50 MHz $1/T_1$ relaxivity values at 25°C, plotted vs integral estimates of the number q of inner-coordinated water molecules for a variety of Gd^{3+} chelate complexes. Except for DTPA and DOTA, the ligands are shown in Figure 3. $DTPA_{OS}$, the 50 MHz outer-sphere result from Figure 1, is taken as the q = 0 intercept. The dashed line indicates the trends. B) The same as (A), except for 0.01 MHz. After Geraldes et al. (1992).

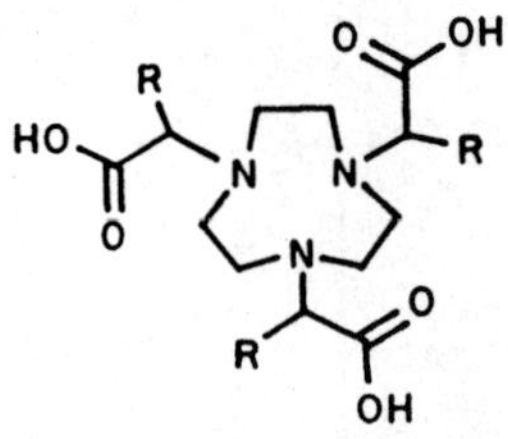

R = H NOTA
R = CH_3 NOTMA

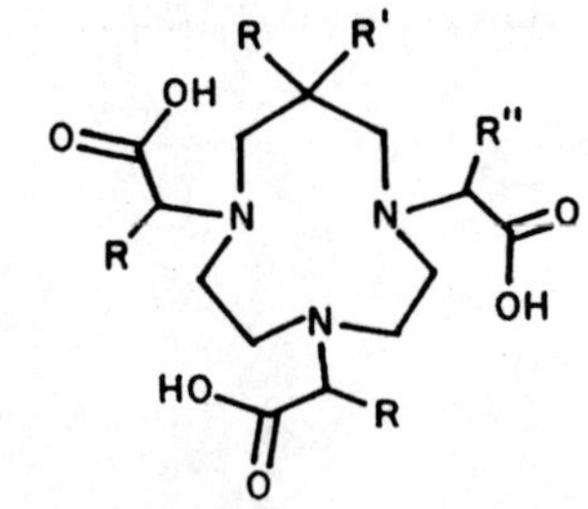

R,R',R" = H DETA
R=CH_3 ; R',R" = H MeDETA
R,R'=CH_3 ; R"=H Me_2 DETA
R,R'=H; R"=CH_3 DETMA

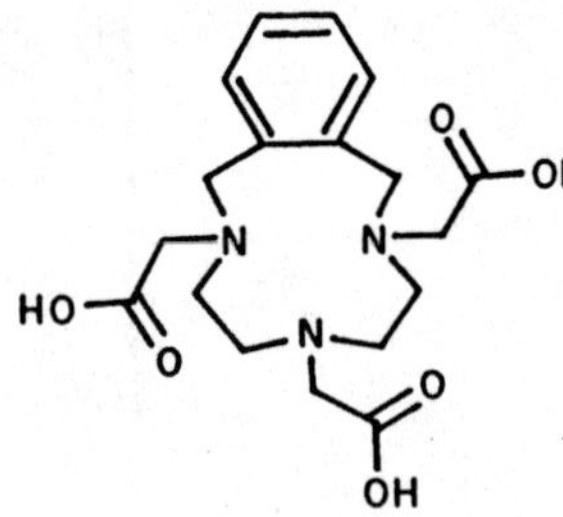

BUNTA

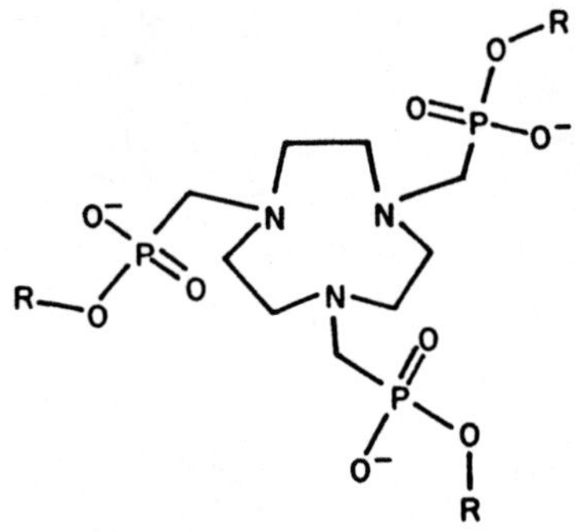

R = CH_3 NOTPMM
R = CH_2CH_3 NOTPME

Figure 3.
Structures of the triazacyclononane ligands relevant to Figure 2. After Geraldes et al. (1992).

estimated q and $1/T_1$ at 50 MHz - apparent in Figure 2A - is now obscured. It was pointed out earlier that the difference between Gd(DTPA) and Gd(DOTA) could be accounted for by the 10-fold larger τ_{S_0} for the macrocyclic Gd(DOTA) (cf Koenig and Brown, 1990). However, the other complexes (Fig. 2B) are all based on macrocyclic ligands; it is clear that a methodical study of the ESR spectra of these complexes is needed to be able to predict (or even estimate well) the relation between τ_S and structure for members of a single structural class of ligands.

IV. MACROMOLECULAR COMPLEXES

Referring to equation 9, we first note that $\tau_{S_0} \sim \tau_R \sim 80$ ps for Gd(DTPA) in water solution. At higher fields, τ_S becomes long quite rapidly, increasing quadratically with B_0 (eq. 9) once $\omega_S \tau_V \geq 1$. Since, for Gd^{3+}, $\tau_M \sim 1$ to 10 ns, τ_E can be made quite large at high fields by lengthening τ_R, potentially increasing the attainable relaxivity 50-fold. This can be accomplished in a number of ways, all involving linkage of the paramagnetic entity to a macromolecular structure; e.g. binding Gd(DTPA) rigidly to a protein 50-fold heavier than Gd(DTPA), ≥ 30 kg mol^{-1}. Figure 4 shows two examples of the general effect, a covalent ternary complex of Gd^{3+}, EDTA (ethylenediaminetetraacetic acid), and bovine serum albumin (Lauffer et al., 1986); and an extended Coulomb association, both specific

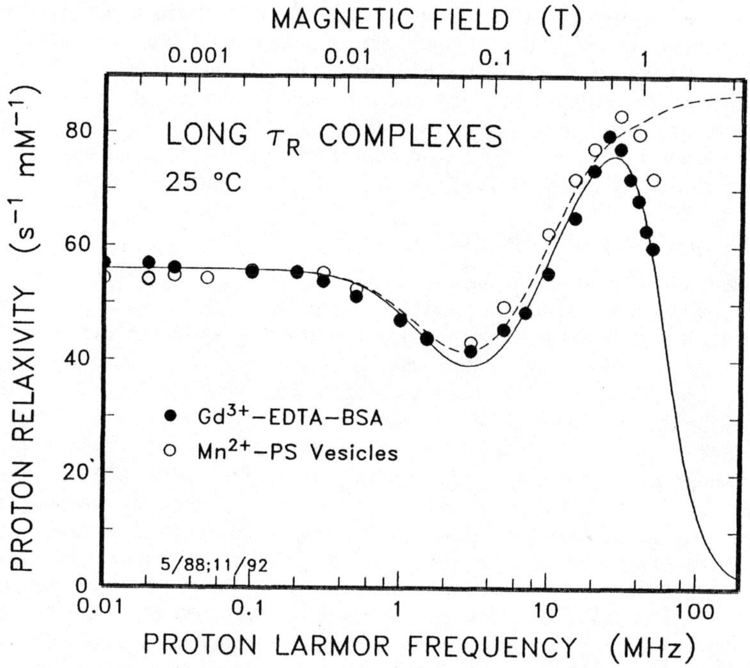

Figure 4.
The data points (●) are $1/T_1$ relaxivity values for a solution of Gd(EDTA) covalently linked to bovine serum albumin, at 25°C. The solid curve derives from a least squares comparison of these data with theory and the dashed curve is the prediction for $1/T_2$ relaxivity. The other data points (O) are for Mn^{2+} electrostatically bound to phosphatidylserine vesicles. (The near equality of the data for the two systems is coincidental.) Results for Gd^{3+} after Lauffer et al. (1986) and for Mn^{2+} after Koenig et al. (1988).

and nonspecific, of Mn^{2+} ions with the lipid-water interface of PS (phosphatidylserine) vesicles (Koenig et al., 1988)(the numerical agreement of the two profiles is coincidental). The solid curve is a fit of inner-sphere theory to the $1/T_1$ data for the Gd^{3+} complex, and the dashed curve the prediction for $1/T_2$ (outer-sphere effects are small and have been ignored). Note the large inner-sphere "secular" value (see below) of $1/T_2$ at high fields, unlike the case for small complexes (Fig. 1).

The major qualitative change, Figure 4 compared with Figure 1, is the appearance of a peak in relaxivity at high fields. Its onset, as field is increased, is caused by the dispersion in electronic relaxation (eq. 9) and the decrease in $1/T_1$ beyond the peak is associated with the dispersion of the bracketted term, $j_{is}(\omega_I)$, of equation 5, which disperses at progressively lower fields as τ_E lengthens. Similar profiles have been observed (unpublished) for many systems: Gd^{3+} ions interacting with polycarboxylates and polyphosphates by Coulomb interactions; binary and ternary complexes of Gd^{3+} or Mn^{2+} with ligands, polypeptides, and macromolecules; and protein complexes with Mn^{2+}, Fe^{3+}, and Gd^{3+}. Recently, the NMRD profiles of a wide variety of Gd^{3+} complexes with lipid vesicles have been reported (Tilcock et al., 1992). There is full theoretical understanding of the form of the dispersion profile in this high-field regime, with long τ_R, but not sufficient knowledge of the magnitude of τ_M, which can determine the maximum relaxivity that can be attained (cf eq. 1). In addition, there are few, if any, reported NMRD measurements of immobilized solute complexes with widely disparate values of τ_S, such as might be expected for macromolecular complexes of Gd(DTPA) and Gd(DOTA).

The importance of slowing rotation by forming, say, ternary hydrophobic complexes between paramagnetic metal ion, ligand, and protein cannot be underestimated. Thus, for example, the relaxivity of Gd(DTPA) in blood could be increased almost two orders of magnitude at imaging fields if the complex were modified to bind at the known hydrophobic surface sites of serum albumin. (Indeed, it was mainly the absence of such a peak in the NMRD profiles of Gd(DTPA) in blood which showed that no interaction existed between this contrast agent and *any* component of blood (Koenig et al., 1986).) Additionally, competitive binding of other molecules at these sites (e.g. bilirubin or a pharmaceutical) would quantitatively alter the observed relaxation rate of any small interactive chelate.

V. LARGE MAGNETIZED SPHERES

The foregoing has emphasized principles and results for small paramagnetic chelate complexes, which have similar $1/T_1$ and $1/T_2$ NMRD profiles. These complexes are most effective (or, at least, have been most used) in the circulatory system, where the unenhanced values of $1/T_1$ and $1/T_2$ are also comparable. Because of this, and because so much of MRI is used for visualizing brain - for which contrast between gray and white matter is greater when comparing T_1- rather than T_2-weighted images (Koenig et al., 1990) - most applications of Gd(DTPA), its analogs, and its extensions, have depended on its increase of $1/T_1$. However, for most tissues, typically $1/T_2 \sim 10\times(1/T_1)$ at MRI fields, and a so-called "T_2-agent" is desirable. This is found in the large solute magnetic particles, generally composed of magnetite stabilized chemically by organic binders and surface coatings. In these systems, relaxation is all outer sphere, and the ideal can be described by equations 10 to 12. Figure 5 shows the expected profiles for magnetite particles of different sizes (Koenig, 1990), with the total iron concentration the same for both samples, 1 mM in $FeO.Fe_2O_3$. The curves were computed using equations 10 and 11, assuming that τ_S is extremely long and can be ignored. For a sample of individual magnetite molecules, the expected dispersion profile is much like the outer-sphere contribution of Gd(DTPA) (Fig.

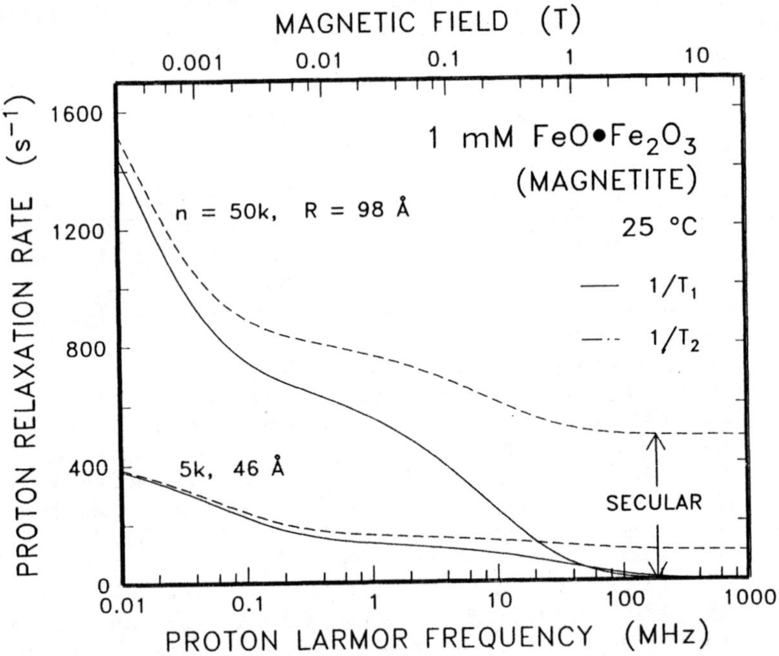

Figure 5.
Theoretical predictions for $1/T_1$ (solid curves) and $1/T_2$ (dashed curves) for solutions of magnetite particles, at 25°C, in one case with 50,000 molecules of $FeO.Fe_2O_3$ (upper curves) and another with 5,000 molecules. The total magnetite content of both samples is the same, corresponding to 1 mM magnetite molecular equivalents. It is assumed that the solute particles are fully magnetized at all fields. After Koenig (1990).

1), but with a low-field relaxivity of only ~$0.25\#s^{-1}$. The advantage of fewer but larger solute particles is clear.

The size of the larger particle (Fig. 5) was chosen to emphasize the two distinct regimes of the NMRD profiles, below and above ~20 MHz. At low fields, $1/T_1$ and $1/T_2$ disperse similarly and have similar values, as expected from equations 10 and 11. At the higher fields, $1/T_1$ disperses essentially to zero, whereas $1/T_2$ becomes constant at its "secular" value, given by the constant in the brackets (eq. 11). One can define the total magnetic dipolar moment μ of the solutes (eqs. 5 and 6) such that

$$\mu^2 = \gamma_S^2 \hbar^2 S(S+1). \tag{13}$$

It is convenient, for larger solute spheres of (field independent) moment μ, to rewrite equations 5, 6, 10, and 11 in terms of the radius R of the solute particles and the volume fraction V_f of the solution that they occupy, and either μ or the equatorial field generated by μ at the solvent interface. The result is, at zero field

$$\frac{1}{T_1} = \frac{1}{T_2} = \frac{16}{27} V_f (\delta\omega)^2 \tau \tag{14}$$

where

$$\delta\omega = \gamma_I \mu / R^3 . \tag{15}$$

Here $\delta\omega$ is the proton Larmor frequency in the equatorial field. Equations 14 and 15 are valid so long as diffusion is rapid in the sense that $(\delta\omega)\tau << 1$, i.e. in the time taken for a molecule to diffuse away from a solute particle, the proton precessional phase shift due to $\delta\omega$ is much less than one cycle. The variable here, $\delta\omega$, is not identical to the contact shift (eq. 3), but they are similar in that each is a measure of the change in longitudinal field experienced by a water proton when "near" the magnetic solute. Indeed, when exchange from an inner-sphere environment is sufficiently rapid, the expressions for both inner- and outer-sphere $1/T_2$ become formally equivalent, as seen by comparing equations 4 and 14. Hence the use of the same symbol for these two different, but analogous, interactions.

The secular contribution to outer-sphere $1/T_2$ (Fig. 5) is often called the "susceptibility" contribution; it can be computed knowing $\delta\omega$ (eq. 15), which is a measure of the change in the longitudinal component of B_0 near a region of discontinuous magnetic susceptibility. It is equivalent to a quasi-classical description - involving the sum of random precessional phase shifts as for the chemical shift term (eq. 3). Both are associated with alterations of $1/T_2$ and not $1/T_1$; hence "T_2-agents". Note, however, that the secular contribution to $1/T_2$ (Fig. 5) is the high-field part of a continuous $1/T_2$ profile which, at lower fields, approaches the $1/T_1$ profile. The field regime of this crossover is determined by the correlation time τ, which is a function of particle size. Thus, the concept of susceptibility agents, or T_2 agents, becomes valid only for relatively large-scale solute structures and higher fields; e.g. above ~20 MHz for 200 Å diameter particles, the size of small viruses. Otherwise, susceptibility effects, then more appropriately called outer-sphere relaxation, contribute to both $1/T_1$ and $1/T_2$.

VI. SUPERPARAMAGNETIC PARTICLES

The profiles (Fig. 5) were computed - as stressed above - for a field-independent moment μ. If the sphere derives its moment from paramagnetism as, for example, does a red blood cell, then the profiles (Fig. 5) must be corrected for this effect, not difficult to do since the correction is linear in B_0. For particles that can exhibit permanent magnetization, this becomes more difficult because of the vagaries of the behavior of small magnetic solid particles at low fields. When possible, and particularly when the particles approach 1000 Å in size, the permanent magnetization of the samples tends to form domains in which the magnetization of each domain is in a different direction. The driving force is the reduction in total magnetic energy generated external to the particles. Opposing this is the energy cost of forming transition regions in the magnetic material from one direction of internal field to another; the existence of "easy" and "hard" directions for magnetization of the solid, and

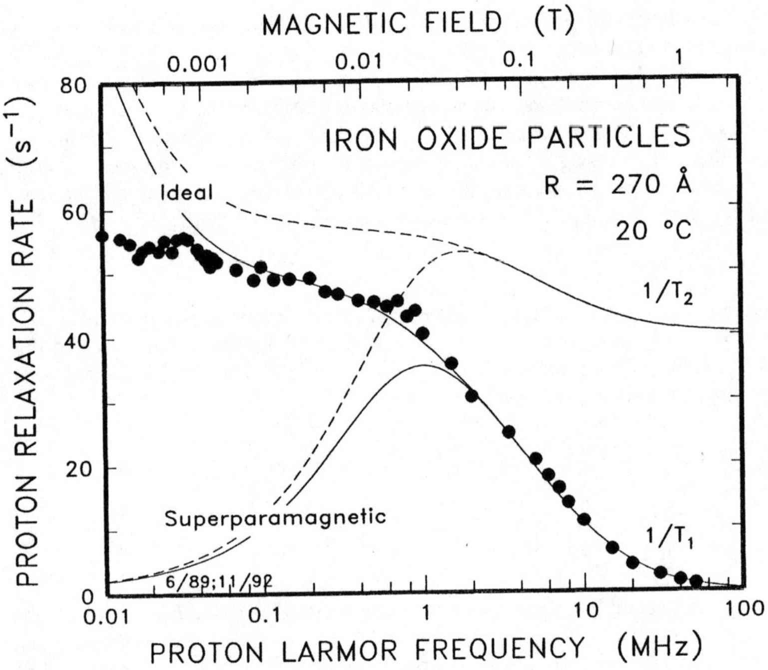

Figure 6.
The points (●) are data for iron oxide particles (magnetite-like) with mean radius 270 Å. The upper set of curves (ideal) was computed assuming that the solute particles are fully magnetized at all fields; the lower set (superparamagnetic) that they have a magnetization curve similar to magnetite. The solid ideal curve is a least squares fit to the data above ~1 MHz, with only the vertical scale adjustable, since R, eq. 15, is known. The three other curves are generated from this one. After Koenig and Brown (1990); data from S. Frank (private communication).

- if the sample is aspheric - large "demagnetizing" forces that depend on geometric details. Generally, all these effects can be overcome by having B_0 greater than some small value ~0.01 T. The result is a retrieval of the total possible permanent magnetization in response to a small applied field; hence the term "superparamagnetism". It is an unfortunate and misleading term, however, since it has nothing to do with the magnetic behavior and contrast enhancement of superparamagnetic particles at imaging fields. At best, this property keeps the dry particulates from aggregating in their containers at low fields (as do iron filings).

Figure 6 shows the NMRD profiles of a typical superparamagnetic sample, and two theoretical fits to the data: one assumes a field-independent moment for the sample, the other assumes zero moment at zero field and typical saturation behavior for the magnetization. It is clear that, for this mixed-valence iron oxide sample, the reality is between the two limits. Nonetheless, the data can be described by the outer-sphere analysis presented here. Clearly, particles of this size behave as T_2-agents above ~10 MHz.

VII. GENERAL REMARKS

Although little has been said about the behavior of the many contrast agents in vivo (except that Gd(DTPA) does not interact with any components of blood), the range of possibilities is limited - in the main - to the phenomena presented here. A major reason is that the water of tissue is little different from pure water in its ability to sample its local environment, generally several cell diameters, within a relaxation time. Thus, for small chelates in tissue, there can be a shift in behavior from that of Figure 1 to that of Figure 4 (or something intermediate) if the rotational mobility of the paramagnetic complex is reduced in vivo. The main complication that can arise is that the agent may accumulate in a small compartment of tissue, generating relaxation rates in that region which are greater than the rate of exchange of water out of that region. In such a case, $1/T_1$ and $1/T_2$ become biexponential, often with the higher rate unobservably large for the usual NMRD (Spiller et al., 1988) or MRI protocols. In this case, the effect of the agent may be missed, or misinterpreted.

For large particulates, the analogous compartmentation in vivo may enhance $1/T_2$ beyond expectations. For example, if sequestered in (small) Kupfer cells, the reticulo-endothelial (RE) component of liver, several separate particles may align in the applied field and make each cell appear - to the external water - as a single, cell-size, particle with a moment given by the sum of those of the sequestered particles, since they would be magnetically aligned by B_0.

The current state is that there is little doubt that the in vivo properties of MRI magnetic contrast agents can be understood from in vitro measurements, and that the theory of both inner- and outer-sphere relaxation is adequate to provide a quantitative explanation of the observations in terms of interactions at the molecular level. What is lacking is a comparable understanding of the relaxation behavior of the magnetic moment of magnetic ions, obtainable from ESR experiments, as well as NMRD data of newer agents in tissue.

Finally, it should be remarked that, in MRI, in vivo changes in $1/T_1$ and $1/T_2$ are ultimately expressed as changes in image contrast. Related to the fact, noted earlier, that $1/T_2 \sim 10\times(1/T_1)$ for tissue in the MRI field range, the established imaging protocols are designed such that reductions in $1/T_1$ or increases in $1/T_2$ both increase image intensity. Moreover, the relationships are nonlinear, so that reliable quantitative estimates of $1/T_1$ and $1/T_2$, and their changes in response to magnetic agents, are needed for a quantitative understanding of image contrast in MRI.

REFERENCES

Bertini, I., and Luchinat, C. (1986). *NMR of Paramagnetic Molecules in Biological Systems*, Benjamin/Cummings Publishing, Menlo Park, CA.

Freed, J. (1978). *J. Chem. Phys.*, *68*: 4034.

Geraldes, C.F.G.C., Brown, R.D., III, Brucher, E., Koenig, S. H., Sherry, A. D., and Spiller, M. (1992). *Magn. Reson. Med.*, *27*: 284.

González, G., Powell, D.H., Tissières, V., and Merbach, A. (1994). *J. Phys. Chem.*, *98*: 53.

Koenig, S.H. (1972). *J. Chem. Phys. 56*: 3188.

Koenig, S.H. (1990). *Acta Radiologica, Suppl. 374*: 17.

Koenig, S.H., and Brown, R.D., III. (1985). *Invest. Radiol.*, *20*: 297.

Koenig, S.H., and Brown, R.D., III. (1987). *NMR Spectroscopy of Cells and Organisms*, Vol. 11 (R.K. Gupta, ed.), CRC Press, Boca Raton, FL, p. 75.

Koenig, S.H., and Brown, R.D., III. (1988). *Invest. Radiol.*, *23*: 495.

Koenig, S.H., and Brown, R.D., III. (1990). *Prog. NMR Spectr.*, *22*: 487.

Koenig, S.H., and Epstein, M. (1975). *J. Chem. Phys.*, *63*: 2279.

Koenig, S.H., and Schillinger, W.E. (1969). *J. Biol. Chem.*, *244*: 3282.

Koenig, S.H., Bryant, R.G., Hallenga, K., and Jacob, G.S. (1978). *Biochemistry*, *17*: 4348.

Koenig, S.H., Brown, R.D., III, and Lindstrom, T.R. (1981). *Biophys. J.*, *34*: 397.

Koenig, S.H., Brown, R.D., III, Adams, D., Emerson, D, and Harrison, C.G. (1984a). *Invest. Radiol.*, *19*: 76.

Koenig, S.H., Baglin, C., Brown, R.D., III, and Brewer, C.F. (1984b). *Magn. Reson. Med.*, *1*: 496.

Koenig S.H., Spiller, M., Brown, R.D., III, and Wolf, G.L. (1986). *J. Magn. Reson.*, *3*: 791.

Koenig, S.H., Brown, R.D., III, Kurland, R., and Ohki, S. (1988). *Magn. Reson. Med.*, *7*: 133.

Koenig, S.H., Brown, R.D., III, Spiller, M., and Lundbom, N. (1990). *Magn. Reson. Med.*, *14*: 482.

Lauffer, R.B., Brady, T.J., Brown, R.D., III, Baglin, C., and Koenig, S.H. (1986). *Magn. Reson. Med.*, *3*: 541.

Spiller, M., Koenig, S.H., Brown, R.D., III, and Wolf, G.L. (1988). *Magn. Reson. Med.*, *8*: 293.

Tilcock, C., Ahkong, Q.F., Koenig, S.H., Brown, R.D., III, Davis, M., and Kabalka, G. (1992). *J. Magn. Reson.*, *27*: 44.

ROLE OF METAL-LIGAND INTERACTIONS IN THE DESIGN OF MRI CONTRAST AGENTS

by

J.F. Desreux and V. Jacques

Laboratoire de Chimie Analytique et Chimie de Coordination,
Université de Liège,
Sart Tilman (B6)
B-4000 Liège
Belgium

I. INTRODUCTION

Medical images can be obtained using a variety of forms of radiation such as radio waves or X-rays. In X-ray computed tomography, tissues are discriminated on the basis of their difference in density because of the interaction between an X-ray beam and orbital electrons. Nuclear magnetic resonance imaging (MRI) recently became a standard clinical technique in numerous hospitals because of its greater diagnostic scope, especially for soft tissues (Morris, 1986). Indeed, the intensity of a MRI signal depends not only on the density of the tissues as in X-ray tomography, but also on parameters that show a much larger dynamic range, such as the relaxation times. The proton is the nucleus of choice in MRI because of its sensitivity and because of its abundance in the body (essentially the protons of water and fat), but MRI images of other nuclei have recently been reported, for instance in the case of ^{23}Na (Kaiser, 1992) or ^{31}P (Morris, 1986). Currently, some of the main applications of MRI are (i) the detection of tumors in the brain (Runge et al., 1991), the liver (Hamm et al., 1992) or the breast (Kaiser, 1992), (ii) acute synovitis in the knee (Hervesomma et al., 1992), (iii) angiography and perfusion measurements (Turner and Keller, 1991), (iv) obstruction of the kidney (Semelka et al., 1990). Research in MRI is being continued at a considerable pace and promises numerous new developments in all fields of medicine.

If the most common MRI pulse sequence is used (spin echo), the intensity S of the MRI signal is given by (Wehrli et al., 1984)

$$S_{\text{spin echo}} = [H]f(v)\left[1 - 2e^{-\frac{(t_r - t_e)/2}{T_1}} + e^{-t_r/T_1}\right] e^{-t_e/T_2} \tag{1}$$

where [H] is the proton density and f(v) is a flow factor. Each tissue is characterized by a longitudinal (or spin-lattice) relaxation time T_1 and a transverse (or spin-spin) relaxation time T_2. The echo time t_e and the repetition time t_r are experimental parameters that can be varied by the radiologist in order to increase the contrast of the images. For instance, a good T_1 contrast is obtained if both t_e and t_r are short, while T_2-weighted images are recorded by lengthening these two parameters. Because the values of the relaxation times depend much more on the structure of the tissues than the spin density, T_1- or T_2-related images show a much better soft tissue contrast than those based primarily on spin density or than those obtained by X-ray computed tomography. Recently, a number of fast scanning techniques have been developed (Frahm et al., 1986). Methods such as FLASH employ radiofrequency pulses that tip the magnetization less than 90 degrees for acquiring more quickly imaging data. The intensity of the MRI signal then also depends on the tip angle. Thus, the radiologist has to find the appropriate settings for several experimental parameters in order to achieve the best contrast between tissues. However, the contrast between normal and malignant tissues is not always large enough, and the clinicians risk missing many lesions or are unable to distinguish between a tumor and the surrounding edema. For instance, meningiomas in the central nervous system feature T_1 and T_2 relaxation times that are similar to those of the normal brain. Even though MRI was originally thought to be a non-invasive technique, it is now realized that the contrast of images must often be increased by the intravenous (iv) injection of compounds that are able to drastically alter the values of the relaxation times. About 25-35% of all MRI images are now taken after the administration of a contrast agent and as many as half of all head studies require the use of a contrast agent (Brasch, 1992).

Contrast agents are substances that contain unpaired electronic spins and thus that have an extremely strong effect on the relaxation times, even if they are present in low concentrations. As indicated in the preceding Chapter 6-D in this book, the interaction between the electronic magnetic moment of a paramagnetic species and the nuclear magnetic moment of the protons of water leads to a shortening of the relaxation times T_1 and T_2. As shown by equation 1, the intensity of the MRI signal increases if T_1 is decreased by the addition of a paramagnetic compound. On the contrary, a decreased value of T_2 brings about a reduced signal intensity. The experimental parameters t_e and t_r are usually chosen so that the presence of a paramagnetic species is revealed by an increase in signal intensity in T_1-related images. Contrast agents are thus adding a new dimension to MRI: the selective distribution of a paramagnetic compound in the body can allow the distinction between normal and diseased tissues. In addition, functional evaluations of organs, for instance the kidney, or analysis of the blood flow can be carried out. Finally, images can be recorded faster because the smaller value of the relaxation times allow a faster collection of the entire spin echo in equation 1.

The MRI contrast agents are divided in two families of compounds (Brasch, 1992): on one hand, the positive enhancers for which T_1 shortenings predominate and thus for which the influence of the unpaired electronic spins is best seen in T_1-weighted images as described above; on the other hand, the negative enhancers that strongly reduce T_2 and thus lead to a decreased signal intensity easily seen in T_2-related images. The first category of contrast agents includes nitroxide radicals as well as metal ions that feature an isotropic distribution of their unpaired electronic spins and thus a long electronic relaxation time. Ferromagnetic particles and metal ions that induce very large paramagnetic NMR shifts belong to the second category. These species cause a loss in phase coherence of the spins of the water protons and thus a shortening of T_2.

An ideal MRI contrast agent should feature several properties that can be antagonistic and that are thus difficult to find in one single compound:

(i) A contrast agent should generate a *maximum increase in relaxation rates*. At least a 10-20 % increase in $1/T_1$ is necessary for an easy detection. The paramagnetic species must thus be in close contact with water molecules.

(ii) A complete *lack of toxicity* is of prime importance. The paramagnetic metal ions that are suitable for MRI are too toxic and they must be complexed by a ligand so as to ensure a complete excretion within a few hours. This implies a high thermodynamic stability of the complex and even more desirably, a high kinetic inertness. Moreover, the *osmolality* of the contrast agent should be low in order to avoid an osmotic shock during an iv injection. Finally, the contrast agents must be *highly purified compounds* and this requirement can sometimes be very difficult to fulfil.

(iii) *Tissue selectivity* is of course an added advantage for a contrast agent: more diluted solutions can be administered and much more informative images are obtained. The ligand of a paramagnetic chelate should thus bear substituents that favor the targeting with a tissue. For instance, a chelate with a hydrophobic group could tend to be retained by the liver. The contrast agents that have been approved so far by the US Food and Drug Administration are non-specific extracellular chelates that allow a better detection of brain tumors only because the latter are highly vascularized.

(iv) The *price* of the ligand can be of practical importance. A standard dose of MRI contrast agent is about 0.2 mmol per kilo of body weight. This represents 5 to 8 g of compound for a 70 kg man.

Central to the design of MRI contrast agents is thus the ability of the metal ion to associate with water molecules and to generate relaxation rate enhancements that are as large as possible. In addition, the metal ion must be strongly chelated by a ligand that should feature substituents that can bind specifically with tissues. These requirements are often contradictory: a high stability is achieved if the ligand in a complex contains several ionizable groups such as carboxylates that strongly interact with a metal ion. However, the complex is then negatively charged and its solutions exhibit a high osmolality. Moreover, the presence of numerous coordinating groups reduce the space available for water molecules and eventually prevents the formation of hydrated species. Finally, chemically modifying the ligand with various groups that must bind specifically with tissues can alter drastically the stability and the relaxation ability of the complex. Thus, the synthesis and the study of the properties of MRI contrast agents represent both an exercise in fundamental coordination chemistry and a key to obtaining more effective compounds for medical applications.

More than $2 billion are spent annually for contrast agents in medicine and MRI should account for a large part of this market in the future. It is thus not surprising that numerous pharmaceutical companies are developing new MRI contrast agents. More new ligands have probably been synthesized during the last 10 years than since the preparation of the first polyaminopolyacetic ligands such as EDTA. Not surprisingly, the use of most ligands is covered by patents.

Because this book is devoted to metal-ligand interactions, the present chapter will be devoted only to metal complexes. Furthermore, particular attention will be given to the complexes of the lanthanides since gadolinium chelates are the most popular contrast agents. No information will be given on ferromagnetic particles that are usually coated with a protective polymeric material but that do not contain metal ligands per se (see preceding Chapter 6-D). The magnetic properties of the paramagnetic particles are discussed in this book. The reader is also referred to several reviews specifically devoted to contrast agents, for instance by Brasch (1992), Desreux and Barthélemy (1988), Koenig and Brown (1990), Kumar and Tweedle (1993), Lauffer (1987), and Tweedle (1992). Finally, it should be noted here that the ligands used in MRI are of interest in other fields of medicine, namely in the use of radiolabelled antibodies for radiotherapy (Parker, 1990; Brechbiel and Gansow, 1991; Deshpande et al., 1990) and in neutron capture therapy (Shih and Brugger, 1992). The chemical structures of some of the ligands mentioned in this review are reproduced in Figures 1-2.

II. PARAMAGNETIC METAL IONS FOR MRI

Seeking a substantial alteration of relaxation times implies the administration of metal ions with high magnetic moments and thus with a large number of unpaired electronic spins. Three metal ions, namely Gd^{3+}, Fe^{3+} and Mn^{2+}, stand for special note because they feature 7 or 5 unpaired spins (for high-spin complexes). These ions are S states in the Russel-Saunders scheme. Their magnetic susceptibility tensor is isotropic and their electronic magnetic moment is only weakly coupled with the lattice. The electronic relaxation time T_{1e} of these ions is thus relatively long, and they are able to strongly enhance the relaxation rates of nuclei that are in their immediate vicinity. This enhancement cannot take place so effectively if T_{1e} is too short. The Gd^{3+} ion has attracted special attention since it has the highest magnetic moment (7.95 Bohr magnetons) of all metallic ions. The use of gadolinium and of the other lanthanides in MRI will be discussed first. The use of the transition metal ions will be dealt with in Part IV.

III. LANTHANIDE COMPLEXES

A. Selection of metal ions

The contribution of the Gd^{3+} ion to the measured longitudinal relaxation rate of an aqueous solution of concentration [Gd] is given by

$$\frac{1}{T_1} = \frac{[Gd]}{55.6} \frac{q_{H_2O}}{T_{1M} + \tau_M} \tag{2}$$

where T_{1M} is the relaxation time of the protons of the water molecules directly bonded to gadolinium and where q_{H_2O} is the number and τ_M the mean residence life time of the coordinated water molecules. The relaxation time, T_{1M}, depends on a correlation time, τ_c, as indicated below:

$$\frac{1}{T_{1M}} = \frac{2}{15} \frac{\gamma_I^2 \beta^2 g^2 S(S+1)}{r^6} \left[\frac{7\tau_c}{1+\omega_s^2\tau_c^2} + \frac{3\tau_c}{1+\omega_I^2\tau_c^2} \right] \tag{3}$$

where γ_I is the magnetogyric ratio of the proton, g is the electronic g-factor, β is the Bohr magneton, r is the distance between the protons of water and the Gd^{3+} ion, ω_S and ω_I are the Larmor angular precession frequencies of the electron spin and of the proton spin respectively ($\omega_S = 658\ \omega_I$), and where $S = 7/2$. The correlation time, τ_c, is related to the electronic relaxation time, T_{1e}, the rotational correlation time of the gadolinium species, τ_R, and the residence life time of the coordinated water molecules, τ_M, by the equation

$$\frac{1}{\tau_c} = \frac{1}{T_{1e}} + \frac{1}{\tau_M} + \frac{1}{\tau_R} \tag{4}$$

The correlation time, τ_c, is dominated by the fastest molecular process. For the Gd^{3+} aquoion and its complexes, the approximate values of the correlation times are $T_{1e} \sim 10^{-9}$ s at a field of 1 T, $\tau_M \sim 10^{-9}$ s and $\tau_R \sim 10^{-10}$ to 10^{-11} s (Koenig and Epstein, 1975; Southwood-Jones et al., 1980). The correlation time, τ_c, of gadolinium complexes is thus primarily determined by τ_R. The above equations also apply to the longitudinal relaxation time of water associated to the other paramagnetic lanthanides, provided the $S(S+1)g^2$ term is replaced by $J(J+1)g_J^2$ where J is the quantum number for the total angular momentum and where g_J is the electronic Landé g-factor. In contrast with Gd^{3+}, the magnetic susceptibility tensor of all the other paramagnetic lanthanide ions is anisotropic and the coupling with the lattice drastically shortens the electronic relaxation time (to less that 10^{-13} s for ions such as Yb^{3+} or Eu^{3+}) so that τ_c is dominated by T_{1e}. The relaxation time, T_{1M}, becomes two to three orders of magnitude higher than in the case of Gd^{3+}, and these lanthanide ions are not suitable T_1 relaxation agents. The division of the lanthanides in two classes also arises in the case of the shift reagents for which Gd^{3+} is a relaxation probe while all the other paramagnetic 4f ions are shift probes (Inagaki and Miyazama, 1981). However, at the border between these two classes of metal ions, one finds ions of intermediate T_{1e} such as Dy^{3+} that induce very large paramagnetic NMR shifts and that also enhance the relaxation rates. The proton spins of water molecules diffusing in the immediate surroundings of Dy^{3+} are submitted to the very large anisotropy of the magnetic susceptibility tensor of this ion and are losing their coherence. The observed transverse relaxation time T_2^* is thus strongly reduced and a loss of spin-echo signal intensity is observed. The Gd^{3+} complexes are thus the preferred positive enhancers while the Dy^{3+} chelates could find a practical use as negative enhancers.

B. Selection of the ligands

The acute iv tolerance LD_{50} of $GdCl_3$ in rats is about 0.5 mmol/kg (Weinmann et al., 1984). The Gd^{3+} ion is thus unsuitable for medical applications, and it must be complexed before iv injection in order to reduce its toxicity and to ensure a rapid excretion from the body. Early in the development of MRI, it was thus suggested to chelate Gd^{3+} with DTPA (see the structures in Fig. 1). With a LD_{50} = 8 to 10 mmol/kg, the tolerance for $GdDTPA^{2-}$ is much higher than for the uncomplexed ion or than for less stable chelates

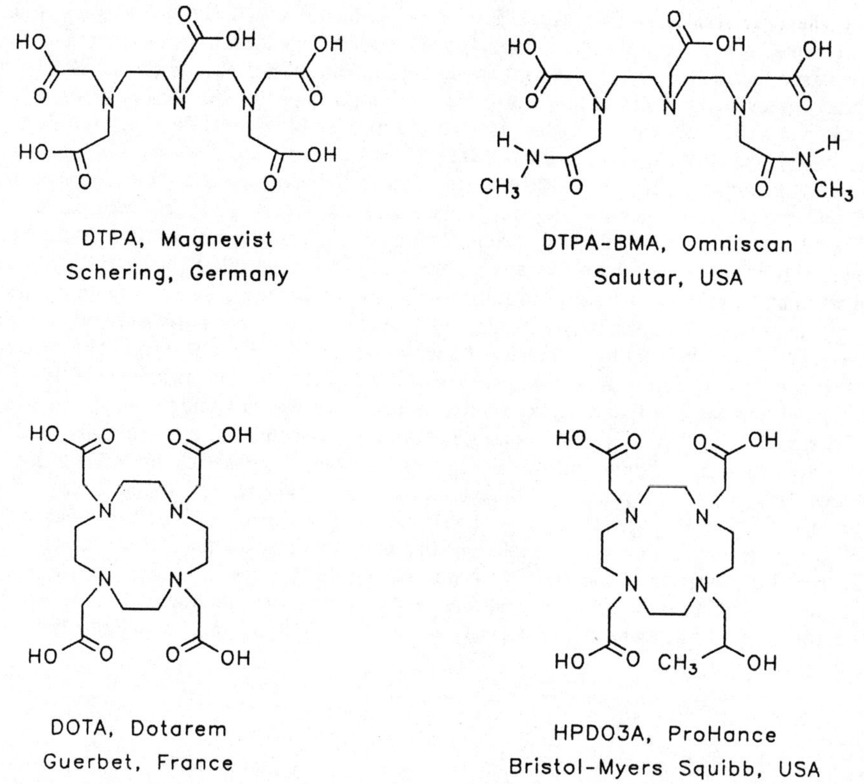

Figure 1

such as $GdEDTA^-$ (LD_{50} = 0.3 mmol/kg) (Weinmann et al., 1990). Solutions of $GdDTPA^{2-}$ are sold under the name Magnevist (Schering, Germany) and have been injected in more than 2 million patients with no severe reactions except for a few problems in the case of patients with a medical history of asthma.

When selecting a strong chelating agent for a lanthanide ion, one should keep in mind that the valence 4f electrons of these elements are buried under filled orbitals and are thus not available for binding. The lanthanides exhibit no directing effect on the geometry of their complexes that are thus highly labile. New ligands have been synthesized to fulfil the requirements of sterically non-demanding ions that feature high coordination numbers with a quasi "spherical" arrangement of donor atoms. Not unexpectedly, macrocyclic polyaza polycarboxylic ligands such as DOTA (Dotarem, Guerbet, France) and its many derivatives have attracted considerable attention. In contrast with the classical linear aminopolycarboxylic ligands such as DTPA, DOTA forms non labile lanthanide chelates that are kinetically inert even at low pH (Desreux, 1980; Wang et al., 1992). The small tetraaza twelve-membered cycle of this ligand seems to impart a high rigidity to the cage it forms with the lanthanide ions and to prevent the protonation of the nitrogen atoms that should take place to achieve the dissociation of its chelates. Moreover, DOTA adopts a square antiprismatic geometry in its complexes (Spirlet et al., 1984). This structure is the least crowded arrangement that can be found for an octacoordinated ion and it could account for the very high thermodynamic stability of the DOTA complexes.

There is of course a price to pay when the toxicity of Gd^{3+} is reduced by chelation: the number of coordinated water molecules q_{H_2O} decreases from 8 for aquo gadolinium to one for $GdDOTA^-$ or $GdDTPA^{2-}$. The relaxivity effect on T_1 is reduced accordingly but it seems that a hydration number close to one is the highest that can be reached while still maintaining a high thermodynamic and kinetic stability (Zhang et al., 1992).

The structures of DOTA and DTPA have been modified in a number of ways so as to fulfil some of the requirements of the "ideal" contrast agent. For instance, the triacetic monoalcoholic ligand HPDO3A forms an electrically neutral gadolinium chelate (ProHance, Bristol-Myers Squibb, USA) that can be injected intravenously with less risks of osmotic shocks. The same objective is achieved with the monoamide derivative of DTA (Sherry et al., 1989) and the DTPA bis-amide (DTPA-BMA, Omniscan, Salutar, USA), but the gadolinium complex of DTPA-BMA is less stable and an excess of the corresponding calcium chelate must be added to the solution of the contrast agent. The acetate arms of DOTA have also been replaced by phosphonate groups (Geraldes et al., 1989). Moreover, a higher rigidity and thus a better kinetic inertness of the macrocyclic ligand DOTA has been achieved by substitution of the acetate arms (Brittain and Desreux, 1984; Kang et al., 1992) or of the tetraaza ring (Schaeffer et al., 1991) with methyl groups. In addition, a higher relaxivity can be obtained by increasing the molar mass of the contrast agent. The rotational correlation time τ_R is the factor that governs the relaxivity at high magnetic fields (20-60 MHz) and thus the addition of bulky groups to a complex that are able to reduce the rate of rotation also brings about a better contrast. However, the solubility of chemically modified complexes in water should remain as high as possible, for instance by substituting the ligands with polyhydroxylated substituents as in ligand 1 in Figure 2 (Aime et al., 1992). Although reliable data are still scarce, it seems that rigid dimeric gadolinium chelates such as 2 could exhibit a longer τ_R correlation time and thus a better relaxivity per metal ion. For practical purposes, polymeric chelates should be as stable as the corresponding monomeric complexes. Because at least three carboxylic groups and one other arm featuring a coordinating group seem to be necessary to ensure stability, it is preferable to link many

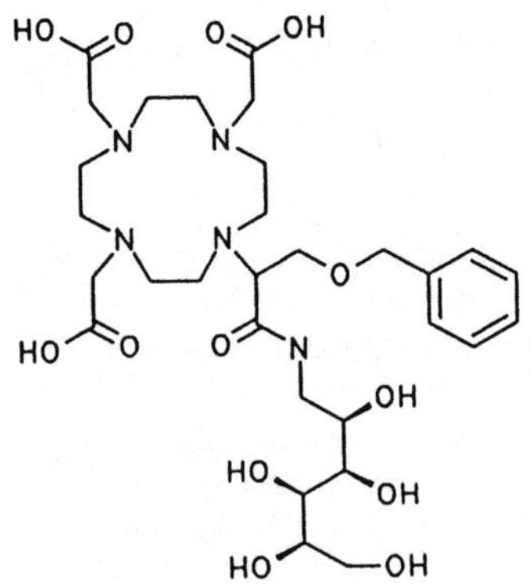

1

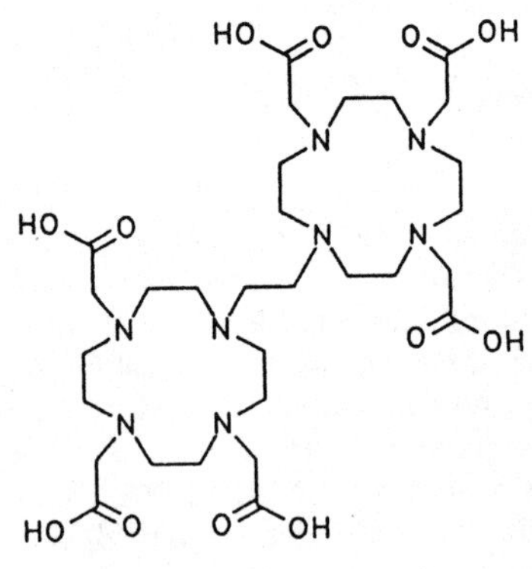

2

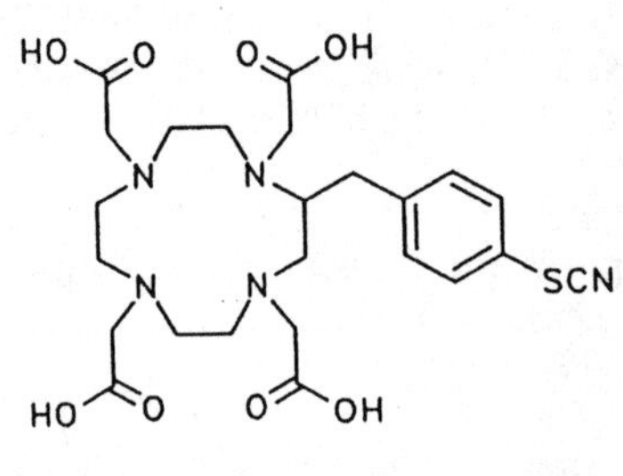

3

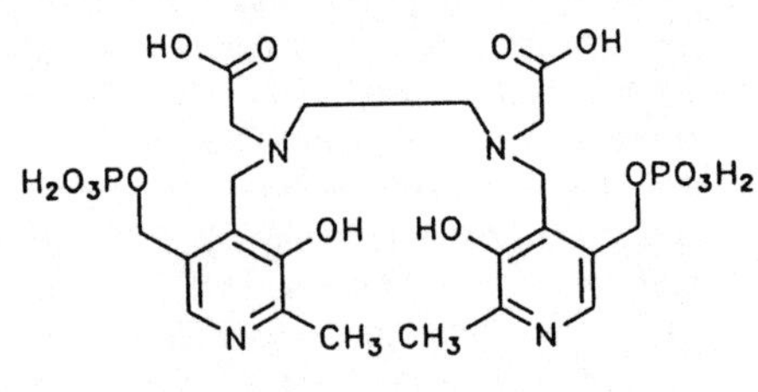

DPDP

Figure 2

gadolinium chelates together or to a polymer with one side arm on one acetate group or on the ethylenic backbone. Although ligands such as 3 are more difficult to prepare, the time spent on organic syntheses should be worth the effort because the complexing properties of the ligands will be barely modified. Other approaches, such as the mixed anhydride method (Krejcarek and Tucker, 1977) or the carbodiimide method (Jacobsen et al., 1986) generate ligands with one of their coordinating groups used as a linkage unit. The stability of the lanthanide complexes is then drastically reduced. Several synthetic procedures for the preparation of ligands such as 3 have been reported recently in the literature (Renn and Meares, 1992; Studer and Meares, 1992; McMurry et al., 1992). These ligands can be coupled to monoclonal antibodies. For the MRI contrast agents, the linkage unit should be as rigid as possible in order to avoid that the polymer backbone and the metal chelates rotate independently and thus exhibit a different τ_R value. Multiple labeling of proteins such as bovine serum albumin with gadolinium chelates yield polymeric water soluble materials that exhibit a high relaxivity around 20 Mhz because the relaxation process is dominated by the electronic relaxation time T_{1e} rather than by the rotational correlation time τ_R that is considerably longer than in the case of monomeric chelates (Niemi et al., 1991; Muller et al., 1990). A macromolecular albumin-GdDTPA contrast agent that cannot diffuse through capillaries because of its size has been used for assessing the blood volume (Ogan et al., 1990).

$GdDTPA^{2-}$ localizes in the extracellular space and shows no specific uptake by hepatocytes. Improvement of the contrast of MRI images of the liver by $GdDTPA^{2-}$ requires a rapid injection followed by fast imaging. A slower injection and a persistent increase of the contrast (> 2 h) are highly desirable. It is hoped that this goal will be achieved with ligands that are substituted with hydrophobic groups attached either to the ethylenic backbone (Weinmann et al., 1991) or to one of the acetate groups of DTPA (Cavagna et al., 1991). So far, the 4-methoxy-and the 4-ethoxybenzyl groups have been used for targeting hepatocytes: the gadolinium chelates of EOB-DTPA (Schering, Germany) and BOPTA (Bracco, Italy) exhibit relatively high hepatobiliary excretion and the latter is undergoing clinical trials in Europe (Saini, 1992). It should be noted here that hepatobiliary agents must be anionic compounds so as to be taken up by the high-capacity non-specific organic anion transporter system that transports bilirubin IXα and bromosulfophtalein (Klaassen and Watkins, 1984). Gadolinium chelates that do not localize in the extracellular space have also been synthesized by attaching two stearyl chains to two of the carboxylic groups of DTPA. These Gd^{3+} chelates can be entrapped into liposomes (Kabalka et al., 1991). These gadolinium-labeled liposomes concentrate either in the hepatocytes or in the reticuloendothelial system, depending on their chemical and physical characteristics.

The search for new contrast agents remains very active and has even been compared recently to a baseball competition (Brasch, 1992). Many aspects of the coordination and relaxation properties of these compounds are not fully understood and one should thus be cautious before giving advice to someone who intends to enter the game. However, it seems that using polyaza cycles that contain either less or more than 12 atoms leads to chelates of greatly reduced stability (Cacheris et al., 1987). One possible exception is the HEHA ligand that has been found to form highly stable lanthanide complexes (Kodama et al., 1991). However, the -3 charge of these chelates probably precludes their use as contrast agents. Ligands featuring propylenic groups rather than ethylenic units should also be avoided because of the much lower stability (Loncin et al., 1986) and the low q_{H_2O} (Bryden and Reilley, 1982) of their lanthanide complexes. Moreover, an excessive crowding of the macrocyle drastically reduces the kinetic inertness of the complexes. For instance, two methyl groups on the same carbon atom in the polyaza cycle of DOTA lead to chelates of little interest for practical purposes. Other ligands such as porphyrins (Lyon et al., 1987), cryptands or Schiff bases (Smith et al., 1989) have not proved useful so far. The complexes formed by these ligands tend to hydrolyze in water. In addition, positively charged chelates are highly toxic because they associate with cell membranes. However, it should be noted that appropriately modified synthetic expanded porphyrins (Sessler et al., 1989) as well as Schiff bases obtained by template reactions (Smith et al., 1989) could yield interesting MRI contrast agents because of the higher hydration number of the encapsulated metal ions.

C. Practical considerations and physicochemical characteristics

No toxicity tests should be performed before a complete assessment of the purity of the ligands and of their complexes. The content of a sample in free gadolinium can be estimated by HPLC with a fluorescence detector (Hagan et al., 1988) or after spiking solutions with radioactive ^{153}Gd (Vora et al., 1986). The total metal content is usually determined by inductively coupled plasma emission spectroscopy (ICP), and the purity of the ligands and their metal derivatives is most often checked by elemental analysis, by HPLC and by fast atom bombardment mass spectrometry (FAB-MS).

The relaxivity at a fixed frequency is easily measured with a NMR spectrometer but the recording of NMR dispersion curves (NMRD) is much more informative as indicated in the preceding chapter in this book. The unknown parameters in equations 2-4 can be estimated by a best-fit interpretation of NMRD curves. If feasible, the direct measurement of these parameters could be a more reliable approach. The hydration numbers q_{H_2O} are deduced from luminescence relaxation times (Horrocks et al., 1979) in the case of the Eu^{3+} and Tb^{3+} complexes. Rotational correlation times τ_R can be estimated from relaxation studies of diamagnetic chelates (Shukla et al., 1991) although with some difficulties while the rate of water exchange τ_M is deduced from ^{17}O high-pressure NMR measurements (Micskei et al., 1992). Finally, electronic relaxation times T_{1e} can be computed from EPR line widths (Geraldes et al., 1992). These measurements have led to the following correlation times: τ_R = 50-100x10^{-12} s, τ_M = 200-300x10^{-9} s and T_{1e} = 150-700x10^{-12} s (at 0.3 T) for $GdDTPA^{2-}$ and $GdDOTA^-$. The relaxivity of $GdDOTA^-$ is about 50% higher than in the case of $GdDTPA^{2-}$ at low fields, presumably because of the highly symmetrical structure of the DOTA ligand. NMR spectroscopy of paramagnetic chelates has yielded a wealth of information on the geometry and the rigidity of lanthanide chelates (Jenkins and Lauffer, 1988; Desreux, 1980). The kinetic properties of these compounds have been investigated by classical techniques (Wang et al., 1992; Kumar et al., 1993). Of particular relevance is the kinetics of dissociation that is acid-assisted and can be exceedingly slow for complexes with sterically demanding macrocyclic ligands.

IV. TRANSITION METAL COMPLEXES

In contrast to lanthanides, the valence electrons of transition metal ions are located in outer d orbitals that are available for bonding. Transition metal ions thus have a strong directing effect on the geometry of their complexes and have a marked preference for a tetrahedral or an octahedral arrangement that is not found in the case of the lanthanides. The Mn^{2+} and Fe^{3+} ions have been selected for MRI because of their high electronic magnetic moment and their long T_{1e} values (vide supra). The highly stable complexes of these ions are usually coordinatively saturated, thus leaving no space for water molecules in their first coordination sphere. The shortening of the relaxation times of water takes place only through an outer sphere mechanism and is thus less pronounced than for gadolinium chelates. Several complexes of ligands containing aromatic groups have been tested for hepatic imaging. Among these, $MnDPDP^{6-}$ (Rocklage et al., 1989) shows a preferential hepatobiliary excretion in rodents. $MnDPDP^{6-}$ is now undergoing clinical trials (Hamm et al., 1992) and was found to feature several desirable properties: the maximum relaxivity enhancement is reached after 15-30 minutes and the contrast enhancement persists for 6 hours. It was suggested that $MnDPDP^{6-}$ is taken up by the hepatocytes because of its resemblance to vitamin B6, but the relaxivity improvement could be due to a partial decomposition of the chelate (Brasch, 1992). Manganese is known to accumulate in the hepatocytes and this could account for the better contrast of the liver images despite the metabolism of the complex. It has been claimed that the release of manganese does not cause serious side effects in patients (Hamm et al., 1992). The conditions for a maximum liver uptake are still poorly known (Klaassen and Watkins, 1984), but a balance between lipophilic and hydrophilic structural features seems to be required. Various attempts at preparing transition metal chelates that exhibit both a high stability and the appropriate structural characteristics have been reported (Lauffer, 1987), and new derivatives will certainly be described in the future.

REFERENCES

Aime, S., Anelli, P.L., Botta, M., Fedeli, F., Grandi, M., Paoli, P., and Uggeri, F. (1992). *Inorg. Chem.*, *31*: 2422.

Brasch, R.C. (1992). *Radiology*, *183*: 1.

Brechbiel, M.W., and Gansow, O.A. (1991). *Bioconjugate Chem.*, *2*: 187.

Brittain, H.G., and Desreux, J.F. (1984). *Inorg. Chem.*, *23*: 4459.

Bryden, C.C., and Reilley, C.N. (1982). *Anal. Chem.*, *54*: 610.

Cacheris, W.P., Nickle, S.K., and Sherry, A.D. (1987). *Inorg. Chem.*, *26*: 958.

Cavagna, F., Dapra, M., Maggioni, F., De Haen, C., and Felder, E. (1991). *Magn. Reson. Med.*, *22*: 329.

Deshpande, S.V., DeNardo, S.J., Kukis, D.L., Moi, M.K., McCall, M.J., Denardo, G.L., and Meares, C.F. (1990). *J. Nucl. Med.*, *31*: 473.

Desreux, J.F. (1980). *Inorg. Chem.*, *19*: 1319.

Desreux, J.F., and Barthélemy, P.P. (1988). *Nucl. Med. Biol.*, *15*: 9.

Frahm, J., Haase, A., and Matthaei, D. (1986). *Magn. Reson. Med.*, *3*: 321.

Geraldes, C.F.G.C., Brown, R.D., III, Cacheris, W.P., Koenig, S.H., and Sherry, A.D. (1989). *Magn. Reson. Med.*, *9*: 94.

Geraldes, C.F.G.C., Brown, R.D., III, Brücher, E., Koenig, S.H., Sherry, A.D., and Spiller, M. (1992). *Magn. Reson. Med.*, *27*: 284.

Hagan, J.J., Taylor, S.C., and Tweedle, M.F. (1988). *Anal. Chem.*, *60*: 514.

Hamm, B., Vogl, T.J., Branding, G., Schnell, B., Taupitz, M., Wolf, K.J., and Lissner, J. (1992). *Radiology*, *182*: 167.

Hervesomma, C.M.P., Sebag, G.H., Prieur, A.M., Bonnerot, V., and Lallemand, D.P. (1992). *Radiology*, *182*: 93.

Horrocks, W.DeW., Jr., and Sudnick, D.R. (1979). *J. Amer. Chem. Soc.*, *101*: 334.

Inagaki, F., and Miyazama, F. (1981). *Progr. NMR Spectr.*, *14*: 67.

Jacobsen, T., Klaveness, J., and Lindberg, B.J. (1986). Eur. Pat. 0 184 899 A1.

Jenkins, B.G., and Lauffer, R.B. (1988). *Inorg. Chem.*, *27*: 4730.

Kabalka, G.W., Davis, M.A., Moss, T.H., Buonocore, E., Hubner, K., Holmberg, E., Maruyama, K., and Huang, L. (1991). *Magn. Reson. Med.*, *19*: 406.

Kaiser, W.A. (1992). *Diagnostic Imaging Int.*, *November/December*: 44.

Kang, I.S., Ranganathan, R.S., Emswiler, J.E., Kumar, K., Gougoutas, J.Z., Malley, M.F., and Tweedle, M.F. (1993). *Inorg. Chem.*, *32*: 2912.

Klaassen, C.D., and Watkins, J.B. (1984). *Pharmacol. Rev.*, *36*: 1.

Kodama, M., Koike, T., Mahatma, A.B., and Kimura, E. (1991). *Inorg. Chem.*, *30*: 1270.

Koenig, S.H., and Brown, R.D. (1990). *Progr. NMR Spectr.*, *22*: 487.

Koenig, S.H., and Epstein, M. (1975). *J. Chem. Phys.*, *63*: 2279.

Krejcarek, G.E., and Tucker, K.L. (1977). *Biochem. Biophys. Res. Commun.*, *77*: 581.

Kumar, K., and Tweedle, M.F. (1993). *Pure Appl. Chem.*, *65*: 515.

Kumar, K., Chang, C.A., and Tweedle, M.F. (1993). *Inorg. Chem.*, *32*: 587.

Lauffer, R.B. (1987). *Chem. Rev.*, *87*: 901.

Loncin, M.F., Desreux, J.F., and Merciny, E. (1986). *Inorg. Chem.*, *25*: 2646.

Lyon, R.C., Faustino, P.J., Cohen, J.S., Katz, A., Mornex, F., Colcher, D., Baglin, C., Koenig, S.H., and Hambright, P. (1987). *Magn. Reson. Med.*, *4*: 24.

McMurry, T.J., Brechbiel, M.W., Kumar, K., and Gansow, O.A. (1992). *Bioconjugate Chem.*, *3*: 108.

Micskei, K., Helm, L., Brücher, E., and Merbach, A.E. (1992). XXIX International Conference on Coordination Chemistry, Lausanne, Switzerland, July 19-24, 1992.

Morris, P.G. (1986). *Nuclear Magnetic Resonance Imaging in Medicine and Biology*, Clarendon Press, Oxford.

Muller, R.N., Vallet, P., Maton, F., Roch, A., Goudemant, J.F., Vanderelst, L., Gillis, P., Peto, S., Moiny, F., and Vanhaverbeke, Y. (1990). *Invest. Radiol.*, *25*: S34.

Niemi, P., Reisto, T., Hemmila, I., and Kormano, M. (1991). *Invest. Radiol.*, *26*: 820.

Ogan, M.D., Schmiedi, U., Moseley, M.E., Grodd, W., Paajanen, H., and Brasch, R.C. (1990). *Invest. Radiol.*, 22: 665.

Parker, D. (1990). *Chem. Soc. Rev.*, *19*: 271.

Renn, O., and Meares, C.F. (1992). *Bioconjugate Chem.*, *3*: 563.

Rocklage, S.M., Cacheris, W.P., Quay, S.C., Hahn, F.E., and Raymond, K.N. (1989). *Inorg. Chem.*, *28*: 477.

Runge, V.M., Bradley, W.G., Brantzawadzki, M.N., Carvlin, M.J., Desimone, D.N., Dean, B.L., Dillon, W.P., Drayer, B.P., Flanders, A.E., Harms, S.E., Haughton, V.M., Howieson, J., Joy, S.E., Kanal, E., Kumar, A.J., Liu, T.H., Lufkin, R.B., Maravilla, K.R., Mezrich, R.S., Mikhael, M.A., Morgan, F.W., Nadel, S.N., Pollei, S.R., Pomeranz, S.J., Price, A.C., Ramsey, R.G., Yuh, W.T.C., and Zelch, J.V. (1991). *Radiology*, *181*: 701.

Saini, S. (1992). *Radiology*, *182*: 12.

Schaeffer, M., Meyer, D., Beaute, S., and Doucet, D. (1991). *Magn. Reson. Med.*, *22*: 238.

Semelka, R.C., Hricak, H., Tomei, E., Floth, A., and Stoller, M. (1990). *Radiology*, *175*: 797.

Sessler, J.L., Murai, T., and Hemmi, G. (1989). *Inorg. Chem.*, *28*: 3390.

Sherry, A.D., Brown, R.D., Geraldes, C.F.G.C., Koenig, S.H., Kuan, K.-T., and Spiller, M. (1989). *Inorg. Chem.*, *28*: 620.

Shih, J.L.A., and Brugger, R.M. (1992). *Med. Phys.*, *19*: 733.

Shukla, R., Zhang, X., and Tweedle, M.F. (1991). *J. Magn. Resonance*, *26*: 5224.

Smith, P.H., Brainard, J.R., Morris, D.E., Jarniven, G.D., and Ryan, R.R. (1989). *J. Amer. Chem. Soc.*, *111*: 7437.

Southwood-Jones, R.V., Earl, W.L., Newman, K.E., and Merbach, A.E. (1980). *J. Chem. Phys.*, *73*: 5909.

Spirlet, M.-R., Rebizant, J., Desreux, J.F., and Loncin, M.F. (1984). *Inorg. Chem.*, *23*: 359.

Studer, M., and Meares, C.F. (1992). *Bioconjugate Chem.*, *3*: 420.

Turner, R., and Keller, P. (1991). *Progr. NMR Spectr.*, *23*: 93.

Tweedle, M.F. (1992). *J. Alloy. Compd.*, *180*: 317.

Vora, M.M., Wukovnig, S., Finn, R.D., Emran, A.M., Boothe, T.E., and Kothari, P.J. (1986). *J. Chromatogr.*, *369*: 187.

Wang, X., Jin, T., Comblin, V., Lopez-Mut, A., Merciny, E., and Desreux, J.F. (1992). *Inorg. Chem.*, *31*: 1095.

Wehrli, F.W., MacFall, J.R., Shutts, D., Breger, R., and Herfkens, R.J. (1984). *J. Comput. Assist. Tomogr.*, *8*: 369.

Weinmann, H.-J., Schuhmann-Giampieri, G., Schmitt-Willich, H., Vogler, H., Frenzel, T., and Gries, H. (1991). *Magn. Reson. Med.*, *22*: 233.

Weinmann, H.J., Brasch, R.C., Press, W.R., and Wesbey, G.E. (1984). *Amer. J. Roentgenol.*, *142*: 619.

Weinmann, H.J., Press, W.R., and Gries, H. (1990). *Invest. Radiol.*, *25*: S49.

Zhang, X., Chang, C.A., Brittain, H.G., Garrison, J.M., Telser, J., and Tweedle, M.F. (1992). *Inorg. Chem.*, *31*: 5597.

EXPERIMENTAL TECHNIQUES FOR METAL ION ANALYSIS IN BIOFLUIDS

by

Ole Jøns and Bente Gammelgaard

Royal Danish School of Pharmacy,
DK-2100 Copenhagen
Denmark

I. INTRODUCTION

Experimental analytical techniques used in bioinorganic chemistry cover all kinds of methods, including methods for quantitative determinations of trace elements but also methods most suitable to give information on the role and influence of metal ions on biomolecule structures. In this chapter, analytical methods used for quantitative determination of metal ions are briefly described. A fundamental discussion of principles of the analytical methods and the modern instruments used can be found in the books of advanced study of instrumental analysis by Skoog and Leary (1992) and Willard et al. (1988).

Analytical techniques and applications to trace element determination in human plasma or serum, which cover the greatest number of biofluid analysis, have recently been reviewed by Versieck and Cornelis (1989).

Analysis of trace metal ion contents in biofluids can be managed with many different methods. In some of the methods, e.g. graphite furnace atomic absorption spectrometry, the sample can be analyzed directly without pretreatment or with only a dilution of the sample prior to determination. Other methods such as voltammetric methods need very careful pretreatment of the sample before measurement. The concentration of metal ions in the different biofluids varies from high levels of e.g. sodium and potassium, through minor levels of iron and zinc to ultratrace levels of e.g. nickel, and different analytical methods have to be used according to different instrumental sensitivity. Reviews of analytical methods used for quantitative trace element analysis of biological materials including discussions on sample handling and pretreatment were recently discussed in a book edited by McKenzie and Smythe (1988).

II. ABSORPTION SPECTROPHOTOMETRY AND FLUOROMETRY

This heading includes methods where chemical substances interact with electromagnetic radiation in the wavelength range 185-750 nm and the transmission or emission of radiation is measured. These methods are the basis of thousands of analytical methods. Relative inexpensive instruments of high quality have been put on the market during the last sixty years and new analytical methods such as chromatography, flow injection and automated analysis continue the use of spectrophotometric and fluorometric based methods. Most organic compounds absorb radiation strongly, especially in the ultraviolet 185-400 nm range, to allow direct determination. Most inorganic species, including metal ions, do not absorb this radiation, hence conversion to an absorbing species through a specific complex formation is demanded.

A. Ultraviolet and visible spectrophotometry

In the spectrophotometer, be it a simple colorimeter or a more sophisticated double-beam instrument, the amount of radiation of a given energy transmitted by the sample is measured. The quantitative use of spectrophotometry is based on Beer's law, which can be written as: $A = -\log T = \varepsilon bC$. Thus, there is a linear relationship between the absorbance (A) measured and the concentration (C) of the substance in solution. ε is the molar absorptivity, which is characteristic of the given species and expresses the sensitivity of the method. The main problem in the analysis of metal ions in biofluids is the low concentration level for most metal ions, and even a complex formation yielding complex species with a high molar absorptivity is not always enough to obtain the sufficient sensitivity. Often time-consuming pretreatment of the sample, which may include a preconcentration step, is involved in the procedure of a spectrophotometric measurement. Determination procedures for nearly all metal ions in biofluids exist, but atomic absorption spectrometry is now the most commonly used method. However, in clinical chemistry (Tietz, 1987), the determination of iron in serum is still in the spectrophotometric automated analyzer program. Iron is separated from serum transferrin by acid treatment, converted to iron(II) by ascorbic acid and dialyzed into a buffer before the complex reaction with 4,7-diphenyl-1,10-phenantroline sulphonate. Some biomolecules containing metal ions such as hemoglobin and vitamin B_{12} have high molar absorptivities and can be determined spectrophotometrically.

B. Fluorometry

In fluorometry, the radiation from excited molecules returning to their ground energy state is measured. It is characteristic in fluorometry that the emitted radiation is less energetic (longer wavelength) than the radiation being absorbed by the molecules. All compounds that absorb ultraviolet or visible radiation have the capability to lose the absorbed energy as fluorescence. The most intense and useful fluorescence is found in compounds containing rigid structures as aromatic rings. Only few metal ions as uranium and some rare earth elements emit fluorescence directly in solution. For metal ion determinations, the ion is therefore combined with an organic ligand forming a fluorescent chelate. Useful reagents for metal ions is given by Guilbault (1973). Selenium has been determined by fluorometry in biological samples, including human blood by Tamari et al. (1986) after wetashing and extraction of the complex formed by selenium(IV) and 2,3-diamino-naphthalene.

Fluorescence spectroscopy has obtained its role in analysis, particularly because the method offers high selectivity and high sensitivity with a large dynamic concentration range in comparison to ultraviolet-visible spectrophotometry. For some elements the working concentration level (10-100 $\mu g\ l^{-1}$) in fluorometry is comparable with graphite furnace atomic absorption spectrometry, but more extensive pretreatment procedures are needed.

III. ATOMIC ABSORPTION AND EMISSION SPECTROMETRY

An exhaustive treatment of atomic absorption and emission spectrometry methods is given by Welz (1985).

A. Flame photometry

Flame photometry is an atomic emission method where the atoms are excited by a flame and the subsequent emission from the excited atoms is measured. The term photometry implicates that the emission lines of the excited atoms are separated by means of filters of fixed wavelengths, contrarily to spectrometry, where gratings are commonly used for the separation. The use of photometers is primarily due to their lower costs.

Flame photometry is practically only used in clinical chemical laboratories for routine determination in biological fluids of sodium and potassium, and occasionally calcium or lithium. The detection limits of the mentioned elements are at the low to the sub- $\mu g\ l^{-1}$ level.

In general, atomic emission methods are considered to be supplementary to atomic absorption methods.

B. Flame atomic absorption spectrometry

In flame atomic absorption spectrometry (FAAS), the flame is used for the atomization of the elements, while the excitation is performed by a hollow cathode lamp which emits radiation of exactly the wavelength required for excitation of the element. The concentration is calculated on basis of the absorbance of the excited element as in ordinary spectrophotometry. The method is a single element technique, each element requiring its own hollow cathode lamp.

Flame atomization is mainly used for determinations at the mg l^{-1} level. Hence, the determination of e.g. calcium, iron and zinc in plasma or calcium and magnesium in urine can be performed directly by the flame technique. The most commonly used flame is an air-acetylene mixture. By variation of fuel-oxidant ratio, most analytical problems can be overcome. FAAS is a fast and simple analytical method with only a few interferences, but requires sample volumes of several millilitres. Detection limits for some elements are given in Table 1. The concentrations in biofluids of normal unexposed individuals of most other elements are too low for the direct determination by FAAS. Consequently, an enrichment procedure has to be performed, e.g. complexation with APDC and solvent extraction into MIBK. However, these procedures involve a serious risk of contamination, and elements at

Table 1

Atomic Spectroscopy Detection Limits in $\mu g\ l^{-1}$ (Perkin-Elmer Sciex) and Approximate Levels of the Elements in Human Serum. (All detection limits refer to elemental standards in aqueous solution).

Element	FAAS	GFAAS*	ICP-AES**	ICP-MS	Serum
Ag	1	0.005	1	0.04	0.2
Al	30	0.04	4	0.1	3
As	20	0.2	20	0.05	3
Ca	1	0.05	0.08	5	200
Cd	0.5	0.003	1	0.02	0.2
Co	6	0.01	2	0.02	0.3
Cr	2	0.01	2	0.02	0.1
Cu	1	0.02	0.9	0.03	1000
Fe	5	0.02	1	1	1000
Hg	200	1	20	0.03	2
K	1	0.02	50	10	150
Li	0.5	0.05	0.9	0.1	0.3
Mg	0.1	0.004	0.08	0.1	40
Mn	1	0.01	0.4	0.04	0.6
Mo	30	0.04	5	0.08	2
Na	0.2	0.05	4	0.06	3300
Ni	4	0.1	4	0.03	0.5
Pb	10	0.05	20	0.02	0.3
Se	100	0.2	60	0.5	80
Sn	20	0.2	40	0.03	0.8
Zn	1	0.01	1	0.08	900

* Based on sample aliquots of 100 μl.

** Instrumental parameters optimized for the individual element. In multielement analyses the values are about the double.

the sub $\mu g\ l^{-1}$ level should not be analyzed this way. Instead, the larger sensitivity of the graphite furnace technique should be exploited.

C. Graphite furnace atomic absorption spectrometry

In graphite furnace atomic absorption spectrometry (GFAAS), the sample is atomized by electrothermal heating in a graphite tube; hence the technique is also abbreviated ETAAS. The sample volume is commonly about 20 μl and the sample is normally introduced automatically. Samples are normally introduced in solution but the application of slurries and solids has been described as well. The heating is performed in several steps including drying, charring and atomization; the total cycle typically taking several minutes. GFAAS is prone to more serious matrix interference problems than FAAS. Very careful temperature programming for each element in various kinds of matrices and addition of suitable matrix modifiers is often needed. The instruments are provided with a background correction system, either a deuterium lamp, which is also used in flame instruments, or the Zeeman-correction system which makes use of magnetic fields. The latter has the advantage that it can compensate for backgrounds up to 2 absorbance units and works in the entire wavelength range.

GFAAS is the overall most commonly used technique for trace elements in biological fluids. The method can be applied directly to most biological fluids as plasma, urine, sweat etc after only a possible dilution of the sample. Whole blood is a more difficult matrix and demands as a rule some pretreatment.

D. Inductively coupled plasma atomic emission spectrometry

In inductively coupled plasma atomic emission spectrometry (ICP-AES), atomization and excitation take place in an argon plasma. The argon plasma is a gaseous mixture containing cations and electrons. Once formed in the plasma, the argon ions are capable of absorbing sufficient power from an external source - here a radio frequency generator - to maintain the temperature at a level at which further ionization sustains the plasma indefinitely. The temperature in the center of the plasma is 6000 to 8000 K. The high temperature together with the inert chemical environment result in a high degree of atomization without solute-volatilization and vapour-phase interferences, thus these types of interferences are very seldom. As oxygen is practically absent in the plasma, elements as the alkaline earths, rare earths and silicon, which are prone to oxide formation in FAAS, are effectively atomized by ICP. The most serious interference in ICP-AES is spectral interferences due to close-lying or broadened lines of different elements. The performance of an instrument is therefore limited by the efficiency of the monochromator. The major advantage of ICP-AES is that it is a multielement technique and several elements can be analyzed per minute. This opens the possibility of surveying the total trace element status of biofluids as urine, blood and serum in a relatively short time compared with the GFAAS method. However, as seen from Table 1, the sensitivity is often not adequate for direct determinations.

E. Inductively coupled plasma mass spectrometry

In recent years, the technique of inductively coupled plasma mass spectrometry (ICP-MS) has been developed. Instrumentation comprises an ICP source for ion production and a mass spectrometer for detection and quantification. In the latter, the ions are separated and analyzed according to their mass-to-charge (m/z) ratios (Beauchemin, 1991). The technique offers the analytical productivity of ICP emission with the detection limits of GFAAS. As seen in Table 1, the detection limits for most elements are two to three orders of magnitude below those of ICP-AES. As the spectra are primarily from singly charged ions and no element have more than 10 isotopes, ICP-MS spectra are simpler than ICP-AES spectra. Different elements may have isotopes having the same nominal mass, but most elements have at least one isotope that is not overlapped by another element. As ICP-MS has the capability of measuring precise isotope ratios, it can be used for stable isotope dilution measurements as described later. The major interference in the method is overlapping of signals by the background spectrum, e.g. ArO^+.

As analysis is performed in a continuous stream, the technique is very suitable for coupling to flow injection systems or chromatographic systems, thus rendering speciation studies possible.

Vanhoe et al. (1990) determined 20 elements in human serum by ICP-MS. Li, Fe, Co, Cu, Zn, Br, Rb, Sr, Mo, and Cs could be determined accurately, while Al, Ti, V, Cr, Mn, Ni, As and Se caused some trouble.

So far, the ICP-MS technique is not a routine method as the instruments are expensive and some development of the applications still have to be made, but the method will probably gain much more interest in the future.

In Table 1, the levels of different trace elements in serum from normal, unexposed humans are given for comparison with the detection limits of the atomic absorption and emission methods.

IV. ELECTROANALYTICAL METHODS

Electroanalytical chemistry provides basis for a group of analytical methods among which only potentiometry with ion-selective electrodes and voltammetry have some application in trace element analysis in biofluids.

A. Potentiometry

In potentiometry, galvanic cell potentials are measured under essentially zero current conditions. The quantitative aspect is that the potential between a working and a reference electrode is proportional to the activity of the species to be determined. Especially the ability to provide information of ion activity rather than concentration makes this method interesting in physiological analysis.

Due to their high concentration in biofluids, the metal ions of interest regarding potentiometric methods are sodium, potassium, magnesium and calcium. The development of new cation-selective membranes shows promising selectivity properties, but improved detection limit capabilities are needed for a more general application. Membrane electrodes for calcium and potassium have already wide use in physiological studies. Miniaturized versions of ion-selective electrodes have become useful for in vivo applications. A survey of ion-selective electrodes available for bioanalytical measurements can be found in a review by Meyerhoff and Collison (1988) and a book of Evans (1987).

B. Voltammetry

In anodic stripping voltammetry, the cell comprises a working mercury electrode and a reference electrode. After a preconcentration step at constant voltage, the diffusion current of the electroactive species as a function of the applied voltage on the mercury electrode allows a quantitative determination of the trace metal ion.

The major problem in direct voltammetric determination of trace metals in biological samples is interference effects caused by adsorption of organic constituents of the sample matrix. Thus, samples should be completely mineralized prior to analysis. As in all trace analysis at very low level, the purity of reagents is important. Otherwise, digestion procedures may introduce contamination exceeding the sample content. If very careful pretreatment procedures are used, stripping voltammetry can determine ultra-trace metal concentrations in biofluids. Voltammetric procedures have mainly been used for the determination of Zn, Cd, Pb, Cu, Ni and Co.

C. Potentiometric stripping analysis (PSA)

Based on the preconcentration of trace metals by potentiostatic deposition on a mercury film electrode and subsequent chemical reoxidation of the amalgamated analyte under well-defined conditions, PSA has been used for e.g. the determination of cadmium and lead in serum and urine samples by Jagner et al. (1981). The sensitivity of the method, which is determined by the preelectrolysis time, is sufficient for metal ion concentrations in biofluids and the method application area is extended beyond metals being amalgamated, but the sample pretreatment procedures can cause problems.

V. STABLE ISOTOPE DILUTION ANALYSIS

Most elements in the periodic system consist of two or more stable isotopes, with identical chemical properties but different atomic masses. In the isotope dilution technique, a known amount of an enriched stable isotope of the analyte to be determined, e.g. ^{58}Fe, is added to the sample, whereby the normal ratio of two isotopes of an element is changed. It is possible to obtain purified (enriched) compounds for the majority of metallic elements from Oak Ridge National Laboratory. A major advantage of isotope dilution methods is that recoveries need not be quantitative since only the ratio of the isotopes is measured. It is an absolute method, no calibration with standards is required. The use of stable isotopes offers the advantages of radioisotopic tracers without exposure of subjects to radioactivity. The method is only applicable to elements with more than one stable isotope and about 17 elements are monoisotopic including some of biological interest, like Na, Mn, Co, Al and As. A review of stable isotope methods and their application to trace metal determination in biological materials is given by Veillon and Alvarez (1983). Different detection methods have been used to determine the isotope composition and are described in the following.

In *neutron activation analysis* the enriched sample is irradiated with neutrons and the induced radioactivity is measured as discussed in the latter section.

Mass spectrometry (MS) is very well suited for stable isotope determinations. MS has a larger sensitivity than most other analytical techniques and its excellent specificity with the characteristic fragmentation pattern gives information of molar mass and structure. After the sample has been transformed to charged ions in gaseous phase the mass spectrometer sorts these ions according to their mass/charge ratio. The isotope composition is obtained and the intensities of the measured ions give a quantitative tool.

In *spark source mass spectrometry* (SSMS), the dry sample is blended with high purity graphite to form an electrode or, due to several interferences from organic ions, the sample is ashed before analysis and an electrode-deposition on a gold wire for sparking is applied. The stable isotope SSMS procedure has been used for certification of biological standard reference materials by NBS, e.g. freeze-dried bovine liver, SRM 1577. Due to both organic and inorganic spectral interferences and contamination problems, the method demands serious experience and has limited use outside certification laboratories. The *thermal (surface) ionization mass spectrometry* (TIMS) procedure is also used primarily as a reference method. It is a very sensitive method with excellent accuracy and precision, but it is time consuming and requires considerable operator knowledge. The method has been used as reference method to the GFAAS method in determination of selenium in plasma by Lewis et al. (1986). Probably the most promising new analytical method for trace element analysis including stable isotope dilution is *inductively coupled plasma mass spectrometry*, which was discussed earlier.

VI. NEUTRON ACTIVATION ANALYSIS

In neutron activation analysis (NAA), the sample is irradiated with thermal neutrons from a nuclear reactor and the resulting radioactivity (α- or β-particles or γ-rays) is measured. The neutrons interact with present nuclei in the sample and convert some of the nuclei into radioactive isotopes. The neutron flux at the reactor can be changed to favor some elements in the sample. The radioactivity measured is related to the elements and their concentration in the sample. An increase in irradiation will increase the sensitivity. In some cases, high neutron fluxes and long radiation times are needed to reach the concentration level in biofluids. NAA is a multielement method. Almost all trace elements in biological materials can be determined simultaneously, although more than one irradiation is often necessary. Advantages in NAA methods are the small risk for contamination of the sample and the possibility of high sensitivity for a number of elements. The disadvantage is obviously the need of a reactor. NAA is used primarily as a reference method, because especially atomic absorption spectrometry methods are available for routine applications. Applications of NAA can be found in several books (Erdtmann and Petri, 1986; Cornelis and Versieck, 1986; Cornelis, 1989). *Instrumental neutron activation analysis* (INAA) is the simplest form of NAA. The radioactivity of the irradiated sampled is measured after a suitable decay time. The elements in biological fluids of interest for INNA are those yielding nuclides with half-lives longer than two weeks such as cobalt, iron, selenium and zinc. In *radiochemical neutron activation analysis* (RNAA) the separation serves to eliminate interferences from other radionuclides in the sample before measurement. Elements with isotopes having shorter half-lives or low concentrations are preferably measured by RNAA, e.g. copper, chromium, manganese, molybdenum, and arsenic. A preconcentration step can be included in the separation.

VII. X-RAY SPECTROMETRY

Two methods of X-ray spectrometry shall be mentioned: XRF and PIXE. In *X-ray fluorescence* the emission of the characteristic X-radiation results from the application of an X-ray beam as exciting radiation. The technique uses either energy-dispersive or wavelength-dispersive instruments. The detection level in XRF is approximately 1 $\mu g\ g^{-1}$ of sample, which is unfavorable to most elements in biofluids. A review of the use of X-ray spectroscopy for analysis of biological materials is given by Valkovic (1980).

Proton-induced X-ray emission spectrometry is a relatively new method. The excitation source consists of particles, normally protons, from an accelerator. PIXE is a multielement method and is available for routine analysis covering all elements from aluminium to uranium, but it is not yet widely used. The spectra produced are complex and need considerable computer handling. The method is non-destructive and requires little sample preparation. This includes as a minimum a freeze-drying to get a solid sample deposited on a thin plastic film. A survey of the method is given by Johansson (1986). The sensitivity is high, the minimum detectable amount is about 10^{-11} g and can be even lower using a microbeam excitation. This means that the trace element level in biofluids can be reached. In an investigation by Maenhaut et al. (1982), 19 elements in human albumin samples were analyzed by PIXE and NAA. Total reflection X-ray fluorescence spectrometry (TXRF) is a new method which has shown promising results regarding detection limits.

VIII. CHROMATOGRAPHIC METHODS

The chromatographic techniques are still primarily used for separation of organic species, but metal ion separation has recently gained considerable interest within e.g. speciation investigations. Gas chromatography can be used for separation of volatile coordination compounds, where ligands like β-diketones are the derivatizing reagents for metal ions. More promising seems high performance liquid chromatography (HPLC) due to the variety of different separation possibilities with ion chromatography in its broadest sense. Conductivity detection is suitable for alkali and alkaline earth metals, but for transition metal ions post-column derivatization with fluorescence or chemiluminescence detection is preferred to obtain sufficient sensitivity. Trace metal ion analysis in biofluids needs more research. Preconcentration procedures (solvent extraction, on-column pre-concentration) have to be developed and HPLC coupling with element specific detectors such as inductively coupled plasma mass spectrometry will continue to facilitate metal ion and speciation studies. A review on chromatography of metals is given by Robards and Worsfold (1991).

REFERENCES

Beauchemin, D. (1991). *Trends Anal. Chem.*, *10*: 71.

Cornelis, R. (1989). *Quantitative Trace Analysis of Biological Materials* (H.A. McKenzie and L.E. Smythe, eds.), Chapter 14, Elsevier Science Publ., Amsterdam, p. 263.

Cornelis, R., and Versieck, J. (1986). *Treatise on Analytical Chemistry*, Part I, Vol. 14 (P.J. Elving, V. Krivan and I.M. Kolthoff, eds.), Wiley, New York, p. 665.

Erdtmann, G., and Petri, H. (1986). *Treatise on Analytical Chemistry*, Part I, Vol. 14 (P.J. Elving, V. Krivan and I.M. Kolthoff, eds.), Wiley, New York, p. 419.

Evans, A. (1987). *Potentiometry and Ion-Selective Electrodes*, Wiley, New York.

Guilbault, G.G. (1973). *Practical Fluorescence. Theory, Methods and Techniques*, Marcel Dekker Inc., New York.

Jagner, D., Josefson, M., and Westerlund, S. (1981). *Anal. Chim. Acta*, *128*: 155.

Johansson, S.A.E. (1986). *Fresenius Z. Anal. Chem.*, *324*: 635.

Lewis, S.A., Hardison, N.W., and Veillon, C. (1986). *Anal. Chem.*, *58*: 1272.

Maenhaut, W., De Reu, L., and Tomza, U. (1982). *Anal. Chim. Acta*, *58*: 301.

McKenzie, H.A., and Smythe, L.E., eds. (1988). *Quantitative Trace Analysis of Biological Materials*, Elsevier Science Publ., Amsterdam.

Meyerhoff, M.E., and Collison, M.E. (1988). *Quantitative Trace Analysis of Biological Materials* (H.A. McKenzie and L.E. Smythe, eds.), Elsevier Science Publ., Amsterdam, p. 241.

Robards, K., and Worsfold, P.J. (1991). *Trends Anal. Chem.*, *10*: 121.

Skoog, D.A., and Leary, J.J. (1992). *Principles of Instrumental Analysis*, 4th ed., Saunders College Publ., Fort Worth.

Tamari, Y., Ohmori, S., and Hiraki, K. (1986). *Clin. Chem.*, *32*: 1464.

Tietz, N., ed. (1987). *Fundamentals of Clinical Chemistry*, 3rd ed., W.B. Saunders Company, Philadelphia.

Valkovic, V. (1980). *Analysis of Biological Material for Trace Elements Using X-Ray Spectroscopy*, CRC Press, Boca Raton, FL.

Vanhoe, H., Vandecasteele, C., Versieck, J., and Dams, R. (1990). *Spec. Publ. R. Soc. Chem.*, *85*: 66.

Veillon, C., and Alvarez, R. (1983). *Metal Ions in Biological Systems*, Vol. 16 (H. Sigel, ed.), Marcel Dekker Inc., p. 103.

Versieck, J., and Cornelis, R. (1989). *Trace Elements in Human Plasma or Serum*, CRC Press, Boca Raton, FL.

Welz, B. (1985). *Atomic Absorption Spectrometry*, 2nd ed., VCH Verlagsgesellschaft mbH, Weinheim.

Willard, H.H., Merritt, L.L., Jr., Dean, J.A., and Settle, F.A., Jr. (1988). *Instrumental Methods of Analysis*, 7th ed., Wadsworth Publ. Comp., Belmont.

SPECIFIC PROBLEMS IN TRACE ELEMENT DETERMINATIONS IN BIOLOGICAL FLUIDS

by

G.V. Iyengar, DSc, PhD

Biomineral Sciences International,
Gaithersburg, MD
U.S.A.

and

V. Iyengar, MD

Georgetown University Medical Centre,
Washington D.C.
U.S.A.

I. INTRODUCTION

From a trace element composition point of view, biological fluids are generally heterogenous media containing suspended or fragmented cells and other proteins or crystalline particles. Fluids such as milk are emulsions. The body produces a variety of fluids required for numerous functions under normal health conditions. These include: aqueous humor, whole blood (includes erythrocytes, leukocytes, plasma, platelets and serum), cerebrospinal fluid, fetal fluids (allantoic and amniotic), intestinal fluids (cecal, duodenal, ileal and jejunal secretions), gastric juice, bile (gallbladder and hepatic), milk (colostrum, transitional and mature), pancreatic juice, pericardial fluids, phlegm, pleural fluids, prelactation secretions of the mammary gland, prostatic fluid, saliva, seminal fluid (seminal plasma and sperm), sputum, sweat, synovial fluid, tears, urine and vaginal secretions. Besides these, transudates and exudates are produced under pathological conditions. Transudate is a non-inflammatory fluid, characterized by a few cells, and contains less than 30 g/l of protein. Exudate consists of the fluid and cells of acute inflammation, containing protein in excess of 30 g/l. The essential features and physiological significance of these biofluids has been recently summarized (Iyengar, 1989).

Body fluids may be grouped under three categories: commonly sought (whole blood, milk, urine), easily accessible but not commonly sought (semen, sweat, tears, sputum, and saliva) and fluids requiring special procedures for collection (amniotic, bile, pancreatic, gastric and cerebrospinal fluid).

This presentation highlights the problems faced in the trace element determination in biological fluids with special reference to multidisciplinary aspects of biomedical specimens, medico-legal (and ethical) implications; presampling factors; contamination by trace elements; losses of trace elements; preservation and storage; sampling and preparatory problems; analysis and quality assurance; and reference parameters and data interpretation.

II. MULTIDISCIPLINARY ASPECTS OF BIOMEDICAL SPECIMENS

A major consideration in any analytical measurement is the "quality" of the sample. In dealing with biological fluids, quality may be divided into two parts: analytical aspects or precautions needed to protect the sample during the course of an analysis, and physiological considerations required to retain the integrity of a specimen before, during and after sampling. An accurate analysis should begin with an uncontaminated sample. Furthermore, it is essential to retain the integrity of the sampled fluids to (i) ensure that the information obtained is compatible with the purpose of the investigation, and (ii) to relate the elemental concentration data to desired biomedical needs and parameters, e.g. to specific cells, cellular components, enzymes or proteins. A range of analytical problems particularly relevant to clinical (Mertz, 1975) and bioenvironmental specimens (Iyengar, 1992) have been outlined.

Biological fluids are susceptible to bacterial growth in unfrozen state, and sampling and preparation should be designed to overcome these difficulties (Anand et al., 1975). Since most biological fluids are suspensions or emulsions, care should be taken to select the phase of interest, or to ensure that the phases are properly mixed if sampled as a whole.

In studies of biological effects of trace elements, besides the determination of their total contents in an organ or its components, it is often necessary to obtain information on the chemical forms in which trace elements occur. Many trace metals are bound to proteins and, therefore, biochemical procedures would be necessary to estimate the fraction bound to organic molecules. Most biochemical procedures can introduce serious extraneous contamination into the processed specimen since the concentrations of several elements of biochemical interest are very low in some clinical specimens such as whole blood, blood serum and urine. Therefore, frequently used materials, e.g. gels and reagents such as buffer solutions, need to be of very high purity.

The biologic validity of a specimen refers to the preservation of certain biological properties such as viability (e.g. cells), motility (e.g. sperm and bacteria) as well as electrolyte and osmotic equilibria in tissues and body fluids. This is essential to meet the requirements of specific biochemical steps, e.g. cell separations, fractionations, enzyme estimations, etc. In practice, it is not a simple task since it involves a systematic approach to the entire problem including the use of several techniques from different scientific disciplines. This is effectively illustrated in sampling platelets (thrombocytes) for trace element analysis (Iyengar et al., 1979). The process involves complicated steps such as obtaining sufficient quantity of the sample, preserving the viability of the isolated platelet fraction, achieving the required cell purity, accounting for the trapped plasma fraction and preventing the problem of sample contamination in the inorganic sense. Solution to these problems requires collective efforts by physicians, biochemists and analysts (Subramanian et al., 1990).

III. MEDICO-LEGAL IMPLICATIONS

Although a physician is justified to obtain tissue and body fluid samples from human subjects (e.g. diagnostic purposes), the end use of the samples is the deciding factor as to what extent the medico-legal implications come into picture. Hence, there are well defined medical restrictions governing human specimen collection. These regulations can be severely restrictive if samples are obtained from living subjects, and greatly differ among different countries. First of all, "informed consent" should be sought from donors to obtain samples such as blood, and for collecting biopsies during surgery. The following information should be provided to the donors before obtaining their consent: procedure to be followed for taking the specimen; description of any attendant discomforts and risks which might reasonably be expected for the individual and the community; an assurance that the data and results will be kept confidential but accessible to the donor; and the donor is at liberty to withdraw from the investigation at any stage if the process involves repeated sampling. Similarly, in case of autopsy sampling, medico-legal restrictions are basically dependent on the country in question, and the existing specifications have to be considered in preparing the protocols. Transboundary situations if prevalent, would probably complicate the situation. However, acceptable norms that have been established for shipping biological reference materials from one country to the other provide useful guidelines.

Medical personnel engaged in specimen collection (e.g. blood) will be required to follow both sterile (medical stipulations) and non-contaminating conditions (trace analysis requirements) if the samples have to be valid for analysis. This involves not only training the medical personnel in trace analysis problems, but also educating the analytical community to observe safety precautions in handling specimens that have potential for infection. For example, the medical personnel should get acquainted with dust-free requirements and use of titanium, teflon or other acceptable non-metallic accessories, while the analysts should learn to work in biohazard hoods (special work benches).

IV. PRESAMPLING FACTORS

Presampling factors address the problems of biological validity of a specimen. Essentially, they are markers of sample viability when biochemical processes affect the analytical integrity of biological specimens. These changes can begin at the in situ stage, and prevail through the sampling phase, and persist even after the samples are delivered to the laboratory for analysis. These events are grouped under biological variations, post mortem changes, internal contaminations (caused by intrinsic and inadvertent sources for specific constituents) and situations such as preferential accumulation of chemicals in selected organs or even within different compartments of an organ (Iyengar, 1982a).

Biological variations include physiological aspects of age, sex, habit, geographical and environmental factors, diet, pregnancy and lactation. Further, circadian rhythm, recent food, fasting, posture and stress during or just before sampling should also be considered.

Post-mortem changes are also of significance to the analyst since they influence the "sample status" as long as the specimens to be sampled remain inside the body. The processes responsible for this are cell swelling, tissue dehydration, imbibition, putrefaction and autolysis (Iyengar, 1981; 1986).

Intrinsic errors in trace element concentrations of specimens may arise due to internal contamination. These are sources that are inherently present in the sample and may falsify the results. For example, normal plasma contains much less hemoglobin in relation to serum which may contain 10 to 20 mg in 100 ml, an equivalent of 350 to 700 ng of Fe per ml, attributable solely to hemolysis (Kasperek et al., 1981). Since naked eye cannot distinguish hemolysis in serum below a certain degree, errors of this kind virtually go unnoticed at the sampling stage. Similarly, medication (e.g. with drugs containing trace elements) and prevalence of subclinical conditions (unidentified health conditions) can induce inexplicable variations in concentrations. These are difficult-to-control situations over which the analyst has little or no control. Careful evaluation of case history is necessary to minimize the impact of these errors in terms of guarded data interpretations. Even habits such as smoking tobacco (elevated Cd levels in blood) and consumption of wine (elevated Pb levels in blood) can have an impact on trace element metabolism as seen from a number of studies documented in the literature (Iyengar, 1989).

In routine clinical samples one has to deal with numerous intrinsic aspects that introduce changes by way of physiologic shifts, accentuating the difficulties in establishing reference values. The following example clarifies the situation. Stress conditions induce water shifts in the circulating blood. Since the red blood cells (RBC) and protein molecules are too large to pass through the water permeable membranes, they undergo changes resulting in higher protein concentrations and elevated hematocrit values. The impact of changes in hematocrit value is best illustrated by considering concentration changes of an element such as Pb which is predominantly bound to RBCs, and shifts in protein content which affect the results for protein-bound elements such as Zn. Since both hematocrit and serum protein concentration can be estimated with high precision, these two parameters are useful in tracing water shifts. Conversely, non-stress conditions can create the opposite effect. The RBCs are also very sensitive to osmolality (total dissolved solids per kg of water) and react by changing shapes. Addition of water increases cell size and vice versa. These size changes are also easy to measure. Since the factors that are responsible for shape changes also initiate concentration changes, e.g. trace metallic elements, one can link them to the wide scatter often observed in the analytical findings in specimens associated with RBCs.

Seasly (1985) has reported the dynamic variations in hematocrit and serum protein concentrations in 200 test subjects. The average daily spread of hematocrit was about 7%; the greatest individual variation in one day was 12.5% and the average 4-day spread between the highest and the lowest values was about 11%. Peaks of hemo-concentrations were observed at meal time. For serum protein (reference value = 70 g per liter), an average drop of 5 g per liter was observed in changing position from standing (which is a form of physical stress) to recumbent (non-stress condition). This difference varied from day to day between 3.5 and 6.5 g per liter (95% confidence limit).

The impact of presampling factors is clearly reflected in sampling body fluids, a process complicated by a host of inherent difficulties. In this context, the "status" of subjects at the time of sampling plays an important role by introducing random changes to the sampled material. For example, Zn in serum changes with age (Kasperek et al., 1977),

Table 1

Parameters Influencing Reference Values: Some Examples *

Sample	Element	Condition	Influence	Remarks
Blood	Cd	Tobacco smoking	Elevated	Remains chronically high in smokers
	Pb	Alcohol consumption	Elevated	Especially wine drinkers
Serum	Zn	Fasting	Elevated	Even overnight fasting
		After normal food	Lowered	During first few hours
		After high Zn food	Elevated	e.g. oysters, liver,
		Low Zn intake	Lowered	Rather sudden effect
		Stress	Elevated	e.g., Standing posture
		Pregnancy	Lowered	Progressive decline
	Cu	Pregnancy	Elevated	Progressive increase
	F	Fasting	Normal	Overnight fasting
		Normal food	Variable	If sampled immediately
		Tea consumption	Elevated	If sampled immediately
		Low F intake	Lowered	Over a few days
	I	High intake e.g. sea foods, iodinated salts	Elevated	Rapid uptake but slow clearance; biological half-life is 2.5 weeks
		Low I intake	Lowered	Over a few days
	As	Fish intake	Elevated	Well absorbed and slowly excreted
Milk	Mn	High input	Elevated	Mn rich cereals
	Fe	Fore-milk	Elevated	Variation due to
		Hind-milk	Elevated	high fat content
	I	Low I intake	Lowered	Over a few days
		High I intake e.g. dietary algae	Elevated	Extremely high levels of I excreted within a short period
Urine	Several elements	Low or high intakes	Variations	Sensitive to fluid intake, requiring 24 h collections

* Information summarized from Iyengar (1982a; 1989)

and pre- and post- prandial situations have profound effects on the elemental concentration profiles in blood serum due to the chelating effects of amino acids (Kasperek et al., 1978). Similarly, certain foods are known to alter the trace element concentration of body fluids: consumption of tea has been shown to induce abrupt changes over a short time in F levels in serum and urine (Carlson et al., 1960); ingestion of fish has been demonstrated to elevate As levels in blood serum over intermediate time intervals (Heydorn et al., 1978); and use of foods such as dietary algae naturally rich in iodine has been documented to increase levels of this element in breast milk over longer periods than blood (Muramatsu et al., 1983). The changes cited are due to the differences in gastrointestinal absorption of various dietary items as well as in residence times of the absorbed fractions.

Minimizing the effect of variations arising from presampling factors in clinical specimens is important but not easy. In Table 1, a few examples are listed which emphasize the impact of physiological status of subjects at the time of sampling. As seen from the Table, a sampling condition suitable for one element may not necessarily be valid for some other element. Importantly, this Table illustrates two situations: the need for an understanding of the metabolic behavior of different elements in the specimen to be sampled and the necessity for adopting compromised but well documented conditions for specimen collection. Indeed, it raises questions on the conclusions drawn from multielement analyses, e.g. serum, for establishing baseline data.

Assuming that the biologic validity of a given specimen is ensured, steps should be initiated to safeguard analytical validity of the specimen. At this stage, prevention of extraneous contamination of the sample is the immediate problem, the magnitude of which is variable depending on the element to be determined. If the sample is not analyzed within a short time, storage of the sample may, depending upon the nature of the sample matrix, create additional difficulties.

V. CONTAMINATION BY TRACE ELEMENTS

"Contamination" is a double-edged concept in sampling body fluids from human subjects. It calls for sterile as well as free of chemical (trace element) contamination. Contamination is a relative concept, the danger of which is related to the concentration of the analyte. For example, aluminum, chromium, manganese and lead are more susceptible to air-borne contamination than are certain other elements. The case of chromium in blood plasma provides a good illustration in this context. The amount of ethylene diamine tetraacetic acid (EDTA) needed (approximately 4 mg) as an anticoagulant for 1 ml of blood, would introduce an equivalent of 1.2 ng of Zn and 0.2 ng of Cr to the resulting plasma, due to the Zn and Cr present as contamination in EDTA. Although this increase would not affect the determination of Zn, for Cr it is unacceptable since the contaminant Cr fraction exceeds 100% of the actual Cr level of the serum samples (Behne and Iyengar, 1986).

It took several years to understand the problems of external contamination of urine before reference values could be established (Iyengar, 1989; Iyengar and Woittiez, 1988). Other crucial factors related to contamination control are the length of storage time before the samples are analyzed, and whether the available technical facilities for storage (e.g. container materials and dust and humidity controlled environment) are adequate for safe storage of the sampled material. A practical solution here is to use tools and containers of highest possible purity, made of materials that do not interfere with the analytical parameters under investigation, e.g. accessories made of titanium, quartz and teflon.

VI. LOSSES OF TRACE ELEMENTS

There are several possibilities of loss of trace elements during sampling as well as sample preparation stages. Adsorption on container walls and tools is a distinct possibility during sampling and storage. Losses due to volatilization, sputtering and chemical conditions during drying and ashing should be prevented by suitably modifying the working procedures. Retention of protein-bound trace elements under moderate drying conditions is in most cases quantitative (Iyengar et al., 1978; 1980), with the exception of methyl-Hg which has been reported to be lost even during freeze drying (Byrne, 1992). Matrix specific problems encountered with various biomaterials during dehydration treatments have been documented (IAEA, 1980).

VII. PRESERVATION AND STORAGE

The long-term storage of biological specimens is an important requirement for various reasons (e.g. for biomonitoring and specimen banking (NAS, 1991)); biological fluids present special problems in this context. Storage, especially from a trace element study point of view, basically implies an overall safe, clean, uncontaminated and non-degraded containment. Use of polyethylene (PE) vials and bottles are acceptable for storing dry powders for several years as very low temperatures are not required. Polycarbonate tubes are adequate for storage of serum and plasma for up to two weeks under refrigeration temperature. These tubes tend to develop cracks under sub-zero conditions, and for storage up to about -20°C, polyethylene tubes are recommended. If deep freezing conditions are desired, PE is not suitable and teflon bags or bottles are recommended. The relative merits of container materials with respect to their suitability for trace analysis work has been evaluated (Iyengar, 1984; IAEA, 1980).

Extensive investigations have been carried out by a Swedish group to understand the storage problems of whole blood for Cd and Pb over a wide range of concentrations in several species (Friberg and Vahter, 1983; Lind et al., 1987). Their results indicate that after 6 years of storage at -20°C in 5 ml polyethylene tubes, the concentration of Pb in blood had decreased by 15%, whereas for the same period, Cd had suffered a loss of about 10%. However, up to 2 years of storage at -20°C no losses were encountered. Similar observations have been recorded for Hg in whole blood (Meranger et al., 1981) and for Se in blood serum (Heydorn and Damsgaard, 1982) especially for the dimethylselenide form. Dahl et al. (1990) have demonstrated that freeze dried blood serum can be stored for over 4 years (1982-86) at -20°C without any problem. After reconstitution, the duration of storage is reduced drastically: one month at -20°C, four days at 2-8°C and only eight hours at room temperature.

In some cases, conventional storage is inadequate for preserving the natural state of organic compounds (e.g. metalloproteins, enzymes). In this context, it should be recognized that preservation relates to the process of retaining morphological and cellular integrity of an organ or tissue. Cryogenic preservation is one such approach which allows the specimens to retain the biochemical identity of the constituents.

VIII. SAMPLING AND PREPARATORY PROBLEMS

A. Blood

The blood is obtained conveniently by venipuncture. Since various considerations limit the widespread use of optimal needles made of alloys such as platinum-rhodium for drawing of blood, disposable type needles and syringes are generally used. Collection of blood in 10 or 20 ml fractions successively using the same needle and discarding the first two or three as rinse fractions is recommended. However, this method is not entirely suitable for elements such as Cr and Mn. Use of a plastic catheter reduces the contamination problem for difficult elements such as Mn and the discomfort to subjects since a single rinse provides a valid sample. Use of teflon catheters is becoming a recognized mode of blood collection for the determination of trace elements.

Preparing neonatal blood samples presents additional problems since only small volumes of sample can be obtained. A technique of handling these samples using a flexible teflon tubing to sample capillary blood from the heel has been described.

Hemolysis must be avoided since elements such as K and Fe which are at higher concentrations in the erythrocytes may affect the serum value depending upon the degree of hemolysis. Visible hemolysis begins at about 50 mg of erythrocytes per 100 ml serum. Use of a dry syringe, slow transfer to a dry test tube and allowing sufficient time for clotting are helpful in avoiding hemolysis (Kasperek et al., 1981).

B. Serum and plasma

The serum should be separated from the clot within one hour. The clotting of blood takes about 15 min. at room temperature and will be delayed when siliconized glassware, teflon or polyethylene containers are used. Blood is fractionated by centrifugation at about 3000 rpm for 10 - 15 min. The contents of the tube should be kept closed until separation to prevent contamination and loss by evaporation. Recentrifugation of the separated serum is sometimes necessary to spin down the residual erythrocytes. One disadvantage with serum samples is that hemolysis is greater than with plasma. In addition clotting releases K from platelets. For this reason K values are lower in plasma than in serum. Plasma can be obtained by immediate centrifugation of heparinized blood, but the presence of anticoagulants is undesirable for trace element analysis. However it is possible to obtain plasma without anticoagulant if fresh blood is immediately centrifuged.

C. Red blood cells

Careful removal of all the serum is necessary to separate the red blood cells (RBCs). However, a certain amount of trapped serum (usually 5 to 8%) is unavoidable and needs correction. Use of a syringe and needle to transfer centrifuged RBCs minimizes the presence of trapped plasma. A check on hematocrit is recommended if RBC values are used to compute whole blood values and vice versa.

D. Human milk

Samples of breast milk are collected either by manual expression or by using plastic breast pumps after cleaning the nipples with distilled water and letting them air dry for a couple of minutes. It is necessary to empty at least one breast completely and mix before collecting the sample, as the mean composition changes during the feed itself, e.g. fat content. Status of the sample, i.e. colostrum, transitional or mature, should be clearly defined since the stage of lactation influences the protein content and therefore the metal concentration (Iyengar, 1982b).

E. Sweat

Sweat is usually collected either by the so-called arm bag method using a polyethylene bag around the arm or from the whole body by collecting the free flowing drops from various points of the body. This is done in specially created sweating environment after the subjects have showered. However, collection of a valid sweat sample is difficult because of the numerous contamination hazards. The sweat collected should be homogenized by rigorous shaking and prepared immediately for the analysis. It is also necessary to centrifuge in order to obtain cell-free sweat. Unpredictable dilutions of the sample may occur since the profuseness of the sweating varies greatly in different parts of the body (Hohnadel et al., 1973).

F. Urine

Random collections of urine have limited use, if any, and therefore, 24 hour collections should be made. This is done after omitting the first void at the start and including the last one at the end of 24 hours using high pressure polyethylene bottles, precleaned by leaching separately with nitric acid and hydrogen peroxide for a few hours followed by liberal rinsing with double distilled water. Following the sample collection, aliquots are taken as soon as possible after vigorous shaking of the contents using pipettes which are rinsed several times with the urine to be transferred. It is preferable to freeze and lyophilize these subsamples (10 - 20 ml) to avoid the interaction of the urine with the container.

IX. ANALYSIS AND QUALITY ASSURANCE

Trace element determination in body fluids is characterized by the extremely small amounts of elements (Table 2) to be determined in the presence of excess quantities of other matrix constituents (Iyengar, 1989). The predominant presence of organic components along with high concentrations of certain inorganic elements complicates the analytical procedures requiring the removal of these interfering components before analytical signals are registered by the measuring instruments. Such analytical problems associated with each trace element and applicability of various analytical techniques for trace element determinations have been discussed on an element by element basis in the folowing monographs (McKenzie and Smythe, 1988; O'Neill et al., 1986).

Table 2

Reference Elemental Concentrations in Adult Human Blood Serum, Urine and Milk*

Element	Whole blood µg/l	Blood serum µg/l	Urine µg/day	Milk µg/l
Aluminum	2 - 8	1 - 5	10 - 100	<1?
Antimony	<0.1 - 3.5	0.01 - 2?	<0.1 - 5?	-
Arsenic	2 - 20	<1 - 5	10 - 30	0.25 - 3
Barium	0.5 - 2.5	-	2 - 4	-
Boron	20 - 50	15 - 45	500 - 3000	<50?
Bromine	3000 - 6000	3000 - 6000	2 - 10(c)	1000 - 2000
Cadmium	0.3 - 1.2(d) 1 - 4 (e)	0.1 - 0.3	1 - 5	<1
Cobalt	5 - 10	0.1 - 0.3	0.5 - 2	0.2 - 0.7
Chromium	<5?	0.1 - 0.2	0.2 - 2	<1.0 - 2.0
Cesium	1.0 - 4.5	1 - 2	5 - 20	1 - 5
Copper	800 - 1100 (a) 1100 - 1400(b)	800 - 1100 (a) 1000 - 1400(b)	30 - 60	200 - 400
Fluorine	200 - 500	20 - 50	500 - 1500	10 - 25
Gold	0.006 - 0.05	0.002-0.08?	0.001-0.6?	-
Iodine	40 - 60	60 - 70	100 - 200	40 - 80
Iron	425 - 500(c)	800 - 1200	100 - 200	300 - 600
Lead	50 - 150	<1	10 - 20	1 - 5
Lithium	0.4 - 1	0.2 - 0.8	25 - 100	<1?
Manganese	8 - 12	0.5 - 1	0.5 - 2	3 - 10
Mercury	2 - 20	<1	5 - 20	1 - 3
Molybdenum	1 - 3	<0.1 - 0.5	20 - 30	1 - 4
Nickel	1 - 5	<1 - 2	2 - 8	10 - 20
Platinum	-	-	<1?	-
Rubidium	2000 - 4000	100 - 300	1000 - 4000	300 - 1200
Scandium	0.002 - 0.12?	0.003-0.1?	<0.001-0.2?	-
Selenium	90 - 130	75 - 120	25 - 50	10 - 25
Silicon	-	300?	3 - 12(c)	-
Thallium	0.15 - 0.60?	0.02-0.34	0.1-1.0	-
Tin	<1	<1	1 - 2	<1 - 2
Uranium	0.001 - 0.005	-	0.004-0.070	-
Vanadium	0.1 - 0.5	0.1 - 1.0	0.2 - 1	0.1 - 0.3
Zinc	6000 - 7000	800 -1100	400 - 600	1000 - 2000

* Data pooled from Iyengar and Woittiez (1988), and Iyengar (1989; 1994).
? uncertain
[a] males
[b] females
[c] mg per litre

Frequent analysis of reference materials is the key for overcoming procedural inconsistencies in establishing quality control in any analytical laboratory. This is the most effective approach to identify results that fluctuate beyond certain set limits (precision) or deviate significantly from true values (accuracy). Use of reference standards helps to trace the discrepancies to reestablish analytical credibility. Ideally, every analyst needs to have available several reference standards that (i) are of a similar matrix type to the samples being analysed, (ii) are at least two in number, and (iii) contain concentrations of the analyte representative of the whole working range that is of interest (i.e. at least one at the low end and one at the high end). These and related analytical quality assurance problems and solutions have been discussed elsewhere (Accuracy in Trace Analysis, 1988).

The following reference materials are available for quality assurance of inorganic analysis in body fluids and products: animal blood (IAEA-A-13); bovine blood (BCR-CRM-194 to 196); bovine serum (NIST-SRM-1598); Pb in blood (NIST-SRM-955 a,b,c,d); human whole blood (NYCO-904-906); serum (KL-146-I to KL-148-II, NYCO 105, 112); urine (KL-110-H to KL-142-II, NYCO 108, NIST-SRM-2670); milk powder (IAEA-A-11, NIST-SRM-1549, BCR-CRM-150, 151 and 063). Information on matrix composition of these standards can be found in the following sources (Cortes Toro et al., 1990; Iyengar, 1989).

X. DATA BASIS, REFERENCE PARAMETERS AND INTERPRETATION

Besides carrying out reliable analytical work, it is also important to express the results on an appropriate basis to facilitate meaningful data interpretation (Iyengar and Behne, 1988). In many cases, the lack of attention to this component has resulted in wrong conclusions being reached, even though the analysis has been carried out correctly. This can always happen when the biological material consists of several components with different element contents, and the ratio of these components differs between the samples.

The results of elemental analysis of whole blood may depend on the percentage of red blood cells in the sample. This should always be considered when the element content is much higher in erythrocytes than in the blood plasma, as for instance in the case of iron and lead. If not, changes in the hematocrit, which can occur as a function of several factors such as pregnancy, age or disease, may lead to false conclusions being drawn about the quantities of these elements in the body.

Errors can also occur by redistribution of elements or water in certain organs due to the interruption of metabolic processes. For instance, significant changes in the weight of the whole organ can occur under post-mortem conditions (Iyengar, 1981), and these can lead to great differences in the results according to whether they are related to the wet or the dry weight of the material. In such cases, it is necessary to know the total weight of the organ as well as its water content, to be able to account for the changes in elemental concentrations of the samples. It is obvious that meaningful data interpretation is not possible under these circumstances.

Many errors of the kind alluded to above can often be avoided if one does not express the analytical value only in the usual way as element amount per volume or weight of the

Table 3

Parameters for Evaluation of Elemental Analysis Data in Biological Fluids*

Specimens	Parameters
Body fluids	
Serum, plasma	Protein, water
Whole blood	Hematocrit
Urine	24 h volume, creatinine
Sweat	Volume
Amniotic fluid	Meconium, cells
Milk	Average daily output
Semen	Sperm count
Bile	Volume
Cells	
RBC, WBC, Platelets,	Cell number and Hemoglobin (for RBC)
Spermatozoa	Viability status

* SG, dry/wet weight ratios, common for all fluids

Table 4

Checklist of Essential Features Related to Elemental Analysis Data of Biological Systems

Specimen	Required information
Plasma	Amount and type of anticoagulant and its chemical purity, body position during sampling
Serum	Body position during sampling
Milk	Days post partum, fat content, specific fraction (hind milk, foremilk) if entire volume not collected
Cells	Viability, (cell age), amount, type and purity of stabilizer used, trapped plasma

sample, but also relates the data to other parameters such as the number of certain cells or the content of a protein, which enable the changes in the tissue composition to be taken into account. In whole blood analysis, for instance, errors due to changes in the hematocrit can be excluded if the amount of hemoglobin, which reflects the number of red blood cells, is used as a base for expressing the element levels.

Some examples of relevant parameters needed to evaluate element analysis data are shown in Table 3. In Table 4, additional features are listed, which should also be checked when elemental data are measured in biological materials.

It is difficult to recommend any particular preference for choosing the most suitable base, as this depends to a large extent on the type of the study and on the problems to be investigated. However, one can improve the present situation by paying adequate attention to document supplemental information on relevant parameters and including them in scientific publications. This would also facilitate a meaningful intercomparison of results from different investigations.

XI. SUMMARY

The complexities involved in sampling biological fluids for trace element research studies require both the analytical and biomedical perceptions. Every effort should be directed at procuring samples that satisfy both biological and analytical validity. The generation of meaningful conclusions from elemental composition data derived from biological systems is a strictly multidisciplinary task, an approach that is vital for the overall success of the investigations. The future efforts towards developing analytical methodologies and reference materials should take into account the multidisciplinary needs of biological trace element investigations. Some of the clinical samples are very valuable since repetitive sampling may not always be possible. Therefore, to cope with the ethical restrictions involved with sampling clinical materials, it is essential that medical sciences personnel appreciate the dangers of sample contamination during collection. This can be achieved only when both the analyst and the physician work in close cooperation.

REFERENCES

Accuracy in Trace Analysis: Goals, Challenges and Accomplishments (1988). *J. Res. Nat. Bur.*, Vol. 93.

Anand, V.D., White, J.M., and Nino, H.V. (1975). *Clin. Chem.*, *21*: 595.

Behne, D., and Iyengar, G.V. (1986). *Analytiker Taschenbuch*, Springer Verlag, Berlin, Vol. 6, p. 237.

Carlson, C.H., Armstrong, W.D., and Singer, L. (1960). *Proc. Soc. Exp. Biol. Med.*, *104*: 235.

Cortes Toro, E., Parr, R.M., and Clements, S.A. (1990). IAEA/RL/128/Vienna, Austria.

Dahl, A., Hald, P., Jorgensen, P., Martinsen, P.I., and Thomassen, Y. (1990). *Fresenius Z. Anal. Chem.*, *338*: 526.

Friberg, M., and Vahter, M. (1983). *Environ. Res.*, *30*: 95.

Heydorn, K., and Damsgaard, E. (1982). *Talanta*, *29*: 1019.

Heydorn, K., Damsgaard, E., Larson, N.A., and Nielsen, B. (1978). Proceedings IAEA Symp. Nuclear Activation Techniques in the Life Sciences, IAEA, Vienna, p. 129.

Hohnadel, D.C., Sunderman, F.W., Nechay, M.W., and McNeely, D. (1973). *Clin. Chem.*, *19*: 1288.

Horwat, M., and Byrne, A.R. (1992). *Analyst*, *117*: 665.

International Atomic Energy Agency (1980). Technical Report Series 197, IAEA, Vienna.

Iyengar, G.V. (1981). *J. Pathol.*, *134*: 173.

Iyengar, G.V. (1982a). *Anal. Chem.*, *54A*: 554.

Iyengar, G.V. (1982b). *The Elemental Composition of Human and Animal Milk*, IAEA TECDOC-269, Vienna.

Iyengar, G.V. (1984). *Quality Assurance in Biomedical Neutron Activation Analysis*, IAEA TECDOC-323, Vienna, p. 83.

Iyengar, G.V. (1986). *NBS J. Res.*, *91*: 67.

Iyengar, G.V. (1989). *Elemental Analysis of Biological Systems*, CRC Press, Boca Raton, FL.

Iyengar, G.V. (1992). *Anal. Sci.*, *8*: 399.

Iyengar G.V. (1994). *Mathematical Modelling for Trace Element Metabolism* (K.N. Subramanian and M. Wastney, eds.), CRC Press, Boca Raton, FL, in press.

Iyengar, G.V., and Behne, D. (1988). *J. Res. Nat. Bur.*, *93*: 326.

Iyengar, G.V., and Woittiez, J.R.W. (1988). *Clin. Chem.*, *34*: 474.

Iyengar, G.V., Kasperek, K., and Feinendegen, L.E. (1978). *Sci. Tot. Environ.*, *10*: 1.

Iyengar, G.V., Borberg, H., Kasperek, K., Kiem, J., Siegers, M., Feinendegen, L.E., and Gross, R. (1979). *Clin. Chem.*, *25*: 699.

Iyengar, G.V., Kasperek, K., and Feinendegen, L.E. (1980). *Analyst*, *105*: 794.

Kasperek, K., Feinendegen, L.E., Lombeck, I., and Bremer, H. (1977). *Eur. J. Pediat.*, *126*, 199.

Kasperek, K., Schicha, H., Siller, V., Feinendegen, L.E., and Hoeck, A. (1978). Proceedings IAEA Symp. Nuclear Activation Techniques in the Life Sciences, IAEA, Vienna, p. 517.

Kasperek, K., Kiem, J., Iyengar, G.V., and Feinendegen, L.E. (1981). *Sci. Tot. Environ.*, *17*: 133.

Lind, B., Elinder, C.G., Friberg, L., Nilsson, B., Svaartengren, M., and Vahter, M. (1987). *Fresenius Z. Anal. Chem.*, *326*: 647.

McKenzie, H.A., and Smythe, L.E., eds. (1988). *Quantitative Trace Analysis of Biological Materials*, Elsevier Science Publishers, Amsterdam.

Meranger, J.C., Hollebone, B.R., and Blanchette, G.A. (1981). *J. Anal. Toxicol.*, *5*: 33.

Mertz, W. (1975). *Clin. Chem.*, *21*: 468.

Muramatsu, Y., Sumiya, M., and Ohmomo, Y. (1983). *Hoken Butsuri*, *18*: 113.

National Academy of Sciences, NAS (1991). *Monitoring Human Tissues for Toxic Substances*, National Academy Press, Washington, DC.

O'Neill, I., Schuller, P., and Fishbein, L., eds. (1986). *Manual of Methods of Analysis on Metals*, Vol. 8, International Agency for Research on Cancer, Geneva.

Seasly, R.H. (1985). 19th Annual Conference on Trace Substances in Environmental Health (D.D. Hemphill, ed.), University of Columbia, MO, p. 314.

Subramanian, K.S., Iyengar, G.V., and Okamoto, K., eds. (1990). *Biological Trace Element Research: Multidisciplinary Perspectives*, ACS Symp. 445, American Chemical Society, Washington, DC.

ION SELECTIVE ELECTRODES AND METAL ION DETERMINATION IN BIOLOGICAL FLUIDS

by

H.G.J. Worth

Director of the Clinical Chemistry Laboratory,
King's Mill Hospital,
Sutton in Ashfield,
Nottinghamshire NG17 4JL
U.K.

I. INTRODUCTION

The use of electrodes in the measurement of low-molar-mass (l.m.m.) species in biological fluids is established for the direct and indirect measurement of a number of analytes of clinical significance, including the measurement of metal ions by ion selective electrodes (ISEs). In practice this has been restricted to sodium, potassium and calcium.

The ion selective electrode is derived from the simple electrode. When a piece of metal is placed in a dilute solution of one of its salts, an electrical potential is established between the solution and the surface of the metal, as shown in Figure 1. In this example, zinc atoms exist in equilibrium with zinc ions in the metal, and an equilibrium is set up between the ions in the metal and those in the solution. In order to achieve this, electrons move between the two phases until the equilibrium is reached, thus establishing an electrical potential. This potential will be dependent upon the position of the metal in the electrochemical series and other factors defined by the Nernst equation

$$E = E^\circ + \frac{RT}{nF} \log_e A$$

where E = potential developed,
E° = specific electrode potential,
R = gas constant,
T = temperature,
n = ionic charge,
F = the Faraday,
A = activity of the ion in solution.

The electrical potential derived in this simple half cell system is related to the activity of the solution and can be considered as a measure of that specific ion in the solution. This forms the basis of the use of ion selective electrodes in measuring specific ions in biological fluids.

Apart from the electrochemical activity of the ion and the activity of the ion in the solution, the potential developed is dependent upon the temperature of the system and the

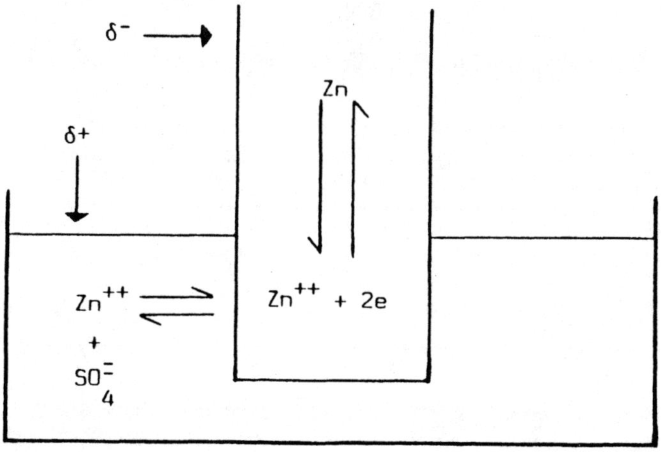

Figure 1.
The simple zinc electrode.

ionic charge of the ion under consideration as defined in the Nernst equation. The specific electrode potential is defined under standard conditions, namely 25°C (298K) and an electrolyte activity of 1 mol dm^{-3}.

Using a simple half cell to measure the activity of a specific ion in a biological fluid requires its electrical potential to be compared with that of a reference electrode. The criterion for a half cell to act as a reference electrode is that it must develop a constant potential under the conditions of operation of the system. The hydrogen electrode is the definitive reference and is defined as consisting of a 1 mol dm^{-3} solution of a strong acid containing an inert platinum electrode saturated by hydrogen at a pressure of one atmosphere at 25°C. By definition, this has an electrode potential of zero. Although a reliable system, it is not practical in the routine measurement of analytes. Alternatively, a like electrode may be used as a reference where it is assumed that any fluctuations in conditions affect both half cells equally. More commonly the glass/calomel electrode is used. This is a complex electrode consisting of solid mercury(I) chloride and metallic mercury in contact with a saturated solution of potassium chloride. It is a good reference electrode as it produces a constant electrical potential varying very little over a wide temperature range.

The reference and the measuring electrodes are maintained as simple half cells, but have to be in electrical contact with each other in order to establish a potential between them. This is achieved by creating an interface which allows movement of electrons and ions but with a minimum of diffusion. This junction will itself have an electrical potential because of the movement of electrons and this is referred to as the junction potential. One of the problems of using ISEs is to ensure that the sum of the junction potentials is not of a significant magnitude compared to the potential developed by the ion to be measured.

The performance of the electrode must ensure that the change in potential can be measured sufficiently precisely for clinical purposes. Table 1 illustrates this by showing the electrical potential that would have to be developed for different ions if there were a change coincidental with their blood plasma reference interval. It also proposes a required clinical precision and shows the equivalent electrical potential. The measurable change in potential for sodium is much smaller than for the other ions if the precision goal is to be achieved. In other words the sensitivity of a sodium electrode has to be greater. This is well recognized in that it is more difficult to maintain a sodium electrode to the required standard than a potassium or calcium electrode or a pH meter.

Blood plasma is a varied mixture of l.m.m. components in a complex macromolecular matrix. Ion selective electrodes, therefore, have to be specific with little or no cross

Table 1

Reference Data and Required Precision with Equivalent Change in Electrical Potential

Ion	Reference range (mmol dm^{-3})	≡	Δ EMF (mV)	Required precision ± (mmol dm^{-3})	≡	EMF ± (mV)
H^+(pH)	7.35-7.45		5.9	0.01		0.59
Na^+	135-145		1.8	0.5		0.09
K^+	3.5-5.0		9.2	0.1		0.6
Ca^{2+}	1.0-1.2		2.3	0.02		0.23

reaction with many components. This is achieved by solid state, liquid ion exchange or gas dialysis membranes. Solid state membranes may consist of glass membranes or crystal lattices. Permeable membranes made from thin films of glass are selective for certain ions. By changing the constitution of the silicates in the glass, it is possible to change the size of the lattice and consequently the selectivity. This is applicable to hydrogen and alkali metals so that glass membranes can be prepared which are selective for hydrogen, lithium, sodium or potassium, although they will allow passage of the ions of a lower molar mass in the series. Thus, a membrane that is selective for sodium allows the transport of hydrogen and lithium. In plasma this does not represent a significant problem. However, a glass membrane that is selective for potassium will also have a cross reactivity for sodium and this is significant as the sodium concentration in plasma is approximately 35-fold that of potassium. Glass membranes form the basis of the hydrogen electrode used in pH meters. This is a glass with a lattice that permits the transfer of hydrogen ions, but no larger ions, and is therefore highly specific.

Crystal lattice membranes rely on natural defects in the lattice which allow the transfer of certain specific ions. This has not been applied to the measurement of metal ions but is used in the measurement of halides and forms the basis of the chloride specific electrode.

Liquid ion exchange membranes consist of an ionizable species in a lyophobic solvent bound by an inert porous matrix on either side, the purpose of the porous matrix is to

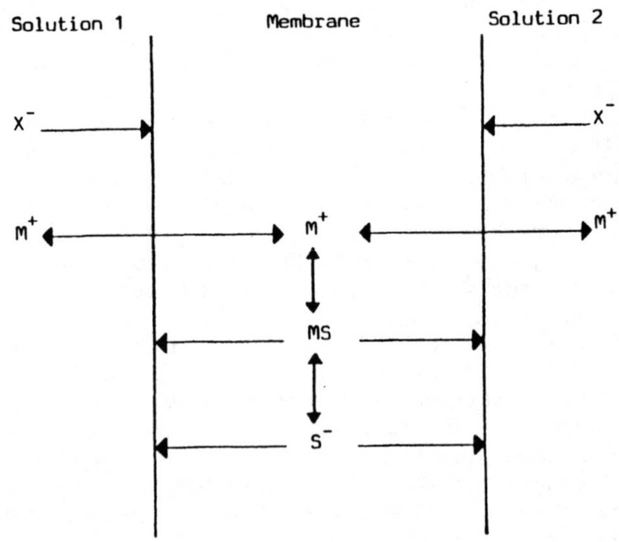

Figure 2.
Liquid ion exchange membrane.

contain the lyophobic solvent and to allow free ionic access. The lyophobic solvent is chosen for its tertiary structure which will enable the transport of certain ions to the exclusion of others. This is demonstrated in Figure 2 where the cation M^+ is preferably transported to the exclusion of the anion X^-. M^+ forms an equilibrium complex with the lyophobic solvent S^-, as well as being in equilibrium with the ion in solution on both sides of the membrane, thus setting up an electrode potential across both sides of the membrane. This is used for measuring calcium and potassium by employing an alkyl phosphate such as di-n-octyl-phenylphosphonate and valinomycin respectively as the lyophobic solvent.

Gas dialysis membranes may be used for selective measurement where gases are derived from components in biological systems, thus affording their indirect measurement. For example polypropylene allows the free transfer of carbon dioxide and ammonia. However, this does not have an application to the measurement of metal ions.

II. MEASUREMENT OF SODIUM AND POTASSIUM

Sodium and potassium are considered together analytically because they have been measured in tandem since the early development of routine clinical chemistry and cannot be assayed by simple colorimetric methods. Their precise measurement is clinically important because they have narrow reference intervals in blood plasma and, in the case of potassium, high tissue levels which may result in marked changes as a consequence of water and electrolyte imbalance associated with many pathological conditions. Variance from the reference interval can have serious and even fatal consequences. The development of emission flame photometry allowed simultaneous measurement and was used exclusively until the introduction of ion selective electrodes.

Emission flame photometry measures the emission of radiation resulting from excited atoms falling back to their ground state and this is related to the concentration of the element according to the Scheibe-Lomakin equation

$$I = Ac^b$$

where
I = emission of radiation,
A and b are constants,
c = concentration.

The introduction of ISEs raises the question of whether there is any difference in measurement of concentration by flame emission photometry and activity by ISE, and if so, which is clinically significant? Concentration is a measure of the number of particles in a system expressed in terms of mass or numbers of molecules in a defined unit volume. Activity, a thermodynamic concept, is a measure of the proportion of particles acting in a manner that is characteristic of that component. The two are related by the equation

$$a = \gamma c$$

where
a = activity,
c = concentration,
γ = activity coefficient.

Strongly electropositive cations may be regarded as fully ionized and therefore their activity coefficient is close to unity, except when the numbers per unit volume become so great that there is an interaction between like ions which causes their full ionic effect and consequently activity to be reduced. This is significant at the levels encountered in blood plasma, the activity coefficients for sodium and potassium being approximately 0.7 and 0.8 respectively. There are other potential causes of differences between activity and concentration in blood plasma, namely: (i) formation of l.m.m. non ionic species, (ii) complexing with macromolecules, (iii) plasma matrix effect.

The formation of l.m.m. non ionic species would reduce activity compared to concentration, although such formation is unlikely with strongly electropositive ions. However, there is some evidence for non ionic complexing with carbonate in plasma (Coleman and Young, 1981), but this is disputed by other workers (Czaban et al., 1982). Similar complexing with macromolecules would have the same effect. There is some evidence of sodium being complexed with blood plasma protein but this effect is probably not significant (Kissel et al., 1982). The final effect is due to the complex matrix of macromolecules in the blood plasma. This has the effect of reducing the available volume of plasma for the l.m.m. species, and is opposite to those effects already considered. The concentration is unchanged by the volume occupied by macromolecules but the activity is increased because the available volume decreases as the quantity of macromolecules increases. This is illustrated in Figure 3. If the matrix volume of a 'normal' plasma with a sodium concentration of 130 mmol dm^{-3} and a potassium concentration of 4.0 mmol dm^{-3} is increased from 10% to 20%, the activity of sodium and potassium would increase to 146.4 and 4.5 mmol dm^{-3} respectively, but the concentration would remain unaltered.

Clearly there is a difference between concentration and activity in pathological situations, but which is clinically correct, and is the difference significant? Chemical (and biochemical) processes comply with the laws of thermodynamics; activity not concentration is a thermodynamic concept; physiological mechanisms must therefore be controlled by the activity not the concentration of their components. Logically, activity not concentration is the scientifically more correct measurement.

Changes in plasma matrix are related to abnormal concentrations of macromolecules, namely proteins or lipoproteins. There are three broad categories of clinical conditions where this may occur, hyperproteinemia such as myelomas, hyperlipidemia such as diabetes and familial hyperlipemias, and through parenteral nutritional feeding. Values as divergent as 40 mmol dm^{-3} have been recorded for plasma sodium. As the model in Figure 3 suggests, the effect on potassium is much less and has not been shown to be clinically significant (Forrest and Shenkin, 1980; Ladenson et al., 1981).

In these situations, *direct* reading ion selective electrodes give the clinically correct result and should be used. However, the majority of clinical chemical analyzers are fitted with either *indirect* reading ion selective electrodes or emission flame photometers, because the rate of throughput is much higher and this is required to cope with the workload of most routine laboratories.

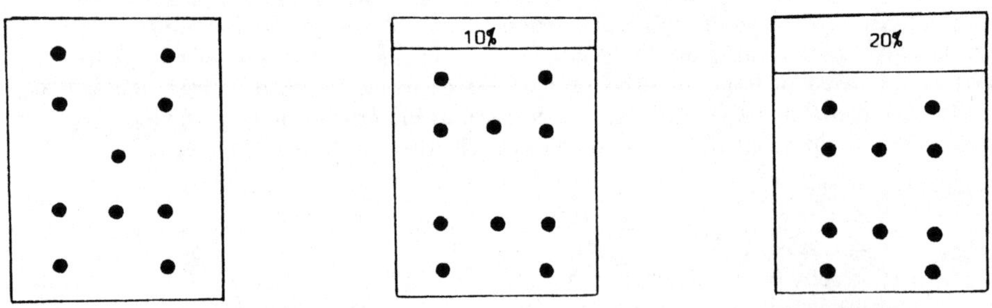

AQUEOUS SOLUTION

Contains 10 parts of a species per unit volume. It will have identical activity and concentration namely 10 parts per unit volume.

'NORMAL' PLASMA

The available circulating volume is reduced by 10%*, the volume occupied by the matrix. Concentration remains the same at 10 per unit volume, but the activity has increased by 10% to 11.1 per unit volume.

*8% is a truer figure, but 10% is convenient for the model.

'ABNORMAL' PLASMA

Matrix volume has been increased to 20% by a pathological abnormality. Concentration remains at 10 per unit volume but activity is now 12.5 per unit volume.

Figure 3.
Plasma matrix effect on activity.

Despite the theoretical advantages of measuring ionized calcium, there is no demonstrative advantage over total calcium measurement in most situations causing calcium homeostatic imbalance. As a result, corrected plasma calcium concentration continues to be used with a few exceptions. Neonates, particularly those delivered prematurely, frequently exhibit hypocalcemia, which is related to an inability to produce adequate levels of parathyroid hormone, and the resultant hypocalcemia is due to reduced ionized calcium.

III. MEASUREMENT OF CALCIUM

The clinical requirements on the measurement of plasma calcium are more stringent than any other analyte commonly measured in blood plasma. If the reference interval is expressed as a percentage of the interval mean, only sodium shows a greater sensitivity. This coupled with the fact that small variations from the limits of the reference interval are clinically significant, may account for why more physical and chemical techniques, including the use of ISEs, have been applied to plasma calcium measurement than any other analyte.

The distribution of calcium in blood plasma is complex. Approximately 50% is present as ionized calcium, 5% unionized, forming an ultrafiltrate component, and the remainder is protein bound, primarily to albumin. The unionized form is present mostly as bicarbonate or organic acid salts. Calcium homeostasis is related to calcium activity and consequently there is a theoretical argument for measuring ionized calcium. It is only in relatively recent years that commercial apparatus has been available and so, with the exception of a few specialist research institutes, blood plasma calcium has been measured as total calcium. The realization of the need to measure calcium activity resulted in the calculation of a derived factor, referred to as corrected calcium. This takes into account the variation in albumin concentration. A number of factors have been used, of which the following is typical:

$$\text{Corrected calcium (mmol dm}^{-3}\text{)} = \text{Measured calcium (mmol dm}^{-3}\text{)} - 0.02\ [\text{albumin (g dm}^{-3}\text{)} - 40]$$

This attempts to correct the plasma calcium concentration to what it would have been if the albumin concentration was 40 g dm^{-3}, the mean for recumbent hospital inpatients. Unlike sodium and potassium, the formation of l.m.m. non ionic species and complexes with macromolecules results in a reference interval for ionized calcium (1.18-1.30 mmol dm^{-3}) that is very different from total plasma calcium (2.10 - 2.70 mmol dm^{-3}).

Technically ionized calcium is not difficult to measure, the electrode is simple and robust using a liquid ion exchange membrane. However, a number of factors affect calcium measurement by electrode. The total plasma ionic strength affects the measured calcium activity, (Siggaard-Andersen et al., 1983), a change of 10 mmol dm^{-3} changes the calcium activity by approximately 1.2%. As the pH increases, the ionized calcium decreases (Fogh-Andersen, 1981). It is important, therefore, to know the pH of the patient's plasma in vivo at the time the sample was collected, and to be sure there is no change in vitro. This requires the sample to be taken under anaerobic conditions. Changes in plasma protein concentration affect ionized calcium measurement, in addition to the effect it has on binding capacity. It is thought that this is a consequence of protein binding to the electrode membrane. For a change in albumin concentration of 10 g dm^{-3}, there can be a mean change of 0.05 mmol dm^{-3} in ionized calcium (Butler et al., 1984). Additionally, erythrocytes have an effect on liquid junction potentials: as the hematocrit increases, so does the liquid junction potential. As the temperature of the sample increases, the amount of protein-bound calcium increases and consequently the ionized fraction decreases (Heining et al., 1983). The converse is also true and, therefore, samples should be measured at patient body temperature.

Despite the theoretical advantages of measuring ionized calcium, there is no demonstrative advantage over total calcium measurement in most situations causing calcium homeostatic imbalance. As a result, corrected plasma calcium concentration continues to be used with a few exceptions. Neonates, particularly those delivered prematurely, frequently exhibit hypocalcemia, which is related to an inability to produced adequate levels of parathyroid hormone, and the resultant hypocalcemia is due to reduced ionized calcium.

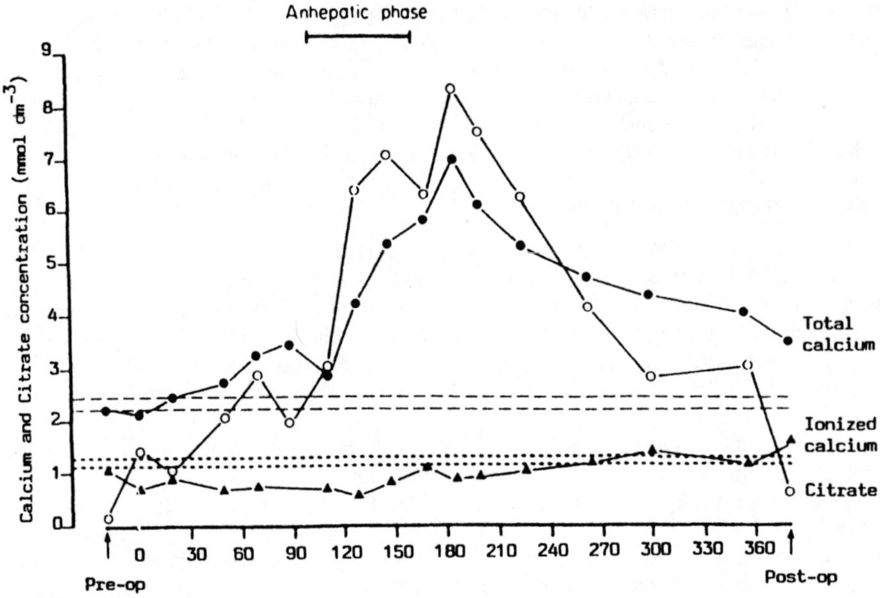

Figure 4.
The effect of citrate infusion on total and ionized plasma calcium during liver transplantation. The upper interrupted horizontal lines indicate the reference range for total calcium, and the lower lines the reference range for ionized calcium.
Reproduced with kind permission from Gray, T.A., and Patterson, C.R. (1988). *Ann. Clin. Biochem.*, *25*: 210.

However, most pediatricians are experienced in interpreting total plasma calcium concentrations and this does not normally produce a problem. Changes in plasma citrate concentration will cause significant changes between ionized calcium and total calcium levels that are not related to the total protein or albumin concentration. This is because calcium citrate is ultrafiltrable but not ionized and will occur as a result of massive blood transfusion or plasmapheresis where there are large infusions of citrate into the patient, in major operative procedures which require a high level of anticoagulation, and the rare condition of reduced citrate metabolism. Operative procedures include coronary arterial bypass graft and transplant surgery where the patient's circulation system has to be bypassed and consequently the blood kept anticoagulated by the addition of citrate. This is illustrated in Figure 4 which shows the citrate infusion required by a liver transplant patient and that the total plasma calcium concentration gives no indication of true calcium homeostasis.

Where coronary aortic bypass graft and transplant operations are carried out, ionized plasma calcium activity needs to be measured and is best carried out in theatre as a side room test by a qualified laboratory technologist. Under these circumstances the quality of results is guaranteed because the measurement is carried out by experienced, qualified staff, and because of the proximity of the equipment to the patient it can be done quickly, avoiding the problems of time delay and transportation.

REFERENCES

Butler, S.J., Payne, R.B., Gunn, I.R., Burns, J., and Paterson, C.R. (1984). *Brit. Med. J.*, *289*: 948.

Coleman, R.L., and Young, C.C. (1981). *Clin. Chem.*, *27*: 1938.

Czaban, J.D., Cormier, A.D., and Legg, K.D. (1982). *Clin. Chem.*, *28*: 1703.

Fogh-Andersen, N. (1981). *Clin. Chem.*, *27*: 1264.

Forrest, A.R.W., and Shenkin, A. (1980). *Lancet*, 2: 1256.

Heining, M.P.D., Band, D.M., and Linton, R.A.F. (1983). *Scand. J. Clin. Lab. Invest.*, *43*: 709.

Kissel, T.R., Sandifer, J.R., and Zumbulyadis, N. (1982). *Clin. Chem.*, *28*: 449.

Ladenson, J.H., Apple, F.S., and Koch, D.D. (1981). *Ann. Intern. Med.*, *95*: 707.

Siggaard-Andersen, O., Thode, J., and Fogh-Andersen, N. (1983). *Scand. J. Clin. Lab. Invest.*, *43 (Suppl. 165)*: 11.

ANALYTICAL TECHNIQUES FOR DRUG DETERMINATION IN BIOLOGICAL FLUIDS

by

Pr. Georges Houin, PhD, and Dr. Sylvie Saivin, PhD,

Unité de Pharmacocinétique Clinique,
Hôpital Purpan,
31059 Toulouse Cedex
France

I. INTRODUCTION

During the last few decades, the development of sensitive, accurate and specific analytical methods for drug measurement in biological fluids has allowed a better understanding of the fate of drugs in the body, particularly in terms of the interactions between the drug and a biological entity. It has constituted the basis for the development of the field of pharmacokinetics (Deutsch et al., 1991).

The overall quality of such studies depends upon careful considerations at each step involved and these can be summarized in two parts: first, the operations required in the collection, storage and preparation of samples for analysis, and second, the choice of the analytical method itself and its optimization in terms of both quality and adaptability for the number of analyses to be performed (Houin, 1991).

Biological sample collection is obviously of critical importance, since it represents the material on the quality of which all the other steps will be based. However, the choice of the analytical method will also depend on the circumstances under which the samples will be collected (administered dose, time for collection after administration, single vs multiple doses, etc) since it precludes the levels to be measured. Analytical considerations include the treatment of the sample and its physical, chemical or immunological analysis, as well as the preparation of standard curves and the mathematical interpretation of the analytical results. Careful control of all these factors will minimize analytical variability, hence increasing the overall quality and specificity of the studies.

Measurements of concentrations may be performed using numerous different techniques. While physicochemical methods are generally chosen because of their universality, in some instances the pharmacological activities of the molecule must be used. Such is the case of antibiotics, for which microbiological methods are commonly employed, and antithrombotic drugs or glycosaminoglycans, for which antithrombotic or anticoagulant properties are used. In these cases, however, it is impossible to distinguish between the activity of the parent drug and/or one or more of its metabolites.

Therefore, this chapter will be divided into three parts: physicochemical methods, immuno-enzymatical methods, and methods specific for some classes of drugs.

II. PREANALYTICAL CONSIDERATIONS

A. Sampling

Of the large number of factors involved, i.e. the site and the method of sampling, the time at which, and tube in which, it takes place only two will be discussed as they can influence interaction studies.

The biochemical composition as well as the drug concentration of a sample may be affected by the position of the subject, supine or sitting, and the duration of the tourniquet application (Fawcett and Wynn, 1960; Eisenberg, 1963; Eisenberg and Wolf, 1967; Husdan et al., 1973; Statland et al., 1974). In particular, the protein concentration of blood is altered by these factors which in turn can affect ligand-protein interactions (Dettli and Spring, 1967; Reidenberg and Kostenbauder, 1969).

Particular attention should be paid to the choice of the tube in which the sample is drawn, especially in regard to the adsorption of drugs to the walls. As a rule, glass tubes are preferable to plastic ones. In the same way, inorganic stabilizers and accelerators may be released from the caps and interfere with the analytical methods and/or affect the binding of drugs to plasma proteins (Cotham and Shand, 1975; Fremstad and Bergerud, 1976; Piafski and Borga, 1976; Wilkinson and Schenker, 1976). The use of commercially available serum separators in collecting tubes should be used with great care, although good results can be obtained (Bailey et al., 1988).

B. Treatment of samples

Obviously, plasma and serum are not interchangeable; they are two distinct fluids. Significant differences have been demonstrated in the assays of various components, including hormones, vitamins, endogenous chemicals and drugs (Ladenson et al., 1974; Lum and Gambino, 1974; Gleispach et al., 1977; Kubasik and Sine, 1978).

Anticoagulants are not equivalent. For example, heparin, the most commonly used anticoagulant, increases lipoprotein lipase activity, inducing an increase in the circulating fatty acids which in turn can displace the binding of drugs to proteins, in particular weakly acidic drugs.

The temperature during the centrifugation may be of critical importance. Such is the case for heat-labile drugs or metabolites. It is also well documented for cyclosporin whose distribution between erythrocytes and plasma is very dependent upon temperature; therefore most studies are made on blood instead of plasma or serum samples.

C. Storage

In most studies, the analysis of the samples occurs at varying times post sampling. The conditions in which the samples will be stored can bring large variabilities from numerous factors. For example, storage at room temperature allows the enzymatic activity to continue. It is then a common situation to freeze the samples at - 20°C, sometimes at - 80°C, to minimize this enzymatic activity; however, protein denaturation can occur, particularly of globulins (Watt, 1976). As a rule, preliminary in vitro experiments should be undertaken and control samples stored under the same conditions in order to verify the integrity of the unknown samples.

III. PHYSICOCHEMICAL METHODOLOGIES

The chemical structure of a drug largely influences the choice of the most suitable procedure for its determination. However, the route and dose of drug administration as well as the pharmacokinetics of the drug govern the ultimate sensitivity and specificity required for the assay. The available methods commonly used are numerous and cover a wide range of quantitation, as listed in Figure 1.

A. Preanalytical treatment

The nature of the preanalytical treatment will depend upon the sensitivity and the specificity to be attained and also on the analytical method employed.

The general aims are to obtain the product in a form that can be analyzed and to clean the samples from interfering molecules as extensively as possible. The various methods used

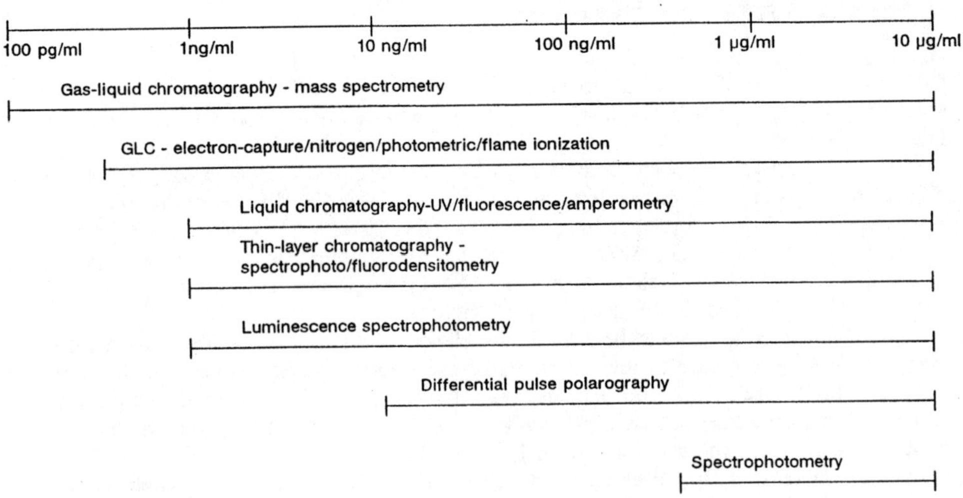

Figure 1.
Practical range of sensitivities of analytical methods. Reprinted from De Silva (1970).

to release the molecules from the cellular structures include physical methods such as heat or ultrasonic treatment. Chemical treatment of the samples is needed to release the compound, either from its binding protein or from its chemical linkage to endogenous compounds such as glucuronic acid. Release of protein binding is achieved, for example, simply by dilution with distilled water, or by means of a protein precipitant such as tungstate or trichloroacetic acid, or with an organic solvent such as acetonitrile, acetone or ethanol. Disruption of covalent linkages involves alkaline hydrolysis of esters or acidic hydrolysis of ethers. Enzymatic hydrolysis can also be used.

Elimination of interfering compounds can be achieved by centrifugation, filtration, precipitation, extraction or chromatography.

Extraction of the drug usually involves an organic extraction at a pH at which the compound is essentially unionized. It may involve a single extraction step or repeated extractions, with or without variation of the pH. This essential step must be optimized, as it influences the recovery, and therefore the sensitivity as well as the variability, of the method. A clean up step and/or a concentration step may be involved, depending on the concentrations of the substances present and the desired sensitivity in relation to that of the detecting instrument. Selective extraction on adsorptive columns can also be employed, incorporating clean up and concentration with a minimum of manipulation. These methods have become more and more popular and give excellent results, although they are in general more difficult to apply to specific requirements. Whatever the chosen solution, the use of radiolabeled drugs may be of great help during these preanalytical studies.

B. Analytical methods

The intrinsic characteristics of the drug, such as its partition characteristics, spectral properties and susceptibility to derivatization, as well as previous experience with analogous compounds, will obviously provide important guidelines in the choice of the suitable analytical method for obtaining the desired sensitivity and specificity.

The choice of techniques is very large and, it is hoped, is increasing continuously, enabling the attainment of optimal specificity and sensitivity. An attempt to summarize the different advantages of the most commonly used methods is described in Table 1.

Obviously, only a rough and subjective choice of an analytical method can be made, differing according to the molecule to be analyzed. Moreover, the choice of the technique may also depend on non-specific considerations, such as the available material, the cost of

Table 1

Advantages of the Most Common Methods Used for Drug Measurement

	HPLC	GLC	GLC-MS	Immunology
Sensitivity	++	++	+++	+++
Accuracy	++	++	+++	++
Reproducibility	+	+	++	++
Specificity	++	++	+++	+
Flexibility	+++	++	+	-
Linearity	++	++	+	-
Matrix effect	+	+	+	+++
Free drug measurement	+	+	+	+

HPLC : High performance liquid chromatography
GLC : Gas liquid chromatography
GLC-MS : Gas liquid chromatography - mass spectrometry

each analysis, the number of samples to be analyzed, and the time necessary to obtain a result vs the degree of emergency. All these factors can be summarized under the general term of "feasibility", which is probably as important as the scientific characteristics of the analysis. The available techniques can be distinguished as non-chromatographic and chromatographic. Of the direct analytical methods, in the main those that are spectrophotometric or fluorimetric are usually employed initially, and have in general adequate sensitivity but low specificity for both endogenous compounds and/or the metabolites. As a consequence, they are used in practice after chromatographic separation. The same applies to polarographic methods, although some specificity may be achieved by virtue of the functional group in the molecule.

IV. IMMUNOLOGICAL METHODS

Immunological methods are based on the interaction between the drug and antibodies. Since the 1980's, advanced technologies are available to produce monoclonal antibodies specific to a well known structure such as a drug or a peptide sequence. Generally, the immuno-enzymatic methods are preferred to the radio-isotopic ones because they are safer and do not need expensive and heavy machines. Several principles are currently involved: EMIT and immuno-enzymo-fluorimetric methods, fluorescence polarization and fluoro-immunologic methods.

The EMIT principle is based on the competition, in the homogenous liquid phase, for the binding to monoclonal antibodies between the drug and the same drug labelled with an enzyme like G6PD. The enzyme-drug complex is only active against the substrate when it is free, and is inactive when it is bound. This method may be totally automatic and very quick (Pesce and Bodourian, 1981; Iosefsohn et al., 1981).

The immuno-enzymo-fluorimetric methods are performed in a biphasic system. It is based on the competition between the analyte drug and the same enzyme-labelled drug, for the binding to specific monoclonal antibodies. These antibodies are bound to a solid phase and the blend of drugs is cleared by radial elution with substrate solution. This substrate is cut by the enzyme carried by the bound drug and this reaction rate is measured by a fluorimetric system.

Fluorescein is a small molecule able to emit polarized rays after excitation. The polarization is a direct function of the molecule size and when the drug is bound to an antibody, its size increases. The principle of the method is the competition between the drug and the fluorescein-labelled drug for their binding to antibodies (Jolley et al., 1981).

In fluoro-immunologic methods, the drug to measure and the same drug labelled by β -galactosyl-umbelliferon are in competition for the binding to antibodies. This binding does not permit β-galactosidase from *Escherichia coli* to cut fluorescent umbelliferon (Burd, 1981); therefore, the fluorescence is proportional to the drug concentration.

V. SPECIFIC METHODS

Specific methods have been developed for some drugs because it has not been possible to measure them with separative methods. For example, before the immuno-enzymatic determination, it was not possible to measure aminoglycosides because HPLC or GC are not applied to glucidic compounds. Another typical example is the measurement of blood heparin, the nature of which does not permit the use of conventional methods.

These specific methods will be described as a function of the activity involved in the measurement.

A. Antibiotic measurements

Before immuno-enzymatic methods, research on antibiotics was carried out by using a very simple idea: measurement of the concentration of an antibiotic may be performed by the determination of its activity against a sensitive bacterial strain. The direct applications are the determination of the activity of a specific antibiotic against a specific strain and the measurement of unknown concentrations after their comparison against a calibration curve.

1. *Efficiency determinations*

To determine the efficiency of antibiotics, a lot of methods are available: the antibiogram, the determination of the minimum inhibitory concentration and minimal bactericidal concentration, the bactericidal power of a biological fluid and even the determination of the efficiency of antibiotics in combination either at the end point or by kinetics. In this chapter, we will only discuss the bactericidal power of biological fluids because other methods are used before the administration of drugs to humans and consequently their measurement is not made in biological fluids.

The bacteriostatic and bactericidal power of a biological fluid such as serum, urine or cerebro-spinal-fluid, is generally used to verify the efficiency of a treatment for a specific patient against his pre-isolated strain. When it is possible, the sampling times correspond to the maximum concentration and just before the next administration. The principle of this measurement is based on the maximum dilution of the fluid able to inhibit the growth of the strain after 18 hours of incubation: this is the bacteriostatic power. In a second manipulation, the survivor germs are counted from tubes where no growth was visible to the naked eye: the bactericidal power is defined as the dilution in which there are less than 0.01 % of survivors. Some technical precautions must be respected: for example the strain must be in exponential phase of growth when it is treated with the antibiotic; this method is not usable for streptogramins; a variation incorporating a solid medium must be used for antibiotics with a high probability of mutations (fosfomycin, rifampicin, sulfamides and trimethoprime) (Courvalin et al., 1985).

2. *Concentration determinations*

Determination of antibiotic concentrations may be performed by a simple method. When the drug, at known or unknown concentrations in a biological fluid, is put on a paper disk or in a hole on agar medium, it will diffuse in the agar. The reference strain has a well-known sensitivity profile. It is added to the agar medium and will grow everywhere if there is not enough antibiotic to inhibit it. The diameter of growth inhibition is a linear function of the logarithm of the antibiotic concentration in the considered biological matrix. It is therefore possible to create calibration curves and to determine unknown concentrations (Courvalin et al., 1985).

This method is very cheap and easy to perform but it presents some disadvantages. First, the calibration curves are very limited when compared with HPLC. The biological matrix may influence the diffusion of the antibiotic, particularly when it contains macro-molecules. This involves the use of the same matrix for samples and calibration curves.

Another problem is the measurement of several antibiotics in the same sample: it requires strains with differential sensitivity profiles and this may be difficult or impossible in some cases.

The most important matter in the measurement of an antibiotic after its administration to an organism is the implicit hypothesis which is done: the relationship between the observed activity and the concentration is strictly linear, even for in vitro measurements. This condition is possible if there is no active metabolite. In the opposite case, it is not possible to know which part of the activity is a function of the antibiotic concentration and which part corresponds to the metabolite. This method may be used to study the pharmacodynamics of antibiotics but not their concentrations and their pharmacokinetic profiles. Even if the drug has no metabolism, the protein binding may certainly influence the diffusion in the agar medium because the diffusion is only possible for the free fraction.

B. Glycosaminoglycans measurements

Glycosaminoglycans are polysaccharidic chains with a wide range of molar masses. Generally, drugs have a small molar mass, less than 1 kg mol^{-1}. Glycosaminoglycans have very high molar mass: 12 kg mol^{-1} for heparin, 25 kg mol^{-1} for dermatan sulfate. Some specific sequences of these polysaccharides (Choay et al., 1983) may interact and catalyze the inhibiting activity of plasma components against coagulation factors. These inhibitors are antithrombin III (AT III) which is able to inhibit factor Xa and factor IIa of coagulation, and heparin cofactor II (HC II) (Tollefsen, 1981) which is only active on factor IIa. The glycosaminoglycans are a large family, but only heparins and dermatan sulfate are used in therapeutics. Briefly, heparin catalyzes the inhibiting activity of AT III on factor Xa and factor IIa (Beguin et al., 1988), and this latter activity is a function of the chain length (Andersonn et al., 1976). Dermatan sulfate only catalyzes the inhibiting activity of HC II on factor IIa (Maimone and Tollefsen, 1988; 1990; Tollefsen et al., 1983).

Glycosaminoglycans are not homogeneous in composition, and their variety of chain lengths and molar masses does not permit the use of classical methods for measuring them. The sole possibility is to determine their different activities. These activities are divided into anticoagulant (due to the inhibition of factor IIa by AT III) and antifactor activities (Buchanan et al., 1985).

1. *Anticoagulant activities*

Anticoagulant activities are classically measured by two tests: the thrombin clotting time and the activated partial thromboplastin time. These two tests are based on the measurement of the time required for the coagulation of plasma in which thrombin or calcium have been added, respectively. These anticoagulant activities require thrombin inhibition after activation of AT III by the drug (Ofosu et al., 1985).

When heparin is present in plasma, its anticoagulant activities increase the clotting time, the logarithmic value of this time being a linear function of the heparin concentration. It is therefore possible to draw a calibration curve from which unknown concentrations can be determined. These methods apply within the narrow range of concentrations generally used in therapeutics. When concentrations are increased, the plasmas become non-coagulable and the test can no longer be used.

These tests are very useful to adapt the dosage regimen of heparin. Low-molar-mass heparins are depolymerized compounds. This chemical modification does not modify their ability to inhibit factor Xa but reduces their ability to inhibit factor IIa; therefore, the anticoagulant activities of these compounds are not possible to determine by these methods (Laurent et al., 1978).

The principle of these methods only permits their use in plasma; therefore, no other matrix can be investigated with these.

2. *Antifactor activities*

As it was specified previously, the natural inhibitors of the coagulation factors are AT III which is effective on factor Xa and IIa, and HC II which is only effective on factor IIa (Dupouy et al., 1988). The catalyzing activities of the glycosaminoglycans used as drugs are shown is Table 2.

Table 2

Glycosaminoglycans Catalyzing Activities Used as Measurement Methods

	Factor Xa		Factor IIa	
Inhibitors	AT III	HC II	AT III	HC II
Heparin	++++	-	++++	+
LMMH	++	-	+	-
Dermatan sulfate	-	-	-	+++

LMMH : Low-molar-mass heparin
ATIII : Antithrombin III
HCII : Heparin coFactor II

The antifactor activities of glycosaminoglycans are all based on the same principle: the coagulation factor and the inhibitor are added in excess in the reaction cup and the limiting factor is the amount of glycosaminoglycan in plasma. The enzymatic reaction occurs and the excess of coagulation factor may interact with a chromogen substrate and generate a colored product which will be measured by absorption spectrometry. The logarithm of the difference of absorption is a linear function of the antifactor activity generated by the glycosaminoglycan; therefore, unknown concentrations may be calculated by reference to a calibration curve. These tests are possible in plasma, urine and buffer solutions. Other methods were described which use clotting measures but they are not dependent on a single factor (Boneu et al., 1991).

REFERENCES

Andersonn, L.O., Barrowcliffe, T.W., Holmer, E., Johnson, E.A. and Sims, G.E. (1976). *Thromb. Res.*, *9*: 575.

Bailey, D.N., Coffee, J.J., and Briggs, J.R. (1988). *Ther. Drug Monit.*, *10*: 352.

Beguin, S., Lindhout, T., and Hemker, H.C. (1988). *Thromb. Haemost.*, *60*: 457.

Boneu, B., Faruel-Bille, V., Pierrejean, D., and Gabaig, A.M. (1991). *Nouv. Rev. Fr. Hématol.*, *33*: 287.

Buchanan, M.R., Boneu, B., Ofosu, F., and Hirsh, J. (1985). *Blood*, *65*: 198.

Burd, J.F. (1981). *Meth. Enzymol.*, *74*: 79.

Choay, J., Petitou, M., Lormeau, J.C., Sinay, P., Casu, B., and Gatti, G. (1983). *Biochem. Biophys. Res. Commun.*, *116*: 492.

Cotham, R.H., and Shand, D. (1975). *Clin. Pharmacol. Ther.*, *18*: 535.

Courvalin, P., Goldstein, F., Philippon, A., and Sirot, J. (1985). *L'antibiogramme*, MPC-Videom, Paris, p. 219.

De Silva, J.A.F. (1970). *Current Concepts in the Pharmaceutical Sciences*, Vol. 1 (J. Swarbrick, ed.), Lea and Febiger, Philadelphia, p. 203.

Dettli, L., Spring, P (1967). *Helv. Med. Acta*, *33*: 291.

Deutsch, T., Ludwig, E., Graber, H., Barre, J., and Houin, G. (1991). *Human Pharmacology* (H. Kummerle, T. Shibuya and J.P. Tillement, eds.), Elsevier Science Publishers, Amsterdam, p. 307.

Dupouy, D., Sie, P., Dol, F., and Boneu, B. (1988). *Thromb. Haemost.*, *60*: 236.

Eisenberg, S. (1963). *Arch. Intern. Med.*, *112*: 544.

Eisenberg, S., and Wolf, P.C. (1967). *Arch. Intern. Med.*, *115*: 17.

Fawcett, J.K., and Wynn, V. (1960). *J. Clin. Pathol.*, *13*: 304.

Fremstad, D., and Bergerud, K. (1976). *Clin. Pharmacol. Ther.*, *20*: 120.

Gleispach, H., Kruetzer, B., Fellier, H., and Brokenstein, B. (1977). *Clin. Chim. Acta*, *80*: 381.

Houin, G. (1991). *Human Pharmacology* (H. Kummerle, T. Shibuya and J.P. Tillement, eds.), Elsevier Science Publishers, Amsterdam, p. 1.

Husdan, H., Rapaport, A., and Locke, S. (1973). *Metabolism*, *22*: 787.

Iosefsohn, M., Boeckx, R.L., and Hicks, J.M. (1981). *Ther. Drug Monit.*, *3*: 365.

Jolley, M.E., Stroupe, .D., Wang, C.H.J., Panas, H.N., Keegan, C.L., Schmidt, R.L., and Schwenzer, Z.S. (1981). *Clin. Chem.*, *27*: 1190.

Kubasik, N.P., and Sine, H.E. (1978). *Clin. Chem.*, *24*: 137.

Ladenson, J.H., Tsai, L.M., Michael, J.M., Kessler, G., and Joist, J.H. (1974). *Amer. J. Clin. Pathol.*, *62*: 545.

Laurent, T.C., Tengblad, A., Thunberg, L., Höök, M., and Lindahl, U. (1978). *Biochem. J.*, *175*: 691.

Lum, G., and Gambino, S.R. (1974). *Amer. J. Clin. Pathol.*, *61*: 108.

Maimone, M.M., and Tollefsen, D.M. (1988). *Biochem. Biophys. Res. Commun.*, *152*: 1056.

Maimone, M.M., and Tollefsen, D.M. (1990). *J. Biol. Chem.*, *265*: 18263.

Ofosu, F.A., Blajchman, M.A., Modi, G.J., Smith, L.M., and Buchanan, M.R. (1985). *Br. J. Haematol.*, *60*: 695.

Pesce, M.A., and Bodourian, S.H. (1981). *Clin. Chem.*, *27*: 1460.

Piafski, K.M., and Borga, O. (1976). *Lancet*, *2*: 963.

Reidenberg, M.M., and Kostenbauder, H.A. (1969). *Metabolism*, *18*: 209.

Statland, B.E., Bokelund, H., and Winkel, P. (1974). *Clin. Chem.*, *20*: 1513.

Tollefsen, D., and Blank, M. (1981). *J. Clin. Invest.*, *68*: 589.

Tollefsen, D., Petska, C., and Monafo, W. (1983). *J. Biol. Chem.*, *258*: 6713.

Watt, J.G. (1976). *Clin. Haematol.*, *5*: 95.

Wilkinson, G.R., and Schenker, S. (1976). *Clin. Pharmacol. Ther.*, *19*: 486.

FRACTIONATION AND CHEMICAL SPECIATION OF TRACE METALS AND METALLOIDS IN BIOLOGICAL FLUIDS

by

P.H.E. Gardiner

Division of Chemistry,
Sheffield Hallam University,
Pond Street
Sheffield S1 1WB
U.K.

I. INTRODUCTION

Trace element-containing species in biological systems can be broadly classified into storage, transport and metabolically active forms. In any biological compartment, more than one type of chemical species of a given element can be found. Attempts at understanding their physiological roles, therefore, require fractionation of the various constituents followed by their characterization and identification.

Regardless of the approach chosen for a particular study, it is essential that the results of the experiments reflect the true situation as found in vivo. The structural and chemical integrity of the species under study must be maintained throughout any physical or chemical manipulation steps. Both chemical and biological contamination which could lead to problems with the interpretation of the results, have to be controlled. Furthermore, the techniques that are chosen for the analysis of the various constituents must have the required sensitivity and specificity.

In this section, the various techniques and methods that have been used for the study of the distribution of metal- and metalloid-containing species in biological fluids will be examined.

II. SOME CONSIDERATIONS IN THE STUDY OF THE DISTRIBUTION OF CHEMICAL SPECIES IN BIOLOGICAL SYSTEMS

The purpose of most investigations into the qualitative and quantitative distribution of chemical species has been to identify the metabolically active forms, the transport and storage species, metabolites of trace element-containing drugs and pollutants, and to find sensitive indicators of the status of trace elements in the organism. Identification of any of these groups of species requires that the appropriate body compartment is chosen for investigation. The chemical species found in readily accessible samples such as blood, urine, feces, sweat, saliva, breast milk, semen and hair, reflect the extracellular pool of the trace elements, and in addition, these species would usually be products of various metabolic processes in different parts of the body. Trace element levels in blood, urine and feces are useful as a measure of recent exposure. Levels in breast milk and semen are relatively constant except in cases of extreme deprivation. However, investigation of the extracellular pool can yield information on the metabolites of trace element-containing drugs and pollutants, their transport forms, and in a few cases, their storage species. In contrast, the search for metabolically active chemical species and sensitive indicators of the status of trace elements in the body is best carried out on cells.

Irrespective of the pool chosen for investigation, the in vivo distribution of the various species depends on the physiological state of the organism at the time the sample was taken. Exercise, medication, fasting, dietary habits, alcohol intake, occupational exposure, circadian rhythm, repeated phlebotomy, pregnancy and stress are all factors that can influence the levels and distribution of various biochemical constituents. Furthermore, the sampling techniques and implements used can alter the nature and prevalence of a number of species. Standardized sample collection procedures, which take into consideration the effects of the above factors, must therefore be adopted. In the absence of standardized procedures, the techniques and methods used should be described in detail so that other investigators can follow the same protocols (Gardiner, 1993).

The need to maintain the structural and chemical integrity of the species has meant that complexing agents and other chemicals that are likely to alter the conformation and destroy the biological activity of the various species should not come into contact with the samples. Hydrolytic enzymes already in the samples must be inhibited if they are perceived to cause degradation of the constituents under study. In order to limit the changes that could occur to relatively unstable species, samples must be processed without too much delay under conditions similar to those found in vivo.

III. SAMPLE COLLECTION, PRE-TREATMENT AND STORAGE

A. Blood serum or plasma

Copper, iron, selenium and zinc are now determined routinely in a number of clinical laboratories. Standard laboratory ware for the collection of blood meant for trace element analysis can be used with the minimum risk of contamination. However, for nickel, chromium, cobalt and manganese, special precautions have to be taken. Some workers (Versieck, 1985; Aitio and Jarvisalo, 1985) have found it necessary to use plastic catheters or polyethylene cannulas in order to prevent contamination of the sample. Particularly in the study of protein-bound fractions, prolonged stasis during blood collection has to be avoided as this affects the distribution of water between compartments.

Once collected, whole blood cannot be used for any meaningful study of the distribution of trace elements since it is composed of a number of compartments, each with its own trace element-containing species profile. For most clinical studies, serum or plasma is used. Anticoagulants added to blood can alter the distribution of trace elements which are loosely bound to proteins or other ligands. Aluminum, chromium, iron, manganese and zinc ions are particularly vulnerable since they have a greater affinity for oxygen-containing ligands such as EDTA, citrate and oxalate. Plasma samples are therefore not suitable for the study of these elements. The use of serum is also not without its drawbacks. Hemolysis is greater when red blood cells are allowed to clot and, as a result, intracellular constituents can leak into serum. Zinc- and iron-containing species and other constituents found at much higher concentrations in erythrocytes can be a source of contamination. However, this may not pose a serious problem if the contaminants are distinct from species found in serum and do not bind to any of the serum constituents. The presence of contaminants, however, puts a greater demand on the resolution of the fractionation techniques.

Blood samples are usually collected into glass, polyethylene or polypropylene containers. Once the red cells clot and the sample is centrifuged, the serum should be transferred into polyethylene, polypropylene or polystyrene containers with tight fitting screw caps. The container material must not selectively adsorb or allow the diffusion of the species through its walls.

Short-term storage, about 3 days, can be carried out at 4°C without discernible changes in the distribution of copper-, iron-, and zinc-containing species. The species of more volatile elements like arsenic, mercury and selenium may have to be stored at lower temperatures. For long-term storage, it is advisable to use a temperature of -20°C or below.

Samples should not be repeatedly thawed and refrozen as this is likely to destroy some constituents. It should be borne in mind that long-term storage could abolish the activity of certain enzymes, and the presence of some species whose identification depends on measuring their enzyme activity may be missed. Chemical species that are oxidized in open air must be stored under nitrogen.

B. Urine

The appearance of a substance in urine depends on its half-life in the body, the nature of its metabolites, and its form and route of intake. It is possible, as has been shown, for arsenic to identify metabolites of an element from dietary sources as distinct from those due to industrial exposure (Johnson and Farmer, 1991; Arbouine and Wilson, 1992). The study of the distribution of the species of a given element can, therefore, be used as a means of distinguishing between different sources of exposure. In this regard, it is important to establish what the likely sources of exposure are, and the time that has elapsed before sample collection. Careful consideration should be given to the variable nature of the amount and type of the urine constituents, and how they affect the distribution of various species of interest.

Potential donors should be made aware of the likelihood of the introduction of extraneous contamination during sample collection. Urine samples should be collected into plain polystyrene containers with screw caps. In contrast to serum, the use of preservatives and stabilizers cannot probably be completely eliminated. Without preservatives, bacterial contamination may destroy some of the species of interest. Furthermore, phosphate, calcium and other constituents are precipitated, thus changing the ionic composition of the sample. This could have an adverse effect on the nature and prevalence of some chemical species. Because the nature and form of the chemical species are important, the choice of preservatives and stabilizers must be made so as to minimize the damage to existing interactions. It is likely that a broad-range preservative or stabilizer may not be appropriate. Instead, these may have to be tailored to the need to preserve the integrity of species of a particular element. Furthermore, their presence must not interfere with subsequent analytical steps. Acidification with hydrochloric acid or glacial acetic acid, a common method of stabilization, could destroy some species and shift the balance between others. Instead of adding thymol or merthiolate as preservative, storage at 4°C is to be preferred. Samples should be filtered through a 0.22 μm filter if bacterial contamination is a problem. Indeed, this should be part of a standard protocol in order to remove epithelial cells, blood and other debris.

C. Intracellular constituents

It is essential that cells used for this type of study be obtained with a sufficient degree of purity, intact and viable if cross-contamination is to be avoided. Fractionation of cells is usually a multistep process which involves the use of a number of reagents. Standard procedures (Peters, 1983; Ford et al., 1990) developed for other purposes may be used for trace element work, provided the blank levels in the reagents of the elements under investigation are acceptable. As a rule, high purity reagents, AnalaR grade or better, should be used. In cases where the presence of reagents that have been used in the fractionation step is found to interfere with subsequent analytical procedures, these have to be removed by either dialysis, ultrafiltration, chromatography or electrophoresis.

A number of procedures have been developed for the separation and determination of manganese (Milne et al., 1990), selenium (Hansson et al., 1989), and zinc (Milne et al., 1985; Goode et al., 1989; Bilbow and Mundal, 1990; Ruz et al., 1992), in cellular components of human blood. Further fractionation is necessary in order to study the intracellular constituents. The cell membranes have to be destroyed and the cell constituents solubilized in an appropriate buffer medium. The medium should have similar pH and ionic strength, and if possible similar redox potential and oxygen tension as that found in the intact cell. The recommendations for storage are the same as those for serum.

D. Other biological fluids

Biological fluids such as amniotic fluid, semen, saliva and milk have been analyzed for trace elements (Heydorn, 1984), and recommendations for the collection of contamination-free samples also apply here. However, the issue of storage, and particularly the effect of storage temperature and the presence of proteolytic enzymes on the integrity of trace element-containing species, are yet to be adequately addressed.

IV. FRACTIONATION AND DETECTION TECHNIQUES FOR TRACE METALS AND METALLOIDS

There are two approaches currently used for the fractionation and detection of trace element-containing species. The first involves either coupling an analytical separation technique directly to an element-selective detection system, or analyzing collected fractions subsequently for constituents of interest. This approach is favored when the amount of material is limited or a rapid method for studying relative changes in the qualitative and quantitative distribution of known species is required. In applying this approach to poorly characterized systems, it is possible to localize fractions containing the species of interest. However, it is difficult to apply any of the classical structure elucidation techniques because of the small quantities (usually less than a microgram) of analyte. Identification is possible only when the fractions can be tested for species thought to be present in the samples and authentic standards are available for comparison. At best, this approach can provide information on the relative molar masses or isoelectric points of the unknown species if their migration during fractionation can be related to these properties.

The second approach involves the complete isolation of a given chemical species from the sample matrix. Characterization is carried out on the pure form of the compound or moiety, and any impurities can make identification difficult. This approach is preferred when additional information which could lead to a better understanding of the biological reactivity of the species is required. Techniques that can be used for structure elucidation and in the study of the reactivity of trace element-containing active centers are listed in Table 1.

Regardless of the fractionation techniques chosen, a number of precautions have to be taken, and these arise out of the need to prevent either the introduction of extraneous chemical contamination into the fractions or the loss of the trace elements from their binding sites. It is essential that the reagents, and parts of equipment which come into contact with the samples, be free of the trace elements under investigation. Recovery values should be reported as a matter of course in order to show that contamination or losses have not occurred. If the chemical species are subjected to chemical reactions, for example derivatization, then it is important that the products be well-defined and the yields high. Side reactions could lead to difficulties with data interpretation. Furthermore, it is important to bear in mind that sample fractionation is not an end in itself but that the fractions are going to be subjected to further analytical steps. Therefore, the reagents chosen for each step have to be compatible with the requirements of subsequent steps, and must not be a source of interference.

Methods and techniques that have been used in the study of the distribution of chemical species in biological samples are summarized in Table 2.

As can be seen from Table 2, a plethora of techniques and methods have been used to study the distribution of chemical species in a number of sample matrices. It seems, however, that limited successes have been achieved in the identification of those chemical species for which prior knowledge does not provide clues as to their identity. For the study of these chemical species, more sophisticated approaches are required.

REFERENCES

Aitio, A., and Jarvisalo, J. (1985). *Ann. Clin. Lab. Sci.*, *15*: 121.

Arbouine, M.W., and Wilson, H.K. (1992). *J. Trace Elem. Electr. Hlth. Dis.*, *6*: 153.

Arnaud, J., André, D., Bouillet, M.C., Kia, D., and Favier, A. (1992). *J. Trace Elem. Electr. Hlth. Dis.*, *6*: 81.

Table 1

Techniques for the Characterization of Trace Element-containing Species

Information	Techniques and methods	Comments
Trace element content	Anodic stripping voltammetry	Complete mineralization of samples is necessary before analysis.
	Atomic Absorption Spectrometry Plasma Emission Spectrometry Inductively Coupled Plasma/Mass Spectrometry	Both on-line and off-line analysis of sample fractions without further pretreatment is possible. However, increased sensitivity of analysis is obtained for some elements when samples are mineralized and converted into hydrides, e.g. antimony, arsenic, bismuth, lead, mercury, selenium, tellurium, and tin.
	Neutron activation analysis	Time-consuming sample preparation required, and long delays, up to several weeks, before results are available.
Identification of volatile thermally stable, low-molar-mass trace element-containing species	Gas Chromatography/Mass Spectrometry Gas Chromatography/Atomic Absorption Spectrometry	Identification is possible if standards are available for comparison. Derivatization can convert relatively non-volatile species into species that are amenable to separation by GC.
Identification of water-soluble, high-molar-mass species	Liquid Chromatographic Techniques/Atomic Spectrometry	Poor chromatographic resolution does make it difficult to identify individual species in complex biological matrices. Multi-fractionation steps may be necessary for unequivocal identification.
The environment around the trace element-containing sites	Electron paramagnetic resonance (EPR) spectroscopy, Magnetic susceptibility, Visible Absorption Spectroscopy, Mössbauer Spectroscopy	Techniques useful in the study of some trace element-containing species.
Characterization of protein and peptides	Circular dichroism Immunochemical methods Sequence analysis Peptide mapping Isoelectric focussing X-ray techniques Biological activity FAB mass spectrometry	The application of all of these techniques and methods can yield useful information only when the fractions are relatively pure.

Table 2

Selected Examples of the Applications of Various Methods and Techniques Used for the Study of the Distribution of Trace Element-containing Species in Biological Fluids

Sample matrix	Fractionation and detection techniques and methods	Detected species	Comments	References
Amniotic fluid	Size exclusion chromatography (SEC), Immunonephelometry and Graphite Furnace Atomic Absorption Spectrometry (GFAAS)	High- and low-molar-mass Cu-, Fe-, and Zn-containing species in human amniotic fluids	Prior knowledge allowed the fractions to be tested for the presence of transferrin, albumin and ceruloplasmin; major iron-, zinc-, and copper-binding proteins, respectively. Identification of the low-molar-mass species was not possible.	Gardiner et al. (1982)
Red blood cells	SEC, GFAAS, Glutathione (GSH-Px) activity	Se-containing fractions in red blood cell lysates.	The presence of the selenium-containing enzyme glutathione peroxide was detected by measuring its activity. Unequivocal identification of the other species was not possible.	Butler et al. (1991)
	Selective uptake of chromates: Cr (VI) instead of Cr (III) by human erythrocytes	Cr (VI) dissolved in erythrocytes	Novel approach to speciation; an example of the use of the discriminating ability of cell membrane to provide information on chemical species.	Neidhart et al. (1990)
	SEC, Isoelectric focussing, (IEF) Carbonic anhydrase activity, GFAAS	Cu- and Zn-bound species in human erythrocytes		Gardiner et al. (1984)
	Extraction, Butylation, Gas Chromatography/Microwave-Induced Plasma Atomic Emission Spectrometry	Methyl-and Inorganic Hg		Bulska et al. (1992)
Breast milk	SEC, Inductively Coupled Plasma Atomic Emission Spectrometry (ICP-AES)	Cu-, Fe-, Zn-containing species in human and cow's milk	Species were identified by comparison with pure standards. Distinct differences in the distribution profiles of the three elements in human and cow's milk were observed.	Brätter et al. (1988)

Table 2 (continued)

Sample matrix	Fractionation and detection techniques and methods	Detected species	Comments	References
	SEC, Differential Pulse Anodic Stripping Voltammetry (DPASV), IEF	Cd species		Michalke and Schramel (1990)
	SEC, IEF, Direct Current Plasma Atomic Emission Spectrometry (DCP)	Zn species		Michalke et al. (1991)
	and			
	SEC and Ultrafiltration	Zn species	A detailed study of the chromatographic factors which affect the distribution of weak metal-ligand complexes.	Arnaud et al. (1992)
Blood serum or plasma	Affinity chromatography, GFAAS, Kinetic immunoturbidimetry	Albumin- and globulin-bound Zn	The methods could be used to study the relative changes in the distribution of zinc in serum in health and disease.	Foote and Delves (1983)
	GFAAS, SEC	Cu- and Zn-protein complexes	Low-molar-mass zinc species identified as associated with the retinol-binding protein complex. Copper is thought to be in the form of a thionein complex.	Chilvers et al. (1984)
	SEC, GFAAS, (GSH-Px) activity	Se-containing proteins	Poor protein resolution.	Borglund and Åkesson (1985)
	SEC, GSH-Px activity Molecular fluorimetry	GSH-Px bound Se	A comparison of the differences in selenium distribution profiles in human and animals is reported.	Beilstein and Wanger (1983)
	Reversed-phase ion-pair chromatography, ICP	Pt species	An example of the use of speciation to study the metabolites of drugs in biological fluids.	De Waal et al. (1987)
	SEC, GFAAS, SDS-PAGE	Al-binding serum proteins		Favarato et al. (1992)

Table 2 (continued)

Sample matrix	Fractionation and detection techniques and methods	Detected species	Comments	References
	Computational approach to chemical speciation	Al and Zn distribution	Useful approach to the study of the distribution of elements, particularly those that form relatively weak metal-ligand complexes which are not amenable to laboratory determination.	Harris and Keen (1989), Venturini and Berthon (1989), Daydé et al. (1990), Harris (1992)
	Anion exchange chromatography, electrophoresis nephelometry	Distribution of ^{51}Cr in plasma	The use of radiotracers circumvents a lot of the problems associated with extraneous chemical contamination.	Borguet et al. (1990)
	SEC, ICP-MS	Cu, Fe, Pb and Zn species	The possibility of using different isotopes to study the distribution of the species of the same element is explored.	Gercken and Barnes (1991)
Semen	SEC, ICP	Cu- and Zn-containing species		Gardiner et al. (1987)
Proteins	SEC, ICP-MS	Zn-containing species and various isotopes: ^{54}Fe, ^{59}Co, ^{67}Zn, ^{88}Sr, ^{114}Cd, ^{127}I and ^{138}Ba	The feasibility of multi-element and multi-isotope determinations of protein fractions is demonstrated.	Owen et al. (1992)
	Reversed-phase chromatography, ICP-MS, Enzymatic digestion	I-containing species in thyroglobulin	A useful approach to the study of the distribution of chemical species in tissue.	Takatera and Watanabe (1993)
Urine	Combined cation-anion exchange Hydride generation atomic absorption spectrometry (HGAAS)	Inorganic arsenic and methylated metabolites	Effects of occupational exposure on the nature and amounts of arsenic species excreted in urine examined.	Farmer and Johnson (1990)
	Capillary gas-liquid chromatography, formation of Methylthioglycolate derivatives, Electron-capture detection	"	Methods could be used for blood and tissue.	Dix et al. (1987)
	Ion chromatography, ICP-MS	Inorganic arsenic {(As (III) and As (V)}, Dimethylarsinic acid and Methylarsonic acid		Sheppard et al. (1992)

Beilstein, M.A., and Whanger, P.D. (1983). *J. Nutr.*, *113*: 2138.

Bilbow, K., and Mundal, H.H. (1990). *Clin. Chem.*, *36*: 1902.

Borglund, M., and Åkesson, B. (1985). *Int. J. Vitam. Nutr. Res.*, *58*: 97.

Borguet, F., Cornelis, R., and Lameire, N. (1990). *Biol. Tr. Elem. Res.*, *26-27*: 449.

Brätter, P., Gercken, B., Rosick, U., and Tomiak, A. (1988). *Trace Element Analytical Chemistry in Medicine and Biology*, Vol. 5 (P. Brätter and P. Schramel, eds.), Walter de Gruyter, Berlin and New York.

Bulska, E., Emteborg, H., Baxter, D.C., Frech, W., Ellingsen, D., and Thomassen, Y. (1992). *Analyst*, *117*: 657.

Butler, J.A., Thomson, C.D., Whanger, P.D., and Robinson, M.F. (1991). *Amer. J. Clin. Nutr.*, *53*: 748.

Chilvers, D.C., Dawson, J.B., Bahreyni-Toosi, M.H., and Hodgkinson, A. (1984). *Analyst*, *109*: 871.

Daydé, S., Filella, M., and Berthon, G. (1990). *J. Inorg. Biochem.*, *38*: 241.

De Waal, W.A.J., Maessen, F.J.M.J., and Kraak, J.C. (1987). *J. Chromatogr.*, *407*: 253.

Dix, K., Cappon, C.J., and Toribara, T.Y. (1987). *J. Chromatogr. Sci.*, *25*: 164.

Farmer, J.G., and Johnson, L.R. (1990). *Brit. J. Ind. Med.*, *47*: 342.

Favarato, M., Mizzen, C.A., Sutherland, M.K., Krishnan, B., Kruck, T.P.A., and McLachlan, D.R.C. (1992). *Clin. Chim. Acta*, *207*: 41.

Foote, J.W., and Delves, H.T. (1983). *Analyst*, *108*: 492.

Ford, T.C., Graham, J., and Rickwood, D. (1990). *Clin. Chim. Acta*, *192*: 115.

Gardiner, P.E., Rosick, E., Rosick, U., Brätter., P., and Kynast, G. (1982). *Clin. Chim. Acta*, *120*: 103.

Gardiner, P.E., Gessner, H., Brätter, P., Stoeppler, M., and Nurnberg, H.W. (1984). *J. Clin. Chem. Clin. Biochem.*, *22*: 159.

Gardiner, P.E., Brätter, P. Gercken, B., and Tomiak, A. (1987). *J. Anal. At. Spectrom.*, *2*: 375.

Gardiner, P.H.E. (1993). *J. Trace Elem. Electr. Hlth. Dis.*, *7*: 1.

Gercken, B., and Barnes, R.M. (1991). *Anal. Chem.*, *63*: 283.

Goode, H.F., Kelleher, J., and Walker, B.E. (1989). *Ann. Clin. Biochem.*, *26*: 89.

Hansson, L., Pettersson, J., Eriksson, L., and Olin, Å. (1989). *Clin. Chem.*, *35*: 537.

Harris, W.R. (1992). *Clin. Chem.*, *38*: 1809.

Harris, W.R., and Keen, C. (1989). *J. Nutr.*, *119*: 1677.

Heydorn, K. (1984). *Neutron Activation Analysis for Clinical Trace Element Analysis*, Vol. I, CRC Press, Boca Raton, Florida.

Johnson, L.R., and Farmer, J.G. (1991). *Bull. Environ. Contam. Toxicol.*, *46*: 53.

Michalke, B., and Schramel, P. (1990). *J. Trace Elem. Electr. Hlth. Dis.*, *4*: 163.

Michalke, B., Munch, D.C., and Schramel, P. (1991). *J. Trace Elem. Electr. Hlth. Dis.*, *5*: 251.

Milne, D.B., Ralston, N.V.C., and Wallwork, J.C. (1985). *Clin. Chem.*, *31*: 65.

Milne, D.B., Sims, R.L., and Ralston, N.V.C. (1990). *Clin. Chem.*, *36*: 450.

Neidhart, B., Herwald, S., Lippmann, Ch., and Straka-Emden, B. (1990). *Fresenius' J. Anal. Chem.*, *337*: 853.

Owen, L.M.W., Crews, H.M., Hutton, R.C., and Walsh, A. (1992). *Analyst*, *117*: 649.

Peters, T.J. (1983). *Methods of Enzymatic Analysis*, 3rd ed. (H.U. Bergmeyer, ed.), Verlag Chemie, Weinheim, Deerfield Beach, Florida and Basel.

Ruz, M., Cavan, K.R., Bettger, W.J., and Gibson, R.S. (1992). *Brit. J. Nutr.*, *68*: 515.

Sheppard, B.S., Caruso, J.A., Heitkemper, D.T., and Wolnik, K.A. (1992). *Analyst*, *117*: 971.

Takatera, K., and Watanabe, T. (1993). *Anal. Chem.*, *65*: 759.

Venturini, M., and Berthon, G. (1989). *J. Inorg. Biochem.*, *37*: 69.

Versieck, J. (1985). *Crit. Rev. Clin. Lab. Sci.*, *22*: 97.

SPECIFIC TECHNIQUES FOR THE DETERMINATION OF BOUND AND FREE METAL IONS IN BIOLOGICAL FLUIDS

by

Henri Faure and Alain Favier

Laboratoire de Biochimie C,
CHU, BP 217
38043 Grenoble Cedex 09
France

I. INTRODUCTION

In biological fluids, the greater part of metal ions is bound to proteins. Actually, only a small fraction of these is bound to low-molar-mass (l.m.m.) compounds, mainly amino acids. Free (hydrated) metal ions exist in biological fluids only at extremely low concentrations, and these concentrations cannot play a significant role in physiological processes.

On the contrary, metals bound to l.m.m. compounds play a major role in many biological and physiological processes like intestinal absorption, cell absorption and renal excretion. In fact, protein-bound fractions are probably also involved in these processes (cells can uptake transferrin-bound iron or zinc, for example), but understanding these processes requires an accurate knowledge of the respective proportions of metal l.m.m. and protein-bound fractions as well as of the delicate equilibria which determine these proportions.

The above reasons explain why determining the relative importance of these two fractions is of interest to biologists. This determination, however, is often difficult for two reasons. First, the concentrations to be measured are very small, which implies that very sensitive and precise methods should be used; second, the equilibria which determine the amount of metal respectively bound to proteins or l.m.m. compounds are easily displaced because of the lability and the weakness of most of the metal-ligand bonds occurring in biological fluids.

The methods used to determine l.m.m. or protein-bound metals are basically the same as those used to analyze proteins in biological fluids. However, these methods require particular attention to ensure that no contamination or retention of metals can occur in the course of the separation procedures. After having performed a separation technique, metals are very often quantified by atomic absorption spectrophotometry. When radioisotopes are used, the detection is performed with scintillation counters.

II. BIOLOGICAL FLUIDS: COMPONENTS WITH POTENTIAL METAL-BINDING PROPERTIES

The composition of biological fluids is extremely complex. Furthermore, it varies with physiological conditions and during disease. We will therefore only review in this section the main biological fluids with significant metal l.m.m. fractions, especially focussing on the ligands involved in these.

A. Blood serum and plasma

Blood is composed of different types of cells and a homogenous fluid, the plasma. In cells, metals are often present at much higher concentrations than in plasma (Faure, 1986). The intracellular physiology of metals totally differs from that in plasma because inside cells metals are biologically active. This may explain why metals are firmly bound to their carriers in intracellular media.

The distinction between plasma and serum lies in the fact that plasma contains fibrinogen and heparin whereas serum does not. Fibrinogen is a protein which (like all proteins) can bind certain amounts of metals. However, this binding is not important enough from the quantitative point of view to significantly modify the distribution of metals in plasma. In contrast, heparins like other anticlotting products (e.g. citrate) can alter the distribution of metals because of their coordinating properties (Faure, 1986; Parrish and Fair, 1981). Heparins are polysulfonated aminoglycans and metals bind to the sulfone function. Moreover, they generally exhibit a large range of molar masses. Heparins thus displace the equilibria between l.m.m. and high-molar-mass (h.m.m.) metal fractions in an unforeseeable manner. This explains why serum, although requiring more time for its preparation (blood must be clotted before collecting serum which must then be centrifugated twice to avoid contamination by red blood cells), is preferred to plasma in metal distribution studies.

1. *Proteins*

Blood plasma and serum contain 72 grams of proteins per liter, which are mainly synthesized by the liver but also by macrophages or immunocompetent cells. These 72 grams are composed of a considerable number of different proteins, but only a few are significantly involved in the distribution of metal ions. As any blood factor likely to influence this distribution, a protein must be sufficiently concentrated to transport a significant amount of metal. Moreover, the affinity of a given protein also can influence the amount of metal carried, but except for specific carriers (e.g. transferrin) and a few proteins which exhibit a very high affinity for particular metals (e.g. metallothioneins for Cu and Zn), plasma proteins can indistinctly bind all metals. Protein affinity for metal ions increases in relation with the percentage of cysteinyl and histidinyl residues. As a result, the amounts of metals bound to proteins are always largely predominant over those bound to l.m.m. compounds. For the same reason, the amounts bound to diverse proteins are roughly proportional to the respective concentrations of these proteins in plasma. So, if we exclude specific carrier proteins, albumin is the major plasma carrier for numerous metals, zinc, copper, manganese, nickel, chromium and vanadium being the most widely known.

Blood protein composition may largely vary in the case of diseases such as myeloma, cirrhosis, hepatitis, and nephrotic syndrome. Consequently, metal distribution in blood is regularly affected by these diseases.

2. *Amino acids*

Blood amino acids are produced by muscle protein catabolism, and by the digestion and intestinal absorption of food proteins. Two pathological situations can increase blood amino acid content: infections that provoke tremendous muscle catabolism, and parenteral feeding because of the high concentrations of amino acids infused in the blood of patients. The distribution of a metal between l.m.m. components and proteins is thus affected by these variations.

In blood, nearly the totality of the l.m.m. fraction of metals is bound to amino acids. The affinity of each amino acid for a given metal can vary considerably from one amino acid to another. The concentrations of the various amino acids present are also very different. In general, therefore, metal ions are preferentially bound to a few amino acids only.

B. Lymph and tissue fluids

Tissue fluids are formed by percolation of blood plasma through the body tissues. Consequently, there are no macromolecules in tissue fluid but only l.m.m. compounds that are for the most part nutrients for the cells which are directly in contact with it.

Lymph is derived from tissue fluids but it contains roughly half the protein concentration of blood because it is also formed by filtration of blood plasma through small capillaries that allow some small proteins to leak through their walls.

Amino acids are major metal carriers in both of these fluids, although lymph proteins transport more important fractions of metals than do l.m.m. compounds.

C. Urine and sweat

In contrast to blood plasma, urine contains almost no protein because proteins cannot filter through the nephron glomerulus, and the small amounts that leak through this microscopical ultrafiltrating organ are then reabsorbed in the proximal tubule. However, researchers working on urinary metals should keep in mind that, in cases of kidney disease, very large amounts of proteins can leak through the glomerulus and therefore be excreted in the urine.

In normal urine, the only ligands significantly binding metal ions are amino acids. These are also reabsorbed in the proximal tubule, but their daily excretion is quantitatively important.

Sweat has roughly the same composition as urine: very small amounts of proteins but higher concentrations of small molecules.

D. Seminal fluid

Seminal fluid contains about 500-fold more zinc than blood plasma (Lafond, 1985), this elevated level being indispensable to fertility.

Proteins are less concentrated in seminal fluid than in blood plasma (1/3), but in comparison amino acids and citric acid are 100-fold more concentrated. Large amounts of metals in seminal fluid are therefore expected to be bound to l.m.m. compounds.

E. Cerebrospinal fluid

A very small amount of proteins is present in the cerebrospinal fluid (0.2 g/l: 300-fold less than in blood plasma), but the concentrations of amino acids and other small molecules which can act as metal ligands are approximately the same as in blood plasma.

F. Milk

The composition of milk greatly differs from one animal species to another (for complete composition tables see Arnaud, 1987). We will focus here on human milk, for which many data are available.

Human milk contains 5.3 to 9 grams of proteins per liter (cow milk: 29-33 g/l), i.e. 7-fold less than blood plasma. Casein exists only in milk where it is much more concentrated than any other protein. Lactoferrin differs from serum transferrin only by a few amino acid residues, but it has a higher affinity for iron.

Concerning the small molecules which are potential ligands for metal ions, human milk contains very high concentrations of amino acids, particularly histidine (1000-fold more than blood plasma). Citric acid is also much more concentrated in human milk than in the plasma.

The composition of human milk explains why a very large proportion of metals is bound to l.m.m. compounds in this fluid.

III. METHODS FOR STUDYING METAL DISTRIBUTION AMONGST PROTEINS

The methods used for discriminating among protein-bound metal fractions are centered on separation techniques. However, investigating metal-protein interactions implies that modifications or displacement of metal ions from their original coordinating sites may arise. The binding capacity of chromatographic beads appears to be correlated with the reticulation degree of the resin (Johnson and Evans, 1980). The first concern of the user of these techniques will therefore be to calculate the recovery percentage of a known solution (see Gardiner et al., 1981), chosen as similar as possible to the biological fluid being studied and containing a known amount of the metal considered. Neglecting this important step would invariably lead to erroneous results. In some situations, chemical treatment of the resin before chromatography may be successfully used to inhibit gel affinity for metals (Lönnerdal and Hofmann, 1981).

Moreover, chromatographic beads are always contaminated with large amounts of metal. They must therefore be cleaned before use. These two problems are not easily solved, especially if metal-protein bonds are weak.

It is also worth pointing out that even a recovery percentage close to 100% does not imply that the relative proportions of metal bound to each fraction at the end of a column necessarily reflect the original distribution of the metal in the fluid investigated. For example, if a h.m.m. protein is a weak ligand of the metal considered, the gel filtration beads can displace the metal from this protein; a smaller protein with a higher affinity for the metal may then pass and recapture the metal from the beads. Under such conditions, the recovery percentage may well be close to 100% but the results seriously biased.

Another important factor is the pH used to separate the molecules in the fluid, which must be as close as possible to that of the original fluid. The stabilities of metal-ligand bonds are pH dependent, and the proportion of metal bound to each fraction may also be erroneous if the buffer used imposes a pH very different from that of the original fluid.

A. Gel filtration and affinity chromatography

The principle of gel filtration chromatography is simple. The sample studied is applied to the column and then pushed through the stationary phase by the mobile phase. Beads retard the progress of the molecules with a size comparable to the mesh size of the network while the others pass faster. Thus, at the end of the column, compounds are collected in an order dependent on the size of their molecules.

Gel filtration chromatography has been the most widely used technique to separate metal-carrier proteins; it has been used either alone (Gardiner et al., 1981; 1984; Chesters and Will, 1981; Baker and Hultquist, 1978) or coupled with affinity chromatography (Failla et al., 1982; Foote and Delves, 1984; Lafond, 1985; 1986; Wirth, 1985). However, gel permeation beads cannot separate wide ranges of compounds all at once, and columns packed with different pore diameter beads are sometimes used sequentially (Lafond, 1986) in order to separate molecules (with the metals bound to these) ranging from 1 kg mol^{-1} up to 900 kg mol^{-1} or more.

Gel filtration has been applied to study the distribution of many metals in many media. It has been used with cold and radioactive metals (Zn: Lafond, 1986; Mn: Scheuhammer and Cherian, 1985; Cu: Weiss and Linder, 1985).

Affinity chromatography is used to selectively bind proteins by specific mechanisms. These proteins are then removed from the affinity gel and the metals they carry can be studied. Cibacron Blue is one of the most popular affinity gels (Foote and Delves, 1982; 1983; 1984; Lafond, 1985; 1986). It selectively binds serum albumin without displacing significant amounts of zinc from this protein. This gel is composed of a dye (Cibacron Blue F3G-A) immobilized on Sepharose CL6B.

Chelex 100 is an ion-exchange resin in which the ion exchangers are iminodiacetic residues. This resin can be used (by competition) to test the strength with which a ligand binds a metal (Weiss and Linder, 1985).

B. Ion exchange chromatography

Ion exchange chromatography is based on the selective binding of molecules to be separated on beads at specific pH or/and ionic strength. A major drawback of this method is that the metal-protein bonds are easily broken by the high affinity of the ion exchangers for metal ions (Gardiner et al., 1981). Moreover, the pH and ionic strength required for this method are often far from physiological conditions. All these factors increase the risk of obtaining results that do not reflect the original distribution of the metal in the biological fluid.

Nevertheless, this method of separation has been often used to identify metal forms in biological fluids (Parisi and Vallee, 1970; Chilvers et al., 1984).

C. Electrophoresis

This technique often uses a radioactive metal as a tracer, which allows the quantification of the metal amount bound to each protein after separation. It was also applied with "cold" copper to study the protein binding of this metal in the plasma of patients suffering from Menkes' disease (Delves, 1976).

Like ion-exchange chromatography, electrophoresis requires non-physiological conditions of pH and ionic strength, and thus suffers the same drawbacks. In addition, the electric field can also break the metal-protein bonds. For these reasons, the results obtained with this method are often different from those found with others (e.g. gel filtration); the number of protein ligands being generally found much higher.

Electrophoresis and ion-exchange chromatography were amongst the first methods used to investigate metal-protein interactions. Polyacrylamide (Lindeman et al., 1978), Pevikon (Boyett and Sullivan, 1970) and agarose gel electrophoresis (Scott and Bradwell, 1983) were used.

Rocket electrophoresis has been used to study zinc-protein interactions (Scott and Bradwell, 1983) in human plasma, but the results obtained with this method are difficult to interpret because radioactive zinc is added after a first separation of plasma proteins. Consequently, no equilibrium between zinc and the different plasma proteins can take place. However, this method is useful to study the absolute affinity of zinc for each protein.

D. Precipitation

Under specific conditions, selective precipitation of proteins is possible and this precipitation sometimes occurs without affecting metal-proteins bonds. This method has been used by Giroux (1975) to measure the fraction of plasma zinc bound to α_2-macroglobulin. This author used a 10% polyethylene glycol 6000 dilution in plasma to precipitate selectively α_2-macroglobulin while the other proteins were not affected. Thus, a simple measurement of total serum zinc before and after this addition permitted to calculate the fraction of zinc bound to α_2-macroglobulin.

E. Ultracentrifugation

Some authors have used ultracentrifugation to separate a protein having a molar mass very different from the others in the same biofluid. After separation, the amounts of metals bound to that protein can be measured (Song and Adham, 1979; Kiilerich and Christiansen, 1984). This method is theoretically very precise and seems to induce only few artifacts, but in practice it is impossible to totally separate the protein chosen and a dilution of the fluid is often required before centrifugation, which may also alter the preexisting equilibria between metal and proteins.

IV. METHODS FOR STUDYING METAL DISTRIBUTION BETWEEN HIGH-MOLAR-MASS AND LOW-MOLAR-MASS COMPOUNDS

A. Ultrafiltration and dialysis

Ultrafiltration uses the sieve principle to separate h.m.m. and l.m.m. compounds when the fluid containing these molecules is forced through a specific membrane. Molecules larger than the cut-off limit are stopped by the membrane, while those with a smaller molar mass can pass through it. If no exchange occurs during this process, a simple measurement in the ultrafiltrate permits to assess the l.m.m. fractions of the metals present in the biofluid. Several manufacturers sell devices designed to ultrafilter the small volumes (1 ml or less) usually at the disposal of the biologist to measure metal l.m.m. fractions in serum (Faure et al., 1990; Henkin et al., 1973; Noubah and Al-Awqati, 1990). Ultrafiltration devices offer the advantage of being versatile enough to be used with a large range of membranes, provided that these have the appropriate shape and dimensions.

Several manufacturers provide membranes exhibiting different cutoffs, but the cutoff range is rarely useful for separating compounds in biological fluids because the particular cutoff of an ultrafiltration membrane never represents a sharp limit under which any smaller molecule can freely pass and above which any larger molecule is stopped with a 100% retention percentage. This phenomenon is easily explained by the pore sizes of the membranes which are always distributed over a large range. For the same reason, membranes are chosen with a specified pore size much larger than the size of the compounds to be measured in the ultrafiltrate (a pore size of 30 kg mol^{-1} is generally used).

Like chromatographic methods, ultrafiltration requires to ensure that the metal being investigated does not bind to the membrane during ultrafiltration and that both the membrane and the ultrafiltration device do not contaminate the fluid to an appreciable extent during the ultrafiltration process. These two phenomena, i.e. adsorption and contamination, usually have negligible effects in the fluids containing high amounts of metal bound to l.m.m. compounds (e.g. zinc in seminal fluid or milk), but they can induce large errors when the l.m.m. fraction of metal is very small (e.g. zinc in blood plasma). Poor precision can significantly alter the results from concentrations of about 0.4 μmol/l. This factor should therefore be carefully assessed.

Ultrafiltration has been widely used to measure metal fractions bound to l.m.m. compounds (Henkin et al., 1973; Yunice et al., 1973; Abu-Hamdam et al., 1981; Whitehouse et al., 1983; Bloxam and Tan, 1984). In serum, the addition of ethylenediaminetetracetic acid (EDTA) before ultrafiltration allows the determination of metals loosely bound to proteins (Faure et al., 1990; Favier and Ruffieux, 1988).

Dialysis is much less used than ultrafiltration because it is more difficult to finalize and it is not adapted to the very small amounts of l.m.m. metal complexes found in serum. It was, however, used in the past by Prasad and Oberleas (1970) to demonstrate the existence of the l.m.m. fraction of serum zinc.

B. Ultracentrifugation

This technique was used to characterize the l.m.m. fraction of copper in human serum (Neuman and Sass-Kortsak, 1967). However, it is time consuming (a separation takes 22 to 24 hours at 114,000 g) and requires a predilution of the samples to achieve a relatively good separation. These reasons explain why ultrafiltration is now often preferred to ultracentrifugation to measure this fraction.

V. CONCLUSION

The above methods allow the characterization of most of the transition metals in biological fluids. There is, however, no general technique. Each metal, biological fluid and even fraction to be investigated (i.e. h.m.m. or l.m.m.) requires a specific technique. The use of radioisotopes after the time required to allow equilibrium can often facilitate these studies, but it is worth noting that metalloproteins (e.g. ceruloplasmin or α_2-macroglobulin) contain significant amounts of specific metals which cannot be exchanged with the added metal. For this reason, the results obtained with radioisotopes must be confronted with results from cold metal studies.

When the fraction to be studied is quantitatively too small, analytical methods are sometimes not sensitive enough to estimate it or may induce artifacts. Zinc in its serum ultrafiltrable fraction is a good example of this problem. In this case, computer simulation (Berthon et al., 1978) represents the only technique to assess the proportion of metal bound to each compound.

ACKNOWLEDGEMENTS

The authors gratefully acknowledge Deryck Jeremy Williams for having reviewed this contribution and for his interesting remarks concerning the manuscript.

REFERENCES

Abu-Hamdam, D.K., Migdal, S.D., Whitehouse, R., Rabani, P., Prasad, A., and Donald, F.D. (1981). *Amer. J. Physiol.*, *241*: F487.

Arnaud, J. (1987). *Etude du Zinc du Lait Maternel*, Thèse d'Etat, Pharmacie, Grenoble.

Baker, B.L., and Hultquist, D.E. (1978). *J. Biol. Chem.*, *253*: 8444.

Berthon, G., May, P.M., and Williams, D.R. (1978). *J. Chem. Soc., Dalton Trans.*, 1433.

Bloxam, D.L., and Tan, J.C.Y. (1984). *Clin. Chim. Acta*, *144*: 81.

Boyett, J.D., and Sullivan, J.F. (1970). *Metabolism, 19*: 148.

Chesters, J.K., and Will, M. (1981). *Brit. J. Nutr., 46*: 111.

Chilvers, D.C., Dawson, J.B., Bahreyni-Toosi, M.H., and Hodkinson, A. (1984). *Analyst, 104*: 871.

Delves, H.T. (1976). *Clin. Chim. Acta, 71*: 495.

Failla, M.L., Van de Veerdonk, M., Morgan, W.T., and Smith, J.C. (1982). *J. Lab. Clin. Med., 100*: 943.

Faure, H. (1986). *Dosage des Fractions Libre et Liées du Zinc Sérique. Intérêt de ces Fractions*, Thèse de 3e Cycle, Pharmacie, Grenoble.

Faure, H., Favier, A., Tripier, M., and Arnaud, J. (1990). *Biol. Tr. Elem. Res., 24*: 25.

Favier, A., and Ruffieux, D. (1988). *Biol. Tr. Elem. Res., 18*: 145.

Foote, J.W., and Delves, H.T. (1982). *Analyst, 107*: 121.

Foote, J.W., and Delves, H.T. (1983). *Analyst, 108*: 492.

Foote, J.W., and Delves, H.T. (1984). *Analyst, 109*: 709.

Gardiner, P.E., Ottaway, J.M., Fell, G.S., and Burns, R.R. (1981). *Anal. Chim. Acta, 124*: 281.

Gardiner, P.E., Gessner, H., Stoeppler, M., and Nurnberg, H.W. (1984). *J. Clin. Chem. Clin. Biochem., 22*: 159.

Giroux, E.L. (1975). *Biochem. Med., 12*: 258.

Henkin, R.I., Schulman, J.D., Schulman, C.B., and Bronzert, D.A. (1973). *J. Pediat., 831*: 831.

Johnson, P.E., and Evans, G.W. (1980). *J. Chromatogr., 188*: 405.

Kiilerich, S., and Christiansen, C. (1984). *Clin. Chim. Acta, 142*: 273.

Lafond, J.L. (1985). *Etude des Ligands du Zinc dans le Liquide Séminal*, Thèse de 3e Cycle, Pharmacie, Grenoble.

Lafond, J.L. (1986). *Biol. Tr. Elem. Res., 9*: 1.

Lindeman, R.D., Baxter, D.J., Yunice, A.A., and Kraikitpatnitch, S. (1978). *Amer. J. Med. Sci., 275*: 17.

Lönnerdal, B., and Hoffman, B. (1981). *Biol. Tr. Elem. Res., 3*: 301.

Neuman, P.Z., and Sass-Kortsak, A. (1967). *J. Clin. Invest., 46*: 646.

Noubah, A.M., and Al-Awqati, M.A. (1990). *Clin. Chem., 36*: 860.

Parisi, A.F., and Vallee, B.L. (1970). *Biochemistry, 9*: 421.

Parrish, R.F., and Fair, W.R. (1981). *Biochem. J., 193*: 407.

Prasad, A.S., and Oberleas, D. (1970). *J. Lab. Clin., 76*: 416.

Scott, B.J., and Bradwell, A.R. (1983). *Clin. Chem., 24*: 629.

Scheuhammer, A.M., and Cherian, M.G. (1985). *Biochim. Biophys. Acta, 840*: 163.

Song, M.K., and Adham, N.F. (1979). *Clin. Chim. Acta, 1150*: 13.

Yunice, A.A., King, R.W., Kraikitpanitch, S., Haygood, C.C., and Lindeman, R.D. (1978). *Amer. J. Physiol., 235*: F40.

Weiss, K.C., and Linder, M.C. (1985). *Amer. J. Physiol., 249*: E77.

Whitehouse, R.C., Prasad, A.S., and Cossack, Z.T. (1983). *Clin. Chem., 29*: 1974.

Wirth, P.L. (1985). *J. Nat. Cancer Inst., 75*: 277.

MEASUREMENT OF COMPLEXATION PROPERTIES OF METAL IONS IN BIOLOGICAL SYSTEMS BY ULTRAFILTRATION

by

Nalini Parthasarathy, Montserrat Filella and Jacques Buffle

Département de Chimie Minérale, Analytique et Appliquée,
Université de Genève,
Quai Ernest Ansermet 30
CH-1211 Genève 4
Switzerland

Because of the vital role played by metal ions in biological processes (see previous chapters), it is important to study the interaction of metals with biological ligands for understanding these processes. One way of proceeding is to determine equilibrium parameters.

I. TECHNIQUES FOR METAL BINDING STUDIES IN BIOLOGICAL SYSTEMS

The equilibrium constant, K, for the reaction involving the interaction of a metal with a ligand is given by equation (2).

$$M + L \rightleftharpoons ML \qquad (1)$$

$$K = [ML]/[M][L] \qquad (2)$$

Complex formation equilibrium may be expressed in several ways (Dwyer and Mellor, 1964). In this chapter equilibrium constants will be described in terms of formation constants as recommended by IUPAC. In order to evaluate binding constants for a simple system it is in theory necessary to measure: (i) the total metal $[M]_t$ and ligand $[L]_t$ concentrations using any method sensitive to total concentration and (ii) free metal ion concentration [M], free ligand concentration [L] or complex concentration [ML] by any method sensitive for one of these species. The same principle is applicable when several well-characterized complexes (e.g. ML, ML_2, ...) are formed between M and L. When L includes a large number of complexing sites interpretation may be complicated (Buffle, 1988 [chapter 5]).

The binding of metal to biological ligands may be measured in several ways. They may be divided into two categories:

(i) direct techniques (without separation) such as potentiometry using glass or ion selective electrodes, voltammetry, UV-visible spectrometry, fluorimetry, NMR, ESR;

(ii) indirect methods (with separation), based on differences either in oxidation-reduction properties (voltammetry) or in size (equilibrium dialysis (ED), ultrafiltration (UF), size exclusion (gel) chromatography).

Direct methods are more accurate but are often applicable only to simple ligands. In complex biological systems or natural aquatic systems, the macromolecules themselves may interfere in the measurement of signals (e.g. due to adsorption on the electrode surface in the case of voltammetry) or the method is not sensitive enough or only applicable to a limited number of metals.

When [M], [L] and [ML] are inaccessible to direct measurement, physical separation methods based on differences in size can be used to separate [M], [L] and [ML]. Of these methods, ED has been most often used because of its simplicity. However, it has the disadvantage that it is very slow (days) (Rudman and Kendal, 1957). In addition to this inconvenience, this long equilibration process may cause problems such as denaturation of proteins or microbial growth (Cheryan, 1986). Since each experiment allows a single measurement of the free metal ion concentration for a given amount of bound metal, several experiments have to be carried out to determine the binding parameters. This implies that relatively large amounts of valuable macromolecules are consumed. The use of flow-dialysis reduces the measurement time but is not as popular as ED (Krause et al., 1991). Gel chromatography (Hummel and Dreyer, 1962; Buffle, 1988; Work and Work, 1969) like ED is a batch method, i.e. a single data point is obtained from one experiment, and several experiments in sequence (using the same column of gel) or in parallel (several identical columns of gel) have to be carried out to obtain a metal-binding isotherm. This is time consuming and requires large amounts of ligands. The amount of ligand used can be reduced provided the formation constants with the metals studied are large (Paulus, 1969). UF, although far less employed than ED for metal-macromolecule complexation measurements, is relatively inexpensive, versatile, rapid, reagent free and may be made a non- (or almost non-) equilibrium perturbing technique. Moreover, it has been shown that UF is theoretically equivalent to ED (Sophianopoulos et al., 1978; Bowers et al., 1984).

In this chapter, the application of UF to the measurement of equilibrium binding data in biological systems will be described and its applications will be illustrated with two selected examples.

II. ULTRAFILTRATION

A. Introduction

The term ultrafiltration is used to describe the separation, on the basis of size at the molecular level, of dissolved species by passage through molecular filters.

These filters are made of cellulose acetate or synthetic polymers of varying polyanion/polycation mixtures. The membrane separation theory as well as membrane fabrications are described in detail in many monographs (Hwang Sun Tak and Kammermeyer, 1975; Meares, 1976; Sourirajan, 1979; Brock, 1983).

Four factors govern the total flux of solute crossing the membrane (Hwang Sun Tak and Kammermeyer, 1975; Meares, 1976; Buffle, 1988):

(i) convective transport of solvent (water) due to hydrostatic pressure ΔP,

(ii) reverse transport of solvent by diffusion due to an osmotic pressure $\Delta\pi$,

(iii) transport of solute by the solvent (convection),

(iv) transport in the same direction as (iii) by diffusion due to concentration gradient.

Effects (i) and (iii) are dominant in UF ($\Delta P >> \Delta\pi$) whereas (ii) and (iv) are more important in ED. The UF measurements are therefore more rapid (minutes) than ED (days) as convection transport is more rapid than diffusion.

The determination of equilibrium formation constants of complexes ML by separation of M from L + ML is commonly used for studying the binding of small molecules to macromolecules (Sophianopoulos and Sophianopoulos, 1985; Cheryan and Saeed, 1989), the complexation of metals by macromolecules in biochemistry and the retention of metals by simple (Bulbula et al., 1975) and polymeric (Tuschall and Brezonik, 1983; Buffle and Staub, 1984) aquatic complexants.

This type of application requires a knowledge of the factors influencing the flux of compounds crossing the membrane. The interpretation of concentrations measured in the filtrate in terms of the complexation occurring in the cell should take into account the trans-membrane fluxes (Fig. 1). The flux of a compound X across a membrane is often represented by the retention coefficient, R, of the membrane (see below).

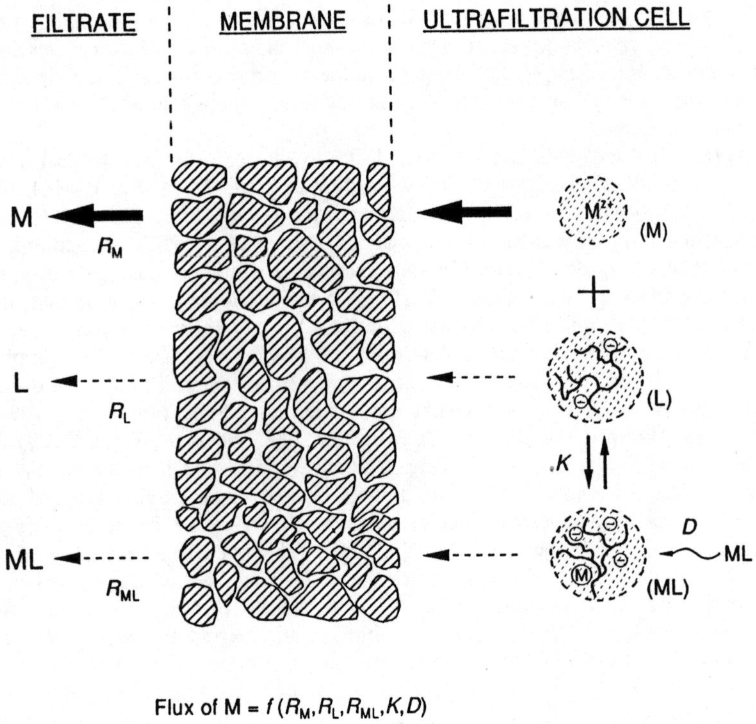

Figure 1.
Schematic representation of passage of M, L and ML through a UF membrane, and the factors which can influence the retention coefficient, R. (Adapted from Buffle (1988) by permission of the copyright holders, Ellis Horwood Limited, Chichester).

B. Membrane separation: principles

The theory of transport processes across membranes is described in Hwang Sun Tak and Kammermeyer (1975), Meares (1976) and Spriggs and Li (1976). A membrane allows the passage of water through all pores, but the passage of solute through only a fraction of them. A theoretical expression for the flux of a solute X, J_X, has been derived assuming the pores to be cylindrical (Buffle, 1988) and is given in equation (3). Transport by diffusion and convection has been taken into account in deriving the following equation.

$$J_X = \frac{(1 - \sigma_X) + D_{m,X} / \lambda.K_m.J_w}{1 + D_{m,X} / \lambda.K_m.J_w} J_w.[X]_c \tag{3}$$

where σ_X is the theoretical retention coefficient of X and differs from experimental retention coefficient R_X (defined below) in that it depends only on the accessibility of pores to X (thus, if X can pass through all the pores indiscriminately, σ_X equals zero), $D_{m,X}$ is the diffusion coefficient of X through the membrane, λ is the average pore length, $K_m = [X]_{m,c}/[X]_c$, $[X]_{m,c}$ and $[X]_c$ are the concentrations of X at the cell side of the internal membrane surface and in the cell, J_w is the water flux. The term $D_{m,X}/\lambda.K_m$, denoted by Δ_X, is often treated as a single parameter because of the difficulty in separating the parameters K_m, λ, and $D_{m,X}$. Δ_X represents the characteristic rate ($cm.s^{-1}$) for diffusional transport of X in the membrane. Note that equation (3) is not applicable in the case where the transport of X is purely diffusion controlled.

The experimental retention coefficient, R_X, is defined by:

$$J_X = (1 - R_X) J_w.[X]_c \tag{4}$$

The experimental retention coefficient can be expressed as:

$$R_X = 1 - \frac{[X]_f}{[X]_c} = \frac{\sigma_X}{1 + \Delta_X / J_w} \tag{5}$$

which shows that $R_X = \sigma_X$ when convection is dominant.

C. Kinetic effect

In the study of metal complexation, the transfer of M across the interface can be considered to result from three processes: diffusion of ML towards the membrane (rate constant = k_D), chemical dissociation of ML (rate constant = k_d) and transport of ML across the membrane (rate constant = k_m), the slowest being the rate limiting (Buffle, 1988). For a ML complex which is retained by the membrane ($k_m >> k_D$) two situations may occur depending on its kinetic properties. If the complex is inert ($k_d \rightarrow 0$), M does not pass through the membrane and the concentration of ML at the surface will be greater than in the solution. In contrast, for a labile complex ($k_d >> k_D$) M bound to L is not retained due to the rapid dissociation of ML and the gradient of ML is the opposite to that when it is retained. If $k_m << k_D$, the concentration gradients in solution are zero since the transport across the membrane is rate limiting. Under these conditions only the filtrate concentration directly represents the cell equilibrium state. Zero concentration gradients imply: (i) a well stirred solution so that $\delta \rightarrow 0$ at the membrane surface and $k_D = D_M/\delta$ is maximum; (ii) the water flux, J_w, is as small as possible (indeed K_m is proportional to J_w (Buffle, 1988)).

The lability and inertness of a complex in UF depends on the ratio between k_m and k_d. For a given value complexation coefficient α_c, complexes are labile if J_w is small ($k_d >> k_m$) and inert if J_w is large ($k_d << k_m$).

III. ULTRAFILTRATION TECHNIQUES

The two most frequently used approaches for binding studies using UF are shown in Figure 2 and described below. In both of them a prior knowledge of retention coefficients of the ligand and of the bound metal ML is required (Meares, 1976; Blatt et al., 1968; Buffle, 1988).

A. Continuous reagent addition method

This method has been widely used for complexation measurements in biochemistry (Yap and Schaffer, 1977). Initially the UF cell contains only a ligand, L, and possibly an electrolyte. This solution is continuously "washed" (Fig. 2) with a reservoir solution containing M of concentration $[M]_r$ and the same electrolyte. The membrane is chosen so as

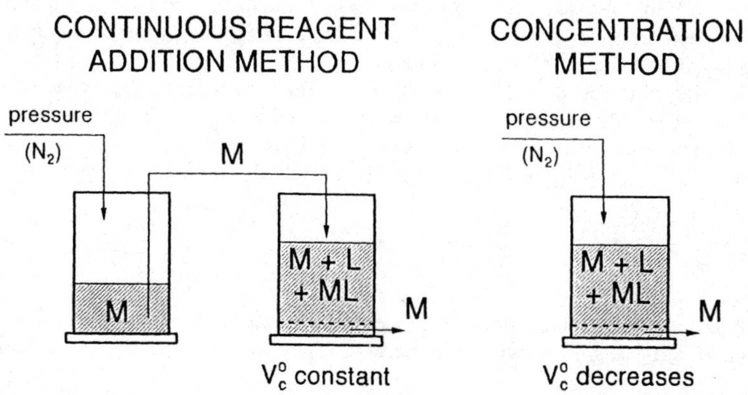

Figure 2.
Principal UF techniques. Membranes shown as broken lines. V_c^o is the initial volume in the cell.

to allow the passage of only M and the electrolyte. The concentration of M in the filtrate ($[M]_f$) is measured as a function of the total filtrate volume (V_f).

If L and ML are totally retained ($R_L = R_{ML} = 1$) and M is not retained ($R_M = 0$), $[M]_c = [M]_f$ where $[M]_c$ is the concentration of M in the cell. Application of the mass balance equation gives:

$$V_f \cdot [M]_r = N_{c,b} + V_c \cdot [M]_c + N_f \tag{6}$$

where $N_{c,b}$, $V_c \cdot [M]_c$ and N_f are the number of moles of M bound to L in the cell, free in the cell, and total in the filtrate after filtration of a volume V_f, respectively. The cell volume, V_c, is constant ($V_c = V_c^o$) and $V_f \cdot [M]_r$ is the total quantity of M introduced into the cell from the reservoir. Equation (6) permits calculation, for all V_f values, of $[M]_c$ and $[ML]_c = N_{c,b}/V_c$.

The continuous reagent addition method is not convenient for studying complexation of metals with low-molar-mass (l.m.m.) ligands as will be discussed in section B below.

B. Concentration method

In the concentration method, small portions of solution containing metal and ligand are filtered through a membrane impermeable to the ligand (Fig. 2). The quantity determined experimentally is the free metal ion concentration. The binding parameters are calculated from the retention coefficient R_X of each solute (M, L and ML) and the mass balances for M and L in the cell and in the filtrate. R_X can be determined by collecting a series of i filtrate fractions and analyzing each of them. From $[X]_{i,f}$ and $[X]_{i,c}$, R_X can be evaluated by

$$R_X = 1 - \frac{[X]_{i,f}}{[X]_{i,c}} \tag{7}$$

Note that a correct value of R_X can be obtained from equation (7) only if the value of the volume of each fraction i is small compared to V_c.

The retention coefficient, R_X, may also be determined by log-linearized forms of the integrated mass balance around the cell. For the concentration method, this expression is (Blatt et al., 1968; Buffle, 1988):

$$R_X = \frac{\ln([X]_c/[X]_c^o)}{\ln(V_c/V_c^o)} \tag{8}$$

where $[X]_c^o$ and V_c^o are the initial cell concentration and volume, and $[X]_c$ is the cell concentration after filtration, the remaining volume being V_c. Equation (8) is valid only if the variation in volume in the cell is less than 20%, i.e. if the volume of the filtered fraction is small compared to the cell volume. It must be noted, however, that if R_X varies during the filtration (for example, by progressive membrane clogging) the R_X calculated by this method will be an average value for the volume filtered. Utilization of equation (7) is therefore preferable, as it allows detection of possible variations in membrane behavior by following R_X during the course of the filtration.

In the case where R_{ML} and R_L are < 1, but the ligand is so large that its physico-chemical properties are not affected by binding with M ($R_{ML} \approx R_L$), Buffle and Staub (1984) have shown that the complexation coefficient in the cell, α_c, can be calculated from the values of $[M]_{f,t}$ (= $[M]_f + [ML]_f$) and $[M]_{c,t}$ (= $[M]_c + [ML]_c$) corresponding to the same filtrate fraction:

$$\frac{[M]_{f,t}}{[M]_{c,t}} = \frac{R_L - R_M}{\alpha_c} + (1 - R_L) \tag{9}$$

$[M]_{f,t}$ and $[M]_{c,t}$ can be readily measured. R_M and R_L can be determined in the absence of either L or M and using equation (7). In the case where $R_L = 1$ and $R_M = 0$, equation (9) becomes

$$\frac{[M]_{f,t}}{[M]_{c,t}} = \frac{1}{\alpha_c} \tag{10}$$

α_c can be calculated from equations (9) or (10) at different values of the ratio $[M]_{f,t}/[M]_{c,t}$ by introducing M into the cell in discontinuous steps and then collecting a filtrate fraction for each addition (Staub et al., 1984).

The continuous reagent addition method is unsuitable for studying metal complexation by l.m.m. ligands owing to long filtration times linked to the use of small pore size membranes (< 1 nm) which must be used so that $R_L \rightarrow 1$ and $R_{ML} \rightarrow 1$. In this case, the concentration method but with filtration of very small fractions, ΔV, of the cell volume, V_c^o, (typically ΔV = 0.1 ml and V_c^o = 50 ml) is preferable. The filtration time is hence reduced and the variations in the cell concentrations are minimal, thus making it a "non equilibrium perturbing" method.

IV. CALCULATION OF THE EQUILIBRIUM CONSTANTS

Metal binding by biological macromolecules is widely analyzed by Scatchard plots (Scatchard, 1949; Scatchard et al., 1950). The Scatchard analysis is a discrete model which results in a straight line if a single type of metal binding site is present:

$$\nu / [M] = n.K - K.\nu \qquad (11)$$

where ν is the degree of site occupation relative to the concentration of ligand (number of moles of metal bound per mole of ligand), K is the formation constant, and n is the number of binding sites per molecule of ligand (Fig. 3). The slope of the line yields K and the number of binding sites is obtained from the intercept. Equation (11) is one of the possible linearization forms of equation (12):

$$\nu = \frac{n.K.[M]}{1 + K.[M]} \qquad (12)$$

This simple equation rarely describes adequately the curvilinear plots of ν/[M] vs ν that are often obtained. At least, a second binding site must be considered, yielding

$$\nu = \frac{n_1.K_1.[M]}{1 + K_1.[M]} + \frac{n_2.K_2.[M]}{1 + K_2.[M]} \qquad (13)$$

Several linearized forms of equation (13) are available. When K_1 and K_2 are very different, two linear segments are obtained. The parameters K_1, K_2, n_1 and n_2 can be determined by combining the mathematical expressions for the ordinates and abscissae and the slopes of these lines (Klotz and Hunston, 1971; Buffle, 1988).

The Scatchard equation has been widely applied to biomolecules such as proteins and nucleic acids. It is not unreasonable to expect that a purified protein might have only one or two distinct sites for binding a metal ion or another substrate. Even in biochemistry,

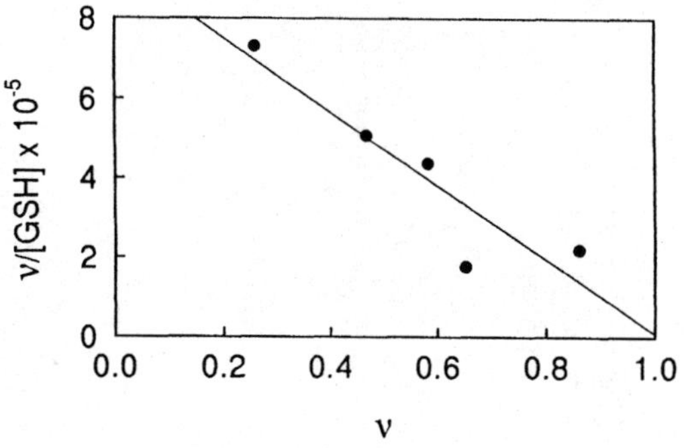

Figure 3.
Scatchard plot of ν/[GSH] vs ν for the binding of GSH to lobster CuMT-III (Reproduced from Brouwer and Brouwer-Hoexum (1991) by permission of the copyright holders, Academic Press, New York).

however, many examples of inappropriate use of the Scatchard equation can be found (Klotz, 1982). This occurs either when the ligand contains many binding sites, or when it undergoes conformational changes or exhibits polyelectrolytic properties as is often the case in macromolecules. The Scatchard treatment then provides a few K_i values which are assumed to correspond to distinct "classes" of metal complexing sites, when in fact these constants are just empirical curve-fitting parameters with no chemical significance.

These systems with many binding sites cannot be described by "average stability constants" and "average ligand concentrations", but must be analyzed by means of continuous distribution complexation functions (Buffle, 1988). See, for instance, the Affinity Spectrum approach (Hunston, 1975; Thakur et al., 1980; Nederlof et al., 1990; van Riemsdijk and Koopal, 1992) and the Differential Equilibrium Function (Gamble, 1970; Gamble et al., 1980; Altmann and Buffle, 1988; Buffle et al., 1990). The former has been applied to biochemical macromolecules whereas the latter has been mostly used to describe fulvic acid behavior. None of these models is entirely satisfactory, but the conceptual approach which is embodied in them is definitely more consistent with the complexity of these substances than the assumptions made when applying any discrete system of data analysis.

V. EXPERIMENTAL PROCEDURES

A. Apparatus

UF cells are commercially available and the selection of cell to use may depend on the conditions required for the system being studied. The standard cell used for binding studies has a pressure inlet valve, a pressure relief valve, a solution compartment (referred to as cell), a magnetic stirrer, a membrane support and a filtrate permeate outlet.

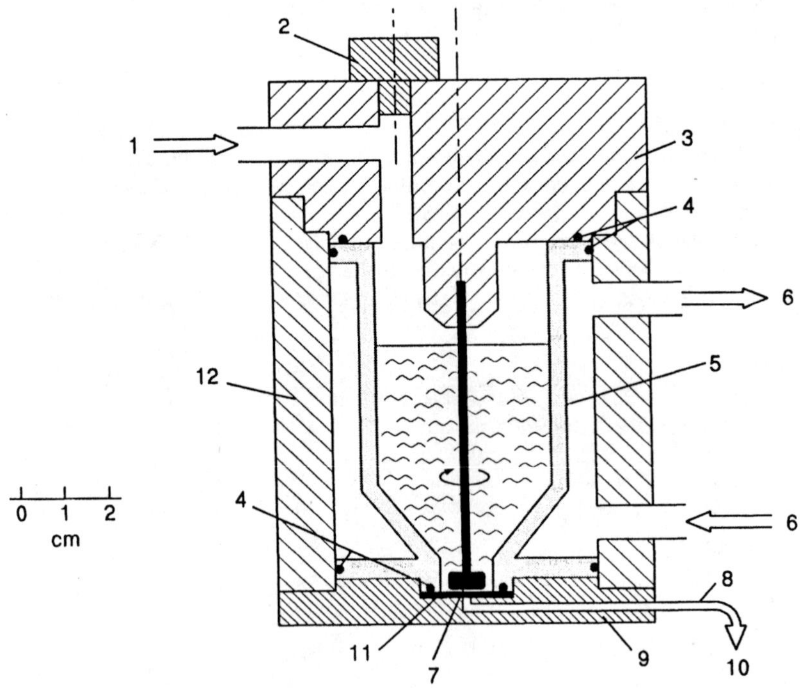

Figure 4.
Diagram of a UF cell with small dead volume: (1) N_2 pressure, (2) wisp, (3) cap, (4) O-ring, (5) Plexiglass cell, (6) circulation of water, (7) magnetic stirrer, (8) capillary outflow, (9) holder, (10) filtrate, (11) membrane on fritted polyethylene. (Adapted from Buffle and Staub (1984) by permission of the copyright holders, American Chemical Society, Washington).

One of the criteria for choosing a cell is that it must have a small void volume so that measurements can be done using small sample volumes. The void volume can be minimized for instance by using outlet tubes with capillary diameters.

Buffle and Staub (1984) designed a special thermostated plexiglass cell of 70 ml cell capacity and a void volume of 50 µl, for metal complexation studies in natural aquatic systems (Fig. 4). This may be also used for conducting binding studies in biological conditions. Cell with smaller cell volumes can also be designed in cases where only a small volume of ligand is available. Cells with other geometries have been designed (Paulus, 1969; Bloxam et al., 1977; Wielders, 1980; Ohshima et al., 1988). Small volume Amicon cells have been adapted to obtain small dead volumes (Sophianopoulos and Sophianopoulos, 1985).

B. Choice of membranes

Membranes should have a well-defined average pore size and a narrow pore size distribution, and these characteristics should be reproducible from membrane to membrane (Buffle et al., 1992). In most cases, the values given by the manufacturers for the nominal pore size of membranes are either the size or molar mass values obtained from curves like those shown in Figure 5, corresponding to 90% rejection by the membrane. As these curves show, the pore size distributions of most filters are such that at least a few percent of the molecules or particles larger than the nominal pore size can pass through the membranes.

The membrane selected should:

(i) be permeable to metal ions but retain the ligand. Commercially available membranes (Amicon, Millipore, Nucleopore) are often used for this purpose. The most often used ones are Amicon PM10 or UM10 with molar mass cutoffs of 10 kg mol^{-1};

(ii) not adsorb either the ligand or the metal ion under study. So, membranes must be tested

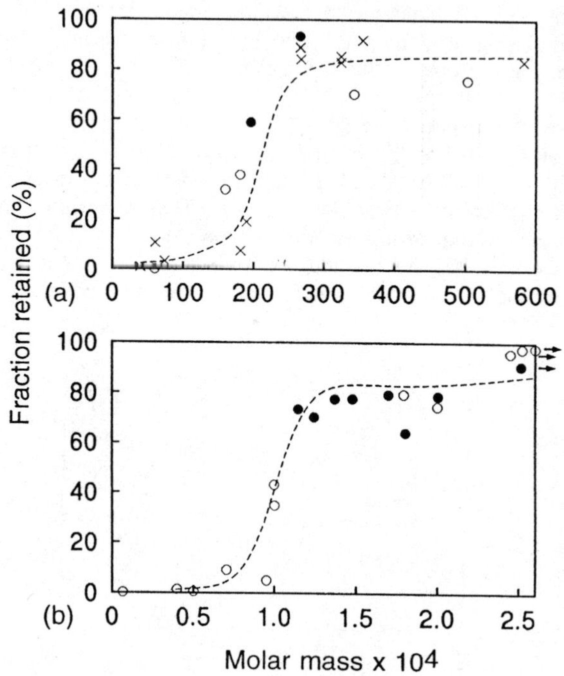

Figure 5.

Calibration of UF membranes with compounds of well-known molar masses. a) Membrane Amicon with nominal cutoff limit of 0.5 kg mol^{-1} (UM05) (Adapted from: Buffle et al. (1978) by permission of the copyright holders, Elsevier Scientific Publishers, Amsterdam). b) Membrane Amicon with nominal cutoff limit of 10 kg mol^{-1} (PM10) (Adapted from: Buffle et al. (1982) by permission of the copyright holders, Birkhäuser Verlag, Basel). The compounds used for calibration are listed in the corresponding references.

prior to complexation studies to ensure that they do not adsorb L or M. This is done by performing the UF experiment and determining R_L and R_M values with the ligand and metal separately under the same conditions as binding studies. The filtrates and retentates are analyzed and if the material balance of the solute is close to the one initially present in the cell, then it indicates that there is no adsorption on the membrane;
(iii) have a small surface area to minimize losses due to adsorption.

C. Metal analysis

Metal ion concentrations in the cell, filtrate and initial solutions may be determined by well known sensitive techniques such as anodic stripping voltammetry, ion selective electrodes, scintillation counters for radiolabelled compounds, atomic absorption spectrometry, EPR, fluorescence, UV-visible spectrometry.

VI. EXAMPLES OF APPLICATION OF ULTRAFILTRATION FOR METAL COMPLEXATION MEASUREMENTS IN BIOLOGICAL SYSTEMS

Table 1 gives a list of applications of UF to metal binding studies and in the subsequent section typical examples are given to illustrate the use of this technique.

A. Binding of calcium to human albumin

Williamson et al. (1978) used the continuous reagent addition method for calculating the binding constant for the calcium-human albumin interaction.

These studies were performed with a Millipore apparatus fitted with PSAC 01310 (M cutoff, 1 kg mol^{-1}) membranes. A solution containing 40 g dm^{-3} of albumin in Tris-HCl buffer (pH = 8) was placed in the cell while 37.8 mmol dm^{-3} calcium in Tris-HCl buffer was taken in the reservoir. Filtrates were collected in scintillation vials. Protein and metal concentrations in the cell and filtrates were analyzed.

A binding isotherm was constructed (calcium bound per mole of albumin vs free calcium concentration) (Fig. 6). The authors found 3 moles of calcium bound per mole of albumin and a formation constant of 200 dm^3 mol^{-1}.

B. Binding of manganese to concanavalin A

Sophianopoulos et al. (1983) investigated the interaction between concanavalin A (Con A), a plant lectin, and Mn and Ca to elucidate the mechanism of the induction of the proper conformation of the saccharide binding site in Con A. They performed their measurements in Amicon cell using PTCG Millipore membranes (M cutoff not given). A solution containing 0.5 mmol dm^{-3} Mn and 1 mmol dm^{-3} Ca and Con A was placed in the cell. Solutions were filtered and small portions of 0.5 ml to 1 ml were collected. The first 0.3 ml

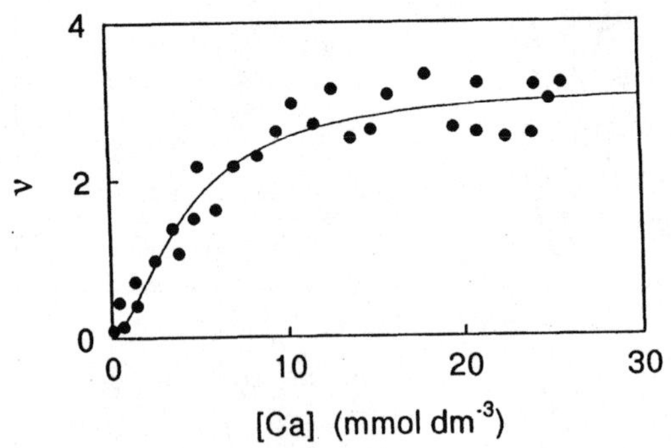

Figure 6.
Binding isotherm for the binding of calcium to human albumin (40 g.dm^{-3}). Experimental points are shown as circles. (Reproduced from Williamson et al. (1978) by permission of the copyright holders, Elsevier/North-Holland Biomedical Press, Amsterdam).

Table 1

Application of UF to Metal Binding Studies in Biological Systems

System	Conditions	Binding parameters	References
A. *Biological fluids*			
Magnesium-bovine serum albumin	Membrane: UM-2 (2)[a] 5-50 g.dm^{-3} albumin; 0.4-8.5 mmol dm^{-3} Ca Medium: 0.25 mol dm^{-3} KCl (pH=7.1) T=8 °C	K=525 dm^3.mol^{-1} [b]	Veloso et al. (1973)
Zinc-bovine serum albumin	Membrane: Millipore XY0 (25)[a] 0-81 μM albumin; 0.2-21 μM Zn pH=3.5-7.52	no values for constants reported	Fiabane and Williams (1978)
Calcium-human albumin	Membrane: PSAC 0310 (1)[a] 40 g.dm^{-3} albumin and Ca conc. varied Medium: Tris buffer (pH=8)	K=200 dm^3.mol^{-1} [b] ν=3[c]	Williamson et al. (1978)
B. *Animal tissues*			
Calcium-protein in rabbit skeletal muscle	Membrane: CF-50 0.1 mg.dm^{-3} troponin; 1-40 μmol dm^{-3} Ca Medium: 50 mol dm^{-3} KCl + 10 mmol dm^{-3} imidazole (pH=7)	K=2.5.10^6 dm^3.mol^{-1}	Fuchs (1972)
Calcium-bovine prothrombin	Membrane: UM-10 (10)[a] 140 μg prothrombin; 1 mmol dm^{-3} Ca Medium: NaCl + acetate buffer (pH=6)	ν=4[c] no values for constants reported	Nelsestuen and Suttie (1972)

cont'd

Table 1 (continued)

System	Conditions	Binding parameters	References
Magnesium-rat brain, liver and kidney	Membrane: UM-2 (2)[a] 0.1-0.13 mmol dm^{-3} Mg Medium: 0.25 mol dm^{-3} KCl (pH=7.1) T=8 °C	kidney: K=727 $kg.mol^{-1}$ [d] liver: K=532 $kg.mol^{-1}$ [d]	Veloso et al. (1973)
Calcium-rabbit hepatic galactoside binding protein	Membrane: UM-10 (10)[a] 0.02-40 mmol dm^{-3} Ca Medium: 0.02 mol dm^{-3} Tris + 0.01 mol dm^{-3} NaCl (pH=7.8)	K=286 $dm^3.mol^{-1}$ [b] K=1613 $dm^3.mol^{-1}$ [b] in the presence of 100 mmol dm^{-3} Mg	Andersen et al. (1982)
Copper-metallothionein - Glutathione	Membrane: YM10 (10)[a] Cu-4 μmol dm^{-3} MT-III; 3 μmol dm^{-3} GSH	K=9.26.10^5 $dm^3.mol^{-1}$ [b] binding sites = 1.01±0.31	Brouwer and Brouwer-Hoexum (1991)
C. Plant proteins			
Manganese-concanavalin A	Membrane: Millipore PTCG (?)[a] 0.2-0.5 mmol dm^{-3} Con A; 0.2-0.5 mmol dm^{-3} Mn. Medium: 0.2 mol dm^{-3} NaCl + 0.05 mol dm^{-3} NaAc buffer (pH=5.2) T=24-25 °C	K=(5±1).10^3 $dm^3.mol^{-1}$ K=(9.1±2.1).10^5 $dm^3.mol^{-1}$ in the presence of 1 mmol dm^{-3} Ca	Sophianopoulos et al. (1978, 1983)
Metal-yeast enolase	Membrane: YM10 (10)[a] 10-40 μmol dm^{-3} enolase; 50-400 μmol dm^{-3} metal. Medium: HEPES buffer (pH=7.5) w/wo substrate (2-phosphoglycerate)	Mg ν=2.3±0.7[c] (SA[e]) Ca ν=2.3±0.5[c] (SA[e]) Ca ν=4.2±0.5[c] (SP[e]) Co ν=3.9±0.5[c] (SA[e]) Co ν=5.6±0.5[c] (SP[e])	Brewer et al. (1983)

[a] Molar mass cutoff in kg mol^{-1}
[b] Recalculated from original data
[c] ν = number of moles of metal bound per mole of ligand
[d] kg weight of tissue per mole of binding sites
[e] SA = substrate absent; SP = substrate present

solution were discarded. This procedure was repeated by adding Con A in discontinuous steps. Con A was varied between 0.2-0.5 mmol dm^{-3}. The metal and Con A concentrations in the filtrate and cell solutions were analyzed by atomic absorption spectrometry and UV spectroscopy, respectively.

The binding constant was determined by a Scatchard plot. The formation constants K of Mn to Con A in the absence and presence of 1 mmol dm^{-3} concentration of Ca at pH = 5.2 were found to be $(5 \pm 1).10^3$ $dm^3\,mol^{-1}$ and $(9.1 \pm 2.1).10^5$ $dm^3\,mol^{-1}$, respectively. At pH = 6.5 Mn bound to Con A with a K of $(7.3 \pm 1.8).10^5$, and the binding affinity was virtually independent of the presence of Ca.

Thus these examples illustrate that UF can be used effectively for metal binding studies in complex media such as biological systems.

VII. ADVANTAGES AND DISADVANTAGES OF ULTRAFILTRATION

The main advantages of UF for metal binding studies are the following: (i) it is a versatile method, (ii) it is rapid, (iii) it is simple, (iv) it can be made almost a non equilibrium perturbing method by filtering volumes much smaller than the total sample volume, (v) it is applicable to a wide range of metals, (vi) it can be easily automated, and (vii) it does not require any addition of reagent.

In contrast, it has the following disadvantages: (i) it requires prior knowledge of the nature of the compound under study, (ii) the pores of the membrane may be clogged reducing the flux of solute particularly at large L concentrations, (iii) adsorption of material on the membrane will hamper the binding measurements, (iv) metal binding with l.m.m. ligands is difficult due to unavailability of membranes with small pore sizes, (v) concentration polarization arising from the build up of ligand concentration at the surface reduces the flux by forming gels. This last problem may be reduced by stirring the sample and by using a low flow rate.

ACKNOWLEDGEMENT

We thank Dr. J. Cox (University of Geneva) for valuable discussion.

List of symbols

$D_{m,X}$	diffusion coefficient of X through the membrane [$cm^2.s^{-1}$]
Δ_X	characteristic rate for diffusional transport of X ($=D_{m,X}/\lambda.K_m$) [$cm.s^{-1}$]
J_w	water flux [$cm.s^{-1}$]
J_X	flux of a solute X
K	equilibrium constant
k_d	dissociation rate constant
k_D	diffusion rate constant
k_m	rate constant for the transfer across the membrane
K_m	distribution coefficient ($= [X]_{m,c}/[X]_c$)
n	number of binding sites per molecule of ligand (Scatchard plot)
$N_{c,b}$	number of moles of M bound to L in the cell
N_f	total number of moles of M in the filtrate after filtration of volume V_f
ΔP	hydrostatic pressure
R_X	experimental retention coefficient of X
V_c	cell volume
V_f	total filtrate volume
V_c^o	initial cell volume
X	solute (M, L, ML)
[X]	free concentration of X
$[X]_c$	concentration of solute X in the cell
$[X]_f$	concentration of solute X in the filtrate
$[X]_{m,c}$	concentration of solute X at the cell side of the internal membrane surface
$[X]^o$	initial concentration of solute X
$[X]_r$	concentration of solute X in the reservoir
$[X]_t$	total concentration of X, irrespective of chemical form
α	complexation coefficient of a metal ion ($=[M]_t/[M]$)

α_c	complexation coefficient of a metal ion in the cell
λ	average pore length
$\Delta\pi$	osmotic pressure
σ_X	theoretical retention coefficient of X
ν	number of moles of metal bound per mole of ligand (Scatchard plot)

Abbreviations used

Con A	concanavalin A
ED	equilibrium dialysis
GSH	glutathione
MT-III	metallothionein isoform class III
UF	ultrafiltration
EPR	electron paramagnetic resonance
ESR	electron spin resonance
NMR	nuclear magnetic resonance
L	ligand or complexing site
M	free, hydrated, metal ion
ML	complex formed between M and L

REFERENCES

Altmann, R.S., and Buffle, J. (1988). *Geochim. Cosmochim. Acta, 52*: 1505.

Andersen, T.T., Freytag, J.W., and Hill, R.L. (1982). *J. Biol. Chem., 257*: 8036.

Blatt, W.F., Robinson, S.M., and Bixler, H.J. (1968). *Anal. Biochem., 26*: 151.

Bloxam, D.L., Hutson, P.H., and Curzon. G. (1977). *Anal. Biochem., 83*: 130.

Bowers, W.F., Fulton, S., and Thompson, J. (1984). *Clin. Pharmacokinet., 9*: 49.

Brewer, J.M., Carreira, L.A., Collins, K.M., Duvall, M.C., Cohen, C., and DerVaitanian, D.V. (1983). *J. Inorg. Biochem., 19*: 255.

Brock, T.D. (1983). *Membrane Filtration*, Sci. Tech. Inc., Madison, WI.

Brouwer, M., and Brouwer-Hoexum, T. (1991). *Arch. Biochem. Biophys., 290*: 207.

Buffle, J. (1988). *Complexation Reactions in Natural Systems. An Analytical Approach*, Ellis-Horwood, Chichester.

Buffle, J., and Staub, C. (1984). *Anal. Chem., 56*: 2837.

Buffle, J., Deladoey, P., and Haerdi, W. (1978). *Anal. Chim. Acta, 101*: 339.

Buffle, J., Deladoey, P., Zumstein, J., and Haerdi, W. (1982). *Schweiz. Z. Hydrol., 44*: 325.

Buffle, J., Altmann, R.S., Filella, M., and Tessier, A. (1990). *Geochim. Cosmochim. Acta, 54*: 1535.

Buffle J., Perret, D., and Newman, M. (1992). *Environmental Particles*, Vol. 1 (J. Buffle and H.P. van Leeuwen, eds.), Lewis Publishers, Boca Raton, Chapter 5.

Bulbula, A., Lindsey, E.E., Archer, R.D., and Lenz, R.W. (1975). *A.I.Ch.E. Symp. Ser., 71*: 105.

Cheryan, M. (1986). *Ultrafiltration Handbook*, Technomic, Lancaster, PA.

Cheryan, M., and Saeed, M. (1989). *J. Food Biochem., 13*: 289.

Dwyer, F.P., and Mellor, D.P., eds. (1964). *Chelating Agents and Metal Chelates*, Academic Press, New York.

Fiabane, A.M., and Williams, D.R. (1978). *J. Inorg. Nucl. Chem., 40*: 1195.

Fuchs, F. (1972). *Int. J. Peptide Protein Res., 4*: 147.

Gamble, D.S. (1970). *Can. J. Chem., 48*: 2662.

Gamble, D.S., Underdown, A.W., and Langford, C.H. (1980). *Anal. Chem., 52*: 1901.

Hummel, J.P., and Dreyer, W.J. (1962). *Biochim. Biophys. Acta, 63*: 530.

Hunston, D.L. (1975). *Anal. Biochem., 63*: 99.

Hwang Sun Tak, and Kammermeyer, K. (1975). *Techniques of Chemistry*, Vol. VII (A. Weissberger, ed.), Wiley, New York.

Klotz, I.M. (1982). *Science (Washington)*, *217*: 1247.

Klotz, I.M., and Hunston, D.L. (1971). *Biochemistry*, *10*: 3065.

Krause, K.-H., Milos, M., Luan-Rilliet, Y., Lew, D.P., and Cox, J.A. (1991). *J. Biol. Chem.*, *266*: 9453.

Meares, P. (1976). *Membrane Separation Processes* (P. Meares, ed.), Elsevier, Amsterdam, Chapter 1.

Nederlof, M.M., van Riemsdijk, W.H., and Koopal, L.K. (1990). *J. Colloid Interface Sci.*, *135*: 410.

Nelsestuen, G.L., and Suttie, J.W. (1972). *Biochemistry*, *11*: 4961.

Ohshima, T., Johno, I., and Kitazawa, S. (1988). *Therapeutic Drug Monitoring*, *10*: 310.

Paulus, H. (1969). *Anal. Biochem.*, *32*: 91.

Riemsdijk, W.H. van, and Koopal, L.K. (1992). *Environmental Particles*, Vol. 1 (J. Buffle and H.P. van Leeuwen, eds.), Lewis Publishers, Boca Raton, Chapter 12.

Rudman, D., and Kendal, F.E. (1957). *J. Clin. Invest.*, *36*: 538.

Scatchard, G. (1949). *Ann. N.Y. Acad. Sci.*, *51*: 660.

Scatchard, G., Scheinberg, I.H., and Armstrong, S.H. (1950). *J. Amer. Chem. Soc.*, *72*: 535.

Sophianopoulos, A.J., and Sophianopoulos, J.A. (1985). *Methods Enzymol.*, *117*: 354.

Sophianopoulos, J.A., Durham, S.J., Sophianopoulos, A.J., Ragsdale, H.L., and Cropper, P.W. (1978). *Arch. Biochem. Biophys*, *187*: 132.

Sophianopoulos, J.A., Sophianopoulos, A.J., and MacMahon, W.C. (1983). *Arch. Biochem. Biophys.*, *223*: 350.

Sourirajan, S., ed. (1979). *Reverse Osmosis and Synthetic Membranes*, National Research Council of Canada, Ottawa.

Spriggs, H.D., and Li, N.N. (1976). *Membrane Separation Processes* (P. Meares, ed.), Elsevier, Amsterdam, Chapter 2.

Staub, C., Buffle, J., and Haerdi, W. (1984). *Anal. Chem.*, *56*: 2843.

Thakur, A.K., Munson, P.J., Hunston, D.L., and Rodbard, D. (1980). *Anal. Biochem.*, *103*: 240.

Tuschall, J.R., and Brezonik, P.L. (1983). *Anal. Chim. Acta*, *149*: 47.

Veloso, D., Guynn, R.W., Oskarsson, M., and Veech, R.L. (1973). *J. Biol. Chem.*, *248*: 4811.

Wielders, J.P.M. (1980). *Anal. Biochem.*, *103*: 394.

Williamson, B.W.A., Strange, R.C., and Percy-Robb, I.W. (1978). *Biochim. Biophys. Acta*, *543*: 397.

Work, T.S., and Work, E. (1969). *Laboratory Techniques in Biochemistry and Molecular Biology*, North Holland, London.

Yap, W., and Schaffer, R. (1977). *Clin. Chem.*, *23*: 986.

MODELLING METAL-LIGAND EQUILIBRIA IN BLOOD PLASMA

by

Peter M. May

School of Mathematical and Physical Sciences,
Murdoch University,
Western Australia 6150
Australia

I. INTRODUCTION

Soon after the first computer programs for dealing with multiple chemical equilibria were developed, they were actively applied by coordination chemists, not only for calculations of laboratory solutions but also for simulating naturally-occurring mixtures of metal ions and ligands. The potential was obvious: such programs could in principle determine the chemical speciation in very complicated systems. Calculating such information without the aid of a computer was, in general, mathematically impracticable; neither could the information be obtained by direct chemical analysis because the complexes are typically too labile and/or present at concentrations too low to be reliably determined.

II. DISTRIBUTION OF METAL IONS AMONGST NATURALLY-OCCURRING LIGANDS

The earliest model of the equilibrium distributions of metals and ligands in a large naturally-occurring system was developed by Sillén (1959; 1963; 1967) for seawater. However, almost concurrently, Perrin began work on a simulation of blood plasma. In a remarkably far-sighted exposition (1965) he described a general approach, based on the simultaneous solution of mass-balance equations, in which he itemized the variety of information which this might provide. He mentioned specifically the quantitative data, such as the concentrations of the free metal ions and of their major complexes but he also included a better understanding of the effects of pH change, metal deficiencies, and the addition of chelating agents such as penicillamine and BAL (2,3-dimercaptopropanol).

Perrin's initial computations were published within a few years (1969). He chose to consider the distribution of copper(II) and zinc(II) amongst a selection of sixteen amino acids. For this purpose, his research group commenced a very broad experimental study of the relevant metal-ligand systems to determine the necessary equilibrium constants. The main outcome of his first model is shown in Table 1A.

Whilst Perrin was clearly aware of the many limitations of his approach, he argued that the various approximations which he had made could be progressively improved "in much the same way that a painting is built up by a succession of strokes of the brush". It is also interesting to note how many of the issues that he raised at the outset have imbued the modelling of blood plasma ever since. On the one hand, he carefully delineated his omissions (of metal-protein binding, of multi-phase equilibria, of compartmental distributions and of kinetic effects) as subjects for future extension of the model. On the

Table 1

Calculated Distributions of Copper and Zinc in Human Blood Plasma

A. Perrin (1969) [135 l.m.m. complexes]

Copper(II)	%	Zinc(II)	%
CuH(His)(Cyst)	33	$Zn(CysH)_2$	21
$Cu(His)(Cys)^-$	29	$Zn(His)^+$	18
$Cu(His)_2$	7	Zn(His)(CysH)	14
Cu(Gln)(His)	7	Zn(His)(Gly)	7

Total Cu(II) = 1.8×10^{-5} M
Calc. $[Cu^{2+}] = 7.0 \times 10^{-12}$ M

Total Zn(II) = 4.6×10^{-5} M
Calc. $[Zn^{2+}] = 4.5 \times 10^{-6}$ M

B. Hallman et al. (1971) [159 l.m.m. complexes]

Copper(II)	%	Zinc(II)	%
$Cu(His)(Cyst)^-$	48	$Zn(Cys)(His)^-$	24
CuH(His)(Cyst)	37	$Zn(His)^+$	21
$Cu(His)_2$	13	$Zn(Cys)_2^{2-}$	16
$Cu(Gln)_2$	0.3	Zn^{2+}	9

Total Cu(II) = 1.1×10^{-6} M
Calc. $[Cu^{2+}] = 3.4 \times 10^{-13}$ M

Total Zn(II) = 4.6×10^{-5} M
Calc. $[Zn^{2+}] = 4.0 \times 10^{-6}$ M

C. Perrin and Agarwal (1973) [217 l.m.m. complexes]

Copper(II)	%	Zinc(II)	%
$Cu(His)(Cyst)^-$	40	$Zn(Cys)(His)^-$	23
CuH(His)(Cyst)	31	$Zn(His)^+$	18
$Cu(His)_2$	11	$Zn(Cys)_2^{2-}$	16

Total Cu(II) = 1.1×10^{-6} M
Calc. $[Cu^{2+}] = 2.7 \times 10^{-13}$ M

Total Zn(II) = 4.6×10^{-5} M
Calc. $[Zn^{2+}] = 3.3 \times 10^{-6}$ M

D. Perrin and Agarwal (1973) [147 l.m.m. complexes]

Copper(II)	%	Zinc(II)	%
$Cu(His)(Cyst)^-$	44	$Zn(Cys)(His)^-$	19
CuH(His)(Cyst)	38	$Zn(Cys)_2^{2-}$	16
$Cu(His)_2$	16	$Zn(His)^+$	13

Total Cu(II) = 1.1×10^{-6} M
Calc. $[Cu^{2+}] = 3.3 \times 10^{-13}$ M

Total Zn(II) = 2.0×10^{-5} M
Calc. $[Zn^{2+}] = 2.0 \times 10^{-6}$ M

E. May et al. (1977) [5000 l.m.m. complexes]

Copper(II)	%	Zinc(II)	%
$Cu(His)(Cyst)^-$	21	$Zn(Cys)(Cit)^{3-}$	43
CuH(His)(Cyst)	17	$Zn(Cys)_2^{2-}$	19
$Cu(His)_2$	11	Zn(Cys)	3
Cu(His)(Thr)	8	$Zn(His)^+$	3

Total Cu(II) = 1.6×10^{-11} M
Fixed $[Cu^{2+}] = 1.0 \times 10^{-18}$ M

Total Zn(II) = 1.8×10^{-7}
Fixed $[Zn^{2+}] = 1.0 \times 10^{-9}$ M

F. Berthon et al. (1978) [5000 l.m.m. complexes]

Zinc(II)	%
$Zn(Cys)_2^{2-}$	40
$Zn(Cys)(His)^-$	24
$Zn(His)^+$	4
Zn(Cys)	3

Fixed $[Zn^{2+}] = 9.0 \times 10^{-9}$ M

G. Berthon et al. (1984) [5000 l.m.m. complexes]

Copper(II)	%
$Cu(His)_2$	19
Cu(His)(Thr)	19
Cu(His)(Ser)	9
Cu(His)(Ala)	7
Cu(His)(Gly)	6
$CuH(His)(Lys)^+$	5
Cu(His)(Val)	5
Cu(His)(Gln)	5
Cu(His)(Leu)	5

Fixed $[Cu^{2+}] = 1.0 \times 10^{-16}$ M

H. Berthon et al. (1986) [10000 l.m.m. complexes]

Copper(II)	%
Cu(His)(Gln)	19
$Cu(His)_2$	15
Cu(His)(Thr)	14
Cu(His)(Ser)	11
Cu(His)(Ala)	5
Cu(His)(Gly)	4
$CuH(His)(Lys)^+$	4
Cu(His)(Val)	4
Cu(His)(Leu)	4
Cu(His)(Asn)	4

Fixed $[Cu^{2+}] = 1.0 \times 10^{-16}$M

I. Brumas et al. (1993) [10000 l.m.m. complexes]

Copper(II)	%
Cu(His)(Gln)	19
$Cu(His)_2$	16
Cu(His)(Thr)	15
Cu(His)(Ser)	8
Cu(His)(Ala)	5
$CuH(His)(Lys)^+$	4
Cu(His)(Gly)	4
Cu(His)(Asn)	4
Cu(His)(Val)	4
Cu(His)(Leu)	4

other, at his very first attempt, he predicted (i) the appearance, in general, of mixed i.e. ternary complexes, (ii) the nexus between Cu(II) and histidinate, (iii) the possible role of electrically-neutral low-molar-mass (l.m.m.) complexes in assisting metals to diffuse more easily than their free ions across cell membranes and (iv) the importance of redox couples such as the oxidized-to-reduced ratio of cystine/cysteine. Time has amply confirmed all of these propositions.

Further studies by Perrin and his group followed. Hallman et al. (1971) published a refinement of the original computation, including the addition of another (inconsequential) amino acid, arginine. The most important change concerned the total copper concentration, which was reduced by more than an order of magnitude, in recognition that the copper bound within ceruloplasmin in plasma would not be available for complexation by the amino acids. (They did not, however, apply the corresponding correction for zinc binding to α_2-macroglobulin since the relevant information was unavailable at that time.) Some significant changes were also made to the mixed complexes, possibly correcting an error in the early statistical estimation of their formation constants. Table 1B gives the distributions computed as a result of the various alterations.

Perrin and Agarwal (1973) then presented a considerably expanded model for Cu(II) and Zn(II) in blood plasma in which they introduced not only five more amino acids but also many more mixed-ligand complexes. However, the results of these calculations differed only marginally, as can be seen by comparing Table 1C with Table 1B.

Of greater import, these authors went on in the same work to consider the competitive binding of metal ions in plasma by peptides and proteins. Although they were able to infer that peptides would mostly be less effective than their monomers, it was evident that some peptides and, especially, some proteins would have a greater avidity for metal ions than the simple amino acids themselves. Unfortunately, there were (and still are) few reliable quantitative studies on reversible metal-binding by proteins. Moreover, it was highly doubtful that any of the equilibrium constants which had been reported were useable. Accordingly, the conclusions which could be drawn from the model in which albumin was first incorporated were only tentative and none has survived the test of time.

Motivated to examine the effect of adding chelating agents such as EDTA to blood, Perrin and Agarwal (1973) also constructed a plasma model for calcium and magnesium in addition to zinc and copper. Since the alkaline earth metal ions are preferentially bound to ligands with many oxygen donor groups, this model encompassed a variety of inorganic and organic ligands in addition to the amino acids. Interestingly, the computed distributions for copper and zinc were not much affected, even though the association of zinc with both α_2-macroglobulin and albumin was now taken into account (cf Table 1D and Table 1C) and a small but significant amount of zinc-citrate binding (3.6%) was found to occur.

The findings of Perrin and Agarwal (1973) for calcium and magnesium are shown in Table 2A. These results have proteins (in this case both globulin and albumin) featuring amongst the most predominant species for the first time. Although these authors considered the possibility that proteins might also be the major binders of zinc, their discussion reveals a distinct preference against such a conclusion. With hindsight, it seems strange that the highly-specific binding by certain plasma proteins for some metal ions and the overwhelming advantage of proteins due to their high effective concentration of donor groups were not credited with greater importance. Nonetheless, the extensive discussion devoted to protein binding in their wide-ranging paper, focussed attention on this issue and this has been a preoccupation in blood plasma modelling ever since.

The importance of protein binding was adopted a priori in the equilibrium model for calcium ion reactions of blood plasma developed by Branegard and Österberg (1974). They assumed that the calcium ions interacted only with the carboxylate groups of the plasma proteins and that a single equilibrium constant was sufficient to simulate the binding function. By including equilibria for CO_2(g) and $CaCO_3$(s), they also introduced a multi-phase speciation calculation of blood plasma for the first time. Their results are given in Table 2B. There are some striking similarities between their model and that of Perrin and Agarwal. Both groups found that the fraction of calcium remaining uncomplexed was reasonably high - somewhere between one third and one half; both groups found that

Table 2

Calculated Distributions of Calcium and Magnesium in Human Blood Plasma

A. Perrin and Agarwal (1973) [147 h.m.m. & l.m.m. complexes]

Calcium(II)	%	Magnesium(II)	%
Ca(globulin)	24	Mg(bicarbonate)	12
Ca(albumin)	16	Mg(phosphate)	18
Ca(bicarbonate)	8	Mg(citrate)	2
Ca(citrate)	2	Mg(lactate)	1

Total Ca(II) = 2.5×10^{-3} M
Calc. $[Ca^{2+}] = 1.2 \times 10^{-3}$ M

Total Mg(II) = 1.1×10^{-3} M
Calc. $[Mg^{2+}] = 9.0 \times 10^{-4}$ M

B. Branegard and Österberg (1974) [5 h.m.m. & l.m.m. complexes]

Calcium(II)	%
Ca(protein)	36
$CaCO_3$(s)	30
Ca(citrate)	3
$CaHPO_4$	1
$CaSO_4$	0.1

Total Ca(II) = 2.5×10^{-3} M; pCO_2 = 44 mm Hg
Calc. $[Ca^{2+}] = 9.0 \times 10^{-4}$ M

C. May et al. (1977) [5000 l.m.m. complexes]

Calcium(II)	%	Magnesium(II)	%
Ca(bicarbonate)	9	Mg(bicarbonate)	6
Ca(citrate)	4	Mg(citrate)	5
Ca(lactate)	3	Mg(carbonate)	2

Total Ca(II) = 1.5×10^{-3} M
Fixed $[Ca^{2+}] = 1.1 \times 10^{-3}$ M

Total Mg(II) = 6.6×10^{-4} M
Fixed $[Mg^{2+}] = 5.2 \times 10^{-4}$ M

approximately 40% of the metal ion was bound to protein; and both groups found (probably coincidentally) that association with citrate accounted for a small but virtually identical amount, about 3%. At the same time, there were also pronounced differences between the two models. Branegard and Österberg concluded that a substantial amount of $CaCO_3(s)$ was formed and perhaps dispersed into particles of colloidal size. No evidence to support this hypothesis has subsequently been forthcoming and it is now generally believed that calcium ions in blood plasma simply exist in a state of supersaturation, stabilized by the presence of large amounts of protein. These authors also omitted bicarbonate from their model whereas Perrin and Agarwal had found its binding to calcium was significant.

Thus far, the computer simulations of metal-ligand equilibria in blood plasma had all been performed using the general computer programs, COMICS (Perrin and Sayce, 1967) and HALTAFALL (Ingri et al., 1967). These programs were written to solve sets of equations that typically might arise with systems of one or two metal ions and one or two ligands. Although they could easily be adapted to deal with a slightly larger number of components, they could not be greatly expanded. In particular, they were both rather limited in respect of the way they stored their data and in respect of their equation-solving algorithms. They could not, therefore, treat the hundreds of components and tens of thousands of complexes which might reasonably be expected to form in a complicated mixture such as blood plasma.

For this reason, a more comprehensive study was initiated by Linder and the present author (May, 1976). A large database of over 1000 equilibrium constants was compiled from the literature for blood plasma conditions (37°C, 150mM NaCl). Two specialized computer progams were then written to handle a large number of possible species. These were called ECCLES (for "Evaluation of Constituent Concentrations in Large Equilibrium Mixtures") and MIX. The former solved the relevant mass balance equations and displayed the resulting species distributions in an ordered way to enable changes in the concentrations of the major complexes to be readily monitored. The latter program, MIX, generated estimates of the equilibrium constants for the many mixed (ternary) complexes that might have been important in the blood plasma mixture but for which no data was available in the literature. As a result, the size of the models for metal-ligand equilibria in blood plasma was substantially increased, ultimately to include 6 - 10 metal ions, over 100 ligands and a set of 10,000 complexes carefully selected to include those most likely to be predominant.

Another notable aspect of this study was the way it dealt with metal-protein binding. Metal ions in the biological fluid were classified into four distinct fractions: those incorporated inertly into metalloproteins and which are non-exchangeable; those which are more loosely bound by other types of protein in a labile equilibrium; those which are complexed by the many l.m.m. ligands in blood plasma; and the free (or aquated) metal ions. Although the absolute concentrations of the metal-complex species were treated as being controlled by protein binding, the percentage distribution of the transition metal ions amongst the l.m.m. ligands was found not to be. This independence meant that errors arising from the omission of metal-protein equilibria were successfully bypassed.

The first distributions published (May et al., 1977) from the enlarged blood plasma model are shown in Table 1E, for copper(II) and zinc(II), and Table 2C, for calcium(II) and magnesium(II). Other conclusions can be summarized as follows. Iron(III) was found to exist almost exclusively as an hydroxy-citrate complex. Bicarbonate species predominated for manganese(II), as they did for calcium(II) and magnesium(II), although the binding appeared weak and a substantial fraction of these cations remained uncomplexed. The importance of histidine in the copper(II) ternary complexes formed with other ligands was confirmed.

The ECCLES and MIX programs have subsequently been widely adopted for calculating metal-ligand distributions in blood plasma. In a succession of studies, most notably by Berthon and co-workers (e.g. Berthon et al., 1978; 1984; 1986), the equilibrium constants which had initially been responsible for promoting complex species to the top of the percentage distribution for each metal ion have been checked experimentally (sometimes by measuring constants which had previously only been estimated and sometimes by repeating the determination). This has caused the calculated distribution of the transition metal ions amongst l.m.m. ligands in blood plasma to change progressively over the years.

The most significant of these reassessments concerned the amino acid ligands involved in the predominant (mostly ternary) complexes of Cu(II) and Zn(II), the outcomes of which are shown in Tables 1F, 1G, 1H and 1I.

A recent and novel application of the ECCLES blood plasma model is that of Harris and Keen (1989). They used this to address a longstanding controversy over the proportional distribution of zinc(II) in plasma between transferrin and albumin. In particular, they focussed on the apparent discrepancy between zinc distributions predicted by protein binding constants and those determined by a wide variety of biochemical separation techniques. As they noted, such a discrepancy is not a minor matter since the possibility that the separation methods were yielding incorrect results, due to unsuspected exchange processes, would cast doubt on numerous studies concerned with the localization of zinc and other transition metals. Fortunately, they found that when anion binding by the proteins is properly included in the calculations, approximately 90% of the labile zinc is predicted to be bound to albumin. This is consistent with the experimental results. Their model suggests that about 1% of the labile zinc in plasma occurs as l.m.m. complexes, with a distribution close to that found by Berthon et al. (1978), shown in Table 1F.

Models for exogenous metal ions in blood plasma using the ECCLES and MIX computer programs have also appeared. The computed distributions for cadmium(II), nickel(II), lead(II) and plutonium(IV) are given in Table 3.

It is abundantly evident from the differences in computed distributions listed in Tables 1 and 2 that there has been little agreement, either between independent research groups or even within the same group over time. The supposed identity of the most predominant l.m.m. metal complexes clearly changes, almost without exception, from one study to the next. In this sense, neither past nor present blood plasma simulations can be regarded as reliable. The reason for this is simply that the equilibrium constants on which the calculations altogether depend, have not been determined accurately enough to yet answer the question which is being posed. Berthon et al. (1984) argue that the calculations are making progress, and hopefully this is the case (Brumas et al., 1993), but there is to date little basis for real confidence, particularly in view of the problems which are still being experienced by those quantifying metal-ligand equilibria in solution (see, for example, May et al., 1988) and

Table 3

Calculated Distributions of Exogenous Metal Ions in Human Blood Plasma

A. Cole et al. (1985) [10000 l.m.m. complexes]

Cadmium(II)	%	Nickel(II)	%
Cd(Cys)	46	$Ni(His)_2$	51
Cd(Cis)	28	$Ni(Cys)(His)^-$	18
$Cd(Cys)(OH)^-$	14	$Ni(Cys)_2^{2-}$	11
$Cd(Cys)_2^{2-}$	8	$Ni(His)^+$	4

B. Duffield et al. (1984) [10000 l.m.m. complexes]

Plutonium(IV)	%
$Pu(citrate)(OH)_2^-$	71
Pu(citrate)(OH)	29

C. May et al. (1977) [5000 l.m.m. complexes]

Lead(II)	%
Pb(Cys)	80
$Pb(Cys)(citrate)^{3-}$	7
$Pb(Cis)^+$	5

those attempting to build these kinds of computer simulation models (May and Murray, 1991).

It would nevertheless be very wrong to conclude that the modelling of metal-ligand equilibria in blood plasma has so far been without value. The identification of the predominant complexes is but one, rather narrow, purpose of the modelling exercise. Far more important is the general insight which modelling provides and which is difficult, if not impossible, to obtain in any other way. This insight arises naturally when the predictions of models are compared with the corresponding experimental investigations and hence used to distinguish good ideas from bad ones. Judged by this criterion, the equilibrium models of trace element complexation in biological fluids have been highly successful.

As mentioned at the beginning of this section, even the earliest models by Perrin led to a range of useful conclusions which still remain valid. These models, and subsequent ones, have proved invaluable in drawing attention to the importance of l.m.m. complexing equilibria in biological systems and to the experimental difficulties which such equilibria often bring about. The chemical principles of this are fully discussed by May and Bulman (1983) and a superbly incisive critique of the failure to take account of complexing equilibria in biological fluids can be found in the paper on zinc(II) concentrations in blood plasma by Magneson et al. (1987).

III. SYNTHETIC CHELATORS IN BLOOD PLASMA

Computer-simulated results for the distribution of metal ions amongst l.m.m. ligands in blood plasma are naturally and conveniently represented as percentages. As employed above, they are useful in discussing the predominant ligands bound to each metal ion and in suggesting which complex species are formed to a significant extent. However, percentage distributions are not at all satisfactory for describing the effects of exogenous chelating agents. There are various problems:

(i) Strong therapeutical chelating agents are calculated to bind nearly 100% of many metal ions even at peak concentrations well below those which have an observable physiological effect. For example, although EDTA produces minimal (or no) cupresis, Perrin and Agarwal (1973) calculated that the addition of EDTA to blood plasma could sequester 98.3% of the total copper(II). The percentage is thus an insensitive indicator of how well the agent is able to complex target metal ions in vivo, particularly in any comparison between different agents.

(ii) Even when the preceding point is discounted, interpreting results presented as percentages is tedious because they always involve a comparison between sets of computed distributions. In other words, the percentage of a metal ion complexed by an exogenous chelator can only be meaningfully understood in terms of the difference between it and other such distributions under comparable circumstances.

(iii) Strictly speaking, the calculated percentage distribution is independent of the extent of metal-protein binding only in "normal" blood plasma. Whilst there are good reasons for thinking that the relative order of binding by a series of introduced agents would not change, the magnitude of differences between them would often do so. This makes the interpretation of any result presented in terms of percentages even more difficult.

For these reasons, it is necessary to find better ways of representing the extent of complexation by exogenous agents in biological fluids.

The method which has been in longest use is based on a concept called the "effective stability constant". This was introduced by Schubert (1955) who recognized that the ability of ligands to bind metal ions in vivo would depend not only on the equilibrium constant of binding between the ligand and the metal ion in question but also on other factors such as the ligand's affinity for protons and for other metal ions which are present in the biological medium. This approach was strikingly successful in showing how the therapeutic effectiveness of EDTA was modified by calcium ion complexation. It is disappointing to note that despite this early insight of Schubert's, numerous instances of comparisons between therapeutic chelating agents have continued to appear in the literature based on the equilibrium constant of the 1:1 metal-ligand complex formation, without due regard to the importance of competition with other metal ions or with protons.

A second method of representing the extent of complexation by chelating agents in blood plasma is the so-called "plasma mobilizing index" or "PMI" proposed by May and Williams (1977). The relative ability of each agent to compete for a metal ion is defined by

$$\text{PMI} = \frac{\text{(total calculated concentration of l.m.m. complex species in the presence of the drug)}}{\text{(total concentration of l.m.m. metal complex species in normal plasma)}}$$

The index is thus a convenient measure of how well the chelator is able to expand the size of the l.m.m. complex fraction. This assumes that the l.m.m. fraction is relatively small and is in a labile equilibrium with a large reservoir of the metal ion from which it can be readily extracted. For metal ions such as copper(II) and zinc(II) the immediate reservoir is albumin, a multifunctional protein carrier which is associated with an overwhelming percentage of these metal ions in plasma. With other metal ions, in particular iron(III), the metal ion may be drawn directly from body sources outside of the plasma, such as the liver. There is now considerable evidence that a labile pool of iron, copper and zinc exists in the cytoplasm of liver cells which is readily exchanged with plasma and at the expense of which the l.m.m. metal complex fraction in plasma may be enlarged.

The third method of expressing the ability of chelating agents to bind particular metal ions in multi-component mixtures is to calculate the equilibrium concentration of the "free" (aquated) metal ion in the relevant solution. This is generally expressed as a negative log or "pM" value. Clearly, the better an agent is able to compete for a target metal ion (overcoming the effects of protons and other metal ions which may be present), the lower the free metal ion concentration at equilibrium will be. Accordingly, a series of chelating agents can be ranked in terms of their relative affinity for a given metal ion by expressing this in terms of increasing pM value.

All three of these methods are founded on some important assumptions. Most fundamentally, they require that every significant competitive reaction is included in the calculations. The parameters being calculated are also highly conditional, i.e. they very much depend on the prevailing concentrations of all the system's components. Often, too much is made of these parameters when their underlying assumptions are not fully justified: it is necessary not only to compare like with like but also to judge the probable effect of the different conditions on any comparisons made between them. Special cognizance must be given to certain rather obvious factors which cannot be adequately taken into account for practical reasons. For instance, the pharmacokinetics of ligand and complex, and their respective lipophilicities, clearly of considerable relevance, yet do not enter into any of the three above-mentioned methods for evaluating therapeutic chelating agents. These deficiencies do not negate the calculated assessments, as is sometimes claimed, but they do limit what can be properly deduced therefrom, as is sometimes overlooked.

In addition to the various limitations inherent in any calculation of chelating agent effectiveness based on equilibrium binding in a single biological medium, the use of effective stability constants suffers from some particular disadvantages. Mainly, this is because they tend to be regarded as analogous to thermodynamic equilibrium constants which obscures their conditional nature and leads to their misuse. Effective stability constants are also difficult to evaluate properly, being less amenable to generalized calculation by computer. Accordingly, they have largely fallen into abeyance in recent years.

On the other hand, as the application of computer simulation methods has spread, the determination of PMI and of pM values has gained increasing acceptance. Nowadays computer models of the metal-ligand interactions in biofluids are readily set up. They provide a comprehensiveness, especially regarding the inclusion of all feasible competitive reactions, which ought to be routinely utilized. They should be regarded as the first of several necessary steps on the way to fully understanding the biological effects of chelating agents in fundamental physico-chemical terms.

In some rather special cases, the biological effects of chelating therapeuticals are dominated by the extent of metal complexation which they exhibit in blood plasma. This happens, for example, with the aminopolycarboxylic acids which, following intravenous injection, are largely unable to cross lipid membranes and are hence effectively confined to the blood plasma compartment until eliminated by glomerular filtration into the urine. Under these circumstances, the decorporation of metal ions induced by the chelating agent closely parallels its ability to compete for the metal ion in plasma; all other considerations have only a minor significance. Accordingly, PMI values are a good predictor of the relative effectiveness of a series of such agents (May and Williams, 1977).

Whilst it is important to recognize the successfulness of these predictions, it is equally necessary to be aware of the limitations outlined above. Most of the chelating agents employed in the clinic are specially effective because they are able to permeate beyond blood plasma into intracellular space. The strength of complexation is thus by no means the sole factor which determines the relative amount of target metal ion which is excreted. Rates of transport of both the ligand into the appropriate body organs and of the resulting metal complex into the bile (or, less often, the urine) also prevail. Unqualified predictions including such kinetic effects must await the development of the comprehensive computer simulation models described in this Part Five.

In the meantime, the results of currently available models must be interpreted with care. This means, in particular, not trying to deduce from them more than is warranted given the underlying assumptions of the calculation. It does not mean that all such results are inherently untrustworthy: the equilibrium calculations can, with care, be carried out reliably and can, in certain circumstances, then be used to draw some conclusions with confidence. Most notable amongst these are the negative ones, i.e. predictions that particular agents will not exert any affect through complexing in vivo with particular metal ions. This is because unforeseen interactions in vivo are unlikely to cause an agent to compete much more successfully in the presence of endogenous ligands than it would do in vitro at the corresponding physiological concentrations. In other words, an ability to overcome the known extent of competition for a metal ion in vivo can reasonably be regarded as the minimum pre-requisite for effectiveness of any chelating agent. Unless the interaction occurs in the first place, other factors are irrelevant. It was on this basis, for example, that Cole et al. (1981) deduced the zinc-related side effects of the anti-tuberculosis agent, ethambutol, were more likely due to a particular metabolic degradation product than to the parent molecule itself.

Provided the models are sufficiently comprehensive, both PMI and pM calculations can provide very reliable measures of how well given agents are able to compete for metal ions in particular biofluids. Interestingly, in respect of the relative degree of complexation by a series of chelating agents, these two parameters tacitly quantify different modelling scenarios. In the case of PMI, it is assumed that the target metal ion is being fully buffered (e.g. by a predominant metal-protein complex) so that the size of its l.m.m. fraction increases on demand. With the pM calculations, on the other hand, it is assumed that the total amount of metal ion available to the chelating agent is fixed, i.e. it can be complexed by the agent only at the expense of other l.m.m. species or by lowering the free (aquated) metal ion concentration. Plainly, the respective calculations represent the two extreme cases which can be conceived. In vivo, there will invariably be some ingress of the metal ion so that the l.m.m. fraction will be at least partly replenished from external sources but this will not be unrestricted and, hence, it will not be enough to compensate fully for the complexation by the chelating agent. Thus, the PMI and pM parameters can be considered to span what actually happens in practice.

IV. CONCLUSION

The computer simulation models described above have explained satisfactorily the relative therapeutic effectiveness of many chelating agents. They also focussed early attention on the critical role played by naturally-occurring l.m.m. species in trace element metabolism; the involvement of such species is now being repeatedly confirmed by a spate of reports in the literature. Furthermore, the models have provided a significant amount of information in their own right. Whilst the precise identity of the predominant complexes must still be regarded as uncertain, there can be no doubt about the special involvement of particular l.m.m. naturally-occurring ligands such as histidinate, cysteinate and citrate. They give a reasonable indication of the extent of complexation by l.m.m. ligands and, hence, of the fraction of labile metal ion pool which can reasonably be attributed to them - although this estimate will undoubtedly be much improved by the development of kinetic models which are currently underway (May and Verhoeven, unpublished work). They have pointed directly to the importance of redox equilibria in determining the physiological fate of both iron and copper (see, for example, Jackson et al., 1978). As a result, nowadays many investigators of transition metal ions in biological systems are taking the necessary experimental precautions which were previously overlooked. (A good example of this is that frequent problems associated early on with the use of metal salt solutions - such as those of chloride or nitrate - are nowadays routinely avoided by instead introducing transition metal ions in a suitably complexed form).

The real benefits of these models then lie in the overview that they provide. As summarized by Perrin and Agarwal (1973), "In seeking in this way to elucidate the roles of metal ions in living organisms it is well to recall Tukey's Principle: 'Far better an approximate answer to the right question, which is often vague, than an exact answer to the wrong question, which can always be made precise'".

REFERENCES

Berthon, G., May, P.M., and Williams, D.R. (1978). *J. Chem. Soc., Dalton Trans.*, 1433.

Berthon, G., Piktas, M., and Blais, M.-J. (1984). *Inorg. Chim. Acta, 93*: 117.

Berthon, G., Hacht, B., Blais, M.-J., and May, P.M. (1986). *Inorg. Chim. Acta, 125*: 219.

Branegard, B., and Österberg, R. (1974). *Clin. Chim. Acta, 54*: 55.

Brumas, V., Alliey, N., and Berthon, G. (1993). *J. Inorg. Biochem., 52*: 287.

Cole, A., May, P.M., and Williams, D.R. (1981). *Agents Actions, 11*: 296.

Cole, A., Furnival, C., Huang, Z.-X., Jones, C.D., May, P.M., Smith, G.L., Whittaker, J., and Williams, D.R. (1985). *Inorg. Chim. Acta, 108*: 165.

Duffield, J.R., May, P.M., and Williams, D.R. (1984). *J. Inorg. Biochem., 20*: 199.

Hallman, P.S., Perrin, D.D., and Watt, A.E. (1971). *Biochem. J., 121*: 549.

Harris, W.R., and Keen, C. (1989). *J. Nutr., 119*: 1677.

Ingri, N., Kakolowicz, W., Sillén, L.G., and Warnqvist, B. (1967). *Talanta, 14*: 1261.

Jackson, G.E., May, P.M., and Williams, D.R. (1978). *FEBS Lett., 90*: 173.

Magneson, G.R., Puvathingal, J.M., and Ray, W.J., Jr. (1987). *J. Biol. Chem., 262*: 11140.

May, P.M. (1976). *Computer Simulation of Metal-Ion Equilibria in Biochemical Systems: Models for Blood Plasma*, M.Sc. Thesis, University of Cape Town.

May, P.M., and Bulman, R.A. (1983). *Prog. Med. Chem., 20*: 225.

May, P.M., and Murray, K. (1991). *Talanta, 38*: 1409.

May, P.M., and Williams, D.R. (1977). *FEBS Lett., 78*: 134.

May, P.M., Linder, P.W., and Williams, D.R. (1977). *J. Chem. Soc., Dalton Trans.*, 588.

May, P.M., Murray, K., and Williams, D.R. (1988). *Talanta, 35*: 825.

Perrin, D.D. (1965). *Nature (London), 206*: 170.

Perrin, D.D. (1969). *Suomen Kemi., 42*: 205.

Perrin, D.D., and Agarwal, R.P. (1973). *Metal Ions in Biological Systems*, Vol. 2 (H. Sigel, ed.), Marcel Dekker, New York, p. 167.

Perrin, D.D., and Sayce, I.G. (1967). *Talanta, 14*: 833.

Schubert, J. (1955). *Annu. Rev. Nucl. Sci., 5*: 369.

Sillén, L.G. (1959). "Oceanography," Proceedings of the International Oceanographic Congress '59, New York, p. 549.

Sillén, L.G. (1963). *Kem. Tidskr., 75*: 161.

Sillén, L.G. (1967). *Chem. Brit., 3*: 291.

SIMULATED DISTRIBUTION OF METAL IONS IN SALIVA

by

Professor David R. Williams, OBE, BSc (Hons), PhD, DSc, EurChem, CChem, FRSC, FRSA,

Dr Clare C. Coombes, BSc (Hons), M Phil, PhD, CChem, MRSC

and Dr Lan Wu, BSc(Hons), MSc, PhD, CChem, MRSC

Department of Chemistry,
University of Wales College of Cardiff,
PO Box 912
Cardiff CF1 3TB
U.K.

I. INTRODUCTION

"Few people outside the field of oral research realise the fascinating range of diverse properties and functions of saliva, or its contributions to the prevention of disease, the promotion of comfort, general health and the pleasures of living"

Professor G. Neil Jenkins (from Edgar and O'Mullane, 1990)

Such statements are a stimulating challenge to the bioinorganic chemist to establish the underlying science so that the benefits and pleasures may be optimized and problems minimized.

Much of this fundamental biochemistry is determined by the chemical speciation which prevails and, for such biofluids, this is below the levels of convenient analysis and so must be simulated using powerful computing techniques. Our researches have used the MINEIR program.

The MINEIR chemical speciation computer program, as developed by Westall et al. (1976) and modified by Schweigruber (1982), and the subsidiary software MINTRATE created by Hall (1989), are used to model the speciation occurring in saliva under equilibrium conditions.

The computer input consists of the concentrations of the known low-molar-mass (l.m.m.) ligands and of the metal ions present. The database compiled contains protonation constants for each of the ligands, formation constants for each metal-ligand interaction judged to be feasible and solubility products for sparingly soluble solids that may potentially precipitate from a combination of any of the components of the solution.

All constants embodied in the database were either measured in our own laboratories or obtained from the literature, or calculated, and in all instances, are believed to be the most reliable available.

II. ADVANTAGES AND DISADVANTAGES ASSOCIATED WITH COMPUTER SPECIATION MODELLING

Computer speciation modelling has many advantages over conventional 'invasive' analytical techniques.

First, the speciation of a component can be computed when the solution is at equilibrium whereas techniques are currently unavailable for analyzing the chemical speciation except for those methods that disturb the delicate labile equilibria present.

Secondly, one can model solutions having component concentrations below those detectable by conventional analytical techniques.

Such simulation is believed to give a true reflection of the relative abundance of the l.m.m. complexes present. However, the exceedingly complex nature of biofluids such as saliva dictates that it is currently impossible to create and execute simulations of all aspects of the biofluids unequivocally reflecting the distribution of all the metal ions involved in all protein-bound and solution states.

Only a rather minor fraction of metal-protein binding has been quantified, the concentrations of some components still being imprecise. Some appropriate formation constants remain unmeasured and those that have been determined are subject to experimental error.

Furthermore, our models are based solely upon thermodynamic considerations and no account is taken of any kinetic effects.

Nevertheless, models have been created which give an informative and useful insight into the l.m.m. complex distribution of metal ions in a range of biofluids.

III. THE SALIVA MODEL

Saliva contains anions, cations, non-electrolytes, amino-acids, proteins, carbohydrates and lipids which flow in waves against, and into, the plaque, having a diurnal tide and varying degrees of intensity (Mandel, 1974).

Saliva is secreted by three pairs of main glands: the parotid, the submandular and the sublingual, and also by the small glands of the mucosa. The most important stimuli for salivary secretion are acid solutions which lead to a maximal secretion of 4 cm^3 of saliva per minute (Davenport, 1982).

The pH of saliva, in the absence of the influence of dietary acids etc, varies throughout the day between 5.5 and 7.8, giving an average of 6.75 according to the conditions, i.e. whether resting or under stimulus (Browley, 1935).

The composition of saliva varies greatly in different individuals and in the same individual under different circumstances (Jenkins, 1977). Therefore, the lists of constituents of saliva shown in Table 1 used in our speciation modelling ought to be regarded as a general guide to the composition of unstimulated saliva. The component concentrations given in the literature were extended by the addition of trace metal-phosphate complexes (Curzon and Cutress, 1983; Hurford, 1989; Wu, 1992; Williams and Wu, 1993). We have utilized this model in two main areas.

A. Calcium speciation in saliva

The major inorganic component of tooth enamel is a mineral similar or identical to hydroxyapatite (HAP, $Ca_5(PO_4)_3OH$). At the most fundamental level, the main reaction involved in enamel de- and re-mineralization can be defined by equation 1.

$$\underset{\text{(HAP)}}{Ca_5(PO_4)_3OH} \underset{\text{Remineralization}}{\overset{\text{Demineralization}}{\rightleftharpoons}} 5Ca^{2+} + 3PO_4^{3-} + OH^- \qquad (1)$$

In practice, however, several different types of calcium phosphate may be precipitated (Table 2) (Nancollas et al., 1979; McCann, 1968). Further, whenever fluoride is present, fluoroapatite (FAP), a mineral with a much larger solubility product than HAP, can be preferentially formed in the remineralization process

$$5Ca^{2+} + 3PO_4^{3-} + F^- \rightleftharpoons \underset{\text{(FAP)}}{Ca_5(PO_4)_3F} \qquad (2)$$

Table 1

Composition of Mixed, Unstimulated Saliva

Metal ions	mol dm^{-3}
Calcium(II)	$1.17x10^{-3}$
Cobalt(II)	$6.30x10^{-7}$
Copper(II)	$3.00x10^{-7}$
Magnesium(II)	$2.69x10^{-4}$
Potassium(I)	$2.05x10^{-2}$
Strontium(II)	$1.50x10^{-7}$
Zinc(II)	$5.00x10^{-6}$
Sodium(I)	$1.00x10^{-2}$
Ligands	**mol dm^{-3}**
Alanate	$1.35x10^{-4}$
Ammonia	$2.63x10^{-3}$
Arginate	$3.80x10^{-4}$
Ascorbate	$8.00x10^{-6}$
Aspartate	$1.10x10^{-5}$
Carbonate	$6.65x10^{-3}$
Chloride	$2.89x10^{-2}$
Citrate	$5.37x10^{-5}$
Cysteinate	$1.00x10^{-5}$
Cystinate	$1.30x10^{-5}$
Fluoride	$7.27x10^{-6}$
Glutaminate	$8.25x10^{-5}$
Glycinate	$1.40x10^{-4}$
Histamine	$1.57x10^{-4}$
Histidinate	$8.50x10^{-5}$
Isoleucinate	$4.20x10^{-5}$
Lactate	$1.35x10^{-6}$
Leucinate	$1.20x10^{-5}$
Lysinate	$5.30x10^{-5}$
Methionate	$5.00x10^{-7}$
Phenylalanate	$9.40x10^{5}$
Phosphate	$3.68x10^{-3}$
Prolinate	$8.00x10^{-5}$
Serinate	$6.30x10^{-5}$
Thiocyanate	$2.13x10^{-3}$
Threoninate	$2.52x10^{-4}$
Tryptophanate	$2.70x10^{-5}$
Tyrosinate	$3.30x10^{-5}$
Valinate	$1.20x10^{-4}$
Urea	$3.01x10^{-3}$
Gustin	$2.43x10^{-6}$
pH	6.75 (5.6-7.6)

Saliva continuously provides calcium and phosphate to surfaces of the teeth in concentrations which not only inhibit the demineralization caused by acids from dental plaque, but also which can encourage some remineralization, or at least halt the early stages of carious lesion formation (enamel demineralization).

Figures 1 and 2 show the major calcium speciation in mixed unstimulated saliva. Since the thermodynamically most stable species, HAP and FAP, form rather slowly and by transformation of more readily precipitated precursors such as octacalcium phosphate (OCP), etc, it is convenient to refer to the model which includes HAP and FAP as the long term model (Fig. 1) and the model which omits HAP and FAP as the short term model (Fig. 2). The long term model may be taken as referring to an equilibrium state of the system, whereas in the short term model kinetic factors inhibiting the direct formation of HAP and FAP predominate (Linder and Little, 1986).

The most cost-effective research progress can be obtained by combining both the long and short term models to give an intermediate model which is closer to reality for a person's normal mouth (Table 3). Structural chemistry suggests that dicalcium phosphate dihydrate (DCPD) and OCP are the most important precursors of HAP and our computer modelling does indeed predict OCP to be the most likely precursor of HAP in mixed, unstimulated saliva at pH = 6.75.

The potential of calcium phosphate solids to precipitate in saliva is represented in our modelling by a Saturation Index (SI) as shown in Figure 3. For a sparingly soluble salt formed by complexing the metal ion, M^{n+}, with an anion, X^{m-}, i.e. M_mX_n the dissolution reaction is:

$$M_mX_n\,(s) \rightleftharpoons mM^{n+}(aq) + nX^{m-}(aq) \tag{3}$$

Table 2

The Phosphate Protonation Constants and Calcium Phosphate Formation Constants Used with Sources (25°C, I=0)

Name	System	log β	Ref.
	$H^+ + PO_4^{3-} \rightleftharpoons HPO_4^{2-}$	12.22 (37°C)	a
	$2H^+ + PO_4^{3-} \rightleftharpoons H_2PO_4^-$	19.34 (37°C)	a
	$3H^+ + PO_4^{3-} \rightleftharpoons H_3PO_4$	21.52 (37°C)	a
	$Ca^{2+} + PO_4^{3-} \rightleftharpoons CaPO_4^-$	6.44 (37°C)	a
	$Ca^{2+} + H^+ + PO_4^{3-} \rightleftharpoons CaHPO_4$	15.04 (37°C)	a
	$Ca^{2+} + 2H^+ + PO_4^{3-} \rightleftharpoons CaH_2PO_4^+$	20.94 (37°C)	a
	$Ca^{2+} + H_2O \rightleftharpoons CaOH^+ + H^+$	1.51	b
(DCPD)	$Ca^{2+} + H^+ + PO_4^{3-} \rightleftharpoons CaHPO_{4(s)}$	18.93	b
(OCP)	$4Ca^{2+} + H^+ + 3PO_4^{3-} \rightleftharpoons Ca_4H(PO_4)_{3(s)}$	46.90	c
(TCP)	$3Ca^{2+} + 2PO_4^{3-} \rightleftharpoons Ca_3(PO_4)_{2(s)}$	28.90	c
(HAP)	$5Ca^{2+} + 3PO_4^{3-} + H_2O \rightleftharpoons Ca_5(PO_4)_3OH_{(s)} + H^+$	44.33	c
(FAP)	$5Ca^{2+} + 3PO_4^{3-} + F^- \rightleftharpoons Ca_5(PO_4)_3F_{(s)}$	59.60	d

Abbreviations: DCPD = Dicalcium phosphate dihydrate; OCP = Octacalcium phosphate; TCP = Tricalcium phosphate; HAP = Hydroxyapatite; FAP = Fluoroapatite.
a = measured in this work
b = Smith and Martell (1976)
c = Nancollas et al. (1979)
d = McCann (1968)

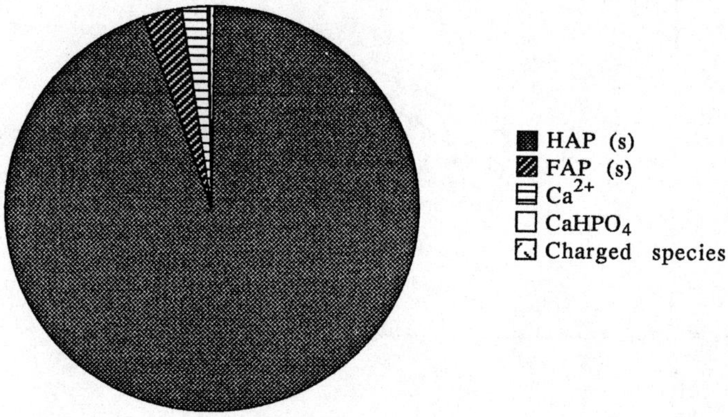

Figure 1.
Calcium speciation in saliva (long term model) at pH = 6.75.

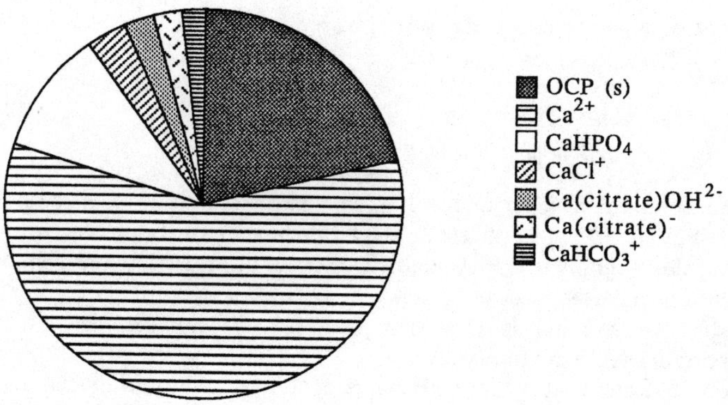

Figure 2.
Calcium speciation in saliva (short term model) at pH = 6.75.

Table 3

Predicted Precipitates at Varying pH and the Effect of Removing HAP and FAP from the Saliva Model

pH	Long term model (all $CaPO_4$ solids)	Short term model (HAP, FAP and TCP omitted)	Intermediate model
5.00	FAP	-	FAP
6.00	FAP and HAP	-	FAP and HAP
6.75	FAP and HAP	OCP	OCP, FAP and HAP
7.00	FAP and HAP	OCP	OCP, FAP and HAP

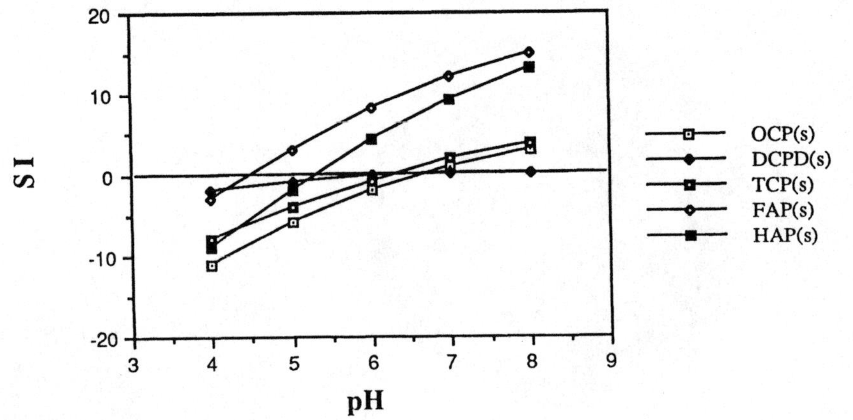

Figure 3.
Saturation Indices in saliva.

and the degree of supersaturation may be expressed in terms of the Saturation Index (SI):

$$SI = \lg (IAP/K_{sp}) \quad (4)$$

where K_{sp} is the solubility product and

$$IAP = (a_M{}^{n+})^m.(a_X{}^{m-})^n \quad (5)$$

If SI>0 the solution is supersaturated. Given sufficient time to encompass the kinetic factors, a precipitate forms. Furthermore, any precipitate already present will not dissolve.

If SI=0, the solution is just at the balance point of saturation and neither precipitation nor dissolution is expected to occur.

If SI<0, the solution is less than saturated. Precipitates do not form and any precipitates present will tend to dissolve.

Figure 3 indicates that below a pH value of 4.5, solubility products are not exceeded, i.e. all saturation indices are less than zero. In the pH range 4.5-5.4, the only possible solid which might be expected to precipitate is FAP, and in the pH range between 5.4 and 6.3, FAP and HAP precipitates are predicted. As the pH value increases, the solubility limits for successive solid phases are reached as the lines cross the horizontal line of SI=0.

B. Distribution of toothpastes in saliva

Much work has been done on the speciation of metal ions in various toothpaste formulations both in the absence and presence of saliva and across a selected range of pH values.

The effects of time on metal ion speciation can also be mimicked by modelling varying ratios of toothpaste and saliva. This serial dilution of toothpaste with saliva reflects what occurs in the mouth immediately upon brushing and at various times afterwards when the toothpaste has been diluted by fresh saliva and by rinsing.

From these results, it is possible to gain information about the efficacy, bioavailability, and hence, the clinical activity of metal ions in the dentifrice when introduced into the mouth. This can lead to increased therapeutic effectiveness and targeting, to improved toothpaste formulations and hence enhance our general understanding of the complex bioinorganic chemistry involved in these media.

ACKNOWLEDGEMENTS

We would like to thank the Science and Engineering Research Council (U.K), Unilever Research and the British Council for funding these researches.

REFERENCES

Browley, R.E. (1935). *J. Dent. Res.*, *15*: 79.

Curzon, M.E.J., and Cutress, T.W. (1983). *Trace Elements and Dental Diseases*, Wright PSG, Bristol, p. 110.

Davenport, H.W. (1982). *Physiology of the Digestive Tract*, Year Book Medical, Chicago.

Edgar, W.M., and O'Mullane, D.M. (1990). *Saliva and Dental Health*, British Dental Association, London.

Hall, S.B. (1989). *MINTRATE* - a set of subsidiary software for database and input management for the operation of MINEIR, Bioinorganic speciation Group, School of Chemistry and Applied Chemistry, UWCC, Cardiff.

Hurford, S.R. (1989). *Speciation and Trace Metals*, Ph.D Thesis, University of Wales, p. 164.

Jenkins, G.N. (1977). *Physiology and Biochemistry of the Mouth*, Blackwell, Oxford.

Linder, P.W., and Little, J.C. (1986). *Inorg. Chim. Acta*, *123*: 137.

Mandel, I.D. (1974). *J. Dent. Res.*, Suppl. to No. 5, *53*: 246.

McCann, H.G. (1968). *Arch. Oral. Biol.*, *13*: 987.

Nancollas, G.H., Amjad, Z., and Koutsoukoz, P. (1979). *Chemical Modelling in Aqueous Systems*, Symposium Series no. 93, American Chemical Society, Washington, USA.

Schweigruber, M. (1982). Swiss Federal Institute - *MINEQL/IR* was developed for Reactor Research and was documented in two Technical Notes, EIR TM-43-82-38 (1982) and EIR AN-43-84-39 (1989).

Smith, R.M., and Martell, A.E. (1976). *Critical Stability Constants*, Vol. 4, Plenum Press, New York and London.

Westall, J.C., Zachary, J.L., and Morel, F.M.M. (1976). *MINEQL - A Computer Program for the Calculation of Chemical Equilibria Composition for Aqueous Systems*, MIT Technical Note No.18, p. 91.

Williams, D.R., and Wu, L. (1993). *Chem. Spec. Bioavail.*, *5*: 107.

Wu, L. (1992). *Chemical Speciation of Calcium and Salicylates in Biofluids*, Ph. D Thesis, University of Wales.

WOUND EXUDATE

by

Dr Anne E. Howells, BSc (Hons), PhD, CChem, MRSC

and Professor David R. Williams, OBE, BSc (Hons), PhD, DSc, EurChem, CChem, FRSC, FRSA

Department of Chemistry,
University of Wales College of Cardiff,
PO Box 912
Cardiff CF1 3TB
U.K.

I. WOUNDS

Wounds originate from various sources, e.g. surgery, accidental injury, the inability of tissue to withstand pressure (e.g. pressure sores) etc. The normal wound healing process incorporates four phases which are the inflammatory, the migratory, the proliferative and the maturation phases (Turner, 1985).

After injury, during the inflammatory phase, a blood clot is formed and plasma flows into the inflamed area. This plasma contains components necessary for repair and for the prevention of infection, e.g. antibodies, components of the complement system, fibrinogen and various leucocytes (white blood corpuscles). Proliferation of epithelial cells also occurs during this phase.

During the migration and proliferation phases, fibroblasts, which will form the new collagen fibrils, and endothelial cells, which will form capillary buds, migrate into the wound area. Epithelial cells continue to proliferate and migrate and a highly vascularized granulation tissue is thus formed.

Finally, during the maturation phase, normal epithelium covers the wound and it closes completely (Turner, 1985).

II. WHAT IS WOUND EXUDATE?

All wounds emit a fluid known as wound exudate. It is an amorphous matrix or 'ground' substance which envelops the cells of the healing wound.

During the inflammatory phase of healing, vasodilation and changes in the endothelium of venules allows blood plasma to flow into the inflamed area (Turner, 1985). Thus, wound exudate, for the first day after injury, has a composition closely resembling blood plasma.

There then follows a period when its composition is altered due to the metabolism of wound cells. It can therefore be said that wound cells exist in an extracellular environment resulting from their metabolism. However, this environment is also partly responsible for the nutrition, growth and development of live cells in this medium.

III. NUTRITIONAL CONSIDERATIONS

Wound healing, in the absence of a whole body nutritional balance, cannot lead to effective repair. Clinical and animal studies have shown the adverse effects exerted on healing by nutrient deficiency, as well as the beneficial effects of repletion (McLaren, 1992).

The wound space is thought to be rather acidotic and hypoxic during the first stages of repair (Silver, 1980). However, instead of inhibiting healing, in practice these conditions are thought to stimulate the repair process. Invariably, the cause of the injury also damages the localized circulation. Thus, when the local circulation is least able to satisfy, the nutritional demands of the healing process seem the greatest (Niinikoski, 1980).

During the first phases of the repair process, wound exudate is responsible for supplying all the nutritional requirements to the wounded area. Re-establishment of the local circulatory system via capillaries, arterioles and venules occurs in the migration and proliferation phases, 48-72 hours after injury (Turner, 1985).

The composition of wound exudate must therefore include essential trace elements, e.g. those of the first transition series of the periodic table (manganese, iron, cobalt, copper and zinc) in addition to carbohydrates, fatty acids, vitamins, etc. It is necessary to keep such trace elements at the recommended daily amounts (RDAs) to optimize healing.

However, even in the non-wounded hospital patient, trace element deficiency can be common due to:

(i) the administration of pharmaceuticals, the majority of which have electron donor groups, which can avidly complex with metal ions causing trace element imbalances and possibly depletion;

(ii) modern diets and convenience foods which tend to be trace element deficient, especially in the elderly;

(iii) trauma, which can cause depletion of certain trace elements, e.g. zinc in hospital patients.

Therefore wounded patients would clearly benefit from supplementation, first to correct the imbalances of (i) and (iii) and secondly, because of the restricted flow of blood to the damaged area. Such supplementation could take place through the wound exudate.

IV. THE COMPOSITION OF WOUND EXUDATE

Unlike other body fluids such as saliva, blood plasma, cerebro-spinal fluid, etc, little work has been done to characterize wound exudate. Limited and scant information is available about its composition. However, there is no knowledge of metal ion concentrations, ionic strength, pH range, etc.

At the University of Wales in Cardiff a well-established chemical speciation model of blood plasma uses the ECCLES program (Evaluation of Constituent Concentrations in Large Equilibrium Systems) (May, 1980). This model has been adapted to simulate wound exudate at 1, 3, 5, 7 and 10 days post-injury by using the wound fluid characterization data of Caldwell (1990) to provide input total low-molar-mass (l.m.m.) ligand concentrations and the blood plasma model of Duffield et al. (1991) to provide total metal ion concentrations (see Tables 1 and 2).

We have used such models to determine the interaction of wound exudate with wound dressings comprised of alginates in order to determine whether the dressings can be used as a source of beneficial metal ions (Howells and Williams, 1993).

We have based our research upon considering the alginate in two ways:

(i) as soluble sodium alginate, formed by ion-exchange between calcium ions in the dressing and the sodium ions of wound exudate. Alginate interactions with the l.m.m. components of wound exudate have been determined;

(ii) as solid alginate dressing, determining whether trace elements could be released from the alginate dressing into the wound during ion-exchange (see Fig. 1).

V. CONCLUSIONS

In conclusion, wound exudate takes an active role in wound healing, particularly with respect to the nutritional status of the individual. More fundamental studies of wound

Table 1

Total Concentrations (mmol dm^{-3}) of the l.m.m. Ligands in Wound Exudate at 1,3,5,7 and 10 Days after Injury (from Caldwell, 1990)

Ligands	Days after injury 1	3	5	7	10
Taurinate	264	464	414	655	1,149
Aspartate	28	79	89	152	271
Hydroxyprolinate	50	49	70	95	93
Threoninate	207	349	391	455	486
Serinate	175	236	212	226	243
Asparaginate	70	138	140	163	178
Glutamate	231	343	507	758	1,123
Glutaminate	530	700	655	703	753
Prolinate	234	323	348	400	406
Glycinate	248	359	366	446	609
Alanate	521	650	741	888	1,017
Citrullinate	138	68	67	82	87
Valanate	271	416	400	445	461
Cysteinate	50	65	60	59	49
Methionate	83	97	91	102	97
Isoleucinate	117	187	170	182	186
Leucinate	196	341	310	332	353
Tyrosinate	88	176	177	169	182
Phenylalanate	88	141	128	134	143
Tryptophanate	64	76	72	79	80
Ornithinate	52	163	199	219	227
Lysinate	387	687	620	625	652
Histidinate	90	131	123	119	114
Arginate	120	161	91	94	69
Lactate	6,500	5,900	5,100	4,100	4,800

Table 2

Total Concentrations of Metal Ions in Blood Plasma (from Duffield et al., 1991)

Metal ions	Concentration (mmol dm^{-3})
Aluminium	9.0×10^{-8}
Calcium	1.5×10^{-3}
Copper(I)	1.9×10^{-8}
Copper(II)	9.4×10^{-11}
Iron(II)	4.7×10^{-11}
Iron(III)	1.2×10^{-11}
Lead	6.6×10^{-11}
Magnesium	6.7×10^{-4}
Manganese	1.8×10^{-12}
Nickel	1.5×10^{-14}
Zinc	1.6×10^{-7}

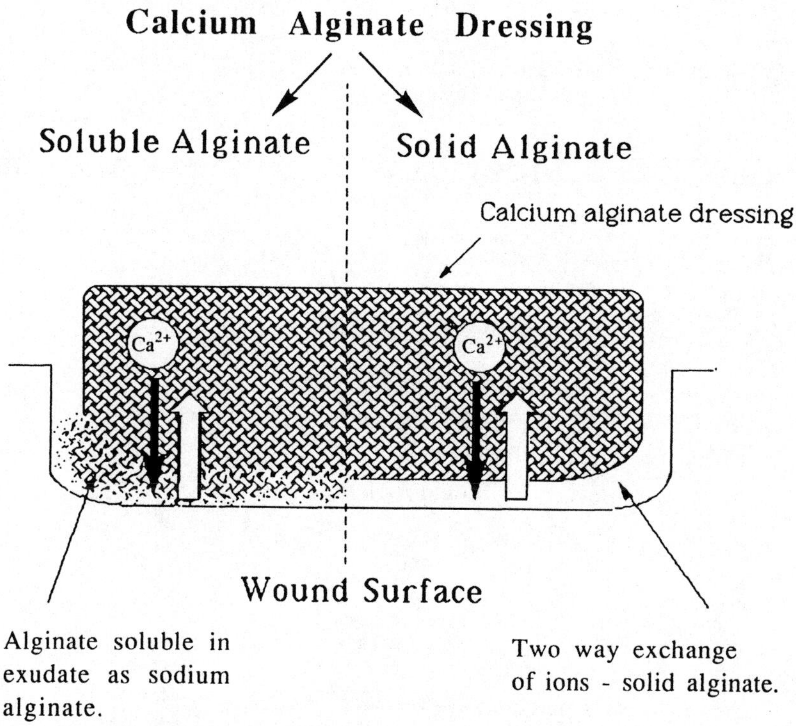

Figure 1.
Alginate interactions with wound exudate.

exudate are necessary, particularly to establish its full composition. In this way it may be possible to supplement trace-element deficient patients topically via the wound exudate, and thus optimize wound healing.

REFERENCES

Caldwell, M.D. (1990). "Clinical and Experimental Approaches to Dermal and Epidermal Repair: Normal and Chronic Wounds", Third International Symposium on Tissue Repair, Miami, Florida.

Duffield, J., Edwards, K., Evans, D.A., Morrish, D.M., Vobe, R.A., and Williams, D.R. (1991). *J. Coord. Chem. (A)*, *23*: 277.

Howells, A.E., and Williams, D.R. (1993). "Alginate Interactions with Trace Metals in Wound Exudate", Proceedings of the 2nd European Conference on Advances in Wound Management, Harrogate, 20-23 October, 1992.

May, P.M. (1980). Ph. D Thesis, University of St. Andrews.

McLaren, S.M.G. (1992). *J. Wound Care*, *1*: 45.

Niinikoski, J. (1980). *Wound Healing and Wound Infection: Theory and Surgical Practice* (T.K. Hunt, ed.), Appleton-Century Crofts, New York, p. 56.

Silver, I.A. (1980). *Wound Healing and Wound Infection: Theory and Surgical Practice* (T.K. Hunt, ed.), Appleton-Century Crofts, New York, p. 11.

Turner, T.D. (1985). *Pharm. Int.*, May, 117.

SPECIATION AS A TOOL FOR CHELATION TREATMENT STRATEGIES

by

Graham E. Jackson

Department of Chemistry,
University of Cape Town,
Rondebosch 7700
South Africa

and

Neil Jarvis

Department of Process Technology,
Atomic Energy Corporation,
Box 582
Pretoria 0001
South Africa

I. INTRODUCTION

The toxic effect of heavy metal poisoning was first reported by Nicander more than 2000 years ago. Pliny the Elder, in 70 AD, knew there were "most pernicious effects" of inhaling mercury. More recently, increased industrialization and its concomitant pollution has increased our levels of exposure to metal ions. Lead, mercury, cadmium and now plutonium are all ever present dangers.

The problems of heavy metal pollution are different from those of organic poisons in that most organic poisons can be detoxified by metabolic or environmental degradation. Thus, cyanide effluent are rendered harmless by the action of sunlight and even DDT is ultimately converted into a less toxic form. Heavy metals, on the other hand, pass on their toxic effects from one organism to another up the food chain. In some instances, the first host even accumulates or converts the metal into a more toxic form, e.g. the "Minimata" disaster in Japan involving high levels of methylmercury in seafood.

Of course life has developed over milieu in the presence of initially toxic metal ions (classified as "detrimental" in this book) which have since become "neutral", or "beneficial", or even "essential" in the course of evolution (see Part Four, Chapter 2-B; see also *Medicine*, Part One, Chapter 1, especially Section C). The advent of industrialization, however, has increased the local concentration and/or converted some of these into more bioavailable and hence toxic forms. The effect of lead poisoning from motor car exhausts is well documented.

With the advent of increased exposure to metal poisons and the clinical recognition of their toxic effects has come the need for treatment-chelation therapy (Jones and Pratt, 1976). Chelate (Greek for "crab's claw") refers to the ability of a ligand to bind to a metal ion at two or more points. Implicit in the term 'chelation therapy' is the recognition that the drug must be a powerful binder of metal ions. The first ligand developed specifically to combat metal poisoning was BAL - British Anti-Lewisite, developed during the 2nd World War as an antidote for arsenic poisoning (Peters et al., 1945). Since then, however, while there have been many attempts, there has been little success with rational ligand design (see Part Four, Chapter 1-B6). Perhaps with the more widespread use of computer simulation of chemical speciation the success rate will increase. In this paper we wish to show how chemical speciation can be used as an aid to rational ligand design. As examples we will look at the design of chelating agents for cadmium, iron and plutonium.

II. DESIGN PRINCIPLES OF CHELATING AGENTS

Because of the nature of metal ion intoxication, the treatment generally falls into two main categories, treatment of chronic poisoning and acute poisoning. In acute poisoning, the body is subjected to a high dose over a short-time period, perhaps only a single dose. The symptoms of poisoning occur rapidly after intoxication and may be quite severe. In this case, if treatment is rapid, the metal ion has to be cleared from the *extra*cellular fluid. Chronic poisoning generally results from a long term, low dose, exposure to the toxic metal. In this case, the symptoms are more progressive and the metal ion has to be cleared from the *intra*cellular space. This makes chronic intoxication far more difficult to detect and treat. In designing a therapeutic chelating agent, then, cognizance has to be taken of the mode of intoxication.

In an elegant paper, Mark Jones (1991) outlined the design of therapeutic chelating agents, highlighting the ligand factors which have to be taken into account (see also Part Four, Chapter 1-B5):

A. Thermodynamic stability

In order to mobilize a metal ion, a chelating agent must be able to form stable complexes with the metal ion under in vivo conditions. Biofluids are highly coordinating and the chelating agent must be able to compete with the endogenous ligands. Coordination must also occur at a specific pH. In general, increasing the basicity of the ligand increases the stability of its metal complex, i.e. often a linear free energy relationship exists between $\log K_{LH}$ and $\log K_{ML}$. However, increasing the basicity may mean that the complex is only formed at higher pH. An example of this is given in the discussion of iron overload where basicity was lowered in order to increase complexation.

B. Selectivity

Biological fluids contain a number of essential metal ions which may be capable of binding to the chelating agent and hence being excreted. This is not such a problem since the essential metal ions are generally present in the diet or, after long-term therapy may be prescribed as dietary supplements. Hence DTPA is generally administered as the calcium salt rather than the free ligand. The main problem with selectivity is that if the ligand binds to an essential metal ion, the effective concentration available for binding to the polluting metal ion is decreased. Ca^{2+} and Zn^{2+} are particularly important here as they are generally present in much higher concentrations than the polluting metal ion. It is for this reason that most attempts at designing therapeutical chelating agents have concentrated on the relative stability of the Ca^{2+} and polluting metal complex. This is an attempt, albeit fairly crude, to take into account in vivo competition.

C. Charge

If the toxic metal ion is to be cleared from intercellular space, a charged complex is desired so that it is rapidly excreted in the urine. Neutral, lipophilic, complexes are to be avoided as they may cause a redistribution of the toxic metal ion into tissue, cf the use of BAL for acute mercury poisoning (Berlin and Lewander, 1965). However, if the metal ion has to be cleared from intracellular space, neutral lipophilic complexes are desired. These should facilitate the passage of the metal ion out of tissue into the intercellular space. From here a second chelating agent could be waiting to form charged complexes with the toxic metal. Before the chelating agent can coordinate to the toxic metal inside the cell, the ligand itself must be able to traverse the cell wall. Here again charge is important, negatively charged ligands being repelled by the phospholipid exterior of the cell membrane. Since chelating agents can in general be protonated, their charge and hence biodistribution will be affected by their in vivo speciation.

In the case of liver deposits of toxic metals, it is possible to form insoluble complexes which are then excreted in the bile. Care must be taken, however, that the excreted metal ion is not reabsorbed from the alimentary canal.

D. Kinetics

For chelation therapy to be effective, the chelate complex must form at an appreciable rate under in vivo conditions. Very often this involves removal of the metal ion from a protein complex. It is for this reason that macrocyclic ligands, which form very stable complexes, are not always successful at complexing metal ions in vivo.

E. Pharmacological factors

Besides complexation factors, effective chelating agents must satisfy a host of pharmacological factors such as toxicity, rate of excretion and resistance to degradation. These are general to all pharmaceuticals and are well documented in the literature.

F. Other factors

The molecular requirements of an effective chelating agent are the same as those for any other pharmaceutical. Biodistribution is controlled by molecular polarity or lipophilicity and molecular size. It has been found that for optimal removal of renal cadmium deposits, a Hansch parameter (Hansch and Leo, 1979) of -3.0 is desirable. Biliary excretion is promoted by increased molecular size, while water soluble low-molar-mass complexes are preferentially excreted in the urine. These and other factors are well documented in the literature.

Of the above factors, the first three are dependent on chemical speciation and so are best evaluated by computer simulation. In the following section we will show how this has been done in the development of chelating agents for cadmium, iron and plutonium.

III. EXAMPLES OF SPECIATION APPLIED TO CHELATION THERAPY

A. Cadmium

Cadmium occurs commonly in base metal ores, fossil fuels and phosphate fertilizers, and as such is commonly present in an industrial environment. The metal is fairly soluble in water, especially at slightly acidic pH values, and so represents a significant threat to health. Tobacco smoke is another possible source of cadmium intoxication. Widespread cadmium intoxication occurred along the Jintsu river in Japan where it became known as itai-itai (Japanese for ouch-ouch). The biological half-life of cadmium is long (~ 20 years) and so it acts as an accumulative poison (Hammond and Foulkes, 1986).

The presence of cadmium in the body stimulates the synthesis of a special cadmium-binding protein, metallothionein. This protein has a high cadmium affinity, making it difficult to mobilize the metal ion once it has become bound. This occurs within 24 h after interperitoneal injection of $CdCl_2$. With time, the cadmium is deposited in the liver and eventually the kidney (Basinger et al., 1981). Thus the treatment of cadmium poisoning is different depending on the time lag between intoxication and treatment.

Using a computer model of blood plasma, Cole et al. (1985) have shown that, upon intoxication, cadmium is distributed amongst several low-molar-mass complexes (Table 1). The predominant complex is the neutral [Cd(cysteinate)] complex which may account for the lack of renal clearance of cadmium from blood plasma. The second most predominant complex [Cd(cystinate)] is also neutral. The affinity of Cd^{2+} for sulphur donor ligands is clearly seen.

Table 1

Predominating Low-molar-mass Cd(II) Complexes in Blood Plasma

Complex formation	Low-molar-mass metal %
Cd(cysteinate)^o	46.4
Cd(cystinate)^o	27.8
Cd(cysteinate)(OH)$^-$	14.3
Cd(cysteinate)$_2$H$^-$	8.4
Cd(cysteinate)$_2^{2-}$	1.9
Cd(cysteinate)$_2$H$_2^o$	0.3
Cd(cysteinate)(histidine)$^-$	0.2
Cd(histidine)$^+$	0.1
Cd(cysteinate)(histidine)H^o	0.1

Jones et al. (1984) have used speciation modelling in the assessment of chelating agents for the removal of cadmium from blood plasma. These authors have used the ECCLES model of blood plasma and the Plasma Mobilizing Index (PMI) to compare the different agents. The results (Fig. 2) compare well with results from animal studies. DTPA is seen to be an efficient mobilizer of Cd^{2+} in vitro, substantiating earlier in vivo animal results. The predominant complex, $[Cd(DTPA)]^{3-}$, 88.7%, is charged however and so DTPA is only able to remove extracellular cadmium (Cantilena and Klaassen, 1981). Unithiol and DMSA, hydrophilic derivatives of BAL, are also only able to mobilize extracellular cadmium (Jones et al., 1978). However, BAL itself is lipid soluble and is effective in mobilizing hepatic deposits of cadmium (Shinobu et al., 1983).

Basinger et al. (1981) tested the efficacy of a series of potential chelating agents in treating acute cadmium intoxication and arrived at the sequence :

$$\text{ZnDTPA} \sim \text{CaDTPA} >> \text{CaTTHA} \sim \text{CaEGTA} \sim \text{CaCDTA} > \text{CaEDTA}$$

This is also the order of thermodynamic stability of the cadmium complexes. The authors noted, however, that a better correlation between efficacy and stability was obtained when comparing not $\log K_{CdL}$ but $\log(K_{CdL}/K_{CaL})$, i.e. when taking into account the relative stability of the Ca^{2+} and Cd^{2+} complexes. The ECCLES blood model automatically takes not only the Ca^{2+} and Cd^{2+} stabilities into account, but also all the other potential coordinating metal ions and ligands. Figure 2 shows the results for a complete simulation of these ligands.

Over the last few years, Hancock and coworkers (Hancock and Martell, 1989) have identified some ligand features which are important in chelation selectivity. In an attempt to design a ligand selective for Pb^{2+} they synthesized the ligand THP-12-aneN_4. While this ligand turned out to be not very selective for Pb^{2+}, it did have a very high selectivity for Cd^{2+}. From the computer simulation results (Fig. 2) we see that THP-12aneN_4 is several orders of magnitude better than DTPA at mobilizing Cd^{2+}. The resulting complex is charged and so the ligand should be able to clear cadmium from extracellular fluid. Preliminary in vivo studies on rats confirm this conclusion (Gulumian et al., 1993).

Results for cadmium clearly show the ability of chemical speciation calculations to explain/predict the observed mobilization of cadmium from extracellular space. Unfortunately, one of the main problems of cadmium intoxication is the rapid deposition of the metal ion in liver and kidney. In order to mobilize cadmium from these stores, a different speciation is required. Not only does the ligand have to enter the storage cell, but also the complex has to exit the cell. BAL, which is fairly lipophilic, has been used to decrease the body burden of cadmium (Cherian, 1980). However, if BAL is given during the acute phase of cadmium poisoning, it can increase the toxicity of the metal by moving it from blood plasma into tissue (Tepperman, 1947).

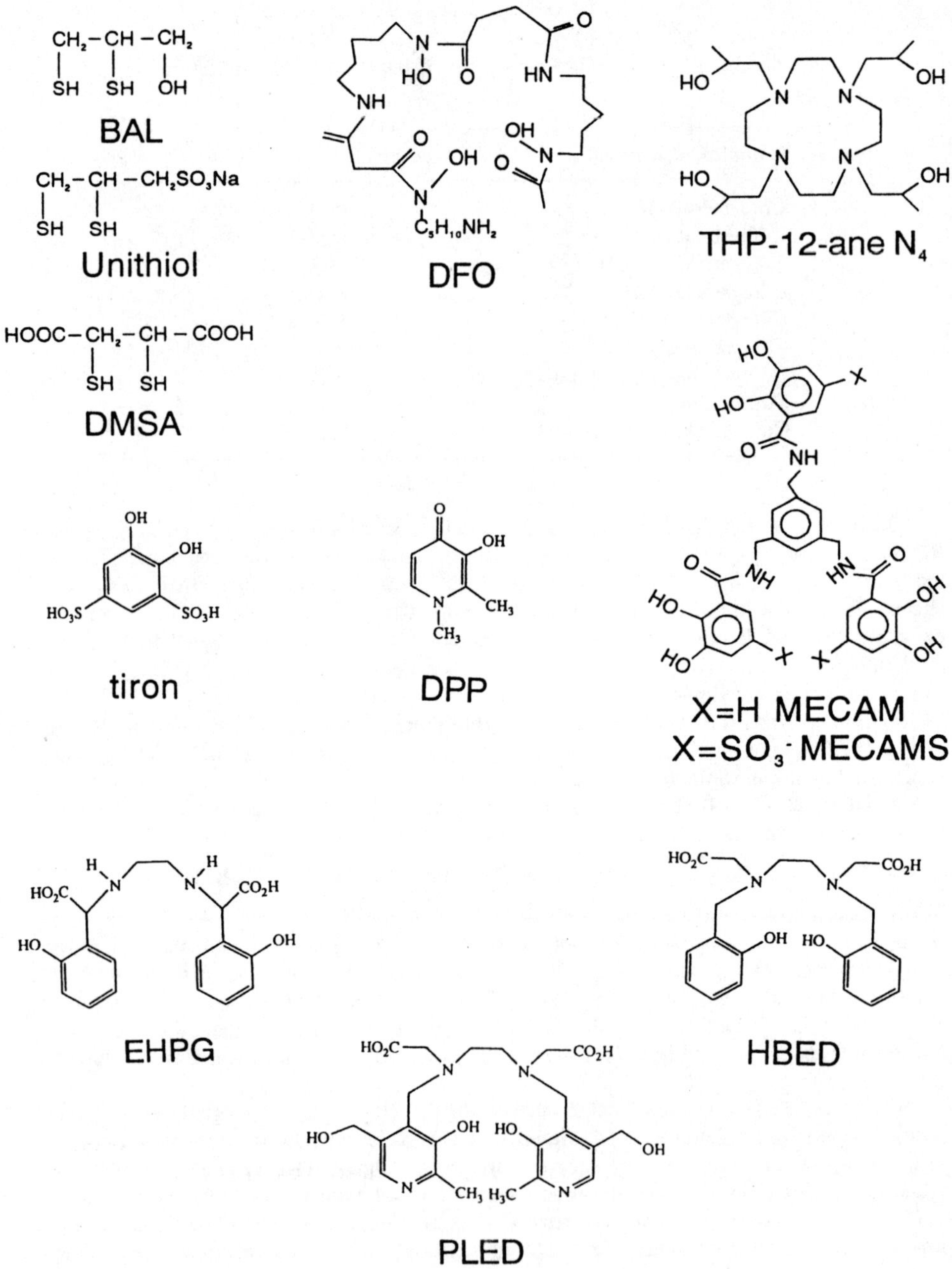

Figure 1.
The ligands.

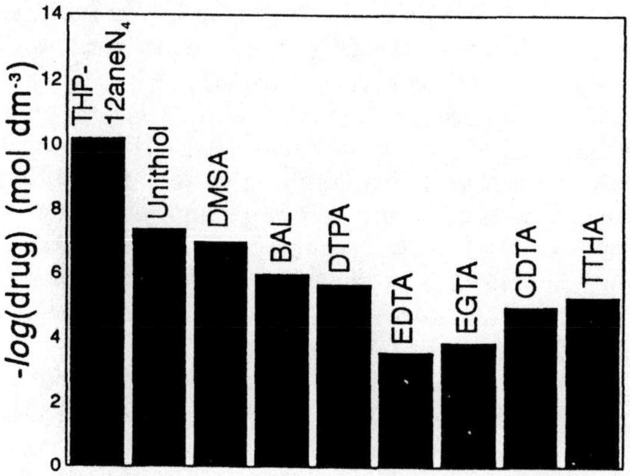

Figure 2.
Concentration of chelating agent needed to increase the low-molar-mass concentration of Cd(II) by a factor of 10.

B. Iron

While cadmium serves as an example of a polluting metal ion which is toxic (classified as "detrimental" in this book; see Part Four, Chapter 2-B), iron serves as an example of an essential element which, if present in excess, is also toxic.

Iron is the most abundant transition metal in man, is essential for oxygen transport and many redox reactions. Most of the metal is found in hemoglobin (65%) or stored in ferritin or hemosiderin (33%) (Bothwell et al., 1979). Iron is transported between different sites by transferrin which, because of its high affinity for Fe^{3+}, maintains a low concentration of free ferric ions in solution. May et al. (1977), using computer simulation, have estimated the concentration of free ferric ions in blood plasma to be $< 10^{-24}$M. This low concentration of free Fe^{3+} is necessary in order to prevent hydrolysis and precipitation of the metal ion.

Because of the low bioavailability of iron and its high biological requirement, humans do not possess an efficient means of excreting iron. In fact, many organisms have evolved means of sequestering iron or increasing its bioavailability. For this reason, the absorption of excessive iron from the diet can be fatal. Actually, one of the most common metal poisonings in children is acute iron intoxication from iron supplement tablets. In adults, acute poisoning is rare but chronic poisoning does occur in patients suffering from certain genetic diseases, idiopathic hemochromatosis, or from oral iron overload. Under normal circumstances, only 5-10% of dietary iron is absorbed. The most convincing example of oral iron overload is provided by the brewing of traditional beer in Africa, where maize is first cooked in iron pots before being left to ferment. Significant quantities of iron are leached from the pots and complex to the carbohydrate in a bioavailable form.

Repeated blood transfusions of patients suffering from certain types of anemia may also lead to iron overload. These repeated transfusions supply iron at a rate of 200 to 250 mg Fe/unit of blood. Upon hemolysis the iron is released and accumulates, particularly in the liver. In some anemias (if not all), there is also an increased absorption of iron. However, it has been found that repeated transfusion does suppress iron absorption.

The treatment of non-anemic iron overload is effectively treated by bloodletting or phlebotomy. However, this treatment is not possible in cases of anemic iron overload. In these cases, a chelating agent is necessary to increase the excretion of iron. The search for a chelating agent which was sufficiently specific for iron was not successful until the discovery of the desferrioxamines. These are hydroxamic siderophores produced by several species of bacteria as iron-sequestering agents. Desferrioxamine B has been used quite successfully in the treatment of iron overload, but it has several disadvantages. It cannot be given orally and so has to be administered by injection which is often painful. Several groups have responded to the challenge of developing new, iron-specific, chelating agents. Raymond et al. (Harris et al., 1980) have based their work on the siderophore, enterobactin, a cyclic triester of (2,3-dihydroxybenzoyl)-l-serine. They have synthesized a series of linear, tripodal and macrocyclic ligands which all have a high affinity for iron. Similarly, Martell et al. (Taliaferro, 1984), have synthesized a series of substituted ethylenediamine ligands which have a high affinity for iron. In both cases, the idea is to produce a ligand which can mobilize iron from transferrin and then to let normal metabolic activity deplete body stores of iron in order to regenerate the transferrin. Alternatively, the chelating agent can be administered at the same time as the blood transfusion. The transfusional iron can then be excreted before it is deposited in the liver. In this way a negative iron balance has been achieved (Seshadri et al., 1974).

Figure 3 shows the relative ability of several chelating agents to remove Fe^{3+} from transferrin. These results have been calculated using computer simulation of blood plasma. It is insufficient just to compare stability constants as account has to be taken of ligand basicity, chelate protonation and competition with endogenous metal ions. In fact, Harris et al. (1980) have designed sulphonated tricatechols which have a lower iron stability but are more effective at coordinating Fe^{3+} at pH 7.4. This is because catechols are such weak acids that, even at pH 7.4, competition with hydrogen ions is important.

The results in Figure 3 show that many of the ligands are thermodynamically able to remove Fe^{3+} from transferrin. EHPG and DPP are relatively modest in their ability to remove Fe^{3+} from transferrin. However, both these ligands have been used to remove iron in vivo, with DPP being hailed as the new orally active challenge to desferrioxamine (Kontoghiorghes, 1991; Gyparaki et al., 1986). The distinction between the different ligands, from a clinical point of view, must rest then on other factors like toxicity, solubility and complexation kinetics.

C. Plutonium

Plutonium is a challenge to the chemical speciationist in that it has a number of available oxidation states, has a very complex aqueous chemistry, is readily hydrolyzed and is hazardous to work with. This latter property restricts the availability of the metal, making

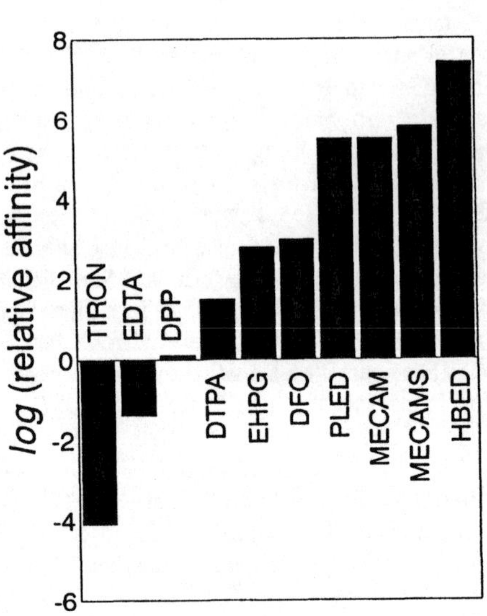

Figure 3.
Relative affinity of a series of chelating agents for iron. The log affinity of transferrin is set at 0.

experimental data scarce and so computer simulation less reliable. However, it also makes computer simulation more essential in order to optimize and minimize the number of experiments.

It has been estimated that with nuclear testing and accidental leakage, some 5000 kg of plutonium have been released into the environment. Plutonium is poorly excreted from the body and most of the assimilated element is deposited in the liver and skeleton. Since the amount of plutonium contamination is likely to be small, the element does not pose a serious chemical threat, but the radiological dangers of the metal deposited in tissue are enormous. Lower limits for plutonium intoxication have now been set (Gerber and Thomas, 1992). Should these be exceeded, therapy to remove plutonium should be commenced.

The evidence to date is that in vivo plutonium exists in the Pu(IV) oxidation state. The metal ion is similar to Fe(III) and is known to be incorporated into ferritin, hemosiderin, and transferrin.

Ligands tested for plutonium decorporation until the early eighties have been listed in a review by Raymond and Smith (1981). Subsequently, Raymond's group have used an interesting "stereognostic" approach in designing modified desferrioxamine B (DFO) ligands. One of these, DFO-HOPO, has been reported to remove 86% of plutonium injected into mice (Freeman, 1985). Presently, DTPA is still the ligand of choice for the treatment of humans contaminated with transuranium elements. This subject has recently been reviewed by Taylor (1991) (see also *Bioinorganic Medicine*, Part Five, Chapter 3-D).

The design of non-toxic, non-carcinogenic, hydroxyalkylamine ligands as selective complexants for larger metal ions has received considerable attention by Hancock and coworkers. These ligands may prove to be important in plutonium chelation therapy (Jarvis and Hancock, 1994). Attractive features are the ligands' ability to suppress hydrolysis as well as selectivity over smaller metal ions or metal ions with a low nitrogen affinity. Thus N,N *bis*(2-hydroxyethyl)glycine (BICIN) was more effective in promoting plutonium excretion than nitrilotriacetate (NTA) because in vivo NTA preferentially coordinates Ca(II) (Foreman, 1953). Recently, formation constants for plutonium complexation with BICIN (Jarvis and Hancock, 1994) and NTA (Nitsche and Becraft, 1992) have been reported.

Attempts to use the blood plasma model of May (May et al., 1977) to predict the in vivo behavior of plutonium have thus far proved unsatisfactory. A study by Duffield et al (1984) using ECCLES was carried out on the ligands EDTA, CDTA, DTPA and Puchel. PMI curves for the polyaminocarboxylic acids were reported. The plutonium database used was based on the similarity between Fe(III) and Pu(IV) and thus Pu(IV) formation constants were estimated by merely adding half a log unit onto Fe(III) constants. In addition, inadequate data on Pu(IV) hydrolysis was available at that time.

A more recent attempt to use ECCLES (Jarvis et al., 1994) to model plutonium speciation in blood plasma has served to highlight problems associated with plutonium speciation modelling in general. When the database of Duffield et al. was updated with recent carbonate and hydrolysis constants obtained by Lierse and Kim (1986), it was found that the carbonate complexes dominated the model. This is not the case in practice. If the carbonate equilibria are rewritten as the hydrolysis of bicarbonate complexes, then the $Pu(OH)_4$ species dominates, with neither DTPA nor transferrin complexation able to compete. This also does not reflect the real situation where a large percentage of plutonium is taken up by the latter. This problem highlights the need to understand the essential role of kinetics in plutonium speciation modelling, especially in relation to transferrin complexation. Empirical observation shows that DFO removes a larger percentage of plutonium than DTPA if administration is prompt. This is presumably due to the lower affinity of DFO for Ca(II). After a few days, treatment becomes ineffective perhaps due to the inability of DFO to compete with transferrin for plutonium. However, DTPA has been shown to be able to mobilize plutonium from transferrin (Duffield and Taylor, 1986) and in practice, it removes moderate amounts of plutonium from animals even after delayed administration.

In summary, much scope remains for the development of speciation modelling for plutonium in vivo. Crucial areas needing to be addressed in the development of software are the updating of formation constant databases, kinetics and the inclusion of solid species.

IV. CONCLUSION

In this paper we have tried to show how chemical speciation can be used to explain/predict clinical observations on chelation therapy. The dangers of metal ion poisoning are ever increasing and so is the need for effective chelating agents. Chemical speciation is a valuable tool to be used in the design of these new chelating agents.

REFERENCES

Basinger, M.A., Jones, M.M., and Shinobu, L.A. (1981). *J. Inorg. Nucl. Chem.*, *43*: 3039.

Berlin, M., and Lewander, T. (1965). *Acta Pharmacol. Toxicol.*, 22: 1.

Bothwell, T.H., Charlton, R.W., Cook, J.D., and Finch, C.A. (1979). *Iron Metabolism in Man*, Blackwell, Oxford.

Cantilena, L.R., and Klaassen, C.D. (1981). *Toxicol. Appl. Pharmacol.*, *58*: 452.

Cherian, M.G. (1980). *J. Toxicol. Environ. Health*, *6*: 379; 393.

Cole, A., Furnival, C., Huang, Z.-X., Jones, D.C., May, P.M., Smith, G.L., Whitaker, J., and Williams, D.R. (1985). *Inorg. Chim. Acta*, *108*: 165.

Duffield, J.R., and Taylor, D.M. (1986). *Handbook of the Physics and Chemistry of the Actinides*, Vol. 3 (A.J. Freeman and C. Keller, eds.), North Holland, Amsterdam.

Duffield, J.R., May, P.M., and Williams, D.R. (1984). *J. Inorg. Biochem.*, *20*: 199.

Foreman, H. (1953). *J. Pharm. Assoc. Sci. Ed.*, *52*: 629.

Freeman, G.E. (1985). *Thesis*, University of California, Berkeley.

Gerber, G.B., and Thomas, R.G. (1992). *Rad. Prot. Dos.*, *41*: 1.

Gulumian, M., Casimiro, E., Linder, P.W., Rama, D.B., and Hancock, R.D. (1993). *Hum. Exp. Toxicol.*, *12*: 247.

Gyparaki, M., Porter, B.J., Huehns, E.R., and Hider, R.C. (1986). *Biochem. Soc. Trans.*, *14*: 181.

Hammond, P.B., and Foulkes, E.C. (1986). *Metal Ions in Biological Systems*, Vol. 20 (H. Sigel, ed.), Marcel Dekker, New York, p. 157.

Hancock, R.D., and Martell, A.E. (1989). *Chem. Rev.*, *89*: 1875.

Hansch, C., and Les, A. (1979). *Substitution Constants for Correlation Analysis in Chemistry and Biology*, John Wiley & Sons, New York.

Harris, W.R., Raymond, K.W., and Weitl, F.L. (1980). *J. Amer. Chem. Soc.*, *103*: 2667.

Jarvis, N.V., and Hancock, R.D. (1994). *Radiochim. Acta*, in press.

Jarvis, N.V., Linder, P.W., and Jackson, G.E. (1994). Unpublished results.

Jones, D.C., Smith, G.L., May, P.M., and Williams, D.R. (1984). *Inorg. Chim. Acta*, *93*: 93.

Jones, M.M. (1991). *J. Coord. Chem.*, *23*: 187.

Jones, M.M, and Pratt, T.H. (1976). *J. Chem. Ed.*, *53*: 342.

Jones, M.M., Weaver, A.D., and Weller, W.L. (1978). *Res. Commun. Chem. Pathol. Pharm.*, *22*: 581.

Kontoghiorghes, G.J. (1991). *Brit. Med. J.*, *303*: 1279.

Lierse, Ch., and Kim, J.I. (1986). *Technische Universität München Report RCM 02286*.

May, P.M., Linder, P.W., and Williams, D.R. (1977). *J. Chem. Soc., Dalton Trans.*, 588.

Nitsche, H., and Becraft, K. (1992). *Transuranium Elements - A Half Century* (L.R. Morss and J. Fuger, eds.), American Chemical Society, Washington, DC.

Peters, R.A., Stocken, L.A., and Thompson, R.H. (1945). *Nature*, *156*: 616.

Raymond, K.N., and Smith, W.L. (1981). *Struct. Bonding*, *43*: 159.

Seshadri, R., Colebatch, J.H., Gordon, P., and Ekert, H. (1974). *Arch. Dis. Child.*, *49*: 621.

Shinobu, L.A., Jones, M.M., Basinger, M.A., Mitchell, W.M., Wendel, D., and Razzuk, A. (1983). *J. Toxicol. Environ. Health*, *12*: 757.

Taliaferro, C.H., Motekaitis, R.J., and Martell, A.E. (1984). *Inorg. Chem.*, *23*: 1188.

Taylor, D.M. (1991). *Handbook of the Physics and Chemistry of the Actinides*, Vol. 6 (A.J. Freeman and C. Keller, eds.), North Holland, Amsterdam.

Tepperman, H.M. (1947). *J. Pharmacol. Exp. Ther.*, *89*: 343.

SPECIATION MODELS IN RELATION TO PHARMACEUTICAL ACTIVITY

by

Montserrat Filella

Département de Chimie Minérale, Analytique et Appliquée,
Université de Genève,
Quai Ernest Ansermet 30
CH-1211 Genève 4
Switzerland

I. INTRODUCTION

A knowledge of metal-ligand speciation has proved to be of great importance in medicine, in particular in the field of trace metal deficiencies and overloads. This chapter has been designed to acquaint the reader with the basic principles of speciation modelling and to provide the conceptual tools needed to evaluate and describe quantitatively metal drug binding phenomena of pharmacological interest.

First, a succinct discussion on the scope and usefulness of modelling in natural sciences is presented in order to provide the reader with the adequate conceptual framework. Since choosing a model first requires choosing a system, that is, a boundary between "in" (system) and "out" (rest of the world), the different compartments in which an organism can be divided when studying the therapeutical action of a drug are briefly presented. The second step for the construction of a model involves choosing the complexity of the "internal" system description. Thus, the basis of the current speciation models is discussed afterwards. The chapter ends with some applications of speciation modelling in the pharmaceutical field and a discussion of their main limitations.

II. MODELLING IN NATURAL SCIENCES

Natural sciences are based on experience and observation and are thus "empirical". Nevertheless, a basic principle of science is that empirical information can be described in terms of mathematical relations. The scientific language is equivalent to the idea of expressing observations by mathematical expressions constructed by a logic formalism. In the most general sense, this procedure can be called "modelling". Nevertheless, it is probably more appropriate to define a model as an imitation of reality which stresses those aspects that are assumed to be important and omits all properties considered to be non-essential (Schwarzenbach et al., 1993).

Models are tools for surmounting the barrier between observing and understanding. The more complicated the situation, the more important is the role of models. Modelling means simplifying intentionally a complex and thus intractable situation. Modelling is a technique which is used in every scientific discipline and by every scientist. A model is not always a complex mathematical system of equations; it can also be a framework of ideas showing relations between system components in a qualitative manner. Table 1 shows different types of models used in natural sciences.

In particular, models can be helpful to the study of natural (biological and environmental) systems for: (i) arrangement of information; (ii) corroboration of hypotheses and formulation of new ones; (iii) sensitivity analysis (exploration of "what if" questions); (iv) design of new experiments; (v) replacement of "true" experiments when full access is impossible, either in time or space, to the phenomena of interest; (vi) prediction. In this respect, prediction is not the only and not even the most important goal of modelling. In recent years, there has been a dramatic increase in the use of the so-called numerical simulation models in science as a means to evaluate large-scale or complex natural processes. In some cases, the predictions generated by these models are considered as a basis for public policy decisions (e.g. global circulation models to predict the behavior of the Earth's climate; geochemical models to evaluate proposed waste disposal sites). This has put an extra emphasis in the predictive value of modelling, often neglecting the fact that the primary value of models is heuristic: models are representations, useful for guiding further study but not susceptible to proof (Oreskes et al., 1994). It is under this perspective that the following should be considered.

III. THE COMPARTMENTAL NATURE OF BODY FLUIDS

To produce a pharmaceutical or therapeutic effect, a drug must reach its site or sites of action in a concentration sufficient to initiate a response. The concentration achieved, while being related to the amount of administered drug, will also depend on the extent and rate at which the drug is absorbed from its site of administration and its distribution by the

Table 1

Types of Models in Natural Sciences
(Imboden, unpublished. With permission)

Conceptual models
- William Harvey (1628) prediction of the existence of capillary blood vessels
- Periodic table of elements
- Alfred Wegener: plate tectonics

Physical (geometrical) models
- Heliocentric planetary model
- Structural models of complex organic molecules (e.g. the double helix of DNA)

Physical (process) models
- All "laws"
- The stagnant boundary model of air/water exchange
- Mass transport by turbulent diffusion

Chemical models
- Thermodynamic models
- Reaction rate models
- Sorption

Biological models
- Population dynamics
- Predator/prey models (Lotka, Volterra)

Statistical (empirical) models
- Any curve fitting

Complex system's models
- Forrester's world dynamics

bloodstream to other parts of the body. The effect of a drug will disappear when the drug is removed from its site of action, either in an unchanged form or after metabolism has taken place giving metabolites which are removed by the process of excretion. Consideration of the compartmental nature of body fluids is essential if a therapeutic effect is to be produced with minimal unwanted effects.

The pattern of drug distribution in the body reflects the interplay of a number of factors relating to permeability, lipid solubility and binding to metal ions and macromolecules. Body fluids can be considered as compartments (percentage of total body weight in brackets): (i) inside cells or intracellular (35%); (ii) outside cells or extracellular: plasma (4%), interstitial fluid that bathes tissues (12%), water in bone and dense connective tissue (4%); (iii) transcellular water present in secretions and lumena (1-5%).

A drug may enter the body by enteral (oral, sublingual, rectal) or by parenteral (intravenous, intramuscular, and subcutaneous injections; inhalation; topical applications) administration. Apart from drugs introduced directly into the systemic circulation by intravenous injection, absorption from the site of administration is essential if the drug is to gain entry to the bloodstream and thus reach its site of action. For the process of absorption to take place, a drug must cross at least one cell membrane. The main mechanisms of transfer of substances across cell membranes are: (i) passive diffusion (direct passage through lipid or aqueous channels of cell membranes down a concentration gradient) and (ii) carrier-aided transfer (either facilitated diffusion or active transport). Most drugs are transferred across membranes by passive diffusion. As well as drug absorption, the processes of distribution, metabolism and excretion likewise involve the passage of drugs or their breakdown products across cell membranes.

When receptors lie on the exterior of cell membranes, diffusion of the drug into the interstitial fluid is sufficient for its effectiveness. For drugs with an intracellular site of action, passage through the cell membranes of the tissue is also necessary. For "small" drugs ($M < 0.600$ kg mol^{-1}), transfer from plasma to the interstitial fluid is rapid, the capillary wall behaving as a leaky sieve. This constitutes an important route for polar compounds, while for lipid-soluble compounds, both this route and diffusion through the actual capillary wall are possible. Uptake by cell membranes is mostly passive and, therefore, lipophilicity is important.

IV. BODY FLUIDS AS MULTICOMPONENT ENVIRONMENTS

The chemical composition of any biological fluid is highly complex; it is possible to find in it most of the elements of the periodic table, plus a vast number of organic components. Innumerable interactions can occur in such a mixture, yielding an enormous number of products often present at very low concentrations. The healthy functioning of the organism is dependent on the continued maintenance of an infinite array of aqueous chemical reactions within certain limits. The evaluation of both the clinically useful and harmful effects of any drug must be considered in such a competitive environment.

It is increasingly being recognized that an understanding of the chemistry and biochemistry of biological systems is only possible with the knowledge of the physical and chemical forms or species in which compounds (either metal ions or organic components) are present. The identification and quantitation of these forms is the topic of speciation. There is a certain confusion concerning the meaning of the term speciation. Rigorously speaking, the term speciation is synonymous with the species distribution of a compound. It has, however, become an ambiguous term because it is often used to designate not only the distribution, but also the group of analytical operations which permit its determination.

A. Effect of pH

Since many drugs are weak electrolytes, they are partly dissociated in solution. The simplest example of the relevance of the speciation concept in the pharmaceutical sciences is provided by the discrimination of a drug into its different forms as a function of pH. The lipid solubility of the neutral form of an organic acid or base is generally several orders of magnitude higher than that of the charged species. The interrelationship between the dissociation constant, pH of the medium and lipid solubility of a drug often plays an important role in determining its pharmaceutical activity since it affects its ability to cross membranes.

The application of these concepts may be traced back to the early part of this century, a typical example being the mesurement of the pK_a values of aspirin. A solution of the weak acid aspirin (pK_a = 3.5) in the stomach at pH = 1 will have more than 99% of the drug in the unionized form. Since this form is lipid soluble, the drug may be expected to be rapidly absorbed in the stomach. It should not be forgotten, however, that absorption is often complicated by other factors. For instance, in this case, two thirds of an orally administered dose of aspirin is being absorbed from the intestine. This is due to the limited time for which the drug is present in the stomach, its higher solubility at intestinal pH, and the larger surface area available for absorption.

Differences in pH between different body compartments may influence drug distribution. For instance, intracellular pH is slightly lower than extracellular pH (approx. 7.0 and 7.4, respectively) and this fact may partly account for the occurrence of relatively higher intracellular concentrations of basic compounds. Similarly, it has been observed that several basic drugs and endogenous amines achieved higher concentration in pulmonary tissue and part of the explanation may be related to the low intracellular pH (6.8 - 6.5) reported for this organ. Alteration of pH gradients as the result of disease should also be considered. Non-steroidal antiinflammatory drugs might accumulate in inflamed areas due to changes in pH at sites of inflammation (Luscombe and Nichols, 1988).

B. Protein binding

The binding of drugs to macromolecules (largely proteins) is usually a reversible process and it affects the concentration of freely diffusible drug available because the high-molar-mass (h.m.m.) complex cannot cross the cell membrane. For most drugs, the free fraction in plasma is essentially constant in practice over the range of therapeutic plasma concentrations. It is generally accepted that the pharmacological activity is related to the concentration of the unbound drug and that this is also the form that is eliminated by drug-metabolizing/excretory organs (Jusko and Gretch, 1976; Øie, 1986).

C. Metal binding

For good health, all the essential metal ions must be maintained at optimum levels. Given that most drugs have electron donor groups, drug administration is very likely to interfere with normal trace element distribution regardless of metal binding being characteristic of the drug's activity or not.

The distribution of metals varies from body compartemnt to body compartment, but in general each metal may be found in four different forms: firmly and inertly complexed to protein, complexed to protein in a labile equilibrium, as low-molar-mass (l.m.m.) complexes, and as free (hydrated) metal ion. The last three are in a state of labile equilibrium. The biological processes involve a lively competition for complexing between different ligands and different metals in vivo.

Understanding the interactions between ligands and metal ions in biological environments is not straightforward. Nevertheless, broad chemical principles for predicting the strength of the binding between a given metal and a given ligand exist and are presented in other chapters of this book. The in vivo stability of a metal complex is affected by its stability constant, hydrogen ion concentration, competition for and by other metal ions and ligands, tendency of the metal to form insoluble hydroxides, and distribution and metabolism of the complex.

The great diversity and the low concentration of most of the species present in the different compartments of any biological system present a formidable challenge to the analytical chemist. Full speciation analysis of a biological system lies far beyond the capabilities of the more sophisticated analytical equipment (Florence, 1989). The reasons are the many thousands of species present, their extremely low concentration, and the lability of the complexes formed.

Although reactions in living systems are never truly at equilibrium, they are often in a steady state that closely approximates equilibrium conditions. The concentrations of the species formed can thus be estimated by thermodynamic calculations. Just as the absorption of aspirin from the stomach may be assessed with a simple manual calculation involving the pK_a value of the ligand, a computer is needed to combine the effects of thousands of equilibrium constants when all interactions are simultaneously considered. These mathematical speciation models will be discussed in detail below.

V. SPECIATION MODELLING IN RELATION TO PHARMACEUTICAL ACTIVITY

Schubert's "effective stability constant" (Schubert, 1955; 1961) constitutes the first attempt to apply thermodynamic calculations to the selection of chelating agents to bind toxic metal ions beyond the mere correlation of the physiological response with the 1:1 metal-ligand formation constant. Schubert's parameter attempted to take into account the most important competitive chemical interactions occurring in plasma (particularly with respect to the proton and some other metal ions), but too many simplifying assumptions were necessary as the number of interactions increased. Many of these difficulties were overcome by the development of computer methods which could incorporate multiple equilibria among biofluid constituents.

A. Theoretical calculation of complex distributions

Thermodynamic speciation models utilize the equilibrium constants or the minimization of the Gibbs free energy, both being thermodynamically interrelated, to determine the distribution of the aqueous species formed in a given system. Both approaches are subjected to the constraints of mass balance and chemical equilibrium. Basically, the mass balance constraint requires that the sum of the uncomplexed element and its derived complexes equals the initial concentration of the element. The equilibrium constraint requires that the most stable (minimum Gibbs free energy) state be computed via the equilibrium constants for all mass action expressions or via Gibbs free energy values for all components and derived species.

In general, the computational modelling of large equilibrium systems uses the formation constant approach. Formation constants are utilized for all of the species which can be considered likely to form. These are then substituted into the mass balance equations for the compartment concentrations and the resulting set of non-linear equations solved simultaneously. Such models require as input data the total (or free) concentrations of the constituent components, accurate values for the metal-ligand formation constants applicable to the temperature and ionic strength of the media under investigation, solubility products for all solids and, where applicable, parameters such as redox potentials.

Pioneer work by Sillén (Sillén, 1961; 1967), Garrels (Garrels, 1960) and Stumm (Stumm and Morgan, 1970) in the geosciences and Perrin (Perrin, 1965; Hallman et al., 1971; Perrin and Agarwal, 1973) in the biological field established the conceptual basis and set the framework for the computer models used today.

B. The ECCLES model

The most widely used computer model in the pharmaceutical field has been ECCLES (Evaluation of Constituent Concentrations in Large Equilibrium Mixtures) (May et al., 1977; Berthon et al., 1978). The ECCLES code and database were developed to simulate the l.m.m. distribution of exogenous metals in human blood plasma. Blood plasma has a reasonably well-defined composition and contains thousands of complex labile equilibrium interactions. The independence of the percentage distribution of the transition metal ions amongst the l.m.m. ligands with respect to metal protein binding allows the omission of metal-protein equilibria in the calculations (May et al., 1976; see also May and Bulman, 1983). The program considers the total concentrations of ligands and the free metal ion concentrations together with all the relevant thermodynamic formation constants to calculate equilibrium concentrations which satisfy all the mass balance equations.

Presently, the ECCLES database contains a set of more than 10,000 formation constants (at physiological conditions of temperature (37 °C) and ionic strength (0.15 mol dm^{-3})). The most relevant complexes thought to be formed in blood plasma have been repeatedly checked experimentally (see Section A1 of this Chapter 3). The assessment of the influence of administered agents on the distribution of metal ions implies the constants for the exogenous ligands under study to be included in the database. This requires critical selection of the necessary formation constants from the literature, their estimation on chemical and/or statistical grounds, or, ideally, their experimental determination. The twin program MIX (see Section A1 of this Chapter 3) includes facilities to generate estimates of the equilibrium constants for the many mixed (ternary) complexes that might be important in the blood plasma mixture but for which no data is available in the literature.

When describing the effects of exogenous chelating agents, the Plasma Mobilizing Index (PMI) concept proposed by May and Williams (1977) has proved very useful. The efficacy of a chelating agent for mobilizing a metal ion from the labile metal-protein complex pool in blood plasma is expressed as the ratio between the total calculated concentration of l.m.m. metal complex species in the presence of the drug and the total concentration of l.m.m. metal complex species in normal plasma. PMI curves where log PMI is plotted vs minus log drug concentration allow visual comparison of the ability of different drugs to mobilize a metal in plasma. An example is shown in section V.A.1.

C. Other speciation computer codes

Computer codes not specifically developed for biofluid simulation have also been used in pharmacological studies. They have been mostly applied to the simulation of body compartments other than blood plasma, mainly the gastrointestinal tract where formation of precipitates (not considered in ECCLES) could be of importance. They range from general complex distribution programs such as Haltafall (Ingri et al., 1967), COMICS (Perrin and Sayce, 1967), SOLGASWATER (Eriksson, 1979), ESTA (May and Murray, 1988) and SPE (Martell and Motekaitis, 1988) to widely-used geochemical models such as MINEQL (Westall et al., 1976). Other programs have been specifically designed for particular applications such as NEUPLOT (Berthon, 1985), which allows the percentage of the neutral fraction of each reactant to be plotted as a function of pH, and TRIPLOT (Berthon, 1987) which does so in three dimensions as a function of the concentration of one of the reactants.

VI. APPLICATIONS

Many drugs contain functional groups capable of binding trace metals in vivo. Nevertheless, before their pharmacological action and/or their adverse reactions can be definitely traced to chelation, it must be ascertained that the affinity of the drugs for metal ions is sufficiently high to overcome the strong competition of endogenous ligands. A significant amount of work has been devoted to the quantitative prediction by computer simulation of the relative capacity of drugs either to induce significant changes in the metabolism of essential metals in vivo or to mobilize toxic metal ions. These studies usually involve: (i) experimental determination of formation constants for all possibly relevant complexes formed under the conditions under investigation, usually by means of potentiometric techniques; (ii) monitoring of the distribution of the l.m.m. pool of essential (i.e. zinc and copper) and detrimental metal ion complexes in the presence of therapeutic levels of the drug by using a computer model such as ECCLES; (iii) calculation of a surrogate variable such as the percentage of neutral complexes in the gastrointestinal tract or the PMI in blood plasma.

A. Blood plasma simulation

Possible studies include the evaluation of both desired and undesired drug effects related to metal binding.

1. *Pharmacological activity*

The relative ability of chelating agents to cause excretion of a given detrimental metal ion plus the concomitant loss of certain essential trace metals have been widely assessed by computer simulation. The ability of current speciation models to screen promising chelating agents for the treatment of metal poisoning and to explain the failure of existing chelating agents is very remarkable. This item is thoroughly analyzed in Section B of this Chapter 3. The main reported applications of speciation modelling in relation to pharmaceutical activity are described in detail in Sections D to G of this Chapter.

Computer simulations may help to identify the compounds causing the pharmacological action. For instance, some anticancer compounds whose biological activity depends on the formation of a metal complex have been studied by computer simulation (Huang et al., 1982; 1985). An interesting group of anti-tumor compounds centers upon ICRF 198 or Raxozane (bisdioxopiperazine). This cyclic diimide is lipophilic in blood plasma thus diffuses through cell membranes. Hydrolysis products of this drug produce EDTA-like molecules that seek out essential metals and so stop cell growth. Nevertheless, there are few clues as to metal identity (Huang et al., 1982).

Berthon and co-workers have applied speciation modelling to the zinc-mediated tissue diffusion of histamine in plasma (Kayali and Berthon, 1980a; 1980b; 1982; Blais et al., 1981; Berthon and Kayali, 1982; Berthon and Germonneau, 1982; Berthon et al., 1987). Dysfunction of histamine release from its storage sites results in various health disorders, particularly, generalized excessive histamine release may cause an anaphylactic shock, with the risk of cardiac arrest. Catabolism of histamine in tissues is prevented by the polarity of the molecule. Computer simulation allowed the identification of the neutral zinc-histamine-cysteine as being the most plausible plasma diffusible ligand (Berthon and Kayali, 1982) and the assessment of different ligand partners for zinc and histamine with the objective of enhancing histamine elimination from plasma.

In some cases, computer simulation has proved useful to reject some hypotheses. For example, the chelation of metal ions had been suggested to play a significant role in the mode of action of isoniazid, but computer simulation results clearly indicates that labile l.m.m. complexes in blood plasma were highly unlikely to account for the pharmacological activity of the drug in vivo (Cole et al., 1983). Similarly, computer modelling helped to reject the hypothesis that a plasma copper-albumin complex is the species excreted when D-penicillamine is administered in the treatment of Wilson's disease (Furnival et al., 1981).

2. *Adverse effects*

The assessment of causality of adverse drug reactions is a very difficult problem. Examples of application of computer simulation to tackle this problem include the evaluation of antihypertensive agents such as hydralazine and prizidilol (Al-Falahi et al., 1984), captopril (Hughes et al., 1985b; Christie et al., 1988), enalapril maleate and lisinopril (Balman et al., 1988a), enalaprilat (Hughes et al., 1985a), and thiopronin (Filella and Williams, 1985); anticonvulsants such as valproic acid (Balman et al., 1988b); antimucolytics such as carboxymethylcysteine (Brumas et al., 1989) and N-acetylcysteine (Brumas et al., 1992); antitubercular agents such as ethambutol (Cole et al., 1981); antitumor agents such as sodium thiazolidine-4-carboxylate and 2-amino-2-thiazoline hydrochloride (Huang et al., 1981); and the histamine H_2-receptor antagonist cimetidine (Akrivos et al., 1984).

Agents which themselves are not chelating agents but which can metabolize into possible chelating agents have also been scrutinized (Cole et al., 1981; Freijanes and Berthon, 1986). Computer simulation allowed Cole et al. (1981) to deduce that the zinc-related side effects of the antitubercular agent ethambutol ((+)-2,2'-(ethylenediimino)-di-1-butanol) (i.e. visual abnormalities) were more likely due to a particular metabolic degradation product (EDBA, 2,2'-(ethylenediimino)-di-butyric acid) than to the parent molecule itself. The PMI curves for ethambutol and EDBA together with those other ligands are shown in Figure 1. The figure shows that EDBA has similar PMI values to those of known pharmaceutical chelating agents such as D-Pen and EDTA. Doses of 25 mg ethambutol per kg body weight which produce an EDBA concentration of 1.25 10^{-6} mol dm^{-3} raise the zinc l.m.m. species concentrations by up to one third, the Zn(EDBA) complexes being neutral. When protracted therapy with essential metal mobilizing ligands is used, topping-up therapy is desirable in order to keep the patient in metal balance. Computer simulation may indicate the extent of metal supplementation required during drug therapy.

B. Gastrointestinal interactions

The most common method of drug administration is by swallowing. Metal imbalances may result from impaired gastrointestinal absorption due to antagonistic interactions with chelators in the gastrointestinal tract. Reciprocally, metal ions may influence drug

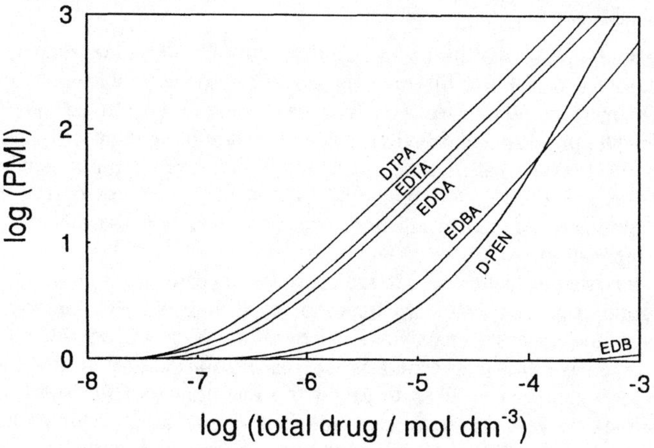

Figure 1.
Curves for log Plasma Mobilizing Index (PMI) *versus* -log drug concentration for zinc(II) (Cole et al., 1981). DTPA = diethylenetriaminepentaacetic acid, EDTA = ethylenediaminetetraacetic acid, EDDA = ethylenediiminodiacetic acid; EDBA = see text; D-PEN = D-penicillamine; EDB = ethambutol (see text).

bioavailability through undesirable associations. As already mentioned, the electrical charge of the species formed in solution constitutes an obvious impediment to their transport across biological membranes. The physico-chemical properties of a drug greatly influence the rate and extent of its absorption. Dietary or administered agents which produce electrically neutral and/or lipophilic complexes help absorption, while h.m.m. ligands that bind metals strongly, and ligands that form electrically charged or insoluble species, hinder it.

An early application of computer modelling concerns the simulation of in vivo conditions for the "colloidal bismuth subcitrate" (CBS) sold as De-Nol (Williams, 1977). This product is very widely used clinically in the treatment of peptic ulcer disease. On acidification of De-Nol to stomach pH, two precipitates (bismuth oxychloride and bismuth citrate) cover the ulcer craters. The protective coating so formed may prevent back-diffusion of H^+ and hence promote reepithelialization.

Potential changes in the bioavailability of essential metals when simultaneously ingested with a certain drug may be easily estimated on the basis of appropriate simulations. Supposing that the metal ion has been liberated from its original matrix, which, under specific conditions such as in presence of soya proteins, may necessitate the use of powerful chelating agents (Makar and Williams, 1977), its subsequent association with anionic substances is then desirable so as to give rise to membrane-diffusible complexes. Several attempts to apply these principles to the oral administration of various metal ions have been reported (Cape et al. 1974; Makar et al., 1976; Roos and Williams, 1977). Particular mention should be made of the application of computer simulation to the development of suitable potential carriers for enhancing magnesium bioavailability (Blaquière and Berthon, 1987).

Recently, risks associated with aluminium bioavailability from its insoluble salts used as antacids or phosphate-binding agents, either alone or in the presence of common dietary acids, have been studied in detail (Venturini and Berthon, 1989; Daydé and Berthon, 1990; Daydé et al., 1990; Duffield et al., 1991). Although this subject is reviewed in Section G of this Chapter 3, these studies are cited here because they constitute an excellent example of how speciation through computer simulation can be applied.

C. Other biofluids

Efforts to expand blood plasma models to other biofluids are scarce (May et al., 1982; Smith et al., 1985; Findlow et al., 1990; Duffield et al., 1991; see also Section A3 of this Chapter 3). Recently, computer simulation has been used to assess the role of zinc (Christie et al., 1991) and tin (Duffield et al., 1990) as therapeutic ingredients of dentrifices.

VII. LIMITATIONS

A clear understanding of the limits of applicability of speciation models and of the uncertainties involved in the use of currently available models is essential to their correct use and to the interpretation of results. The key point to be stressed is that chemical thermodynamics just provides information on the direction of spontaneous change and that, as with any thermodynamic calculation, speciation results can obviously not be taken as a proof of the formation of a given compound. The main limitations of this approach are listed below. Some are common to all thermodynamic models while others are specific to this particular application.

(i) *Lack of thermodynamic data*. Results are critically dependent upon the species included in the computations and it is thus very important to include all possible species. In cases where formation constants are unavailable for species which are considered likely to be formed, it is recommended that estimates be made from trends in the literature. The resulting speciation can then be used to assess the importance of the species in question. Should the complex prove to be significant, the formation constant must be experimentally determined. In any case, the most reliable thermodynamic property is always the one obtained by the most direct path, i.e. the one closest to the actual measurement.

(ii) *Low quality of available thermodynamic data*. Caution should be exercised when acquiring values for formation constants from the literature since in many cases the reported constants are of questionable validity. In addition, in most situations, literature values will be applicable to conditions of temperature and ionic strength other than those under consideration and must be corrected to the conditions of interest.

(iii) *Complexation kinetics is not considered*. The hypothesis that thermodynamic equilibrium is reached is not always justified. Hydrated metal ions found in biological systems are highly labile, but this is not the case when they occupy binding sites in macromolecules. Kinetics of reaction of metal ion complexes of serum constituents such as transport proteins like albumin and transferrin with the administered drugs in plasma may be slow. For example, it is well known that the slow reaction of Desferrioxamine B (DFO) with iron bound to transferrin severely limits the ability of this compound to mobilize iron (Peter, 1985).

(iv) *Only l.m.m. complexing equilibria are considered*. This is because labile metal-protein interactions are still too poorly characterized in a thermodynamic sense. This is not a serious limitation since l.m.m. species act as important intermediates in the transport of metal ions between macromolecules and across biological membranes. In the ECCLES approach, the assumption is made that the free metal-ion concentration is perfectly buffered by the protein-metal interaction. This approach requires good estimations of the free metal ion concentrations in plasma which are not always available.

(v) *Little attention is paid to the effects of redox equilibria and solubility*. The partition of some metals and ligands among their oxidation states (e.g. copper and cysteine) are not yet adequately characterized.

(vi) *The defined system is not a closed system*. Most of the drugs employed (and their metal complexes) are able to permeate blood plasma into intracellular space and they do not remain confined to the blood plasma compartment. Changing ligand binding by proteins and drug metabolism are also possible. Some attempts to include drug kinetics have been made (Jones and May, 1987). These authors conclude that all the kinetic factors considered, when comparing chelating agents relative efficacies, tend to reduce the efficacy of the chelating agents compared with that which might be expected on the basis of thermodynamic considerations alone. These models need to be taken much further (see Section I of this Chapter 3).

(vii) *The multicompartmental nature of biological systems is not adequately taken into account*. Most of the reported studies refer only to blood plasma and gastrointestinal fluid. No attempt has yet been made to couple speciation and transport models, presumably because of the high conceptual and mathematical complexity involved.

VIII. CONCLUSIONS

The preceding discussion is not intended to leave the reader with the feeling that little can be done with speciation modelling in the pharmaceutical field. On the contrary, computer modelling is always useful in refining knowledge and testing hypotheses. As pointed out in the introduction, prediction is not the only, and not even the most important, goal of modelling. Experience in speciation calculations applied to pharmaceutical development clearly shows that their enormous potentiality for hypothesis testing is probably their best quality.

Over the past several decades, speciation models and resultant codes have evolved greatly in areas where their application is more generalized than in biomedical sciences. Advancements include evaluating model sensitivities, improvement of the thermodynamic and kinetic databases, and the introduction of organic macromolecules to existing codes. The resulting changes have greatly enhanced the flexibility and utility of most aqueous geochemical codes. Unfortunately, a parallel development has not been observed in the biomedical field. There is, in principle, no reason why better models cannot be developed. It is to be hoped that the situation will change in the near future. Particularly, if it is considered that, at present, a great deal of experimental evidence has accumulated in support of the idea that trace element balance by itself can be used to combat disease. Unfortunately, this possibility has been up to now largely unexplored. Computer simulation could prove an indispensable tool to realize the full potential of this approach.

REFERENCES

Akrivos, F., Blais, M.-J., Hoffelt, J., and Berthon, G. (1984). *Agents Actions, 15*: 649.

Al-Falahi, H., May, P.M., Roe, A.M., Slater, R.A., Trott, W.J., and Williams, D.R. (1984). *Agents Actions, 14*: 113.

Balman, J.A., Christie, G.L., Duffield, J.R., and Williams, D.R. (1988a). *Inorg. Chim. Acta, 152*: 81.

Balman, J.A., Christie, G.L., Duffield, J.R., and Williams, D.R. (1988b). *Inorg. Chim. Acta, 151*: 169.

Berthon, G. (1985). NEUPLOT. Unpublished program.

Berthon, G. (1987). TRIPLOT. Unpublished program.

Berthon, G., and Germonneau, P. (1982). *Agents Actions, 12*: 619.

Berthon, G., and Kayali, A. (1982). *Agents Actions, 12*: 398.

Berthon, G., May, P.M., and Williams, D.R. (1978). *J. Chem. Soc., Dalton Trans.*, 1433.

Berthon, G., Varsamidis, A., Blaquière, C., and Rigal, D. (1987). *Agents Actions, 22*: 231.

Blais, M.-J., Kayali, A., and Berthon, G. (1981). *Inorg. Chim. Acta, 56*: 5.

Blaquière, C., and Berthon, G. (1987). *Inorg. Chim. Acta, 135*: 179.

Brumas, V., Venturini, M., Filella, M., and Berthon, G. (1989). *J. Inorg. Biochem., 37*: 309.

Brumas, V., Hacht, B., Filella, M., and Berthon, G. (1992). *Agents Actions, 36*: 278.

Cape, J.N., Cook, D.H., and Williams, D.R. (1974). *J. Chem. Soc., Dalton Trans.*, 1849.

Christie, G.L., Hughes, M.A., Rees, S.B., and Williams, D.R. (1988). *Inorg. Chim. Acta, 151*: 215.

Christie, G.L., Cummins, D., Duffield, J.R., Hurford, S.R., Morris, C.R., Riley, P.I., Vesey, J.A., and Williams, D.R. (1991). *J. Inorg. Biochem., 42*: 273.

Cole, A., May, P.M., and Williams, D.R. (1981). *Agents Actions, 11*: 296.

Cole, A., May, P.M., and Williams, D.R. (1983). *Agents Actions, 13*: 91.

Daydé, S., and Berthon, G. (1990). *Food Add. Contam., 7 Suppl. 1*: S155.

Daydé, S., Filella, M., and Berthon, G. (1990). *J. Inorg. Biochem., 38*: 241.

Duffield, J.R., Morris, C.R., Morrish, D.M., Vesey, J.A., and Williams, D.R. (1990). *Tin-based Antitumour Drugs* (M. Gielen, ed.), NATO ASI Series, Ser. H, 37 (Tin-Based Antitumour Drugs), Springer-Verlag, Berlin, p. 147.

Duffield, J.R., Edwards, K., Evans, D.A., Morrish, D.M., Vobe, R.A., and Williams, D.R. (1991). *J. Coord. Chem.*, *23*: 277.

Eriksson, G. (1979). *Anal. Chim. Acta*, *112*: 375.

Filella, M., and Williams, D.R. (1985). *Inorg. Chim. Acta*, *106*: 49.

Findlow, J.A., Duffield, J.R., and Williams, D.R. (1990). *Chem. Spec. Bioavail.*, 2: 3.

Florence, T.M. (1989). *Trace Element Speciation: Analytical Methods and Problems* (G.E. Batley, ed.), C.R.C. Press, Boca Raton FL, p. 319.

Freijanes, E., and Berthon, G. (1986). *Inorg. Chim. Acta*, *124*: 141.

Furnival, C., May, P.M., and Williams, D.R. (1981). *Trace Elements in the Pathogenesis and Treatment of Inflammations* (K.D. Rainsford and K. Brune, eds.), Birkhäuser Verlag, Basel, p. 241.

Garrels, R.M. (1960). *Mineral Equilibria at Low Temperature and Pressure*, Harper and Row, New York.

Hallman, P.S., Perrin, D.D., and Watt, A.E. (1971). *Biochem. J.*, *121*: 549.

Huang, Z.-X., May, P.M., Williams, D.R., and Gosálvez, M. (1981). *Inorg. Chim. Acta*, *56*: 41.

Huang, Z.-X., May, P.M., Quinlan, K.M., Williams, D.R., and Creighton, A.M. (1982). *Agents Actions*, *12*: 536.

Huang, Z.-X., Creighton, A.M., and Williams, D.R. (1985). *Inorg. Chim. Acta*, *107*: L29.

Hughes, M.A., Smith, G.L., and Williams, D.R. (1985a). *Inorg. Chim. Acta*, *107*: L11.

Hughes, M.A., Smith, G.L., and Williams, D.R. (1985b). *Inorg. Chim. Acta*, *107*: 247.

Ingri, N., Kakolowicz, W., Sillén, L.G., and Warnqvist, B. (1967). *Talanta*, *14*: 1261.

Jones, M.M., and May, P.M. (1987). *Inorg. Chim. Acta*, *138*: 67.

Jusko, W.J., and Gretch, M. (1976). *Drug Metab. Rev.*, *5*: 43.

Kayali, A., and Berthon, G. (1980a). *J. Chim. Phys.*, *77*: 333.

Kayali, A., and Berthon, G. (1980b). *J. Chem. Soc., Dalton Trans.*, 2374.

Kayali, A., and Berthon, G. (1982). *Polyhedron*, *1*: 371.

Luscombe, D.K., and Nichols, P.J. (1988). *Smith and Williams' Introduction to the Principles of Drug Design*, 2nd ed., Wright, London, p. 1.

Makar, G.K.R., and Williams, D.R. (1977). *J. Inorg. Nucl. Chem.*, *39*: 201.

Makar, G.K.R., Touche, M.L.D., and Williams, D.R. (1976). *J. Chem. Soc., Dalton Trans.*, 1016.

Martell, A.E., and Motekaitis, R.J. (1988). *Determination and Use of Stability Constants*, VCH, New York.

May, P.M., and Bulman, R.A. (1983). *Progress in Medicinal Chemistry*, Vol. 20 (G.P. Ellis and G.B. West, eds.), Elsevier, Amsterdam, p. 225.

May, P.M., and Murray, K. (1988). *Talanta*, *35*: 825.

May, P.M., and Williams, D.R. (1977). *FEBS Lett.*, *78*: 134.

May, P.M., Linder, P.W., and Williams, D.R. (1976). *Experientia*, *32*: 1492.

May, P.M., Linder, P.W., and Williams, D.R. (1977). *J. Chem. Soc., Dalton Trans.*, 588.

May, P.M., Smith, G.L., and Williams, D.R. (1982). *J. Nutr.*, *112*: 1990.

Øie, S. (1986). *J. Clin. Pharmacol.*, *26*: 583.

Oreskes, N., Shrader-Frechette, K., and Belitz, K. (1994). *Science*, *263*: 641.

Perrin, D.D. (1965). *Nature*, *206*: 170.

Perrin, D.D., and Agarwal, R.P. (1973). *Metal Ions in Biological Systems*, Vol. 2 (H. Sigel, ed.), Marcel Dekker, New York, p. 168.

Perrin, D.D., and Sayce, I.G. (1967). *Talanta*, *14*: 833.

Peter, H.H. (1985). *J. Pharmacol. (Paris)*, *16*: 365.

Roos, J.T.H., and Williams, D.R. (1977). *J. Inorg. Nucl. Chem.*, *39*: 367.

Schubert, J. (1955). *Ann. Rev. Nucl. Sci.*, *5*: 369.

Schubert, J. (1961). *Fed. Proc.*, *10 Suppl.*: 220.

Schwarzenbach, R.P., Gschwend, P.M., and Imboden, D.M. (1993). *Environmental Organic Chemistry*, Wiley, New York.

Sillén, L.G. (1961). *Oceanography* (M. Sears, ed.), Amer. Assoc. Adv. Sci., Washington; Bailey, London, p. 550.

Sillén, L.G. (1967). *Equilibrium Concepts in Natural Water Systems* (R.F. Gould, ed.), ACS, Washington DC, p. 45.

Smith, G.L., Rees, S.B., and Williams, D.R. (1985). *Polyhedron*, *4*: 713.

Stumm, W., and Morgan, J.J. (1970). *Aquatic Chemistry*, 1st ed., Wiley-Interscience, New York.

Venturini, M., and Berthon, G. (1989). *J. Inorg. Biochem.*, *37*: 69.

Westall, J.C., Zachary, J.L., and Morel, F.M.M. (1976). *MINEQL, a Computer Program for the Calculation of Chemical Equilibrium Compositions of Aqueous Systems*, Tech. Note 18, Department of Civil Engineering, Massachusetts Institute of Technology.

Williams, D.R. (1977). *J. Inorg. Nucl. Chem.*, *39*: 711.

SPECIATION APPROACH TO THE DESIGN OF ANTIINFLAMMATORY COPPER COMPLEXES

by

Graham E. Jackson

Department of Chemistry,
University of Cape Town,
Rondebosch 7700
South Africa

I. INTRODUCTION

For many years, drug designers have adopted the black box approach to ligand design in which a myriad of compounds were screened, without a justifiable theoretical basis, for their biological activity in the hope of finding one active compound. The less understood the target disease the more prevalent this approach became. Very few drugs have been designed from chemical theory.

In more recent times, because of cost and time factors, structure activity relationships (QSAR) have been employed in an attempt to home in on desired drug characteristics (Ramsden et al., 1990). This approach requires no knowledge of the mechanism of action of the drug but requires experimental activity and physico-chemical data on the compounds. If the mechanism of action at the molecular level is known, molecular modelling tools can be used to design active molecules and to simulate drug-receptor interactions. Neither of these methods, however, have been successful where labile metal complexes are concerned because the active component may bear no relation to the administered drug. An alternative approach to this problem, speciation design, has developed out of speciation modelling of biological systems.

In this regard, much of the work has centered around chelation therapy, i.e. the use of chelate ligands to decrease the body burden of toxic metal ions (Jones, 1983). Initial attempts at the design of therapeutic chelating agents for toxic metal ions involved their selection on the basis of the stability of the metal chelate complex. However, it was soon realized that it was the ability of the ligand to bind metal ions under physiological conditions that was important. This led to the development of an 'effective stability constant', which tried to take into account the most important competitive chemical equilibria, particularly with respect to the proton (Schubert, 1955). While the simplicity of this approach was and still is attractive, it suffers from a lack of generality and the calculations became unwieldy as the number of interactions increase. It is for this reason that computer simulation methods which could incorporate, in a relatively easy way, multiple equilibria, were developed. These methods have been described in detail in Chapter 3.

II. SPECIATION MODELLING

Speciation modelling of blood plasma was pioneered by Perrin in a series of papers (Perrin, 1965; 1969; Hallman et al., 1971). These early models were restricted to one or two metal ions with a limited number of ligands. Notwithstanding their restrictive nature, the success of these early models led to the development of more comprehensive models of biofluids, many of which have already been described in previous chapters.

The essential feature of any speciation model is to define a series of chemical equilibria, which represent all the equilibria possible for the system under investigation. This series of equilibria, together with the concentration of each component in the system, constitute the database of the model. Many computer programs now exist which can interrogate this database and solve for the individual species concentrations. Multiple equilibria (May et al., 1977) adsorption and precipitation (Morel and Morgan, 1972; Westall et al., 1975; Allison et al., 1990) can all be taken into account.

The advantage of speciation modelling is that it gives some insight into what is actually happening in solution and why. It also allows questions like "what if?" to be asked. In blood plasma, for example, we could ask the question "would EDTA promote the excretion of Cu^{2+}?". The results (Table 1) show that at an EDTA concentration of 10^{-6} mol dm^{-3} copper excretion would not be affected because most of the EDTA would be complexed by Ca^{2+} and Zn^{2+}. This result would be very difficult to predict by inspection of the relevant equilibrium constants as concentration and pH play a vital role. It is not until 10^{-5}mol dm^{-3} EDTA that significant changes to the copper speciation become apparent with the formation of the $[Cu(EDTA)]^{2-}$ complex. This complex is charged and so is likely to be excreted via the kidneys.

One limitation of present day speciation modelling is that it does not take into account the kinetics of the situation. Thus it is assumed that the system is at equilibrium, a situation which is rarely achieved in biofluids. Certain complexes are kinetically inert and so, even though it is predicted that at equilibrium the metal ion would be redistributed amongst the available ligands, in practice the complex is excreted before this can happen. Some attempts have been made to correct this deficiency (Jones and May, 1987; Jackson et al., 1990) but as yet no model exists which explicitly includes kinetics.

Since speciation modelling is a quick and relatively easy method of predicting the in vivo fate of an administered ligand or complex, it is an ideal tool for the design of new

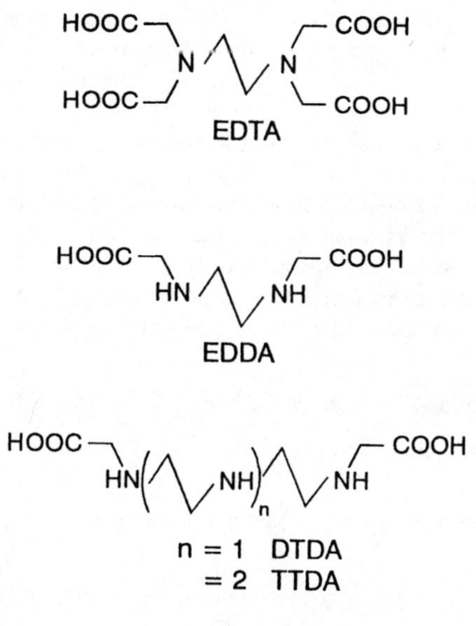

Figure 1.
Structural formulae of ligands used in the discussion.

Table 1

Calculated Low-molar-mass Species Distribution of EDTA (1.0×10^{-6} mol dm^{-3}) upon Addition to Blood Plasma

Species	log β	Concentration (mol dm^{-3})
$[Ca(EDTA)]^{2-}$	9.36	9.88×10^{-7}
$[Mg(EDTA)]^{2-}$	7.76	1.14×10^{-8}
$[Pb(EDTA)]^{2-}$	16.70	1.9×10^{-10}
$[Zn(EDTA)]^{2-}$	14.6	1.6×10^{-10}
$[Fe(EDTA)]^{2-}$	13.5	1.2×10^{-10}
$[Cu(EDTA)]^{2-}$	17.6	1.5×10^{-15}

Constitution of blood plasma taken from May et al. (1977)

drugs. The process is one of systematically varying the metal ion coordinating characteristics of the ligand and calculating, using an appropriate model, the effect these changes have on the metal ion speciation. In this way, the ligand can be varied until the optimal metal ion speciation is obtained. In the case of chelation therapy this may be the formation of charged complexes for renal excretion, while in another case neutral, lipophilic complexes may be required. An example of speciation modelling as applied to the design of copper-based antiinflammatory drug is given in a latter section.

In biofluids, there is the problem that the metal ion may bind to proteins. In this case, it is necessary to test whether the ligand is able to sequester the metal ion from the protein or whether the protein will remove the metal ion from any administered complex. While it is difficult to explicitly take into account the influence of protein binding upon metal ion speciation, some progress has been made using the plasma mobilizing index (May and Williams, 1977).

A. Plasma Mobilizing Index (PMI)

The absolute concentration of low-molar-mass (l.m.m.) metal complexes depend on metal-protein equilibria. However, in terms of simulating these equilibria, the metal-protein interactions are ill-defined and hence they are difficult to include in the calculation. Fortunately the total metal ion concentration is very much less than the total protein concentration and the free metal ion concentration is very much lower than the concentration of the metal-protein complex. This means that until a substantial portion of the metal is stripped off the protein, its free concentration is buffered and remains constant. Using this buffered free concentration for a metal ion it is possible to calculate the total concentration of l.m.m. metal ion species in solution. The ability of different drugs (chelating agents) to move a metal ion from the protein bound fraction to the l.m.m. form can then be expressed as a Plasma Mobilizing Index (PMI), defined by May and Williams (1977):

$$\text{PMI} = \frac{\text{Total concentration of l.m.m. metal complex species in the presence of the drug}}{\text{Total concentration of l.m.m. metal complex species in normal plasma}}$$

III. DESIGN OF COPPER-BASED ANTIINFLAMMATORY DRUGS

A. Copper in vivo

Copper was first shown to be an essential metal ion by Hart et al. (1928) who showed that it was required for haemoglobin synthesis. Since then, copper has been shown to be essential for a myriad of cell functions (Underwood, 1971). Copper is unevenly distributed throughout the body, the levels being homeostatically maintained by adsorption, excretion

Table 2

Mean Concentration (ppm) of Cu(II) in Tissues and Fluids

Bile	500
Blood (total)	1.0
(plasma)	1.2
(serum)	1.9
Brain	370
Gallbladder	750
Intestine	180-300
Gastric juice	0.1-0.4
Kidney	270
Liver	680
Muscle	85
Urine	40 ppm

Source: data from Sorenson (1974)

and storage. Typical concentration ranges for selected tissues are given in Table 2. Of particular relevance to the present exercise is the concentration of copper in blood serum. Of the total plasma copper, the main fraction is irreversibly bound to ceruloplasmin (Gubler et al., 1952) while the remaining copper is reversibly bound to serum albumin and transcuprein (Wirth and Linder, 1985). A small portion of the non-ceruloplasmin copper is distributed amongst l.m.m. complexes. It is this l.m.m. fraction of copper which is postulated to be involved in transport of the metal across membranes and between biologically active sites (Underwood, 1971; Caruthers et al., 1966; Österberg, 1974). Unfortunately, the diversity and concentration of these l.m.m. complexes make their detection and analysis very difficult if not impossible with present day techniques. Hence the only way in which some idea of the composition of this fraction can be obtained is by computer simulation. Table 3 contains a list of the predicted most predominant l.m.m.

Table 3

Calculated Percentage Distribution of Cu(II) among its most Predominant l.m.m. Complexes in Blood Plasma (Brumas et al., 1993)

Complex species[a]	log β	% of total l.m.m. Cu(II)
[Cu(his)(gln)]	16.70	19.2
$[Cu(his)_2]$	17.50	15.5
[Cu(his)(thr)]	17.03	14.7
[Cu(his)(ser)]	16.97	7.5
[Cu(his)(ala)]	17.00	5.4
$[Cu(his)(lys)H]^+$	27.05	4.4
[Cu(his)(gly)]	16.94	4.3
[Cu(his)(asn)]	16.81	4.2
[Cu(his)(val)]	16.93	3.9
[Cu(his)(leu)]	17.18	3.7

[a] Symbols: his = histidinate, gln = glutaminate, thr = threoninate, ser = serinate, ala = alaninate, lys = lysinate, gly = glycinate, asn = asparaginate, val = valinate, leu = leucinate

complexes present in normal blood plasma (Brumas et al., 1993). As can be seen, most of these complexes are mixed, ternary complexes of histidine and another amino acid. Most of the complexes are uncharged and hence in the form most likely to be transported into tissue (Neumann and Silverberg, 1966).

In patients with active inflammation, it has been found that total plasma copper is elevated, but that it returns to normal with remission. It appears that copper is mobilized from the liver in response to the onset of inflammation both as ceruloplasmin copper and non-ceruloplasmin copper. Results have also shown that there is a significant increase in total copper in the synovial fluid of inflamed joints. Most of this copper is bound to ceruloplasmin but a second fraction, suggested to be albumin-bound has also been found (Bonebrake et al., 1968).

B. Rheumatoid arthritis. Justification for copper administration

Rheumatoid arthritis (RA) is a chronic inflammation of the joints and muscles which is not directly related to infection or injury. The disease is progressive and incapacitating due to the resulting ankylosis. It is one of the more important connective tissue diseases affecting some 5% of the population of the Western world (May and Williams, 1981). It has been said "...everything is known about rheumatoid arthritis except its cause, natural history and treatment. It is one of the perversities of the human condition that the rarer the disease, the more clearly it seems to be understood" (Williams, 1974).

Copper in the form of copper bracelets and copper-rich diets (shell-fish, nuts, cider vinegar) has been used in the treatment of arthritic conditions for many centuries. Sorenson and others have reviewed the role of copper in the treatment of inflammation (Sorenson, 1978; 1982; Milanino and Velo, 1981) and indeed there is overwhelming evidence that a direct connection between RA and copper exists. Historically copper complexes have been used successfully in the treatment of arthritic conditions. It has been suggested that the acceptance of this form of therapy was impeded by the development of steroidal and non-steroidal antiinflammatory drugs. Recently, however, Sorenson has suggested that in contrast to other antiinflammatory agents the antiinflammatory activity of copper complexes is via their ability to promote tissue-repair processes (see *Bioinorganic Medicine*, Part Five, Chap. 1-D). For this reason copper complexes deserve serious consideration. Second, many antiarthritic drugs have to be discontinued because they cause gastrointestinal irritation. On the other hand, many copper complexes have been shown to have antiulcer activity (Sorenson, 1982).

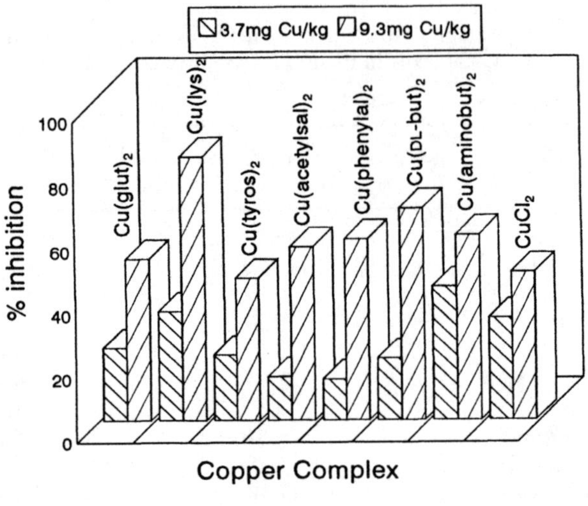

Figure 2.
Percent inhibition of carrageenan induced foot oedema following subcutaneous injection of different copper complexes (data taken from Jackson et al., 1978).

Sorenson (1974; 1976) has tested the antiinflammatory activity of a variety of copper complexes using a rat carrageenan foot edema model, and reported substantial reduction in inflammation. More recently (Jackson et al., 1978a), we have repeated some of these experiments and found (Fig. 2) that the response is independent of the complex used but is dependent on the dose administered. At a moderate dose of 9.3 mg copper/kg a 60% reduction in inflammation is obtained.

C. Design strategy

The design of copper-based antiinflammatory drugs is based on the assumption that (i) their therapeutic effect arises from an increase in the total labile l.m.m. pool of copper in the inflamed tissue and (ii) that this labile pool may be increased by an increase in the tissue-permeable neutral fraction of copper in plasma.

Scheme 1 outlines the different strategies which may be employed in order to bring about an increase in the tissue permeable fraction of copper in blood plasma. Firstly, since a significant percentage of the exchangeable plasma copper is reversibly bound to serum albumin, powerful chelators may be used to move this copper into a l.m.m. form. It has been postulated that some of the acidic antiinflammatory drugs are able to liberate copper indirectly from serum albumin by distorting the protein molecule (Fiabane and Williams, 1978). On the other hand, penicillamine - which is currently used in the treatment of RA - is postulated to liberate copper from other sources, notably ceruloplasmin (Jackson et al., 1978b).

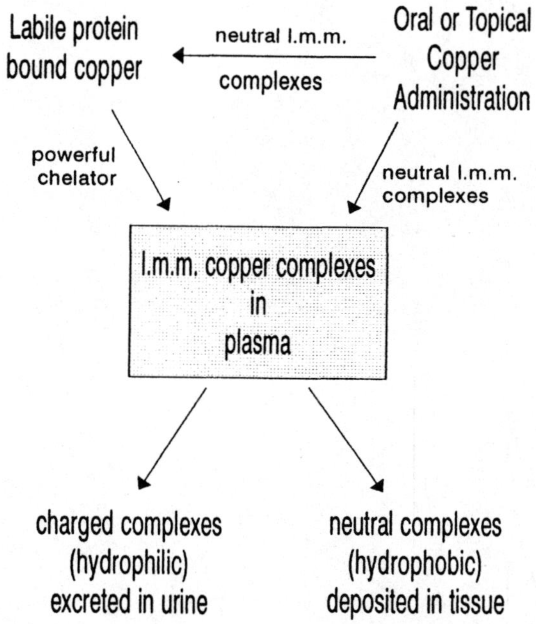

Scheme 1.
Different routes for increasing labile copper in tissue.

A second method of increasing the l.m.m. concentration of copper in plasma is by exogenous administration. For long-term therapy, this would have to be via oral or topical administration, as intravenous or subcutaneous injection is too inconvenient and problematical. The limit for intravenous injection of copper is <1 mg dm^{-1} while an oral dose of 1g (May and Williams, 1981) is tolerated. Also, subcutaneous injection of copper can lead to local edema. In designing a ligand for oral copper administration, speciation modelling can be used to maximize the concentration of assimilable (neutral, lipophilic) copper complexes in intestinal fluids. These complexes would have to be able to withstand the acidic conditions of the stomach as well as the alkaline conditions of the intestine. The

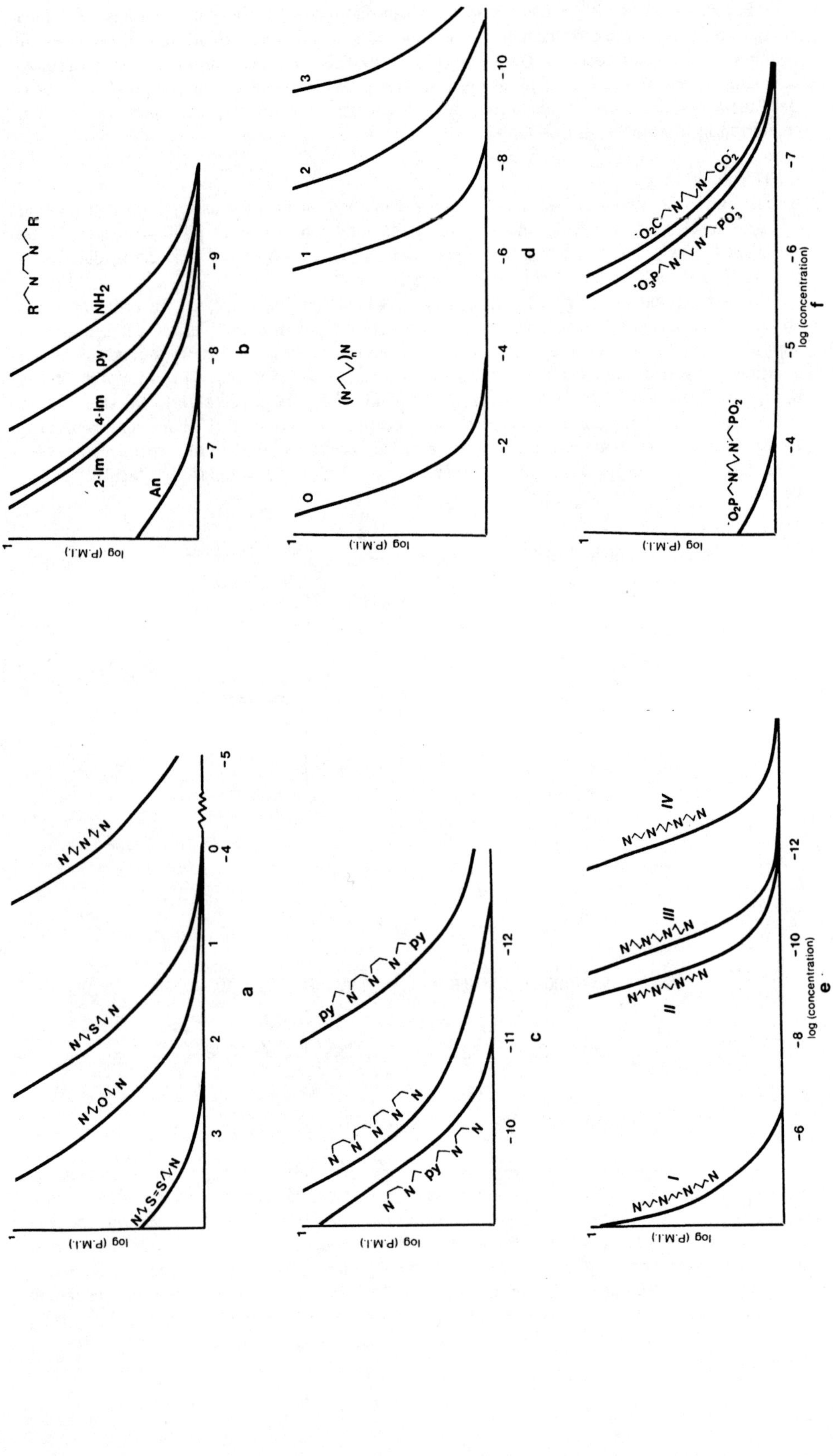

log (P.M.I.)
log (concentration)
a
b
c
d
e
f

complexes would also have to withstand metal binding by the mucosal cells which homeostatically regulate copper absorption (Owen, 1965; Marceau et al., 1970). On the other hand, the complexes must not be so inert that they are not able to release the copper to the labile plasma pool once they have been absorbed. This applies particularly to the use of macrocyclic ligands to complex the copper.

More recently attention has focused on dermal absorption of copper. Here the difficulties of oral absorption are circumvented. Experiments have shown that the overall resistance to dermal absorption is characteristic of passive Fickian diffusion (Scheuplein and Blank, 1973). Compounds that are readily absorbed have a balance between water solubility and lipid solubility, but molecules which are hydrophilic are often more permeative than anticipated (Houk and Guy, 1988). Because copper complexes are generally hydrophilic, it becomes easier then to design a copper pharmaceutical for dermal absorption. Indeed, copper salts have been shown to have antiinflammatory activity following topical application (Beveridge et al., 1980).

D. Example of speciation modelling applied to ligand design

Speciation modelling can be used to simulate the effect of structural changes to a ligand upon its ability to mobilize copper in vivo. In this example, the following changes are considered.

1. *Donor atom*

Copper(II) is considered to be borderline/soft on the HSAB scale while Zn(II) and Ca(II), its main competitors in vivo, are both HSAB hard (Pearson, 1963). Since soft metal ions prefer to bond to soft donor atoms while hard metal ions prefer to bond to hard donor atoms, the use of nitrogen as opposed to oxygen as a donor atom in a ligand should favor coordination of Cu(II) as opposed to Zn(II) or Ca(II). Figure 3a shows the effect of changing the ligand donor atom upon the plasma mobilizing efficiency of the ligand. As expected, the nitrogen analogue is far more efficient (10^7 times) at mobilizing copper than the oxygen analogue. Perhaps unexpectedly the soft sulphur donor does not mobilize copper as well as its nitrogen analogue. This is because of the relative strengths of these two bases, the nitrogen being the strong base. Note that the thiol group, which is known to form very stable copper complexes, has not been considered because it is notoriously toxic and undergoes reductive chelation with copper.

2. *Donor strength*

The strength of nitrogen donor ligands is measured in terms of their basicity. In order to examine the effect of donor atom basicity on copper mobilization, a comparison of *N,N'* disubstituted ethylenediamines can be made. Figure 3b shows that as the basicity of the nitrogen donor atom is increased, so the PMI increases.

However, structural constraints also have to be considered when designing a ligand. Figure 3c shows that as the position of pyridine within a ligand is changed, so the ability of the ligand to mobilize copper changes. Thus, despite the amine group being a stronger donor than pyridine, *N,N'*-bis(2-pyridylmethyl)-1,3-diaminopropane is a factor of 10 more effective at mobilizing copper than tetraethylenepentamine.

Figure 3 (*turned*).

The effect of different ligand features upon plasma mobilization of Cu^{2+}. For clarity only the coordinating atoms are shown in the structural formulae of the ligands: (a) donor atom; (b) donor strength, Am = 2-aminobenzyl, 2-Im = 2-imidazolyl, 4-Im = 4-imidazolyl, py = 2-pyridyl; (c) structural constraints; (d) number of donor atoms; (e) chelate ring size; (f) anionic groups. Reproduced with permission of Inorg. Chim. Acta (Jackson, 1988).

3. *Number of donor atoms*
Clearly, as the number of donor atoms on the ligand increases, so the stability of the copper complexes should increase. Figure 3d shows that this has a concomitant effect on the mobilization of copper in vivo. There is a large increase in mobilization in going from mono- to bi-dentate. This reflects the increased stability of the copper complex of these ligands relative to the Ca^{2+} and Zn^{2+} complexes. Further increases in the denticity of the ligand result in improved copper mobilization, but above a coordination number of 4 there is no significant increase in mobilizing ability. This is because copper complexes are generally Jahn-Teller distorted and so addition of the fifth donor group to an axial position on the metal ion is not very favorable.

4. *Ring size*
The stability of inorganic chelate complexes is affected by the size of the chelate rings formed upon coordination. The effect of this is shown in Figure 3e using quadradentate ligands as examples. Ligand I forms three 6-membered rings upon coordination and, as expected, is far less effective at mobilizing copper than II which forms one 5- and two 6-membered chelate rings upon coordination. Ligand III forms three 5-membered rings upon coordination and is 1000-fold more efficient at mobilizing copper. However, the best mobilization of copper is shown by ligand IV which forms a 5,6,5-membered ring system. Subsequent calculations have shown that this mobilization results from a decreased ring strain within the complex (Jackson et al., 1991).

5. *Anionic groups*
Since the objective is to form neutral Cu^{2+} complexes, the ligand must be dianionic. Figure 3f shows the effect of different anionic groups upon copper mobilization. There appears to be little to choose between the phosphate, carboxylate and phenolate as a source of negative charge. However, the phosphate is dianionic and so the phosphate mono ester should be considered as the source of negative charge.

IV. RESULTS

A. Intravenous injection

Based on speciation design results, an ideal ligand for binding copper in vivo should be a linear dicarboxylate, diphenolate or dialkylphosphate-substituted polyamine. The results (Jackson and Kelly, 1990; Jackson et al., 1981) for three such ligands EDDA, TTDA and DTDA are shown in Figure 4. Computer simulation shows that substantial mobilization of protein-bound copper is predicted to occur in plasma using TTDA or DTDA, a concentration of 10^{-6} mol dm^{-3} resulting in a 10-fold increase in l.m.m. copper. These l.m.m. complexes are predominantly neutral and hence meet our criteria for tissue deposition and antiinflammatory activity. On the other hand, as expected, EDDA is unable

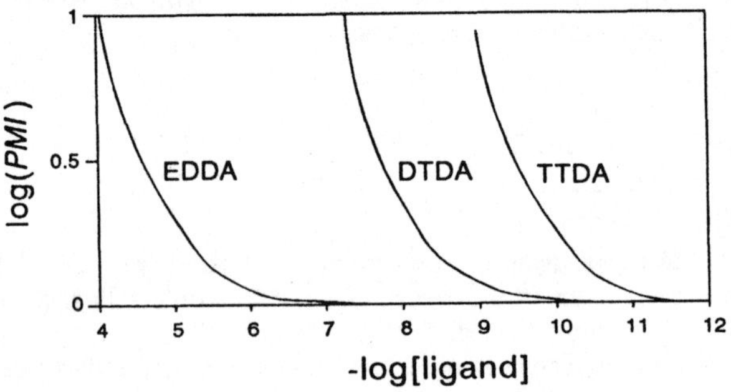

Figure 4.
Copper PMI curves for TTDA, DTDA and EDDA.

Table 4

Tissue Distribution of ^{67}Cu Labelled Complexes of TTDA, DTDA and $CuCl_2$ in White Mice, 6h and 24h Post Injection

Complex	Time (h)	% Counts/g of tissue					
		Carcass	Urine	Blood	Kidney	Liver	Muscle
[^{67}Cu(TTDA)]	6.0	26.4	64.3	1.3	3.6	9.7	0.04
	24.0	21.0	72.9	1.4	2.3	4.5	0.02
[^{67}Cu(DTDA)]	6.0	18.6	70.9	0.15	2.3	17.7	0.0
	24.0	18.0	71.8	0.53	1.6	10.1	0.0
$^{67}CuCl_2$	6.0	43.1	21.1	4.0	7.0	36.2	0.8
	24.0	41.5	34.4	3.6	6.0	23.4	0.4

to compete effectively for Cu(II) in plasma, its mobilization being suppressed by competition with zinc and calcium.

Experimental in vivo tissue distribution results (Table 4) confirm the computer simulation predictions. In mice, the [Cu(TTDA)] complex is distributed throughout the body. This distribution is totally different to that of $CuCl_2$. The normal route for copper excretion is via the liver with only a small fraction being eliminated in the urine. This is seen to be the case for $CuCl_2$. The [Cu(TTDA)] and [Cu(DTDA)] complexes, however, are eliminated mainly in the urine and substantially less copper is excreted via the liver. As predicted by speciation modelling, both TTDA and DTDA are able to prevent loss of copper to serum albumin. This is shown by the low levels of radioactive copper in the blood 6h post injection.

While TTDA and DTDA are able to cause a redistribution of copper in vivo, the results are still disappointing in that the complexes are rapidly eliminated in the urine. This is presumably because the complexes are too hydrophilic and are therefore not reabsorbed from the urine.

B. Oral administration

While for RA topical administration of copper complexes may be preferred, oral administration is still a viable alternative, particularly in view of the antiulcer activity of many copper complexes (Sorenson, 1982). Thus a dual protocol involving both non-steroidal antiinflammatory drugs (NSAIDs) and copper antiinflammatory drugs is envisaged.

For maximal oral absorption, neutral lipophilic complexes should be formed under conditions typical of the intestinal lumen. The predicted speciation of copper/TTDA and copper/EDDA under these conditions is shown in Figure 5 as a function of pH. The results show that for TTDA, only at higher pH values is the neutral [Cu(TTDA)] complex formed. At pH values typical of the stomach, the charged complexes $[Cu(TTDA)H]^+$ and $[Cu(TTDA)H_2]^{2+}$ predominate. This indicates that TTDA would promote the absorption of copper in the small intestine but suppress absorption of copper in the stomach. On the other hand, with EDDA the neutral [Cu(EDDA)] complex predominates at much lower pH values and so both gastric and intestinal absorption is possible.

These results, together with the plasma mobilization results, indicate that oral administration of either Cu/TTDA or Cu/EDDA should be beneficial in the treatment of inflammation. In the case of EDDA, the activity is due to an increase in the copper absorption while in the case of TTDA there is both an increase in oral absorption and a redistribution of endogenous copper into a l.m.m. form.

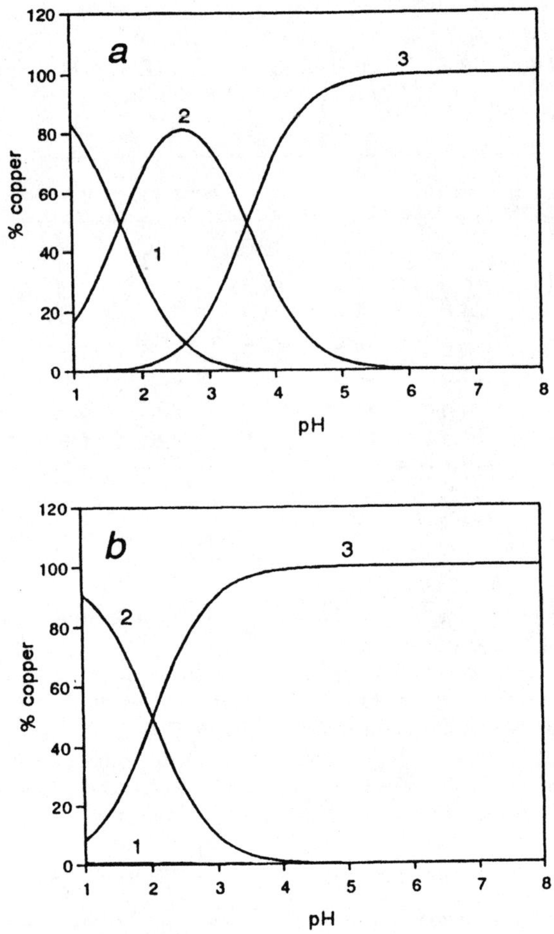

Figure 5.
Species distribution of 1.0 mmol dm^{-3} Cu(II) in the presence of (a) 1.0 mmol dm^{-3} TTDA, 1 = $[Cu(TTDA)H_2]^{2+}$, 2 = $[Cu(TTDA)H]^+$, 3 = [Cu(TTDA)]; and (b) 1.0 mmol dm^{-3} EDDA, 1=$[Cu(EDDA)H]^+$, 2 = $[Cu(H_2O)_6]^{2+}$, 3 = [Cu(EDDA)].

V. CONCLUSION

Computer simulation as applied to speciation modelling is a useful method which can be applied to the design of therapeutic chelating agents in general and copper antiinflammatory agents in particular. The approach provides a quick, inexpensive method of prescreening potential drugs before they are subjected to testing in animal studies. Furthermore, the results are such that an insight into the mode of action of a ligand at the molecular level is obtained. This allows one to design more effective drugs based on chemical principles rather than on intuitive guess-work.

REFERENCES

Allison, J.D., Brown, D.S., and Novo-Gradac, K.J. (1990). MINTEQA2/PRODEFA2, Environmental Research Laboratory, U.S. Environmental Protection Agency, Athens, Georgia.

Beveridge, S.J., Walker, V.R., and Whitehouse, M.W. (1980). *J. Pharm. Pharmacol.*, *32*: 425.

Bonebrake, R.A., McCall, J.T., Hunder, G., and Polley, H. (1968). *Arthritis Rheum.*, *11*: 95.

Brumas, V., Alliey, N., and Berthon, G. (1993). *J. Inorg. Biochem.*, *52:* 287.

Caruthers, M.E., Hobbs, C.B., and Warren, R.L. (1966). *J. Clin. Pathol.*, *19*: 498.

Fiabane, A., and Williams, D.R. (1978). *J. Inorg. Nucl. Chem.*, *40*: 1195.

Gubler, C.J., Lakey, M.E., Cartwright, G.E., and Wintrobe, M.M. (1952). *Amer. J. Physiol.*, *171*: 652.

Hallman, P.S., Perrin, D.D., and Watt, A.E. (1971). *Biochem. J.*, *121*: 549.

Hart, E.D., Steenbock, H., Waddell, J., and Elvenjeri, C.A. (1928). *J. Biol. Chem.*, *77*: 797.

Houk, J., and Guy, R. (1988). *Chem. Rev.*, *88*: 455.

Jackson, G.E., and Kelly, M.J. (1988). *Inorg. Chim. Acta*, *152*: 215.

Jackson, G.E., and Kelly, M.J. (1990). *J. Chem. Soc., Dalton Trans.*, 1889.

Jackson, G.E., May, P.M., and Williams, D.R. (1978a). *J. Inorg. Nucl. Chem.*, *40*: 1189.

Jackson, G.E., May, P.M., and Williams, D.R. (1978b). *FEBS Lett.*, *90*: 173.

Jackson, G.E., May, P.M., and Williams, D.R. (1981). *J. Inorg. Nucl. Chem.*, *43*: 825.

Jackson, G.E., Wynchank, S., and Woudenberg, M. (1990). *Magn. Reson. Med.*, *16*: 57.

Jackson, G.E., Linder, P.W., and Voyé, A. (1991). *Polyhedron*, *10*: 883.

Jones, M.M. (1983). *Metal Ions in Biological Systems*, Vol. 16 (H. Sigel, ed.), Marcel Dekker, New York, p. 47.

Jones, M.M., and May, P.M. (1987). *Inorg. Chim. Acta*, *138*: 67.

Marceau, N., Aspin, N., and Sass-Kortsak, A. (1970). *Amer. J. Physiol.*, *218*: 377.

May, P.M., and Williams, D.R. (1977). *FEBS Lett.*, *78*: 134.

May, P.M., and Williams, D.R. (1981). *Metal Ions in Biological Systems*, Vol. 12 (H. Sigel, ed.), Marcel Dekker, New York, p. 283.

May, P.M., Linder, P.W., and Williams, D.R. (1977). *J. Chem. Soc., Dalton Trans.*, 588.

Milanino, R., and Velo, G.P. (1981). *Trace Element in the Pathogenesis and Treatment of Inflammation* (K. Rainsford, K. Brune and M. Whitehouse, eds.), Birkhäuser Verlag, Basel, p. 209.

Morel, F., and Morgan, J. (1972). *Environ. Sci. Tech.*, *6*: 58.

Neumann, P.Z., and Silverberg, M. (1966). *Nature*, *210*: 414.

Österberg, R. (1974). *Coord. Chem. Rev.*, *12*: 309.

Owen, C. (1965). *Amer. J. Physiol.*, *209*: 900.

Pearson, R. (1963). *J. Chem. Educ.*, *45*: 581.

Perrin, D.D. (1965). *Nature*, *206*: 170.

Perrin, D.D. (1969). *Suomen Kem.*, *42*: 205.

Ramsden, C.A., Hansch, C., Sammes, P.G., Taylor, J.B., and Drayton, C.J., eds. (1990). *Comprehensive Medicinal Chemistry*, Vol. 4, Pergamon Press, New York.

Scheuplein, R., and Blank, I. (1973). *J. Invest. Dermatol.*, *60*: 286.

Schubert, J. (1955). *Annu. Rev. Nucl. Sci.*, *5*: 369.

Sorenson, J.R.J. (1974). *Trace Substances in Environmental Health*, Vol. 8 (D.D. Hemphill, ed.), Univ. Missouri-N.S.F., p. 305.

Sorenson, J.R.J. (1976). *J. Med. Chem.*, *19*: 135.

Sorenson, J.R.J. (1978). *Prog. Med. Chem.*, *15*: 211.

Sorenson, J.R.J. (1982). *Metal Ions in Biological Systems*, Vol. 14 (H. Sigel, ed.), Marcel Dekker, New York, p. 77.

Underwood, E.J. (1971). *Trace Elements in Human and Animal Nutrition*, Academic Press, New York.

Westall, J., Zachary, J., and Morel, F. (1975). *Massachusetts Institute of Technology*, Technical Note No 18.

Williams, R.C. (1974). *Major Problems in Medicine*, Vol. 4 (L.H. Smit, ed.), Saunders, Philadelphia.

Wirth, P.L., and Linder, M.C. (1985). *J. Nat. Cancer Inst.*, Vol. 75, p. 277.

COMPUTER-BASED STUDIES ON TRACE METAL EXCRETION AND NEEDS IN TOTAL PARENTERAL NUTRITION

by

Guy Berthon

INSERM U 305,
Equipe "Bioréactifs: Spéciation et Biodisponibilité",
38 rue des Trente-six Ponts
31400 Toulouse
France

I. INTRODUCTION

In severe forms of gastrointestinal disease, part or all of a digestive organ may be so damaged that it can no longer fulfil its biological functions, this failure necessitating for the patient to be fed artificially. Depending on the nature of the organ that is affected, administration of the nutritive fluid may be enteral (whenever the absorption of the nutrients is still effective), or parenteral (when small bowel functions are too seriously impaired).

Parenteral nutrition has long been used in the short term for improving the clinical status of depleted patients temporarily, for example before or after surgery. More recently, technical progress in the placement and maintenance of catheters as well as in the (aseptic) handling of nutritive solutions has allowed to extend this kind of therapy for several months, sometimes even years, hence its designation as long-term total parenteral nutrition (TPN) - and, for stable patients, home parenteral nutrition (HPN) (Fleming, 1989).

As far as protein, glucose and lipid fractions of the nutritive solutions are concerned, extending the duration of TPN does not pose any special problem. The composition of nutritive solutions in these macronutrients is well-known, and nutritionists have long had available a choice of mixtures of different compositions adapted to the therapeutic objectives dictated by the patient's status. In contrast, the situation regarding the micronutrient composition of these solutions is less certain and depends on many factors given below. For trace elements in particular, balance studies that would be necessary to assess the apparent* needs of each patient (Jacobson and Wester, 1977; Phillips and Garnys, 1981; Shulman, 1989) cannot be performed on a large scale, and therefore symptoms due to essential and beneficial trace metal deficiencies, especially copper (Karpel and Peden, 1972; Dunlap et al., 1974; Vilter et al., 1974), zinc (Kay and Tasman-Jones, 1975; Kay et al., 1976; Arakawa et al., 1976; Okada et al., 1976; Tucker et al., 1976; Weisman et al., 1976) and chromium (Jeejeebhoy et al., 1975; Freund et al., 1979; Brown et al., 1986) were reported to parallel long-term TPN therapy (McKenzie, 1977; McClain, 1981).

* Apparent in that they do not indicate whether individual tissues have been maintained or have been adequately repleted with the nutrient under study (Shenkin, 1986).

Much attention was paid to this issue in the late seventies, and guidelines for essential trace element preparations for parenteral use were established by a panel of experts for the AMA Department of Foods and Nutrition (Shils et al., 1979), which were reviewed recently (Greene et al., 1988). Based on these recommendations, most hospital formularies supply by default a standard mixture that closely approximates the corresponding doses. Yet, individual supplementation of appropriate quantities of trace elements is imperative in the management of patients on TPN (Frankel, 1993).

Such individual trace metal requirements are difficult to establish, however. Theoretically, the maintenance doses to be infused should correspond to those amounts that would exit from the hepatic vein in the patient with normal digestion and absorption and after first passage through the liver (Fleming, 1989). Obviously, these doses cannot be deduced from quantitative studies of oral intake and extent of absorption (Frankel, 1993). Trace element parenteral requirements are conditioned by several factors: (i) pre-existent patient's nutritional status, (ii) losses in stool and urine, and (iii) nutrient bioavailability (Fleming, 1989). The fact that the infusion of the nutritive mixture is in itself a cause of important metal losses during TPN (Fleming et al., 1976; Solomons et al., 1976; Jacobson and Wester, 1977; McKenzie, 1977; Askari et al., 1979; Matuchansky et al., 1979; Wolman et al., 1979; Shike et al., 1981; Bengoa et al., 1983) greatly complicates the problem (see also *Bioinorganic Medicine*, Part Three, Chapter 5-B); this effect is due to the capacity of some components of nutritive solutions to induce metal excretion individually (Freeman et al., 1975; Henkin et al., 1972; Freeman and Taylor, 1977; Yunice et al., 1978), the excretion route depending on the nature of the metal (Wolman et al., 1979; Shike et al., 1981).

On the other hand, TPN has developed in a controlled environment favorable to the study of metal balances and their metabolic influences (Wolman et al., 1979). It also constitutes a simplified model of nutrition that lends itself well to speciation calculations. In particular, the individual effects of components of nutritive solutions on the low-molar-mass (l.m.m.) fractions of trace metals such as zinc and copper in blood plasma - hence on the fate of these metals during TPN - have been simulated (Berthon, 1983; Berthon et al., 1984a). Furthermore, the metal quantities that should be added to each nutritive mixture to compensate for the losses it specifically induces can also be obtained from appropriate calculations, and a few examples of such predictions have already been published (Berthon, 1984b; Blaquière and Berthon, 1987).

This short review gives an account of the works done in this context and outlines their possible clinical applications.

II. TRACE METAL DEFICIENCIES DURING TOTAL PARENTERAL NUTRITION

A. Zinc

The most spectacular symptom due to TPN-induced metal depletion is certainly the skin disease similar to acrodermatitis enteropathica that is observed during zinc deficiency states (see *Bioinorganic Medicine*, Part Four, Chapter 2).

The whole syndrome characterizing zinc depletion was first reported by Kay and Tasman-Jones (1975): 2-5 weeks after the commencement of high-caloric infusion feeding, the patients became apathetic and depressed, and developed a facial rash, diarrhea and alopecia at a time when plasma-zinc levels were very low. Dermatitis is the most commonly reported clinical manifestation (Kay and Tasman-Jones, 1975; Kay et al., 1976; Arakawa et al., 1976; Okada et al., 1976; McClain, 1981). The rash characteristically starts as a moist eczematoid area in the nasolabial folds and progresses rapidly to widespread lesions (see Ladefoged and Jarnum, 1983 and references therein). In all cases, the administration of zinc rapidly reverses the skin lesions (usually healed within 10 days), the diarrhea and the mental depression. Zinc depletion is especially prejudicial to surgical patients as the metal is an essential factor in wound healing and tissue repair (Pories et al., 1976). For details on other consequences of zinc deficiency (growth retardation, impaired immunity, impairment of insulin response, etc.), see *Bioinorganic Medicine*, Part Two, Chapters 2 and 3-A6, and Part Three, Chapters 5-B, 5-C, and 6-D.

B. Copper

Although inducing less spectacular symptoms, copper deficiency was the first trace metal deficiency described during TPN, originally in infants (Karpel and Peden, 1972) then in adults (Dunlap et al., 1974; Vilter et al., 1974). Appearing after several months of hyperalimentation, the depletion syndrome is characterized by anemia, leukopenia and neutropenia, with severe bone demineralization in the infants. In these, oral copper supplementation prompted complete recovery in 3 weeks (Karpel and Peden, 1972). For more details on the effects of copper deficiency, see *Bioinorganic Medicine*, Part Two, Chapter 3-A5, and Part Three, Chapters 5-B, 5-C, and 6-D.

C. Other trace metals

Chromium deficiency was first described by Jeejeebhoy et al. (1975; 1977) on a patient who had been receiving TPN for 3½ years. The symptoms of weight loss, peripheral neuropathy and impaired glucose tolerance were reversed on chromium administration. Two other cases were reported since then with similar profiles (Freund et al., 1979; Brown et al., 1986), the shortest period of time until appearance of clinical deficiency being five months.

A case of manganese deficiency was first suspected in a subject on a 17-week experimental diet from which manganese was later discovered to be absent (Doisy, 1972). More recently, several subjects on a 39-day manganese-deficient diet developed a symptom of itching rash which disappeared in 2-3 days with repletion (Friedman, 1987).

Molybdenum deficiency was reported to affect a patient with Crohn's disease and short-bowel syndrome after 12 months of TPN (Abumrad et al., 1981). The observed defect in sulfur amino acid metabolism was reversed and the clinical condition improved following treatment with ammonia molybdate.

Selenium is also a trace element whose status is altered by TPN (Van Rij et al., 1979; Fleming et al., 1984), but, as a non-metal, is out of the scope of this review. The only point that should be stressed here is that selenium supplementation must be considered carefully given the narrow margin of safety of this element (Vanderhoof and Antonson, 1986).

III. ORIGIN OF TRACE METAL DEFICIENCIES DURING TOTAL PARENTERAL NUTRITION

A. Non TPN-specific causes of metal depletion

Trace metal deficiencies observed during TPN may have different causes that are not directly due to the mode of administration of the nutrients:

(i) The absence or presence in insufficient amount of trace metals in the nutritive mixtures administered. Metal ions are present as contaminants in commercially available solutions, sometimes in significant amounts (Hoffmann and Ashby, 1976; Hauer and Kaminski, 1978). However, while this contamination may introduce non-negligible levels of detrimental metals into the patient's blood (e.g. aluminum; see e.g. Sedman et al., 1985), it can by no means satisfy the patient's needs in essential and beneficial elements.

(ii) Increased utilization, especially during anabolic periods of gain of weight. There is an avid demand of zinc during new protein formation, and in particular new enzyme synthesis, in the anabolic phase (Kay and Tasman-Jones, 1975). Hyperzincuria is observed from the institution of therapy, but plasma zinc only falls and overt clinical zinc deficiency appears at a time of sustained weight gain and anabolism (Kay and Tasman-Jones, 1975; Matuchansky et al., 1979).

(iii) Losses due to the disease itself, in case of fistula or diarrhea for example (Wolman et al., 1979; Ladefoged and Jarnum, 1983).

(iv) Losses due to stresses or trauma in case of surgery or burns (Pories et al., 1976). Whereas normal urinary excretions of both zinc and copper are independent of diet, the body's response to sudden, severe injury upsets this homeostasis: the mean loss for each element was reported to be 2-3 times greater for patients than for normal volunteers (Askari et al., 1979). The peak losses (2-10 times greater than normal for zinc, and 3-10 times for copper, on a daily basis) reflect the severity of trauma and emphasize the metabolic response to stress. There is a general correlation between zinc and nitrogen excretions. The origin of zinc losses would be direct damage to tissue cells resulting in intercellular losses, but also the catabolism of muscle mass which follows severe injury (Askari et al., 1979; Burbrink Moser et al., 1985).

B. TPN-specific losses

Cases of metal extra losses specifically induced by TPN administration itself are documented. For calcium for example, the acid load that results from amino acid infusion would contribute to the hypercalciuria observed in TPN and has been suggested as a possible cause of metabolic bone disease (Bengoa et al., 1983; Berkelhammer et al., 1988).

Zinc and copper plasma levels studied by Solomons et al. (1976) were shown to decrease during TPN administration, this tendency being reversed as soon as oral diet was restored. The authors suggested, however, that the deficiency of trace metals in the parenteral fluids was the major factor of this effect, redistribution and increased utilization also being contributory. Other results obtained by Matuchansky et al. (1979) are more characteristic of the specificity of TPN in metal losses. A comparison was made of the effects of parenteral and enteral administration of mixtures of similar composition on plasma and urine levels of zinc and copper: relative to enteral nutrition, TPN tended to decrease both zinc and copper plasma levels (not shown here), but whereas zinc urinary excretion was clearly increased (Fig. 1A), urinary copper was reduced instead (Fig. 1B), suggesting increased copper excretion through another emunctory, especially bile.

Regarding the processes underlying these observations, several attempts have been made to examine different parameters separately. All of these hinge upon the fact that a small fraction of trace metals exists in blood plasma in the form of l.m.m. complexes, as experimentally demonstrated for copper (Neumann and Sass-Kortsak, 1967) and zinc (Prasad and Oberleas, 1968; 1970), and that this fraction is enlarged under the effect of TPN administration. For example, Henkin et al. (1972) established a direct relationship between histidine infusion and hyperzincuria in patients with scleroderma as well as normal volunteers. Attention was later called to the influence on zinc urinary excretion of amine-sugar compounds present in TPN mixtures after heat sterilization of either protein hydrolysates or amino acid solutions with dextrose (Freeman et al., 1975). More recently, urinary zinc excretion was found to be significantly elevated in newborn infants receiving TPN during high-dose histidine and cysteine infusion periods (Zlotkin, 1989).

In rats, histidine was shown to increase zinc urinary excretion whether administered orally or parenterally (Freeman and Taylor, 1977), as did cysteine-fortified diet (Hsu and Smith, 1983). A detailed study on the relative influence of three amino acids upon zinc urinary excretion in dogs found it in the order cysteine > histidine > glycine (Yunice et al., 1978) (Fig. 2), which was in line with the fact that cysteine and histidine had previously been recognized as the only important amino acid ligands of zinc in plasma (Giroux and Henkin, 1972).

Giroux and Prakash (1977) tried to correlate the capacity of amino acids to induce trace metal excretion with the stability constants of their apparently main complexes, but unsuccessfully, confirming the complexity of the interactions at the origin of the metal excretion process. The appearance of the ECCLES computer program the same year (May et al., 1977; see Section A1 of this Chapter 3) constituted a real breakthrough in the investigation of these individual ligand effects.

IV. SPECIATION STUDIES RELATED TO TRACE METAL METABOLISM DURING TPN

A. Simulated effects of TPN on metal plasma distribution

The first study using ECCLES (Berthon et al., 1980) analyzed the effect of the infusion of a TPN mixture based on a study solution (Azonutril 25®, Roger Bellon Lab., Paris) on the distribution of zinc, copper and manganese in the plasma of three patients. For the amino acid concentrations directly measured in patients' plasma as well as for a model imitating the infusion process, nothing changed in the distribution *order* of the main complexes of the three metals. The difference in *percentages* observed for the real model was due to the fact that the histidine concentration in the TPN patients was lower than normal.

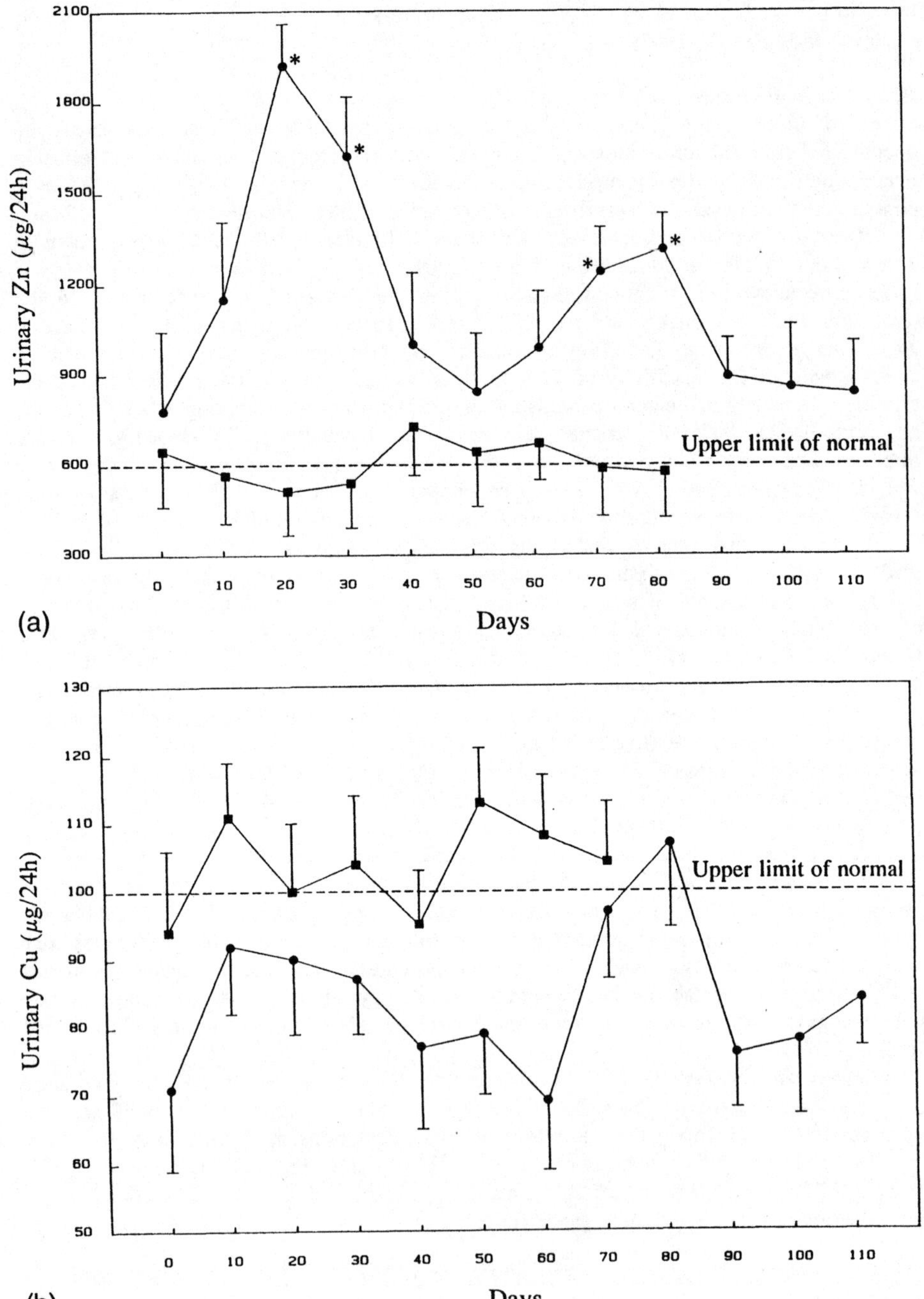

Figure 1.
A. Changes in urine zinc levels of patients during the enteral (■, 29 patients) or parenteral (●, 36 patients) administration of nutritive mixtures of similar composition (from Matuchansky et al., 1979).
B. Changes in urine copper levels of patients during the enteral (■, 29 patients) or parenteral (●, 36 patients) administration of nutritive mixtures of similar composition (from Matuchansky et al., 1979).

∗ p < 0.01 vs enteral.

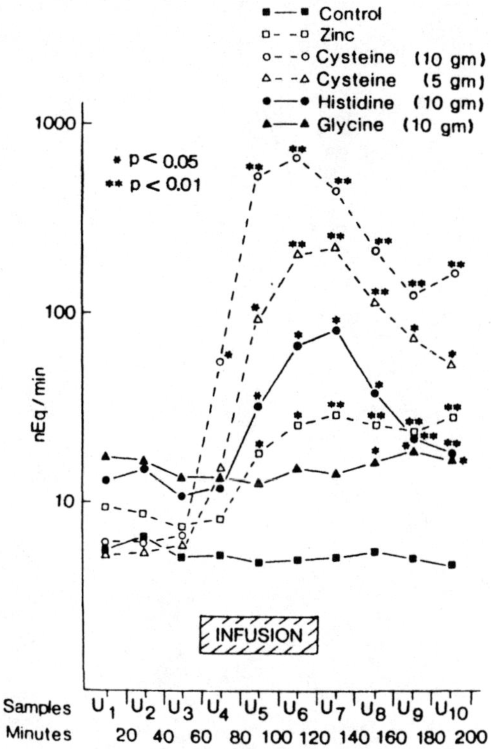

Figure 2.
Relative influence of cysteine, histidine and glycine on zinc urinary excretion in dogs (from Yunice et al., 1978).

More revealing was the comparative study simulating the infusion of increasing amounts of each amino acid separately. P.M.I.** logarithmic curves relative to zinc perfectly corroborated the findings of Yunice et al. (1978): (i) the effect of cysteine on the mobilization of zinc into its l.m.m. fraction appeared by far the most important, (ii) histidine ranked in second position, (iii) glycine and glycine-like amino acids had almost a negligible influence.

Judged the most effective model of representation of the mobilizing effect in practical terms, these P.M.I. curves were then regularly improved as the constants of zinc-amino acid complexes featuring on top of the distribution were progressively checked (Alemdaroglu and Berthon, 1981a; 1981b; 1981c), and were later extended to copper (Berthon, 1983; Berthon et al., 1984a; 1984b). For zinc, the effect of the whole TPN mixture (based on the study solution Azonutril 25®) turned out to be almost equivalent to the sum of the cysteine and histidine individual effects (Berthon, 1983), confirming the leading role of these two amino acids in zinc metabolism (Fig. 3). The fact that the main zinc complexes in plasma during the simulated infusion of the TPN mixture, i.e. $Zn(Cys)_2$ and Zn(Cys)(His), were electrically charged and hence filtrable by the kidneys, substantiated the excessive urinary losses of zinc observed (Fig. 1A) (Matuchansky et al., 1979). For copper, the most important amino acid was found to be histidine, glycine and its homologues being much less effective ligands (Berthon et al., 1984a) (Fig. 4). All important complexes in the copper

** P.M.I. (Plasma Mobilizing Index) has been defined as the ratio by which the l.m.m. fraction of a metal ion is increased in the presence of a drug with respect to normal plasma (May and Williams, 1977).

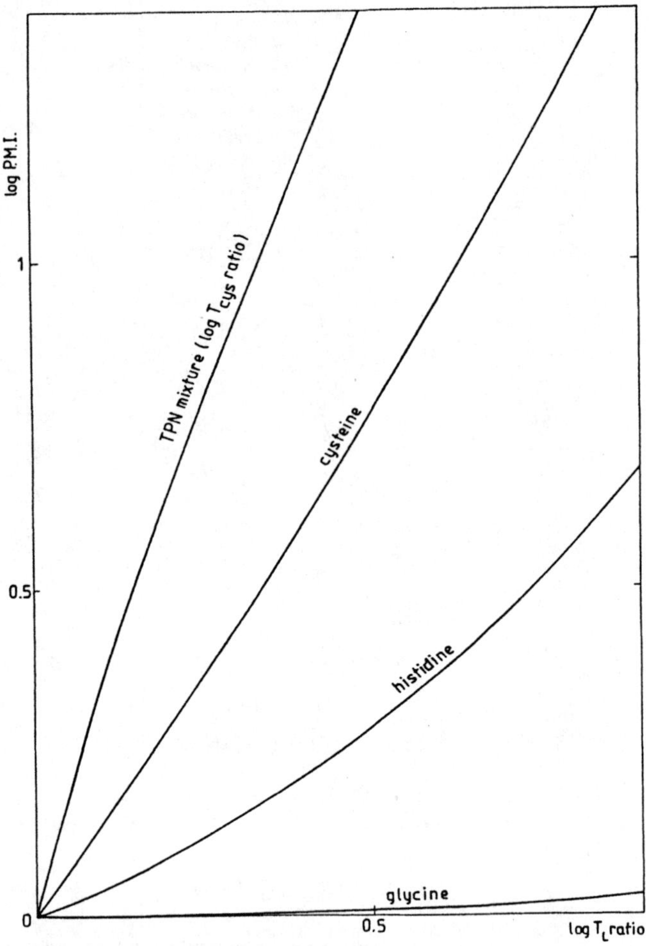

Figure 3.
Simulated influence of the separate infusion of cysteine, histidine, glycine, and of the whole TPN mixture based on Azonutril 25®, on the low-molar-mass fraction of zinc in blood plasma (from Berthon, 1983).

distribution ($Cu(His)_2$, Cu(His)(Thr), etc) were electrically neutral, which was consistent with the hypothesized predominant biliary excretion of copper (see Section III.B above).

B. Evaluation of metal doses compensating for TPN-induced losses

1. *Metal doses for a given nutritive mixture*

In parallel to the studies relative to metal excretions, ECCLES simulations were run to calculate the metal total concentrations to be added to the nutritive mixture under study (based on Azonutril 25®) in order to compensate for the TPN-induced losses. Based on the principle that mixing two solutions with identical free component concentrations avoids a redistribution of these components among their complex species, it was proposed that nearly ideal total concentrations of each metal ion in the TPN mixture could be obtained by simulation using as input data free metal ion concentrations pertaining in normal plasma (Berthon et al., 1980; 1981).

These calculations were performed for calcium (used as a reference as its required daily dose was pretty well known), magnesium, zinc, copper and manganese (Berthon et al., 1980; 1981). A daily dose of 507mg of calcium was found, to be compared with 500mg

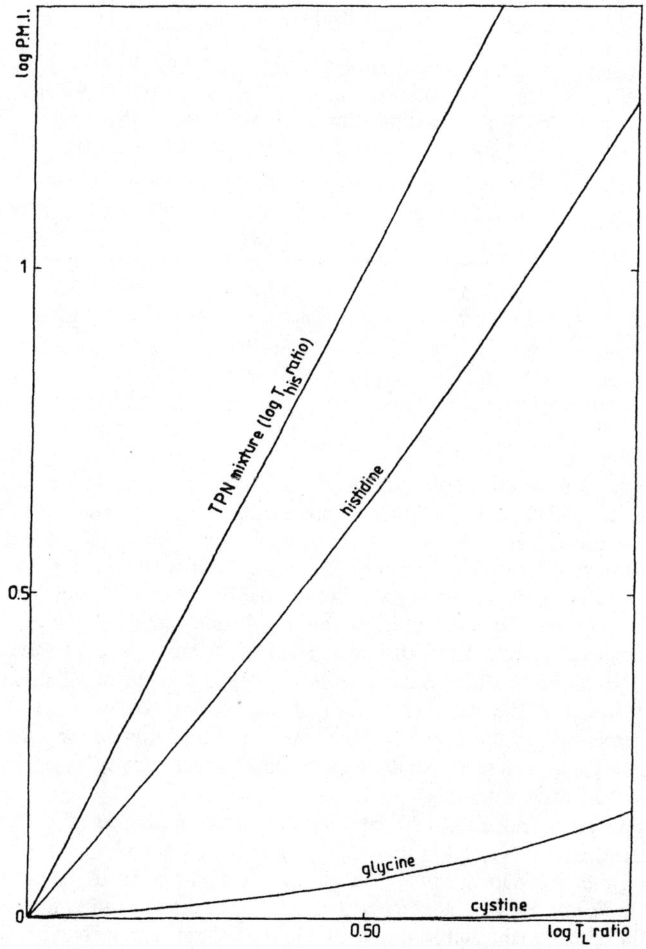

Figure 4.
Simulated influence of the separate infusion of histidine, glycine, cystine, and of the whole TPN mixture based on Azonutril 25®, on the low-molar-mass fraction of copper in blood plasma (from Berthon et al., 1984a).

actually administered. This very satisfactory agreement was interpreted as arising from the plasma free calcium concentration directly measured with ion-sensitive electrodes. For magnesium, the predicted daily dose lay between 200 and 400mg whereas the real quantity administered was about 450mg. For zinc, the daily dose was calculated between 25 and 50mg. For copper it was found equal to 1.3mg, and was virtually negligible for manganese. As these preliminary doses were found of the correct order of magnitude (Berthon et al., 1980; see also *Medicine*, Part Three, Chapter 1), the approach was considered valid, and further works were undertaken to improve the accuracy of the predictions (Alemdaroglu and Berthon, 1981a; 1981b; 1981c; Berthon et al., 1984a; 1984b). Table 1 shows the final results relative to the nutritive solution considered. The corresponding doses were close to the required daily allowances of these metals (see *Medicine*, Part Three, Chapter 1).

2. *Influence of cysteine on zinc predicted doses*
As cysteine was confirmed as the main amino acid ligand of zinc, further studies were done on the effect of its concentration in the nutritive mixture on the predicted daily dose of zinc (Blaquière and Berthon, 1987).

Table 1

Free and Total Concentrations, and Predicted Doses for Calcium, Magnesium, Zinc and Copper in the Nutritive Solution Based on Azonutril 25® (from Berthon et al., 1984b; see composition in Berthon et al., 1980)

Metal ion	Free concentration (mol dm^{-3})	Total concentration (mol dm^{-3})	Predicted dose (mg/day)
Ca^{2+}	1.13×10^{-3}	5.47×10^{-3}	566
Mg^{2+}	5.2×10^{-4}	4.63×10^{-3}	290
Zn^{2+}	1.0×10^{-9}	1.30×10^{-4}	21.9
Cu^{2+}	1.0×10^{-16}	9.96×10^{-6}	1.6

The TPN solution studied (based on Azonutril 25®, mixture I in Table 2) was taken as a reference for two other mixtures whose composition was the same as the first except for the source of amino acids: one was richer in cysteine/cystine (Hyperamine®, Bruneau Lab.), the other was poorer (Trophysan®, Egic Lab.). For each of these two mixtures, two cases were considered: first, amino acid solutions Hyperamine® and Trophysan® were considered to be introduced into the TPN solution in the same volume as Azonutril 25®; then, to attenuate the effect of the absolute extent of nitrogen in each mixture, they were considered to be added in such volumes as the nitrogen content of all three mixtures was finally equivalent. For the sake of a comparison, copper doses were also calculated. Considerable variations of the zinc dose were found (Table 2), even for equivalent nitrogen amounts. Although to a lesser extent, copper doses were also affected by the different compositions of the three solutions.

In order to get a closer insight into the individual influence of cysteine in a given nutritive mixture, its concentration (together with that of cystine) was made to vary in the solution based on Azonutril 25®. Considerable variations were again found in zinc predicted doses (Table 3). It is also clear, however, that the dose virtually stops decreasing below a certain cysteine concentration, histidine then behaving as a "buffer" ligand.

Table 2

Variation of Zinc Predicted Doses as a Function of the Cysteine Content of Different TPN Mixtures (from Blaquière and Berthon, 1987)

Mixtures	Concentrations (mol dm^{-3})		Predicted doses (mg)	
	Zn	Cu	Zn	Cu
I	1.297×10^{-4}	1.005×10^{-5}	21.88	1.65
II	3.597×10^{-4}	2.133×10^{-5}	60.66	3.50
III	1.542×10^{-4}	1.418×10^{-5}	0.26	0.02
IV	1.542×10^{-4}	1.509×10^{-5}	43.57	2.47
V	1.542×10^{-4}	1.207×10^{-5}	2.87	0.20

I = Mixture based on Azonutril 25®
II = Mixture based on Hyperamine® (volume equivalent to I)
III = Mixture based on Trophysan® (volume equivalent to I)
IV = Mixture based on Hyperamine® (nitrogen amount equivalent to I)
V = Mixture based on Trophysan® (nitrogen amount equivalent to I)

Table 3

Variation of the Predicted Dose of Zinc as a Function of the Cysteine Concentration in a Given Mixture (from Blaquière and Berthon, 1987)

Cys (mol dm^{-3})	log R	Zn (mol dm^{-3})	dose (mg)	A (%)	B(%)	C(%)
3.732×10^{-6}	-2.0	4.972×10^{-5}	8.38	0.0	0.7	44.1
1.180×10^{-5}	-1.5	5.130×10^{-5}	8.65	0.0	2.2	42.8
3.732×10^{-5}	-1.0	5.635×10^{-5}	9.50	0.2	6.4	38.9
1.180×10^{-4}	-0.5	7.285×10^{-5}	12.29	1.7	15.3	29.9
3.732×10^{-4}	0.0	1.297×10^{-4}	21.88	8.5	25.8	16.6
1.180×10^{-3}	0.5	3.436×10^{-4}	57.95	25.7	27.0	6.1
3.732×10^{-3}	1.0	1.197×10^{-3}	201.95	46.6	18.7	1.6
1.180×10^{-2}	1.5	4.535×10^{-3}	764.93	60.5	10.1	0.4
3.732×10^{-2}	2.0	1.697×10^{-2}	2862.07	64.7	4.7	0.1

A = $Zn(Cys)_2$, B = Zn(Cys)(His), C = $Zn(His)_2$
R = ratio by which the cysteine concentration varies in the mixture

VI. CONCLUSION

Many sets of guidelines for the use of vitamins and essential trace elements in TPN have appeared in the literature (see Shenkin, 1986). The AMA Expert Panel (Shils et al., 1979) recommended that preparations of individual trace elements should be available; specifically recommended were individual preparations of Zn, Cu, Cr and Mn, in order to allow maximum flexibility so that greater amounts of one element could be provided when necessary without causing overprovision of the others. Shenkin (1986) concluded that a mixture of elements could be used in the majority of patients, and that for those who may have a particular requirement for an individual element, adapted supplements should be given in addition to the multi-element mixture. An adapted version of the AMA guidelines by the ASCN (Greene et al., 1988) recommends the use of a Zn single-entity solution for hospitalized patients but that of a multi-element solution for Zn, Cu, Se, Cr, Mn, Mo and I for stable patients on long-term TPN in order to reduce costs and diminish the risk of error from multiple additions.

As it has now been demonstrated that specific TPN-induced extra-losses of trace metals are due to particular components of nutritive mixtures, it seems reasonable to recommend that new multi-element solutions be selectively adapted to the current amino acid solutions commercially available. Computer-aided speciation may greatly help to optimize these future adaptations. The only limitations to the present approach are (i) the degree of reliability of the formation constants used in the calculations, a number of which has to be determined (or redetermined) as a function of the preliminary complex distributions for each nutritive mixture, and (ii) the accuracy of the free metal ion concentrations (particularly zinc and copper) in normal blood plasma. The values of May et al. (1977) used in the studies reported here seemed a reasonable choice as these values are slightly inferior to those calculated by Agarwal and Perrin (1976). However, more accurate values are desirable to evaluate absolute quantities of trace metals to be routinely added to commercial mixtures.

ACKNOWLEDGEMENT

Warm thanks are due to Pr K.N. Jeejeebhoy and Pr C. Matuchansky for their kind reviewing of the manuscript.

REFERENCES

Abumrad, N.N., Schneider, A.J., Steel, D., and Rogers, L.S. (1981). *Amer. J. Clin. Nutr.*, *34*: 2551.

Agarwal, R.P., and Perrin, D.D. (1976). *Agents Actions*, *6*: 667.

Alemdaroglu, T., and Berthon, G. (1981a). *Bioelectrochem. Bioenerg.*, *8*: 49.

Alemdaroglu, T., and Berthon, G. (1981b). *Inorg. Chim. Acta*, *56*: 51.

Alemdaroglu, T., and Berthon, G. (1981c). *Inorg. Chim. Acta*, *56*: 115.

Arakawa, T., Tamura, T., Igarashi, Y., Suzuki, H., and Sandstead, H.H. (1976). *Amer. J. Clin. Nutr.*, *29*: 197.

Askari, A., Long, C.L., and Blakemore, W.S. (1979). *J. Parent. Ent. Nutr.*, *3*: 151.

Bengoa, J.M., Sitrin, M.D., Wood, R.J., and Rosenberg, I.H. (1983). *Amer. J. Clin. Nutr.*, *38*: 264.

Berkelhammer, C.H., Wood, R.J., and Sitrin, M.D. (1988). *Amer. J. Clin. Nutr.*, *48*: 1482.

Berthon, G. (1983). *Inorg. Chim. Acta*, *79*: 46.

Berthon, G., Matuchansky, C., and May, P.M. (1980). *J. Inorg. Biochem.*, *13*: 63.

Berthon, G., Matuchansky, C., and May, P.M. (1981). *Experientia*, *37*: 735.

Berthon, G., Blais, M.-J., Piktas, M., and Houngbossa, K. (1984a). *J. Inorg. Biochem.*, *20*: 113.

Berthon, G., Piktas, M., and Blais, M.-J. (1984b). *Inorg. Chim. Acta*, *93*: 117.

Blaquière, C., and Berthon, G. (1987). *Le Zinc en Médecine et Biologie* (A. Favier, J. Arnaud and H. Faure, eds.), Editions Médicales Internationales, Paris, p. 285.

Brown, R.O., Forloines-Lynn, S., Cross, R.E., and Heizer, W.D. (1986). *Digest. Dis. Sci.*, *31*: 661.

Burbrink Moser, P., Borel, J., Majerus, T., and Anderson, R.A. (1985). *Nutr. Res.*, *5*: 253.

Doisy, E.A. (1972). *Trace Substances in Environmental Health - VI* (D.D. Hemphill, ed.), University of Missouri Press, Columbia, p. 193.

Dunlap, W.M., James, G.W., and Hume, D.M. (1974). *Ann. Intern. Med.*, *80*: 470.

Fleming, C.R. (1989). *Amer. J. Clin. Nutr.*, *49*: 573.

Fleming, C.R., Hodges, R.E., and Hurley, L.S. (1976). *Amer. J. Clin. Nutr.*, *29*: 70.

Fleming, C.R., McCall, J.T., O'Brien, J.F., Forsman, R.W., Ilstrup, D.M., and Petz, J. (1984). *J. Parent. Ent. Nutr.*, *8*: 258.

Frankel, D.A. (1993). *Nutr. Res.*, *13*: 583.

Freeman, R.M., and Taylor, P.R. (1977). *Amer. J. Clin. Nutr.*, *30*: 523.

Freeman, J.B., Stegink, L.D., Meyer, P.D., Fry, L.K., and Denbenstein, L. (1975). *J. Biol. Res.*, *18*: 463.

Freund, H., Atamian, S., and Fischer, J.E. (1979). *J. Amer. Med. Assoc.*, *241*: 496.

Friedman, B.J. (1987). *J. Nutr.*, *117*: 133.

Giroux, E.L., and Henkin, R.J. (1972). *Biochim. Biophys. Acta*, *273*: 64.

Giroux, E.L., and Prakash, N.J. (1977). *J. Pharm. Sci.*, *66*: 391.

Greene, H.L., Hambidge, K.M., Schauler, R., and Tsang, R.C. (1988). *Amer. J. Clin. Nutr.*, *48*: 1324.

Hauer, E.C., and Kaminski, M.V., Jr. (1978). *Amer. J. Clin. Nutr.*, *31*: 264.

Henkin, R.I., Keiser, H.R., and Bronzert, D. (1972). *J. Clin. Invest.*, *51*: 44a.

Hoffmann, R.P., and Ashby, D.M. (1976). *Drug Intell. Clin. Pharm.*, *10*: 74.

Hsu, J.M., and Smith, J.C. (1983). *J. Nutr.*, *113*: 2171.

Jacobson, S., and Wester, P.O. (1977). *Brit. J. Nutr.*, *37*: 107.

Jeejeebhoy, K.N., Shu, R., Marliss, E.B., Greenberg, G.R., and Robertson, A.B. (1975). *Clin. Res.*, *23*: 636A.

Jeejeebhoy, K.N., Shu, R., Marliss, E.B., Greenberg, G.R., and Robertson, A.B. (1977). *Amer. J. Clin. Nutr.*, *30*: 531.

Karpel, J.T., and Peden, V.H. (1972). *J. Pediat.*, *80*: 32.

Kay, R.G., and Tasman-Jones, C.T. (1975). *Lancet*, *2*: 605.

Kay, R.G., Tasman-Jones, C.T., Pybus, J., Whiting, R., and Black, H. (1976). *Ann. Surg.*, *183*: 331.

Ladefoged, K., and Jarnum, S. (1983). *Metal Ions in Biological Systems*, Vol. 15 (H. Sigel, ed.), Marcel Dekker, New York and Basel, p. 415.

Matuchansky, C., Druart, F., Aries, J., and Guillard, O. (1979). Proceedings of the 1st European Congress on Parenteral and Enteral Nutrition, Stockholm, September 2-5, p. 7.5.

May, P.M., and Williams, D.R. (1977). *FEBS Lett.*, *78*: 134.

May, P.M., Linder, P.W., and Williams, D.R. (1977). *J. Chem. Soc., Dalton Trans.*, 577.

McClain, C.J. (1981). *J. Parent. Ent. Nutr.*, *5*: 424.

McKenzie, J.M. (1977). *Tr. Elem. Hum. Anim. Health Dis. N.-Z.*, 59.

Neumann, P.Z., and Sass-Kortsak, A. (1967). *J. Clin. Invest.*, *46*: 646.

Okada, A., Takagi, Y, Itakura, T, Satani, M., Manabe, H., Iida, Y., Tanigaki, T., Iwasaki, M., and Kasahara, N. (1976). *Surgery*, *80*: 629. (reprinted in (1989) *Nutrition*, *5*: 11.)

Phillips, G.D., and Garnys, V.P. (1981). *J. Parent. Ent. Nutr.*, *5*: 11.

Pories, W.J., Mansour, E.G., Plecha, F.R., Flynn, A., and Strain, W.H. (1976). *Trace Elements in Human Health and Disease*, Vol. 1 (A.S. Prasad and D. Oberleas, eds.), Academic Press, New York, p. 115.

Prasad, A.S., and Oberleas, D. (1968). *J. Lab. Clin. Med.*, *72*: 1006.

Prasad, A.S., and Oberleas, D. (1970). *J. Lab. Clin. Med.*, *76*: 416.

Sedman, A.B., Klein, G.L., Merritt, R.J., Miller, N.L., Weber, K.O., Gill, W.L., Anand, H., and Alfrey, A.C. (1985). *N. Engl. J. Med.*, *312*: 1337.

Shenkin, A. (1986). *Proc. Nutr. Soc.*, *45*: 383.

Shike, M., Roulet, M., Kurian, R., Whitwell, J., Stewart, S., and Jeejeebhoy, K.N. (1981). *Gastroenterology*, *81*: 290.

Shils, M.E., Burke, A.W., Greene, H.L., Jeejeebhoy, K.N., Prasad, A.S., and Sandstead, H.H. (1979). *J. Amer. Med. Assoc.*, *241*: 2051.

Shulman, R.J. (1989). *Amer. J. Clin. Nutr.*, *49*: 879.

Solomons, N.W., Layden, T.J., Rosenberg, I.H., Khactu, K.V., and Sandstead, H.H. (1976). *Amer. J. Physiol.*, *70*: 1022.

Tucker, S.B., Schroeter, A.A., Brown, P.W., and McCall, J.T. (1976). *J. Amer. Med. Assoc.*, *235*: 2399.

Vanderhoof, J.A., and Antonson, D.L. (1986). *Total Parenteral Nutrition, Utilization, Complications, and Pathophysiological Considerations* (E. Lebenthal, ed.), Raven Press, New York, p. 109.

Van Rij, A.M., Thompson, C.R., McKenzie, J.M., and Robinson, M.F. (1979). *Amer. J. Clin. Nutr.*, *32*: 2076.

Vilter, R.W., Bozian, R.C., Hess, E.V., Zellner, D.C., and Petering, H.G. (1974). *N. Engl. J. Med.*, *291*: 188.

Weissman, K., Hjorth, N., and Fischer, A. (1976). *Clin. Exp. Dermatol.*, *1*: 237.

Wolman, S.L., Anderson, G.H., Marliss, E.B., and Jeejeebhoy, K.N. (1979). *Gastroenterology*, *76*: 458.

Yunice, A.A., King, R.W., Kraikitpanich, S., Haygood, C.C., and Lindeman, R.D. (1978). *Amer. J. Physiol.*, *235*: F40.

Zlotkin, S.H. (1989). *J. Pediat.*, *114*: 859.

SPECIATION STUDIES IN RELATION TO THE BIOAVAILABILITY AND DRUG ACTIVITY OF TETRACYCLINES

by

Guy Berthon and Luc Lambs

INSERM U 305,
Equipe "Bioréactifs: Spéciation et Biodisponibilité",
38 rue des Trente-six Ponts
31400 Toulouse
France

I. INTRODUCTION

Like many pharmaceuticals (see Part Four, Chapter 4; also *Medicine*, Volume 2, Part Five, Chapter 4), the tetracycline (TC) molecule and its active derivatives (Scheme 1) contain several electron-donor groups likely to coordinate metal ions in vivo. In his pioneering work in view of a possible similarity between tetracyclines and 8-hydroxyquinoline in respect of metal potentiation (Albert, 1953; Albert and Rees, 1956), Albert (Albert and Rees, 1956) noted that "In any event, knowledge of the metal-binding properties of these drugs (was) valuable because in the human body they (could) not fail to combine with

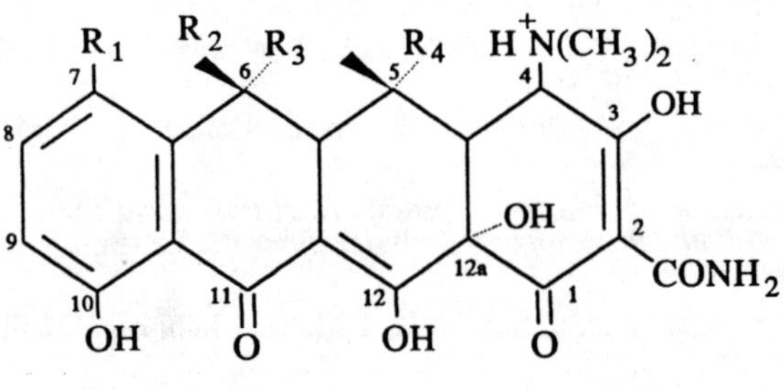

	R_1	R_2	R_3	R_4	
Tetracycline (TC)	H	OH	CH_3	H	
Oxytetracycline (OTC)	H	OH	CH_3	OH	(5α-oxy-TC)
Doxycycline (DOX)	H	H	CH_3	OH	(6-deoxy-OTC)
Minocycline (MIN)	NH_2	H	H	H	(7-amino-6-deoxy-6-demethyl-TC)
Methacycline (MTC)	H	$-CH_2-$		OH	(6-methylene-OTC)
Chlorotetracycline (CTC)	Cl	OH	CH_3	H	(7-chloro-TC)
Demeclocycline (DMC)	Cl	OH	H	H	(6-demethyl-CTC)

Scheme 1

metallic ions". Some twenty years later, Newman and Frank (1976) pointed out that "From an antimicrobial standpoint, the tetracycline complexes with Mg^{2+} and Ca^{2+} (were) probably the most important, because the concentrations of these cations and the large formation constants of their complexes with tetracycline (dictated) that in vivo TC would exist as one of these two complexes". In the meantime, Kohn (1961) had been the first to report on the possible mediation of metal ions in the binding of tetracyclines to biological macromolecular sites with which these drugs could not associate by themselves.

As is frequent in Science, the essentials of our present knowledge on metal-tetracycline interactions in vivo were already contained in these early views, even if in many respects reality should eventually reveal different from what was anticipated at that time. For example, it would now seem more appropriate in this context to distinguish between drug bioavailability and therapeutic activity. Even though metal-drug interactions may influence both, the extent to which they do so depends on several factors that condition the strength of the bonds at different physiological sites: (i) the pH, which determines the donating center involved as well as its degree of coordinating capacity, (ii) the formation constants pertaining to this donating center, and (iii) the metal and drug concentrations and concentration ratios. For a drug like TC, whose numerous potential O- and/or N-containing binding sites may contribute to coordinate different metal ions with roughly comparable stabilities, the concentration factor intuitively appears as determining in regard to bioavailability. For example, if a ML complex with calcium is "only" 10^6 times less stable than its copper homologue whereas, in blood plasma, free $[Ca^{2+}]$ is about 10^{13} more concentrated than $[Cu^{2+}]$, it seems logical to anticipate that calcium will play a greater role than copper towards TC bioavailability. Nonetheless, if the reactivity of TC with a redox metal like copper is to be taken as a possible key factor for its therapeutic action (see *Medicine*, Part Five, Chapter 4-D), the free concentration of drug remains sufficiently high with respect to the total concentration of metal for such an interaction to occur.

The great complexity of the TC molecule and the influence of substituents on its coordinating capacity were also beyond approach at the time of the above studies. The stoichiometries of metal-tetracycline complexes that were then characterized were limited to ML_n and $M(LH)_n$ types, and the consideration of all L, LH and LH_2 forms of the ligands in complexation has been relatively recent (Brion et al., 1981). Also recent has been the assignment of the conformations assumed by the TC molecule and its main complexes as a function of pH, a direct result of the better knowledge of the involved equilibria (see later).

These advances were made possible by the huge development of computer programs for the analysis of metal complex equilibria in the seventies (see Part Three, Chapter 1-A3)*. In parallel, a decisive progress was made in the field of computer simulation models, the most recent of these taking into account virtually all the concentration factors that may influence metal and drug distributions in the main biofluids. In conjunction with the data collected on tetracycline action in relation to metal ions, the use of these models allowed a closer insight into metal-drug interactions at the molecular level.

Finally, the recent discovery of the role of metal-induced free radicals in the potentiation of organic drugs has also changed the context in which the therapeutic activity of antibiotics was considered, re-centering the interest of researchers on the interactions of these drugs with redox metal ions (see *Bioinorganic Medicine*, Part Five, Chapter 4).

In this context, this short review summarizes the different aspects of metal-tetracycline interactions, with special emphasis on the use of computer-aided speciation for the comprehension of the roles of metal ions in the bioavailability and activity of these drugs.

II. METAL ION - TETRACYCLINE INTERACTIONS IN VIVO

A. Metal ion - tetracycline interactions in hard tissues

The strong affinity of tetracyclines for mineralizing systems such as teeth and bone has long been reported (Filippi and Mela, 1957; Bevelander and Cohlan, 1962; Cohlan et al., 1963;

---

* In particular, the appearance of MINIQUAD (Sabatini et al., 1974) and of PSEUDOPLOT (Corrie et al., 1975), an adapted version of the former HALTAFALL (Ingri et al., 1967) used to plot pseudo-experimental formation curves from different constant sets, represented a real breakthrough in complex species selection.

Bevelander, 1964) and is well documented (Skinner and Nalbandian, 1975; Kerley and Kollar, 1978; Lambrou et al., 1977 and references therein). Given the rate of mineralization in the early stages of life, infants and children are especially affected. Long-term administration of tetracyclines from prenatal life through childhood results in enamel discoloration and hyperplasia of tooth crowns (Kerley and Kollar, 1978). As little as 1g/day TC for 3 days during the third trimester can produce yellow staining of deciduous teeth (Cohlan, 1977). Reduced growth rate of long bones of premature children receiving high doses of tetracyclines has also been observed (Cohlan et al., 1963).

Investigations on mouse embryonic long bones in culture (Kaitila et al., 1970) have shown that the deleterious impact of TC on osteogenesis is two-fold:
(i) At low drug levels, i.e. 1-10 μg/ml, TC has a markedly depressing effect on bone mineralization. As was shown by Perrin (1965) from molecular models, TC can bind to calcium atoms of bone apatite without steric strains. The inhibition of calcification by TC could thus be a direct blocking effect of the further growth of a mineral crystal (Kaitila et al., 1970). On the other hand, in vitro studies of the incorporation of TC into enamel (Lambrou et al., 1977) have shown that this phenomenon occurs only during mineralization or remineralization. The drug deposition seems to require the formation of a calcium-TC complex of ionic character in the calcifying solution, which then precipitates with the anionic phosphates of the interface. Incidentally, the calcium-TC fluorescent complex has been used as a biological tag in developing tissues (Hoerman, 1975).
(ii) At higher concentrations, impairment of osteogenesis would be due to both inhibited mineralization *and* depressed matrix formation (Kaitila et al., 1970). Tetracycline affects collagen synthesis by inhibiting both the incorporation of proline into protocollagen and the conversion of protocollagen to collagen. The latter inhibition is probably due to an interaction of TC with ferrous iron, which is required in the hydroxylation of proline by protocollagen hydroxylase (Halme et al., 1969).

B. Metal ion - tetracycline interactions in relation to drug bioavailability

The first factor to determine the therapeutic response of a drug is its bioavailability at the receptor site. For orally administered pharmaceuticals, this bioavailability involves both the absorption and distribution processes.

1. *Gastrointestinal absorption*

Like most drugs (Gudzinowicz et al., 1984), tetracyclines are absorbed from the gastrointestinal tract by passive diffusion (Colaizzi and Klink, 1969). As lipid solubility, essential to membrane penetration, mainly derives from electrical neutrality, the electron-donor groups of tetracyclines have a strong influence on their absorption through their interactions with proton and metal ions.

The first effect of metal-containing compounds to consider with regard to drug absorption is that due to Al-, Ca-, and Mg-based antacids, whose induced elevation of pH may affect drug solubility. As a matter of fact, tetracycline absorption is considerably reduced in the presence of such antacids (Khalil et al., 1977; Mapp and McCarthy, 1976; Schumacher and Linn, 1978; see also *Bioinorganic Medicine*, Part Six, Chapter 2-A).

Metal complexation also has a determining influence on tetracycline absorption. Among the metal ions involved, iron (both as Fe^{2+} and Fe^{3+}) has often been mentioned (Neuvonen et al., 1970; Neuvonen and Turakka, 1974; D'Arcy et al., 1976; Poiger and Schlatter, 1979; D'Arcy, 1985), with considerable quantitative differences from one iron compound to another. In particular, the stability of the iron complex administered together with the antibiotic is determining for the effect observed (Neuvonen and Turakka, 1974): the more stable the iron complex, the lesser effect on drug absorption. Interestingly, the order of activity of different iron salts was found to be the same as that of their own intestinal absorption (Neuvonen and Turakka, 1974). This seems to be in relation with the fact that too stable complexes in the gastrointestinal fluid cannot release their metal ions to the endogenous ligands present at the interface of the membrane (see Part Four, Chapter 1-B). Reciprocally, TC also impairs the absorption of iron (Neuvonen et al., 1975).

That both antibiotic and metal ion administered simultaneously interfere with the gastrointestinal absorption of each other has also been demonstrated for zinc:

coadministration of zinc with the antibiotic markedly reduces the absorption of TC (Mapp and McCarthy, 1976; Andersson et al., 1976; Pentillä et al., 1975) and of doxycycline, although to a lesser extent (Pentillä et al., 1975). On the other hand, impaired zinc absorption by oral TC has been observed in rats (Weismann et al., 1978).

Calcium also affects TC absorption (Poiger and Schlatter, 1979). Experiments on rat small intestine have shown increased disappearance of TC from recirculating solution in the presence of calcium, the maximum effect being reached at a calcium-to-TC ratio around 2 (Kakemi et al., 1968a). A correlation was noted between the disappearance percent and the partition ratio of TC between isoamyl alcohol and water in the presence of calcium, indicating increased lipid solubility. In spite of this, the antibiotic was demonstrated to be less favorably absorbed in the presence of calcium in vivo (Kakemi et al., 1968a), because of excessively strong binding to the gastrointestinal mucosa (Kakemi et al., 1968b).

The antagonistic roles of magnesium (Khalil et al., 1977) and even bismuth (Mapp and McCarthy, 1976) against tetracycline absorption have also been mentioned. All these observations have led to recommend the administration of tetracycline antibiotics on an empty stomach (Poiger and Schlatter, 1979).

2. *Body distribution*

Comparative studies of total and "free" (from protein binding) concentrations of various tetracyclines in serum implicitely suggest a significant role for metal ions in drug body distribution. For example, the remarkably high serum concentrations of methacycline (MTC) cannot be explained by its strong interaction with serum protein alone (Schach von Wittenau and Yeary, 1963). No direct quantitative data, however, has ever been produced, and lipid solubility of the free drug molecule was considered as the only criterion for its distribution between tissues and interstitial fluid (Schach von Wittenau and Delahunt, 1966).

C. Metal ion - tetracycline interactions in relation to drug antibacterial action

Experimental data on this topic are diverse and often controversial. Metal ions were first considered as antagonists of the intrinsic antibacterial action of tetracyclines. Thus, Fe^{3+}, Mn^{2+} and Mg^{2+} ions were shown to reduce the inhibitory action of TC against *Pseudomonas aeruginosa* (Weinberg, 1954). Calcium and magnesium were observed to limit the accumulation of TC in *Bacillus cereus* cells (Connamacher et al., 1967), magnesium also reducing the ability of the drug to bind to isolated t-RNA (Day, 1966). More recently, in an in vitro test of the influence of metal ions on the action of six tetracyclines against *Klebsiellae*, all drugs were strongly neutralized by Fe^{3+}, three being moderately neutralized by Cu^{2+}, two by Mg^{2+}, and one by Ca^{2+} (Miles and Maskell, 1986). The antagonistic action of TC and ferric ion in vivo has also been reported (Miles and Maskell, 1985). These results may be related to a direct influence of redox metal ions on the antibacterial action of tetracyclines (see below).

A positive role for metal ions was suggested by the ability of a zinc-TC complex to bind single- and double-stranded DNA (Pato, 1977), a result in line with Kohn's (1961) demonstration that TC could bind DNA in the presence of zinc, calcium, manganese, or magnesium, in decreasing order of effectiveness. In contrast, Mikelens and Levinson (1978) found that TC complexes with Cu^{2+}, Ni^{2+} and Co^{2+} could bind DNA, but that those with Zn^{2+}, Fe^{2+}, Fe^{3+}, Mn^{2+}, Mg^{2+}, and Ca^{2+} could not. The possible mediation of Mg^{2+} in the binding of TC to *E. Coli* ribosomes has also been debated (White and Cantor, 1971; Tritton, 1977; Martin, 1979). By fluorescence, Dokter and Magnuson (1974) demonstrated that 7-chloro-tetracycline (CTC) was actively transported into *Staphylococcus aureus* by Ca^{2+} or Mg^{2+} ions located inside the membrane at a site inaccessible to EDTA.

Da Silva and Dias (1973) suggested that the bacteriostatic action of tetracyclines was linked with the formation of mixed-metal complexes, consisting mainly in the inhibition of protein and nucleic acid synthesis. More importantly, Quinlan and Gutteridge (1988) have recently shown that tetracyclines caused the degradation of carbohydrate in the presence of a ferric salt at pH 7.4, and that in the presence of copper salts they could substantially damage linear duplex DNA, hydroxyl radicals being involved in both cases. In connection with this, the above-mentioned antagonistic action of cupric (and ferric) ions with respect to tetracycline antibacterial activity (Miles and Maskell, 1985; 1986) may be related to the recently disclosed inhibitory effect of Cu^{2+} ions towards Fe^{2+}- and Cu^{+}-driven Fenton reactions (Maestre et al., 1992; 1994).

III. TETRACYCLINE COMPLEX EQUILIBRIA

The main objective of this chapter is not to critically review the formation constants of proton and metal complexes of tetracyclines, nor to discuss in detail the assignments of corresponding bonds, which have already been examined (Martin, 1985). However, the reliability of the predictions made from computer-aided speciation hinges upon the true selection of formation constants on which relevant simulation models are based, and the distribution of electrical charges within a molecule or an ion is critical to its bioavailability. Related basic information is thus worth recalling in this respect.

A. Proton-tetracycline equilibria

The TC molecule (Scheme 1) contains three protonation centers: the C4 dimethylamino group (I), the C10-C12 phenolic diketone system (II) (most presumably O12; Martin (1985)), and the C1-C3 tricarbonyl methane moiety (III) (O3; Martin (1985)) in the increasing acidity order in aqueous solution (Leeson et al., 1963). However, this order is subject to change with the medium: the lower the dielectric constant, the less predominant the protonation of I over II (Martin, 1985) - for example, the protonation of II is favored in 50% methanol (Martin, 1985).

A second point is worth considering: the conformation of the TC molecule changes with its protonation state. In basic aqueous solutions, the dianion adopts a form known as conformation A (Mitscher et al., 1968; 1972) or "extended" conformation (Lambs et al., 1988a), in which the C1-C3 and amide carbons lie above the roughly accomodated BCD ring plan, hydrogen bonding being expected to take place between N4 and $OH12_a$ atoms. In neutral to acidic aqueous media, a similar structure prevails in which the A ring is somewhat twisted to relieve the steric crowding between the bulky solvated C4 dimethylammonium and $C12_a$ hydroxyl groups, with the $N4H^{+}$-O3 hydrogen bond characterized on solid being presumably maintained. This form is referred to as conformation B (Mitscher et al., 1968; 1972) or "twisted" conformation (Lambs et al., 1988a) - in contrast to the former "extended" designation by Martin (1985).

The aqueous zwitterionic TC, which assumes "twisted" conformation B seen in hydrated crystals (Stezowski, 1976), is in equilibrium with the unionized free base in conformation E (Mitscher et al., 1968; 1972) or "folded" conformation (Martin, 1985) characterized from anhydrous crystals (Stezowski, 1976; Prevo and Stezowski, 1977; Hughes et al., 1979). This equilibrium may reveal important in that speciation calculations do not take into account microconstant analyses (see Leeson et al., 1963 and Martin, 1985). It was contended (Hughes et al., 1979) that "the nonionized form of the free base is the important lipid-soluble tetracycline species", but this statement was considered excessive in the light of more recent findings establishing the prime role of calcium and magnesium complexes in tetracycline bioavailability (Lambs et al., 1984; see below). For details on tetracycline conformations, see Lambs et al. (1988a; 1988b) and references therein.

B. Metal-tetracycline equilibria

This review being concerned with the role of tetracycline complex equilibria in vivo, the following section exclusively deals with determinations relative to the physiological medium. For discussions of other data, see Martin (1985).

1. *Calcium and magnesium*

As pointed out by Martin (1985), the variations observed from one tetracycline to another in their calcium and magnesium complexation schemes (Table 1) seemed initially difficult to understand. A careful circular dichroism (CD) analysis of the different conformations adopted by tetracyclines in their various complexes in solution revealed very helpful in unravelling the complexity of these observations (Lambs et al., 1988a; 1988b).

For example, it was surprising at first sight that, in contrast with oxytetracycline (OTC) and doxycycline (DOX), no ML complex formed in the calcium-TC system whereas

Table 1

Formation Constants for Calcium and Magnesium Complexes of Tetracyclines under Physiological Conditions (for abbreviations, see text).

Ligand	Calcium			Magnesium		
	Complex	logβ	Reference	Complex	logβ	Reference
TC	ML_2H_2	25.540	Brion et al.	MLH	12.657	Berthon et al.
	ML_2H	17.618	(1981)	ML_2H_2	25.275	(1983)
	ML_2	8.731		ML_2H	17.597	
	M_2L	8.671		ML_2	8.698	
				M_2L	7.740	
OTC	ML_2H_2	24.625	Brion et al.	ML_2H_2	24.095	Berthon et al.
	ML_2H	16.654	(1981)	ML_2H	17.423	(1983)
	ML_2	8.385		ML_2	9.560	
	ML	4.462		ML	4.874	
	M_2L	7.884		M_2L	8.346	
				M_2LH	14.970	
DOX	MLH	13.058	Brion et al.	MLH	12.988	Berthon et al.
	ML_2H_2	25.263	(1981)	ML_2H_2	25.559	(1983)
	ML	5.600		ML_2H	17.420	
	M_2L	8.885		M_2L	8.546	
MIN	MLH	14.183	Brion et al.	MLH	13.088	Berthon et al.
	ML_2H_2	27.228	(1981)	ML_2H_2	26.728	(1983)
	ML	5.981		ML_2H	17.905	
				ML	5.886	
				MLH_2	18.391	
				ML_2H_3	32.323	
				M_2LH	15.824	
CTC	MLH_2	17.984	Lambs et al.	ML_2H_2	23.995	Lambs et al.
	ML_2H_2	25.230	(1985)	MLH	11.515	(1985)
DMC	ML_2H_2	24.269	Lambs et al.	ML_2H_2	23.592	Lambs et al.
	ML_2H	16.355	(1985)	ML_2H	15.416	(1985)
	ML	4.569		MLH	11.852	
	M_2L	8.343		M_2L	7.605	
MTC	ML_2H_2	24.860	Lambs and	ML_2H_2	25.196	Lambs and
	ML	5.306	Berthon (1988)	ML_2H	17.233	Berthon (1988)
				MLH	12.373	
				ML	5.142	
DSC	MLH	12.718	Lambs et al.	MLH	13.074	Lambs et al.
	ML_2H_2	26.376	(1988a)	ML_2H_2	26.566	(1988a)
	ML_2H	19.550		ML_2H	19.325	
	ML_2	9.942		ML_2	9.307	
	ML	5.159		ML	5.495	
DTC	ML	4.061	Lambs et al.	ML	4.744	Lambs et al.
	ML_2	5.913	(1988a)	ML_2	7.642	(1988a)

M_2L did (Brion et al., 1981). The observation of unchanged CD spectra relative to TC in the presence of calcium up to pH 9 revealed that no coordination can occur as long as the dimethylammonium group remains protonated (Lambs et al., 1988b). Then, as it dissociates, a Ca^{2+} ion binds through N4 and $O12_a$, which "locks" the molecule in conformation A and renders O12 and O1 atoms (which pointed in different directions in conformation B) accessible to a second Ca^{2+} ion. Thus, both Ca^{2+} ions bind simultaneously as the dimethylammonium group dissociates.

To help discriminate among the potential metal-binding sites of active tetracyclines, calcium and magnesium complexes with two simpler parent substances, 4-dedimethylamino-tetracycline (DTC) and 6-desoxy-6-demethyl-tetracycline (DSC), were investigated. In contrast with the largely accepted principle that the first Ca^{2+} or Mg^{2+} ion to bind a tetracycline as the pH is raised would do so through its C10-C12 donor site (see references in Lambs et al., 1988a), both stability constant analysis and CD spectra suggested the N4 atom as first coordinating site for calcium (with $O12_a$) and magnesium (with O3) in their 1:1 complexes, the C10-C12 system being also involved in 1:2 M:L species.

General rules eventually emerged from the various observations made (Berthon, 1989):
- tetracyclines preferentially bind Ca^{2+} through their N4-$O12_a$ atoms (TC, DSC, CTC: conformation A), or N4-O5 when an OH group is present in C5 (DOX: conformation C), via six-membered chelate rings;
- in contrast, the smaller Mg^{2+} is preferentially chelated through N4-O3 atoms (TC, DSC, CTC: conformation B; DOX: conformation C) via five-membered rings.

However, these general tendencies suffer exceptions for even closely related derivatives. In CTC for example, the Cl7 substituent not only exerts its classical electron-withdrawing effects, but also interacts with the 6-α-methyl-substituent, inducing a twisting in the BCD ring system so that the pseudoboat form of the C ring is almost totally reversed at basic pH (D-like conformation). More details are available in Lambs et al. (1988a; 1988b).

The distinct binding modes of tetracyclines with calcium and magnesium as well as the occurrence of binuclear complexes with these two metals suggested the possible formation of mixed-metal ternary species, which was confirmed experimentally (Lambs et al., 1984).

2. *Copper*

Tetracyclines form very stable complexes with copper (Brion et al., 1986; Lambs and Berthon, 1988; Jezowska-Bojczuk et al., 1993). The main characteristic of the Cu^{2+} ion is its capacity to bind to their monoanionic form (i.e. LH_2^-; or LH_3^- for minocycline). This phenomenon, common to all bioactive tetracyclines, is reminiscent of the binding of copper to carboxylate in glycine-like amino acids (Berthon et al., 1984 and references therein).

The involvement of O3 as first copper coordination center as tetracyclines start deprotonating has been confirmed through various spectroscopies (Jezowska-Bojczuk et al., 1993). When the pH is raised, MLH substitutes for MLH_2 with chelation taking place at the O10-O12 system as the OH12 group dissociates. The ultimate deprotonation of the C4 dimethylammonium group allows the formation of ML with a change of the binding mode in favor of N4-$OH12_a$ centers (conformation A). As to 1:2 M:L complexes, except for ML_2H_4 which is bound in the MLH_2 manner, all others (ML_2H_2, ML_2H, ML_2) presumably imply alternate bonds at the N4-$O12_a$ pinch and the O10-O12 juncture.

3. *Zinc*

Zinc forms significantly less stable complexes than copper with tetracyclines (Brion et al., 1985; Lambs and Berthon, 1988). In particular, the bonds involved in MLH_2 complexes are weaker by some three orders of magnitude (logβ differences between MLH_2 and LH_2, found near 5 for copper, are only around 2 for zinc). In absolute terms, a zinc-tetracycline bond is only about one log unit more stable than its homologue with calcium or magnesium.

IV. SPECIATION STUDIES

A. Blood plasma investigations

1. *Influence of metal ions on drug distribution*

Until the early eighties, tetracyclines were considered by pharmacologists to be distributed into two fractions in blood plasma: (i) a protein-bound pool acting as storage buffer, and (ii) the molecule in the free state (i.e. active), hence the exclusive use of partition coefficient considerations for assessing drug bioavailability (Schach von Wittenau and Yeary, 1963; Schach von Wittenau and Delahunt, 1966).

The first speciation studies relative to the distribution of four tetracyclines (TC, OTC, DOX and MIN) in blood plasma, which took into account interactions with calcium only, clearly demonstrated that more than 95% of the fraction of antibiotic previously considered as free was actually comprised of calcium complexes (Brion et al., 1981). The use of drug pharmacokinetic data within 24h following injection also clearly indicated that, given the high metal-to-drug ratios throughout, the time factor could be neglected.

A subsequent study devoted to magnesium interactions with the same antibiotics (Berthon et al., 1983) led to the conclusion that in blood plasma tetracyclines were predominantly bound to calcium but that the influence of magnesium was far from negligible, the relative proportions of drug bound to each metal depending on the nature of the antibiotic. Another important conclusion of that study was that, due to the high metal-to-ligand ratios pertaining to plasma, binuclear complexes of both metals, especially calcium, were predominant in drug distributions, suggesting the possible occurrence of mixed-metal complexes. These were effectively shown to take a significant part in the plasma distribution of OTC, TC and MIN (Table 2) (Lambs et al., 1984).

Considering the predominance of these electrically charged homo- or hetero-binuclear complexes in drug distributions (Table 2) whereas neutral species only are lipid soluble, it appeared that blocking one of the two coordination centers of calcium and/or magnesium might, by reducing the extent of its charged fraction, allow the drug a better tissue affinity. The C10-C12 site being presumably involved in the coordination of both metals, and its electron density being under the control of substituents in the C5-C9 region, the effect of the chloro group at C7 was investigated (Lambs et al., 1985).

Of the two derivatives chosen for that study (i.e. CTC and 6-demethyl-7-chloro-tetracycline, DMC), CTC effectively proved to be a rather poor ligand of calcium and magnesium, its protonated species being even predominant at plasma therapeutic levels. As a result, its percentage of neutral species was the highest of all the tetracyclines investigated at that time (Table 2). Surprisingly, the coordinating behavior of DMC was almost not affected (Lambs et al., 1985) (see Table 1).

In MTC, the electron-attracting methylene group in C6 is also expected to reduce the affinity of the C10-C12 system for metal ions. The investigation of corresponding complex equilibria (Lambs and Berthon, 1988) confirmed the nonexistence of binuclear species for both calcium and magnesium (Table 1). As a result, electrically neutral complexes with the two metals proved to represent about 80% of MTC not bound to proteins, which is by far the highest neutral fraction found for a tetracycline in plasma (Table 2).

The above result is in line with several clinical observations. In humans, MTC is more slowly cleared by the kidneys than TC (Kunin, 1962). The more marked serum binding of MTC advocated to interpret this effect was effectively confirmed later in humans and dogs, and 'cannot be explained by its strong interaction with serum proteins alone' (Schach von Wittenau and Yeary, 1963). According to the latter authors, 'the distribution of a tetracycline is ... primarily influenced by its affinity for body tissues': in other words, the comparatively low extent of the fraction of MTC not bound to proteins would derive from both protein binding *and* tissue affinity. Logically, the half-life of MTC was found to be the highest of all the tetracyclines examined (Schach von Wittenau and Yeary, 1963).

As for zinc and copper, their free concentrations are far too low, especially with regard to calcium and magnesium, for them to exert any quantifiable effect on plasma tetracycline distributions (Brion et al., 1985; Brion et al., 1986; Lambs and Berthon, 1988).

2. *Influence of drugs on metal distribution*

Compared with the section above, the situation concerning metal ions is reversed: calcium and magnesium concentrations are too high for their distributions to be affected by tetracyclines. In contrast, zinc and copper distributions may a priori be influenced by these. In fact, this influence is limited and never becomes significant at usual therapeutic levels (Brion et al., 1985; Brion et al., 1986; Lambs and Berthon, 1988).

Table 2

Simulated Plasma Distributions of Tetracyclines into their Complexes with Proton, Calcium and Magnesium at Therapeutic Level

Antibiotic	Species	logβ	Percentage	Electrical charge	Reference
TC	Ca_2L	8.67	62.4	+2	Lambs et al. (1984)
	CaMgL	8.46	17.8	+2	
	MgLH	12.66	9.8	+1	
	LH	9.05	4.7	-1	
	LH_2	16.32	3.5	0	
OTC	CaMgL	8.31	30.7	+2	Lambs et al. (1984)
	Ca_2L	7.88	25.2	+2	
	Mg_2L	8.35	15.4	+2	
	MgL	4.87	10.0	0	
	CaL	4.46	8.4	0	
	LH	8.66	4.7	-1	
	Mg_2LH	14.97	2.6	+3	
	LH_2	15.77	2.4	0	
DOX	Ca_2L	8.89	43.0	+2	Lambs et al. (1984)
	CaLH	13.06	22.5	+1	
	CaL	5.60	19.7	0	
	MgLH	12.99	8.8	+1	
	Mg_2L	8.55	4.2	+2	
MIN	CaLH	14.18	71.0	+1	Lambs et al. (1984)
	CaL	5.98	11.2	0	
	CaMgLH	16.59	9.4	+3	
	MgL	5.89	4.1	0	
	MgLH	13.09	2.6	+1	
CTC	LH	8.7	31.6	-1	Lambs et al. (1985)
	CaL_2H_2	25.23	15.4	0	
	LH_2	15.71	13.0	0	
	MgLH	11.52	10.8	+1	
	LTA-CaLH	13.91	10.3	0	
	PO_4-$CaLH_2$	25.57	3.5	-1	
	$CaLH_2$	17.98	2.7	+2	
DMC	Ca_2L	8.34	74.3	+2	Lambs et al. (1985)
	CaL	4.57	11.1	0	
	LH	8.69	5.2	-1	
	MgLH	11.85	3.9	+1	
	Mg_2L	7.60	2.9	+2	
MTC	CaL	5.31	57.2	0	Lambs and Berthon (1988)
	MgL	5.14	18.1	0	
	MgLH	12.37	12.2	+1	
	LH	8.73	5.4	-1	
	LH_2	15.81	2.6	0	

$C_{drug} = 1.0\ 10^{-5}$ mol dm^{-3}, $C_{Ca} = 1.13\ 10^{-3}$ mol dm^{-3}, $C_{Mg} = 5.2\ 10^{-4}$ mol dm^{-3}
Species whose percentage is inferior to 2% are omitted (LTA = lactate, PO_4 = phosphate).

B. Gastrointestinal interactions

1. *Calcium*

Given its high dietary levels (see *Medicine*, Part Three, Chapter 1), calcium is logically expected to exert an important effect on the absorption of tetracyclines. Calcium complexation effectively influences their distribution in the gastrointestinal fluid, e.g. TC and OTC as from pH 5 and DOX and MIN above pH 4 (Brion et al., 1981). In particular, two neutral species (ML_2H_2 and/or ML) are likely to improve drug absorption. As demonstrated by Kakemi et al. (1968a; 1968b), however, either or both of these complexes are sequestered by the intestinal membrane, presumably through mixed-ligand complexation (Poiger and Schlatter, 1979), which actually impairs drug absorption.

2. *Magnesium*

The effect of magnesium on tetracycline absorption is expected to be comparable to that of calcium, complex stabilities (Berthon et al., 1983) and dietary concentrations (*Medicine*, Part Three, Chapter 1) being of the same order of magnitude for the two metals.

3. *Zinc*

Metal and drug concentrations chosen by pharmacologists to assess the influence of zinc on the absorption of TC (Pentillä et al., 1975; Andersson et al., 1976; Mapp and McCarthy, 1976) and DOX (Pentillä et al., 1975) were used to simulate the distribution of these drugs within the gastrointestinal pH range (Brion et al., 1985). The results obtained corroborate the clinical observations. In particular, the lesser influence of zinc on DOX than on TC at the different drug concentrations used by Pentillä et al. (1975) was interpreted. Figure 1 shows the simulated effect of zinc on the potential bioavailability of several tetracyclines.

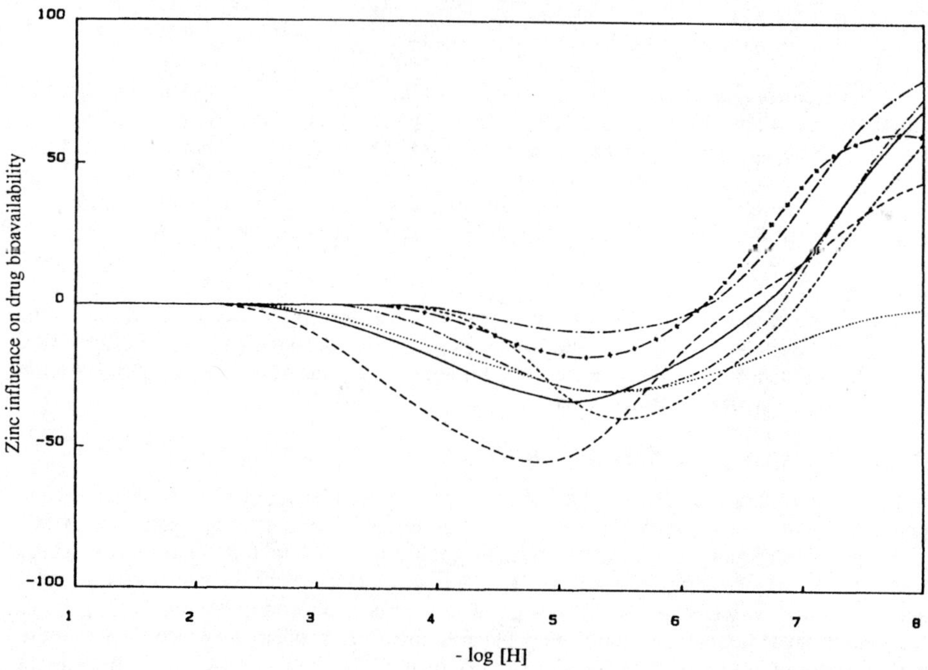

Figure 1.

Relative influence of zinc on the neutral fraction of antibiotic in the gastrointestinal tract. C_{Zn}= 0.003442 mol dm^{-3} and C_{drug}= 0.0052 mol dm^{-3}. The symbols are as follows: TC ——, OTC ......, DOX – – – –, MIN -------, CTC —·—·—, DMC –+–+–, MTC —··—··— . From Brion et al. (1985), and Lambs and Berthon (1988).

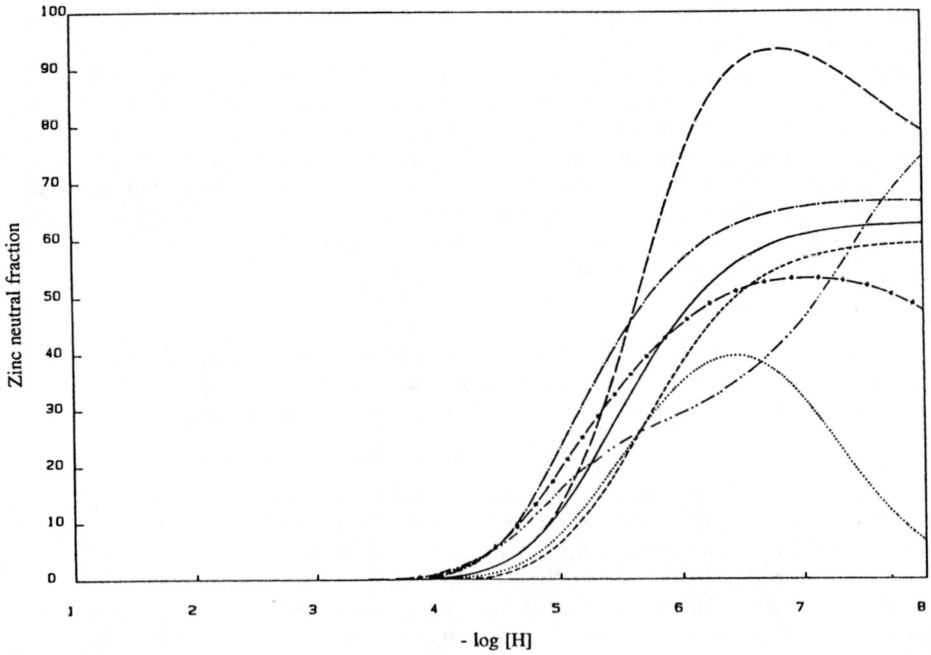

Figure 2.
Influence of tetracyclines on the neutral fraction of zinc in the gastrointestinal tract. Concentrations and symbols as in Figure 1. From Brion et al. (1985), and Lambs and Berthon (1988).

The reciprocal influence of tetracyclines on zinc absorption was also analyzed under the same conditions (Brion et al., 1985). Any impairment of zinc absorption when both metal and drug are simultaneously administered to a fasting subject should be ruled out (Fig. 2). On the contrary, the important neutral fraction of zinc induced by all drugs suggests a favorable influence of these on zinc absorption. This does not confirm the antagonistic effect evidenced in rats under different conditions (Weismann et al., 1978).

4. *Copper*
The influence of copper on drug potential absorption was also simulated (Brion et al., 1986; Lambs and Berthon, 1988). As could be expected from the metal-to-drug concentration ratio, this influence proved negligible. For this very reason, however, copper absorption is expected to be favored by all tetracyclines.

V. CONCLUSION AND FUTURE PROSPECTS

In the context of this Chapter 3, the results reported here illustrate one more aspect of the possible use of computer-aided speciation in drug design. The particular usefulness of this technique pointed out here is its multivalent applicability: (i) to the identification of the particular forms of a drug that are likely to be active in vivo in terms of bioavailability and/or therapeutic action; (ii) to the design of appropriate experimental conditions (pH and reactant concentrations) to study the physico-chemical properties which make these particular forms active; (iii) to test the validity of the initiatives that can be taken, e.g. in respect of molecular structure changes, to improve the efficiency of the drug. In this way, speciation can help to refine the more classical notion of structure/activity relationship.

This review is mainly concerned with the bioavailability of tetracyclines. Similar studies could be devoted to the processes that condition the therapeutic activity of these drugs. For instance, whereas copper tetracycline complexes are apparently of little relevance to the metabolism of either parent reactant in extracellular biofluids, intracellular roles for these species are much more likely. Copper tetracycline interactions have been recently implicated in DNA degradation through free radical production (see *Bioinorganic Medicine*, Part Five, Chapter 4-D). Substantial damage to the DNA molecule was observed following its incubation with a cupric salt and TC or any of its OTC, DOX, and CTC analogs, whereas antibiotics alone had no effect (Quinlan and Gutteridge, 1988). This is consistent with the observation that copper-TC complexes can bind DNA whereas TC itself cannot (Mikelens and Levinson, 1978). It has, therefore, been recently suggested that, like calcium and magnesium in the transport of tetracyclines, copper can act as a cofactor of their antibiotic activity: the structural flexibility of copper binary complexes within the three distinct donor sites of bioactive tetracyclines is expected to favor mixed-ligand coordination with bacterial nucleic acids; it is presumably through such ternary complexes that copper can induce the attack of free radicals known to damage these nucleic acids (Jezowska-Bojczuk et al., 1993). No doubt that speciation will find a new field of application in this context in the foreseeable future.

ACKNOWLEDGEMENT

Warm thanks are due to Professor Bruce Martin for his kind reviewing of the manuscript.

REFERENCES

Albert, A. (1953). *Nature, 172*: 201.

Albert, A., and Rees, C.W. (1956). *Nature, 177*: 433.

Andersson, K.E., Bratt, L., Dencker, H., Kamme, C., and Lanner, E. (1976). *Eur. J. Clin. Pharmacol., 10*: 59.

Berthon, G. (1989). Proceedings of the IInd Symposium on Inorganic Biochemistry and Molecular Biophysics, Wroclaw-Karpacz, May 27- June 2, p. 20.

Berthon, G., Brion, M., and Lambs, L. (1983). *J. Inorg. Biochem., 19*: 1.

Berthon, G., Piktas, M., and Blais, M.-J. (1984). *Inorg. Chim. Acta, 93*: 117.

Bevelander, G. (1964). *Adv. Oral Biol., 1*: 205.

Bevelander, G., and Cohlan, S.Q. (1962). *Biol. Neonat., 4*: 365.

Brion, M., Berthon, G., and Fourtillan, J.-B. (1981). *Inorg. Chim. Acta, 55*: 47.

Brion, M., Lambs, L., and Berthon, G. (1985). *Agents Actions, 17*: 229.

Brion, M., Lambs, L., and Berthon, G. (1986). *Inorg. Chim. Acta, 123*: 61.

Cohlan, S.Q. (1977). *Teratology, 15*: 127.

Cohlan, S.Q., Bevelander, G., and Tiamsic, T. (1963). *Amer. J. Dis. Child., 105*: 453.

Colaizzi, J.L., and Klink, P.R. (1969). *J. Pharm. Sci., 58*: 1184.

Connamacher, R.H., Mandel, H.G., and Hahn, F.E. (1967). *Mol. Pharmacol., 3*: 586.

Corrie, A.M., Makar, G.K.R., Touche, M.L.D., and Williams, D.R. (1975). *J. Chem. Soc., Dalton Trans.*, 105.

D'Arcy, P.F. (1985). *Pharm. Int.*, 222.

D'Arcy, P.F., Muhyiddin, H.A., and McElnay, J. (1976). *J. Pharm. Pharmacol., 28 Suppl.*: 33P.

Da Silva, J.J.R.F., and Dias, M.H.M. (1973). *Rev. Port. Quim., 15*: 1.

Day, L.E. (1966). *J. Bacteriol., 91*: 1917.

Dokter, M.E., and Magnuson, J.A. (1974). *J. Supramol. Struct.*, 2: 32.

Filippi, B., and Mela, V. (1957). *Minerva Chir., 12*: 106.

Gudzinowicz, B.J., Younkin, B.T., Jr., and Gudzinowicz, M.J. (1984). *Drugs and the Pharmaceutical Sciences*, Vol. 26, Marcel Dekker, Inc., New York and Basel.

Halme, J., Kivirikko, K.I., Kaitila, I., and Saxén, L. (1969). *Biochem. Pharmacol., 18*: 827.

Hoerman, K.C. (1975). *J. Dent. Res., 54*: B131.

Hughes, L.J., Stezowski, J.J., and Hughes, R.E. (1979). *J. Amer. Chem. Soc.*, *101*: 7655.

Ingri, N., Kakolowicz, W., Sillén, L.G., and Warnqvist, B. (1967). *Talanta*, *14*: 1261.

Jezowska-Bojczuk, M., Lambs, L., Kozlowski, H., and Berthon, G. (1993). *Inorg. Chem.*, *32*: 428.

Kaitila, I., Wartiovaara, J., Laitinen, O., and Saxén, L. (1970). *J. Embryol. Exp. Morph.*, *23*: 185.

Kakemi, K., Sezaki, H., Ogata, H., and Nadai, T. (1968a). *Chem. Pharm. Bull.*, *16*: 2200.

Kakemi, K., Sezaki, H., Hayashi, M., and Nadai, T. (1968b). *Chem. Pharm. Bull.*, *16*: 2206.

Kerley, M.A., and Kollar, E.J. (1978). *J. Exp. Zool.*, *203*: 89.

Khalil, S.A., Daabis, N.A., Naggar, V.F., and Wafik, M. (1977). *Pharmazie*, *32*: 519.

Kohn, K.W. (1961). *Nature*, *191*: 1156.

Kunin, C.M. (1962). *Proc. Soc. Exp. Biol. N.Y.*, *110*: 311.

Lambrou, D.B., Tahos, B.S., and Lambrou, K.D. (1977). *J. Dent. Res.*, *56*: 1527.

Lambs, L., and Berthon, G. (1988). *Inorg. Chim. Acta*, *151*: 33.

Lambs, L., Brion, M., and Berthon, G. (1984). *Agents Actions*, *14*: 743.

Lambs, L., Brion, M., and Berthon, G. (1985). *Inorg. Chim. Acta*, *106*: 151.

Lambs, L., Venturini, M., Decock-Le Révérend, B., Kozlowski, H., and Berthon, G. (1988a). *J. Inorg. Biochem.*, *33*: 193.

Lambs, L., Decock-Le Révérend, B., Kozlowski, H., and Berthon, G. (1988b). *Inorg. Chem.*, *27*: 3001.

Leeson, L.J., Koneger, J.E., and Nash, R.A. (1963). *Tetrahedron Lett.*, *18*: 1155.

Maestre, P., Lambs, L., Thouvenot, J.P., and Berthon, G. (1992). *Free Rad. Res. Commun.*, *15*: 305.

Maestre, P., Lambs, L., Thouvenot, J.P., and Berthon, G. (1994). *Free Rad. Res. Commun.*, in press.

Mapp, R.K., and McCarthy, T.J. (1976). *S. Afr. Med. J.*, *50*: 1829.

Martin, S.R. (1979). *Biophys. Chem.*, *10*: 319.

Martin, R.B. (1985). *Metal Ions in Biological Systems*, Vol. 19 (H. Sigel, ed.), Marcel Dekker, Inc., New York and Basel, p. 19.

Mikelens, P., and Levinson, W. (1978). *Bioinorg. Chem.*, *9*: 421.

Miles, A.A., and Maskell, J.P. (1985). *J. Med. Microbiol.*, *20*: 17.

Miles, A.A., and Maskell, J.P. (1986). *J. Antimicrob. Chemother.*, *17*: 481.

Mitscher, L.A., Slater-Eng, B., and Sokoloski, T.D. (1968). *Antimicrob. Agents Chemother.*, 78.

Mitscher, L.A., Bonacci, A.C., and Sokoloski, T.D. (1972). *Antimicrob. Agents Chemother.*, 2: 66.

Neuvonen, P.J., and Turakka, H. (1974). *Europ. J. Clin. Pharmacol.*, *7*: 357.

Neuvonen, P.J., Gothoni, G., Hackman, R., and Björksten, K. (1970). *Brit. Med. J.*, *4*: 532.

Neuvonen, P.J., Pentikaïnen, P.J., and Gothoni, G. (1975). *Brit. J. Clin. Pharmacol.*, 2: 94.

Newman, E.C., and Frank, C.W. (1976). *J. Pharm. Sci.*, *65*: 1728.

Pato, M.L. (1977). *Antimicrob. Agents Chemother.*, *11*: 318.

Pentillä, O., Hurme, H., and Neuvonen, P.J. (1975). *Europ. J. Clin. Pharmacol.*, *9*: 131.

Perrin, D.D. (1965). *Nature*, *208*: 787.

Poiger, H., and Schlatter, C. (1979). *Naunyn-Schmiedeberg's Arch. Pharmacol.*, *306*: 89.

Prevo, R., and Stezowski, J.J. (1977). *J. Amer. Chem. Soc.*, *99*: 1117.

Quinlan, G.J., and Gutteridge, J.M.C. (1988). *Free Rad. Biol. Med.*, *5*: 341.

Sabatini, A., Vacca, A., and Gans, P. (1974). *Talanta*, *21*: 53.

Schach von Wittenau, M., and Yeary, R. (1963). *J. Pharmacol. Exp. Ther.*, *140*: 258.

Schach von Wittenau, M., and Delahunt, C.S. (1966). *J. Pharmacol. Exp. Ther.*, *152*: 164.

Schumacher, G.E., and Linn, E.E. (1978). *J. Pharm. Sci.*, *67*: 1717.

Skinner, H.C.W., and Nalbandian, J. (1975). *Yale J. Biol. Med.*, *48*: 377.

Stezowski, J.J. (1976). *J. Amer. Chem. Soc.*, *98*: 6012.

Tritton, T.R. (1977). *Biochemistry*, *16*: 4133.

Weinberg, E.D. (1954). *Antibiot. Chemother.*, *4*: 35.

Weismann, K., Knudsen, L., and Høyer, H. (1978). *Arch. Dermatol. Res.*, *263*: 135.

White, J.P., and Cantor, C.R. (1971). *J. Mol. Biol.*, *58*: 397.

SPECIATION STUDIES IN RELATION TO ALUMINUM BIOAVAILABILITY

by

Chris Orvig

Department of Chemistry,
University of British Columbia,
2036 Main Mall
Vancouver, British Columbia V6T 1Z1
Canada

and

Guy Berthon

INSERM U 305,
Equipe "Bioréactifs: Spéciation et Biodisponibilité",
38 rue des Trente-six Ponts
31400 Toulouse
France

I. INTRODUCTION

Among the metals that occur in living species, aluminum is particular in many respects. First, although being the third in abundance of the elements constituting the earth's crust (7-8%, after O and Si) and the first metal among these, it has, paradoxically, no known biological role. Its constant sequestration in soils in the form of insoluble salts (oxides, hydroxides and silicates) throughout evolution presumably explains this. Second, aluminum has a very strong tendency to hydrolyze. The strong competition of the hydroxide anion with other ligands for binding aluminum throughout virtually the whole physiological pH range has a critical influence on the chemistry of Al in vivo; the slow kinetics of the hydrolysis process also greatly complicates the study of aqueous Al complexation in vitro. Finally, electrically neutral Al complexes have a strong tendency to precipitate (e.g. $Al(OH)_3$, $AlPO_4$). This has often led to soluble forms being ignored in discussions concerning the distribution of Al in vivo, although these forms are potentially significant in biological fluids. This also makes the experimental study of their stabilities a real challenge.

Thus, the investigation of Al complexes in solution is not an easy task, and it is not surprising that most of the available speciation data pertinent to living systems are controversial. This is rather unfortunate since aluminum, long considered as "neutral"

(Sorenson et al., 1974) according to the metal classification used in this book (see Volume 2, Part One, Chapter C), is in fact a "detrimental" element whose implication in neurological disorders of several kinds (see *Bioinorganic Medicine*, Part Four, Chapter 1-C1), combined with its increasing availability from the environment, is raising growing concern (see *Bioinorganic Medicine*, Part Four, Chapter 5-B).

The occurrence of increasing levels of aluminum in food and beverages, often together with appreciable amounts of dietary acids likely to favor its absorption (see e.g. Partridge et al., 1989; Domingo et al., 1993), makes it urgent to attain a better understanding of the chemical reactions which can influence its absorption process. Likewise, a better knowledge of the distribution of aluminum in blood plasma, with respect to which the influence of exogenous ligands could be tested, is a prerequisite for the design of new sequestering agents for the treatment of acute Al intoxications (see *Medicine*, Part Five, Chapter 3-B).

In this context, as a fundamental background to the observations developed in *Medicine* (see above references), the objective of this review is to summarize the essentials of the chemistry of aluminum in vivo and to sum up the information available regarding the "hot" and constantly moving topic of Al speciation in the main biological fluids.

II. AQUEOUS ALUMINUM COORDINATION CHEMISTRY

A. Hydrolysis

The low standard reduction potential ($Al^{3+} + 3e^- \rightarrow Al$, E° = -1.67 V) dictates that +3 is the only oxidation state for aluminum in water (Martin, 1988). In the absence of effective competing ligands, the highly charged and small Al^{3+} ion (r = 0.53 Å) is easily hydrolyzed; the solubility of $Al(OH)_3$ has a minimum of about 10^{-8} M at pH 6.

The species in aqueous solution are pH dependent. In very acidic media (pH < 3), Al^{3+} remains unhydrolyzed; the main mononuclear species $Al(H_2O)_6{}^{3+}$ contains an octahedral array of water molecules around the ion and undergoes proton transfer reactions instantaneously. As the pH increases, mononuclear species such as $[Al(OH)(H_2O)_5]^{2+}$, $[Al(OH)_2(H_2O)_4]^+$ and $Al(OH)_3$ are formed, as well as polynuclear species if the Al concentration is high enough (see e.g. Fig. 1 at 25°C). Most of the studies at 25°C concur, indicating the existence of $[Al_2(OH)_2(H_2O)_8]^{4+}$, $[Al_3(OH)_4(H_2O)_{10}]^{5+}$ and $[Al_{13}O_4(OH)_{24}(H_2O)_{12}]^{7+}$ (Baes and Mesmer, 1976; Bertsch and Anderson, 1989). Other polynuclear species of the $[Al_{2n}(OH)_{5n}]^{n+}$ type have been suggested (Charlet et al., 1984). The difficulty in studying Al hydrolysis by glass electrode potentiometry lies in the very

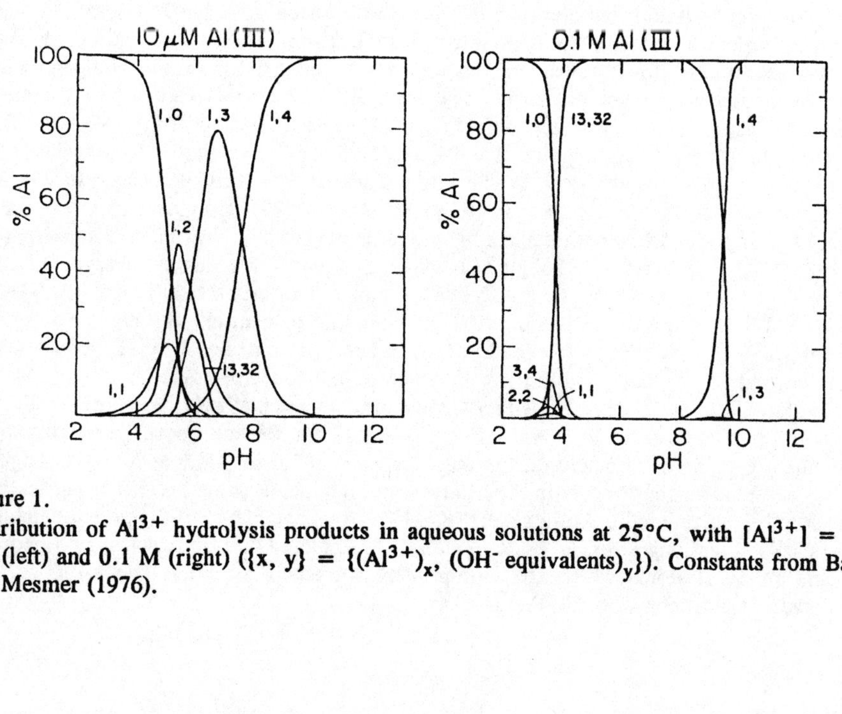

Figure 1.
Distribution of Al^{3+} hydrolysis products in aqueous solutions at 25°C, with $[Al^{3+}]$ = 10 μM (left) and 0.1 M (right) ({x, y} = {$(Al^{3+})_x$, $(OH^-$ equivalents$)_y$}). Constants from Baes and Mesmer (1976).

slow drift of the potential which never seems to reach true equilibrium. The variety of ways of considering this problem are probably at the origin of the discrepancies observed. Temperature may also be an additional factor; as the only investigation at 37°C refers to "kinetic" determinations following a specific protocol (Venturini and Berthon, 1987), corresponding results apply only in those specific conditions. $Al(OH)_3$ is amphoteric and readily dissolves under alkaline conditions to give the predominant Al-containing product in basic solution $Al(OH)_4^-$ (regardless of concentration). Examples of distributions of hydrolysis products in dilute (10 μM) and concentrated (0.1 M) aluminum solutions at 25 °C are shown in Figure 1.

The hydrolysis of Al^{3+} under varying conditions and in the presence of very weakly coordinating ligands has been extensively studied and has recently been reviewed in detail (Bertsch, 1989). It is a controversial topic of great chemical interest because of the variation in the hydrolysis behavior of Al^{3+} with factors such as the concentration and the nature of the other ions present, and because of the centrally important role hydrolysis takes in any consideration of in vivo Al chemistry, especially speciation studies which can only be as accurate as the model adopted.

B. Ligand binding

Ligands compete with OH^- for binding Al^{3+} to various extents. Two governing factors are the rate at which aluminum reacts with a given ligand (kinetics) and the strength of the resulting complex (thermodynamics).

Kinetically, Al^{3+} is generally regarded as a "sluggishly labile" metal ion. Intermolecular reactions in water (ligand substitution, H_2O exchange) are generally slow, although the few examples of intramolecular exchange which have been studied vary considerably depending on the steric bulk of the ligand (Gennaro et al., 1983). The complicated hydrolysis and polymerization of the Al^{3+} ion, combined with the lack of a UV-visible print to monitor reaction progress, severely hamper the study of its aqueous reaction rates. Aluminum solvent exchange is relatively slow, with first-order rate constants of up to 10^4 s^{-1}, which is, according to Langford and Gray (1966), defined as class III in the classification of metal ion rates; the self-exchange of $Al(H_2O)_6{}^{3+}$ with H_2O, whose rate constant is 1.29 ± 0.04 s^{-1} at 25°C, follows a dissociative interchange (I_d) pathway (Hugi-Cleary et al., 1985). The exchange of bound ligands at metal ion centers roughly follows the aquo exchange rate. Exchange reactions of Al complexes in aqueous solution would thus be I_d in nature (Burgess, 1978), consistent with the small size of Al^{3+} requiring a ligand to dissociate before another may coordinate. The hydrolyzed species $[Al(OH)(H_2O)_5]^{2+}$ actually undergoes anation reactions faster than the hexaaquo ion $[Al(H_2O)_6]^{3+}$ by more than three orders of magnitude (Crumbliss and Garrison, 1988). It must be stressed, however, that these kinetic considerations are not accounted for in the computer-aided speciation models with which this review is concerned, since these have been based entirely on thermodynamic data so far (Jones and May, 1987; see also Section I of this Chapter 3).

From the thermodynamic point of view, the hard Al^{3+} ion prefers to associate with hard bases (see Part One, Chapter 3) - hence its tendency to hydrolyze. Thus, Al^{3+} has a very strong affinity for fluoride (Driscoll and Schecher, 1989). However, as the F^- anion cannot be incorporated into multidentate ligands, it is oxygen (the next most electronegative element) which is found in strongly Al-chelating ligands. Negatively charged organic functional groups give rise to very stable Al complexes (Martell and Smith, 1974-1989; Taylor, 1987; Nordström and May, 1989) in the acidic to neutral pH range. No ligand, however, can compete with hydroxide in the very basic pH range (pH > 11.5), but this is largely beyond the physiological limits with which we are concerned here.

The stability of complexes with O^--containing ligands closely correlates with the affinity of the metal ion for hydroxide (Evers et al., 1989). Because most ligands are weak acids, there is a fierce competition for these between H^+ and Al^{3+} at low pH. Highest Al^{3+}-ligand binding thus occurs in the neutral to weakly acidic range, which is particularly relevant to biological media. More generally, Al^{3+} forms complexes of the greatest stability with ligands containing strongly basic O atoms (phenolate O^-, carboxylates, phosphonate O^-), and amine N atoms which are appropriately spaced to form five-membered or six-membered metallacycles (e.g. EDTA, DTPA).

III. ALUMINUM COORDINATION CHEMISTRY IN VIVO

Within a fairly constant environmental level, the ecological age of a metal determines its behavior with respect to living systems. Above this level, concentration becomes determining (see *Bioinorganic Medicine*, Part One, Chapter 1). For any essential, beneficial, or even neutral metal—in the classification used in this book (see *Medicine*, Part One, Chapter 1-C)—this implies that only a contamination (well) above the usual environmental level will constitute a real threat to the living organisms in contact with it (Venchikov, 1960; see also *Medicine*, Part One, Chapter 1, and Part Four, Chapter 1). On the contrary, for a detrimental metal like aluminum, for which no homeostatic regulatory mechanism has been elaborated during evolution, *any* contamination level may be hazardous. In particular, its absorption and retention are under the control of the ligands with which it may interact.

Some of these ligands are endogenous. Although there is virtually no way to influence their roles with respect to Al in vivo, a better knowledge of these roles is highly desirable. Other ligands are exogenous. A knowledge of their influence on Al bioavailability may lead to effective action against Al absorption and retention (by limiting the ingestion of ligands which enhance, or by increasing that of ligands which reduce, this bioavailability, or by administering agents which favor Al excretion). In particular, prevention measures are desirable for patients who are administered aluminum salts for therapeutic purposes. The great similarity in chemistry between Al^{3+} and Fe^{3+} ions is of underlying importance in these considerations (Crumbliss and Garrison, 1988; Fatemi et al., 1991).

A. Endogenous ligands

Transferrin, whose sites are only 30% occupied by Fe^{3+}, is the major high-molar-mass ligand for Al^{3+} in blood plasma (Cochran et al., 1984; Martin, 1986a; 1988). First thought to be significant (Gardiner et al., 1984; Bertholf et al., 1984), the role of albumin in this respect was later shown to be quantitatively negligible (Martin, 1986a; 1988; Venturini and Berthon, 1989). A new Al carrier, provisionally called albindin, has recently been identified in human serum (Favarato et al., 1992a). The main low-molar-mass (l.m.m.) ligands for Al in blood plasma are citrate and phosphate, whose importance relative to each other is still debated (Martin, 1986a; 1988; Daydé et al., 1990; Duffield et al., 1991a; Harris, 1992). Citrate must be the most extensively investigated (and controversial) ligand for aluminum in aqueous solution. Its active role in oral Al uptake in rats and humans (Slanina et al., 1984; 1985; 1986; Partridge et al., 1989; Molitoris et al., 1989) has largely been the cause of this particular interest. The involvement of citrate in Al metabolism, however, characterizes it as a dietary (exogenous) ligand; its case will thus be developed in the following section.

In brain tissue, concentrations of citrate (~2 μM) and transferrin are very low. Other small molecules thus become possible Al^{3+} ligands. Among these, catecholamines are neurotransmitters which perfuse into tissue and bind aluminum strongly (Weinland and Denzel, 1914; Öhman and Sjöberg, 1983a; Kiss et al., 1989). Trace amounts of Al^{3+} may disrupt neurochemical processes through this binding (Martin, 1988). As Al^{3+} prefers oxygen donors, its coordination to catecholamines occurs at the diphenolato moiety, whether the ammonium group is protonated or not (Kiss et al., 1989; see also Part Three, Chapter 2-C). However, according to Kiss et al. (1989), large excesses of catecholamines are required for them to compete effectively for Al^{3+} with citrate and ATP (3000-fold and 400-fold excesses, respectively). If Al^{3+} can penetrate red cells, 2,3-diphosphoglycerate (its most likely l.m.m. ligand inside these cells) would bind it strongly enough to equalize Al concentrations on both sides of the cell membrane (Sóvágó et al., 1990).

Phosphate is a ligand of considerable importance in biological systems; delineation of the complexes formed with Al^{3+} have long been hampered by the insolubility of $AlPO_4$ and much effort has been expended on the study of this system without a definitive result having been achieved (Hsu, 1968; Karlik et al., 1982; 1983; Arp and Meyer, 1985; Jackson and Voyi, 1988; Martin, 1988; Wilson et al., 1989; Daydé et al., 1990; Duffield et al., 1991a; Harris, 1992). The binding of Al^{3+} by nucleosides and nucleotides has been of considerable interest in the last decade because of the increasing importance of Al bioinorganic chemistry (Bock and Ash, 1980; Karlik et al., 1980; 1983; Laussac and Commenges, 1983; Martin, 1986b; Dyrssen et al., 1987; Jackson and Voyi, 1987; Kiss et al., 1991). Although it has been suggested that Al can bind to both base nitrogens and phosphate backbone oxygens, quantitative work shows the binding to be almost completely to the latter, in agreement with the general trends for ligand binding obvious from the multitude of stability constant determinations (Martell and Smith, 1974-1989). ADP and ATP chelate Al^{3+} to form complexes of greater stability than nucleoside phosphates with only one terminal phosphate such as AMP (Jackson and Voyi, 1987; Kiss et al., 1991).

Polycarboxylate, hydroxycarboxylate and aminocarboxylate ligands are of special interest in aqueous Al chemistry in vivo since (i) most show strong affinity towards aluminum, (ii) most of their Al complexes are water soluble, and (iii) some exist in the human body (in particular those involved in the Krebs cycle). However, as their relevance to Al metabolism is more attributable to their quality as exogenous ligands, they will be, like citrate, examined in the next section.

Several inorganic ligands such as sulfate, thiocyanate and hydrogen carbonate can also form complexes with aluminum in vivo, but not nearly as strongly as does fluoride.

B. Exogenous ligands

The Al-citrate system has been extensively investigated in diverse chemical (Jackson, 1982; Öhman and Sjöberg, 1983b; Lopez-Quintela et al., 1984; Motekaitis and Martell, 1984; Gregor and Powell, 1986; Martin, 1986b; Öhman, 1988; Venturini and Berthon, 1989; Marques, 1991) as well as physiological (Molitoris et al., 1989; van Ginkel et al., 1990) aspects. The neutral ML complex which forms near pH 2-4 (Martin, 1986b; 1988; Venturini and Berthon, 1989) is considered to be the culprit in human Al-citrate loading studies (Slanina et al., 1984; 1985; Slanina et al., 1986; Weber and Berstad, 1986; Molitoris et al., 1989; Walker et al., 1990); simultaneous administration of Al-containing drugs and citrus fruits or juices should thus be avoided. Citrate administered separately, however, favors Al excretion (Domingo et al., 1986; Ecelbarger and Greger, 1991; Lote et al., 1992a). This effect also has been interpreted by simulation of the corresponding equilibria (Domingo et al., 1988a; Venturini and Berthon, 1989; Daydé et al., 1989).

Other dietary acids are important modifiers of Al metabolism. Like citric acid, ascorbic, gluconic, lactic, malic, oxalic and tartaric acids elevate the pH of precipitation above 8 during the titration of 0.1 M aluminum chloride by 0.1 N hydroxide (Partridge et al., 1989). Concurrent ingestion of lactate and Al can result in developmental toxicity in mice (Colomina et al., 1992). Ascorbic acid also enhances gastrointestinal absorption of Al in rats and humans (Domingo et al., 1991). On the other hand, malic, malonic and salicylic acids have been described as possible antidotes for acute Al toxicity (Domingo et al., 1986; 1988b; Llobet et al., 1987). Likewise, ascorbate enhances Al excretion in rabbits (Fulton and Jeffery, 1990).

The reference ligand for Al detoxification is desferrioxamine (DFO). Its beneficial effect lies in its capacity to mobilize Al stores in tissues, increasing the total plasma level of Al as well as its ultrafiltrable fraction (Domingo, 1989). DFO significantly decreases the proportion of Al bound to transferrin/albumin but increases that bound to albindin (Favarato et al., 1992b). Despite its widespread use, DFO therapy suffers from serious drawbacks: (i) its lack of Al specificity and concurrent sequestration of Fe, (ii) its uncomfortable parenteral route of administration, (iii) its important side effects, and (iv) its high cost. The search for new chelating drugs has thus been active during the last decade. In addition to DFO and the above-mentioned acids often found in appreciable amounts in normal human blood plasma, synthetic carboxylic acids (NTA, EDTA, DTPA, CDTA, etc) and other types of chelating agents (hydroxamic acids, pyridinones, 2,3-dihydroxybenzoic acid (DHBA), 3,4-LICAMS, tiron, etc) have been quantitatively investigated and tested in vitro (Rajan et al., 1981; Yokel and Kostenbauder, 1987; Martell, 1989; Domingo, 1989; Martell et al., 1990; Harris and Sheldon, 1990; Yokel et al., 1991; see also *Medicine*, Part Five, Chapter 3-B).

In the absence of direct experimental speciation data (see e.g. Gardiner et al., 1984), studies of the aqueous coordination chemistry of Al with endogenous as well as exogenous ligands, coupled with speciation modelling, have proven to be a strong approach to elucidate the parameters which influence Al bioavailability. As the declared objective of this Handbook implies, the short review of the results obtained with these techniques will be limited to human biology and medicine. However, it is worth noting that these also apply in the environmental problems - due to the Al release following soil acidification - which are at the origin of the Al threat to humans (Tam and Williams, 1986). See for example the comprehensive work done by the Umeå group in that field (Marklund et al., 1989; Marklund and Öhman, 1990a; 1990b; 1990c; Öhman, 1991).

IV. SPECIATION STUDIES AND BIOFLUIDS

As already expressed (see previous Sections in this Chapter 3), the reliability of any simulation model depends on that of the thermodynamic data of which it is constituted. This consideration carries a particular weight in the case of aluminum, because of the difficulty in studying its complexes. The frequent lack of agreement on the main species of many Al-ligand systems logically leads to conflicting models. Other factors like temperature, ionic strength, and kinetic protocols used in constant determinations may add more confusion. A good example of this is the Al-citrate system: (i) the $M_3L_3H_{-4}$ complex, which requires hours to form at 25 °C (Öhman and Sjöberg, 1983b; Öhman, 1988), was observed at equilibrium within 2 min at 37 °C (Venturini and Berthon, 1989); (ii) the same complex was judged unlikely to occur in a biological milieu (Martin, 1986a) while it can reach nearly 100 % of Al for common concentrations of Al and citrate in the small intestine in the absence of solid $Al(OH)_3$ (Öhman and Sjöberg, 1983b; Öhman, 1988) and about 70 % in its presence (Venturini and Berthon, 1989); (iii) constants calculated from "zero-time" extrapolated data were used to calculate a simplified Al plasma equilibrium model (Öhman, 1988); (iv) the detection of some hydrolytic species (Öhman, 1988) has been suggested to be due to metastable equilibria (Martell et al., 1990); and yet (v) it is estimated that "there is reasonable agreement on the overall strength of citrate binding under physiological conditions"! (Harris, 1992).

Difficulties in distribution modelling are also inherent in the choice of analytical data for the biofluid considered. Gastrointestinal (g.i.) fluid and blood plasma are the two biofluids of greatest pertinence for the absorption and biodistribution of aluminum.

The basic difficulty with g.i. fluid is that the diversity of the nutrients ingested over a population of individuals renders it impossible to establish a standardized diet composition, and the pH along the g.i. tract varies from about 2 to 8. It is thus commonly accepted practice to represent the distribution of a given element or substance in an empty stomach, with the percentages of the different species varying as a function of pH. This graphical representation greatly facilitates the analysis of the results and has the advantage of being extendable to three dimensions if, for example, one wants to assess the influence of a ligand on the absorption of a metal in the presence of another ligand (Blaquière and Berthon, 1987). The common characteristics necessary for Al absorption being solubility and neutral charge, these are flagged in searching the results of such calculations.

Blood plasma is a more complicated problem. Beside the possible use of "effective stability constants" (which are difficult to evaluate properly), the notion of "pM" to assess the potential mobilizing capacity of a ligand, or very simple models (Clevette and Orvig, 1990), ECCLES is the model of choice for blood plasma investigations (see also Section A1 in this Chapter 3). The critical advantage of ECCLES is the possibility it offers of using free metal concentrations, which bypasses the difficult problem of the lack of thermodynamic constants for metal-protein associations (May et al., 1976). Unfortunately, this advantage is not a priori determining with aluminum. Unlike essential or beneficial elements, aluminum suffers no constant regulation of its free concentration in plasma, and the extent of its buffering by proteins (mainly transferrin; see above) is not totally independent of its total concentration or that of l.m.m. ligands. This problem can be circumvented, however (Venturini and Berthon, 1989). Another advantage of ECCLES is the great number of species that can be accommodated simultaneously. Given the preference of Al^{3+} for oxygen donors, it is important to take into account the interactions of its preferred ligands with the more concentrated Ca^{2+}, Mg^{2+}, and even Mn^{2+} ions in the calculations.

A. Gastrointestinal fluid

The factors which govern gastrointestinal uptake of Al are still poorly understood. It was originally considered that the g.i. tract constituted a formidable barrier against absorption

(Kaehny et al., 1977; Alfrey, 1984). This belief was based on the fact that Al accumulation in the tissues of individuals with normal renal function is very low, despite widespread exposure (cookware, antacids, food additives, drinking water, etc). In fact, no specific protection exists against Al absorption, as there is against excessive intakes of essential metals (see *Medicine*, Part Three, Chapter 2). Before the recent intrusion of Al into the environment due to industrial activity, its uptake from the g.i. tract was naturally limited by its low occurrence in the food chain and the poor solubility of its ingested salts (Martin, 1986b). This, combined with efficient renal clearance, maintained the Al body burden below limits of overt noxiousness (Sorenson et al., 1974), and if it had not been for the discovery of dialysis encephalopathy (Alfrey et al., 1976), Al toxicity would probably have gone unnoticed for some time yet. Recent advances in Al bioinorganic chemistry have permitted some of the processes in which aluminum may take part along the g.i. tract to be outlined.

First, the pH in the stomach is low enough (< 3) to solubilize part or all of ingested Al (depending on the transit rate) and to dissociate Al^{3+} ions even in the presence of ligands - with the likely exception of oxalate (Venturini et al., 1987). It has therefore been suggested that Al absorption primarily occurs in the stomach or the proximal duodenum (Kaehny et al., 1977; Alfrey, 1986). However, the Al^{3+} ion does not benefit in the stomach from any "solvent drag" mechanism, such as that involved in Mg^{2+} uptake from the small intestine (see *Medicine*, Part Three, Chapter 2-F2), and it reprecipitates in the form of the trihydroxide when it reaches the milder pH conditions of the duodenum. The presence of phosphate in the diet (1g a day on average; Ciba-Geigy, 1981) also helps to limit the extent of soluble forms of Al even at relatively acidic pH.

This protection, however, is not effective in all circumstances since the simultaneous ingestion of Al and dietary acids may alter it (Partridge et al., 1989; Molitoris et al., 1989; Lote and Saunders, 1991). The anionic form(s) of some of these acids may act through two distinct processes to induce Al absorption: (i) the Al complexes they form may be stable enough to dissolve a significant amount of the metal ion ingested, and (ii) some of these complexes are electrically neutral and can pass somewhat freely (depending on their lipophilicity) through the g.i. membrane. It may also be that the acid does not form any Al neutral complex, but once the Al salt is dissolved, the Al^{3+} ions can be coordinated into neutral complexes by other nutrients in the g.i. tract. This risk increases with the amount of aluminum ingested, as in the case of treatment with Al-containing drugs. Computer-aided speciation can help to understand these processes, provided reliable formation and solubility product constants are available.

For example, the analysis of the influence of dietary acids on the pH range over which $Al(OH)_3$ and $AlPO_4$ can totally dissolve (Berthon and Daydé, 1992) confirms the experimental observations made by Partridge et al. (1989) on the effect of some of these acids on $Al(OH)_3$ precipitation. Figure 2 shows the distribution of 0.05 M aluminum in the presence of 0.0312 M citrate, which corresponds to 270 mg of aluminum (a single dose of an antacid like Maalox) administered with 40 cm^3 of lemon juice in 200 cm^3 of water. Under these conditions, the percentage of the neutral complex $AlCta^\circ$ reaches 30% near pH 2.5, which represents some 80 mg of Al in an absorbable form. Fortunately, this effect of citrate on Al absorption is partly compensated for by its tendency to increase Al excretion from plasma (see next section). Tartaric acid is still more effective than citric acid in mobilizing Al into an absorbable form in the g.i. tract. As Figure 3 shows, a concentration of 0.02 M of tartrate (approximately the amount present in a glass of wine diluted to 200 cm^3 in water) can form up to 35% of Al as a neutral complex (almost 100 mg of Al) near pH 4, with still significant percentages from pH 2 to 6. The apparent risk with tartrate is all the more important as, unlike citrate, it exerts no compensatory effect on Al excretion at the plasma level. However, tartrate is much more sensitive than citrate to the competition of phosphate, which limits the formation of soluble neutral species, provided the phosphate concentration is at least as great as that of aluminum (Daydé and Berthon, 1992). In the case of the administration of Al-containing drugs, it is in fact only so when therapeutic aluminum is administered as $AlPO_4$. The situation is comparable with succinate (Venturini, M., Alliey, N., and Berthon, G., unpublished results).

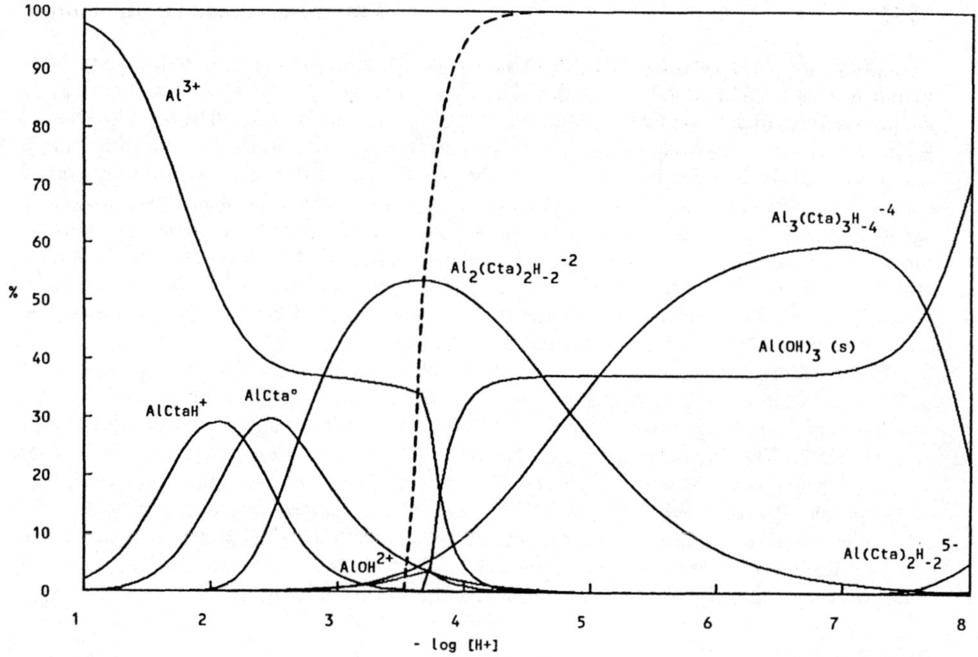

Figure 2.
Distribution profiles of Al complexes with citrate in the g.i. fluid. Reactant concentrations are $C_{Al} = 0.05$ M and $C_{Cta} = 0.0312$ M. The broken line represents the percentage of solid $Al(OH)_3$ in the absence of citrate. From Berthon and Daydé (1992).

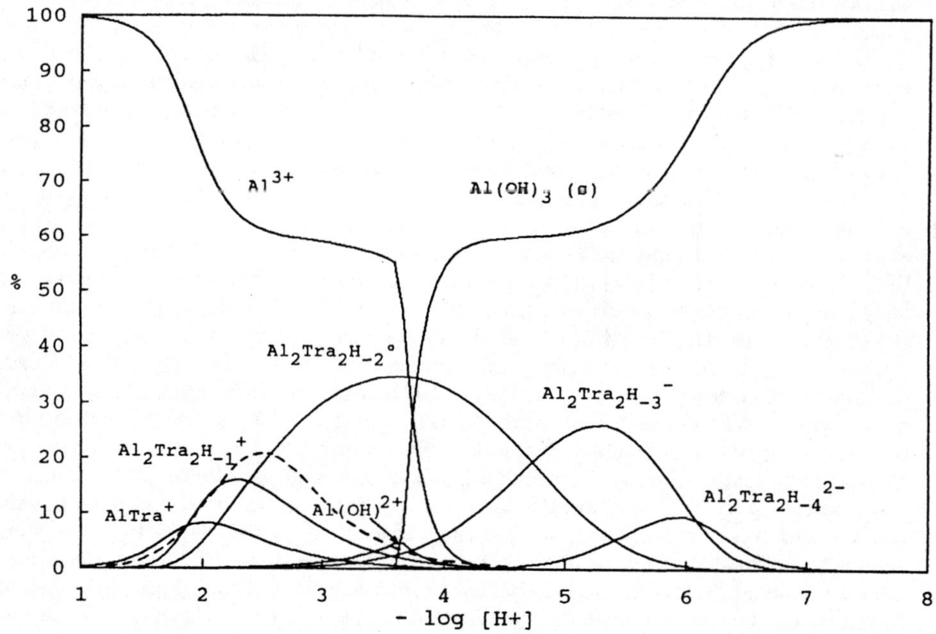

Figure 3.
Distribution profiles of Al complexes with tartrate in the g.i. fluid. Reactant concentrations are $C_{Al} = 0.05$ M and $C_{Tra} = 0.02$ M. The broken line represents the percentage of the $AlCta^{\circ}$ complex under identical conditions ($C_{Al} = 0.05$ M and $C_{Cta} = 0.02$ M). From Daydé and Berthon (1992).

Concerning the potential relative toxicities of Al-containing drugs, it has long been established that $AlPO_4$ is safer than $Al(OH)_3$ (Kaehny et al., 1977). This was attributed to the poorer solubility of the former. Yet, while $AlPO_4$ was shown to be virtually insoluble at acidic pH in vitro, it proved more soluble than $Al(OH)_3$ under alkaline conditions, which led to the above-mentioned hypothesis of the absorption of aluminum in the acidic region of the g.i. tract (Kaehny et al., 1977). The better safety of aluminum phosphate as a protective agent of the g.i. wall has recently been attributed to the combination of the coating protection of solid $AlPO_4$ with that of the non-absorbable $Al_2PO_4(OH)_2$ complex (Berthon and Daydé, 1992). However, as the validity of this interpretation lies on stability constants which have recently been questioned (Duffield et al., 1991a; Harris, 1992), it should be considered with caution until the situation is clarified.

Distributions of aluminum have also been calculated in the presence of phosphate at different concentrations and Al:phosphate concentration ratios by Duffield et al. (1991a). The constants used by these authors for these calculations have also been questioned (Harris, 1992). The Al-phosphate system is a highly controversial one. It has even been said that "the most likely pathogenic route of aluminum (was) via the phosphate complex" (Jackson and Qubeka, 1987). Incidentally, Jackson has also examined a gastrointestinal speciation model to explain the biodistribution of tris(maltolato)Al(III) and suggested that this neurologically interesting complex should be absorbed intact (Jackson and Steyn, 1991) - this can only occur if it were to pass the stomach and reform in the small intestine.

B. Blood plasma

Blood plasma is the biofluid which has been studied most extensively because of its physiological importance. The pH (7.4), ionic strength (μ = 0.15 M NaCl), and the concentrations of its contents are tightly buffered, making it (apparently) easier to model. However, the interdependent effects due to the complexity of its chemical composition and the impossibility to determine thermodynamic formation constants for metal-protein complexes seriously complicate the problem. The main variables in existing Al speciation models of blood plasma are (i) the number of metals and ligands which can be accommodated, (ii) the consideration of either total or free Al concentrations, and (iii) the selection of the stability constants used. Here again, the complexes into which aluminum is distributed are differentiated by charge; those which are cationic or anionic tend to be excreted via the renal system, while those which are electrically neutral induce tissue retention before they are ultimately excreted via the hepatobiliary system. The half-life of Al in plasma has been found to vary, depending on species and experiment, from one hour to two days, with a much longer $t_{1/2}$ for dialysis patients (Wilhelm et al., 1990).

The number of reactants to be included in the model depends on the capacities of the program used. The way the ECCLES program has been conceived (May et al., 1977) enables the user to incorporate as many of these as necessary. Also, as stressed above, ECCLES allows a choice between total and free metal concentrations as input data. Among the groups of authors of the different models published (Daydé et al., 1990; Clevette and Orvig, 1990; Jackson, 1990; Duffield et al., 1991a; Harris, 1992), however, all but one (Daydé et al., 1990) have used the total concentration - even those running ECCLES (Jackson, 1990; Duffield et al., 1991a; Harris, 1992). Actually, both approaches were used by one group (Venturini and Berthon, 1989; Daydé et al., 1990), which brought to light two interesting points. Albumin was quantitatively confirmed to be a negligible ligand for Al in plasma, and the free Al^{3+} concentration was shown to vary linearly with the total concentration up to 10^{-16} M; above this limit, the free concentration decreased more rapidly than the total concentration, due to the growing influence of the $[Al_2PO_4(OH)_2]^+$ complex. Although the stability constant for the latter complex has been questioned since then (Duffield et al., 1991a; Harris, 1992), this prediction that the fraction of protein-bound Al decreases with increasing plasma Al concentration has been validated experimentally in rats (Van Ginkel et al., 1990; Lote et al., 1992b).

Since all groups have based their calculations on the total concentration of Al (solely or not), it is interesting to compare the values obtained for the fraction of protein-bound aluminum (Table 1). These results are roughly divided into two groups: 57% (Daydé et al., 1990) and 63% (Jackson, 1990) on the one hand, and 83% (Clevette and Orvig, 1990),

Table 1

Simulated Distributions of l.m.m. Aluminum in Blood Plasma

Protein-bound Al	Al l.m.m. fraction	References
57%	$Al(OH)_3$ [51%] $AlPO_4$ [41.5%] $Al_2PO_4(OH)_2$ [7.2%]	Daydé et al. (1990)
83%	Citrate [98%] Hydroxide [2%] (phosphate not considered)	Clevette and Orvig (1990)
63%	$AlPO_4$ [62%] AlCta(OH) [23%] $AlPO_4CtaH$ [10%]	Jackson (1990)
80%	$AlCta(OH)_2$ [94%] AlCta(OH) [3%] $AlPO_4$ [1.5%] $AlOxa(OH)_2$ [1.4%]	Duffield et al. (1991)
81%	$AlPO_4(OH)$ [80%] AlCta(OH) [10%] $Al(OH)_3$ [4%] $Al(OH)_4$ [3%] $AlPO_4$ [2%]	Harris (1992)

80% (Duffield et al., 1991a) and 81% (Harris, 1992) on the other hand, whereas a value of $54 \pm 7\%$ has been determined experimentally on healthy individuals (Rahman et al., 1985). As the conditional constants used for the mono- and di-Al transferrin complexes ($K_1 = 10^{12.9}$ and $K_2 = 10^{12.3}$ (Martin et al., 1987), or $K_1 = 10^{13.5}$ and $K_2 = 10^{12.5}$ (Harris and Sheldon, 1990)) and Al-albumin ($K = 10^{12.1}$ (Bertholf et al., 1984)) are common to all these authors, the discrepancies found are due to differences in the selection of the sets of constants for l.m.m. complexes.

The selection of appropriate constants is a difficult and subjective matter, and great variations are found from one group to the other in the different above-mentioned models (Daydé et al., 1990; Clevette and Orvig, 1990; Jackson, 1990; Duffield et al., 1991a; Harris, 1992); the variation in the resulting l.m.m. Al distributions (Table 1) reflects this.

The model obtained by Daydé et al. (1990) highlights the importance of the two poorly soluble Al salts, trihydroxide and monophosphate, which must exist in soluble forms in blood plasma (Table 1).* The constant for $Al(OH)_3$ has been determined under specified kinetic conditions (Venturini and Berthon, 1987). The principle of that strategy was to obtain approximate data, rather than no data at all. Of course, the clear disadvantage of such an approach is that it can always be questioned from a purely thermodynamic point of view. The situation with $AlPO_4$ is even more difficult. Many attempts have been made to determine formation constants for the Al-phosphate system (Hsu, 1968; Arp and Meyer, 1985; Jackson and Voyi, 1988; Wilson et al., 1989; Daydé et al., 1990; Duffield et al., 1991a). The constants used by Daydé et al. (1990) have been called into question based on the argument that corresponding experiments were pursued in the presence of a precipitate of $AlPO_4$ because "autotitration equipment was employed" (Duffield et al., 1991a). Actually

*It is noteworthy that considering AlCta(OH) (with a constant of $10^{5.25}$ for $AlCtaH_{-1}$ as rejected as a second choice by Venturini and Berthon (1989)) instead of $Al_2Cta_2(OH)_2$ in the calculations would lead to 30% of AlCtaOH at the expense of other species shown in Table 1.

no such equipment was used and if there was misapprehension of the experimental reality, it was human in origin. In new experiments, precipitation has effectively been observed above pH around 3 (Alliey, N., and Berthon, G., unpublished results), a significant difference from the previous study (Daydé, 1990). Also, the stoichiometries of the complexes present within this range are different from those (apparently wrongly) characterized previously (Daydé et al., 1990) and which were confirmed later by Duffield et al. (1991a). If there is effectively a flaw in the experiments reported by Daydé et al. (1990) as suggested by Duffield et al. (1991a), not only are the constants questionable, but the species stoichiometries too. Moreover, it appears very hazardous in that case to simulate Al distribution at pH 7.4 with constants determined within the 2-3 pH range.

Concerning its applications, the model used by Daydé et al. (1990) was the first to interpret the positive effect of citrate on Al urinary excretion (Venturini and Berthon, 1989) observed in many animal studies (Domingo et al., 1986; 1988a; 1988b; Ecelbarger and Greger, 1991; Lote et al., 1992a).

A fairly simple model has been developed at pH 7.4, using 100 mM citrate and 50 mM vacant sites on transferrin (Clevette et al., 1989; Clevette and Orvig, 1990), to evaluate 28 ligands as Al^{3+} binders. Applications of speciation computer modelling to animal studies were also discussed. It was found that a 3-hydroxy-4-pyridinone was more effective at complexing Al than EDTA, maltol or catechol (Clevette and Orvig, 1990). This model was useful in a comparative sense but is very primitive by the standards of ECCLES program and blood plasma model (May et al., 1977; Berthon et al., 1986) which have been used by most investigators.

Jackson (1990) used the ECCLES model to simulate the distribution of plasma Al in the presence and absence of transferrin. A number of constants were selected relative to complexes of naturally occurring acids (malonate, cysteinate, citrate, etc) and small inorganic ligands (fluoride, carbonate, etc). The results (Table 1) are critically dependent on the phosphate constants used. Incidentally, Jackson used constants estimated from the literature rather than his own (Jackson and Voyi, 1988). The plasma mobilizing indices (PMIs) for various potential in vivo chelating agents were also evaluated, with DFO proving to be the most effective chelator for plasma Al^{3+}. The speciation of 4 mM tris(maltolato)Al(III) in blood was also examined with this model (Jackson and Steyn, 1991) and found to show less than 1 % intact complex with the Al^{3+} having been removed to transferrin and citrate species - in agreement with previous work (Clevette and Orvig, 1990) - as well as to a phosphate complex.

Duffield et al. (1991a), also using ECCLES, have found l.m.m. Al to be predominantly present as $AlCta(OH)_2$ (Table 1). The constants used for phosphate in this model have been indirectly evoked above. That of the $AlCta(OH)_2$ complex has been questioned (Martell et al., 1990; Harris, 1992); found by Öhman (1988) at 25°C but not by Venturini and Berthon (1989) under identical conditions, it needs confirmation.

Based on all the above works and also using ECCLES, the model recently developed by Harris (1992) represents a critical synthesis of the preceding models. The constants used have been selected after a comparison of different sets of literature values for a given ligand, with the result that composite sets were constructed and used for citrate, hydroxide, and phosphate. Taking into account the insolubility of $AlPO_4$ above pH 4, Harris used linear free energy relationships to estimate a formation constant for the $AlPO_4$ complex, and also for its hydrolyzed counterpart, $AlPO_4(OH)$, which reveals largely predominant (Table 1). There is, however, no experimental evidence of the existence of this species.

In addition to the above models, a study of aluminosilicate stabilities under blood plasma conditions has been used to posit an explanation for the formation of aluminosilicates in the senile plaques of Alzheimer's disease (Martin, 1990). A blood plasma model per se was not used, but the free Al^{3+} concentrations allowed by six aluminosilicates were compared with those allowed by transferrin and citrate to postulate that supersaturation of solutions containing Al bound to phosphorylated proteins and bathed in excess silicic acid might cause the formation of an amorphous aluminosilicate which could act as a precursor to the forms found.

The usefulness of speciation modelling to the real world problem of Al in blood and particularly to the effectiveness of chelating agents in Al-related diseases is only now beginning to be shown (Harris, 1992). Dialysis encephalopathy and osteomalacia have both been treated using a variety of chelating agents (see III.B). As described above, some work has been done using speciation modelling to understand the relative effects of these chelating agents on Al balance; however, with the threat of increased Al loading of people, the importance of speciation modelling to a better understanding of the in vivo coordination chemistry of Al will continue to grow. As Table 1 shows, much work remains to be done before general agreement can be obtained on plasma Al speciation.

C. Other biofluids

Williams and colleagues have investigated Al speciation in mammalian milks and in parenteral nutrition fluids using ECCLES and the Cardiff thermodynamic database; these seem to be the only studies of their kind but are harbingers of the direction that this research will take as models become more sophisticated and life-like (Findlow et al., 1990; Duffield et al., 1991a; 1991b).

V. CONCLUSION

The influence of dietary and environmental constituents on Al biodistribution is complex. There is a lack of easily accessible isotopes of Al, and analytical techniques such as atomic absorption give information on localization, not speciation. Speciation modelling studies allow researchers the opportunity to correlate Al biodistribution with coordination chemistry but also to discriminate between the biological movement of Al. With this in view, the elaboration of a reliable model of distribution of Al in blood plasma is an urgent necessity.

ACKNOWLEDGMENT

We are indebted to the Natural Sciences and Engineering Research Council of Canada and to INSERM for their continuing support of our Al work.

REFERENCES

Alfrey, A.C. (1984). *N. Engl. J. Med.*, *310*: 1113.

Alfrey, A.C. (1986). *Kidney Int.*, *29*: S-8.

Alfrey, A.C., LeGendre, G.R., and Kaehny, W.D. (1976). *N. Engl. J. Med.*, *294*: 184.

Arp, P.A., and Meyer, W.L. (1985). *Can. J. Chem.*, *63*: 3357.

Baes, C.F.J., and Mesmer, R.E. (1976). *The Hydrolysis of Cations*, Wiley, New York, New York.

Bertholf, R.L., Wills, M.R., and Savory, J. (1984). *Biochem. Biophys. Res. Commun.*, *125*: 1020.

Berthon, G., and Daydé, S. (1992). *J. Amer. Coll. Nutr.*, *11*: 340.

Berthon, G., Hacht, B., Blais, M.-J., and May, P.M. (1986). *Inorg. Chim. Acta*, *125*: 219.

Bertsch, P.M. (1989). *The Environmental Chemistry of Aluminum* (G. Sposito, ed.), CRC Press, Boca Raton, FL, p. 87.

Bertsch, P.M., and Anderson, M.A. (1989). *Anal. Chem.*, *61*: 535.

Blaquière, C., and Berthon, G. (1987). *Inorg. Chim. Acta*, *135*: 179.

Bock, J.L., and Ash, D.E. (1980). *J. Inorg. Biochem.*, *13*: 105.

Burgess, J. (1978). *Metal Ions in Solution*, Ellis Horwood, Chichester.

Charlet, P., Deloume, J.P., Duc, G., and Thomas-David, G; (1984). *Bull. Soc. Chim. Fr., Part I*: 222.

Ciba-Geigy (1981). *Geigy Scientific Tables*, Vol. 1, Ciba-Geigy Ltd., Basel.

Clevette, D.J., and Orvig, C. (1990). *Polyhedron*, *9*: 151.

Clevette, D.J., Nelson, W.O., Nordin, A., Orvig, C., and Sjöberg, S. (1989). *Inorg. Chem.*, *28*: 2079.

Cochran, M., Patterson, D., Coates, J.H., and Coates, P.T.H. (1984). *Trace Element - Analytical Chemistry in Medicine and Biology*, Vol. 3 (P. Brätter and P. Schramel, eds.), Walter de Gruyter & Co., Berlin, p. 311.

Colomina, M.T., Gomez, M., Domingo, J.L., Llobet, J.M., and Corbella, J. (1992). *Res. Commun. Chem. Pathol. Pharm.*, *77*: 95.

Crumbliss, A.L., and Garrison, J.M. (1988). *Comments Inorg. Chem.*, *8*: 1.

Daydé, S. (1990). PhD Thesis, Université Paul Sabatier, Toulouse, 1990.

Daydé, S., and Berthon, G. (1992). *Metal Compounds in Environment and Life* (E. Merian and W. Haerdi, eds.), Science and Technology Letters, Northwood, U.K., p. 401.

Daydé, S., Venturini, M., and Berthon, G. (1989). *J. Inorg. Biochem.*, *36*: 348.

Daydé, S., Filella, M., and Berthon, G. (1990). *J. Inorg. Biochem.*, *38*: 241.

Domingo, J.L. (1989). *Clin. Toxicol.*, *27*: 355.

Domingo, J.L., Llobet, J.M., Gomez, M, and Corbella, J. (1986). *Res. Commun. Chem. Pathol. Pharm.*, *53*: 93.

Domingo, J.L., Gomez, M., Llobet, J.M., and Corbella, J. (1988a). *Lancet*, *ii*: 1362.

Domingo, J.L., Gomez, M., Llobet, J.M., and Corbella, J. (1988b). *Hum. Toxicol.*, *7*: 259.

Domingo, J.L., Gomez, M., Llobet, J.M., and Richart, C. (1991). *Lancet*, *338*: 1467.

Domingo, J.L., Gomez, M., Sanchez, D.J., Llobet, J.M., and Corbella, J. (1993). *Res. Commun. Chem. Pathol. Pharm.*, *79*: 377.

Driscoll, C.T., and Schecher, W.D. (1989). *Aluminum and Health: A Critical Review* (H.J. Gitelman, ed.), Marcel Dekker, New York, p. 27.

Duffield, J.R., Edwards, K., Evans, D.A., Morrish, D.M., Vobe, R.A., and Williams, D.R. (1991a). *J. Coord. Chem.*, *23*: 277.

Duffield, J.R., Hall, S.B., Williams, D.R., and Barnett, M.I. (1991b). *Prog. Med. Chem.*, *28*: 175.

Dyrssen, D., Haraldsson, C., Nyberg, E., and Wedborg, M. (1987). *J. Inorg. Biochem.*, *29*: 67.

Ecelbarger, C.A., and Greger, J.L. (1991). *J. Nutr.*, *121*: 1755.

Evers, A., Hancock, R.D., Martell, A.E., and Motekaitis, R.J. (1989). *Inorg. Chem.*, *28*: 2189.

Fatemi, S.J.A., Kadir, F.H.A., Williamson, D.J., and Moore, G.A. (1991). *Adv. Inorg. Chem.*, *36*: 409.

Favarato, M., Mizzen, C.A., and McLachlan, D.R. (1992a). *J. Chromatogr.*, *576*: 271.

Favarato, M., Mizzen, C.A., Sutherland, M.K., Krishnan, B., Kruck, T.P.A., and McLachlan, D.R. (1992b). *Clin. Chim. Acta*, *207*: 41.

Findlow, J.A., Duffield, J.R., and Williams, D.R. (1990). *Chem. Speciation Bioavail.*, *2*: 3.

Fulton, B., and Jeffery, E.H. (1990). *Fund. Appl. Toxicol.*, *14*: 788.

Gardiner, P.E., Stoeppler, M., and Nürnberg, H.W. (1984). *Trace Element - Analytical Chemistry in Medicine and Biology*, Vol. 3 (P. Brätter and P. Schramel, eds.), Walter de Gruyter & Co., Berlin, p. 299.

Gennaro, M.C., Mirti, P., and Casalino, C. (1983). *Polyhedron*, *2*: 13.

Gregor, J.E., and Powell, H.K.J. (1986). *Aust. J. Chem.*, *39*: 1851.

Harris, W.R. (1992). *Clin. Chem.*, *38*: 1809.

Harris, W.R., and Sheldon, J. (1990). *Inorg. Chem.*, *29*: 119.

Hsu, P.H. (1968). *Adv. Chem.*, *73*: 115.

Hugi-Cleary, D., Helm, L., and Merbach, A.E. (1985). *Helv. Chim. Acta*, *68*: 545.

Jackson, G.E. (1982). *S. Afr. J. Chem.*, *35*: 89.

Jackson, G.E. (1990). *Polyhedron*, *9*: 163.

Jackson, G.E., and Qubeka, V.V.K. (1987). "Aluminium in Biofluids - Chelation by Phosphates", Proceedings of the XXV ICCC, Nanjing, p. 685.

Jackson, G.E., and Singh, M. (1990). *Polyhedron*, *9*: 505.

Jackson, G.E., and Steyn, C.B. (1991). *S. Afr. J. Chem.*, *44*: 110.

Jackson, G.E., and Voyi, K.V.V. (1987). *Polyhedron, 6*: 2095.

Jackson, G.E., and Voyi, K.V.V. (1988). *S. Afr. J. Chem., 41*: 17.

Jones, M.M., and May, P.M. (1987). *Inorg. Chim. Acta, 138*: 67.

Kaehny, W.D., Hegg, A.P., and Alfrey, A.C. (1977). *N. Engl. J. Med., 296*: 1389.

Karlik, S.J., Eichhorn, G.L., Lewis, P.N., and Crapper, D.R. (1980). *Biochemistry, 19*: 5991.

Karlik, S.J., Elgavish, G.A., Pillai, R.P., and Eichhorn, G.L. (1982). *J. Mag. Res., 49*: 164.

Karlik, S.J., Elgavish, G.A., and Eichhorn, G.L. (1983). *J. Amer. Chem. Soc., 105*: 602.

Kiss, T., Sóvágó, I., and Martin, R.B. (1989). *J. Amer. Chem. Soc., 111*: 3611.

Kiss, T., Sóvágó, I., and Martin, R.B. (1991). *Inorg. Chem., 30*: 2130.

Langford, C.H., and Gray, H.B. (1966). *Ligand Substitution Processes*, W.A. Benjamin, Reading, MA.

Laussac, J.-P., and Commenges, G. (1983). *Nouv. J. Chim., 7*: 579.

Llobet, J.M., Domingo, J.L., Gomez, M., Tomas, J.M., and Corbella, J. (1987). *Pharmacol. Toxicol., 60*: 280.

Lopez-Quintela, M.A., Knoche, W., and Veith, J. (1984). *J. Chem. Soc., Faraday Trans., 80*: 2313.

Lote, C.J., and Saunders, H. (1991). *Clin. Sci., 81*: 289.

Lote, C.J., Saunders, H., Wood, J.A., and Spencer, A.J. (1992a). *Clin. Sci., 83*: 431.

Lote, C.J., Wood, J.A., and Saunders, H.C. (1992b). *Clin. Sci., 82*: 13.

Marklund, E., and Öhman, L.-O. (1990a). *J. Chem. Soc., Dalton Trans.*, 755.

Marklund, E., and Öhman, L.-O. (1990b). *Acta Chem. Scand., 44*: 228.

Marklund, E., and Öhman, L.-O. (1990c). *Acta Chem. Scand., 44*: 353.

Marklund, E., Öhman, L.-O., and Sjöberg, S. (1989). *Acta Chem. Scand., 43*: 641.

Marques, H.M. (1991). *J. Inorg. Biochem., 41*: 187.

Martell, A.E. (1989). *Biol. Tr. Elem. Res., 21*: 295.

Martell, A.E., and Smith, R.M. (1974-1989). *Critical Stability Constants*, Vols. 1-6, Plenum Press, New York.

Martell, A.E., Motekaitis, R.J., and Smith, R.M. (1990). *Polyhedron, 9*: 171.

Martin, R.B. (1986a). *Clin. Chem., 32*: 1797.

Martin, R.B. (1986b). *J. Inorg. Biochem., 28*: 181.

Martin, R.B. (1988). *Metal Ions in Biological Systems*, Vol. 24 (H. Sigel and A. Sigel, eds.), Marcel Dekker, New York, p. 1.

Martin, R.B. (1990). *Polyhedron, 9*: 193.

Martin, R.B., Savory, J., Brown, S., Bertholf, R.L., and Wills, M.R. (1987). *Clin. Chem., 33*: 405.

May, P.M., Linder, P.W., and Williams, D.R. (1976). *Experientia, 32*: 1492.

May, P.M., Linder, P.W., and Williams, D.R. (1977). *J. Chem. Soc., Dalton Trans.*, 588.

McCarthy, J.T., Milliner, D.S., and Johnson, W.J. (1990). *Quart. J. Med., 74 (New Ser.)*: 257.

Molitoris, B.A., Froment, D.A., MacKenzie, T.A., Huffer, W.H., and Alfrey, A.C. (1989). *Kidney Int., 36*: 949.

Motekaitis, R.J., and Martell, A.E. (1984). *Inorg. Chem., 23*: 18.

Nordström, D.K., and May, H.M. (1989). *The Environmental Chemistry of Aluminum* (G. Sposito, ed.), CRC Press, Boca Raton, FL, p. 29.

Öhman, L.-O. (1988). *Inorg. Chem., 27*: 2565.

Öhman, L.-O. (1991). *Acta Chem. Scand., 45*: 258.

Öhman, L.-O., and Sjöberg, S. (1983a). *Polyhedron, 2*: 1329.

Öhman, L.-O., and Sjöberg, S. (1983b). *J. Chem. Soc., Dalton Trans.*, 2513.

Partridge, N.A., Regnier, F.E., White, J.L., and Hem, S.L. (1989). *Kidney Int.*, *35*: 1413.

Rahman, H., Skillen, A.W., Channon, S.M., Ward, M.K., and Kerr, D.N.S. (1985). *Clin. Chem.*, *31*: 311.

Rajan, K.S., Mainer, S., Rajan, N.L., and Davis, J.M. (1981). *J. Inorg. Biochem.*, *14*: 339.

Slanina, P., Falkeborn, Y., Frech, W., and Cedergren, A. (1984). *Food Chem. Toxicol.*, 22: 391.

Slanina, P., Frech, W., Bernhardson, A., Cedergren, A., and Mattson, P. (1985). *Acta Pharmacol. Toxicol.*, *56*: 331.

Slanina, P., Frech, W., Ekstrom, L.-G., Lööf, L., Slorach, S., and Cedergren, A. (1986). *Clin. Chem.*, *32*: 539.

Sorenson, J.R.J., Campbell, I.R., Tepper, L.B., and Lingg, R.D. (1974). *Environ. Health Perspect.*, *8*: 3.

Sóvágó, I., Kiss, T., and Martin, R.B. (1990). *Polyhedron*, *9*: 189.

Tam, S.C., and Williams, R.J.P. (1986). *J. Inorg. Biochem.*, *26*: 35.

Taylor, M.J. (1987). *Comprehensive Coordination Chemistry*, Vol. 3 (G. Wilkinson, R.D. Gillard and J.A. McCleverty, eds.), Pergamon Press, Oxford, p. 106.

van Ginkel, M.F., van der Voet, G.B., van Eijk, H.G., and de Wolff, F.A. (1990). *J. Clin. Chem. Clin. Biochem.*, *28*: 459.

Venchikov, A.I. (1960). *Voprosy Pitanyia*, *19*: 3.

Venturini, M., and Berthon, G. (1987). *J. Chem. Soc., Dalton Trans.*, 1145.

Venturini, M., and Berthon, G. (1989). *J. Inorg. Biochem.*, *37*: 69.

Venturini, M., Berthon, G., and May, P.M. (1987). *Rec. Trav. Chim. Pays-Bas*, *106*: 406.

Walker, J.A., Sherman, R.A., and Cody, R.P. (1990). *Arch. Intern. Med.*, *150*: 2037.

Weber, R., and Berstad, A. (1986). *Eur. J. Clin. Invest.*, *16*: 428.

Weinland, R.F., and Denzel, W. (1914). *Ber.*, *47*: 737.

Wilhelm, M., Jäger, D.E., and Ohnesorge, F.K. (1990). *Pharmacol. Toxicol.*, *66*: 4.

Wilson, M.A., Collin, P.J., and Akitt, J.W. (1989). *Anal. Chem.*, *61*: 1253.

Yokel, R.A., and Kostenbauder, H.B. (1987). *Toxicol. Appl. Pharmacol.*, *91*: 281.

Yokel, R.A., Datta, A.K., and Jackson, E.G. (1991). *J. Pharmacol. Exp. Ther.*, *257*: 100.

COMPUTER-AIDED APPROACH TO IMAGING CONTRAST AGENTS

by

Graham E. Jackson

Department of Chemistry,
University of Cape Town,
Rondebosch 7700
South Africa

I. INTRODUCTION

In recent years a new medical speciality, *medical imaging*, has rapidly evolved from traditional X-ray imaging. This new speciality encompasses additional imaging techniques like computer tomography, thermography, ultrasonic, radionuclide, and nuclear magnetic imaging. In particular, the latter two imaging techniques rely on contrast agents for their tissue discrimination, and awareness is now developing of the role that chemical speciation plays in the efficacy of these agents. Biodistribution, excretion, and toxicity are all intimately related to the in vivo speciation of an administered contrast agent. A knowledge of chemical speciation can help to explain existing clinical observations, alert the clinician to new possible side effects and help predict iatrogenic effects of contrast agents. Ultimately, chemical speciation is a powerful tool in the design of new and more effective contrast agents.

II. PROPERTIES OF A CONTRAST AGENT

Besides the physical characteristics which allow its easy detection, a clinically useful contrast agent must meet a number of criteria:

(i) Chemical and physical stability. For reproducible results, it is essential that the chemical integrity of the contrast agent be maintained prior to administration. At the same time, the contrast agent must be physically stable and not precipitate, coagulate or flocculate during storage.

(ii) Target specificity and localization. The contrast agent must have good target tissue specificity so as to reduce background activity. At the same time, a high target tissue localization is necessary in order to limit the required doses.

(iii) Low toxicity. At the concentrations needed for tissue or organ function visualization, the contrast agent must have no adverse effects on the patient. In terms of using contrast agents in magnetic resonance and X-ray imaging, this is often the limiting factor. For radionuclide imaging, the injected doses of radioisotope are usually very low (pmol dm^{-3} range) and so chemical toxicity is not normally a problem. However, in this case the radiation dose that the patient receives is important.

(iv) Reasonable clearance rate. In order for a contrast agent to be clinically useful it must have a reasonable clearance rate or biological half life. This is in order, not only to reduce the toxicity of the agent, but also to enable alternative or repeat diagnostic procedures to be carried out.

(v) Easy preparation. Contrast agents are often supplied in kit form, with final preparation taking place just prior to administration. Clearly if preparation involves a complicated synthetic and/or purification procedure, this will detract from the use of the contrast agent.

In all of the above criteria, chemical speciation plays a vital role and so speciation modeling can be used to advantage.

III. FACTORS DETERMINING BIODISTRIBUTION OF CONTRAST AGENTS

The physical and chemical properties of a contrast agent are partly responsible for its biodistribution. The charge, size and lipophilicity of the contrast agent will affect its membrane permeability, excretion and biodistribution. The contrast agent may bind to blood protein or blood cells; it may enter an active transport mechanism or bind to cellular receptors; it may interact with the reticuloendothelial system or undergo phagocytotic absorption. In addition, metabolism of the contrast agent may occur.

Tissue localization of a contrast agent can occur as a result of local alteration in physiology due to disease. Indeed, the value of a contrast agent is in its ability to respond to these differences in physiology and hence visualize the disease states themselves.

One of the prime uses of contrast agents in medical imaging is the early detection of neoplastic disease and delineation of tumorous growths. More difficult is the differentiation between benign and malignant tumors.

A. Mechanisms of tumor uptake

Essential to the enhancement of tumor signal in any imaging technique is the accumulation of the contrast agent within the tumor. This may be accomplished via several mechanisms (Larson et al., 1975).

1. *Agents as substrates*

The rationale behind this method of tumor localization is the increased metabolic activity of tumor cells resulting in a large turnover rate of a suitably labelled substrate. Thus, [^{32}P]phosphate has been used to detect malignant tumors with increased DNA synthesis (Hagler et al., 1970).

2. *Tumor-associated antigens*

Labelling of tumor-specific antigens, while holding enormous promise, has only had limited success (Lauffer, 1991). Recently, Bair et al. (1991) have used a ^{99m}Tc-labelled anti NCA-95 antibody to detect prosthetic valve endocarditis.

3. *Increased blood flow and vascularity*

Most tumor scanning pharmaceuticals owe their localization not to binding to specific metabolites or antigens but to an increased residency time within the tumor. This may come about via increased tumor blood flow or via altered vascularity of tumor tissue. Increased blood volume, increased membrane permeability due to larger pore size, pinocytosis of neoplastic cells, and decreased lymphatic drainage have all been listed as factors contributing to contrast agent localization (Larson et al., 1975).

IV. RADIOPHARMACEUTICALS

Radionuclide imaging is the general term used to describe the procedure in nuclear medicine in which radioactive material is administered and its distribution in the human body detected and displayed. This procedure can be carried out to detect abnormalities in tissue and organ function, the type of radioactive material, radiopharmaceutical, employed being a function of the clinical problem. Radiopharmaceuticals are used at pmol dm^{-3} concentrations and so their in vivo speciation is difficult, if not impossible, to study via conventional analytical techniques. On the other hand, using non-radioactive isotopes, the chemical thermodynamics of these metal ions can be studied and their in vivo speciation calculated using modeling techniques.

A. Gallium

[^{67}Ga], as gallium citrate, has been widely used in the diagnosis of neoplastic disorders, particularly lymphomas and Hodgkin's disease (Edwards and Hayes, 1970). A wide variety of benign disorders have also been visualized using [^{67}Ga], the ion being accumulated in inflammatory conditions (Boxen and Ballinger, 1991). Doses of 2-10 mCi (1-100pmol dm^{-3}) are commonly injected intravenously with imaging taking place 24 to 72h post injection.

The most important oxidation state of gallium is +3 and because of its high charge and small ionic radius (0.62 Å) it is extensively hydrolyzed in solution. For this reason [^{67}Ga] is usually supplied as the citrate complex or with a high background acid concentration. At pH 7.4 the free concentration of Ga^{3+} is limited to 10^{-16} mol dm^{-3} by hydroxide precipitation. At the same time Ga^{3+} forms very insoluble phosphate complexes which also limits the plasma concentration of free Ga^{3+}. Hence it is possible to predict that Ga(III) would be poorly absorbed after oral, subcutaneous, or intramuscular administration. Clinically the radionuclide has to be given via intravenous or intraperitoneal injection.

Recently computer models of blood plasma have been developed which include Ga(III) (Jackson, 1990; Clevette and Orvig, 1990). These models show that, at concentrations typical of normal diagnostic procedures, the metal ion is bound almost exclusively to transferrin (Tf), the Fe^{3+} transport protein. In the absence of Tf binding, or if the concentration of Ga(III) exceeds the available concentration of Tf binding sites, then the predominant low-molar-mass (l.m.m.) complexes are hydrolyzed phosphate and citrate complexes (Table 1). As the incorporation of the metal ion into Tf is likely to be slow, the simulation conditions including and excluding Tf binding represent two extremes of the real situation, immediately after administration and at equilibrium (after several hours). Clinically it has been shown that initially ~15% of the injected dose of Ga(III) is excreted in the urine and that the remaining Ga(III) is concentrated in the liver (5%), kidney (2%), skeleton (including marrow) (24%), and other soft tissue (34%) (Nelson et al., 1972). These clinical observations are in accord with the simulation results which predict that initially the metal ion would be distributed in plasma as charged (readily excretable) l.m.m. complexes, but that ultimately the metal ion would be incorporated into Tf. The high affinity of Ga(III) for phosphate is consistent with its localization in bone.

[^{67}Ga]gallium citrate is commonly used to visualize neoplastic disorders. It has been proposed (Hoffer, 1980) that gallium localizes as a l.m.m. complex or is mediated by Tf

Table 1

Calculated Species Distribution of Ga^{3+} Complexes (mol dm^{-3}) in Blood Plasma in (a) Absence of Tf Binding, (b) the Presence of 50 μmol dm^{-3} Tf Binding sites, (c) + 0.014 mol dm^{-3} Ca^{2+}, and (d) pH 6.4. Total [Ga^{3+}] = 3.6×10^{-10} mol dm^{-3}

Species	(a)	(b)	(c)	(d)
$[Ga(OH)_4]^-$	3.5×10^{-10}	2.7×10^{-13}	2.6×10^{-13}	2.7×10^{-14}
$[Ga(OH)_3]$	1.4×10^{-11}	1.1×10^{-14}	1.0×10^{-14}	1.1×10^{-14}
$[Ga(cit)(OH)]^-$	3.3×10^{-12}	2.5×10^{-15}	4.7×10^{-16}	2.5×10^{-13}
$[Ga(OH)_2]^+$	1.4×10^{-14}	1.1×10^{-17}	-	-
$[Ga(cit)(HPO_4)]^{2-}$	1.7×10^{-15}	7.8×10^{-18}	-	-
[Ga(Tf)]	-	3.6×10^{-10}	3.6×10^{-10}	3.6×10^{-10}
$[Ga_2(Tf)]$	-	6.5×10^{-16}	5.7×10^{-16}	2.4×10^{-16}
$[Ga]^{3+}_{free}$	1.3×10^{-22}	9.0×10^{-26}	1.0×10^{-25}	1.6×10^{-22}

cit = citrate

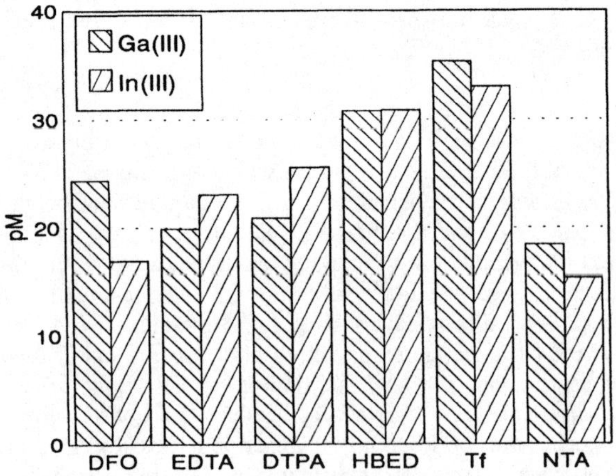

Figure 1.
Calculated free metal ion concentration (pM = -log[M]) for selected ligands with Ga(III) and In(III). Source: Clevette and Orvig, 1990. DFO = desferrioxamine-B; HBED = *N,N'*-bis(*o*-hydroxylbenzyl)ethylenediamine-*N,N'*-diacetate; Tf = transferrin; NTA = nitrilotriacetate.

specific receptors on cell membranes, the gallium being bound to the Tf. Furthermore, a local decrease in pH or the high concentration of intracellular Ca^{2+} displaces Ga(III) from Tf fixing it within the cell. Speciation calculations (Table 1) show that Tf-mediated uptake of gallium is a viable mechanism as most of the metal ion circulating in the plasma is bound to this protein. On the other hand a 10-fold increase in calcium concentration does not affect gallium speciation (Table 1) and, while a decrease in pH from 7.4 to 6.4 has a substantial effect on the l.m.m. speciation of gallium, there is no change in the concentration of [Ga(Tf)].

B. Indium

The solution chemistry of indium is very similar to that of gallium, both being group 13 (IIIA) metals. Above pH 3.4, indium tends to precipitate as the hydroxide, $[In(OH)_3]$, or phosphate, $[In(PO_4)]$, complex. Precipitation of indium hydroxide can, however, be prevented by using a chelating agent to bind the metal ion or by working at low pH. Typically 0.05 M HCl is used to elute ^{113m}In from its generator, giving a 10^{-8} M solution of ^{113m}In. This eluate can be used unaltered as a blood pool scanning agent (Adatepe et al., 1971; Goodwin et al., 1973) or complexed to other organic ligands in order to achieve a different biodistribution.

Clevette and Orvig (1990) have published an abbreviated model of blood plasma which allows the in vivo chelation of indium by a range of common chelating agents to be calculated. Figure 1 shows the free metal concentration for a selected group of ligands with indium and gallium. Under the conditions of the calculation, in the presence of DTPA or EDTA, there is 10^3 - 10^4 less free metal ion if the metal is In(III) than if it is Ga(III). This may account for the fact that only limited success has been achieved using ^{67}Ga-labelled DTPA coupled to proteins (Meares, 1986) while more success has been achieved using $^{111}In^{3+}$-labelled DTPA and EDTA protein conjugates (Scheinberg et al., 1982; Hnatowich et al., 1983). However, computer speciation calculations show that both metal ions are ultimately lost to Tf and so the slow decomposition kinetics relative to the duration of the diagnostic procedure must play a role. Similar results have been obtained with $^{67}Cu^{2+}$ labelled DTPA and EDTA protein conjugates (Cole et al., 1986). Here, however, the computer model predicts that the metal ion is lost to serum albumin.

V. MAGNETIC RESONANCE IMAGING

Magnetic resonance imaging (MRI) has rapidly evolved from an experimental to a clinical very useful modality. Excellent tissue contrast is readily obtained without the use of contrast agents. However, more recently the difficulty in differentiating certain lesions from adjacent edema and fibrosis, the need for faster imaging techniques, and the study of tissue or organ function has led to the development of MRI contrast agents.

Modern MRI images are essentially T_1 or T_2-weighted two and three dimensional maps of proton density. In general paramagnetic metal ions are used as contrast agents, in which the unpaired electron spin affects tissue contrast. Gadolinium(III) (S = 7/2) is the metal of choice with gadopentetic acid ($[Gd(DTPA)]^{2-}$) being licensed for use in Europe and the United States.

A. Gadolinium in blood plasma

Jackson et al. (1990) and Jackson and Du Toit (1991) have developed a computer model of blood plasma which includes Gd(III) equilibria and have applied this model to an analysis of MRI relaxivity (Jackson and du Toit, 1992). The model is based on the blood plasma model of May et al. (1977) (Berthon et al., 1986) to which the appropriate constants for Gd(III) have been added. This model, and others, have been described in detail elsewhere in this book. A similar model of blood plasma which includes Gd(III) has been published by Gennaro et al.(1990).

Figure 2 shows the calculated l.m.m. species distribution of Gd(III) as a function of metal ion concentration. The graphs show that at a low concentration of Gd(III) (<2 x 10^{-4} mol dm^{-3}) the metal ion is bound almost exclusively as the $[Gd(citrate)(OH)]^-$ (69.1%) and

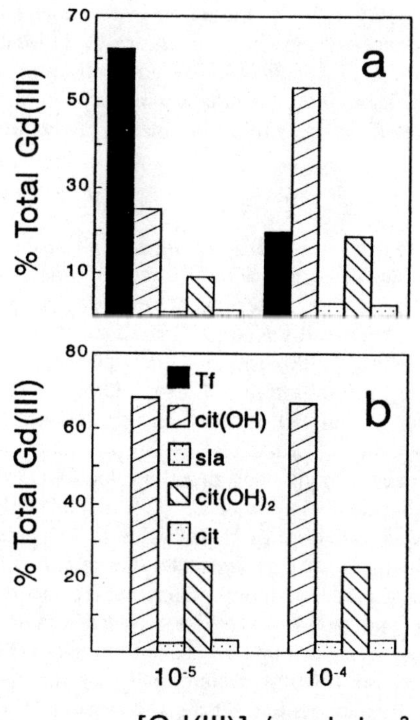

Figure 2.
Calculated low-molar-mass species distribution of Gd(III) in blood plasma as a function of total metal concentration. (a) including Tf binding, (b) excluding Tf binding. The coordinated ligands and their stoichiometry are as indicated; Tf = transferrin; cit = citrate; sla = salicylate.

$[Gd(citrate)(OH)_2]^{2-}$ (24.4%) complexes. As the concentration of the metal ion is increased (LD_{50} $GdCl_3$ ~6 mmol dm^{-3}) so the amount of available citrate is exceeded and a new species distribution is obtained.

In the above simulation, the interaction between Gd(III) and in vivo protein metal binders was ignored. Metal uptake by proteins is often slow compared to the biological half-life of the metal and so in practice two conditions exist, one in which no protein binding occurs (immediately after administration) and another when equilibrium with the protein has been established. Unfortunately, simulation programs which include the effect of kinetics have yet to be developed.

The introduction of Gd(III)-Tf binding into the speciation model results in the species distribution shown in Figure 2. At a total Gd(III) concentration less than 10^{-5} mol dm^{-3}, the metal ion is bound predominantly as the $[Gd(Tf)HCO_3]$ complex, while at a concentration of 10^{-4} mol dm^{-3}, Gd(III) is bound to citrate ($[Gd(citrate)OH]^-$, 53%) and Tf ($[Gd(Tf)HCO_3]$, 20%).

The effect of Gd(III) on the species distribution of in vivo metal ions is shown in Figure 3. Cu(II) is unaffected by the presence of Gd(III) but significant changes occur in the speciation of Zn(II), Ca(II) and Fe(III). These predicted results are in accord with clinical observations that $GdCl_3$ can cause elevated levels of iron in serum and inhibit blood clotting (Sax, 1984). It would be interesting to monitor changes in zinc excretion upon Gd(III) administration.

B. DTPA and EDTA

Of interest to the clinician is the effect of DTPA, the currently used ligand for preparing MRI contrast agents, and other chelating ligands upon the in vivo species distribution of Gd(III) (Fig. 4). In the absence of Tf, binding DTPA and EDTA lower the concentration of free metal ion by a factor of 10 - 10^7, the reduction in free metal ion concentration being higher, the higher the drug concentration. In the presence of Tf binding, only at nanomolar concentrations, in the case of DTPA, and mmolar concentrations, in the case of EDTA, does Tf binding become significant. These results are in accord with the clinical observation that $[Gd(DTPA)]^{2-}$ is excreted intact while the metal ion is lost from $[Gd(EDTA)]^-$ (Weinmann et al., 1984).

C. Iatrogenic interactions

In general, contrast agents are used to visualize changes in physiology as a result of disease. However, the possibility exists that drugs administered to combat disease may also affect the biodistribution of a contrast agent. An excellent review of drug-induced changes to the biodistribution of radiotracers is given by Hodges (1987).

Recently, Hattner and White (1990) reported the interference of $[Gd(DTPA)]^{2-}$ upon the biodistribution of $[^{67}Ga]$gallium citrate. These authors noted that an injection of $[^{67}Ga]$gallium citrate 4h after a $[Gd(DTPA)]^{2-}$ enhanced MRI scan resulted in unexpected renal excretion of the radioisotope. Hattner and White postulate that some dissociation of the $[Gd(DTPA)]^{2-}$ complex occurs in vivo and that the released Gd^{3+} gives rise to a carrier effect on the ^{67}Ga biodistribution. Wiggins et al. (1991) reject this argument, suggesting instead that the change in biodistribution of ^{67}Ga is due to the presence of excess DTPA, found in all commercial formulations of gadopentate. These two different explanations can be evaluated using a speciation model of blood plasma which incorporates both metal ions. The results (Table 2) show that neither DTPA nor $[Gd(DTPA)]^{2-}$ are able to affect the speciation of Ga(III). Indeed, even at mmolar concentrations, DTPA does not affect the speciation of Ga(III), the ligand binding preferentially to the much more concentrated Ca(II). However, Gd(III) on its own does affect the speciation of Ga(III). At 10^{-4} M Gd(III) there is a 100-fold increase in the concentration of $[Ga(OH)_4]^-$ and a new mixed-metal Tf complex is formed. These changes in speciation are consistent with the observed changes in the biodistribution of the radioisotope.

While the above change was detrimental to the diagnostic procedure, other metal ions may cause changes which improve the clinical picture. Thus Fe(III), if administered 24h after a tracer injection of $[^{67}Ga]$gallium citrate, lowers the whole-body retention of $[^{67}Ga]$

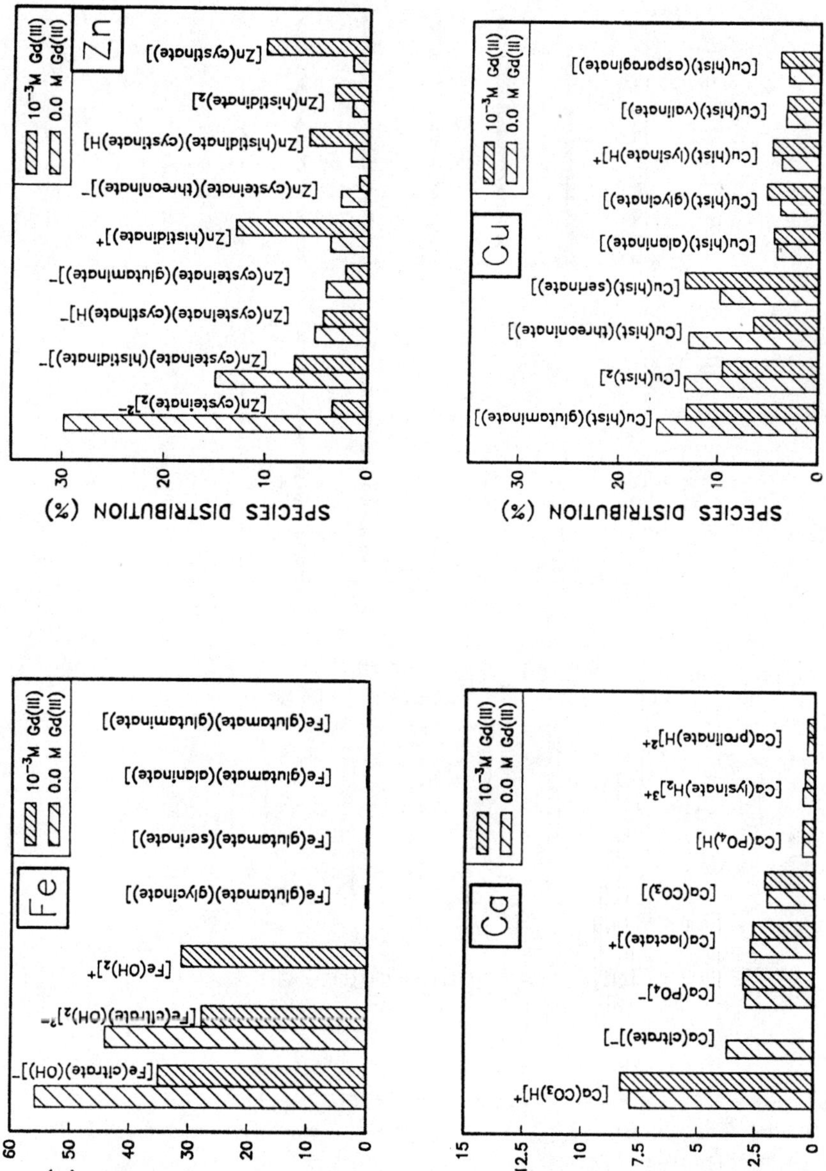

Figure 3.
Effect of Gd(III) on the low-molar-mass species distribution of Cu(II), Zn(II), Ca(II) and Fe(III). (Courtesy of G.E. Jackson et al. (1990), Academic Press, Duluth).

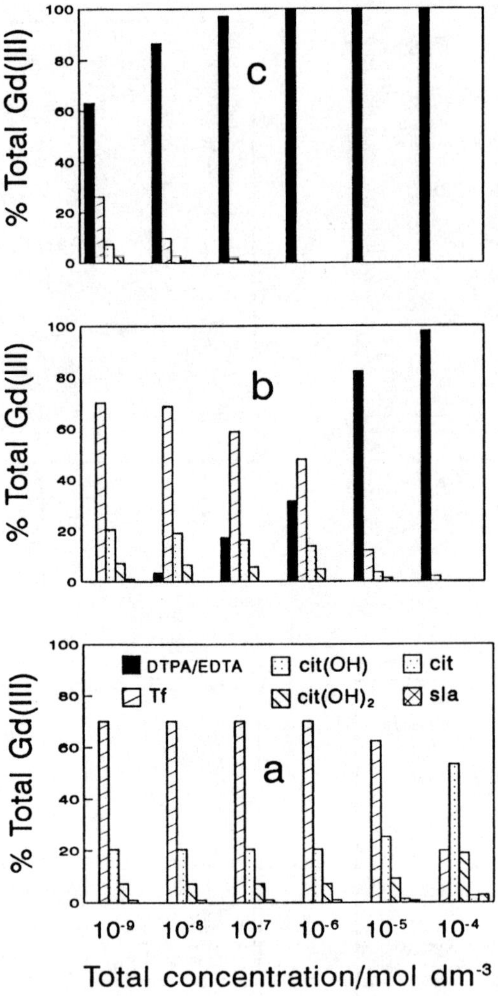

Figure 4.

Effect of contrast agent upon the in vivo speciation of Gd(III). (a) Gd^{3+}; (b) $[Gd(EDTA)]^{-}$; (c) $[Gd(DTPA)]^{2-}$. The coordinated ligands and their stoichiometry are as indicated. Tf = transferrin; cit = citrate; sla = salicylate.

Table 2

Calculated Species Distribution of Ga(III) Complexes in (a) Blood Plasma, (b) + 9.4×10^{-9} mol dm^{-3} DTPA, (c) + 9.4×10^{-3} mol dm^{-3} $[Gd(DTPA)]^{2-}$, (d) + 3.8×10^{-4} mol dm^{-3} Gd(III). Total Ga(III) = 3.75×10^{-10} mol dm^{-3} and Total Citrate = 3.4×10^{-4} mol dm^{-3}

Species	(a), (b), (c)		(d)	
	Conc. (mol dm^{-3})	Total Ga^{3+} (%)	Conc. (mol dm^{-3})	Total Ga^{3+} (%)
$[Ga(Tf)HCO_3]$	3.8×10^{-9}	99.9	5.5×10^{-10}	14.7
$[GdGa(Tf)(HCO_3)_2]$	-	-	3.0×10^{-9}	80.1
$[Ga(OH)_4]^-$	1.3×10^{-13}	0.1	1.3×10^{-10}	4.8
$[Ga_2(Tf)(HCO3)_2]$	3.1×10^{-14}	0.0	6.7×10^{-13}	0.0
$[Ga(OH)_3]$	5.4×10^{-14}	0.0	5.4×10^{-12}	0.2
$[Ga(citrate)(OH)]^-$	1.3×10^{-14}	0.0	2.1×10^{-13}	0.0
$[Ga(DTPA)]^{2-}$	3.4×10^{-14}	0.0^b	-	-
$[Ga]^{3+}_{free}$	5.0×10^{-25}	0.0	5.0×10^{-23}	0.0

Source: Jackson and Byrne (1994)

resulting in lower background activity. Speciation modeling shows that the Fe(III) displaces Ga(III) from its Tf-binding site thereby increasing its excretion and lowering the background activity. The [Ga(Tf)] already accumulated at the lesion site is less accessible to the Fe(III) and so improved contrast is obtained.

VI. CONCLUSION

Medical imaging is a valuable tool in the armory of the clinical diagnostician. Increasingly contrast agents are being used to improve the quality and versatility of these imaging modalities. Chemical speciation plays a central role in the biodistribution, excretion, and toxicity of contrast agents and hence is crucial to their efficacy.

REFERENCES

Adatepe, M.H., Welch, M., Evens, R.G., and Potchen, E.J. (1971). *Amer. J. Roentgenol. Radium Ther. Nucl. Med.*, *112*: 701.

Bair, H.J., Becker, W., Volkholz, H.J., and Wolf, F. (1991). *Nuklearmedizin*, *30*: 149.

Berthon, G., Hacht, B., Blais, M.-J., and May, P.M. (1986). *Inorg. Chim. Acta*, *125*: 219.

Boxen, I., and Ballinger, J.R. (1991). *Curr. Opin. Radiol.*, *3*: 840.

Clevette, D.J., and Orvig, C. (1990). *Polyhedron*, *9*: 151.

Cole, W., DeNardo, S., Meares, C., McCall, M., DeNardo, G., Epstein, A., O'Brien, H., and Moi, M. (1986). *Nucl. Med. Biol.*, *13*: 363.

Edwards, C.L., and Hayes, R.L. (1970). *JAMA*, *212*: 1182.

Gennaro, M.C., Aime, S., Santucci, E., Causa, M., and De Stefano, C. (1990). *Anal. Chim. Acta*, *233*: 85.

Goodwin, D.A., Song, C.H., Finston, R., and Martin, P. (1973). *Radiology*, *108*: 91.

Hagler, W.S., Jarrett, W.H., and Humphrey, W.T. (1970). *Arch. Ophthalmol.*, *83*: 548.

Hattner, R.S., and White, D.L. (1990). *J. Nucl. Med.*, *31*: 1844.

Hnatowich, D., Layne, W., Childs, R., Lanteigne, D., Davis, M., Griffin, T., and Doherty, P. (1983). *Science*, *220*: 613.

Hodges, R. (1987). *Essentials of Nuclear Medicine Science* (W. Hladik, G. Saha, K. Study and B. Friedman, eds.) Williams & Wilkins, Baltimore, Chapter 13.

Hoffer, P. (1980). *J. Nucl. Med.*, *21*: 282.

Jackson, G.E. (1990). *Polyhedron*, *9*: 163.

Jackson, G.E., and Du Toit, L. (1991). *J. Chem. Soc., Dalton Trans.*, 1463.

Jackson, G.E., and Du Toit, L. (1992). *Polyhedron*, 11: 1315.

Jackson, G.E., and Byrne, M.J. (1994). *Int. J. Nucl. Med. Biol.*, submitted.

Jackson, G.E., Wynchank, S., and Woudenberg, M. (1990). *Magn. Reson. Med.*, *16*: 57.

Larson, S.M., Milder, M.S., and Johnston, G.S. (1975). *Radiopharmaceuticals* (G. Subramanian, B. Rhodes, J. Cooper and V. Sodd, eds.), The Society of Nuclear Medicine, New York, p. 413.

Lauffer, R.B. (1991). *Magn. Reson. Med.*, 22: 339.

May, P.M., Williams, D.R., and Linder, P.W. (1977). *J. Chem. Soc., Dalton Trans.*, 588.

Meares, C. (1986). *Nucl. Med. Biol.*, *13*: 311.

Nelson, B.M., Hayes, R.L., Edwards, C.L., Kniseley, R.M., and Andrews, G.A. (1972). *J. Nucl. Med.*, *13*: 92.

Sax, N.I. (1984). *Dangerous Properties of Industrial Materials*, 5th ed., Van Nostrand-Reinhold, New York.

Scheinberg, D., Strand, M., and Gansow, O. (1982). *Science*, *215*: 1511.

Weinmann, H.J., Brasch, R.C., Press, W.R., and Weskey, G.F. (1984). *Amer. J. Roentgenol.*, *142*: 619.

Wiggins, J., Goldstein, H., and Weinmann, H. (1991). *J. Nucl Med.*, *32*: 1830.

FUTURE TRENDS

by

Peter M. May

School of Mathematical and Physical Sciences,
Murdoch University,
Western Australia 6150
Australia

I. INTRODUCTION

It is always difficult and usually risky for those who are "persuaded" to attempt predictions of the future. The future has an unkind habit of doing the unexpected. In this section, therefore, I will try to avoid speculation and stick instead to developments currently taking place at the forefront of research into the quantification of metal-ligand interactions in biological fluids. This will indicate the areas which, in my opinion, are likely to impact most and, by implication, suggest what lies immediately ahead.

I will commence with a description of what is happening presently in the general domain of computing. Whilst developments of this kind are clearly not specific to the subject of this Handbook, there can be no doubt whatsoever that they are about to revolutionize how metal-ligand interactions are studied and, especially, how such knowledge can be applied to biochemical and medical problems.

I will then go on to discuss two of the most difficult matters which are now holding back computer simulations of metal-ligand equilibria in biofluids. Their importance lies in trying to build more comprehensive models of how metal ions behave in biological systems. The first concerns the better description of how metal ions interact with biological macromolecules and the second is associated with modelling the kinetics of metal ion transport through many physiological compartments, each containing many potential complexing agents.

II. ADVANCES IN COMPUTER TECHNOLOGY

Computer simulation established itself as an important technique for understanding trace element interactions in vivo in the late 1960s, quite early on in the scientific application of computers, and its use has grown rapidly ever since. This, of course, reflects the power of computers to deal with highly complicated systems such as are typical in biology. The huge number of possible interactions created by all the permutations of metal and ligand in biological fluids simply cannot be evaluated thoroughly in any other way. On the other hand, compared with developments in the computer modelling of astrophysical phenomena and of weather forecasting, simulations of biological systems are sadly underdeveloped. The fact that many of the individual reactions can be described in relatively simple chemical terms makes computer modelling of these large systems an entirely realistic long-term goal in a wide variety of different contexts. However, despite their obvious potential and their early introduction, computer methods in this area remain far from being fully exploited. This section discusses why this is and how it seems likely to change in future.

Perhaps the most fundamental reason why the study of metal-ligand interactions in biological fluids has not embraced computer technology as avidly as it might have done arises from the difficulty inherent in any multidisciplinary research. Our academic institutions have still to evolve truly effective training programmes that cross traditional disciplinary boundaries or even integrate new with old areas of study. The problem is particulary acute in the present setting since an exceptionally broad span of disciplines is involved. These cover not only chemistry and biology; in addition to a sound knowledge of the physiological and medical implications of the work, skills are also required in mathematics and computer science.

Another reason has been the somewhat ambivalent attitude adopted by many scientists towards computer-assisted methodology. As Marsili (1990) has observed, computers are accepted as powerful tools so long as they perform in "subordinate" ways but are regarded less positively if their role extends beyond simple "number cruching". Machines that make decisions or determine strategy are seen as something of a threat, as evidenced by the excessive delight which often accompanies any perceived lack of commonsense by the computer.

Neither of these reasons is fundamental; nor are they likely to persist indefinitely. Whilst there are also some technical obstacles which have held back the development of more sophisticated models of metal-ligand interactions in biological systems, it is evident that a profound change in attitude will not be long in coming. For one thing, familiarity with computers, nowadays even in schoolchildren, is increasing dramatically. For another, it is becoming increasingly difficult to compete effectively in science (as in other areas) without computer aid. There is a general realization that computers are now a prerequisite in almost all cutting-edge research, which is already starting to impact on specific areas such as are covered by this Handbook.

It is important to appreciate the magnitude of the developments in computer technology which are soon to be forthcoming. Few would be unaware of the great improvements which have occurred over the preceding couple of decades but, so dramatic have these been, it might be felt that the rate of progress is now due to decline. This is not the case. Machines for specific CPU-intensive tasks will soon be capable of a teraflop (10^{12} floating point operations per second). Ready access to supercomputers is becoming commonplace; the vector processor at the Australian National University, for example, can already provide a peak speed of over a gigaflop and over a gigabyte of memory, free of charge, purely on the basis of academic merit. Desktop workstations with hundreds of times the speed of yesterday's mainframes and with gigabytes of disk storage are already available in the price bracket of personal computers. So-called "multicomputers", based on massively parallel computer architecture, are set to appear in the next few years. The development of systematic methods that realize the enormous potential of this parallel processing for all kinds of program will ensure the exponential growth of computer power continues unabated (Denning and Tichy, 1990). Ten-fold increases in computer processing power can be reasonably anticipated every four years or so, at least until the turn of the century.

The implications of such improvements in computer technology are awesome. In many areas of science, breakthroughs can be confidently predicted, either because the additional computer power permits tasks to be undertaken that are currently just too large or because improvements in general problem-solving by computational techniques will systematically provide new insight. It is, of course, impossible to foresee all the details of the progress which will result but one can well imagine that the impact will be profound and widely-embracing. In the remainder of this section, the specific developments which seem to the author most likely to alter our knowledge of metal-ligand interactions in biological fluids will be considered.

A. Chemometrics

The methodology of chemical data analysis, particularly in respect of computer usage, has developed so rapidly in recent times that it has acquired its own special name, "chemometrics". The two main objectives are (i) to optimize experiments and analytical procedures so that one obtains data with the maximum possible information content, and (ii) then to process this data in ways that yield the best possible numerical results. These general computational approaches clearly have much to offer wherever metal-ligand interactions are studied quantitatively. The only significant factor limiting the application of chemometrics to this area is the dearth of investigators who are properly qualified in respect of both disciplines. The multivariate nature of biological systems in general and the pronounced tendency for correlations to beset phenomena associated with metal-ligand interactions makes the application of chemometric techniques especially necessary. The various types of factor analysis, cluster analysis, sampling theory, and multidimensional optimization are the main chemometric methods to note (Jurs, 1990).

B. Molecular graphics

Molecular graphics is the visual representation of various kinds of surface which can be imagined at the molecular level. These surfaces can be contours of electronic charge density, molecular orbitals, electrostatic potentials, van der Waals radii, areas accessible to solvents, etc (Mezey, 1990). Their representation by computer has become a very important element in the interpretation of molecular properties and processes. Three-dimensional representations, nowadays often in full technicolor, are in particular most useful for identifying reactive sites of molecules, conformational rearrangements and the effects of atomic size and volume. Whilst an all-too-loud fanfare announcing the possibilities of molecular graphics techniques in understanding the mechanism of action of enzymes has inevitably led to an anti-climax, this should not detract from the undoubted benefits of being able to visualize routinely and easily what is happening to the atoms involved in chemical transformations. The author hazards a guess that the application of molecular graphics methodology to the study of interfacial phenomena will become one of the high fashions of chemistry in this decade.

C. Computer simulation of molecular energetics

The calculation of molecular dynamics has now progressed to the extent that the relative energies of various species postulated to exist in solution can be estimated with some confidence. Major advances have been made through the development of semi-empirical molecular mechanics, force-field, and molecular orbital models which, unlike ab initio quantum chemistry calculations, are parametrized using real data. Consequently, it is now feasible to determine activity coefficients, pair correlation functions, internal energies and other thermodynamic functions theoretically. For example, state-of-the-art molecular simulations can now predict various free energies of interaction (Damewood et al., 1990) and relative free energy differences can be obtained even more reliably due to the cancellation of certain errors (Straatsma and McCammon, 1992). Of particular interest is the accuracy of the recent calculation of the hydration free energies of Fe^{3+} and Fe^{2+} (Urban and Underwood, 1990). Whilst such successes are presently limited to rather simple systems, it is evident that straightforward extensions of these computational methods coupled with more computer power will lead to ever greater complexity in the problems being tackled and solved. It will not be too long before the application of reaction dynamics and the ab initio calculation of rate constants will be feasible for many processes (Hirst, 1990), including those of biochemical importance. Given the sparsity of such information, especially variations due to experimental conditions, the ability to calculate rates of reaction in this way will revolutionize computer simulations of transition element physiology.

D. Applications of artificial intelligence

No area of computing is more likely to impact in the near future on the development of scientific ideas more than that of artificial intelligence. As the complexity of the systems which we are trying to understand grows ever more rapidly, there is an obvious need for computing tools to help manage the multitude of relevant interrelationships involved in, say, biological systems. It is clear from the considerable success of so-called "expert systems" that computer-assisted interpretations of chemical phenomena can be usefully invoked both as a research tool (for those with the necessary experience) and as a learning-aid (for those who are new to the subject). Expert systems are decision-making computer programs which assist in areas requiring specialist knowledge. They have proved particularly effective in cases where the domain is intellectually demanding but narrow. They have emerged in fields as diverse as geological prospecting, manufacturing design and medical diagnosis. In chemistry, the pioneering DENDRAL project (Buchanan et al., 1976) has revolutionized the interpretation of certain spectra and much progress has been made in the artificial design of organic synthetic pathways. There can be little doubt that the development of such techniques in artificial intelligence coupled with the unleashing of the prodigious amounts of computer power that this requires will be one of the main features of science over the next decade or two. Whilst few of the consequences of this will be specific to those investigating metal-ligand interactions in biofluids, the general effect will be profound.

E. Chemical databases

Online access to information in the chemical literature is now well established as an indispensible research tool. Chemical knowledge is said to be doubling every five years or thereabouts. Certainly, the number of chemical publications is increasing rapidly: several appearing per minute (Hassanaly and Dou, 1986). This remarkable growth makes coping properly with the acquisition of relevant material impossible without a computerized approach. The availability of literature databases such as provided by the American Chemical Society's Chemical Abstract Service on large host computers accessible by electronic network has accordingly become commonplace. It doesn't require much imagination to appreciate that the role of these automated information providers will continue to expand. This means that the slow step, i.e. one involving people, will shift from gathering the information to rationalizing it. Hence, it can be confidently predicted that fresh applications of the computer will soon flourish. In particular, advances will be greatest wherever the context permits data to be stored not only in large quantity but also in a form that makes it amenable to subsequent automatic processing. Two examples from the realm of chemistry stand out. The first concerns information about molecular structures. Secondly, there is the vast, accumulated body of rules and data required for the general simulation of chemical processes. In both cases, there is an obvious need for databases which can be interpreted and manipulated without further human intervention. To accomplish this, much effort will have to be expended on capturing, assessing and classifying the knowledge in the first place. However, the rewards for doing so will surely be spectacular. Not only will our access to all the information be greatly improved, but interesting new chemical relationships will inevitably be discovered as specially designed computer programs search for and quantify collective properties not otherwise apparent (May and Murray, 1991b).

III. NEW APPROACHES TO METAL ION-MACROMOLECULE INTERACTIONS

Macromolecules (like proteins, polysaccharides and fulvic/humic acids) have a significant, often paramount, role in determining the chemical fate of metal ions in biological systems (Betti and Papoff, 1988). Conversely, as this Handbook partially documents, metal ion binding has many, very important functions in biology. Unfortunately, the chemical interactions between macromolecules and metal ions in multicomponent biological milieux, such as blood plasma, cannot at present be satisfactorily characterized. Quantitative descriptions of the binding behavior are especially problematical. Since macromolecules often complex a major fraction of the metal ions present in biofluids, this is a major reason why our understanding of, and hence our ability to control and modify, the biological effects of metal ions remain severely limited.

Serious difficulties arise with the theoretical description of how cations become attached to any ligand with a large number of chemical binding sites (Dixon, 1992). In essence, this is mainly because the interaction depends not only on the functional groups at which complexation occurs but also on the state of ionization of surrounding functional groups. The number of possible states rises combinatorially with the number of surrounding ionizable groups so that the strength of binding at any particular site soon becomes a highly complicated function of the prevailing conditions, such as the solution pH. In addition, there are often multiple sites available on macromolecules at which binding can take place, each with a huge variety of possible complex stoichiometries (especially in respect of the number of protons associated with the complex). For these various reasons, it is simply not possible to express the relevant complexation reactions in terms of proper equilibrium constants and, hence, the valid thermodynamic characterization of these interactions is put far beyond any realistic prospect of experimental measurement.

Numerous attempts have been made to simplify the mathematical treatment of metal ion-macromolecule interactions but these have, in large measure, only served to yield empirical descriptions of the interactions which are highly dependent on the conditions under which they were measured, and thus of little or no applicability elsewhere. It is surprising to find how often this dependence is neglected in the literature. A singular constant for metal binding to a multifunctional complexing agent such as a protein is virtually without value except when it is used to back-calculate the particular measurement on which it was originally based under the identical conditions operating at the time.

Recently, however, it has been shown (Buffle and Altmann, 1987; Altmann and Buffle, 1988) how fulvic acid, at least, should be treated mathematically to avoid the artefactual results which so abound in the literature. Based on a differential equilibrium function derived by Gamble et al. (1980), this approach permits both a better representation of the experimental information and a better molecular interpretation of the heterogeneous complexation which typifies this kind of "macromolecular" system. It is reasonable to think that in principle other polyfunctional and polyelectrolytic macromolecules (including proteins, nucleic acids, and polysaccharides) would be amenable to a similar analysis, notwithstanding the distinct differences in their metal-binding profiles.

A particular interest arising from the approach of Buffle and Altmann is the possibility it creates of properly linking high- and low-molar-mass speciation calculations. This is presently being implemented as an extension of the capabilities of JESS (for Joint Expert Speciation System), a large software package currently being developed by us to deal with various kinds of chemical partitioning in solution (May and Murray, 1991a). The idea is to marry the functions describing metal-macromolecule binding to the conventional procedures for calculating low-molar-mass distributions. It has not previously been possible to incorporate terms representing metal-macromolecule complexes in the mass balance equations which are solved to determine speciation distributions because of the problems with multifunctional binding groups described earlier and for several other reasons (e.g. the concentration of the molecule can often not be defined in terms of moles and there may be sudden changes in behavior corresponding to switches in macromolecule conformation). However, since the Buffle and Altmann functions can be used to calculate the total amount of metal bound to the macromolecule, the overall species distribution can be calculated by an iterative method in which the separate high- and low-molar-mass fractions are evaluated successively. In practice, whether this would be successful depends on the way the interactive procedure is set up: to achieve a convergence, it is necessary that the two parts of the calculation do not counteract each other.

Our proposed method of ensuring convergence is based on the so-called "Surrogate Functions" which have been employed with success for other purposes within JESS (May and Murray, 1993). A set of empirical functions will be developed to approximate the metal-binding behavior of the macromolecular fraction in a form that is suitable for direct insertion into the mass balance equations. These would be parametrized by evaluating the degree of metal binding over an appropriate range of conditions from experimental data, or as represented by Buffle and Altmann's relationships. This information would mean that the effects of the macromolecular equilibria were sufficiently taken into account when the mass balance equations are solved to achieve calculational stability. As the iterative calculation approaches the solution, the range of conditions covered by the surrogate functions will be reduced progressively so that at convergence the terms representing metal binding by the macromolecular fraction becomes exactly correct.

It is interesting to note that this proposed approach can readily accommodate changes in the functions describing the macromolecular metal binding. Indeed, it can even be used to interface the low-molar-mass calculations directly with experimental data. The ease of managing additional effects, say of binding to the macromolecule by secondary ions or of conformational changes, is an especially attractive feature of the surrogate function method. In these and other respects, there are strong parallels between the way JESS will handle metal-macromolecule binding and the various different kinds of approach which can be used to model adsorption phenomena (May and Murray, unpublished work).

IV. KINETIC FACTORS INFLUENCING METAL-LIGAND INTERACTIONS IN VIVO

Without doubt, the most worrying deficiency of existing models describing the interaction of metals and ligands in vivo is the widespread neglect of kinetic effects. All of the major simulations of complexing reactions that occur in biological fluids are essentially thermodynamic in nature. This is true of all the blood plasma models and of other calculations of trace element distributions in vivo described in this Chapter 3, as well as for urine (e.g. Linder and Little, 1986). It is also true even of the CPMI calculations by Jones and May (1987) which attempt to take into account certain effects such as the pharmacokinetics of chelating drugs, the redistribution of metal ions between serum and other organs and the rates of the complex-forming reactions. It is simply not yet possible to incorporate all the kinetic factors which are important in determining the complexing behavior of metal ions and ligands in the whole physiological environment. Kinetic simulations, such as those of Jones (1985) are, given the current state of the art, necessarily severe oversimplifications. The need is to amalgamate all available knowledge into a single model encompassing the vast number of relevant thermodynamic equilibria in vivo and every conceivable rate-limiting process, even indirect ones such as metalloprotein synthesis.

That this ambitious goal has not yet even been attempted is due almost entirely to the practical limit which currently exists regarding the size of system which can be solved by setting up the appropriate differential equations. Not only is the number of processes completely overwhelming but also there are large differences in the rate constants which ultimately preclude numerical solution.

However, progress in understanding and thus remedying these difficulties is now being made. A project is presently underway in the author's laboratory (Verhoeven, unpublished work) which involves the careful division of the calculation into several parts so that the equilibrium and kinetic equations can be solved separately. The considerable advantage of this is that all the thermodynamic equilibria can be treated in the conventional way (i.e. in terms of mass balance equations) and that, as a corollary, they do not need to be included directly into the differential equations. This code will be incorporated into the JESS modelling package in the near future. Whilst the number of kinetic processes that can be considered simultaneously is still limited, it is possible to identify the most important ones and restrict the model to these. In this way, the kinetic and equilibrium reactions can be effectively coupled together and models of the metal-ligand interactions in vivo made orders of magnitude more comprehensive than has hitherto been possible.

Of course, the trick to this approach is to ensure that the separate thermodynamic and kinetic parts of the calculation do not counteract each other, leading to non-convergence. The problem is similar to that encountered with the partitioning of equilibrium calculations in order that the effects of metal-macromolecule binding can be accommodated. It transpires that it can be overcome in much the same way, based on a special type of surrogate function approach (Verhoeven, unpublished work).

As a result of these developments, large computer simulation models of the species distribution and kinetics of trace metals in whole-body mammalian systems will shortly become available. These models will be much more capable of rationalizing the immense amount of physiological data which has been collected experimentally over recent decades. This process of rationalization should itself greatly improve our understanding by exposing inconsistencies in experimental results and thereby uncovering artefactual results or deficiencies in the model. Even more importantly, though, these models will grow rapidly in their predictive capabilities as their parameters become better and better established. This greater knowledge of how metal ions in a physiological environment are expected to behave will impact in many different ways, but none more exciting than the prospect of scanning large numbers of potential therapeutic agents to determine, in advance of animal experiments, how any agent is likely to interact with metal ions in vivo.

REFERENCES

Altmann, R.S., and Buffle, J. (1988). *Geochim. Cosmochim. Acta, 52*: 1505.

Betti, M., and Papoff, P. (1988). *Crit. Rev. Anal. Chem., 19*: 271.

Buchanan, B.G., Smith, D.H., White, W.C., Gritter, R.J., Feigenbaum, E.A., Lederberg, J., and Djerassi, C. (1976). *J. Amer. Chem. Soc., 98*: 6168.

Buffle, J., and Altmann, R.S. (1987). *Aquatic Surface Chemistry* (W. Stumm, ed.), John Wiley & Sons, N.Y., p. 351.

Damewood, J.R., Kumpf, R.A., Muhlbauer, W.C.F., Urban, J.J., and Eksterowicz, J.E. (1990). *J. Phys. Chem., 94*: 6619.

Denning, P.J., and Tichy, W.F. (1990). *Science, 250*: 1217.

Dixon, H.B.F. (1992). *Essays Biochem., 27*: 161.

Gamble, D.S., Underdown, A.W., and Langford, C.H. (1980). *Anal. Chem., 52*: 1901.

Hassanaly, P., and Dou, H. (1986). *Computer Aids to Chemistry* (G. Vernin and M. Chanon, eds.), Ellis Horwood, Chichester, p. 334.

Hirst, D.M. (1990). *A Computational Approach to Chemistry*, Blackwell, Oxford, p. 204.

Jones, M.M. (1985). *Inorg. Chim. Acta, 107*: 235.

Jones, M.M., and May, P.M. (1987). *Inorg. Chim. Acta, 138*: 67.

Jurs, P.C. (1990). *Reviews in Computational Chemistry* (K.B. Lipkowitz and D.B. Boyd, eds.), VCH Publishers Inc., New York, p. 169.

Linder, P.W., and Little, J.C. (1986). *Inorg. Chim. Acta, 123*: 137.

Marsili, M. (1990). *Computer Chemistry*, CRC Press, Boca Raton, Florida, p. 1.

May, P.M., and Murray, K. (1991a). *Talanta, 38*: 1409.

May, P.M., and Murray, K. (1991b). *Talanta, 38*: 1419.

May, P.M., and Murray, K. (1993). *Talanta, 40*: 819.

Mezey, P.G. (1990). *Reviews in Computational Chemistry* (K.B. Lipkowitz and D.B. Boyd, eds.), VCH Publishers Inc., New York, p. 265.

Straatsma, T.P., and McCammon, J.A. (1992). *Annu. Rev. Phys. Chem., 43*: 407.

Urban, J.J., and Underwood, J.R. (1990). *J. Chem. Soc. Chem. Commun.*, 1636.

INDEX TO VOLUMES 1 AND 2

Volume 1: 1–724
Volume 2: 725–1297

ABOUT THE EDITOR

Guy Berthon is Directeur de Recherche at the Centre National de la Recherche Scientifique and Head of a research group in Research Unit 305 of the Institut National de la Santé et de la Recherche Médicale, Toulouse, France. The author or coauthor of more than 125 scientific articles, Dr. Berthon is a former editorial-board member of *Thermochimica Acta.* His research interests focus on zinc and copper dosage in total parenteral nutrition, the role of metal ions in the bioavailability of tetracyclines, the speciation of histamine in blood plasma, the speciation of aluminum in vivo, and the use of copper as a catalyst in nonsteroidal anti-inflammatory drugs. Dr. Berthon received the Docteur ès Sciences Physiques degree (1970) from the University of Poitiers, France.